Mathematical Analysis, Optimization, Approximation and Applications

Series on Computers and Operations Research

ISSN: 1793-7973

Series Editor: Panos M. Pardalos *(University of Florida, USA)*

Published

Vol. 12 *Mathematical Analysis, Optimization, Approximation and Applications*
edited by P. M. Pardalos and T. M. Rassias

Vol. 11 *Mathematical Analysis, Differential Equations and Applications*
edited by T. M. Rassias and P. M. Pardalos

Vol. 10 *Analysis, Cryptography and Information Science*
edited by N. J. Daras, P. M. Pardalos and M. Th. Rassias

Vol. 9 *Analysis, Geometry, Nonlinear Optimization and Applications*
edited by P. M. Pardalos and T. M. Rassias

Vol. 8 *Network Design and Optimization for Smart Cities*
edited by K. Gakis and P. M. Pardalos

Vol. 7 *Computer Aided Methods in Optimal Design and Operations*
edited by I. D. L. Bogle and J. Zilinskas

Vol. 6 *Recent Advances in Data Mining of Enterprise Data: Algorithms and Applications*
edited by T. W. Liao and E. Triantaphyllou

Vol. 5 *Application of Quantitative Techniques for the Prediction of Bank Acquisition Targets*
by F. Pasiouras, S. K. Tanna and C. Zopounidis

Vol. 4 *Theory and Algorithms for Cooperative Systems*
edited by D. Grundel, R. Murphey and P. M. Pardalos

Vol. 3 *Marketing Trends for Organic Food in the 21st Century*
edited by G. Baourakis

Vol. 2 *Supply Chain and Finance*
edited by P. M. Pardalos, A. Migdalas and G. Baourakis

Vol. 1 *Optimization and Optimal Control*
edited by P. M. Pardalos, I. Tseveendorj and R. Enkhbat

Series on Computers and Operations Research Vol. 12

Mathematical Analysis, Optimization, Approximation and Applications

Editors

Panos M Pardalos
University of Florida, USA

Themistocles M Rassias
National Technical University of Athens, Greece

World Scientific

NEW JERSEY • LONDON • SINGAPORE • GENEVA • BEIJING • SHANGHAI • TAIPEI • CHENNAI

Published by

World Scientific Publishing Co. Pte. Ltd.
5 Toh Tuck Link, Singapore 596224
USA office: 27 Warren Street, Suite 401-402, Hackensack, NJ 07601
UK office: 57 Shelton Street, Covent Garden, London WC2H 9HE

Library of Congress Control Number: 2024028344

British Library Cataloguing-in-Publication Data
A catalogue record for this book is available from the British Library.

Series on Computers and Operations Research — Vol. 12
MATHEMATICAL ANALYSIS, OPTIMIZATION, APPROXIMATION AND APPLICATIONS

ISBN 978-981-12-9299-6 (hardcover)
ISBN 978-981-12-9300-9 (ebook for institutions)
ISBN 978-981-12-9301-6 (ebook for individuals)

For any available supplementary material, please visit
https://www.worldscientific.com/worldscibooks/10.1142/13833#t=suppl

Desk Editors: Soundararajan Raghuraman/Steven Patt

Typeset by Stallion Press
Email: enquiries@stallionpress.com

https://doi.org/10.1142/9789811293009_fmatter

Preface

In this book, one can find both research and survey papers presenting results in a broad spectrum of subjects in pure and applied mathematics. The content is devoted to various subjects in which mathematical analysis plays an integral role, such as in approximation theory, optimization, and their applications. In more detail, topics treated within this book include relator spaces, Sobolev spaces, Banach spaces, locally convex spaces, integral operators, Szasz–Mirakyan operators, stochastic properties of circuit chains, boundary value problems, functional equations, Ulam stability, analytic inequalities, variational inequalities, orthogonal polynomials, Kuratowski's topology, and Lyusternik–Schnirelmann theory. Furthermore, this book presents applications of the theory and techniques investigated to supply chain network models, cardiac hemodynamics, and various game theoretic problems.

This book will be useful to graduate students and advanced research scientists in various branches of pure and applied mathematics in which methods of analysis are applied. Effort was made for the content to constitute a reference source for researchers in mathematics, physics, and engineering.

We would like to express our thanks to all contributing authors for participating in this publication. We are very thankful as well to the staff of World Scientific Publishing for their very valuable assistance throughout the preparation of this book.

https://doi.org/10.1142/9789811293009_fmatter

About the Editors

Panos M. Pardalos serves as Professor Emeritus of industrial and systems engineering at the University of Florida. Additionally, he is the Paul and Heidi Brown Preeminent Professor of industrial and systems engineering. He is also an affiliated faculty member of the Computer and Information Science Department, the Hellenic Studies Center, and the biomedical engineering program, as well as the Director of the Center for Applied Optimization. Pardalos is a world-leading expert in global and combinatorial optimization. His recent research interests include network design problems, optimization in telecommunications, e-commerce, data mining, biomedical applications and massive computing.

Themistocles M. Rassias is a Professor at the National Technical University of Athens, Greece. He has published more than 300 papers, 10 research books and 45 edited volumes in research Mathematics as well as four textbooks in Mathematics (in Greek) for university students. He serves as a member of the Editorial Board of several international mathematical journals. His work extends over several fields of mathematical analysis. It includes nonlinear functional analysis, functional equations, approximation theory, analysis on manifolds, calculus of variations, inequalities, metric geometry and their applications. He has contributed a number of results in the stability of minimal submanifolds, in the solution of Ulam's Problem for approximate homomorphisms in Banach spaces, in the theory of isometric mappings in metric spaces and in complex analysis (Poincaré's inequality and harmonic mappings).

https://doi.org/10.1142/9789811293009_fmatter

Contents

https://doi.org/10.1142/9789811293009_0001

Chapter 1

Extensions of the Cosine-Sine Functional Equation

O. Ajebbar[*,‡] and E. Elqorachi[†,§]

[*] *Department of Mathematics and Computer Sciences,*
Sultan Moulay Slimane University, Multidisciplinary faculty,
Beni Mellal, Morocco

[†] *Department of Mathematics,*
Ibn Zohr University, Faculty of Sciences,
Agadir, Morocco

[‡] *omar-ajb@hotmail.com*

[§] *elqorachi@hotmail.com*

The aim of this chapter is to give extensions of the cosine-sine functional equation.

1. Introduction

1.1. *Connection to the literature*

In this chapter, we solve the following functional equations:

$$f(xy) = g_1(x)h_1(y) + g(x)h_2(y) + \chi(x)h(y),\ x, y \in G, \tag{1}$$

on a group G, where g is the average of two different characters μ_1 and μ_2 of G and χ is a character of G such that $\mu_1 \neq \chi$ and $\mu_2 \neq \chi$;

$$f(xy) = g(x)h(y) + \sum_{j=1}^{N} \chi_j(x)h_j(y),\ x, y \in G, \tag{2}$$

where $\chi_1, \ldots, \chi_N$ are N distinct characters of a group G;

$$f(xy) = g_1(x)h_1(y) + g(x)h_2(y) + \chi(x)A(x)h(y), \quad x, y \in G, \tag{3}$$

where g is the average of two given nonzero multiplicative functions μ and χ on a monoid G and A is an additive function on G such that $\chi A \neq 0$;

$$f(xy) = g_1(x)h_1(y) + \mu(x)h_2(y) + \chi(x)A(x)h(y),\ x, y \in G, \tag{4}$$

where μ and χ are two different characters of a group G and $A : G \to \mathbb{C}$ is a nonzero additive function on G. To solve the functional equations (1), (2), (3), and (4), we need to solve respectively the corresponding functional equations

$$f(xy) = f(x)h_1(y) + g(x)h_2(y) + \chi(x)h(y),\ x, y \in G, \tag{5}$$

$$f(xy) = f(x)h(y) + \sum_{j=1}^{N} \chi_j(x)h_j(y),\ x, y \in G, \tag{6}$$

$$f(xy) = f(x)h_1(y) + g(x)h_2(y) + \chi(x)A(x)h(y),\ x, y \in G, \tag{7}$$

and

$$f(xy) = f(x)h_1(y) + \mu(x)h_2(y) + \chi(x)A(x)h(y),\ x, y \in G. \tag{8}$$

We note that (1), (2), (3), (4), (5), (6), (7), and (8) are examples of Levi-Civita's functional equation

$$f(xy) = \sum_{j=1}^{N} g_j(x)h_j(y),\ x, y \in G,$$

on which lots of work has been done. Székelyhidi [17, Section 10] computes for any abelian group $(G, +)$ the solution of

$$f(x + y) = \sum_{j=1}^{N} g_j(x)h_j(y),\ x, y \in G,$$

under the assumption that each of the sets $\{g_1, g_2, \ldots, g_N\}$ and $\{h_1, h_2, \ldots, h_N\}$ is linearly independent. The structure of any solution of Levi-Civita's functional equation on monoids was described by using matrix coefficients of the right regular representations [15, Theorem 5.2]. In Ref. [3, pp. 25–27], Vincze's equation, which is a particular case of our functional equations, was solved. The most recent results about Levi-Civita's functional equation on abelian monoid are obtained by Ebanks [9] and Ebanks [10]. We refer also to Refs. [1, 13–15, 18].

If $h = 0$, then the functional equations (1) and (3) become

$$f(xy) = g_1(x)h_1(y) + g(x)h_2(y),$$

which was recently solved by Ebanks [11], Stetkær [16], and Ebanks and Stetkær [12].

If $h = 0$, $f = g_1 = h_2$, and $h_1 = g$, then Eqs. (1) and (3) reduce to the sine addition law

$$f(xy) = f(x)g(y) + g(x)f(y)$$

for which solutions are known on semigroups, see Ebanks [6] and Ebanks [7].

Chung, Kannappan, and Ng [5] solved the functional equation

$$f(xy) = f(x)g(y) + f(y)g(x) + h(x)h(y), \quad x, y \in G,$$

where G is a group. Recently, the results in Ref. [5] were extended by the authors [2] to semigroups generated by their squares. Ebanks [8] find the solutions on a large class of semigroups that contains the semigroups generated by their squares.

We relate the solutions of the functional equations (1), (5), (2), (6), (4), and (8) to those of a functional equation of type

$$f(xy) = f(x)\chi_1(y) + \chi_2(x)f(y),$$

where χ_1 and χ_2 are two characters of G, which was solved by Stetkær [16].

1.2. *Motivation*

In Ref. [11], Ebanks solved the functional equation

$$f(xy) = k(x)M(y) + g(x)h(y),\ x, y \in S, \tag{9}$$

for four unknown central functions f, g, h, k on certain semigroups S, where M is a fixed multiplicative function on S.

In Ref. [16], Stetkær solved, by removing the restriction of the solutions, the functional equation

$$f(xy) = f(x)\chi_1(y) + \chi_2(x)f(y),\ x, y \in G \tag{10}$$

and its pexiderized version

$$f(xy) = g_1(x)h_1(y) + \chi_2(x)h_2(y),\ x, y \in G \tag{11}$$

on a group G, where χ_1 and χ_2 are fixed characters of G.

In Ref. [12], Ebanks and Stetkær studied the more general variant of (10)

$$f(xy) = g_1(x)h_1(y) + g(x)h_2(y), \tag{12}$$

for four unknown f, g_1, h_1, h_2 on a monoid, where g is a linear combination of $n \geq 2$ distinct multiplicative functions.

Recently, Belfakih and Elqorachi [4] solved the functional equation

$$f(xy) = g(x)h(y) + \sum_{j=1}^{n} g_j(x)h_j(y),\ x, y \in M$$

on a monoid M, where g_j are linear combinations of at least two distinct nonzero multiplicative functions on M, and f, g, h, h_j are the unknown functions.

Let G be a group. Let μ and χ be two different characters of G and let $A : G \to \mathbb{C}$ be a nonzero additive function. Let $c_1, c_2 \in \mathbb{C}$ be two constants satisfying $1 + c_1c_2^2 = 0$. We check by elementary computations that the functions $f, g, k : G \to \mathbb{C}$ defined by $f := -c_1\mu + c_1\chi - c_1c_2\chi A$, $g := \frac{\mu+\chi}{2} - \frac{c_2}{2}\chi A$, and $k := \chi A$ satisfy the functional equation

$$f(xy) = f(x)g(y) + g(x)f(y) + k(x)k(y)$$

on the group G, which implies, by simple computations, that the triple (f, g, h) is a solution of the functional equation

$$f(xy) = f(x)g(y) + \frac{\mu(x) + \chi(x)}{2} f(y) + \chi(x)A(x)h(y)$$

for all $x, y \in G$, where $h := -\frac{1}{2}c_2 f + \chi A$ and $h \neq \chi A$ because $c_1, c_2 \in \mathbb{C} \setminus \{0\}$. Moreover, for a given character m of G, we check easily that the quadruple $\left(m, \frac{m+\mu+\chi}{2} + \chi A, 2m, -2m, -2m\right)$ satisfies the functional equation

$$f(xy) = g_1(x)h_1(y) + \frac{\mu(x) + \chi(x)}{2} h_2(y) + \chi(x)A(x)h(y),$$

while the functions $f := \mu$, $g_1 := -\mu + \chi A$, $h_1 := \chi$, $h_2 := \mu + \chi$, and $h := -\chi$ satisfy the functional equation

$$f(xy) = g_1(x)h_1(y) + \mu(x)h_2(y) + \chi(x)A(x)h(y),$$

with $g_1 \neq f$, $h_2 \neq f$, $h \neq 0$, and $h \neq \chi A$.

The papers cited above motivate us to deal with the functional equations (1), (2), (3), (4), (5), (6), (7), and (8), which are not covered by the cited works.

1.3. *About our results*

In this chapter, we solve the functional equations (1), (2), (3), (4), (5), (6), (7), and (8), which are not in the literature. We express the solutions of the functional equations (1), (2), (3), and (4) respectively in term of the solutions of (5), (6), (7), and (8). We relate some categories of solutions to the solutions of the functional equation $f(xy) = f(x)\chi_1(y) + \chi_2(x)f(y)$, where χ_1 and χ_2 are two characters of the group G, which was recently solved by Stetkær [16]. Our results are extensions of those of Refs. [16] and [12].

There is no connection between the functional equations (1), (2), (3), and (4) nor between (5), (6), (7), and (8), but the reasonings for the functional equations with additive function, i.e., (3), (4), (7), and (8), are similar to those for the functional equations that do not contain the additive function, i.e., (1), (2), (5), and (6). This is due to fact that the resolution of this functional equation

(1) makes use of the linear independence between some unknown functions and the given arguments in equations, namely, multiplicative functions, additive functions, or characters and
(2) relates some of them to the functional equation $f(xy) = f(x)\chi_1(y) + \chi_2(x)f(y)$, where χ_1 and χ_2 are two characters of the group G.

The organization of this chapter is as follows. In the following section, we give some definitions and notations. In Section 3, we give preliminary results that we need in this chapter. In Sections 4, 5, 6, and 7, we prove our main results and give some examples.

Our methods are elementary algebraic manipulations (no analysis or geometry come into play).

2. Set Up and Terminology

Definition 1. Let $f : S \to \mathbb{C}$ be a function on a semigroup S. We say that
f is central if $f(xy) = f(yx)$ for all $x, y \in G$,
f is additive if $f(xy) = f(x) + f(y)$ for all $x, y \in S$,
f is multiplicative if $f(xy) = f(x)f(y)$ for all $x, y \in S$.
If $S = G$ is a group, then we say that f is a character of G if f is a homomorphism of G into $\mathbb{C}^*$, where $\mathbb{C}^*$ denotes the multiplicative group of nonzero complex numbers.

If G is a group, then $[G, G]$ denotes the commutator of G, i.e., the subgroup of G generated by the set of commutators $[x, y] := xyx^{-1}y^{-1}$, $x, y \in G$.

If χ_1 and χ_2 are two characters of a group G, we denote by φ_{χ_1,χ_2} a solution of the functional equation

$$f(xy) = f(x)\chi_1(y) + \chi_2(x)f(y),\ x, y \in G.$$

Blanket assumption: Throughout this chapter, G denotes a group or a monoid with identity element e.

3. Preliminaries

In this section, we give the essential tools that we need in this chapter.

Proposition 2 ([15, Theorem 3.18(a)]). *Let S be a semigroup and $n \in \mathbb{N}$. Let $\chi, \chi_2, \ldots, \chi_n : S \to \mathbb{C}$ be n different multiplicative functions and let $a_1, a_2, \ldots, a_n \in \mathbb{C}$.*

If $a_1\chi + a_2\chi_2 + \cdots + a_n\chi_n = 0$, then $a_1\chi = a_2\chi_2 = \cdots = a_n\chi_n = 0$, in particular, any set of nonzero multiplicative functions is linearly independent.

Lemma 3 ([2, pp. 1122–1123]). *Let $a : S \to \mathbb{C}$ be an additive function and $m : S \to \mathbb{C}$ be a multiplicative function on a semigroup S.*

If $ma = \Sigma_{j=1}^{N} c_j\mu_j$, where $c_j \in \mathbb{C}$ and $\mu_j : S \to \mathbb{C}$ is multiplicative for each $j = 1, 2, \ldots, N$, then $ma = 0$.

Let χ_1 and χ_2 be characters of G. The following result, which is taken from Ref. [16, Proposition 4, Theorem 11], solves the functional equation

$$f(xy) = f(x)\chi_1(y) + \chi_2(x)f(y), \; x, y \in G. \tag{13}$$

Theorem 4 ([16, p. 5, pp. 8–9]). *Let G be a group.*

(1) *If $\chi_1 = \chi_2$, then the solutions of* (13) *are the functions $f = \chi_1 A$, where $A : G \to \mathbb{C}$ is an additive function on G.*
(2) *If $\chi_1 \neq \chi_2$, then the solutions of* (13) *are the functions*

$$f(x) = \alpha(\chi_1(x) - \chi_2(x)) + A([y_0, x])\chi_1(x), \; x \in G,$$

where $y_0 \in G$ such that $\chi_1(y_0) \neq \chi_2(y_0)$, α ranges over $\mathbb{C}$, and $A : [G, G] \to \mathbb{C}$ ranges over the additive functions on $[G, G]$ with the transformation property

$$A(xcx^{-1}) = \frac{\chi_2(x)}{\chi_1(x)} A(c) \quad \textit{for all} \quad x \in G \quad \textit{and} \quad c \in [G, G].$$

Remark 5. According to Ref. [16, Proposition 4], if $\chi_1 \neq \chi_2$, the central solutions of (13) are the functions $f = \alpha(\chi_1 - \chi_2)$, where α ranges over $\mathbb{C}$.

Lemma 6. *Let G be a group. Let (f, g_1, h_1, h_2, h) be a solution of the functional equation* (1)*:*

(1) *If $h_1 = 0$, then $f = h = b\chi$, $h_2 = 0$, and g_1 is arbitrary, where $b \in \mathbb{C}$ is a constant.*
(2) *If $h_1 \neq 0$ and the set $\{f, \mu_1, \mu_2, \chi\}$ is linearly dependent, then the set $\{g_1, \mu_1, \mu_2, \chi\}$ is also linearly dependent.*

Proof. (1) Assume that $h_1 = 0$. Then g_1 is arbitrary. By putting $y = e$ in (1), we get that $f = a\mu_1 + a\mu_2 + b\chi$, where $a := \frac{h_2(e)}{2}$ and $b := h(e)$. Substituting this back into (1), we obtain

$$\begin{aligned} &a\mu_1(y)\mu_1(x) + a\mu_2(y)\mu_2(x) + b\chi(y)\chi(x) \\ &= \frac{h_2(y)}{2}\mu_1(x) + \frac{h_2(y)}{2}\mu_2(x) + h(y)\chi(x), \end{aligned}$$

for all $x, y \in G$. Since the characters μ_1, μ_2, and χ are different, the set $\{\mu_1, \mu_2, \chi\}$ is linearly independent. Then we get, from the identity above, that $\frac{h_2}{2} = a\mu_1 = a\mu_2$ and $h = b\chi$. So $a = 0$ because $\mu_1 \neq \mu_2$. Hence, $f = h = b\chi$ and $h_2 = 0$.

(2) Assume that $h_1 \neq 0$ and $\{f, \mu_1, \mu_2, \chi\}$ is linearly dependent. Since the set $\{\mu_1, \mu_2, \chi\}$ is linearly independent, there exist $\alpha_1, \alpha_2, \alpha_3 \in \mathbb{C}$ such that $f = \alpha_1\mu_1 + \alpha_2\mu_2 + \alpha_3\chi$. So

$$f(xy) = \alpha_1\mu_1(x)\mu_1(y) + \alpha_2\mu_2(x)\mu_2(y) + \alpha_3\chi(x)\chi(y)$$

for all $x, y \in G$, from which we get, taking (1) into account, that

$$\begin{aligned} g_1(x)h_1(y) = {} & [\alpha_1\mu_1(y) - \frac{1}{2}h_2(y)]\mu_1(x) + [\alpha_2\mu_2(y) - \frac{1}{2}h_2(y)]\mu_2(x) \\ & + [\alpha_3\chi(y) - h(y)]\chi(x), \end{aligned}$$

for all $x, y \in G$. As $h_1 \neq 0$, we deduce from the identity above, by choosing $y_0 \in G$ such that $h_1(y_0) \neq 0$, that the set $\{g_1, \mu_1, \mu_2, \chi\}$ is linearly dependent. □

Proposition 7. *Let G be a monoid. For $f = 0$, the solutions of the functional equation* (3) *can be listed as follows:*

(1) *g_1 arbitrary and $h_1 = h_2 = h = 0$.*
(2) *$g_1 = a\frac{\mu+\chi}{2} + b\chi A$, $h_1 \neq 0$ arbitrary, $h_2 = -ah_1$, and $h = -bh_1$, where a and b range over $\mathbb{C}$.*

Proof. (1) Assume that $h_1 = 0$. Then g_1 is arbitrary. Let $y \in G$ be arbitrary. Since $f = 0$, the functional equation (3) implies

$$\frac{h_2(y)}{2}\mu(x) + \frac{h_2(y)}{2}\chi(x) = -h(y)\chi(x)A(x),$$

for all $x \in G$. So, the set $\{\mu, \chi, \chi A\}$ being linearly independent, $h_2(y) = h(y) = 0$ and y arbitrary, we deduce that $h_2 = h = 0$.

(2) Assume that $h_1 \neq 0$. Taking $x = e$ in (3) and seeing that $\mu(e) = \chi(e) = 1$ and $A(e) = 0$, we get that $g_1(e)h_1(y) + h_2(y) = 0$ for all $y \in G$. So

$$h_2 = -ah_1, \tag{14}$$

where $a := g_1(e)$. Substituting (15) back into (3), we get that

$$\left(g_1(x) - a\frac{\mu(x) + \chi(x)}{2}\right) h_1(y) + h(y)\chi(x)A(x) = 0$$

for all $x, y \in G$. Choosing $y_0 \in G$ such that $h_1(y_0) \neq 0$, we deduce from the last functional equation that

$$g_1 = a\frac{\mu + \chi}{2} + b\chi A, \tag{15}$$

where $b := -\frac{h(y_0)}{h_1(y_0)}$.

Substituting (15) and (15) back into (3), we get that

$$(bh_1(y) + h(y))\chi(x)A(x) = 0$$

for all $x, y \in G$. Since $\chi A \neq 0$, we derive that

$$h = -bh_1.$$

Conversely, it is easy to check that the formulas of (1) and (2) define solutions of (3) for $f = 0$. □

Proposition 8. *Let G be a group and let $\alpha \in \mathbb{C}$ be a constant. For $f = \alpha\mu$, the solutions of the functional equation (4) can be listed as follows:*

(1) g_1 *arbitrary,* $h_1 = 0$, $h_2 = \alpha\mu$, *and* $h = 0$;
(2) $g_1 = a\mu + b\chi\, A$, $h_1 \neq 0$ *arbitrary,* $h_2 = -ah_1 + \alpha\mu$, *and* $h = -bh_1$, *where* a *and* b *range over* $\mathbb{C}$.

Proof. By proceeding as in the proof of Proposition 7. □

4. Solutions of Eq. (5) and Eq. (1)

In this section, μ_1, μ_2, and χ are different characters of a group G, and $g := \frac{\mu_1+\mu_2}{2}$.

4.1. *Solutions of Eq. (5)*

In this section, we solve the functional equation (5), i.e.,

$$f(xy) = f(x)h_1(y) + g(x)h_2(y) + \chi(x)h(y),\ x, y \in G,$$

where $f, h_1, h_2, h : G \to \mathbb{C}$ are the unknown functions to be determined.

Proposition 9 solves the functional equation (5) when the set $\{f, \mu_1, \mu_2, \chi\}$ is linearly independent.

Proposition 9. *Let* (f, h_1, h_2, h) *be a solution of (5) such that the set* $\{f, \mu_1, \mu_2, \chi\}$ *is linearly independent. Then*

$$f = c\mu + \varphi_{\mu,\chi},\ h_1 = \mu,\ h_2 = 0 \text{ and } h = \varphi_{\mu,\chi},$$

where μ *is a character of* G *and* $c \in \mathbb{C}$ *is a constant.*

Proof. Let $x, y, z \in G$ be arbitrary. We compute $f(xyz)$ as $f(x(yz))$ and then as $f((xy)z)$. Using Eq. (5) and that $g = \frac{\mu_1+\mu_2}{2}$, we obtain

$$f(x(yz)) = f(x)h_1(yz) + \frac{1}{2}\mu_1(x)h_2(yz) + \frac{1}{2}\mu_2(x)h_2(yz) + \chi(x)h(yz). \tag{16}$$

On the other hand, by applying Eq. (5) to the pair (xy, z), we get that

$$\begin{aligned} f((xy)z) &= f(xy)h_1(z) + g(xy)h_2(z) + \chi(xy)h(z) \\ &= \left[f(x)h_1(y) + \frac{1}{2}\mu_1(x)h_2(y) + \frac{1}{2}\mu_2(x)h_2(y) + \chi(x)h(y)\right] \\ &\quad \times h_1(z) + \frac{1}{2}\mu_1(x)\mu_1(y)h_2(z) + \frac{1}{2}\mu_2(x)\mu_2(y)h_2(z) \\ &\quad + \chi(x)\chi(y)h(z). \end{aligned}$$

So

$$\begin{aligned} f((xy)z) = f(x)h_1(y)h_1(z) &+ \frac{1}{2}\mu_1(x)[h_2(y)h_1(z) + \mu_1(y)h_2(z)] \\ &+ \frac{1}{2}\mu_2(x)[h_2(y)h_1(z) + \mu_2(y)h_2(z)] \\ &+ \chi(x)[h(y)h_1(z) + \chi(y)h(z)]. \end{aligned} \tag{17}$$

As the set $\{f, \mu_1, \mu_2, \chi\}$ is linearly independent and x, y, z are arbitrary, we derive from (16) and (17) that (h_1, h_2, h) is a solution of the following functional equations as a system:

$$h_1(xy) = h_1(x)h_1(y), \tag{18}$$

$$h_2(xy) = h_2(x)h_1(y) + \mu_1(x)h_2(y), \tag{19}$$

$$h_2(xy) = h_2(x)h_1(y) + \mu_2(x)h_2(y), \tag{20}$$

and

$$h(xy) = h(x)h_1(y) + \chi(x)h(y) \tag{21}$$

for all $x, y \in G$. The functional equation (18) implies that $h_1 = \mu$, where $\mu : G \to \mathbb{C}$ is a multiplicative function. By subtracting (19) from (20), we get that $[\mu_1(x) - \mu_2(x)]h_2(y) = 0$ for all $x, y \in G$. Since $\mu_1 \neq \mu_2$, we have $h_2(y) = 0$. So, y being arbitrary, we deduce that $h_2 = 0$.

If $\mu = 0$, then $h_1 = 0$ and the functional equation (21) reduces to $h(xy) = \chi(x)h(y)$ for all $x, y \in G$, which yields, by putting $y = e$, that $h = \alpha\chi$, where $\alpha := h(e)$. Thus, the functional equation (5) becomes $f(xy) = \alpha\chi(xy)$ for all $x, y \in G$. This implies, by putting $y = e$, that $f = \alpha\chi$, which contradicts the linear independence of the set $\{f, \mu_1, \mu_2, \chi\}$.

Hence, $\mu \neq 0$ and then μ is a character of G. So, the functional equation (21) implies that $h = \varphi_{\mu,\chi}$. So the functional equation (5) reduces to $f(xy) = f(x)\mu(y) + \chi(x)\varphi_{\mu,\chi}(y)$ for all $x, y \in G$, from which we get, by putting $x = e$ and $c = f(e)$, that $f = c\mu + \varphi_{\mu,\chi}$. □

Theorem 10. *The solutions (f, h_1, h_2, h) of (5) are as follows:*

(1) $f = c\mu + \varphi_{\mu,\chi}$, $h_1 = \mu$, $h_2 = 0$, *and* $h = \varphi_{\mu,\chi}$, *where μ is a character of G and $c \in \mathbb{C}$ is a constant;*

(2) $f = c\chi$, h_1 *is arbitrary*, $h_2 = 0$, *and* $h = c(\chi - h_1)$, *where $c \in \mathbb{C}$ is a constant;*

(3)

$$f = a\mu_1 + b\mu_2 + c\chi, \; h_1 = \frac{1}{a-b}(a\mu_1 - b\mu_2),$$

$$h_2 = -\frac{2ab}{a-b}(\mu_1 - \mu_2), \; h = \frac{c}{a-b}(-a\mu_1 + b\mu_2 + (a-b)\chi),$$

where $a, b, c \in \mathbb{C}$ are constants such that $a \neq b$.

Proof. It easy to check that the formulas in parts (1)–(3) define solutions of the functional equation (5), so what is left is to show that any solution (f, h_1, h_2, h) fits into (1)–(3).

If the set $\{f, \mu_1, \mu_2, \chi\}$ is linearly independent, we get, by Proposition 9, the part (1) of Theorem 19.

In the remainder of the proof, we assume that $\{f, \mu_1, \mu_2, \chi\}$ is linearly dependent. Recall that the set $\{\mu_1, \mu_2, \chi\}$ is linearly independent because μ_1, μ_2, and χ are different characters of G. Then there exists a triple $(a, b, c) \in \mathbb{C}^3$ such that

$$f = a\mu_1 + b\mu_2 + c\chi. \tag{22}$$

Substituting this back into (5), we get by a small computation that

$$\begin{aligned} &a\mu_1(y)\mu_1(x) + b\mu_2(y)\mu_2(x) + c\chi(y)\chi(x) \\ &= \left[ah_1(y) + \frac{1}{2}h_2(y)\right]\mu_1(x) \\ &\quad + \left[bh_1(y) + \frac{1}{2}h_2(y)\right]\mu_2(x) + [ch_1(y) + h(y)]\chi(x) \end{aligned}$$

for all $x, y \in G$. Since μ_1, μ_2, and χ are different characters of G, we derive by Proposition 2 that (h_1, h_2, h) is a solution of the following identities as a system:

$$ah_1 + \frac{1}{2}h_2 = a\mu_1, \tag{23}$$

$$bh_1 + \frac{1}{2}h_2 = b\mu_2, \tag{24}$$

and

$$ch_1 + h = c\chi. \tag{25}$$

When we subtract (24) from (23), we obtain

$$(a - b)h_1 = a\mu_1 - b\mu_2. \tag{26}$$

If $a = b$, the identity (26) reduces to $a(\mu_1 - \mu_2) = 0$. Hence, $a = b = 0$ because $\mu_1 \neq \mu_2$. So, taking (22) and (25) into account, we get that

$$f = c\chi,\ h_1 \text{ is arbitrary},\ h_2 = 0 \text{ and } h = c(\chi - h_1),$$

which is solution (2).

If $a \neq b$, identity (26) implies that $h_1 = \frac{1}{a-b}(a\mu_1 - b\mu_2)$. Substituting this back into (23) and (25), we derive that

$$h_2 = 2a(\mu - h_1) = 2a\left[\mu_1 - \frac{1}{a-b}(a\mu_1 - b\mu_2)\right] = -\frac{2ab}{a-b}(\mu_1 - \mu_2)$$

and

$$h = c(\chi - h_1) = c\left[\chi - \frac{1}{a-b}(a\mu_1 - b\mu_2)\right] = \frac{c}{a-b}(-a\mu_1 + b\mu_2 + (a-b)\chi).$$

The solution occurs in part (3). This completes the proof. □

4.2. *Solutions of Eq.* (1)

In Theorem 11, we solve the functional equation (1), i.e.,

$$f(xy) = g_1(x)h_1(y) + g(x)h_2(y) + \chi(x)h(y), \quad x, y \in G,$$

where $f, g_1, h_1, h_2, h : G \to \mathbb{C}$ are the unknown functions to be determined.

Theorem 11. *The solutions (f, g_1, h_1, h_2, h) of (1) can be listed as follows:*

(1) $f = b\chi$, g_1 *is arbitrary,* $h_1 = h_2 = 0$, *and* $h = b\chi$, *where* $b \in \mathbb{C}$ *is a constant;*

(2) $f = b\chi$, $g_1 = \alpha\mu_1 + \alpha\mu_2 + \gamma\chi$, $h_1 \neq 0$ *is arbitrary,* $h_2 = -2\alpha h_1$, *and* $h = b\chi - \gamma h_1$, *where* $b \in \mathbb{C} \setminus \{0\}$ *and* $\alpha, \gamma \in \mathbb{C}$ *are constants;*

(3)

$$f = (d_1 - d_2)\left(\frac{1}{2}a_1\mu_1 - \frac{1}{2}a_2\mu_2 - \frac{1}{2}a_3\chi\right),$$

$$g_1 = \frac{1}{2}d_1\mu_1 + \frac{1}{2}d_2\mu_2 + d_3\chi,$$

$$h_1 = a_1\mu_1 + a_2\mu_2,\ h_2 = -a_1d_2\mu_1 - a_2d_1\mu_2,\ \text{and}$$

$$h = -a_1d_3\mu_1 - a_2d_3\mu_2 - \frac{1}{2}a_3(d_1 - d_2)\chi,$$

where $a_1, a_2, a_3, d_1, d_2, d_3 \in \mathbb{C}$ *are constants such that* $d_1 \neq d_2$*;*

(4) $f = a_1b\mu + b\varphi_{\mu,\chi}$, $g_1 = -\frac{1}{2}a_2\mu_1 - \frac{1}{2}a_2\mu_2 - a_3\chi + a_1\mu + \varphi_{\mu,\chi}$, $h_1 = b\mu$, $h_2 = a_2b\mu$, *and* $h = a_3b\mu + b\varphi_{\mu,\chi}$, *where* $b \in \mathbb{C} \setminus \{0\}, a_1, a_2, a_3 \in \mathbb{C}$ *are constants and* μ *is a character of* G*;*

(5) $f = a_1b\chi$, $g_1 = -\frac{1}{2}a_1\mu_1 - \frac{1}{2}a_2\mu_2 + (a_1 - a_2)\chi$, h_1 *is arbitrary,*

$h_2 = a_2h_1$, *and* $h = a_1b\chi - (a_1 - a_2)h_1$,

where $b \in \mathbb{C} \setminus \{0\}$ *and* a_1, a_2, $a_3 \in \mathbb{C}$ *are constants.*

Proof. We check by elementary computations that if f, g_1, h_1, h_2, and h are of the forms (1)–(5), then (f, g_1, h_1, h_2, h) is a solution of (1), so left is that any solution (f, g_1, h_1, h_2, h) of (1) fits into (1)–(5). By putting $y = e$ in (1), we get that

$$f = h_1(e)g_1 + \frac{h_2(e)}{2}\mu_1 + \frac{h_2(e)}{2}\mu_2 + h(e)\chi. \tag{27}$$

We split the discussion into the subcases $h_1(e) = 0$ and $h_1(e) \neq 0$.

Case 1: Suppose $h_1(e) = 0$. Then (27) gives

$$f = a\mu_1 + a\mu_2 + b\chi, \tag{28}$$

where $a := \frac{h_2(e)}{2}$ and $b := h(e)$. Hence, the set $\{f, \mu_1, \mu_2, \chi\}$ is linearly dependent.

If $h_1 = 0$, then, according to Lemma 6(1), we get that g_1 is arbitrary, $h_2 = 0$, and $f = h = b\chi$, which is solution (1).

If $h_1 \neq 0$, then, by Lemma 6(2), the set $\{g_1, \mu_1, \mu_2, \chi\}$ is linearly dependent. Since the set $\{\mu_1, \mu_2, \chi\}$ is linearly independent, there exists a triple $(\alpha, \beta, \gamma) \in \mathbb{C}^3$ such that

$$g_1 = \alpha\mu_1 + \beta\mu_2 + \gamma\chi. \tag{29}$$

By substituting (28) and (29) in (1), we derive by simple computations that

$$\begin{aligned} &a\mu_1(y)\mu_1(x) + a\mu_2(y)\mu_2(x) + b\chi(y)\chi(x) \\ &= \left[\alpha h_1(y) + \frac{1}{2}h_2(y)\right]\mu_1(x) \\ &\quad + \left[\beta h_1(y) + \frac{1}{2}h_2(y)\right]\mu_2(x) + [\gamma h_1(y) + h(y)]\chi(x) \end{aligned}$$

for all $x, y \in G$. So, the set $\{\mu_1, \mu_2, \chi\}$ being linearly independent, we derive from the identity above that (h_1, h_2, h) satisfies the following identities as a system:

$$\alpha h_1 + \frac{1}{2}h_2 = a\mu_1, \tag{30}$$

$$\beta h_1 + \frac{1}{2}h_2 = a\mu_2, \tag{31}$$

and

$$\gamma h_1 + h = b\chi. \tag{32}$$

By subtracting (31) from (30), we get that

$$(\alpha - \beta)h_1 = a(\mu_1 - \mu_2). \tag{33}$$

If $\alpha = \beta$, then (33) implies that $a = 0$ because $\mu_1 \neq \mu_2$. Hence, from (28), (29), (30), and (33), we get that

$$\begin{aligned} &f = b\chi,\ g_1 = \alpha\mu_1 + \alpha\mu_2 + \gamma\chi,\ h_1 \neq 0 \text{ is arbitrary}, \\ &h_2 = -2\alpha h_1, \text{ and } h = b\chi - \gamma h_1. \end{aligned}$$

This is solution (2).

If $\alpha \neq \beta$, then we obtain, from (33), that

$$h_1 = \frac{a}{\alpha - \beta}(\mu_1 - \mu_2). \tag{34}$$

By substituting (34) in (30) and in (32), we get that

$$h_2 = -\frac{2\beta a}{\alpha - \beta}\mu_1 + \frac{2\alpha a}{\alpha - \beta}\mu_2 \tag{35}$$

and

$$h = -\frac{\gamma a}{\alpha - \beta}\mu_1 + \frac{\gamma a}{\alpha - \beta}\mu_2 + b\chi. \tag{36}$$

Let

$$d_1 := 2\alpha,\ d_2 := 2\beta,\ d_3 := \gamma$$

$$a_1 := \frac{2a}{d_1 - d_2},\ a_2 := -\frac{2a}{d_1 - d_2},\ \text{and}\ a_3 := -\frac{2b}{d_1 - d_2}.$$

Note that $d_1 \neq d_2$ because $\alpha - \beta \neq 0$. By using (28), (29), (34), (35), and (36), we obtain by small computations

$$f = (d_1 - d_2)\left(\frac{1}{2}a_1\mu_1 - \frac{1}{2}a_2\mu_2 - \frac{1}{2}a_3\chi\right),\ g_1 = \frac{1}{2}d_1\mu_1 + \frac{1}{2}d_2\mu_2 + d_3\chi,$$

$$h_1 = a_1\mu_1 + a_2\mu_2,$$

$$h_2 = -a_1 d_2\mu_1 - a_2 d_1\mu_2,$$

and

$$h = -a_1 d_3\mu_1 - a_2 d_3\mu_2 - \frac{1}{2}a_3(d_1 - d_2)\chi,$$

which is solution (3).

Case 2: Suppose $h_1(e) \neq 0$. Then from (27), we get that

$$g_1 = \frac{1}{h_1(e)}f - \frac{h_2(e)}{h_1(e)}g - \frac{h(e)}{h_1(e)}\chi. \tag{37}$$

Substituting this back into (1), we obtain

$$f(xy) = \left[\frac{1}{h_1(e)}f(x) - \frac{h_2(e)}{h_1(e)}g(x) - \frac{h(e)}{h_1(e)}\chi(x)\right] h_1(y)+g(x)h_2(y)+\chi(x)h(y)$$

for all $x, y \in G$. So

$$f(xy) = f(x)\left(\frac{1}{h_1(e)}h_1\right)(y)+\left(h_2 - \frac{h_2(e)}{h_1(e)}h_1\right)(y)g(x)+\chi(x)\left(h - \frac{h(e)}{h_1(e)}h_1\right)(y)$$

for all $x, y \in G$, i.e., the quadruple $(f, h_1/h_1(e), h_2 - \frac{h_2(e)}{h_1(e)}h_1, h - \frac{h(e)}{h_1(e)}h_1)$ is a solution of the functional equation (5). So, according to Theorem 19, that quadruple falls into three categories:

(i) $f = c\mu + \varphi_{\mu,\chi}$, $h_1/h_1(e) = \mu$, $h_2 - \frac{h_2(e)}{h_1(e)}h_1 = 0$, $h - \frac{h(e)}{h_1(e)}h_1 = \varphi_{\mu,\chi}$, where μ is a character of G and $c \in \mathbb{C}$ is a constant. We define complex constants by $b := h_1(e) \in \mathbb{C} \setminus \{0\}$, $a_1 := \frac{c}{h_1(e)}$, $a_2 := \frac{h_2(e)}{h_1(e)}$, and $a_3 := \frac{h(e)}{h_1(e)}$. So, the identities above and (37) yield, by writing $\varphi_{\mu,\chi}$ instead of $\frac{1}{b}\varphi_{\mu,\chi}$, that

$$f = a_1 b\mu + b\varphi_{\mu,\chi},\ g_1 = -\frac{1}{2}a_2\mu_1 - \frac{1}{2}a_2\mu_2 - a_3\chi + a_1\mu + \varphi_{\mu,\chi},$$
$$h_1 = b\mu,\ h_2 = a_2 b\mu \text{ and } h = a_3 b\mu + b\varphi_{\mu,\chi},$$

which is solution (4).

(ii)

$$f = c\chi,\ h_1/h_1(e) \text{ arbitrary},\ h_2 - \frac{h_2(e)}{h_1(e)}h_1 = 0, \text{ and}$$
$$h - \frac{h(e)}{h_1(e)}h_1 = c\left(\chi - \frac{1}{h_1(e)}h_1\right),$$

where $c \in \mathbb{C}$ is a constant.

Defining complex constants $b := h_1(e) \in \mathbb{C} \setminus \{0\}$, $a_1 := \frac{c}{h_1(e)} \in \mathbb{C}$, $a_2 := \frac{h_2(e)}{h_1(e)}$, and $a_3 := \frac{h(e)}{h_1(e)}$, we obtain, by a small computation, from the identities above and (37), that

$$f = a_1 b\chi,\ g_1 = -\frac{1}{2}a_1\mu_1 - \frac{1}{2}a_2\mu_2 + (a_1 - a_2)\chi,\ h_1 \text{ is arbitrary},$$
$$h_2 = a_2 h_1,\ \text{and } h = a_1 b\chi - (a_1 - a_2)h_1,$$

which is solution (5).
(iii)

$$f = a\mu_1 + b\mu_2 + c\chi,\ h_1/h_1(e) = \frac{1}{a-b}(a\mu_1 - b\mu_2),$$
$$h_2 - \frac{h_2(e)}{h_1(e)}h_1 = -\frac{2ab}{a-b}(\mu_1 - \mu_2),$$
$$h - \frac{h(e)}{h_1(e)}h_1 = \frac{c}{a-b}(-a\mu_1 + b\mu_2 + (a-b)\chi),$$

where $a, b, c \in \mathbb{C}$ such that $a \neq b$. Using (37) and the identities above, we get that

$$\begin{aligned} g_1 &= \frac{2a - h_2(e)}{2h_1(e)}\mu_1 + \frac{2b - h_2(e)}{2h_1(e)}\mu_2 + \frac{c - h(e)}{h_1(e)}\chi, \\ h_1 &= \frac{ah_1(e)}{a-b}\mu_1 - \frac{bh_1(e)}{a-b}\mu_2, \\ h_2 &= -\frac{a(2b - h_2(e))}{a-b}\mu_1 + \frac{b(2a - h_2(e))}{a-b}\mu_2, \\ h &= -\frac{a(c - h(e))}{a-b}\mu_1 + \frac{b(c - h(e))}{a-b}\mu_2 + c\chi. \end{aligned} \tag{38}$$

Defining

$$a_1 := \frac{ah_1(e)}{a-b},\ a_2 := -\frac{bh_1(e)}{a-b},\ a_3 := -\frac{ch_1(e)}{a-b}$$
$$d_1 := \frac{2a - h_2(e)}{h_1(e)},\ d_2 := \frac{2b - h_2(e)}{h_1(e)},\ \text{and } d_3 := \frac{c - h(e)}{h_1(e)}.$$

We have $a = \frac{1}{2}a_1(d_1 - d_2)$, $b = -\frac{1}{2}(d_1 - d_2)$, and $c = -\frac{1}{2}a_3(d_1 - d_3)$. Sine $a \neq b$, we get $d_1 \neq d_2$. Moreover, the identities $f = a\mu_1 + b\mu_2 + c\chi$

and (38) read

$$f=(d_1-d_2)\left(\frac{1}{2}a_1\mu_1-\frac{1}{2}a_2\mu_2-\frac{1}{2}a_3\chi\right),\ g_1=\frac{1}{2}d_1\mu_1+\frac{1}{2}d_2\mu_2+d_3\chi,$$

$$h_1=a_1\mu_1+a_2\mu_2,\ h_2=-a_1d_2\mu_1-a_2d_1\mu_2,$$

and

$$h=-a_1d_3\mu_1-a_2d_3\mu_2-\frac{1}{2}a_3(d_1-d_2)\chi,$$

which is solution (3). This completes the proof of Theorem 11. □

From Case 1 of the proof of Theorem 11, we derive that if $f, g_1, h_1, h_2, h : G \to \mathbb{C}$ satisfy the functional equation (1) and $h_1(e) = 0$, then f and g_1 are abelian, or f, g_1, h_1, h_2, and h are abelian.

According to Ref. [16, Proposition 5], if $\mu \neq \chi$ and G is an abelian group, then the function $\varphi_{\mu,\chi}$, in the formulas of solutions of form (4) in Theorem 11, take the form $\varphi_{\mu,\chi} = c(\mu - \chi)$, where c ranges over $\mathbb{C}$. On groups that need not be abelian, the expression of $\varphi_{\mu,\chi}$ is given by Theorem 4.

Example 12. If we choice $b - \gamma = 0$ in Theorem 11(2), or $a_3 = d_3 = 0$ in Theorem 11(3), or $a_3 = 0$ and $\varphi_{\mu,\chi} = 0$ in Theorem 11(4), we get that $h = 0$, then the functional equation (1) reduces to

$$f(xy) = g_1h_1(y) + g(x)h_2(y),\ x, y \in G$$

which was solved in Ref. [12, Theorem 8]:
(1) For $b = \gamma = 0$ in Theorem 11(2), we get, by putting $c = 2\alpha$, that

$f = 0$, $g_1 = \alpha\mu_1+\alpha\mu_2 = cg$, $h_1 \neq 0$ is arbitrary, $h_2 = -ah_1$, and $h = -bh_1$,

which is the solution obtained in Ref. [12, Theorem 8(d)].
(2) For $a_3 = d_3 = 0$ in Theorem 11(3), we obtain

$$f=(d_1-d_2)\left(\frac{1}{2}a_1\mu-\frac{1}{2}a_2\chi\right),\ g_1=\frac{1}{2}d_1\mu+\frac{1}{2}d_2\chi,$$

and

$$h_1=a_1\mu+a_2\chi,\ h_2=-a_1d_2\mu-a_2d_1\chi,$$

where a_1, a_2, d_1, $d_2 \in \mathbb{C}$ are constants. Furthermore, for $f \neq 0$, we have $d_1 \neq d_2$ and $(a_1, a_2) \neq (0, 0)$. The solution was obtained in

Ref. [12, Proposition 6(b), Theorem 8(b)].
(3) For $a_3 = 0$ and $\varphi_{\mu,\chi} = 0$ in Theorem 11(4),

$$f = a_1 b\mu,\ g_1 = a_1\mu - a_2 g,\ h_1 = b\mu \text{ and } h_2 = a_2 b\mu,$$

where $a_1,\ a_2 \in \mathbb{C}$ are constants. Furthermore, for $f \neq 0$, we have $a_1,\ b \in \mathbb{C}\setminus\{0\}$. The solution was obtained in Ref. [12, Theorem 8(a)].

Example 13. By taking $d_1 = -d_2$, $a_1 = a_2 = \frac{1}{2}$, and $a_3 = d_3 = 0$ in Theorem 11(3), we get that

$$f = g_1 = h_2 = d_1\frac{\mu_1 - \mu_2}{2} \quad \text{and} \quad h_1 = \frac{\mu_1+\mu_2}{2}, \quad \text{where } d_1 \in \mathbb{C}\setminus\{0\},$$

which is the solution of the classic sine addition law obtained in Ref. [15, Theorem 4.1(c)].

5. Solutions of Eq. (6) and Eq. (2)

In this section, $\chi_1, \ldots, \chi_N$ are N distinct characters of a group G.

5.1. *Solutions of Eq.* (6)

In Theorem 14, we solve the functional equation (6), i.e.,

$$f(xy) = f(x)h(y) + \sum_{j=1}^{N} \chi_j(x)h_j(y),\ x, y \in G,$$

where $f,\ h,\ h_1, \ldots,\ h_N : G \to \mathbb{C}$ are the unknown functions to be determined.

Theorem 14. *The solutions* $(f,\ h,\ h_1, \ldots,\ h_N)$ *of* (6) *are as follows:*

(1)

$$f = \sum_{j=1}^{N} \alpha_j\chi_j,\ h \text{ arbitrary and } h_j = \alpha_j\chi_j - \alpha_j h$$

for all $j = 1, \ldots, N$, *where* $\alpha_1, \ldots, \alpha_N \in \mathbb{C}$ *are constants;*

(2)

$$f = a\chi + \sum_{j=1}^{N} \varphi_{\chi,\chi_j},\ h = \chi \textit{ and } h_j = \varphi_{\chi,\chi_j}$$

for all $j = 1, \ldots, N$, *where* χ *is a character of* G *and* $a \in \mathbb{C}$ *is a constant.*

Proof. Elementary computations show that if $f, h, h_1, \ldots, h_N$ are of forms (1)–(2), then $(f, h, h_1, \ldots, h_N)$ is a solution of (6), so left is that any solution $(f, h, h_1, \ldots, h_N)$ of (6) fits into (1)–(2). We consider two cases:

Case 1: Suppose that the set $\{f, \chi_1, \ldots, \chi_N\}$ is linearly dependent. Since $\chi_1, \ldots, \chi_N$ are different characters, then, according to Proposition 2, the set $\{\chi_1, \ldots, \chi_N\}$ is linearly independent. So there exist $\alpha_1, \ldots, \alpha_N \in \mathbb{C}$ such that

$$f = \sum_{j=1}^{N} \alpha_j \chi_j.$$

Substituting this in (6), we get that

$$\sum_{j=1}^{N} \alpha_j \chi_j(y)\chi_j(x) = \sum_{j=1}^{N} (\alpha_j h(y) + h_j(y))\chi_j(x)$$

for all $x, y \in G$. So, by using the linear independence of $\{\chi_1, \ldots, \chi_N\}$, we obtain $\alpha_j h(y) + h_j(y) = \alpha_j \chi_j(y)$ for all $y \in G$ and all $j = 1, \ldots, N$. Hence, $h_j = \alpha_j \chi_j - \alpha_j h$ for all $j = 1, \ldots, N$, with h arbitrary. The result occurs in part (1).

Case 2: Suppose that the set $\{f, \chi_1, \ldots, \chi_N\}$ is linearly independent.

Let $x, y, z \in G$ be arbitrary. We compute $f(zxy)$ by using the associativity of the operation of G. We have

$$f(z(xy)) = f(z)h(xy) + \sum_{j=1}^{N} \chi_j(z)h_j(xy). \tag{39}$$

On the other hand, by using (6), and that $\chi_1, \ldots, \chi_N$ are multiplicative, we obtain

$$\begin{aligned} f((zx)y) &= f(zx)h(y) + \sum_{j=1}^{N} \chi_j(zx)h_j(y) \\ &= \left[f(z)h(x) + \sum_{j=1}^{N} \chi_j(z)h_j(x) \right] h(y) + \sum_{j=1}^{N} \chi_j(zx)h_j(y) \\ &= f(z)h(x)h(y) + \sum_{j=1}^{N} \chi_j(z)[h_j(x)h(y) + \chi_j(x)h_j(y)]. \end{aligned} \tag{40}$$

Since the set $\{f, \chi_1, \ldots, \chi_N\}$ is linearly independent and x, y, z are arbitrary, we derive from (39) and (40) that

$$h(xy) = h(x)h(y) \tag{41}$$

and

$$h_j(xy) = h_j(x)h(y) + \chi_j(x)h_j(y) \tag{42}$$

for all $x, y \in G$ and all $j = 1, \ldots, N$. The functional equation (41) says that $\chi := h$ is a multiplicative function of G. If there exists $x_0 \in G$ such that $\chi(x_0) = 0$, then $h = \chi = 0$. So, by putting $y = e$ in (6), we get that $f = \sum_{j=1}^{N} h_j(e)\chi$, which contradicts that the set $\{f, \chi_1, \ldots, \chi_N\}$ is linearly independent. Thus, χ is a character of G and the functional equations (43) become

$$h_j(xy) = h_j(x)\chi(y) + \chi_j(x)h_j(y) \tag{43}$$

for all $x, y \in G$ and all $j = 1, \ldots, N$, i.e., $h_j = \varphi_{\chi, \chi_j}$ for all $j = 1, \ldots, N$.

To find f, we put $x = e$ in (6) which yields that $f = a\chi + \sum_{j=1}^{N} \varphi_{\chi, \chi_j}$, where $a := f(e)$. The result occurs in part (2). This completes the proof of Theorem 14. $\square$

5.2. *Solutions of Eq.* (2)

In Theorem 15, we solve the functional equation (2), i.e.,

$$f(xy) = g(x)h(y) + \sum_{j=1}^{N} \chi_j(x)h_j(y),\ x, y \in G,$$

where f, g, h, $h_1, \ldots, h_N : G \to \mathbb{C}$ are the unknown functions to be determined.

Theorem 15. *The solutions* $(f, g, h, h_1, \ldots, h_N)$ *of* (2) *are as follows:*

(1)

$$f = \sum_{j=1}^{N} a_j\chi_j,\ g \text{ arbitrary },h = 0 \text{ and } h_j = a_j\chi_j$$

for all $j = 1, \ldots, N$, *where* $a_1, \ldots, a_N \in \mathbb{C}$ *are constants;*

(2)

$$f = \sum_{j=1}^{N} a_j\chi_j,\ g = \sum_{j=1}^{N} \beta_j\chi_j,\ h \neq 0 \text{ arbitrary and } h_j = a_j\chi_j - \beta_j h$$

for all $j = 1, \ldots, N$, *where* $a_1, \ldots, a_N, \beta_1, \ldots, \beta_N \in \mathbb{C}$ *are constants;*

(3)

$$f = \alpha b\chi + \sum_{j=1}^{N} b\varphi_{\chi,\chi_j},\ g = \alpha\chi - \sum_{j=1}^{N} a_j\chi_j + \sum_{j=1}^{N} \varphi_{\chi,\chi_j},$$

$$h = b\chi \text{ and } h_j = a_j b\chi + b\varphi_{\chi,\chi_j}$$

for all $j = 1, \ldots, N$, *where* χ *is a character of* G, *and* $\alpha, a_1, \ldots, a_N \in \mathbb{C}$ *and* $b \in \mathbb{C} \setminus \{0\}$ *are constants.*

Proof. We check by elementary computations that if f, g, h, $h_1, \ldots, h_N$ are of forms (1)–(3), then $(f, g, h, h_1, \ldots, h_N)$ is a solution of (2), so left is that any solution $(f, g, h, h_1, \ldots, h_N)$ of (2) fits into (1)–(3).

We discuss two cases according to whether the set $\{f, \chi_1, \ldots, \chi_N\}$ is linearly independent or not.

Case 1: Suppose that the set $\{f, \chi_1, \ldots, \chi_N\}$ is linearly dependent. As seen earlier, the set $\{\chi_1, \ldots, \chi_N\}$ is linearly independent. Then there exist $a_1, \ldots, a_N \in \mathbb{C}$ such that

$$f = \sum_{j=1}^{N} a_j \chi_j. \tag{44}$$

There are two subcases to consider:

Subcase 1.1: Suppose that $h = 0$. Then g is arbitrary, and taking (44) into account, the functional equation (2) implies that

$$\sum_{j=1}^{N} a_j \chi_j(y)\chi_j(x) = \sum_{j=1}^{N} h_j(y)\chi_j(x)$$

for all $x, y \in G$ and all $j = 1, \ldots, N$. So, the set $\{\chi_1, \ldots, \chi_N\}$ being linearly independent, and x and y arbitrary, we get, according to Proposition 2, that $h_j = a_j \chi_j$ for all $j = 1, \ldots, N$. The result occurs in part (1).

Subcase 1.2: Suppose that $h \neq 0$. Then, we derive from Eq. (2) that there exist constants $\beta_1, \ldots, \beta_N \in \mathbb{C}$ such that

$$g = \sum_{j=1}^{N} \beta_j \chi_j. \tag{45}$$

By substituting (45) in (2), we obtain, by a small computation, that

$$\sum_{j=1}^{N} \chi_j(x)[a_j \chi_j(y) - \beta_j h(y) - h_j(y)] = 0$$

for all $x, y \in G$. Since the set $\{\chi_1, \ldots, \chi_N\}$ is linearly independent, and x and y arbitrary, we get, according to Proposition 2, that

$$h_j = a_j \chi_j - \beta_j h \tag{46}$$

for all $j = 1, \ldots, N$. The identities (44), (45), and (46), with $h \neq 0$ arbitrary, constitute result (2).

Case 2: Suppose that the set $\{f, \chi_1, \dots, \chi_N\}$ is linearly independent. By putting $y = e$ in the functional equation (2), we get that

$$f = h(e)g + \sum_{j=1}^{N} h_j(e)\chi_j. \tag{47}$$

Since the set $\{f, \chi_1, \dots, \chi_N\}$ is linearly independent, identity (47) imposes that $h(e) \neq 0$, which implies that

$$g = \frac{1}{h(e)} f - \sum_{j=1}^{N} \frac{h_j(e)}{h(e)} \chi_j. \tag{48}$$

Substituting this back into (2), we get, by elementary computation, that

$$f(xy) = f(x)\frac{h(y)}{h(e)} + \sum_{j=1}^{N} \left[h_j(y) - \frac{h_j(e)}{h(e)} h(y) \right] \chi_j(x) = 0$$

for all $x, y \in G$. Then the functions $f, h/h(e), h_1 - \frac{h_1(e)}{h(c)} h, \dots, h_N - \frac{h_N(c)}{h(e)} h$ satisfy the functional equation (6). Hence, according to Theorem 14 and seeing that the set $\{f, \chi_1, \dots, \chi_N\}$ is linearly independent, we derive that

$$f = a\chi + \sum_{j=1}^{N} \varphi_{\chi,\chi_j}, \; h/h(e) = \chi \text{ and } h_j - \frac{h_j(e)}{h(e)} h = \varphi_{\chi,\chi_j}$$

for all $j = 1, \dots, N$, where χ is a character of G and $a \in \mathbb{C}$ is a constant. Using this and identity (48), we get that

$$f = a\chi + \sum_{j=1}^{N} \varphi_{\chi,\chi_j},$$

$$g = \frac{a}{h(e)} \chi + \sum_{j=1}^{N} \frac{1}{h(e)} \varphi_{\chi,\chi_j} - \sum_{j=1}^{N} \frac{h_j(e)}{h(e)} \chi_j,$$

$$h = h(e)\chi$$

and

$$h_j = h_j(e)\chi + \varphi_{\chi,\chi_j}$$

for all $j = 1, \ldots, N$. Defining complex constants

$$b := h(e) \neq 0, \; \alpha := \frac{a}{h(e)} \quad \text{and} \quad a_j := \frac{h_j(e)}{h(e)},$$

and writing φ_{χ,χ_j} instead of $\frac{1}{b}\varphi_{\chi,\chi_j}$ for all $j = 1, \ldots, N$, the formulas above read

$$f = \alpha b\chi + \sum_{j=1}^{N} b\varphi_{\chi,\chi_j}, \; g = \alpha\chi - \sum_{j=1}^{N} a_j\chi_j + \sum_{j=1}^{N} \varphi_{\chi,\chi_j},$$

$$h = b\chi, \text{ and } h_j = a_j b\chi + b\varphi_{\chi,\chi_j}$$

for all $j = 1, \ldots, N$. The result occurs in part (3). This completes the proof. □

We close this section by relating our results to two functional equations in the literature. The proof of Theorem 15 reveals that if the set $\{f, \chi_1, \ldots, \chi_N\}$ is linearly independent, then the solutions $(f, g, h, h_1, \ldots, h_N)$ of the functional equation (2) are of form (3) in Theorem 15. This is an extension of Ref. [16, Theorem 14] as Example 16 says.

Example 17 relates Theorem 15(3) to the simple generalization of Prexider's functional equation

$$f(xy) = g_1(x)\chi(y) + \chi(x)h_1(y), \; x, y \in G,$$

where χ is a character of G (see Ref. [16, Proposition 15]).

Example 16. If we choice $a_j = 0$ and $\varphi_{\chi,\chi_j} = 0$ for all $j = 2, \ldots, N$ in part (3) of Theorem 15, we get that $h_j = 0$ for all $j = 2, \ldots, N$. With the notations $c_1 := \alpha b$, $c_2 := a_1 b - c_1$, and $f_{\chi,\chi_1} := c_1(\chi - \chi_1) + b\varphi_{\chi,\chi_1}$, we get that

$$f = c_1\mu + f_{\chi,\chi_1}, \; g = (f_{\chi,\chi_1} - c_2\mu)/b, \; h = b\chi, \quad \text{and} \quad h_1 = c_1\chi_1 + c_2\chi + f_{\chi,\chi_1},$$

where $b \in \mathbb{C} \setminus \{0\}$ and $c_1, c_2 \in \mathbb{C}$ are constants, and $f_{\chi,\chi_1} : G \to \mathbb{C}$ satisfies the functional equation

$$f_{\chi,\chi_1}(xy) = f_{\chi,\chi_1}(x)\chi(y) + \chi_1(x)f_{\chi,\chi_1}(y), \; x, y \in G,$$

which is the solution obtained in Ref. [16, Theorem 14].

Example 17. If we take $b = 1$, $\chi = \chi_1$, $a_j = 0$, and $\varphi_{\chi,\chi_j} = 0$ for all $j = 2, \dots, N$ in part (3) of Theorem 15, we get that $h_j = 0$ for all $j = 2, \dots, N$ and $\varphi_{\chi,\chi_1} = \varphi_{\chi,\chi} = \chi A$, where A is an additive function on G. With the notations $c_1 := \alpha - a_1$ and $c_2 := a_1$, we obtain

$$f = (c_1 + c_2 + A)\chi, \; g = (c_1 + A)\chi \quad \text{and} \quad h_1 = (c_2 + A)\chi,$$

where $c_1, c_2 \in \mathbb{C}$ are constants, which is the solution obtained in Ref. [16, Proposition 15].

6. Solutions of Eq. (7) and Eq. (3)

Throughout this section, G denotes a monoid with identity element e, $\mu, \chi : G \to \mathbb{C}$ different nonzero multiplicative functions, and $A : G \to \mathbb{C}$ an additive function such that $\chi A \neq 0$ and $g := \frac{\mu+\chi}{2}$.

6.1. *Solutions of Eq. (7)*

Proposition 18 solves the functional equation (7), i.e.,

$$f(xy) = f(x)h_1(y) + g(x)h_2(y) + \chi(x)A(x)h(y), \; x, y \in G,$$

under the assumption that the set $\{f, \mu, \chi, \chi A\}$ is linearly independent and $g = \frac{\mu+\chi}{2}$.

Proposition 18. *The solutions (f, h_1, h_2, h) of (7) such that the set $\{f, \mu, \chi, \chi A\}$ is linearly independent are as follows:*

$$f = cm, \; h_1 = m, \; h_2 = h = 0,$$

where m is a nonzero multiplicative function and $c \in \mathbb{C} \setminus \{0\}$ is a constant such that $m \neq \mu$ and $m \neq \chi$.

Proof. It is easy to check that the formulas in Proposition 18 define solutions of Eq. (7), so left is that any solution (f, h_1, h_2, h) is of that form.

Let $x, y, z \in G$ be arbitrary. First, we compute $f(xyz)$ as $f(x(yz))$ and then as $f((xy)z)$. Using Eq. (7) and that $g = \frac{\mu+\chi}{2}$, we obtain

$$f(x(yz)) = f(x)h_1(yz) + \frac{1}{2}\mu(x)h_2(yz) + \frac{1}{2}\chi(x)h_2(yz) + \chi(x)A(x)h(yz). \tag{49}$$

On the other hand, by making use that μ and χ are multiplicative and that A is additive, we get from Eq. (7) that

$$\begin{aligned} f((xy)z) &= f(xy)h_1(z) + g(xy)h_2(z) + \chi(xy)A(xy)h(z) \\ &= \left[f(x)h_1(y) + \frac{1}{2}\mu(x)h_2(y) + \frac{1}{2}\chi(x)h_2(y) + \chi(x)A(x)h(y)\right] \\ &\quad \times h_1(z) + \frac{1}{2}\mu(x)\mu(y)h_2(z) + \frac{1}{2}\chi(x)\chi(y)h_2(z) \\ &\quad + \chi(x)\chi(y)A(y)h(z) + \chi(x)A(x)\chi(y)h(z). \end{aligned}$$

So

$$\begin{aligned} f((xy)z) = f(x)h_1(y)h_1(z) &+ \frac{1}{2}\mu(x)\left[h_2(y)h_1(z) + \mu(y)h_2(z)\right] \\ &+ \frac{1}{2}\chi(x)\left[h_2(y)h_1(z) + \chi(y)h_2(z) + 2\chi(y)A(y)h(z)\right] \\ &+ \chi(x)A(x)\left[h(y)h_1(z) + \chi(y)h(z)\right]. \end{aligned} \tag{50}$$

As the set $\{f, \mu, \chi, \chi A\}$ is linearly independent and x, y, z are arbitrary, we derive from (49) and (50) that (h_1, h_2, h) is a solution of the following functional equations as a system:

$$h_1(xy) = h_1(x)h_1(y), \tag{51}$$

$$h_2(xy) = h_2(x)h_1(y) + \mu(x)h_2(y), \tag{52}$$

and

$$h_2(xy) = h_2(x)h_1(y) + \chi(x)h_2(y) + 2\chi(x)A(x)h(y) \tag{53}$$

for all $x, y \in G$. The functional equation (51) implies that $h_1 = m$, where $m : G \to \mathbb{C}$ is a multiplicative function. From (52) and (53), we get that $[\mu(x) - \chi(x)]h_2(y) = 2\chi(x)A(x)h(y)$ for all $x, y \in G$, from which we deduce that $h(y) = h_2(y) = 0$ for all $y \in G$, and then $h_2 = h = 0$. Hence, the functional equation (7) reduces to $f(xy) = f(x)m(y)$ for all $x, y \in G$, which yields, by putting $x = e$, that $f = cm$, where $c := f(e)$. If $c = 0$ or $m \in \{0, \mu, \chi\}$, then the set $\{f, \mu, \chi, \chi A\}$ is linearly dependent, contradicting the hypothesis. Thus, $c \in \mathbb{C} \setminus \{0\}$, $m \neq 0$, $m \neq \mu$, and $m \neq \chi$. □

Theorem 19. *The solutions (f, h_1, h_2, h) of (7) are as follows:*

(1) $f = cm$, $h_1 = m$, $h_2 = h = 0$, *where m is a nonzero multiplicative function and $c \in \mathbb{C} \setminus \{0\}$ is a constant such that $m \neq \mu$ and $m \neq \chi$;*
(2) $f = 0$, h_1 *is arbitrary, and* $h_2 = h = 0$*;*
(3)

$$f = a\mu + b\chi + c\chi A,\ h_1 = \frac{1}{a-b}(a\mu - b\chi - c\chi A),$$

$$h_2 = \frac{2a}{a-b}(-b\mu + b\chi + c\chi A),\ h = \frac{c}{a-b}(-a\mu + a\chi + c\chi A),$$

where $a, b, c \in \mathbb{C}$ are constants such that $(a, b, c) \neq (0, 0, 0)$ and $a \neq b$.

Proof. (1) If the set $\{f, \mu, \chi, \chi A\}$ is linearly independent, we get, by Proposition 18, part (1) of Theorem 19.

In the remainder of the proof, we assume that $\{f, \mu, \chi, \chi A\}$ is linearly dependent.

(2) If $f = 0$, then h_1 is arbitrary and the functional equation (7) implies that

$$\frac{h_2(y)}{2}\mu(x) + \frac{h_2(y)}{2}\chi(x) + h(y)(\chi A)(x) = 0$$

for all $x, y \in G$. So $h_2 = h = 0$. In what follows, we assume that $f \neq 0$.

(3) By using Lemma 3, we get that the set $\{\mu, \chi, \chi A\}$ is linearly independent. So, seeing that $f \neq 0$, there exists a triple $(a, b, c) \in \mathbb{C}^3 \setminus \{(0, 0, 0)\}$ such that $f = a\mu + b\chi + c\chi A$. Substituting this back

into (7), we get by a small computation that

$$\begin{aligned}(a\mu(y))\mu(x) + [b\chi(y) + c\chi(y)A(y)]\,\chi(x) + (c\chi(y))(\chi\,A)(x)\\ = \left[ah_1(y) + \frac{1}{2}h_2(y)\right]\mu(x) + \left[ah_1(y) + \frac{1}{2}h_2(y)\right]\chi(x)\\ + [ch_1(y) + h(y)]\,(\chi\,A)(x)\end{aligned}$$

for all $x, y \in G$. We derive that (h_1, h_2, h) is a solution of the following identities as a system:

$$ah_1 + \frac{1}{2}h_2 = a\mu, \tag{54}$$

$$bh_1 + \frac{1}{2}h_2 = b\chi + c\chi\,A, \tag{55}$$

and

$$ch_1 + h = c\chi. \tag{56}$$

When we subtract (55) from (54), we obtain

$$(a - b)h_1 = a\mu - b\chi - c\chi\,A. \tag{57}$$

We have $a \neq b$. Indeed, if $a = b$, the identity (57) reduces to $a\mu - b\chi - c\chi\,A = 0$. Hence, $a = b = c = 0$, which contradicts the assumption on f. Hence, the identity (57) implies that $h_1 = \frac{1}{a-b}(a\mu - b\chi - c\chi\,A)$. Substituting this back into (54) and (56), we derive that

$$h_2 = 2a(\mu - h_1) = 2a\left[\mu - \frac{1}{a-b}(a\mu - b\chi - c\chi\,A)\right] = \frac{2a}{a-b}(-b\mu + b\chi + c\chi\,A)$$

and

$$h = c(\chi - h_1) = c\left[\chi - \frac{1}{a-b}(a\mu - b\chi - c\chi\,A)\right] = \frac{c}{a-b}(-a\mu + a\chi + c\chi\,A).$$

Conversely, simple computations show that the formulas in (1)–(3) define solutions of (7). □

6.2. *Solutions of Eq.* (3)

In Theorem 20, we solve the functional equation (3), i.e.,

$$f(xy) = g_1(x)h_1(y) + g(x)h_2(y) + \chi(x)A(x)h(y), \quad x, y \in G,$$

where $f, g_1, h_1, h_2, h : G \to \mathbb{C}$ are the unknown functions to be determined.

Theorem 20. *The solutions* (f, g_1, h_1, h_2, h) *of* (3) *can be listed as follows:*

(1) $f = 0$, g_1 *is arbitrary, and* $h_1 = h_2 = h = 0$*;*

(2) $f = 0$, $g_1 = a\dfrac{\mu + \chi}{2} + b\chi A$, $h_1 \neq 0$ *is arbitrary,* $h_2 = -ah_1$*, and* $h = -bh_1$*, where* $a, b \in \mathbb{C}$ *are constants;*

(3)
$$f = (d_1 - d_2)\left(\frac{1}{2}a_1\mu - \frac{1}{2}a_2\chi - \frac{1}{2}a_3\chi A\right),\ g_1 = \frac{1}{2}d_1\mu + \frac{1}{2}d_2\chi + d_3\chi A,$$

$$h_1 = a_1\mu + a_2\chi + a_3\chi A,\ h_2 = -a_1d_2\mu - a_2d_1\chi - a_3d_1\chi A,$$

and

$$h = -a_1d_3\mu + \left(-a_2d_3 - \frac{1}{2}a_3d_1 + \frac{1}{2}a_3d_2\right)\chi - a_3d_3\chi A,$$

where $a_1, a_2, a_3, d_1, d_2, d_3 \in \mathbb{C}$ *are constants such that* $d_1 \neq d_2$ *and* $(a_1, a_2, a_3) \neq (0, 0, 0)$*;*

(4) $f = a_1\alpha m$, $g_1 = \alpha m - a_2\dfrac{\mu + \chi}{2} - a_3\chi A$, $h_1 = a_1 m$, $h_2 = a_1a_2m$, *and* $h = a_1a_3m$*, where* $\alpha, a_1 \in \mathbb{C}\setminus\{0\}$ *and* $a_2, a_3 \in \mathbb{C}$ *are constants, and* $m : G \to \mathbb{C}$ *is a nonzero multiplicative function.*

Proof. We check by elementary computations that if f, g_1, h_1, h_2, and h are of forms (1)–(4), then (f, g_1, h_1, h_2, h) is a solution of (3), so left is that any solution (f, g_1, h_1, h_2, h) of (3) fits into (1)–(4). There are two cases to consider:

Case 1: Suppose $f = 0$. By Proposition 7, we obtain solutions (1) and (2).

Case 2: Suppose $f \neq 0$. Taking $y = e$ in (3), we get that

$$f = h_1(e)g_1 + \frac{h_2(e)}{2}\mu + \frac{h_2(e)}{2}\chi + h(e)\chi A. \tag{58}$$

We split the discussion into the subcases $h_1(e) = 0$ and $h_1(e) \neq 0$:

Subcases 2.1: Suppose $h_1(e) = 0$. Then (58) gives

$$f = a\mu + a\chi + b\chi A, \tag{59}$$

where $a := \frac{h_2(e)}{2}$, $b := h(e)$, and $(a,b) \neq (0,0)$. Hence, the set $\{f, \mu, \chi, \chi A\}$ is linearly dependent. As $f \neq 0$, we get, according to Lemma 6(1), that $h_1 \neq 0$. So, by Lemma 6(2), the set $\{g_1, \mu, \chi, \chi A\}$ is linearly dependent. Since the set $\{\mu, \chi, \chi A\}$ is linearly independent, there exists a triple $(\alpha, \beta, \gamma) \in \mathbb{C}^3$ such that

$$g_1 = \alpha\mu + \beta\chi + \gamma\chi A. \tag{60}$$

By substituting (59) and (60) in (3), we derive by simple computations that

$$\begin{aligned}(a\mu(y))\mu(x) &+ [a\chi(y) + b\chi(y)A(y)]\chi(x) + (b\chi(y))(\chi A)(x)\\ &= \left[\alpha h_1(y) + \frac{1}{2}h_2(y)\right]\mu(x) + \left[\beta h_1(y) + \frac{1}{2}h_2(y)\right]\chi(x)\\ &\quad + [\gamma h_1(y) + h(y)]\,(\chi A)(x)\end{aligned}$$

for all $x, y \in G$. So, the set $\{\mu, \chi, \chi A\}$ being linearly independent, we derive from the identity above that (h_1, h_2, h) satisfies the following identities as a system:

$$\alpha h_1 + \frac{1}{2}h_2 = a\mu,$$

$$\beta h_1 + \frac{1}{2}h_2 = a\chi + b\chi A,$$

and

$$\gamma h_1 + h = b\chi.$$

The first and the second identities imply that $(\alpha - \beta)h_1 = a\mu - a\chi - b\chi A$. Since $(a,b) \neq (0,0)$ and $h_1 \neq 0$, we have $\alpha - \beta \neq 0$, so

$$h_1 = \frac{a}{\alpha - \beta}\mu - \frac{a}{\alpha - \beta}\chi - \frac{b}{\alpha - \beta}\chi A. \tag{61}$$

Substituting this in the first identity, we get that

$$h_2 = 2a\mu - \frac{2\alpha a}{\alpha - \beta}\mu + \frac{2\alpha a}{\alpha - \beta}\chi + \frac{2\alpha b}{\alpha - \beta}\chi A,$$

hence,

$$h_2 = -\frac{2\beta a}{\alpha - \beta}\mu + \frac{2\alpha a}{\alpha - \beta}\chi + \frac{2\alpha b}{\alpha - \beta}\chi\, A. \tag{62}$$

The third identity implies that

$$h = b\chi - \frac{a}{\alpha - \beta}\mu - \frac{a}{\alpha - \beta}\chi - \frac{b}{\alpha - \beta}\chi\, A,$$

so

$$h = -\frac{\gamma a}{\alpha - \beta}\mu + \left[b + \frac{\gamma a}{\alpha - \beta}\right]\chi + \frac{\gamma b}{\alpha - \beta}\chi\, A. \tag{63}$$

Let

$$d_1 := 2\alpha,\ d_2 := 2\beta,\ d_3 := \gamma$$

$$a_1 := \frac{2a}{d_1 - d_2},\ a_2 := -\frac{2a}{d_1 - d_2},\ \text{and}\ a_3 := -\frac{2b}{d_1 - d_2}.$$

Note that $d_1 \neq d_2$ and $(a_1, a_2, a_3) \neq (0, 0, 0)$ because $\alpha - \beta \neq 0$ and $(a, b) \neq (0, 0)$. From (59), (60), (61), and (62), we get that

$$f = (d_1 - d_2)\left(\frac{1}{2}a_1\mu - \frac{1}{2}a_2\chi - \frac{1}{2}a_3\chi\, A\right),\ g_1 = \frac{1}{2}d_1\mu + \frac{1}{2}d_2\chi + d_3\chi\, A,$$

$$h_1 = a_1\mu + a_2\chi + a_3\chi\, A,$$

and

$$h_2 = -a_1 d_2\mu - a_2 d_1\chi - a_3 d_1\chi\, A.$$

From (63), we obtain

$$\begin{aligned} h &= -a_1 d_3\mu + \left[-\frac{1}{2}(d_1 - d_2)a_3 + a_1 d_3\right]\chi - a_3 d_3\chi\, A \\ &= -a_1 d_3\mu + \left(-a_2 d_3 - \frac{1}{2}a_3 d_1 + \frac{1}{2}a_3 d_2\right)\chi - a_3 d_3\chi\, A, \end{aligned}$$

which is solution (3).

Subcases 2.2: Suppose $h_1(e) \neq 0$. Then, from (58), we get that

$$g_1 = \frac{1}{h_1(e)} f - \frac{h_2(e)}{h_1(e)} g - \frac{h(e)}{h_1(e)} \chi A. \tag{64}$$

Substituting this back into (3), we obtain

$$f(xy) = \left[\frac{1}{h_1(e)} f(x) - \frac{h_2(e)}{h_1(e)} g(x) - \frac{h(e)}{h_1(e)} \chi(x) A(x)\right] h_1(y) + g(x) h_2(y) + \chi(x) A(x) h(y)$$

for all $x, y \in G$. So

$$f(xy) = f(x) \left(\frac{1}{h_1(e)} h_1\right)(y) + \left(h_2 - \frac{h_2(e)}{h_1(e)} h_1\right)(y) g(x) + \chi(x) A(x) \left(h - \frac{h(e)}{h_1(e)} h_1\right)(y)$$

for all $x, y \in G$, i.e., the quadruple $\left(f, h_1/h_1(e), h_2 - \frac{h_2(e)}{h_1(e)} h_1, h - \frac{h(e)}{h_1(e)} h_1\right)$ is a solution of the functional equation (7), so, according to Theorem 19, it falls into two categories:

(i) $f = cm$, $h_1/h_1(e) = m$, $h_2 - \frac{h_2(e)}{h_1(e)} h_1 = 0$, $h - \frac{h(e)}{h_1(e)} h_1 = 0$,

where m is a nonzero multiplicative function and $c \in \mathbb{C} \setminus \{0\}$ is a constant such that $m \neq \mu$ and $m \neq \chi$. We define complex constants by $\alpha := \frac{c}{h_1(e)} \in \mathbb{C} \setminus \{0\}$, $a_1 := h_1(e) \in \mathbb{C} \setminus \{0\}$, $a_2 := \frac{h_2(e)}{h_1(e)}$, and $a_3 := \frac{h(e)}{h_1(e)}$. So, the identities above and (64) yield that
$f = a_1 \alpha m$, $g_1 = \alpha m - a_2 \frac{\mu + \chi}{2} - a_3 \chi A$, $h_1 = a_1 m$, $h_2 = a_1 a_2 m$, and $h = a_1 a_3 m$, which is solution (4).

(ii)

$$f = a\mu + b\chi + c\chi A, \quad h_1/h_1(e) = \frac{1}{a - b}(a\mu - b\chi - c\chi A),$$

$$h_2 - \frac{h_2(e)}{h_1(e)} h_1 = \frac{2a}{a - b}(-b\mu + b\chi + c\chi A),$$

$$h - \frac{h(e)}{h_1(e)} h_1 = \frac{c}{a - b}(-a\mu + a\chi + c\chi A),$$

where $a, b, c \in \mathbb{C}$ such that $(a, b, c) \neq (0, 0, 0)$ and $a \neq b$. Taking (64) into account and using the identities above, we get, by simple computations, that

$$\begin{aligned} f &= a\mu + b\chi + c\chi A, \\ g_1 &= \frac{2a - h_2(e)}{2h_1(e)}\mu + \frac{2b - h_2(e)}{2h_1(e)}\chi + \frac{c - h(e)}{h_1(e)}\chi A, \\ h_1 &= \frac{ah_1(e)}{a-b}\mu - \frac{bh_1(e)}{a-b}\chi - \frac{ch_1(e)}{a-b}\chi A, \\ h_2 &= -\frac{a(2b - h_2(e))}{a-b}\mu + \frac{b(2a - h_2(e))}{a-b}\chi + \frac{c(2a - h_2(e))}{a-b}\chi A, \\ h &= -\frac{a(c - h(e))}{a-b}\mu + (c + \frac{b(c - h(e))}{a-b})\chi + \frac{c(c - h(e))}{a-b}\chi A. \end{aligned} \tag{65}$$

Defining

$$a_1 := \frac{ah_1(e)}{a-b},\ a_2 := -\frac{bh_1(e)}{a-b},\ a_3 := -\frac{ch_1(e)}{a-b},$$
$$d_1 := \frac{2a - h_2(e)}{h_1(e)},\ d_2 := \frac{2b - h_2(e)}{h_1(e)},\ \text{and } d_3 := \frac{c - h(e)}{h_1(e)},$$

we have $d_1 \neq d_2$ because $a \neq b$, $(a_1, a_2, a_3) \neq (0, 0, 0)$ because $(a, b, c) \neq (0, 0, 0)$, $c = -\frac{1}{2}a_3d_1 + \frac{1}{2}a_3d_2$, then (65) reads

$$f = (d_1 - d_2)(\frac{1}{2}a_1\mu - \frac{1}{2}a_2\chi - \frac{1}{2}a_3\chi A),\ g_1 = \frac{1}{2}d_1\mu + \frac{1}{2}d_2\chi + d_3\chi A,$$
$$h_1 = a_1\mu + a_2\chi + a_3\chi A,\ h_2 = -a_1d_2\mu - a_2d_1\chi - a_3d_1\chi A,$$

and

$$h = -a_1d_3\mu + \left(-a_2d_3 - \frac{1}{2}a_3d_1 + \frac{1}{2}a_3d_2\right)\chi - a_3d_3\chi A,$$

which is solution (3). □

Remark 21. If $f, g_1, h_1, h_2, h : G \to \mathbb{C}$ satisfy the functional equation (3), the formulas in parts (3) and (4) of Theorem 20 reveal that if $f \neq 0$, the functions f, g_1, h_1, h_2, and h are abelian.

Example 22. If we choose $a_3 = d_3 = 0$ in Theorem 20(3) or $a_3 = 0$ in Theorem 20(4), we get that $h = 0$, then the functional equation (3) reduces to $f(xy) = g_1 h_1(y) + g(x)h_2(y)$, $x, y \in G$, which was solved in Ref. [12, Theorem 8], and we obtain the following:
(a) For $a_3 = d_3 = 0$ in Theorem 20(3),

$$f = (d_1 - d_2)\left(\frac{1}{2}a_1\mu - \frac{1}{2}a_2\chi\right),\ g_1 = \frac{1}{2}d_1\mu + \frac{1}{2}d_2\chi$$

and

$$h_1 = a_1\mu + a_2\chi,\ h_2 = -a_1 d_2\mu - a_2 d_1\chi,$$

which is the solution obtained in Ref. [12, Theorem 8(b)].
(b) For $a_3 = 0$ in Theorem 20(4),

$$f = a_1\alpha m,\ g_1 = \alpha m - a_2\frac{\mu+\chi}{2},\ h_1 = a_1 m, \quad \text{and} \quad h_2 = a_1 a_2 m,$$

where $\alpha, a_1, a_2 \in \mathbb{C}$ are constants and $m : G \to \mathbb{C}$ is a nonzero multiplicative function such that $\alpha \neq 0$ and $a_1 \neq 0$, which is the solution obtained in Ref. [12, Theorem 8(a)].

Example 23. By taking $d_1 = -d_2$, $a_1 = a_2 = \frac{1}{2}$, and $a_3 = d_3 = 0$ in Theorem 20(3), we get that

$$f = g_1 = h_2 = d_1\frac{\mu - \chi}{2} \quad \text{and} \quad h_2 = \frac{\mu+\chi}{2}, \quad \text{where} \quad d_1 \in \mathbb{C}\setminus\{0\},$$

which is the solution of the classic sine addition law in case $\mu \neq \chi$ obtained in Ref. [15, Theorem 4.1(c)].

7. Solutions of Eq. (8) and Eq. (4)

Throughout this section, G denotes a group with identity element e, $\mu, \chi : G \to \mathbb{C}^*$ different characters, and $A : G \to \mathbb{C}$ a nonzero additive function. We deal with the functional equations (8) and (4).

7.1. *Solutions of Eq.* (8)

In Theorem 24, we solve the functional equation (8), i.e.,

$$f(xy) = f(x)h_1(y) + \mu(x)h_2(y) + \chi(x)A(x)h(y),\ x, y \in G,$$

where f, h_1, h_2, h are the unknown functions to be determined.

Theorem 24. *The solutions* (f, h_1, h_2, h) *of* (8) *are as follows:*

(1) $f = a\mu$, h_1 *arbitrary,* $h_2 = a\mu - ah_1$, *and* $h = 0$, *where* a *ranges over* $\mathbb{C}$;
(2) $f = a\mu_1 + \varphi_{\mu_1,\mu}$, $h_1 = \mu_1$, $h_2 = \varphi_{\mu_1,\mu}$, *and* $h = 0$, *where* μ_1 *is a character of* G *and* a *ranges over* $\mathbb{C}$;
(3) $f = a\mu + b\chi + b\alpha\chi A$, $h_1 = \chi + \alpha\chi A$, $h_2 = a\mu - a\chi - a\alpha\chi A$, *and* $h = -\alpha^2 b\chi A$, *where* a *and* α *range over* $\mathbb{C}$, *and* b *over* $\mathbb{C} \setminus \{0\}$.

Proof. Elementary computations show that if f, h_1, h_2, and h are of forms (1)–(3), then (f, h_1, h_2, h) is a solution of (8), so left is that any solution (f, h_1, h_2, h) of (8) fits into (1)–(3).

If $f = 0$, then h_1 is arbitrary, and the functional equation (8) becomes $\mu(x)h_2(y) + \chi(x)A(x)h(y) = 0$ for all $x, y \in G$. Since the set $\{\mu, \chi A\}$ is linearly independent, we get that $h_2 = h = 0$, which occurs in (1) for $a = 0$. In the rest of the proof, we assume that $f \neq 0$. We consider two cases:

Case 1: Suppose that the set $\{f, \mu, \chi, \chi A\}$ is linearly independent. Let $x, y, z \in G$ be arbitrary. We compute $f(zxy)$ by using the associativity of the operation of G. We have

$$f(z(xy)) = f(z)h_1(xy) + \mu(z)h_2(xy) + \chi(z)A(z)h(xy). \tag{66}$$

On the other hand, by using (8), that μ and χ are multiplicative and A is additive, we get that

$$\begin{aligned} f((zx)y) &= f(zx)h_1(y) + \mu(zx)h_2(y) + \chi(zx)A(zx)h(y) \\ &= [f(z)h_1(x) + \mu(z)h_2(x) + \chi(z)A(z)h(x)]h_1(y) \\ &\quad + \mu(zx)h_2(y) + \chi(zx)A(zx)h(y) \\ &= f(z)h_1(x)h_1(y) + \mu(z)[h_2(x)h_1(y) + \mu(x)h_2(y)] \\ &\quad + \chi(z)\chi(x)A(x)h(y) + \chi(z)A(z)[h(x)h_1(y) + \chi(x)h(y)]. \end{aligned} \tag{67}$$

Since the set $\{f,\mu,\chi,\chi A\}$ is linearly independent and x,y,z are arbitrary, we derive from (66) and (67) that

$$h_1(xy)=h_1(x)h_1(y), \tag{68}$$

$$h_2(xy)=h_2(x)h_1(y)+\mu(x)h_2(y), \tag{69}$$

$$\chi(x)A(x)h(y)=0, \tag{70}$$

and

$$h(xy)=h(x)h_1(y)+\chi(x)h(y) \tag{71}$$

for all $x,y\in G$. The functional equation (68) says that $\mu_1:=h_1$ is a multiplicative function on G. If there exists $x_0\in G$ such that $\mu_1(x_0)=0$, then $\mu_1=0$. So, by putting $y=e$ in (8), we get that $f=h_2(e)\mu+h(e)\chi A$, which contradicts that the set $\{f,\mu,\chi,\chi A\}$ is linearly independent. Thus, μ_1 is a character of G.

Since χ is a character of G and A is an additive function on G, we derive from (70) that $h=0$.

Taking into account that $h_1=\mu_1$, the functional equation (69) becomes $h_2(xy)=h_2(x)\mu_1(y)+\mu(x)h_2(y)$ for all $x,y\in G$, i.e., $h_2=\varphi_{\mu_1,\mu}$.

To find f, we put $x=e$ in (8) which yields that $f=a\mu_1+\varphi_{\mu_1,\mu}$, where $a:=f(e)$. The result occurs in part (2).

Case 2: Suppose that the set $\{f,\mu,\chi,\chi A\}$ is linearly dependent. Since $\{\mu,\chi,\chi A\}$ is linearly independent and $f\neq 0$, there exists a triple $(a,b,c)\in\mathbb{C}^3\setminus\{(0,0,0)\}$ such that

$$f=a\mu+b\chi+c\chi A. \tag{72}$$

Combining (8) and (72), we get by a small computation that

$$f(xy)=(ah_1(y)+h_2(y))\mu(x)+bh_1(y)\chi(x)+(ch_1(y)+h(y))\chi(x)A(x) \tag{73}$$

for all $x,y\in G$.

On the other hand, (72) gives

$$f(xy)=a\mu(y)\mu(x)+(b\chi(y)+c\chi(y)A(y))\chi(x)+c\chi(y)\chi(x)A(x) \tag{74}$$

for all $x,y\in G$. Since the set $\{\mu,\chi,\chi A\}$ is linearly independent, we get from (73) and (74) that

$$ah_1+h_2=a\mu, \tag{75}$$

$$bh_1=b\chi+c\chi A, \tag{76}$$

and

$$ch_1 + h = c\chi. \tag{77}$$

If $b = 0$, then we get from (76) that $c = 0$. So, (72) and (77) imply that $f = a\mu$ and $h = 0$. From (75), we obtain $h_2 = a\mu - ah_1$ with h_1 arbitrary. The result occurs in part (1).

If $b \neq 0$, then we derive from (75), (76), and (77) that

$$h_1 = \chi + \frac{c}{b}\chi A, \; h_2 = a\mu - a\chi - \frac{ac}{b}\chi A, \; h = -\frac{c^2}{b}\chi A.$$

In (72) and the identities above, we put $\alpha = \frac{c}{b}$, which yields that

$$f = a\mu + b\chi + b\alpha\chi A, \; h_1 = \chi + \alpha\chi A, \; h_2 = a\mu - a\chi - a\alpha\chi A, \; h = -\alpha^2 b\chi A.$$

The result occurs in part (3). □

7.2. *Solutions of Eq.* (4)

In Theorem 25, we solve the functional equation (4), i.e.,

$$f(xy) = g_1(x)h_1(y) + \mu(x)h_2(y) + \chi(x)A(x)h(y), \; x, y \in G,$$

where f, g_1, h_1, h_2, h are the unknown functions to be determined.

Theorem 25. *The solutions (f, g_1, h_1, h_2, h) of (4) are as follows:*

(1) $f = a\mu$, *g_1 arbitrary,* $h_1 = 0$, $h_2 = a\mu$, *and* $h = 0$, *where a ranges over $\mathbb{C}$;*
(2) $f = a\mu$, $g_1 = -b\mu + c\chi A$, $h_1 \neq 0$ *arbitrary,* $h_2 = a\mu + bh_1$, *and* $h = -ch_1$, *where a, b, and c range over $\mathbb{C}$;*
(3) $f = a\mu + b\chi A$, $g_1 = \alpha\mu + \beta\chi + \gamma\chi A$, $h_1 = c\chi A$, $h_2 = a\mu - \alpha c\chi A$, *and* $h = b\chi - \gamma c\chi A$, *where a, α, β, γ, b, and c are constants such that $\beta bc \neq 0$ and $b = \beta c$;*
(4) $f = a_1\mu + \varphi_{\mu_1,\mu}$, $g_1 = (\varphi_{\mu_1,\mu} - a_2\mu)/a + a_3\chi A$, $h_1 = a\mu_1$, $h_2 = a_1\mu + a_2\mu_1 + \varphi_{\mu_1,\mu}$, *and* $h = -aa_3\mu_1$, *where μ_1 is a character of G, a_1, a_2, and a_3 range over $\mathbb{C}$, and a over $\mathbb{C} \setminus \{0\}$;*
(5) $f = a\mu + ca_2\chi + \alpha ca_2\chi A$, $g_1 = a_1\mu + a_2\chi + a_3\chi A$, $h_1 = c\chi + \alpha c\chi A$, $h_2 = a\mu - ca_1\chi - \alpha ca_1\chi A$, *and* $h = c(\alpha a_2 - a_3)\chi - \alpha ca_3\chi A$, *where α, a, a_1, and a_3 range over $\mathbb{C}$, and c and a_2 over $\mathbb{C} \setminus \{0\}$.*

Proof. We check by elementary computations that if f, g_1, h_1, h_2, and h are of forms (1)–(5), then (f, g_1, h_1, h_2, h) is a solution of (4), so left is that any solution (f, g_1, h_1, h_2, h) of (4) fits into (1)–(5). There are two cases to consider:

Case 1: f is proportional to μ. By Proposition 8, we get solutions (1) and (2).

Case 2: f is not proportional to μ. By putting $y = e$ in the functional equation (4), we get that

$$f = h_1(e)g_1 + h_2(e)\mu + h(e)\chi A. \tag{78}$$

We consider two subcases according to whether $h_1(e) = 0$ or not:

Subcase 2.1: $h_1(e) = 0$. Then (78) becomes

$$f = a\mu + b\chi A, \tag{79}$$

where $a := h_2(e)$ and $b := h(e)$. So the set $\{f, \mu, \chi, \chi A\}$ is linearly dependent. As f is not proportional to μ, we derive, according to Lemma 6(1), that $h_1 \neq 0$. So, by Lemma 6(2), the set $\{g_1, \mu, \chi, \chi A\}$ is linearly dependent. As the set $\{\mu, \chi, \chi A\}$ is linearly independent, we deduce that there exists a triple $(\alpha, \beta, \gamma) \in \mathbb{C}^3$ such that

$$g_1 = \alpha\mu + \beta\chi + \gamma\chi A. \tag{80}$$

Let $y \in G$ be arbitrary. By substituting (79) and (80) in (4), we obtain

$$\begin{aligned} &a\mu(x)\mu(y) + b\chi(x)\chi(y)A(y) + b\chi(x)A(x)\chi(y) \\ &\quad = (\alpha\mu(x) + \beta\chi(x) + \gamma\chi(x)A(x))h_1(y) + \mu(x)h_2(y) + \chi(x)A(x)h(y) \\ &\quad = \mu(x)(\alpha h_1(y) + h_2(y)) + \chi(x)(\beta h_1(y)) + \chi(x)A(x)(\gamma h_1(y) + h(y)) \end{aligned}$$

for all $x \in G$. Since the set $\{\mu, \chi, \chi A\}$ is linearly independent, we derive from the identity above that $\alpha h_1(y)+h_2(y) = a\mu(y)$, $\beta h_1(y) = b\chi(y)A(y)$, and $\gamma h_1(y) + h(y) = b\chi(y)$. So, y being arbitrary, we deduce that

$$\alpha h_1 + h_2 = a\mu,\ \beta h_1 = b\chi A \quad \text{and} \quad \gamma h_1 + h = b\chi. \tag{81}$$

Since f is not proportional to μ, we get from (79) that $b \neq 0$. Hence, the second identity in (81) implies that $\beta \neq 0$ because $\chi A \neq 0$. So, by putting $c = \frac{b}{\beta}$, we derive from (81) that

$$h_1 = c\chi A,\ h_2 = a\mu - \alpha c\chi A \quad \text{and} \quad h = b\chi - \gamma c\chi A.$$

The result occurs in part (3).

Subcase 2.2: $h_1(e) \neq 0$. Here we get from (78) that

$$g_1 = \frac{1}{h_1(e)} f - \frac{h_2(e)}{h_1(e)} \mu - \frac{h(e)}{h_1(e)} \chi A. \tag{82}$$

Substituting this back into (4), we get, by elementary computation, that

$$f(xy) = f(x)\frac{h_1(y)}{h_1(e)} + \mu(x)\left(h_2(y) - \frac{h_2(e)}{h_1(e)} h_1(y)\right) + \chi(x)A(x)(h(y) - \frac{h(e)}{h_1(e)} h_1(y))$$

for all $x, y \in G$. Then the quadruple $(f, h_1/h_1(e), h_2 - \frac{h_2(e)}{h_1(e)} h_1, h - \frac{h(e)}{h_1(e)} h_1)$ satisfies the functional equation (8), hence, according to Theorem 24 and seeing that f is not proportional to μ, that quadruple falls into two categories:

(i) $f = a_1\mu_1 + \varphi_{\mu_1,\mu}$, $h_1/h_1(e) = \mu_1$, $h_2 - \frac{h_2(e)}{h_1(e)} h_1 = \varphi_{\mu_1,\mu}$, and $h - \frac{h(e)}{h_1(e)} h_1 = 0$, where μ_1 is a character of G and a_1 ranges over $\mathbb{C}$. Then,

$$f = a_1\mu_1 + \varphi_{\mu_1,\mu},\ h_1 = h_1(e)\mu_1,\ h_2 = h_2(e)\mu_1 + \varphi_{\mu_1,\mu}, \quad \text{and} \quad h = h(e)\mu_1.$$

So, (82) gives

$$g_1 = (a_1\mu_1 + \varphi_{\mu_1,\mu} - h_2(e)\mu)/h_1(e) - \frac{h(e)}{h_1(e)} \chi A.$$

Defining

$$a := h_1(e) \neq 0,\ a_2 := h_2(e) - a_1, \quad \text{and} \quad a_3 := -\frac{h(e)}{h_1(e)}$$

and writing $\varphi_{\mu_1,\mu}$ instead of $a_1(\mu_1 - \mu) + \varphi_{\mu_1,\mu}$, the formulas above read

$$f = a_1\mu + \varphi_{\mu_1,\mu},\ g_1 = (\varphi_{\mu_1,\mu} - a_2\mu)/a + a_3\chi A,\ h_1 = a\mu_1,$$
$$h_2 = a_1\mu + a_2\mu_1 + \varphi_{\mu_1,\mu}, \quad \text{and} \quad h = -aa_3\mu_1.$$

Moreover, since f is not proportional to μ, $\varphi_{\mu_1,\mu}$ is nonzero. The result obtained occurs in part (4).
(ii) $f = a\mu + b\chi + \alpha b\chi A$, $h_1/h_1(e) = \chi + \alpha\chi A$, $h_2 - \frac{h_2(e)}{h_1(e)}h_1 = a\mu - a\chi - \alpha a\chi A$, and $h - \frac{h(e)}{h_1(e)}h_1 = -\alpha^2 b\chi A$, where a and α range over $\mathbb{C}$, and b over $\mathbb{C}\setminus\{0\}$. Then,

$$h_1 = h_1(e)\chi + \alpha h_1(e)\chi A,\ h_2 = a\mu + (h_2(e) - a)\chi + \alpha(h_2(e) - a)\chi A,\\ \text{and}\quad h = h(e)\chi + \alpha(h(e) - \alpha b)\chi A. \tag{83}$$

So, by using (82), we get that

$$g_1 = \frac{a - h_2(e)}{h_1(e)}\mu + \frac{b}{h_1(e)}\chi + \frac{\alpha b - h(e)}{h_1(e)}\chi A. \tag{84}$$

Defining $c := h_1(e) \neq 0$, $a_1 := \frac{a-h_2(e)}{h_1(e)}$, $a_2 := \frac{b}{h_1(e)} \neq 0$, and $a_3 := \frac{\alpha b - h(e)}{h_1(e)}$, we get, by using the formula of f above, (83), and (84), that

$$f = a\mu + ca_2\chi + \alpha ca_2\chi A,\ g_1 = a_1\mu + a_2\chi + a_3\chi A,\ h_1 = c\chi + \alpha c\chi A,\\ h_2 = a\mu - ca_1\chi - \alpha ca_1\chi A,\quad \text{and}\quad h = c(\alpha a_2 - a_3)\chi - \alpha ca_3\chi A.$$

The result occurs in part (5). □

Example 26. If we choice $a_3 = 0$ in part (4) of Theorem 25, we get that $h = 0$ and

$$f = a_1\mu + \varphi_{\mu_1,\mu},\ g_1 = (\varphi_{\mu_1,\mu} - a_2\mu)/a,\ h_1 = a\mu_1 \quad \text{and}\\ h_2 = a_1\mu + a_2\mu_1 + \varphi_{\mu_1,\mu},$$

with the same constraints on μ_1, $\varphi_{\mu_1,\mu}$, a, a_1, and a_2, which is the solution obtained in Ref. [16, Theorem 14].

Example 27. If we choice $\alpha = a_3 = 0$ in part (5), we get that $h = 0$ and

$$f = a\mu + ca_2\chi,\ g_1 = a_1\mu + a_2\chi,\ h_1 = c\chi, \quad \text{and} \quad h_2 = a\mu - ca_1.$$

With the notations $c_1 := a + cc_2$, $c_2 := -ca_1 - ca_2$, $\lambda := ca_2 \neq 0$, and $\widetilde{f} := \lambda(\chi - \mu)$, the formulas above read

$$f = c_1\mu + \widetilde{f},\ g_1 = \left(\widetilde{f} - c_2\mu\right)/c,\ h_1 = c\chi, \quad \text{and} \quad h_2 = c_1\mu + c_2\chi + \widetilde{f},$$

which is a solution obtained in Ref. [16, Theorem 14] corresponding to the nonzero central solutions $\widetilde{f} = \lambda(\chi - \mu)$ of the functional equation $\widetilde{f}(xy) = \widetilde{f}(x)\chi(y) + \mu(x)\widetilde{f}(y)$, $x, y \in G$.

We close with non-examples. On finite monoids or groups, the only additive function is $A = 0$, so the functional equations (3) and (4) become $f(xy) = g_1(x)h_1(y) + g(x)h_2(y)$, $x, y \in G$, where G is a monoid, g is the average of two distinct nonzero multiplicative functions μ_1 and μ_2 on G, and

$$f(xy) = g_1(x)h_1(y) + \mu(x)h_2(y), \quad x, y \in G, \tag{85}$$

where G is a group and μ is a character of G, which was recently solved in Refs. [12] and [15].

On the other hand, on perfect groups, i.e., groups G such that $G = [G, G]$, the only character is $\chi = 1$, so the function $g = \frac{\mu_1+\mu_2}{2}$ cannot exist. As examples of perfect groups, we cite connected groups and semisimple Lie groups like $SO(n)$, $SL(n, \mathbb{R})$, and $SL(n, \mathbb{C})$ for $n \geq 2$.

References

[1] Aczél, J. *Lectures on functional equations and their applications. Mathematics in Sciences and Engineering.* Academic Press, New York-London, 19 (1966).

[2] Ajebbar, O. and Elqorachi, E. The Cosine-Sine functional equation on a semigroup with an involutive automorphism, *Aequ. Math.* 91(6), 1115–1146 (2017).

[3] Andreescu, T., Boreico, I., Mushkarov, O. and Nicolov, N. *Topics in functional equations.* 2nd Edition. XYZ Press, LLC (2015).

[4] Belfakih, K. and Elqorachi, E. A note on the Levi-Civita functional equation, *Aequ. Math.* 93, 1275–1291 (2017).

[5] Chung, J.K., Kannappan, Pl. and Ng, C.T. A generalization of the cosine-sine functional equation on groups, *Linear Algebra Appl.* 66, 259–277 (1985).

[6] Ebanks, B. Around the sine addition law and d'Alembert's equation on semi- groups, *Results Math.* 77(11), 14 (2022).

[7] Ebanks, B. The cosine and sine addition and subtraction formulas on semigroups, *Acta Math. Hungar.* 165, 337–354 (2021).

[8] Ebanks, B. The cosine-sine functional equation on semigroups, *Annales Math. Silesianae.* 36(1), 30–52 (2022).

[9] Ebanks, B. A Levi-Cevita equation on monoids, Two ways, *Annales Math. Silesianae.* 36(2), 151–166 (2022), DOI: 10.2478/amsil-2022-0009.

[10] Ebanks, B., Some Levi-Civita functional equations on semigroups, *Results Math.* 77(154), 19 (2022).

[11] Ebanks, B. An extension of the sine addition formula on groups and semigroups, *Publ. Math. Debrecen.* 93 (1–2), 9–27 (2018).

[12] Ebanks, B. and Stetkær, H. Extensions of the sine addition formula on monoids, *Results Math.* 73, 13 (2018). https://doi.org/10.1007/s00025-018-0880-z.

[13] Levi-Civita, T. Sulle funzioni che ammettono una formula d'addizione del tipo$f(x+y) = \sum_{i=1}^{n} X_i(x)Y_i(y)$, *(Italian) Rom. Acc. L. Rend.* (5) 22(2), 181–183, (1913).

[14] Shulman, E. Group representations and stability of functional equations, *J. London Math. Soc.* 54, 111–120 (1996).

[15] Stetkær, H. *Functional Equations on Groups.* World Scientific, Singapore, (2013).

[16] Stetkær, H. Extensions of the sine addition law on groups, *Aequ. Math.* 93, 467–484, (2019).

[17] L. Székelyhidi, *Convolution Type Functional Equations on Topological Abelian Groups.* World Scientific Publishing Co., Inc., Teaneck, NJ (1991).

[18] L. Székelyhidi, *On the Levi-Civita functional equation.* (Berichte der Mathematisch-Statistischen Sektion in der Forschungsgesellschaft Joanneum, 301. Forschungszentrum Graz, Mathematisch-Statistische Sektion, Graz, 1988).

https://doi.org/10.1142/9789811293009_0002

Chapter 2

No Ulam Stability, but Modified Ulam Stability of First-Order q-Difference Equations with Periodic Coefficient

Douglas R. Anderson

Department of Mathematics
Concordia College
Moorhead, MN 56562 USA
andersod@cord.edu

First, we show that first-order linear q-difference equations with 1-periodic or 2-periodic coefficients are not Ulam stable. Second, we establish the best (minimum) constant for modified Ulam stability of first-order linear q-difference equations with a 1-periodic coefficient. Third, we establish the best (minimum) constant for modified Ulam stability of first-order linear q-difference equations with a 2-periodic coefficient and give several examples.

1. Introduction

Ulam [34] initiated a new approach to stability concerning linear operators and approximately linear operators, which was partially addressed by Hyers [18] and generalized by Rassias [31]. From this beginning, Ulam stability, or Hyers–Ulam stability, or Hyers–Ulam–Rassias stability, has had a robust growth into the context of operators, functional equations, differential equations, difference and q-difference equations, and dynamic equations on time scales. For Ulam stability and functional equations, some representative papers include Abdollahpour and Rassias [1], Abdollahpour, Aghayaria, and Rassias [2], Elqorachi and Rassias [15], Jung, Popa, and Rassias [20], Jung and Rassias [21], Lee, Jung, and Rassias [22]. Works such as Brillouët–Belluot, Brzdęk, and Ciepliński [12] offer a nice topical survey of the literature. Other, recent works include Brzdęk, Popa, Raşa, and Xu [13]. Specific to Ulam stability in the discrete domain, Popa had some of the earliest papers [28, 29]. Contemporary works include András and Mészáros [8], Brzdęk and Wójcik [14], Hua, Li, and Feng [17], Jung and Nam [19], Nam [23–25], Shen [32, 33], Rasouli, Abbaszadeh, and Eshaghi [30], and Anderson and Onitsuka [3–6], covering Ulam stability via recurrences, difference equations, quantum equations, or dynamic equations on time scales, respectively.

Very little work has been done in the area of Ulam stability and q-difference equations with periodic coefficients, for $q > 1$. We employ the definition of periodicity introduced by Bohner and Chieochan [9]; see also Bohner, Mesquita, and Streipert [10]. In what follows, we start by showing that under the usual definition of Ulam stability, the q-difference equation with 1-periodic coefficient is not Ulam stable. Due to that, we define what modified Ulam stability is in the context of first-order q-difference equations with a periodic coefficient and establish parameter values in terms of the periodic coefficient for which the equations exhibit modified Ulam stability. In the case of modified Ulam stability, best constants will be found in the sense of the minimum constant needed for an approximate solution (perturbation) to track close to a specific solution of the said equation. We begin with the easier case of a 1-periodic coefficient, followed by the case of a 2-periodic coefficient. These results set the stage for researchers exploring second- and higher-order q-difference equations with periodicity in the coefficients.

Remark 1. Ulam stability for non-autonomous quantum equations (q-difference equations) has been studied by Anderson and Onitsuka [6]. However, as we see in the following, if the non-autonomous coefficient is periodic in the sense of Refs. [9, 10], the results do not apply, as the periodic coefficient does not meet the criteria established in Ref. [6, Proposition 2.3].

2. Best Constant for First-Order q-Equations with Period-1 Coefficient

Let $q > 1$, and define the discrete set $\mathbb{T} = q^{\mathbb{N}_0} := \{1, q, q^2, q^3, \ldots\}$. In this section, we consider on $\mathbb{T}$ the Ulam stability and modified Ulam stability of the first-order linear homogeneous q-difference equation with period-one coefficient:

$$\mathcal{D}_q z(t) - p(t)z(t) = 0, \qquad \mathcal{D}_q z(t) := \frac{z(qt) - z(t)}{(q-1)t}, \tag{1}$$

where $p : \mathbb{T} \to \mathbb{C}$ is 1-periodic [9, Definition 3.1] if

$$p(t) = qp\,(qt), \qquad \forall t \in \mathbb{T}. \tag{2}$$

Let $p(1) = p_1$ for $p_1 \in \mathbb{C}\backslash\{\frac{-1}{q-1}\}$. Through iteration on (2), one sees that p is given by

$$p(t) := \frac{p(1)}{t} = \frac{p_1}{t}. \tag{3}$$

From (3), we see that $\liminf |p(t)| = 0$, so Ref. [6, Proposition 2.3] does not apply here.

Remark 2 (Exponential function). For $q > 1$, let $p(1) = p_1$ for $p_1 \in \mathbb{C}\backslash\{\frac{-1}{q-1}\}$. Set

$$e_p(t) := \prod_{j=0}^{\log_q t - 1} \left(1 + (q-1)q^j p(q^j)\right), \quad \text{where} \quad \prod_{\ell=0}^{-1} f(\ell) \equiv 1.$$

By the 1-periodic nature of p given in (2) and (3), we have

$$e_p(t) = [1 + (q-1)p_1]^{\log_q t}. \tag{4}$$

Note that the condition $p_1 \in \mathbb{C}\backslash\{\frac{-1}{q-1}\}$ is really a regressivity condition, since $p(1) = p_1 \neq \frac{-1}{q-1}$ is equivalent to $1 + \mu(1)p_1 \neq 0$. See Ref. [11, Definition 2.25] for regressivity on general time scales.

For notational convenience throughout the remainder of this section, define the base of the exponential function to be

$$A := 1 + (q-1)p_1. \tag{5}$$

Remark 3. Equation (1) and the results of this section can be easily extended to the Cayley q-difference equation

$$\mathcal{D}_q z(t) - p(t)\left\langle z(t)\right\rangle_\beta = 0, \quad \left\langle z(t)\right\rangle_\beta := \beta z(qt) + (1-\beta)z(t), \quad 0 \le \beta \le 1.$$

In this more general case, the exponential function and the constant A are expressed via

$$e_p(t) := A^{\log_q t} \quad \text{and} \quad A = \frac{1 + (1-\beta)(q-1)p_1}{1 - \beta(q-1)p_1},$$

respectively. The regressivity conditions become $1 + (1-\beta)(q-1)p_1 \neq 0 \neq 1 - \beta(q-1)p_1$. If $\beta = 0$, we recover (1).

Definition 4 (Ulam and modified Ulam stability). Equation (1) has Ulam stability if and only if there exists a constant $L > 0$ such that, for arbitrary $\varepsilon > 0$, if a function $\psi : \mathbb{T} \to \mathbb{C}$ satisfies

$$|\mathcal{D}_q \psi(t) - p(t)\psi(t)| \le \varepsilon, \qquad t \in \mathbb{T}, \tag{6}$$

then there exists a solution $z : \mathbb{T} \to \mathbb{C}$ of (1) such that $|\psi(t) - z(t)| \le L\varepsilon$ for all $t \in \mathbb{T}$. Equation (1) has modified Ulam stability if and only if there exists a constant $L > 0$ such that, for arbitrary $\varepsilon > 0$, if a function $\psi : \mathbb{T} \to \mathbb{C}$ satisfies

$$|\mathcal{D}_q \psi(t) - p(t)\psi(t)| \le \frac{\varepsilon}{(q-1)t}, \qquad t \in \mathbb{T}, \tag{7}$$

then there exists a solution $z : \mathbb{T} \to \mathbb{C}$ of (1) such that $|\psi(t) - z(t)| \le L\varepsilon$ for all $t \in \mathbb{T}$. In either case, such a constant L is called a Ulam stability constant for (1) on $\mathbb{T}$.

Remark 5. In the definition of modified Ulam stability above, in particular in (7), we typically find just the constant ε on the right-hand side of the inequality, as in (6). According to Ref. [10, Theorem 5.1], the role of constant functions in the classical sense is assumed by 1-periodic functions on isolated time scales, where a function f is 1-periodic on $\mathbb{T}$ if and only if there exists a real constant c such that $f(t) = \frac{c}{\mu(t)}$ for all $t \in \mathbb{T}$. For $\mathbb{T} = q^{\mathbb{N}_0}$, as in this chapter, we point out that $\mu(t) = (q-1)t$. Therefore, the function $f(t) = \frac{\varepsilon}{(q-1)t}$ in (7) is a 1-periodic function for all $t \in \mathbb{T}$.

Theorem 6 (No Ulam stability). *Assume $q > 1$, p is as in* (3), *$p_1 \in \mathbb{C}\backslash\{\frac{-1}{q-1}\}$, and A is as in* (5). *Then,* (1) *is not Ulam stable.*

Proof. Since $\liminf |p(t)| = 0$ for p given in (2), the Ulam stability result [6, Proposition 2.3] for non-autonomous quantum equations does not apply here, as that result requires the non-autonomous coefficient function p to satisfy $\liminf |p(t)| > 0$. Let $\varepsilon > 0$ be a fixed arbitrary constant throughout the proof, which we split according to the following cases:

(1) if $|A| = q$, then (1) is not Ulam stable;
(2) if $|A| \neq q$, then (1) is not Ulam stable.

(i) Suppose $|A| = q$, for A given in (5). By the definition of Ulam stability given in Definition 4, we need to find a function $\psi : \mathbb{T} \to \mathbb{C}$ fulfilling (6), such that there is no solution z of (1) satisfying $|\psi(t) - z(t)| \leq L\varepsilon$ for all $t \in \mathbb{T}$, for any constant $L \in \mathbb{R}$. For a function ψ to fulfill (6), there exists a (perturbation) function $\eta : \mathbb{T} \to \mathbb{C}$ such that

$$\mathcal{D}_q\psi(t) - p(t)\psi(t) = \eta(t), \quad |\eta(t)| \leq \varepsilon \tag{8}$$

for all $t \in \mathbb{T}$. Let

$$\eta(t) = \frac{\varepsilon A^{\log_q t}}{|A|^{\log_q t}} = \frac{\varepsilon A^{\log_q t}}{t}$$

so that $|\eta(t)| = \varepsilon$ for all $t \in \mathbb{T}$, as $|A| = q$ in this case (i). The variation of constants formula [11, Theorem 2.77] applied to (8) yields

$$\psi(t) = \psi(1)e_p(t) + e_p(t) \sum_{\tau=0}^{\log_q t-1} \frac{(q-1)q^\tau \eta(q^\tau)}{e_p\left(q^{\tau+1}\right)},$$

$$\text{assuming} \quad \sum_{\tau=0}^{-1} f(\tau) \equiv 0. \tag{9}$$

With the η given above, and noting that $e_p(t) = A^{\log_q t}$ in this case, we have from (9) that

$$\psi(t) = A^{\log_q t} \sum_{\tau=0}^{\log_q t-1} \frac{(q-1)q^\tau \varepsilon A^\tau}{q^\tau A^{\tau+1}} = (q-1)\varepsilon \log_q t A^{\log_q t-1}$$

by taking $\psi(1) = 0$. Then, ψ satisfies (8) for all $t \in \mathbb{T}$, but

$$\begin{aligned}|\psi(t) - z(t)| &= \left|(q-1)\varepsilon \log_q t A^{\log_q t-1} - cA^{\log_q t}\right| \\ &= t\left|(q-1)\varepsilon A^{-1} \log_q t - c\right| \longrightarrow \infty\end{aligned}$$

as $t \to \infty$ for any choice of $c \in \mathbb{C}$. Thus, in this case of $|A| = q > 1$, equation (1) is not Ulam stable.

(ii) Suppose $|A| \neq q$, for A given in (5). Utilizing (9) with $\eta(t) \equiv \varepsilon$ for all $t \in \mathbb{T}$,

$$\begin{aligned}\psi(t) = A^{\log_q t} \sum_{\tau=0}^{\log_q t-1} \frac{(q-1)q^\tau \varepsilon}{A^{\tau+1}} &= \frac{(q-1)\varepsilon\left(A^{\log_q t} - q^{\log_q t}\right)}{A-q} \\ &= \frac{(q-1)\varepsilon\left(A^{\log_q t} - t\right)}{A-q},\end{aligned}$$

by again taking $\psi(1) = 0$ in (9). Then, ψ satisfies (8) for all $t \in \mathbb{T}$. If $|A| > q$, then

$$\begin{aligned}|\psi(t) - z(t)| &= \left|\frac{(q-1)\varepsilon\left(A^{\log_q t} - t\right)}{A-q} - cA^{\log_q t}\right| \\ &= \left|\left(\frac{(q-1)\varepsilon}{A-q} - c\right) A^{\log_q t} - \frac{(q-1)\varepsilon t}{A-q}\right|, \quad t = q^n \\ &= |A|^n \left|\left(\frac{(q-1)\varepsilon}{A-q} - c\right) - \frac{(q-1)\varepsilon q^n}{(A-q)A^n}\right| \longrightarrow \infty\end{aligned}$$

as $t \to \infty$ for any choice of $c \in \mathbb{C}$. If $|A| < q$, then

$$\begin{aligned}|\psi(t) - z(t)| &= \left| \left(\frac{(q-1)\varepsilon}{A-q} - c \right) A^{\log_q t} - \frac{(q-1)\varepsilon t}{A-q} \right|, \quad t = q^n \\ &= q^n \left| \left(\frac{(q-1)\varepsilon}{A-q} - c \right) \left(\frac{A}{q} \right)^n - \frac{(q-1)\varepsilon}{A-q} \right| \longrightarrow \infty\end{aligned}$$

as $t \to \infty$ for any choice of $c \in \mathbb{C}$. Thus, in either case of $|A| \neq q$, equation (1) is not Ulam stable. Overall, equation (1) has no Ulam stability in the case that the coefficient function p is 1-periodic. □

Remark 7. As a consequence of Theorem 6, we see that (1) with 1-periodic coefficient function p is always unstable in the Ulam sense. This is an important observation for readers concerned with perturbations of (1). By using the definition of modified Ulam stability given in Definition 4, however, the stronger assumption on the perturbation η given there, essentially that it decays according to the graininess μ, does allow modified stability results, as we see in the remainder of this chapter.

Theorem 8 (Modified Ulam stability). *Assume $q > 1$, p is given by (2) with $p_1 \in \mathbb{C}\backslash\left\{\frac{-1}{q-1}\right\}$, and e_p is given by (4). Let A be as in (5), $\varepsilon > 0$ be a fixed arbitrary constant, and let the function $\psi : \mathbb{T} \to \mathbb{C}$ satisfy inequality (7). One of the following holds:*

(1) *If $|A| > 1$, then equation (1) is modified Ulam stable, $\lim_{t\to\infty} \frac{\psi(t)}{e_p(t)}$ exists, and the function z given by*

$$z(t) := \left(\lim_{t\to\infty} \frac{\psi(t)}{e_p(t)} \right) e_p(t)$$

is the unique solution of (1) with $|\psi(t) - z(t)| \leq L_1\varepsilon$ for all $t \in \mathbb{T}$, where

$$L_1 := \frac{1}{|A| - 1} \tag{10}$$

is the minimum Ulam stability constant for (1).

(2) *If $|A| = 1$, then (1) is not modified Ulam stable.*

(3) *If* $|A| < 1$, *then equation* (1) *is modified Ulam stable, and any solution* z *of* (1) *with*

$$|\psi(1) - z(1)| < \varepsilon\left(\frac{1}{1-|A|}\right)$$

satisfies

$$|\psi(t) - z(t)| < \varepsilon\left(\frac{1}{1-|A|}\right)$$

for all $t \in \mathbb{T}$.

Proof. Fix $q > 1$, and refer to the definition of modified Ulam stability in Definition 4. Assume $p_1 \in \mathbb{C}\backslash\{\frac{-1}{q-1}\}$. Let $\varepsilon > 0$ be an arbitrary constant. In the cases of the proof that follow, since we are assuming $\psi : \mathbb{T} \to \mathbb{C}$ is a function fulfilling (7), there exists a function $\eta : \mathbb{T} \to \mathbb{C}$ such that

$$\mathcal{D}_q\psi(t) - p(t)\psi(t) = \eta(t), \quad |\eta(t)| \leq \frac{\varepsilon}{(q-1)t} \tag{11}$$

for all $t \in \mathbb{T}$.

(i) First, we consider the case $|A| = |1 + (q-1)p_1| > 1$. From (9), we have

$$\left|\frac{\psi(t)}{e_p(t)}\right| \leq |\psi(1)| + \varepsilon \sum_{\tau=0}^{\log_q t - 1} \frac{1}{|e_p(q^{\tau+1})|},$$

as $|\eta(q^\tau)| \leq \frac{\varepsilon}{(q-1)q^\tau}$ for all $t = q^\tau \in \mathbb{T}$ by the assumption in (11). Note that

$$\sum_{\tau=0}^{\infty} \frac{1}{|e_p(q^{\tau+1})|} = \frac{1}{|A|} + \frac{1}{|A|^2} + \frac{1}{|A|^3} + \cdots = \frac{1}{|A|-1}$$

converges as a geometric series in this case of $1 < |A|$. Thus, ψ can be rewritten as

$$\psi(t) = \left[\psi(1) + \sum_{\tau=0}^{\infty} \frac{(q-1)q^\tau\eta(q^\tau)}{e_p(q^{\tau+1})}\right] e_p(t) - e_p(t) \sum_{\tau=\log_q t}^{\infty} \frac{(q-1)q^\tau\eta(q^\tau)}{e_p(q^{\tau+1})}, \tag{12}$$

where

$$z_1 := \psi(1) + \sum_{\tau=0}^{\infty} \frac{(q-1)q^\tau \eta(q^\tau)}{e_p\left(q^{\tau+1}\right)} \in \mathbb{C}$$

exists as a finite number due to the definition of e_p in (4), and the assumption $|A| > 1$. Clearly

$$z(t) := z_1 e_p(t), \quad t \in \mathbb{T}$$

is a solution of (1), and

$$\lim_{t\to\infty} \frac{\psi(t)}{e_p(t)} = \psi(1) + \sum_{\tau=0}^{\infty} \frac{(q-1)q^\tau \eta(q^\tau)}{e_p\left(q^{\tau+1}\right)} = z_1$$

exists. Consequently,

$$z(t) = \left(\lim_{t\to\infty} \frac{\psi(t)}{e_p(t)}\right) e_p(t),$$

and

$$\begin{aligned} |\psi(t) - z(t)| &= \left| -e_p(t) \sum_{\tau=\log_q t}^{\infty} \frac{(q-1)q^\tau \eta(q^\tau)}{e_p\left(q^{\tau+1}\right)} \right| \\ &\leq \varepsilon |e_p(t)| \sum_{\tau=\log_q t}^{\infty} \frac{1}{\left|e_p\left(q^{\tau+1}\right)\right|} = \varepsilon \left(\frac{1}{|A|-1}\right) \end{aligned}$$

holds for all $t \in \mathbb{T}$. We have again used a geometric series and the fact that $|A| > 1$. Therefore, (1) has modified Ulam stability with Ulam constant L_1 given by (10).

Next, we show that in case (i), z given above is the unique solution of (1) such that $|\psi(t) - z(t)| \leq L_1\varepsilon$ for all $t \in \mathbb{T}$. Suppose $\psi : \mathbb{T} \to \mathbb{C}$ satisfies (7) for some $\varepsilon > 0$. Further, suppose that $z_1, z_2 : \mathbb{T} \to \mathbb{C}$ are two different solutions of (1) such that $|\psi(t) - z_k(t)| \leq L_1\varepsilon$ for all $t \in \mathbb{T}$, for $k = 1, 2$. Then we have for constants $c_k \in \mathbb{C}$ that

$$z_k(t) = c_k e_p(t), \quad c_1 \neq c_2$$

and

$$|e_p(t)||c_1 - c_2| = |z_1(t) - z_2(t)| \leq |z_1(t) - \psi(t)| + |\psi(t) - z_2(t)| \leq 2L_1\varepsilon.$$

If we let $t \to \infty$, then $\infty < 2L_1\varepsilon$, a contradiction. As a result, z is the unique solution of (1) such that $|\psi(t) - z(t)| \leq L_1\varepsilon$ for all $t \in \mathbb{T}$.

Finally, we show that in case (i), L_1 in (10) is the best (minimum) Ulam constant possible. In (11), if

$$\eta(t) = \frac{\varepsilon e_p(qt)}{\mu(t)|e_p(qt)|} \tag{13}$$

for $t \in \mathbb{T}$, then (9) and (12) imply the function $\psi_* : \mathbb{T} \to \mathbb{C}$ given by

$$\psi_*(t) := \left(\psi_*(1) + \sum_{\tau=0}^{\infty} \frac{\varepsilon}{|e_p(q^{\tau+1})|}\right) e_p(t) - e_p(t) \sum_{\tau=\log_q t}^{\infty} \frac{\varepsilon}{|e_p(q^{\tau+1})|}$$

satisfies the equality

$$|\mathcal{D}_q\psi_*(t) - p(t)\psi_*(t)| = \left|\frac{\varepsilon e_p(qt)}{\mu(t)|e_p(qt)|}\right| = \frac{\varepsilon}{\mu(t)}, \qquad t \in \mathbb{T}.$$

As

$$z(t) := \left(\psi_*(1) + \sum_{\tau=0}^{\infty} \frac{\varepsilon}{|e_p(q^{\tau+1})|}\right) e_p(t)$$

is a solution of (1),

$$|\psi_*(t) - z(t)| = \varepsilon|e_p(t)| \sum_{\tau=\log_q t}^{\infty} \frac{1}{|e_p(q^{\tau+1})|} = \varepsilon\left(\frac{1}{|A|-1}\right) = L_1\varepsilon$$

holds for all $t \in \mathbb{T}$. As a result, the entirety of the claims in (i) holds. ♠

(ii) We claim that if $|A| = |1 + (q-1)p_1| = 1$, then (1) is not modified Ulam stable. In this case, we have

$$|e_p(t)| = |A|^{\log_q t}$$

from (4) and the assumption $|A| = 1$. If $\varepsilon > 0$ is arbitrary, let η satisfy (13), and let $\psi^\dagger$ be defined by

$$\psi^\dagger(t) := e_p(t) \sum_{\tau=0}^{\log_q t-1} \frac{\mu(q^\tau)\eta(q^\tau)}{e_p(q^{\tau+1})} = e_p(t) \sum_{\tau=0}^{\log_q t-1} \frac{\varepsilon}{|e_p(q^{\tau+1})|}, \quad t \in \mathbb{T}.$$

Then, $\psi^\dagger$ satisfies the equality

$$\left|\mathcal{D}_q\psi^\dagger(t)-p(t)\psi^\dagger(t)\right|=|\eta(t)|=\left|\frac{\varepsilon e_p(qt)}{\mu(t)|e_p(qt)|}\right|=\frac{\varepsilon}{\mu(t)},\qquad t\in\mathbb{T}.$$

Since $z(t)=ce_p(t)$ is the general solution of (1), then

$$\left|\psi^\dagger(t)-z(t)\right|=\left|\sum_{\tau=0}^{\log_q t-1}\frac{\varepsilon}{|e_p(q^{\tau+1})|}-c\right||e_p(t)|=\left|\sum_{\tau=0}^{\log_q t-1}\varepsilon-c\right|\to\infty$$

as $t\to\infty$ for $t\in\mathbb{T}$ and for any $c\in\mathbb{C}$. Consequently, in this case, (1) is not modified Ulam stable on $\mathbb{T}$. ♠

(iii) Now assume $0<|1+(q-1)p_1|<1$, that is, $0<|A|<1$. As before, we know that a function ψ that satisfies (7) takes the form

$$\psi(t)=\psi(1)e_p(t)+e_p(t)\sum_{\tau=0}^{\log_q t-1}\frac{\mu(q^\tau)\eta(q^\tau)}{e_p(q^{\tau+1})}$$

by (9). Let z be any solution of (1) satisfying the inequality

$$|\psi(1)-z(1)|<\varepsilon\left(\frac{1}{1-|A|}\right).$$

It follows that z has the form

$$z(t)=z(1)e_p(t),\quad t\in\mathbb{T},$$

and we have

$$\psi(t)-z(t)=\big(\psi(1)-z(1)\big)e_p(t)+e_p(t)\sum_{\tau=0}^{\log_q t-1}\frac{\mu(q^\tau)\eta(q^\tau)}{e_p(q^{\tau+1})}.$$

Consequently, we have

$$\begin{aligned}|\psi(t)-z(t)|&\le|\psi(1)-z(1)||e_p(t)|+\varepsilon|e_p(t)|\sum_{\tau=0}^{\log_q t-1}\frac{1}{|e_p(q^{\tau+1})|}\\&<\varepsilon\left(\frac{1}{1-|A|}\right)|e_p(t)|+\varepsilon|e_p(t)|\sum_{\tau=0}^{\log_q t-1}\frac{1}{|e_p(q^{\tau+1})|}.\end{aligned}$$

From (4), we note for $n \in \mathbb{N}$ that $|e_p(t)| = |A|^n$ if $t = q^n$, and

$$|e_p(t)| \sum_{\tau=0}^{\log_q t-1} \frac{1}{|e_p(q^{\tau+1})|} = \frac{1-|A|^n}{1-|A|}.$$

Combining the above calculations, we have

$$|\psi(t) - z(t)| < \varepsilon \left(\frac{|A|^n + 1 - |A|^n}{1-|A|} \right) = \frac{\varepsilon}{1-|A|}$$

for all $t \in \mathbb{T}$. Therefore, (iii) holds and the proof is complete. □

3. Best Constant for First-Order q-Equations with Period-2 Coefficient

In this section, we consider on $\mathbb{T} = q^{\mathbb{N}_0}$ for $q > 1$ the modified Ulam stability of the first-order linear homogeneous q-difference equation with 2-periodic coefficient

$$\mathcal{D}_q z(t) - p(t)z(t) = 0, \qquad \mathcal{D}_q z(t) := \frac{z(qt) - z(t)}{(q-1)t}, \tag{14}$$

where $p : \mathbb{T} \to \mathbb{C}$ is 2-periodic [9, Definition 3.1] if

$$p(t) = q^2 p(q^2 t), \qquad \forall t \in \mathbb{T}. \tag{15}$$

Let $p(1) = p_1$ and $p(q) = p_q$, for $p_1 \in \mathbb{C}\backslash\{\frac{-1}{q-1}\}$ and $p_q \in \mathbb{C}\backslash\{\frac{-1}{q(q-1)}\}$. Through iteration on (15), one sees that p is given by

$$p(t) := \frac{1}{t} \begin{cases} p_1 & \text{if} \quad \log_q t \equiv 0 \pmod 2, \\ qp_q & \text{if} \quad \log_q t \equiv 1 \pmod 2. \end{cases} \tag{16}$$

Remark 9 (Exponential function). For $q > 1$, let $p(1) = p_1$ and $p(q) = p_q$, for $p_1 \in \mathbb{C}\backslash\{\frac{-1}{q-1}\}$ and $p_q \in \mathbb{C}\backslash\{\frac{-1}{q(q-1)}\}$. Set

$$e_p(t) := \prod_{j=0}^{\log_q t-1} (1 + (q-1)q^j p(q^j)), \quad \text{where} \quad \prod_{j=0}^{-1} f(j) \equiv 1. \quad (17)$$

By the 2-periodic nature of p given in (15) and (16), we have

$$e_p(t) = \begin{cases} [1 + (q-1)p_1]^{\frac{1}{2}\log_q t} \\ [1 + (q-1)qp_q]^{\frac{1}{2}\log_q t} \end{cases} \text{if} \quad \log_q t \equiv 0 \pmod 2, \\ \begin{cases} [1 + (q-1)p_1]^{\frac{1}{2}(\log_q t+1)} \\ [1 + (q-1)qp_q]^{\frac{1}{2}(\log_q t-1)} \end{cases} \text{if} \quad \log_q t \equiv 1 \pmod 2. \quad (18)$$

From (16), we see that $\liminf |p(t)| = 0$, so Ref. [6, Proposition 2.3] does not apply here. Note that the conditions $p_1 \in \mathbb{C}\backslash\{\frac{-1}{q-1}\}$ and $p_q \in \mathbb{C}\backslash\{\frac{-1}{q(q-1)}\}$ are really regressivity conditions, since $p(1) = p_1 \neq \frac{-1}{q-1}$ is equivalent to $1 + \mu(1)p_1 \neq 0$, and $p(q) = p_q \neq \frac{-1}{q(q-1)}$ is equivalent to $1 + \mu(q)p_q \neq 0$. See Ref. [11, Definition 2.25] for regressivity on general time scales.

For notational convenience throughout the remainder of this section, define the key constants as

$$A := 1 + (q-1)p_1, \qquad B := 1 + q(q-1)p_q. \quad (19)$$

It is left to the reader to check that e_p indeed satisfies (14), with $e_p(1) = 1$.

The results in this section are different from the periodic results for constant step size by Anderson, Onitsuka, and Rassias [7] and from the results by Fukutaka and Onitsuka [16] given for first-order homogeneous linear differential equations with a periodic coefficient, due to the interesting implications of the definition of periodicity on isolated time scales and our modified Ulam stability definition.

Theorem 10 (No Ulam stability 2-periodic). *Assume* $q > 1$, $p_1 \in \mathbb{C}\backslash\{\frac{-1}{q-1}\}$, $p_q \in \mathbb{C}\backslash\{\frac{-1}{q(q-1)}\}$. *Let* p *be as in* (16), *and* A, B *be as in* (19). *Then,* (14) *is not Ulam stable.*

Proof. Since $\liminf |p(t)| = 0$ for p given in (16), the Ulam stability result [6, Proposition 2.3] for non-autonomous quantum equations does not apply here, as that result requires the non-autonomous coefficient function p to satisfy $\liminf |p(t)| > 0$. The assumptions on p_1, p_q imply that A, B as in (19) satisfy $A, B \neq 0$. Let $\varepsilon > 0$ be a fixed arbitrary constant throughout the proof, which we split according to the following cases:

(1) if $|AB| = q^2$, then (14) is not Ulam stable;
(2) if $|AB| \neq q^2$, then (14) is not Ulam stable.

(i) Suppose $|AB| = q^2$, for A, B given in (19). By the definition of Ulam stability given in Definition 4, we need to find a function $\psi : \mathbb{T} \to \mathbb{C}$ fulfilling (6), such that there is no solution z of (14) satisfying $|\psi(t) - z(t)| \leq L\varepsilon$ for all $t \in \mathbb{T}$, for any constant $L \in \mathbb{R}$. For a function ψ to fulfill (6), there exists a function $\eta : \mathbb{T} \to \mathbb{C}$ such that

$$\mathcal{D}_q\psi(t) - p(t)\psi(t) = \eta(t), \quad |\eta(t)| \leq \varepsilon \tag{20}$$

for all $t \in \mathbb{T}$. Let

$$\eta(t) = \frac{\varepsilon e_p(qt)}{|e_p(qt)|} \tag{21}$$

so that $|\eta(t)| = \varepsilon$ for all $t \in \mathbb{T}$. The variation of constants formula [11, Theorem 2.77] applied to (20) yields

$$\psi(t) = \psi(1)e_p(t) + e_p(t) \sum_{\tau=0}^{\log_q t-1} \frac{(q-1)q^\tau \eta(q^\tau)}{e_p\left(q^{\tau+1}\right)},$$

$$\text{assuming} \quad \sum_{\tau=0}^{-1} f(\tau) \equiv 0. \tag{22}$$

Note that for $|AB| = q^2$,

$$|e_p(t)| = t \begin{cases} 1 & \text{if} \quad \log_q t \equiv 0 \pmod 2 \\ \frac{|A|}{q} & \text{if} \quad \log_q t \equiv 1 \pmod 2, \end{cases}$$

and, with the η given above, we have from (22) and $|AB| = q^2$ that

$$\begin{aligned} \psi(t) &= \varepsilon(q-1)e_p(t) \sum_{\tau=0}^{\log_q t-1} \frac{q^\tau}{|e_p(q^{\tau+1})|} \\ &= \varepsilon(q-1)e_p(t) \begin{cases} \frac{1}{2}\left(\frac{1}{|A|} + \frac{1}{q}\right) \log_q t & \text{if} \quad t \equiv 0 \pmod 2 \\ \left(\frac{\log_q t+1}{2|A|} + \frac{\log_q t-1}{2q}\right) & \text{if} \quad t \equiv 1 \pmod 2 \end{cases} \end{aligned}$$

by taking $\psi(1) = 0$. Then, ψ satisfies (8) for all $t \in \mathbb{T}$, but

$$\begin{aligned} |\psi(t) - z(t)| &= \left| \varepsilon(q-1)e_p(t) \sum_{\tau=0}^{\log_q t-1} \frac{q^\tau}{|e_p(q^{\tau+1})|} - ce_p(t) \right| \\ &= t \begin{cases} \left| \frac{\varepsilon(q-1)}{2}\left(\frac{1}{|A|} + \frac{1}{q}\right) \log_q t - c \right| & \text{if} \quad t \equiv 0 \pmod 2 \\ \frac{|A|}{q}\left| \frac{\varepsilon(q-1)}{2}\left(\frac{\log_q t+1}{|A|} + \frac{\log_q t-1}{q}\right) - c \right| & \text{if} \quad t \equiv 1 \pmod 2 \end{cases} \\ &\longrightarrow \infty \end{aligned}$$

as $t \to \infty$ for any choice of $c \in \mathbb{C}$. Thus, in this case of $|AB| = q^2$, equation (14) is not Ulam stable.

(ii) Suppose $|AB| \neq q^2$, for A, B given in (19). Utilizing (22) with $\eta(t)$ given by (21) for all $t \in \mathbb{T}$,

$$\psi(t) = \varepsilon(q-1)e_p(t) \sum_{\tau=0}^{\log_q t-1} \frac{q^\tau}{|e_p(q^{\tau+1})|},$$

by again taking $\psi(1) = 0$. Then, ψ satisfies (8) for all $t \in \mathbb{T}$. If $t = q^{2n}$ and $|AB| > q^2$, then

$$\begin{aligned}
|\psi(t) - z(t)| &= \left|\psi\left(q^{2n}\right) - z\left(q^{2n}\right)\right| \\
&= \left|\frac{\varepsilon(q-1)(q+|B|)\left(\frac{AB}{|AB|}\right)^n}{|AB| - q^2}\left(|AB|^n - q^{2n}\right) - c(AB)^n\right| \\
&= \left|\left(\frac{\varepsilon(q-1)(q+|B|)}{|AB| - q^2} - c\right)(AB)^n\right. \\
&\qquad \left. - \left(\frac{\varepsilon(q-1)(q+|B|)}{|AB| - q^2}\right)\left(\frac{AB}{|AB|}\right)^n q^{2n}\right| \\
&= |AB|^n \left|\left(\frac{\varepsilon(q-1)(q+|B|)}{|AB| - q^2} - c\right)\right. \\
&\qquad \left. - \left(\frac{\varepsilon(q-1)(q+|B|)}{|AB| - q^2}\right)\left(\frac{q^2}{|AB|}\right)^n\right| \longrightarrow \infty
\end{aligned}$$

as $t \to \infty$ $(n \to \infty)$ for any choice of $c \in \mathbb{C}$. If $t = q^{2n}$ and $|AB| < q^2$, then

$$\begin{aligned}
|\psi(t) - z(t)| &= \left|\left(\frac{\varepsilon(q-1)(q+|B|)}{|AB| - q^2} - c\right)(AB)^n\right. \\
&\qquad \left. - \left(\frac{\varepsilon(q-1)(q+|B|)}{|AB| - q^2}\right)\left(\frac{AB}{|AB|}\right)^n q^{2n}\right| \\
&= t\left|\left(\frac{\varepsilon(q-1)(q+|B|)}{|AB| - q^2} - c\right)\left(\frac{|AB|}{q^2}\right)^n\right. \\
&\qquad \left. - \left(\frac{\varepsilon(q-1)(q+|B|)}{|AB| - q^2}\right)\right| \longrightarrow \infty
\end{aligned}$$

as $t \to \infty$ $(n \to \infty)$ for any choice of $c \in \mathbb{C}$. Thus, if $t = q^{2n}$ and $|AB| \neq q^2$, equation (14) is not Ulam stable. If $t = q^{2n+1}$ and

$|AB| > q^2$, then

$$
\begin{aligned}
|\psi(t) - z(t)| &= \left|\psi\left(q^{2n+1}\right) - z\left(q^{2n+1}\right)\right| \\
&= \left|\frac{\varepsilon(q-1)}{|AB|-q^2}\left((q+|B|)|A|^{n+1}|B|^n - (q+|A|)q^{2n+1}\right)\right. \\
&\qquad \left. - c|A|^{n+1}|B|^n\right| \\
&= \left|\left(\frac{\varepsilon(q-1)(q+|B|)}{|AB|-q^2} - c\right)|A|^{n+1}|B|^n\right. \\
&\qquad \left. - \left(\frac{\varepsilon(q-1)(q+|A|)}{|AB|-q^2}\right)q^{2n+1}\right| \\
&= |A|^{n+1}|B|^n\left|\left(\frac{\varepsilon(q-1)(q+|B|)}{|AB|-q^2} - c\right)\right. \\
&\qquad \left. - \left(\frac{\varepsilon(q-1)(q+|A|)}{|AB|-q^2}\right)\left(\frac{q^{2n+1}}{|A|^{n+1}|B|^n}\right)\right| \longrightarrow \infty
\end{aligned}
$$

as $t \to \infty$ $(n \to \infty)$ for any choice of $c \in \mathbb{C}$. If $t = q^{2n+1}$ and $|AB| < q^2$, then

$$
\begin{aligned}
|\psi(t) - z(t)| &= \left|\left(\frac{\varepsilon(q-1)(q+|B|)}{|AB|-q^2} - c\right)|A|^{n+1}|B|^n\right. \\
&\qquad \left. - \left(\frac{\varepsilon(q-1)(q+|A|)}{|AB|-q^2}\right)q^{2n+1}\right| \\
&= t\left|\left(\frac{\varepsilon(q-1)(q+|B|)}{|AB|-q^2} - c\right)\frac{|A|^{n+1}|B|^n}{q^{2n+1}}\right. \\
&\qquad \left. - \left(\frac{\varepsilon(q-1)(q+|A|)}{|AB|-q^2}\right)\right| \longrightarrow \infty
\end{aligned}
$$

as $t \to \infty$ for any choice of $c \in \mathbb{C}$. Thus, if $t = q^{2n+1}$ and $|AB| \neq q^2$, equation (14) is not Ulam stable. Overall, equation (14) has no Ulam stability in the case that the coefficient function p is 2-periodic. □

Theorem 11. *Assume $q > 1$, $p_1 \in \mathbb{C}\backslash\{\frac{-1}{q-1}\}$, $p_q \in \mathbb{C}\backslash\{\frac{-1}{q(q-1)}\}$. Let A, B be as in (19). Let $\varepsilon > 0$ be a fixed arbitrary constant, and let the* *function $\psi : \mathbb{T} \to \mathbb{C}$ satisfy inequality* (7)*. Then one of the following holds, where e_p is given in* (18)*:*

(1) *If $|AB| > 1$, then* (14) *is modified Ulam stable, $\lim_{t\to\infty} \frac{\psi(t)}{e_p(t)}$ exists, and the function z given by*

$$z(t) := \left(\lim_{t\to\infty} \frac{\psi(t)}{e_p(t)} \right) e_p(t)$$

is the unique solution of (14) *with $|\psi(t) - z(t)| \leq L_1\varepsilon$ for all $t \in \mathbb{T}$, where*

$$L_2 := \max \left\{ \frac{|A| + 1}{|AB| - 1}, \frac{|B| + 1}{|AB| - 1} \right\} \tag{23}$$

is the minimum Ulam stability constant for (14)*.*

(2) *If $|AB| = 1$, then* (14) *is not modified Ulam stable.*

(3) *If $|AB| < 1$, then* (14) *is modified Ulam stable, and any solution z of* (14) *with*

$$|\psi(1) - z(1)| < \varepsilon \left(\frac{|B| + 1}{1 - |AB|} \right)$$

satisfies

$$|\psi(t) - z(t)| < \varepsilon \max \left\{ \frac{|A| + 1}{1 - |AB|}, \frac{|B| + 1}{1 - |AB|} \right\}$$

for all $t \in \mathbb{T}$.

Proof. Fix $q > 1$. Assume $p_1 \in \mathbb{C}\backslash\{\frac{-1}{q-1}\}$ and $p_q \in \mathbb{C}\backslash\{\frac{-1}{q(q-1)}\}$. Let $\varepsilon > 0$ be an arbitrary constant. In the cases of the proof that follow, since we are assuming $\psi : \mathbb{T} \to \mathbb{C}$ is a function fulfilling (7), there exists a function $\eta : \mathbb{T} \to \mathbb{C}$ such that

$$\mathcal{D}_q\psi(t) - p(t)\psi(t) = \eta(t), \quad |\eta(t)| \leq \frac{\varepsilon}{(q-1)t} \tag{24}$$

for all $t \in \mathbb{T}$.

(i) First, we consider the case $|AB| = |1 + (q-1)p_1||1 + (q-1)qp_q| > 1$. The variation of constants formula [11, Theorem 2.77] applied to (24) yields

$$\psi(t) = \psi(1)e_p(t) + e_p(t) \sum_{\tau=0}^{\log_q t - 1} \frac{(q-1)q^\tau \eta(q^\tau)}{e_p(q^{\tau+1})},$$

$$\text{assuming} \quad \sum_{\tau=0}^{-1} f(\tau) \equiv 0. \tag{25}$$

From (25), we have

$$\left|\frac{\psi(t)}{e_p(t)}\right| \le |\psi(1)| + \varepsilon \sum_{\tau=0}^{\log_q t - 1} \frac{1}{|e_p(q^{\tau+1})|},$$

as $|\eta(q^\tau)| \le \frac{\varepsilon}{(q-1)q^\tau}$ for all $t = q^\tau \in \mathbb{T}$ by the assumption in (24). Note that

$$\begin{aligned}
\sum_{\tau=0}^{\infty} \frac{1}{|e_p(q^{\tau+1})|} &= \frac{1}{|A|} + \frac{1}{|AB|} + \frac{1}{|A||AB|} + \frac{1}{|AB|^2} \\
&\quad + \frac{1}{|A||AB|^2} + \frac{1}{|AB|^3} + \cdots \\
&= \left(\frac{1}{|A|} + \frac{1}{|AB|}\right)\left[1 + \frac{1}{|AB|} + \frac{1}{|AB|^2} + \cdots\right] \\
&= \left(\frac{1}{|A|} + \frac{1}{|AB|}\right)\left(\frac{|AB|}{|AB| - 1}\right)
\end{aligned}$$

converges as a geometric series in this case of $1 < |AB|$. Thus, ψ can be rewritten as

$$\psi(t) = \left[\psi(1) + \sum_{\tau=0}^{\infty} \frac{(q-1)q^\tau \eta(q^\tau)}{e_p(q^{\tau+1})}\right] e_p(t) - e_p(t) \sum_{\tau=\log_q t}^{\infty} \frac{(q-1)q^\tau \eta(q^\tau)}{e_p(q^{\tau+1})}, \tag{26}$$

where

$$z_1 := \psi(1) + \sum_{\tau=0}^{\infty} \frac{(q-1)q^\tau \eta(q^\tau)}{e_p(q^{\tau+1})} \in \mathbb{C}$$

exists as a finite number due to the definition of e_p in (18), and the assumption $|AB| > 1$. Clearly,

$$z(t) := z_1 e_p(t), \quad t \in \mathbb{T}$$

is a solution of (14), and

$$\lim_{t\to\infty} \frac{\psi(t)}{e_p(t)} = \psi(1) + \sum_{\tau=0}^{\infty} \frac{(q-1)q^\tau \eta(q^\tau)}{e_p\left(q^{\tau+1}\right)} = z_1$$

exists. Consequently,

$$z(t) = \left(\lim_{t\to\infty} \frac{\psi(t)}{e_p(t)}\right) e_p(t),$$

and

$$\begin{aligned} |\psi(t) - z(t)| &= \left| -e_p(t) \sum_{\tau=\log_q t}^{\infty} \frac{(q-1)q^\tau \eta(q^\tau)}{e_p\left(q^{\tau+1}\right)} \right| \\ &\leq \varepsilon |e_p(t)| \sum_{\tau=\log_q t}^{\infty} \frac{1}{\left|e_p\left(q^{\tau+1}\right)\right|} \\ &= \varepsilon \begin{cases} \left(\frac{1}{|A|} + \frac{1}{|AB|}\right)\left(\frac{|AB|}{|AB|-1}\right) & \text{if} \quad \log_q t \equiv 0 \pmod 2 \\ \left(\frac{1}{|B|} + \frac{1}{|AB|}\right)\left(\frac{|AB|}{|AB|-1}\right) & \text{if} \quad \log_q t \equiv 1 \pmod 2 \end{cases} \end{aligned}$$

holds for all $t \in \mathbb{T}$. We have again used a geometric series and the fact that $|AB| > 1$. Therefore, (14) has modified Ulam stability with modified Ulam constant L_2 given by (23). The proof that z is the unique solution of (14) such that $|\psi(t) - z(t)| \leq L_2\varepsilon$ for all $t \in \mathbb{T}$ is omitted, since it is the same as the one given earlier in Theorem 8.

Finally, we show that in case (i), L_2 in (23) is the best (minimum) modified Ulam constant possible. In (24), if $\eta(t)$ satisfies (13) for

$t \in \mathbb{T}$, then (25) and (26) imply the function $\psi_* : \mathbb{T} \to \mathbb{C}$ given by

$$\psi_*(t) := \left(\psi_*(1) + \sum_{\tau=0}^{\infty} \frac{\varepsilon}{|e_p(q^{\tau+1})|} \right) e_p(t) - e_p(t) \sum_{\tau=\log_q t}^{\infty} \frac{\varepsilon}{|e_p(q^{\tau+1})|}$$

satisfies the equality

$$|\mathcal{D}_q \psi_*(t) - p(t)\psi_*(t)| = \frac{\varepsilon}{\mu(t)}, \qquad t \in \mathbb{T}.$$

As

$$z(t) := \left(\psi_*(1) + \sum_{\tau=0}^{\infty} \frac{\varepsilon}{|e_p(q^{\tau+1})|} \right) e_p(t)$$

is a solution of (14),

$$\begin{aligned} |\psi_*(t) - z(t)| &= \varepsilon |e_p(t)| \sum_{\tau=\log_q t}^{\infty} \frac{1}{|e_p(q^{\tau+1})|} \\ &= \varepsilon \begin{cases} \left(\frac{1}{|A|} + \frac{1}{|AB|}\right)\left(\frac{|AB|}{|AB|-1}\right) & \text{if} \quad \log_q t \equiv 0 \pmod 2 \\ \left(\frac{1}{|B|} + \frac{1}{|AB|}\right)\left(\frac{|AB|}{|AB|-1}\right) & \text{if} \quad \log_q t \equiv 1 \pmod 2 \end{cases} \\ &\le L_2 \varepsilon \end{aligned}$$

holds for all $t \in \mathbb{T}$. As a result, the entirety of the claims in (i) holds. ♠

(ii) We claim that if $|AB| = |1 + (q-1)p_1||1 + (q-1)qp_q| = 1$, then (14) is not modified Ulam stable. In this case, we have

$$|e_p(t)| = |A|^{\log_q t \pmod 2}$$

from (18) and the assumption $|AB| = 1$. If $\varepsilon > 0$ is arbitrary, let η satisfy (13), and let $\psi^\dagger$ be defined by

$$\psi^\dagger(t) := e_p(t) \sum_{\tau=0}^{\log_q t - 1} \frac{\mu(q^\tau)\eta(q^\tau)}{e_p(q^{\tau+1})} = e_p(t) \sum_{\tau=0}^{\log_q t - 1} \frac{\varepsilon}{|e_p(q^{\tau+1})|}, \qquad t \in \mathbb{T}.$$

Then, $\psi^\dagger$ satisfies the equality

$$\left| \mathcal{D}_q \psi^\dagger(t) - p(t)\psi^\dagger(t) \right| = |\eta(t)| = \left| \frac{\varepsilon e_p(qt)}{\mu(t)|e_p(qt)|} \right| = \frac{\varepsilon}{\mu(t)}, \qquad t \in \mathbb{T}.$$

Since $z(t) = ce_p(t)$ is the general solution of (14), then

$$\left|\psi^\dagger(t) - z(t)\right| = \left|\sum_{\tau=0}^{\log_q t-1} \frac{\varepsilon}{|e_p(q^{\tau+1})|} - c\right| |e_p(t)|$$
$$= \left|\sum_{\tau=0}^{\log_q t-1} \frac{\varepsilon}{|A|^{(\tau+1) \pmod 2}} - c\right| |A|^{\log_q t \pmod 2} \to \infty$$

as $t \to \infty$ for $t \in \mathbb{T}$ and for any $c \in \mathbb{C}$ because the series diverges to infinity. Consequently, in this case, (14) is not modified Ulam stable on $\mathbb{T}$. ♠

(iii) Now assume $0 < |1 + (q-1)p_1||1 + (q-1)qp_q| < 1$, that is, $0 < |AB| < 1$. As before, we know that a function ψ that satisfies (7) takes the form

$$\psi(t) = \psi(1)e_p(t) + e_p(t) \sum_{\tau=0}^{\log_q t-1} \frac{\mu(q^\tau)\eta(q^\tau)}{e_p(q^{\tau+1})}$$

by (25). Let z be any solution of (14) satisfying the inequality

$$|\psi(1) - z(1)| < \varepsilon\left(\frac{|B|+1}{1-|AB|}\right).$$

It follows that z has the form

$$z(t) = z(1)e_p(t), \quad t \in \mathbb{T},$$

and we have

$$\psi(t) - z(t) = \big(\psi(1) - z(1)\big)e_p(t) + e_p(t) \sum_{\tau=0}^{\log_q t-1} \frac{\mu(q^\tau)\eta(q^\tau)}{e_p(q^{\tau+1})}.$$

Consequently, we have

$$|\psi(t) - z(t)| \le |\psi(1) - z(1)||e_p(t)| + \varepsilon|e_p(t)| \sum_{\tau=0}^{\log_q t-1} \frac{1}{|e_p(q^{\tau+1})|}$$
$$< \varepsilon\left(\frac{|B|+1}{1-|AB|}\right)|e_p(t)| + \varepsilon|e_p(t)| \sum_{\tau=0}^{\log_q t-1} \frac{1}{|e_p(q^{\tau+1})|}.$$

From (18), we note for $n \in \mathbb{N}$ that

$$|e_p(t)| = \begin{cases} |AB|^n & \text{if} \quad t = q^{2n} \\ |A|^n|B|^{n-1} & \text{if} \quad t = q^{2n-1} \end{cases}$$

and

$$|e_p(t)| \sum_{\tau=0}^{\log_q t-1} \frac{1}{|e_p(q^{\tau+1})|} = \begin{cases} \left(\frac{1-|AB|^n+|B|-|A|^n|B|^{n+1}}{1-|AB|}\right) & \text{if} \quad t = q^{2n} \\ \left(\frac{1-|AB|^n+|A|-|A|^n|B|^{n-1}}{1-|AB|}\right) & \text{if} \quad t = q^{2n-1}. \end{cases}$$

Combining the above calculations, we have

$$\begin{aligned} |\psi(t) - z(t)| &< \varepsilon \begin{cases} \frac{|A|^n|B|^{n+1}+|AB|^n+1-|AB|^n+|B|-|A|^n|B|^{n+1}}{1-|AB|} \\ \frac{|AB|^n+|A|^n|B|^{n-1}+1-|AB|^n+|A|-|A|^n|B|^{n-1}}{1-|AB|} \end{cases} \\ &= \varepsilon \begin{cases} \frac{|B|+1}{1-|AB|} & \text{if} \quad \log_q t \equiv 0 \pmod 2 \\ \frac{|A|+1}{1-|AB|} & \text{if} \quad \log_q t \equiv 1 \pmod 2 \end{cases} \\ &\leq \varepsilon \max\left\{\frac{|A|+1}{1-|AB|}, \frac{|B|+1}{1-|AB|}\right\} \end{aligned}$$

for all $t \in \mathbb{T}$. Therefore, (iii) holds and the proof is complete. □

As a direct consequence of Theorem 11, we immediately have the following result.

Theorem 12. *Assume $q > 1$, $p_1 \in \mathbb{C}\backslash\{\frac{-1}{q-1}\}$, $p_q \in \mathbb{C}\backslash\{\frac{-1}{q(q-1)}\}$. Let A, B be as in* (19), *with $|AB| \neq 1$. Then, the* 2*-periodic equation* (14) *has modified Ulam stability with Ulam stability constant*

$$L_0 := \max\left\{\frac{|A|+1}{||AB|-1|}, \frac{|B|+1}{||AB|-1|}\right\} \tag{27}$$

on $\mathbb{T}$. Moreover, if $|AB| = |1+(q-1)p_1||1+(q-1)qp_q| > 1$, then L_0 is the best possible (minimum) Ulam stability constant for (14).

Remark 13. It is known [3, Lemma 2.4] that the q-difference equation with constant coefficient $w \in \mathbb{C}$ represented as

$$\mathcal{D}_q z(t) - wz(t) = 0 \tag{28}$$

is not Ulam stable on $\mathbb{T}$ when $w = 0$. If $w = 0$, then (28) is not modified Ulam stable either. This follows from Theorem 8 (ii). ♠

Remark 14. Considering Remark 13, it may be a natural follow-up to ask the following question: If the 2-periodic coefficient function $p = p(t)$ has infinitely many zeros, does (14) have modified Ulam stability on $\mathbb{T}$? To find out, set $p_1 = \frac{c}{q-1}$ for $c \in \mathbb{C} \setminus \{-2, -1, 0\}$ and $p_q = 0$, and consider the function

$$p_*(t) := \frac{1}{t}\begin{cases} \frac{c}{q-1} & \text{if} \quad \log_q t \equiv 0 \pmod 2, \\ 0 & \text{if} \quad \log_q t \equiv 1 \pmod 2 \end{cases}$$

for $t \in \mathbb{T}$. In other words, p_* is 2-periodic and equals zero infinitely often. Clearly, we see that $0 \neq |1 + c| \neq 1$ hold. The following result is obtained from Theorem 12, where $A = 1 + c$ and $B = 1$. Our main equation with $p = p_*$ has modified Ulam stability with Ulam constant L_0 given by (27). In this case, the Ulam constant L_0 is given by

$$L_0 = \begin{cases} \frac{|1+c|+1}{|1+c|-1} & \text{if} \quad |1 + c| > 1, \\ \frac{2}{1-|1+c|} & \text{if} \quad |1 + c| < 1. \end{cases}$$

In addition, $\frac{|1+c|+1}{|1+c|-1}$ is the minimum (best) constant for Ulam stability when $|1 + c| > 1$. ♠

4. Conclusion and Future Directions

In this chapter, we proved that first-order linear homogeneous q-difference equations with 1-periodic or 2-periodic coefficient functions are not Ulam stable. We also proved, however, that the same equation with 1-periodic or 2-periodic coefficient function is modified Ulam stable, and in some cases, we were able to obtain the best Ulam stability constant. In the future, we will explore the case for an n-periodic coefficient function, for both q-difference equations and general isolated time scales.

References

[1] Abdollahpour, M.R. and Rassias, M. Th. Hyers-Ulam stability of hypergeometric differential equations, *Aequ. Mathe.*, 93(4), 691–698 (2019).

[2] Abdollahpour, M.R., Aghayaria, R. and Rassias, M. Th. Hyers-Ulam stability of associated Laguerre differential equations in a subclass of analytic functions, *J. Math. Anal. Appl.*, 437, 605–612 (2016).

[3] Anderson, D.R. and Onitsuka, M. Hyers–Ulam stability for quantum equations, *Aequ. Mathe.*, (2020) DOI: 10.1007/s00010-020-00734-1.

[4] Anderson, D.R. and Onitsuka, M. Hyers–Ulam stability for quantum equations of Euler type, *Discrete Dyn. Nature Soc.*, 5626481, 10 (2020), https://doi.org/10.1155/2020/5626481.

[5] Anderson, D.R. and Onitsuka, M. Hyers-Ulam stability for Cayley quantum equations and its application to h-difference equations, *Mediterranean J. Math.*, 18(168), 13 (2021).

[6] Anderson, D.R. and Onitsuka, M. Ulam stability for nonautonomous quantum equations, *J. Inequalities Appl.*, 2021(161) (2021) https://doi.org/10.1186/s13660-021-02699-4.

[7] Anderson, D.R., Onitsuka, M. and Rassias, J.M. Best constant for Ulam stability of first-order h-difference equations with periodic coefficient, *J. Math. Anal. Appl.*, 491, 124363 (2020). https://doi.org/10.1016/j.jmaa.2020.124363.

[8] András, S. and Mészáros, A.R. Ulam–Hyers stability of dynamic equations on time scales via Picard operators, *Appl. Math. Computation*, 219, 4853–4864 (2013).

[9] Bohner, M. and Chieochan, R. Floquet theory for q-difference equations, *Sarajevo J. Math.* 8(21), 355–366 (2012).

[10] Bohner, M., Mesquita, J. and Streipert, S. Periodicity on isolated time scales, *Mathematische Nachrichten,* 295, 259–280 (2022) DOI: 10.1002/mana.201900360.

[11] Bohner, M. and Peterson, A. *Dynamic Equations on Time Scales, An Introduction with Applications*, Birkhäuser, Boston, (2001).

[12] Brillouët–Belluot, N., Brzdęk, J. and Ciepliński, K. On some recent developments in Ulam's type stability, *Abstract Appl. Anal.*, 2012(716936), 41 (2012).

[13] Brzdęk, J., Popa, D., Raşa, I. and Xu, B. *Ulam Stability of Operators*, A volume in *Mathematical Analysis and its Applications*, Academic Press, London (2018).

[14] Brzdęk, J. and Wójcik, P. On approximate solutions of some difference equations, *Bull. Australian Math. Soc.*, 95(3), 476–481 (2017).

[15] Elqorachi, E. and Rassias, M. Th. Generalized Hyers-Ulam stability of trigonometric functional equations, *Mathematics*, 6(5), 83 (2018) https://doi.org/10.3390/math6050083.

[16] Fukutaka, R. and Onitsuka, M. Best constant in Hyers–Ulam stability of first-order homogeneous linear differential equations with a periodic coefficient, *J. Math. Anal. Appl.*, 473(2), 1432–1446 (2019).

[17] Hua, L., Li, Y. and Feng, J. On Hyers–Ulam stability of dynamic integral equation on time scales, *Mathematica Aeterna*, 4(6), 559–571 (2014).

[18] Hyers, D.H. On the stability of the linear functional equation, *Proc. Nat. Acad. Sci.* U.S.A., 27, 222–224, (1941).

[19] Jung, S.-M. and Nam, Y.W. Hyers–Ulam stability of Pielou logistic difference equation, *J. Nonlinear Sci. Appl.*, 10, 3115–3122 (2017).

[20] Jung, S.-M., Popa, D. and Rassias, M. Th. On the stability of the linear functional equation in a single variable on complete metric groups, *J. Global Optimization*, 59, 165–171 (2014).

[21] Jung, S.-M. and Rassias, M. Th. A linear functional equation of third order associated to the Fibonacci numbers, *Abstract and Applied Analysis*, 2014(137468), 7 (2014),

[22] Lee, Y.-H., Jung, S.-M. and Rassias, M. Th. Uniqueness theorems on functional inequalities concerning cubic-quadratic-additive equation, *J. Math. Inequalities*, 12(1), 43–61 (2018).

[23] Nam, Y.W. Hyers–Ulam stability of hyperbolic Möbius difference equation, *Filomat*, 32(13), 4555–4575 (2018) https://doi.org/10.2298/FIL1813555N.

[24] Nam, Y.W. Hyers–Ulam stability of elliptic Möbius difference equation, *Cogent Math. & Stat.*, 5(1), 1–9 (2018).

[25] Nam, Y.W. Hyers–Ulam stability of loxodromic Möbius difference equation, *Appl. Math. Comput.*, 356(1), 119–136 (2019) https://doi.org/10.1016/j.amc.2019.03.033.

[26] Onitsuka, M. Influence of the stepsize on Hyers–Ulam stability of first-order homogeneous linear difference equations, *Int. J. Difference Equ.*, 12(2), 281–302 (2017).

[27] Onitsuka, M. Hyers–Ulam stability of second-order nonhomogeneous linear difference equations with a constant step size, *J. Comput. Anal. Appl.*, 28(1), 152–165 (2020).

[28] Popa, D. Hyers–Ulam stability of the linear recurrence with constant coefficients, *Adv. Differ. Equ.*, 2005, 407076 (2005).

[29] Popa, D. Hyers–Ulam–Rassias stability of a linear recurrence, *J. Math. Anal. Appl.*, 309, 591–597 (2005).

[30] Rasouli, H., Abbaszadeh, S. and Eshaghi, M. Approximately linear recurrences, *J. Appl. Anal.*, 24(1), 81–85 (2018).

[31] Rassias, Th. M. On the stability of linear mapping in Banach spaces, *Proc. Amer. Math. Soc.*, 72, 297–300 (1978).

[32] Shen, Y.H. The Ulam stability of first order linear dynamic equations on time scales, *Results Math.*, 72(4), 1881–1895 (2017).

[33] Shen, Y.H. and Li, Y.J. Hyers–Ulam stability of first order nonhomogeneous linear dynamic equations on time scales, *Comm. Math. Research* 35(2), 139–148 (2019). DOI: 10.13447/j.1674-5647.2019.02.05.

[34] Ulam, S.M. *A Collection of the Mathematical Problems.* Interscience, New York, (1960).

https://doi.org/10.1142/9789811293009_0003

Chapter 3

Generalized Fractional Hermite–Hadamard–Noor Inequalities Via n-Polynomial Preinvex Functions

Muhammad Uzair Awan[*,§], Muhammad Ubaid Ullah[*,¶], Michael Th. Rassias[†,‖], Sadia Talib[*,#], Muhammad Aslam Noor[‡,**] and Khalida Inayat Noor[‡,††]

[*] *Department of Mathematics, Government College University, Faisalabad, Pakistan*
[†] *Institute of Mathematics, University of Zurich, 8057, Zurich, Switzerland*
[‡] *Department of Mathematics, COMSATS University Islamabad, Islamabad, Pakistan*
[§] *awan.uzair@gmail.com*
[¶] *mubaidullah07@gmail.com*
[#] *sadiatalib2015@gmail.com*
[‖] *michail.rassias@math.uzh.ch*
[**] *noormaslam@gmail.com*
[††] *khalidan@gmail.com*

We derive a new variant of fractional Hermite–Hadamard–Noor-type inequality via n-polynomial preinvex functions involving the general form of fractional integrals. Furthermore, a new auxiliary result

[*] Corresponding author

and associated estimates of Hermite–Hadamard–Noor inequality via n-polynomial preinvex functions are obtained. We also discuss special cases which show that our results are quite unifying.

1. Introduction and Preliminaries

A set C is said to be a convex set if

$$(1-\tau)\flat_1 + \tau\flat_2 \in C, \qquad \forall \flat_1, \flat_2 \in C.$$

Similarly, a real-valued function $\Upsilon : C \subseteq \mathbb{R} \to \mathbb{R}$ is said to be convex function if

$$\Upsilon((1-\tau)\flat_1 + \tau\flat_2) \leq (1-\tau)\Upsilon(\flat_1) + \tau\Upsilon(\flat_2)$$

holds for all $\flat_1, \flat_2 \in C$ and $\tau \in [0,1]$. Furthermore, the functions satisfying this inequality in reverse direction are known as concave functions.

The classical concept of convexity has been extended and generalized in different directions using novel and innovative ideas. For example, Mititelu [11] introduced invex sets as follows:

Definition 1. A set $\mathcal{C}_\varsigma$ is said to be an invex set with respect to the bifunction $\varsigma : \mathcal{C}_\varsigma \times \mathcal{C}_\varsigma \to \mathbb{R}$, if

$$\flat_1 + \tau\varsigma(\flat_2, \flat_1) \in \mathcal{C}_\varsigma \qquad \forall \flat_1, \flat_2 \in \mathcal{C}_\varsigma.$$

For more details on invexity property of s set, see Ref. [6].

Weir and Mond [19] introduced the notion of preinvex functions, as follows:

Definition 2. A function $\Upsilon : \mathcal{C}_\varsigma \to \mathbb{R}$ is said to be a preinvex function with respect to the bifunction $\varsigma : \mathcal{C}_\varsigma \times \mathcal{C}_\varsigma \to \mathbb{R}$, if

$$\Upsilon(\flat_1 + \tau\varsigma(\flat_2, \flat_1)) \leq (1-\tau)\Upsilon(\flat_1) + \tau\Upsilon(\flat_2), \quad \forall \flat_1, \flat_2 \in \mathcal{C}_\varsigma, \tau \in [0,1].$$

Recently, Toplu *et al.* [18], introduced the notion of n-polynomial convex functions as follows:

Definition 3. Let $n \in N$. A real-valued function $\Upsilon : C \to \mathbb{R}$ is said to be n-polynomial convex function if

$$\Upsilon((1-\tau)\flat_1 + \tau\flat_2) \leq \frac{1}{n}\sum_{s=1}^{n}(1-\tau^s)\Upsilon(\flat_1) + \frac{1}{n}\sum_{s=1}^{n}(1-(1-\tau)^s)\Upsilon(\flat_2),$$

for all $\flat_1, \flat_2 \in C$ and $\tau \in [0,1]$.

Very recently, Wu *et al.* [20] defined the class of n-polynomial preinvex functions as follows:

Definition 4. Let $n \in N$. A real-valued function $\Upsilon : \mathcal{C}_\zeta \to \mathbb{R}$ is said to be n-polynomial preinvex function with respect to bifunction $\zeta : \mathcal{C}_\zeta \times \mathcal{C}_\zeta \to \mathbb{R}$ if

$$\Upsilon(\flat_1 + \tau\zeta(\flat_2, \flat_1)) \leq \frac{1}{n}\sum_{s=1}^{n}(1-\tau^s)\Upsilon(\flat_1) + \frac{1}{n}\sum_{s=1}^{n}(1-(1-\tau)^s)\Upsilon(\flat_2),$$

for all $\flat_1, \flat_2 \in \mathcal{C}_\zeta$ and $\tau \in [0,1]$.

For some recent details on n-polynomial convexity property, its generalizations, and associated inequalities, see Refs. [1,2,4].

Although convexity attracted many researchers due to its applications, another aspect of convexity which is also very fascinating is its relationship with the theory of inequalities. Many inequalities known to us are direct consequences of the applications of convex functions.

A well-known inequality that is Hermite–Hadamard inequality is one of the most significant results pertaining to the convex functions. This result reads as follows:

Theorem 5. *Let $\Upsilon : C \subseteq \mathbb{R} \to \mathbb{R}$ be a real valued convex function, then*

$$\Upsilon\left(\frac{\flat_1 + \flat_2}{2}\right) \leq \frac{1}{\flat_2 - \flat_1}\int_{\flat_1}^{\flat_2} \Upsilon(x)\mathrm{d}x \leq \frac{\Upsilon(\flat_1) + \Upsilon(\flat_2)}{2}.$$

With the assumption that $\zeta(.,.)$ satisfies Condition C, Noor [14] obtained a new version of Hermite–Hadamard's inequality using the class of preinvex functions. Before presenting the result of Noor, let us recall Condition C, which was introduced and studied by Mohan and Neogy [12].

Condition C. A set $C_\zeta \subset \mathbb{R}$ is said to be an invex set with respect to bifunction $\zeta(.,.)$ if and only if for any $x, y \in C_\zeta$ and $\tau \in [0,1]$, we have

(1) $\zeta(x, x + \tau\xi(y,x)) = -\tau\zeta(y,x)$,
(2) $\zeta(y, x + \tau\xi(y,x)) = (1-\tau)\zeta(y,x)$.

Note that for any $x, y \in C_\zeta$, $\tau_1, \tau_2 \in [0,1]$ and from Condition C, we can deduce

$$\zeta(x + \tau_2\zeta(y,x), x + \tau_1\zeta(y,x)) = (\tau_2 - \tau_1)\zeta(y,x).$$

Theorem 6. *Let $\Upsilon : C_\zeta \to \mathbb{R}$ be a preinvex function. If $\zeta(.,.)$ satisfies Condition C, then for $\zeta(\flat_2, \flat_1) > 0$, we have*

$$\Upsilon\left(\frac{2\flat_1 + \zeta(\flat_2,\flat_1)}{2}\right) \leq \frac{1}{\zeta(\flat_2,\flat_1)}\int_{\flat_1}^{\flat_1+\zeta(\flat_2,\flat_1)} \Upsilon(x)\mathrm{d}x \leq \frac{\Upsilon(\flat_1) + \Upsilon(\flat_2)}{2}.$$

We call the above result as Hermite–Hadamard–Noor inequality. This result reduces to classical Hermite–Hadamard inequality by taking $\zeta(\flat_2, \flat_1) = \flat_2 - \flat_1$.

Using the concepts of Riemann–Liouville fractional integrals, Sarikaya *et al.* [16] obtained the fractional version of Hermite–Hadamard inequality. Before we discuss the result of Sarikaya *et al.* [16], let us recall the definition of Riemann–Liouville fractional integrals.

The classical Riemann–Liouville fractional integrals are defined as follows:

Definition 7 ([10]). Let $\Upsilon \in L_1[\flat_1, \flat_2]$. Then, the Riemann–Liouville integrals $J^\alpha_{\flat_1+}\Upsilon$ and $J^\alpha_{\flat_2^-}\Upsilon$ of order $\alpha > 0$ with $\flat_1 \geq 0$ are defined by

$$J^\alpha_{\flat_1^+}\Upsilon(x) = \frac{1}{\Gamma(\alpha)}\int_{\flat_1}^{x} (x - v)^{\alpha-1}\Upsilon(v)\mathrm{d}v, \quad x > \flat_1,$$

and

$$J^\alpha_{\flat_2^-}\Upsilon(x) = \frac{1}{\Gamma(\alpha)}\int_{x}^{\flat_2} (v - x)^{\alpha-1}\Upsilon(v)\mathrm{d}v, \quad x < \flat_2,$$

where

$$\Gamma(x) = \int_0^\infty e^{-v}v^{x-1}\mathrm{d}v$$

is the well-known gamma function.

Theorem 8. *Let $\Upsilon : C \subseteq \mathbb{R} \to \mathbb{R}$ be a positive function, if Υ is convex function on $[\flat_1, \flat_2]$, then*

$$\Upsilon\left(\frac{\flat_1+\flat_2}{2}\right) \leq \frac{\Gamma(\alpha+1)}{2(\flat_2-\flat_1)^{\alpha}}[J^{\alpha}_{\flat_1+}\Upsilon(\flat_1)+J^{\alpha}_{\flat_2-}\Upsilon(\flat_2)] \leq \frac{\Upsilon(\flat_1)+\Upsilon(\flat_2)}{2}$$

with $\alpha > 0$.

Iscan [5] obtained fractional analogue of Hermite–Hadamard–Noor-type inequality using the class of preinvex functions.

Recently, Sarikaya and Ertugral [15] introduced a general class of left and right fractional integrals as follows:

Let $\psi : [0,\infty) \to [0,\infty)$ be a function satisfying the following conditions:

(1) $\int_0^1 \frac{\psi(\tau)}{\tau} d\tau < \infty$,

(2) $\frac{1}{M_1} \leq \frac{\psi(s)}{\psi(r)} \leq M_1$ for $\frac{1}{2} \leq \frac{s}{r} \leq 2$,

(3) $\frac{\psi(r)}{r^2} \leq M_2 \frac{\psi(s)}{s^2}$ for $s \leq r$,

(4) $\left|\frac{\psi(r)}{r^2} - \frac{\psi(s)}{s^2}\right| \leq M_3 |r-s| \frac{\psi(r)}{r^2}$ for $\frac{1}{2} \leq \frac{s}{r} \leq 2$,

where M_1, M_2, and M_3 are independent of $r, s > 0$. Under the assumptions of ψ, the left- and right-sided generalized fractional integrals are

$$_{\flat_1+}I_{\psi}\Upsilon(x) = \int_{\flat_1}^{x} \frac{\psi(x-\tau)}{x-\tau}\Upsilon(\tau)d\tau, \qquad x > \flat_1, \tag{1}$$

$$_{\flat_2-}I_{\psi}\Upsilon(x) = \int_{x}^{\flat_2} \frac{\psi(\tau-x)}{\tau-x}\Upsilon(\tau)d\tau, \quad x < \flat_2. \tag{2}$$

These fractional integrals are the generalization of some well-known fractional integrals like Riemann–Liouville fractional integrals [10], κ-Riemann–Liouville fractional integrals [13], Katugampola fractional integrals [8], conformable fractional integrals [17], etc.

(1) If we substitute $\psi(\tau) = \tau$ in operators (1) and (2), we obtain Riemann integrals.
(2) If we substitute $\psi(\tau) = \frac{\tau^{\alpha}}{\Gamma(\alpha)}$ in operators (1) and (2), we obtain Riemann–Liouville fractional integrals. For details, see Ref. [10].

(3) If we substitute $\psi(\tau) = \frac{\tau^{\frac{\alpha}{\kappa}}}{\kappa\Gamma_\kappa(\alpha)}$ in operators (1) and (2), we obtain κ-Riemann–Liouville fractional integrals. For details, see Ref. [13].
(4) If we substitute $\psi(\tau) = \tau(x-\tau)^{\alpha-1}$ in operators (1) and (2), we obtain conformable fractional integrals which are defined by Khalil *et al.* [9].
(5) If we substitute $\psi(\tau) = \frac{\tau}{\alpha}\exp\left(-\frac{1-\alpha}{\alpha}\tau\right)$ in operators (1) and (2), we obtain left-sided and right-sided fractional integrals with exponential kernel for $\alpha \in (0,1)$ which were defined in Ref. [7].

For further details, see Ref. [15].

Motivated by the research work of Sarikaya and Ertugral, [15] we in this chapter use the general class of fractional integrals and obtain some new generalized fractional estimates of Hermite–Hadamard–Noor-type inequalities involving n-polynomial preinvex functions.

2. Hermite–Hadamard–Noor-type Inequality Involving General Fractional Integrals

In this section, we derive a new Hermite–Hadamard–Noor-type inequality using the class of n-polynomial preinvex functions. For the sake of simplicity, let us denote by

$$\Delta(y) = \int_0^y \frac{\psi(\zeta(\flat_2,\flat_1)u)}{u}\mathrm{d}u,$$

$$\nabla(y) = \int_y^1 \frac{\psi(\zeta(\flat_2,\flat_1)u)}{u}\mathrm{d}u.$$

Theorem 9. *Let $\Upsilon : [\flat_1, \flat_1 + \zeta(\flat_2,\flat_1)] \to \mathbb{R}$ be n-polynomial preinvex function with $0 < \zeta(\flat_2,\flat_1)$. Further, suppose that $\zeta(.,.)$ satisfies Condition C, then*

$$\begin{aligned}
&\frac{n}{n+2^{-n}-1}\Upsilon\left(\frac{2\flat_1+\zeta(\flat_2,\flat_1)}{2}\right)\Delta(1)\\
&\leq_{(\flat_1+\zeta(\flat_2,\flat_1))^-} I_\psi\Upsilon(\flat_1) +_{\flat_1^+} I_\psi\Upsilon(\flat_1+\zeta(\flat_2,\flat_1))\\
&\leq \frac{\Upsilon(\flat_1)+\Upsilon(\flat_2)}{n}\int_0^1\sum_{s=1}^n[2-\tau^s-(1-\tau)^s]\frac{\psi(\zeta(\flat_2,\flat_1)\tau)}{\tau}\mathrm{d}\tau. \quad (3)
\end{aligned}$$

Proof. Since Υ is a n-polynomial preinvex function, then by using Condition C, we have

$$\begin{aligned}
&\Upsilon\left(\frac{2\flat_1+\zeta(\flat_2,\flat_1)}{2}\right)\\
&\quad=\Upsilon\left(\flat_1+\tau\zeta(\flat_2,\flat_1)+\frac{1}{2}\zeta\left(\flat_1+(1-\tau)\zeta(\flat_2,\flat_1),\flat_1+\tau\zeta(\flat_2,\flat_1)\right)\right)\\
&\quad\le\frac{1}{n}\sum_{s=1}^{n}\left(1-\frac{1}{2^s}\right)\left[\Upsilon(\flat_1+\tau\,\zeta(\flat_2,\flat_1))+\Upsilon(\flat_1+(1-\tau)\zeta(\flat_2,\flat_1))\right]\\
&\quad=\frac{n+2^{-n}-1}{n}(\Upsilon(x)+\Upsilon(y)),
\end{aligned}\tag{4}$$

where $x=\flat_1+\tau\zeta(\flat_2,\flat_1)$ and $y=\flat_1+(1-\tau)\zeta(\flat_2,\flat_1)$.

Multiplying both sides of (4) by $\frac{\psi(\zeta(\flat_2,\flat_1)\tau)}{\tau}$, then integrating with respect to τ on $[0,1]$, we get

$$\begin{aligned}
&\frac{n}{n+2^{-n}-1}\Upsilon\left(\frac{2\flat_1+\zeta(\flat_2,\flat_1)}{2}\right)\Delta(1)\\
&\quad\le\int_0^1\frac{\psi(\zeta(\flat_2,\flat_1)\tau)}{\tau}\Upsilon(\flat_1+\tau\zeta(\flat_2,\flat_1))d\tau\\
&\qquad+\int_0^1\frac{\psi(\zeta(\flat_2,\flat_1)\tau)}{\tau}\Upsilon(\flat_1+(1-\tau)\zeta(\flat_2,\flat_1))d\tau\\
&\quad=\int_{\flat_1}^{\flat_1+\zeta(\flat_2,\flat_1)}\frac{\psi(x-\flat_1)}{x-\flat_1}\Upsilon(x)dx\\
&\qquad+\int_{\flat_1}^{\flat_1+\zeta(\flat_2,\flat_1)}\frac{\psi(\flat_1+\zeta(\flat_2,\flat_1)-x)}{\flat_1+\zeta(\flat_2,\flat_1)-x}\Upsilon(x)dx\\
&\quad={}_{(\flat_1+\zeta(\flat_2,\flat_1))^-}I_\psi\Upsilon(\flat_1)+{}_{\flat_1^+}I_\psi\Upsilon(\flat_1+\zeta(\flat_2,\flat_1)).
\end{aligned}\tag{5}$$

Also, we know that

$$\begin{aligned}
&\Upsilon\left(\flat_1+\tau\zeta(\flat_2,\flat_1)\right)+\Upsilon(\flat_1+(1-\tau\zeta(\flat_2,\flat_1)))\\
&\quad\le\frac{1}{n}\sum_{s=1}^{n}[1-\tau^s]\Upsilon(\flat_1)+\frac{1}{n}\sum_{s=1}^{n}[1-(1-\tau)^s]\Upsilon(\flat_2)
\end{aligned}$$

$$+\frac{1}{n}\sum_{s=1}^{n}[1-(1-\tau)^{s}]\Upsilon(\flat_1)+\frac{1}{n}\sum_{s=1}^{n}[1-\tau^{s}]\Upsilon(\flat_2)$$
$$=\frac{\Upsilon(\flat_1)+\Upsilon(\flat_2)}{n}\sum_{s=1}^{n}[2-\tau^{s}-(1-\tau)^{s}]. \tag{6}$$

Multiplying both sides of (6) by $\frac{\psi(\zeta(\flat_2,\flat_1)\tau)}{\tau}$, then integrating with respect to τ on $[0,1]$, we get

$$\begin{aligned}&{}_{(\flat_1+\zeta(\flat_2,\flat_1))^-}I_{\psi}\Upsilon(\flat_1)+{}_{\flat_1^+}I_{\psi}\Upsilon(\flat_1+\zeta(\flat_2,\flat_1))\\&\le\frac{\Upsilon(\flat_1)+\Upsilon(\flat_2)}{n}\int_0^1\sum_{s=1}^{n}[2-\tau^{s}-(1-\tau)^{s}]\frac{\psi(\zeta(\flat_2,\flat_1)\tau)}{\tau}\mathrm{d}\tau.\end{aligned} \tag{7}$$

Combining (5) and (7), we get

$$\begin{aligned}&\frac{n}{n+2^{-n}-1}\Upsilon\left(\frac{2\flat_1+\zeta(\flat_2,\flat_1)}{2}\right)\Delta(1)\\&\le{}_{(\flat_1+\zeta(\flat_2,\flat_1))^-}I_{\psi}\Upsilon(\flat_1)+{}_{\flat_1^+}I_{\psi}\Upsilon(\flat_1+\zeta(\flat_2,\flat_1))\\&\le\frac{\Upsilon(\flat_1)+\Upsilon(\flat_2)}{n}\int_0^1\sum_{s=1}^{n}[2-\tau^{s}-(1-\tau)^{s}]\frac{\psi(\zeta(\flat_2,\flat_1)\tau)}{\tau}\mathrm{d}\tau.\end{aligned}$$

This completes the proof. □

Following are some special cases of Theorem 9.

Remark 10. If all the assumptions of Theorem 9 hold, then for $\psi(\tau)=\tau$, inequality (3) reduces to

$$\begin{aligned}\frac{1}{2}\left(\frac{n}{n+2^{-n}-1}\right)\Upsilon\left(\frac{2\flat_1+\zeta(\flat_2,\flat_1)}{2}\right)&\le\frac{1}{\zeta(\flat_2,\flat_1)}\int_{\flat_1}^{\flat_1+\zeta(\flat_2,\flat_1)}\Upsilon(x)\mathrm{d}x\\&\le\frac{\Upsilon(\flat_1)+\Upsilon(\flat_2)}{n}\sum_{s=1}^{n}\frac{s}{s+1},\end{aligned}$$

which is the Hermite–Hadamard-type inequality for n-polynomial preinvex functions.

Corollary 11. *If all the assumptions of Theorem 9 hold, then for* $\psi(\tau) = \frac{\tau^\alpha}{\Gamma(\alpha)}$, *we have*

$$\begin{aligned}
&\frac{n}{n+2^{-n}-1}\Upsilon\left(\frac{2\flat_1+\zeta(\flat_2,\flat_1)}{2}\right)\\
&\quad\leq \frac{\Gamma(\alpha+1)}{[\zeta(\flat_2,\flat_1)]^\alpha}\left[I^\alpha_{(\flat_1+\zeta(\flat_2,\flat_1))^-}\Upsilon(\flat_1)+I^\alpha_{\flat_1^+}\Upsilon(\flat_1+\zeta(\flat_2,\flat_1))\right]\\
&\quad\leq \frac{\alpha[\Upsilon(\flat_1)+\Upsilon(\flat_2)]}{n}\sum_{s=1}^{n}\left(\frac{2s+\alpha-\alpha(\alpha+s)\beta(\alpha,s+1)}{\alpha(\alpha+s)}\right).
\end{aligned}$$

Corollary 12. *If all the assumptions of Theorem 9 hold, then for* $\psi(\tau) = \frac{\tau^{\frac{\alpha}{k}}}{k\Gamma_k(\alpha)}$, *we have*

$$\begin{aligned}
&\frac{n}{n+2^{-n}-1}\Upsilon\left(\frac{2\flat_1+\zeta(\flat_2,\flat_1)}{2}\right)\\
&\quad\leq \frac{\alpha\Gamma_k(\alpha)}{[\zeta(\flat_2,\flat_1)]^{\frac{\alpha}{k}}}\left[I^\alpha_{(\flat_1+\zeta(\flat_2,\flat_1))^-,k}\Upsilon(\flat_1)+I^\alpha_{\flat_1^+,k}\Upsilon(\flat_1+\zeta(\flat_2,\flat_1))\right]\\
&\quad\leq \frac{\alpha(\Upsilon(\flat_1)+\Upsilon(\flat_2))}{kn}\sum_{s=1}^{n}\left(\frac{2k}{\alpha}-\frac{k}{ks+\alpha}-k\beta_k\left(\alpha,k(s+1)\right)\right).
\end{aligned}$$

3. Key Lemmas

We now derive significant auxiliary results, which will help us in obtaining our next results.

Lemma 13. *Let the function* $\Upsilon : [\flat_1, \flat_1+\zeta(\flat_2,\flat_1)] \to \mathbb{R}$ *be differentiable on* $(\flat_1, \flat_1+\zeta(\flat_2,\flat_1))$ *with* $\zeta(\flat_2,\flat_1) > 0$. *If* $\Upsilon' \in L[\flat_1, \flat_1+\zeta(\flat_2,\flat_1)]$, *then*

$$\begin{aligned}
&\frac{\Upsilon(\flat_1)+\Upsilon(\flat_1+\zeta(\flat_2,\flat_1))}{2}-\frac{1}{2\Delta(1)}\Big[{}_{\flat_1^+}I_\psi\,\Upsilon(\flat_1+\zeta(\flat_2,\flat_1))\\
&\quad+{}_{(\flat_1+\zeta(\flat_2,\flat_1))^-}I_\psi\,\Upsilon(\flat_1)\Big]=\frac{\zeta(\flat_2,\flat_1)}{2\Delta(1)}\int_0^1\Delta(\tau)[\Upsilon'(\flat_1+\tau\zeta(\flat_2,\flat_1))\\
&\qquad+\Upsilon'(\flat_1+(1-\tau)\zeta(\flat_2,\flat_1))]d\tau.
\end{aligned}$$

Proof. Integrating by parts, we have

$$\begin{aligned}
K_1 &= \int_0^1 \left[\int_0^\tau \frac{\psi(\zeta(\flat_2,\flat_1)u)}{u}\mathrm{d}u\right] \Upsilon'(\flat_1+\tau\zeta(\flat_2,\flat_1))\mathrm{d}\tau \\
&= -\frac{\Upsilon(\flat_1)}{\zeta(\flat_2,\flat_1)}\int_0^1 \frac{\psi(\zeta(\flat_2,\flat_1)u)}{u}\mathrm{d}u \\
&\quad + \frac{1}{\zeta(\flat_2,\flat_1)}\int_0^1 \frac{\psi(\zeta(\flat_2,\flat_1)u)}{u}\Upsilon(\flat_1+(1-\tau)\zeta(\flat_2,\flat_1))\mathrm{d}\tau \\
&= -\frac{\Upsilon(\flat_1)}{\zeta(\flat_2,\flat_1)}\int_0^1 \frac{\psi(\zeta(\flat_2,\flat_1)u)}{u}\mathrm{d}u \\
&\quad + \frac{1}{\zeta(\flat_2,\flat_1)}\int_{\flat_1}^{\flat_1+\zeta(\flat_2,\flat_1)} \frac{\psi(\flat_1+\zeta(\flat_2,\flat_1)-x)}{\flat_1+\zeta(\flat_2,\flat_1)-x}\Upsilon(x)\mathrm{d}x \\
&= -\frac{\Upsilon(\flat_1)}{\zeta(\flat_2,\flat_1)}\int_0^1 \frac{\psi(\zeta(\flat_2,\flat_1)u)}{u}\mathrm{d}u + \frac{1}{\zeta(\flat_2,\flat_1)}\,{}_{\flat_1^+}I_\psi\,\Upsilon(\flat_1+\zeta(\flat_2,\flat_1)).
\end{aligned} \tag{8}$$

Similarly,

$$\begin{aligned}
K_2 &= \int_0^1 \left[\int_0^\tau \frac{\psi(\zeta(\flat_2,\flat_1)u)}{u}\mathrm{d}u\right] \Upsilon'(\flat_1+(1-\tau)\zeta(\flat_2,\flat_1))\mathrm{d}\tau \\
&= \frac{\Upsilon(\flat_1+\zeta(\flat_2,\flat_1))}{\zeta(\flat_2,\flat_1)}\int_0^1 \frac{\psi(\zeta(\flat_2,\flat_1)u)}{u}\mathrm{d}u \\
&\quad - \frac{1}{\zeta(\flat_2,\flat_1)}\,{}_{(\flat_1+\zeta(\flat_2,\flat_1))^-}I_\psi\,\Upsilon(\flat_1).
\end{aligned} \tag{9}$$

Subtracting (8) from (9), we get

$$\begin{aligned}
&\frac{\Upsilon(\flat_1)+\Upsilon(\flat_1+\zeta(\flat_2,\flat_1))}{2} \\
&\quad - \frac{1}{2\Delta(1)}\left[{}_{\flat_1^+}I_\psi\,\Upsilon(\flat_1+\zeta(\flat_2,\flat_1)) + {}_{(\flat_1+\zeta(\flat_2,\flat_1))^-}I_\psi\,\Upsilon(\flat_1)\right] \\
&= \frac{\zeta(\flat_2,\flat_1)}{2\Delta(1)}\int_0^1 \Delta(\tau)[\Upsilon'(\flat_1+\tau\zeta(\flat_2,\flat_1)) + \Upsilon'(\flat_1+(1-\tau)\zeta(\flat_2,\flat_1))]\mathrm{d}\tau,
\end{aligned}$$

which is required. □

We now discuss some special cases of Lemma 13.

Remark 14. If all the assumptions of Lemma 13 hold, then for $\psi(\tau) = \tau$, we have

$$\begin{aligned}
&\frac{\Upsilon(\flat_1) + \Upsilon(\flat_1 + \zeta(\flat_2, \flat_1))}{2} - \frac{1}{\zeta(\flat_2, \flat_1)} \int_{\flat_1}^{\flat_1 + \zeta(\flat_2, \flat_1)} \Upsilon(x) \mathrm{d}x \\
&\quad = \frac{\zeta(\flat_2, \flat_1)}{2} \int_0^1 (1 - 2\tau) \Upsilon'(\flat_1 + (1 - \tau)\zeta(\flat_2, \flat_1)) \mathrm{d}\tau,
\end{aligned}$$

which is given by Ref. [3].

Corollary 15 ([5]). *If all the assumptions of Lemma* 13 *hold, then for* $\psi(\tau) = \frac{\tau^\alpha}{\Gamma(\alpha)}$, *we have*

$$\begin{aligned}
&\frac{\Upsilon(\flat_1) + \Upsilon(\flat_1 + \zeta(\flat_2, \flat_1))}{2} \\
&\quad - \frac{\Gamma(\alpha + 1)}{2[\zeta(\flat_2, \flat_1)]^\alpha} \left[I_{\flat_1^+}^\alpha \Upsilon(\flat_1 + \zeta(\flat_2, \flat_1)) + I_{(\flat_1 + \zeta(\flat_2, \flat_1))^-}^\alpha \Upsilon(\flat_1) \right] \\
&= \frac{\zeta(\flat_2, \flat_1)}{2} \int_0^1 [(1 - \tau)^\alpha - \tau^\alpha] \Upsilon'(\flat_1 + (1 - \tau)\zeta(\flat_2, \flat_1)) \mathrm{d}\tau.
\end{aligned}$$

Corollary 16. *If all the assumptions of Lemma* 13 *hold, then for* $\psi(\tau) = \frac{\tau^{\frac{\alpha}{k}}}{k\Gamma_k(\alpha)}$, *we have*

$$\begin{aligned}
&\frac{\Upsilon(\flat_1) + \Upsilon(\flat_1 + \zeta(\flat_2, \flat_1))}{2} \\
&\quad - \frac{\Gamma_k(\alpha + k)}{2[\zeta(\flat_2, \flat_1)]^{\frac{\alpha}{k}}} \left[I_{\flat_1^+, k}^\alpha \Upsilon(\flat_1 + \zeta(\flat_2, \flat_1)) + I_{(\flat_1 + \zeta(\flat_2, \flat_1))^-, k}^\alpha \Upsilon(\flat_1) \right] \\
&= \frac{\zeta(\flat_2, \flat_1)}{2} \int_0^1 \left[(1 - \tau)^{\frac{\alpha}{k}} - \tau^{\frac{\alpha}{k}} \right] \Upsilon'(\flat_1 + (1 - \tau)\zeta(\flat_2, \flat_1)) \mathrm{d}\tau.
\end{aligned}$$

Corollary 17. *If all the assumptions of Lemma 13 hold, then for* $\psi(\tau) = \frac{\tau}{\alpha} e^{\left(-\frac{1-\alpha}{\alpha}\tau\right)}$, *we have*

$$\begin{aligned}
&\frac{\Upsilon(\flat_1) + \Upsilon(\flat_1 + \zeta(\flat_2, \flat_1))}{2} \\
&\quad - \frac{1-\alpha}{2\left(1 - e^{-A}\right)} \left[I^{\alpha}_{\flat_1^+} \Upsilon(\flat_1 + \zeta(\flat_2, \flat_1)) + I^{\alpha}_{(\flat_1 + \zeta(\flat_2, \flat_1))^-} \Upsilon(\flat_1) \right] \\
&= \frac{\zeta(\flat_2, \flat_1)}{2\left(1 - e^{-A}\right)} \int_0^1 \left[e^{-A\tau} - e^{-A(1-\tau)} \right] \Upsilon'(\flat_1 + (1-\tau)\zeta(\flat_2, \flat_1)) \mathrm{d}\tau,
\end{aligned}$$

where $A = \frac{1-\alpha}{\alpha}\zeta(\flat_2, \flat_1)$.

Lemma 18. *Let the function* $\Upsilon : [\flat_1, \flat_1 + \zeta(\flat_2, \flat_1)] \to \mathbb{R}$ *be differentiable on* $(\flat_1, \flat_1 + \zeta(\flat_2, \flat_1))$ *with* $\zeta(\flat_2, \flat_1) > 0$. *If* $\Upsilon' \in L[\flat_1, \flat_1 + \zeta(\flat_2, \flat_1)]$, *then*

$$\begin{aligned}
&\Upsilon\left(\frac{2\flat_1 + \zeta(\flat_2, \flat_1)}{2} \right) \\
&\quad - \frac{1}{2\Delta(1)} \left[{}_{\flat_1^+}I_{\psi} \Upsilon(\flat_1 + \zeta(\flat_2, \flat_1)) + {}_{(\flat_1 + \zeta(\flat_2, \flat_1))^-} I_{\psi} \Upsilon(\flat_1) \right] \\
&= \frac{\zeta(\flat_2, \flat_1)}{2\Delta(1)} \sum_{i=1}^{4} J_i,
\end{aligned}$$

where

$$\begin{aligned}
J_1 &= \int_0^{\frac{1}{2}} \Delta(\tau)\Upsilon'(\flat_1 + \tau\zeta(\flat_2, \flat_1))\mathrm{d}\tau, \\
J_2 &= \int_0^{\frac{1}{2}} -\Delta(\tau)\Upsilon'(\flat_1 + (1-\tau)\zeta(\flat_2, \flat_1))\mathrm{d}\tau, \\
J_3 &= \int_{\frac{1}{2}}^{1} -\nabla(\tau)\Upsilon'(\flat_1 + \tau\zeta(\flat_2, \flat_1))\mathrm{d}\tau, \\
J_4 &= \int_{\frac{1}{2}}^{1} \nabla(\tau)\Upsilon'(\flat_1 + (1-\tau)\zeta(\flat_2, \flat_1))\mathrm{d}\tau.
\end{aligned}$$

Proof. Integrating by parts, we have

$$J_1 = \int_0^{\frac{1}{2}} \Delta(\tau)\Upsilon'(\flat_1 + \tau\zeta(\flat_2, \flat_1))d\tau$$

$$= \frac{1}{\zeta(\flat_2, \flat_1)} \left[\Upsilon\left(\frac{2\flat_1 + \zeta(\flat_2, \flat_1)}{2}\right) \int_0^{\frac{1}{2}} \frac{\psi[\zeta(\flat_2, \flat_1)u]}{u} du \right.$$

$$\left. - \int_0^{\frac{1}{2}} \frac{\psi[\zeta(\flat_2, \flat_1)\tau]}{\tau} \Upsilon(\flat_1 + \tau\zeta(\flat_2, \flat_1))d\tau\right].$$

Similarly,

$$J_2 = \int_0^{\frac{1}{2}} -\Delta(\tau)\Upsilon'(\flat_1 + (1-\tau)\zeta(\flat_2, \flat_1))d\tau$$

$$= \frac{1}{\zeta(\flat_2, \flat_1)} \left[\Upsilon\left(\frac{2\flat_1 + \zeta(\flat_2, \flat_1)}{2}\right) \int_0^{\frac{1}{2}} \frac{\psi[\zeta(\flat_2, \flat_1)u]}{u} du \right.$$

$$\left. - \int_0^{\frac{1}{2}} \frac{\psi[\zeta(\flat_2, \flat_1)\tau]}{\tau} \Upsilon(\flat_1 + (1-\tau)\zeta(\flat_2, \flat_1))d\tau\right],$$

$$J_3 = \int_{\frac{1}{2}}^1 -\nabla(\tau)\Upsilon'(\flat_1 + \tau\zeta(\flat_2, \flat_1))d\tau$$

$$= \frac{1}{\zeta(\flat_2, \flat_1)} \left[\Upsilon\left(\frac{2\flat_1 + \zeta(\flat_2, \flat_1)}{2}\right) \int_{\frac{1}{2}}^1 \frac{\psi[\zeta(\flat_2, \flat_1)u]}{u} du \right.$$

$$\left. - \int_{\frac{1}{2}}^1 \frac{\psi[\zeta(\flat_2, \flat_1)\tau]}{\tau} \Upsilon(\flat_1 + \tau\zeta(\flat_2, \flat_1))d\tau\right],$$

$$J_4 = \int_{\frac{1}{2}}^1 -\nabla(\tau)\Upsilon'(\flat_1 + (1-\tau)\zeta(\flat_2, \flat_1))d\tau$$

$$= \frac{1}{\zeta(\flat_2, \flat_1)} \left[\Upsilon\left(\frac{2\flat_1 + \zeta(\flat_2, \flat_1)}{2}\right) \int_{\frac{1}{2}}^1 \frac{\psi[\zeta(\flat_2, \flat_1)u]}{u} du \right.$$

$$\left. - \int_{\frac{1}{2}}^1 \frac{\psi[\zeta(\flat_2, \flat_1)\tau]}{\tau} \Upsilon(\flat_1 + (1-\tau)\zeta(\flat_2, \flat_1))d\tau\right].$$

Adding J_1, J_2, J_3, and J_4, then multiplying both sides by $\frac{\zeta(\flat_2,\flat_1)}{2\Delta(1)}$, we get the required result. □

Following are some special cases of Lemma 18.

Corollary 19. *If all the assumptions of Lemma* 18 *hold, then for* $\psi(\tau)=\tau$, *we have*

$$\frac{1}{\zeta(\flat_2,\flat_1)}\int_{\flat_1}^{\flat_1+\zeta(\flat_2,\flat_1)}\Upsilon(x)\mathrm{d}x-\Upsilon\left(\frac{2\flat_1+\zeta(\flat_2,\flat_1)}{2}\right)$$
$$=\zeta(\flat_2,\flat_1)\left[\int_0^{\frac{1}{2}}\tau\Upsilon'(+(1-\tau)\zeta(\flat_2,\flat_1))\mathrm{d}\tau\right.$$
$$\left.+\int_{\frac{1}{2}}^{1}(\tau-1)\Upsilon'(\flat_1+(1-\tau)\zeta(\flat_2,\flat_1))\mathrm{d}\tau\right].$$

Corollary 20. *If all the assumptions of Lemma* 18 *hold, then for* $\psi(\tau)=\frac{\tau^\alpha}{\Gamma(\alpha)}$, *we have*

$$\Upsilon\left(\frac{2\flat_1+\zeta(\flat_2,\flat_1)}{2}\right)$$
$$-\frac{\Gamma(\alpha+1)}{2[\zeta(\flat_2,\flat_1)]^\alpha}\left[I^\alpha_{\flat_1^+}\Upsilon(\flat_1+\zeta(\flat_2,\flat_1))+I^\alpha_{(\flat_1+\zeta(\flat_2,\flat_1))^-}\Upsilon(\flat_1)\right]$$
$$=\frac{\zeta(\flat_2,\flat_1)}{2}[K_1+K_2+K_3+K_4],$$

where

$$K_1=\int_0^{\frac{1}{2}}\tau^\alpha\Upsilon'(\flat_1+\tau\zeta(\flat_2,\flat_1))\mathrm{d}\tau,$$
$$K_2=\int_0^{\frac{1}{2}}-\tau^\alpha\Upsilon'(\flat_1+(1-\tau)\zeta(\flat_2,\flat_1))\mathrm{d}\tau,$$
$$K_3=\int_{\frac{1}{2}}^{1}(\tau^\alpha-1)\Upsilon'(\flat_1+\tau\zeta(\flat_2,\flat_1))\mathrm{d}\tau,$$
$$K_4=\int_{\frac{1}{2}}^{1}(1-\tau^\alpha)\Upsilon'(\flat_1+(1-\tau)\zeta(\flat_2,\flat_1))\mathrm{d}\tau.$$

Corollary 21. *If all the assumptions of Lemma 18 hold, then for* $\psi(\tau)=\frac{\tau^{\frac{\alpha}{k}}}{k\Gamma_k(\alpha)}$*, we have*

$$\begin{aligned}\Upsilon&\left(\frac{2\flat_1+\zeta(\flat_2,\flat_1)}{2}\right)\\&-\frac{\Gamma(\alpha+1)}{2[\zeta(\flat_2,\flat_1)]^{\frac{\alpha}{k}}}\left[I^{\alpha}_{\flat_1^+,k}\Upsilon(\flat_1+\zeta(\flat_2,\flat_1))+I^{\alpha}_{(\flat_1+\zeta(\flat_2,\flat_1))^-,k}\Upsilon(\flat_1)\right]\\=&\frac{\zeta(\flat_2,\flat_1)}{2}[S_1+S_2+S_3+S_4],\end{aligned}$$

where

$$\begin{aligned}S_1&=\int_0^{\frac{1}{2}}\tau^{\frac{\alpha}{k}}\Upsilon'(\flat_1+\tau\zeta(\flat_2,\flat_1))\mathrm{d}\tau,\\S_2&=\int_0^{\frac{1}{2}}-\tau^{\frac{\alpha}{k}}\Upsilon'(\flat_1+(1-\tau)\zeta(\flat_2,\flat_1))\mathrm{d}\tau,\\S_3&=\int_{\frac{1}{2}}^1(\tau^{\frac{\alpha}{k}}-1)\Upsilon'(\flat_1+\tau\zeta(\flat_2,\flat_1))\mathrm{d}\tau,\\S_4&=\int_{\frac{1}{2}}^1(1-\tau^{\frac{\alpha}{k}})\Upsilon'(\flat_1+(1-\tau)\zeta(\flat_2,\flat_1))\mathrm{d}\tau.\end{aligned}$$

4. New Inequalities for General Fractional Integrals

Using Lemma 13 as an auxiliary result, we in this section obtain the right estimates of Hermite–Hadamard–Noor-type inequality for the functions whose first-order derivative's absolute values are n-polynomial preinvex functions.

Theorem 22. *Let* $\Upsilon:[\flat_1,\flat_1+\zeta(\flat_2,\flat_1)]\to\mathbb{R}$ *be a differentiable mapping on* $(\flat_1,\flat_1+\zeta(\flat_2,\flat_1))$ *with* $0<\zeta(\flat_2,\flat_1)$*. If* $|\Upsilon'|$ *is an* n*-polynomial preinvex function, then*

$$\Bigg|\frac{\Upsilon(\flat_1)+\Upsilon(\flat_1+\zeta(\flat_2,\flat_1))}{2}\\-\frac{1}{2\Delta(1)}\left[{}_{\flat_1^+}I_\psi\,\Upsilon(\flat_1+\zeta(\flat_2,\flat_1))+{}_{(\flat_1+\zeta(\flat_2,\flat_1))^-}I_\psi\,\Upsilon(\flat_1)\right]\Bigg|$$

$$\leq \frac{\zeta(\flat_2,\flat_1)}{n\Delta(1)}\sum_{s=1}^{n}\int_0^1(1-\tau^s)\left|\int_0^{1-\tau}\frac{\psi(\zeta(\flat_2,\flat_1)u)}{u}\mathrm{d}u - \int_0^{\tau}\frac{\psi(\zeta(\flat_2,\flat_1)u)}{u}\mathrm{d}u\right|\frac{|\Upsilon'(\flat_1)|+|\Upsilon'(\flat_2)|}{2}\mathrm{d}\tau. \quad (10)$$

Proof. Lemma 13 and n-polynomial preinvexity property of $|\Upsilon'|$ imply that

$$\begin{aligned}
&\left|\frac{\Upsilon(\flat_1)+\Upsilon(\flat_1+\zeta(\flat_2,\flat_1))}{2}\right.\\
&\quad\left. - \frac{1}{2\Delta(1)}\left[{}_{\flat_1^+}I_\psi\,\Upsilon(\flat_1+\zeta(\flat_2,\flat_1)) + {}_{(\flat_1+\zeta(\flat_2,\flat_1))^-}I_\psi\,\Upsilon(\flat_1)\right]\right|\\
&\leq \frac{\zeta(\flat_2,\flat_1)}{2\Delta(1)}\int_0^1\left[\left|\int_0^{1-\tau}\frac{\psi(\zeta(\flat_2,\flat_1)u)}{u}\mathrm{d}u - \int_0^{\tau}\frac{\psi(\zeta(\flat_2,\flat_1)u)}{u}\mathrm{d}u\right|\right]\\
&\quad\times\left|\Upsilon'(\flat_1+(1-\tau)\zeta(\flat_2,,\flat_1))\right|\mathrm{d}\tau\\
&\leq \frac{\zeta(\flat_2,\flat_1)}{2\Delta(1)}\frac{1}{n}\sum_{s=1}^{n}\int_0^1(1-(1-\tau)^s)\\
&\qquad\times\left|\int_0^{1-\tau}\frac{\psi(\zeta(\flat_2,\flat_1)u)}{u}\mathrm{d}u - \int_0^{\tau}\frac{\psi(\zeta(\flat_2,\flat_1)u)}{u}\mathrm{d}u\right||\Upsilon'(\flat_1)|\mathrm{d}\tau\\
&\quad+\frac{\zeta(\flat_2,\flat_1)}{2\Delta(1)}\frac{1}{n}\sum_{s=1}^{n}\int_0^1(1-\tau^s)\\
&\qquad\times\left|\int_0^{1-\tau}\frac{\psi(\zeta(\flat_2,\flat_1)u)}{u}\mathrm{d}u - \int_0^{\tau}\frac{\psi(\zeta(\flat_2,\flat_1)u)}{u}\mathrm{d}u\right||\Upsilon'(\flat_2)|\mathrm{d}\tau.
\end{aligned}$$

On using change of variable technique, we have

$$\begin{aligned}
&\left|\frac{\Upsilon(\flat_1)+\Upsilon(\flat_1+\zeta(\flat_2,\flat_1))}{2}\right.\\
&\quad\left.-\frac{1}{2\Delta(1)}\left[{}_{\flat_1^+}I_\psi\,\Upsilon(\flat_1+\zeta(\flat_2,\flat_1)) + {}_{(\flat_1+\zeta(\flat_2,\flat_1))^-}I_\psi\,\Upsilon(\flat_1)\right]\right|
\end{aligned}$$

$$\leq \frac{\zeta(\flat_2,\flat_1)}{\Delta(1)}\frac{1}{n}\sum_{s=1}^{n}\int_0^1(1-\tau^s)\left|\int_0^{1-\tau}\frac{\psi(\zeta(\flat_2,\flat_1)u)}{u}\mathrm{d}u\right.$$
$$\left.-\int_0^{\tau}\frac{\psi(\zeta(\flat_2,\flat_1)u)}{u}\mathrm{d}u\right|\frac{|\Upsilon'(\flat_1)|+|\Upsilon'(\flat_2)|}{2}\mathrm{d}\tau,$$

which is required. □

We now discuss some special cases of Theorem 22.

Corollary 23. *If all the assumptions of Theorem 22 hold, if* $\psi(\tau)=\tau$, *then we have*

$$\left|\frac{\Upsilon(\flat_1)+\Upsilon(\flat_1+\zeta(\flat_2,\flat_1))}{2}-\frac{1}{\zeta(\flat_2,\flat_1)}\int_{\flat_1}^{\flat_1+\zeta(\flat_2,\flat_1)}\Upsilon(x)\mathrm{d}x\right|$$
$$\leq\frac{\zeta(\flat_2,\flat_1)}{n}\sum_{s=1}^{n}\left(\frac{1}{2}-\frac{1+s.2^s}{2^s(s+1)(s+2)}\right)\left(\frac{|\Upsilon'(\flat_1)|+|\Upsilon'(\flat_2)|}{2}\right).$$

Corollary 24. *If all the assumptions of Theorem 22 hold, if* $\psi(\tau)=\frac{\tau^\alpha}{\Gamma(\alpha)}$, *then we have*

$$\left|\frac{\Upsilon(\flat_1)+\Upsilon(\flat_1+\zeta(\flat_2,\flat_1))}{2}\right.$$
$$\left.-\frac{\Gamma(\alpha+1)}{2[\zeta(\flat_2,\flat_1)]^\alpha}\left[I^\alpha_{(\flat_1+\zeta(\flat_2,\flat_1))^-}\Upsilon(\flat_1)+I^\alpha_{\flat_1^+}\Upsilon(\flat_1+\zeta(\flat_2,\flat_1))\right]\right|$$
$$\leq\zeta(\flat_2,\flat_1)\frac{1}{n}\sum_{s=1}^{n}M(\alpha,s)\frac{|\Upsilon'(\flat_1)|+|\Upsilon'(\flat_2)|}{2},$$

where

$$M(\alpha,s)=\frac{2^{\alpha+1}-2}{2^\alpha(\alpha+1)}+\frac{1-2^{\alpha+s}}{2^{\alpha+s}(\alpha+s+1)}$$
$$+\frac{1}{2^{\alpha+1}}\beta(\alpha+1,1)_2\mathcal{F}_1\left(-s,\alpha+1,\alpha+2;\frac{1}{2}\right)$$
$$-\frac{1}{2^{\alpha+s+1}}\beta(1,s+1)_2\mathcal{F}_1(-\alpha,1,s+2;-1).$$

Corollary 25. *If all the assumptions of Theorem 22 hold, if* $\psi(\tau) = \frac{\tau^{\frac{\alpha}{k}}}{k\Gamma_k(\alpha)}$*, then we have*

$$\begin{aligned}
&\left|\frac{\Upsilon(\flat_1) + \Upsilon(\flat_1 + \zeta(\flat_2, \flat_1))}{2}\right. \\
&\qquad \left. - \frac{\Gamma_k(\alpha + k)}{2[\zeta(\flat_2, \flat_1)]^{\frac{\alpha}{k}}} \left[J^{\alpha}_{\flat_1^+, k}\Upsilon(\flat_1 + \zeta(\flat_2, \flat_1)) + J^{\alpha}_{(\flat_1+\zeta(\flat_2,\flat_1))^-, k}\Upsilon(\flat_1)\right]\right| \\
&\quad \leq \frac{\zeta(\flat_2, \flat_1)}{n} \sum_{s=1}^{n} N(\alpha, s, k) \frac{|\Upsilon'(\flat_1)| + |\Upsilon'(\flat_2)|}{2},
\end{aligned}$$

where

$$\begin{aligned}
N(\alpha, s, k) &= \frac{k(2^{\frac{\alpha}{k}+1} - 2)}{2^{\frac{\alpha}{k}}(\alpha + k)} + \frac{k(1 - 2^{\frac{\alpha}{k}+s})}{2^{\frac{\alpha}{k}+s}(\alpha + sk + k)} \\
&\quad + \frac{kB_k(\alpha + k, k)}{2^{\frac{\alpha}{k}+1}} {}_2\mathcal{F}_{1,k}\left(-ks, \alpha + k, \alpha + 2k, \frac{1}{2k}\right) \\
&\quad - \frac{k}{2^{s+\frac{\alpha}{k}+1}} \beta_k(k, k(s+1)) {}_2\mathcal{F}_{1,k}\left(-\alpha, k, k(s+2); -\frac{1}{k}\right).
\end{aligned}$$

Theorem 26. *Let* $\Upsilon : [\flat_1, \flat_1 + \zeta(\flat_2, \flat_1)] \to \mathbb{R}$ *be a differentiable mapping on* $(\flat_1, \flat_1 + \zeta(\flat_2, \flat_1))$ *with* $0 < \zeta(\flat_2, \flat_1)$*. If* $|\Upsilon'|^q$ *is an* n*-polynomial preinvex function, then*

$$\begin{aligned}
&\left|\frac{\Upsilon(\flat_1) + \Upsilon(\flat_1 + \zeta(\flat_2, \flat_1))}{2}\right. \\
&\qquad \left. - \frac{1}{2\Delta(1)} \left[{}_{\flat_1^+}I_{\psi}\, \Upsilon(\flat_1 + \zeta(\flat_2, \flat_1)) + {}_{(\flat_1+\zeta(\flat_2,\flat_1))^-}\, I_{\psi}\, \Upsilon(\flat_1)\right]\right| \\
&\leq \frac{\zeta(\flat_2, \flat_1)}{2\Delta(1)} \left(\int_0^1 \left|\int_0^{1-\tau} \frac{\psi(\zeta(\flat_2, \flat_1)u)}{u} \mathrm{d}u - \int_0^{\tau} \frac{\psi(\zeta(\flat_2, \flat_1)u)}{u} \mathrm{d}u\right|^p \mathrm{d}\tau\right)^{\frac{1}{p}} \\
&\quad \times \left(\frac{1}{n}\sum_{s=1}^{n} \int_0^1 (1 - (1-\tau)^s)|\Upsilon'(\flat_1|^q \mathrm{d}\tau + \frac{1}{n}\sum_{s=1}^{n} \int_0^1 (1 - \tau^s)|\Upsilon'(\flat_2)|^q \mathrm{d}\tau\right)^{\frac{1}{q}}.
\end{aligned} \tag{11}$$

Proof. Using Lemma 13 and well-known Hölder's inequality, we have

$$\left|\frac{\Upsilon(\flat_1)+\Upsilon(\flat_1+\zeta(\flat_2,\flat_1))}{2}\right.$$
$$\left.-\frac{1}{2\Delta(1)}\left[{}_{\flat_1^+}I_\psi\,\Upsilon(\flat_1+\zeta(\flat_2,\flat_1))+{}_{(\flat_1+\zeta(\flat_2,\flat_1))^-}I_\psi\,\Upsilon(\flat_1)\right]\right|$$
$$\leq\frac{\zeta(\flat_2,\flat_1)}{2\Delta(1)}\left(\int_0^1\left|\int_0^{1-\tau}\frac{\psi(\zeta(\flat_2,\flat_1)u)}{u}\mathrm{d}u-\int_0^{\tau}\frac{\psi(\zeta(\flat_2,\flat_1)u)}{u}\mathrm{d}u\right|^p\mathrm{d}\tau\right)^{\frac{1}{p}}$$
$$\times\left(\int_0^1|\Upsilon'(\flat_1+(1-\tau)\zeta(\flat_2,\flat_1))|^q\mathrm{d}\tau\right)^{\frac{1}{q}}$$
$$\leq\frac{\zeta(\flat_2,\flat_1)}{2\Delta(1)}\left(\int_0^1\left|\int_0^{1-\tau}\frac{\psi(\zeta(\flat_2,\flat_1)u)}{u}\mathrm{d}u-\int_0^{\tau}\frac{\psi(\zeta(\flat_2,\flat_1)u)}{u}\mathrm{d}u\right|^p\mathrm{d}\tau\right)^{\frac{1}{p}}$$
$$\times\left(\frac{1}{n}\sum_{s=1}^{n}\int_0^1(1-(1-\tau)^s)|\Upsilon'(\flat_1|^q\mathrm{d}\tau+\frac{1}{n}\sum_{s=1}^{n}\int_0^1(1-\tau^s)|\Upsilon'(\flat_2)|^q\mathrm{d}\tau\right)^{\frac{1}{q}},$$

which is required. □

We now present some special cases of Theorem 26.

Corollary 27. *If all the assumptions of Theorem 26 hold, if* $\psi(\tau)=\tau$, *then we have*

$$\left|\frac{\Upsilon(\flat_1)+\Upsilon(\flat_1+\zeta(\flat_2,\flat_1))}{2}-\frac{1}{\zeta(\flat_2,\flat_1)}\int_{\flat_1}^{\flat_1+\zeta(\flat_2,\flat_1)}\Upsilon(x)\mathrm{d}x\right|$$
$$\leq\frac{\zeta(\flat_2,\flat_1)}{2(p+1)^{\frac{1}{p}}}\left(\frac{1}{n}\sum_{s=1}^{n}\frac{s}{s+1}(|\Upsilon'(\flat_1)|^q+|\Upsilon'(\flat_2)|^q)\right)^{\frac{1}{q}}.$$

Corollary 28. *If all the assumptions of Theorem 26 hold, if* $\psi(\tau) = \frac{\tau^{\alpha}}{\Gamma(\alpha)}$, *then we have*

$$\left|\frac{\Upsilon(\flat_1)+\Upsilon(\flat_1+\zeta(\flat_2,\flat_1))}{2} - \frac{\Gamma(\alpha+1)}{2[\zeta(\flat_2,\flat_1)]^{\alpha}}\left[I^{\alpha}_{(\flat_1+\zeta(\flat_2,\flat_1))-}\Upsilon(\flat_1)+I^{\alpha}_{\flat_1+}\Upsilon(\flat_1+\zeta(\flat_2,\flat_1))\right]\right| \leq \frac{\zeta(\flat_2,\flat_1)}{2(\alpha p+1)^{\frac{1}{p}}}\left[\frac{1}{n}\sum_{s=1}^{n}\frac{s}{s+1}(|\Upsilon'(\flat_1)|^q+|\Upsilon'(\flat_2)|^q)\right]^{\frac{1}{q}}.$$

Corollary 29. *If all the assumptions of Theorem 26 hold, if* $\psi(\tau) = \frac{\tau^{\frac{\alpha}{k}}}{k\Gamma_k(\alpha)}$, *then we have*

$$\left|\frac{\Upsilon(\flat_1)+\Upsilon(\flat_1+\zeta(\flat_2,\flat_1))}{2} - \frac{\Gamma_k(\alpha+k)}{2[\zeta(\flat_2,\flat_1)]^{\frac{\alpha}{k}}}\left[J^{\alpha}_{\flat_1+,k}\Upsilon(\flat_1+\zeta(\flat_2,\flat_1))+J^{\alpha}_{(\flat_1+\zeta(\flat_2,\flat_1))-,k}\Upsilon(\flat_1)\right]\right| \leq \frac{\zeta(\flat_2,\flat_1)k^{\frac{1}{p}}}{2(\alpha p+k)^{\frac{1}{p}}}\left[\frac{1}{n}\sum_{s=1}^{n}\frac{s}{s+1}(|\Upsilon'(\flat_1)|^q+|\Upsilon'(\flat_2)|^q)\right]^{\frac{1}{q}}.$$

5. Midpoint-type Inequalities via General Fractional Integrals

Theorem 30. *Let* $\Upsilon : [\flat_1, \flat_1+\zeta(\flat_2,\flat_1)] \to \mathbb{R}$ *be a differentiable mapping on* $(\flat_1, \flat_1+\zeta(\flat_2,\flat_1))$ *with* $0 < \zeta(\flat_2,\flat_1)$. *If* $|\Upsilon'|$ *is an* n*-polynomial preinvex function, then*

$$\left|\Upsilon\left(\frac{2\flat_1+\zeta(\flat_2,\flat_1)}{2}\right) - \frac{1}{2\Delta(1)}\left[{}_{\flat_1+}I_{\psi}\,\Upsilon(\flat_1+\zeta(\flat_2,\flat_1))+{}_{(\flat_1+\zeta(\flat_2,\flat_1))-}I_{\psi}\,\Upsilon(\flat_1)\right]\right|$$

$$\leq \frac{\zeta(\flat_2,\flat_1)}{2\Delta(1)}\left[\frac{|\Upsilon'(\flat_1)|+|\Upsilon'(\flat_2)|}{n}\int_0^{\frac{1}{2}}|\Delta(\tau)|\sum_{s=1}^{n}[2-\tau^s-(1-\tau)^s]\mathrm{d}\tau\right.$$

$$\left.+\frac{|\Upsilon'(\flat_1)|+|\Upsilon'(\flat_2)|}{n}\int_{\frac{1}{2}}^{1}|\nabla(\tau)|\sum_{s=1}^{n}[2-\tau^s-(1-\tau)^s]\mathrm{d}\tau\right].$$

Proof. Using Lemma 18 and n-polynomial preinvexity property of $|\Upsilon'|$, we have

$$\left|\Upsilon\left(\frac{2\flat_1+\zeta(\flat_2,\flat_1)}{2}\right)\right.$$

$$\left.-\frac{1}{2\Delta(1)}\left[{}_{\flat_1^+}I_{\psi}\,\Upsilon(\flat_1+\zeta(\flat_2,\flat_1))+{}_{(\flat_1+\zeta(\flat_2,\flat_1))^-}\,I_{\psi}\,\Upsilon(\flat_1)\right]\right|$$

$$\leq \frac{\zeta(\flat_2,\flat_1)}{2\Delta(1)}\left[\int_0^{\frac{1}{2}}|\Delta(\tau)||\Upsilon'(\flat_1+\tau\zeta(\flat_2,\flat_1))|\mathrm{d}\tau\right.$$

$$+\int_0^{\frac{1}{2}}|\Delta(\tau)||\Upsilon'(\flat_1+(1-\tau)\zeta(\flat_2,\flat_1))|\mathrm{d}\tau$$

$$+\int_{\frac{1}{2}}^{1}|\nabla(\tau)||\Upsilon'(\flat_1+\tau\zeta(\flat_2,\flat_1))|\mathrm{d}\tau$$

$$\left.+\int_{\frac{1}{2}}^{1}|\nabla(\tau)||\Upsilon'(\flat_1+(1-\tau)\zeta(\flat_2,\flat_1))|\mathrm{d}\tau\right]$$

$$\leq \frac{\zeta(\flat_2,\flat_1)}{2\Delta(1)}\left\{\int_0^{\frac{1}{2}}|\Delta(\tau)|\left[\frac{1}{n}\sum_{s=1}^{n}(1-\tau^s)|\Upsilon'(\flat_1)|\right.\right.$$

$$\left.+\frac{1}{n}\sum_{s=1}^{n}(1-(1-\tau)^s)|\Upsilon'(\flat_2)|\right]\mathrm{d}\tau$$

$$+\int_0^{\frac{1}{2}}|\Delta(\tau)|\left[\frac{1}{n}\sum_{s=1}^{n}(1-(1-\tau)^s)|\Upsilon'(\flat_1)|\right.$$

$$\left.+\frac{1}{n}\sum_{s=1}^{n}(1-\tau^s)|\Upsilon'(\flat_2)|\right]\mathrm{d}\tau$$

$$+\int_{\frac{1}{2}}^{1}|\nabla(\tau)|\left[\frac{1}{n}\sum_{s=1}^{n}(1-\tau^s)|\Upsilon'(\flat_1)|\right.$$
$$\left.+\frac{1}{n}\sum_{s=1}^{n}(1-(1-\tau)^s|\Upsilon'(\flat_2)|)\right]\mathrm{d}\tau$$
$$+\int_{\frac{1}{2}}^{1}|\nabla(\tau)|\left[\frac{1}{n}\sum_{s=1}^{n}(1-(1-\tau)^s)|\Upsilon'(\flat_1)|\right.$$
$$\left.\left.+\frac{1}{n}\sum_{s=1}^{n}(1-\tau^s)|\Upsilon'(\flat_2)|\right]\mathrm{d}\tau\right\}$$
$$=\frac{\zeta(\flat_2,\flat_1)}{2n\Delta(1)}\int_0^{\frac{1}{2}}|\Delta(\tau)|\sum_{s=1}^{n}[2-\tau^s-(1-\tau)^s](|\Upsilon'(\flat_1)|+|\Upsilon'(\flat_2)|)\mathrm{d}\tau$$
$$+\frac{\zeta(\flat_2,\flat_1)}{2n\Delta(1)}\int_{\frac{1}{2}}^{1}|\nabla(\tau)|\sum_{s=1}^{n}[2-\tau^s-(1-\tau)^s](|\Upsilon'(\flat_1)|+|\Upsilon'(\flat_2)|)\mathrm{d}\tau.$$

This completes the proof. □

We now discuss some special cases of Theorem 30.

Corollary 31. *If all the assumptions of Theorem* 30 *hold, then for* $\psi(\tau)=\tau$, *we have*

$$\left|\Upsilon\left(\frac{2\flat_1+\zeta(\flat_2,\flat_1)}{2}\right)-\frac{1}{\zeta(\flat_2,\flat_1)}\int_{\flat_1}^{\flat_1+\zeta(\flat_2,\flat_1)}\Upsilon(x)\mathrm{d}x\right|$$
$$\leq\frac{\zeta(\flat_2,\flat_1)}{n}\sum_{s=1}^{n}\left[\frac{2^{s-1}(s+1)(s+2)+1-2^{s+1}}{2^{s+1}(s+1(s+2))}\right](|\Upsilon'(\flat_1)|+|\Upsilon'(\flat_2)|).$$

Remark 32. If all the assumptions of Theorem 30 hold, then for $\psi(\tau)=\tau$ and $n=1$, we have

$$\left|\Upsilon\left(\frac{2\flat_1+\zeta(\flat_2,\flat_1)}{2}\right)-\frac{1}{\zeta(\flat_2,\flat_1)}\int_{\flat_1}^{\flat_1+\zeta(\flat_2,\flat_1)}\Upsilon(x)\mathrm{d}x\right|$$
$$\leq\frac{\zeta(\flat_2,\flat_1)}{8}[|\Upsilon'(\flat_1)|+|\Upsilon'(\flat_2)|].$$

Corollary 33. *If all the assumptions of Theorem 30 hold, then for* $\psi(\tau) = \frac{\tau^{\alpha}}{\Gamma(\alpha)}$, *we have*

$$\left|\Upsilon\left(\frac{2\flat_1 + \zeta(\flat_2, \flat_1)}{2}\right) - \frac{\Gamma(\alpha+1)}{2[\zeta(\flat_2, \flat_1)]^{\alpha}}\left[I^{\alpha}_{\flat_1^+}\Upsilon(\flat_1 + \zeta(\flat_2, \flat_1)) + I^{\alpha}_{(\flat_1+\zeta(\flat_2,\flat_1))^-}\Upsilon(\flat_1)\right]\right| \leq \frac{\zeta(\flat_2, \flat_1)}{2n}\sum_{s=1}^{n} H(\alpha, s)(|\Upsilon'(\flat_1)| + |\Upsilon'(\flat_2)|),$$

where

$$H(\alpha, s) = \frac{1}{2^{\alpha-1}(\alpha+1)} - \frac{1}{2^{\alpha+s}(\alpha+s+1)} - \frac{2}{\alpha+1} + \frac{1}{\alpha+s+1} - \frac{1}{s+1} - \frac{1}{2^{\alpha+s+1}}\beta(1, \alpha+1)\,{}_2\mathcal{F}_1(-s, 1, \alpha+2; -1) + \frac{\beta(s+1, 1)}{2^{s+1}}\,{}_2\mathcal{F}_1\left(\alpha, s+1, s+2; \frac{1}{2}\right) + 1.$$

Corollary 34. *If all the assumptions of Theorem 30 hold, then for* $\psi(\tau) = \frac{\tau^{\frac{\alpha}{k}}}{k\Gamma_k(\alpha)}$, *we have*

$$\left|\Upsilon\left(\frac{2\flat_1 + \zeta(\flat_2, \flat_1)}{2}\right) - \frac{\Gamma(\alpha+k)}{2[\zeta(\flat_2, \flat_1)]^{\frac{\alpha}{k}}}\left[I^{\alpha}_{\flat_1^+, k}\Upsilon(\flat_1 + \zeta(\flat_2, \flat_1)) + I^{\alpha}_{(\flat_1+\zeta(\flat_2,\flat_1))^-, k}\Upsilon(\flat_1)\right]\right| \leq \frac{\zeta(\flat_2, \flat_1)}{2n}\sum_{s=1}^{n}\left[\frac{4k(1-2^{\frac{\alpha}{k}})}{2^{\frac{\alpha}{k}+1}(\alpha+k)} - \frac{2k - k2^{\frac{\alpha}{k}+s+1}}{2^{\frac{\alpha}{k}+s+1}(\alpha+sk+k)} - \frac{1}{s+1} + 1 - \frac{k\beta_k(k, \alpha+k)}{2^{\frac{\alpha}{k}+s+1}}\,{}_2\mathcal{F}_{1,k}\left(-ks, k, \alpha+2k, \frac{-1}{k}\right) + \frac{k\beta_k(k(s+1), k)}{2^{s+1}}\,{}_2\mathcal{F}_{1,k}\left(\alpha, k(s+1), k(s+2); \frac{1}{2k}\right)\right] \times (|\Upsilon'(\flat_1)| + |\Upsilon'(\flat_2)|).$$

Theorem 35. *Let* $\Upsilon : [\flat_1, \flat_1 + \varsigma(\flat_2, \flat_1)] \rightarrow \mathbb{R}$ *be a differentiable mapping on* $(\flat_1, \flat_1 + \varsigma(\flat_2, \flat_1))$ *with* $0 < \varsigma(\flat_2, \flat_1)$. *If* $|\Upsilon'|^q$ *is an* n*-polynomial preinvex function for all* $q > 1$, *then*

$$\begin{aligned}
&\left|\Upsilon\left(\frac{2\flat_1 + \varsigma(\flat_2, \flat_1)}{2}\right)\right. \\
&\qquad \left. - \frac{1}{2\Delta(1)}\left[{}_{\flat_1^+}I_{\psi}\,\Upsilon(\flat_1 + \varsigma(\flat_2, \flat_1)) + {}_{(\flat_1+\varsigma(\flat_2,\flat_1))^-}I_{\psi}\,\Upsilon(\flat_1)\right]\right| \\
&\leq \frac{\varsigma(\flat_2, \flat_1)C^*}{2\Delta(1)} \\
&\qquad \times \left[\left(\frac{1}{n}\sum_{s=1}^{n}\frac{2^s(s+1) - 2^{s+1} + 1}{2^{s+1}(s+1)}|\Upsilon'(\flat_1)^q| + \frac{1}{n}\sum_{s=1}^{n}\frac{2^s(s+1) - 1}{2^{s+1}(s+1)}|\Upsilon'(\flat_2)|^q\right)^{\frac{1}{q}}\right. \\
&\qquad \left. + \left(\frac{1}{n}\sum_{s=1}^{n}\frac{2^s(s+1) - 2^{s+1} + 1}{2^{s+1}(s+1)}|\Upsilon'(\flat_2)^q| + \frac{1}{n}\sum_{s=1}^{n}\frac{2^s(s+1) - 1}{2^{s+1}(s+1)}|\Upsilon'(\flat_1)|^q\right)^{\frac{1}{q}}\right],
\end{aligned}$$

where

$$C^* = \left(\int_0^{\frac{1}{2}}|\Delta(\tau)|^p \mathrm{d}\tau\right)^{\frac{1}{p}} + \left(\int_{\frac{1}{2}}^{1}|\nabla(\tau)|^p \mathrm{d}\tau\right)^{\frac{1}{p}}.$$

Proof. Using Lemma 18, power mean inequality, and n-polynomial preinvexity property of $|\Upsilon'|^q$, we have

$$\begin{aligned}
&\left|\Upsilon\left(\frac{2\flat_1 + \varsigma(\flat_2, \flat_1)}{2}\right)\right. \\
&\qquad \left. - \frac{1}{2\Delta(1)}\left[{}_{\flat_1^+}I_{\psi}\,\Upsilon(\flat_1 + \varsigma(\flat_2, \flat_1)) + {}_{(\flat_1+\varsigma(\flat_2,\flat_1))^-}I_{\psi}\,\Upsilon(\flat_1)\right]\right| \\
&\leq \frac{\varsigma(\flat_2, \flat_1)}{2\Delta(1)}\left[\left(\int_0^{\frac{1}{2}}|\Delta(\tau)|^p \mathrm{d}\tau\right)^{\frac{1}{p}}\left(\int_0^{\frac{1}{2}}|\Upsilon'(\flat_1 + \tau\varsigma(\flat_2, \flat_1))|^q \mathrm{d}\tau\right)^{\frac{1}{q}}\right. \\
&\qquad + \left(\int_0^{\frac{1}{2}}|\Delta(\tau)|^p \mathrm{d}\tau\right)^{\frac{1}{p}}\left(\int_0^{\frac{1}{2}}|\Upsilon'(\flat_1 + (1-\tau)\varsigma(\flat_2, \flat_1))|^q \mathrm{d}\tau\right)^{\frac{1}{q}}
\end{aligned}$$

$$+\left(\int_{\frac{1}{2}}^{1}|\nabla(\tau)|^{p}\mathrm{d}\tau\right)^{\frac{1}{p}}\left(\int_{\frac{1}{2}}^{1}|\Upsilon'(\flat_1+\tau\zeta(\flat_2,\flat_1))|^{q}\mathrm{d}\tau\right)^{\frac{1}{q}}$$

$$+\left(\int_{\frac{1}{2}}^{1}|\nabla(\tau)|^{p}\mathrm{d}\tau\right)^{\frac{1}{p}}\left(\int_{\frac{1}{2}}^{1}|\Upsilon'(\flat_1+(1-\tau)\zeta(\flat_2,\flat_1))|^{q}\mathrm{d}\tau\right)^{\frac{1}{q}}\Bigg]$$

$$\leq\frac{\zeta(\flat_2,\flat_1)}{2\Delta(1)}\left[\left(\int_0^{\frac{1}{2}}|\Delta(\tau)|^{p}\mathrm{d}\tau\right)^{\frac{1}{p}}\right.$$

$$\times\left\{\left(\frac{1}{n}\sum_{s=1}^{n}\int_0^{\frac{1}{2}}(1-(1-\tau)^s)\,|\Upsilon'(\flat_1)|^q+\frac{1}{n}\sum_{s=1}^{n}\int_0^{\frac{1}{2}}(1-\tau^s)|\Upsilon'(\flat_2)|^q\right)^{\frac{1}{q}}\right.$$

$$\left.+\left(\frac{1}{n}\sum_{s=1}^{n}\int_0^{\frac{1}{2}}(1-(1-\tau)^s)\,|\Upsilon'(\flat_2)|^q+\frac{1}{n}\sum_{s=1}^{n}\int_0^{\frac{1}{2}}(1-\tau^s)|\Upsilon'(\flat_1)|^q\right)^{\frac{1}{q}}\right\}$$

$$+\left(\int_{\frac{1}{2}}^{1}|\nabla(\tau)|^{p}\mathrm{d}\tau\right)^{\frac{1}{p}}$$

$$\times\left\{\left(\frac{1}{n}\sum_{s=1}^{n}\int_{\frac{1}{2}}^{1}(1-(1-\tau)^s)\,|\Upsilon'(\flat_1)|^q+\frac{1}{n}\sum_{s=1}^{n}\int_{\frac{1}{2}}^{1}(1-\tau^s)|\Upsilon'(\flat_2)|^q\right)^{\frac{1}{q}}\right.$$

$$\left.+\left(\frac{1}{n}\sum_{s=1}^{n}\int_{\frac{1}{2}}^{1}(1-(1-\tau)^s)\,|\Upsilon'(\flat_2)|^q+\frac{1}{n}\sum_{s=1}^{n}\int_{\frac{1}{2}}^{1}(1-\tau^s)|\Upsilon'(\flat_1)|^q\right)^{\frac{1}{q}}\right\}$$

$$=\frac{\zeta(\flat_2,\flat_1)C^*}{2\Delta(1)}$$

$$\times\left[\left(\frac{1}{n}\sum_{s=1}^{n}\frac{2^s(s+1)-2^{s+1}+1}{2^{s+1}(s+1)}|\Upsilon'(\flat_1)^q|+\frac{1}{n}\sum_{s=1}^{n}\frac{2^s(s+1)-1}{2^{s+1}(s+1)}|\Upsilon'(\flat_2)|^q\right)^{\frac{1}{q}}\right.$$

$$\left.+\left(\frac{1}{n}\sum_{s=1}^{n}\frac{2^s(s+1)-2^{s+1}+1}{2^{s+1}(s+1)}|\Upsilon'(\flat_2)^q|+\frac{1}{n}\sum_{s=1}^{n}\frac{2^s(s+1)-1}{2^{s+1}(s+1)}|\Upsilon'(\flat_1)|^q\right)^{\frac{1}{q}}\right].$$

The proof is completed. □

Following are some special cases of Theorem 35:

Corollary 36. *If all the assumptions of Theorem 35 hold, then for* $\psi(\tau)=\tau$*, we have*

$$\begin{aligned}
&\left|\Upsilon\left(\frac{2\flat_1+\varsigma(\flat_2,\flat_1)}{2}\right)-\frac{1}{\varsigma(\flat_2,\flat_1)}\int_{\flat_1}^{\flat_1+\varsigma(\flat_2,\flat_1)}\Upsilon(x)\mathrm{d}x\right|\\
&\leq\frac{\varsigma(\flat_2,\flat_1)}{2^{1+\frac{1}{p}}(p+1)^{\frac{1}{p}}}\\
&\quad\times\left[\left(\frac{1}{n}\sum_{s=1}^{n}\frac{2^s(s+1)-2^{s+1}+1}{2^{s+1}(s+1)}|\Upsilon'(\flat_1)^q|+\frac{1}{n}\sum_{s=1}^{n}\frac{2^s(s+1)-1}{2^{s+1}(s+1)}|\Upsilon'(\flat_2)|^q\right)^{\frac{1}{q}}\right.\\
&\quad\left.+\left(\frac{1}{n}\sum_{s=1}^{n}\frac{2^s(s+1)-2^{s+1}+1}{2^{s+1}(s+1)}|\Upsilon'(\flat_2)^q|+\frac{1}{n}\sum_{s=1}^{n}\frac{2^s(s+1)-1}{2^{s+1}(s+1)}|\Upsilon'(\flat_1)|^q\right)^{\frac{1}{q}}\right].
\end{aligned}$$

Corollary 37. *If all the assumptions of Theorem 35 hold, then for* $\psi(\tau)=\frac{\tau^\alpha}{\Gamma(\alpha)}$*, we have*

$$\begin{aligned}
&\left|\Upsilon\left(\frac{2\flat_1+\varsigma(\flat_2,\flat_1)}{2}\right)\right.\\
&\quad\left.-\frac{\Gamma(\alpha+1)}{2[\varsigma(\flat_2,\flat_1)]^\alpha}\left[I_{\flat_1+}^{\alpha}\Upsilon(\flat_1+\varsigma(\flat_2,\flat_1))+I_{(\flat_1+\varsigma(\flat_2,\flat_1))-}^{\alpha}\Upsilon(\flat_1)\right]\right|\\
&\leq\frac{\varsigma(\flat_2,\flat_1)}{(\alpha p+1)^{\frac{1}{p}}2^\alpha}\\
&\quad\times\left[\left(\frac{1}{n}\sum_{s=1}^{n}\frac{2^s(s+1)-2^{s+1}+1}{2^{s+1}(s+1)}|\Upsilon'(\flat_1)^q|+\frac{1}{n}\sum_{s=1}^{n}\frac{2^s(s+1)-1}{2^{s+1}(s+1)}|\Upsilon'(\flat_2)|^q\right)^{\frac{1}{q}}\right.\\
&\quad\left.+\left(\frac{1}{n}\sum_{s=1}^{n}\frac{2^s(s+1)-2^{s+1}+1}{2^{s+1}(s+1)}|\Upsilon'(\flat_2)^q|+\frac{1}{n}\sum_{s=1}^{n}\frac{2^s(s+1)-1}{2^{s+1}(s+1)}|\Upsilon'(\flat_1)|^q\right)^{\frac{1}{q}}\right].
\end{aligned}$$

Corollary 38. *If all the assumptions of Theorem 35 hold, then for* $\psi(\tau)=\frac{\tau^{\frac{\alpha}{k}}}{k\Gamma_k(\alpha)}$*, we have*

$$\begin{aligned}&\left|\Upsilon\left(\frac{2\flat_1+\zeta(\flat_2,\flat_1)}{2}\right)\right.\\&\left.\quad-\frac{\Gamma(\alpha+k)}{2[\zeta(\flat_2,\flat_1)]^{\frac{\alpha}{k}}}\left[I^{\alpha}_{\flat_1^+,k}\Upsilon(\flat_1+\zeta(\flat_2,\flat_1))+I^{\alpha}_{(\flat_1+\zeta(\flat_2,\flat_1))^-,k}\Upsilon(\flat_1)\right]\right|\\&\leq\frac{\zeta(\flat_2,\flat_1)}{(\frac{\alpha}{k}p+1)^{\frac{1}{p}}2^{\frac{\alpha}{k}}}\\&\quad\times\left[\left(\frac{1}{n}\sum_{s=1}^{n}\frac{2^s(s+1)-2^{s+1}+1}{2^{s+1}(s+1)}|\Upsilon'(\flat_1)^q|+\frac{1}{n}\sum_{s=1}^{n}\frac{2^s(s+1)-1}{2^{s+1}(s+1)}|\Upsilon'(\flat_2)|^q\right)^{\frac{1}{q}}\right.\\&\quad\left.+\left(\frac{1}{n}\sum_{s=1}^{n}\frac{2^s(s+1)-2^{s+1}+1}{2^{s+1}(s+1)}|\Upsilon'(\flat_2)^q|+\frac{1}{n}\sum_{s=1}^{n}\frac{2^s(s+1)-1}{2^{s+1}(s+1)}|\Upsilon'(\flat_1)|^q\right)^{\frac{1}{q}}\right].\end{aligned}$$

Acknowledgments

The authors are thankful to Prof Dr. Themistocles M. Rassias and to the anonymous referee for their valuable comments and suggestions. This research was supported by HEC Pakistan under project 8081/Punjab/NRPU/R&D/HEC/2017.

References

[1] Awan, M.U., Akhtar, N., Iftikhar, S., Noor, M.A. and Chu, Y.-M. New Hermite-Hadamard type inequalities for n-polynomial harmonically convex functions, *J. Inequal. Appl.*, 2020(125), 1–12 (2020).

[2] Awan, M.U., Talib, S., Noor, M.A., Chu, Y.-M. and Noor, K.I. Some trapezium–like inequalities involving functions having strongly n-polynomial preinvexity property of higher order, *J. Func. Spaces*, 2020(9154139), 9 (2020).

[3] Barani, A., Ghazanfari, A.G. and Dragomir, S.S. Hermite-Hadamard inequality for functions whose derivatives absolute values are preinvex, *J. Inequal. Appl.*, 2012(247), 1–9 (2012).

[4] Deng, Y.-P., Awan, M.U., Talib, S., Noor, M.A., Noor, K.I., Mohammed, P.O. and Wu, S.H. Inequalities for estimations of integrals related to higher–order strongly n-polynomial preinvex functions, *J. Math.*, 2020(8841356), 12 (2020).

[5] Iscan, I. Hermite–Hadamard's Inequalities for preinvex function via fractional integrals and related fractional inequalities, *American J. Math. Anal.*, 1(3), 33–38 (2013).

[6] Israel, A.-B. and Mond, B. What is invexity? *The ANZIAM Journal.* 28(1), 1–9 (1986).

[7] Kirane, M. and Torebek, B.T. Hermite–Hadamard, Hermite–Hadamard–Fejer, Dragomir–Agarwal and Pachpatte type inequalities for convex functions via fractional integrals, arXiv:1701.00092.

[8] Katugampola, U.N. New approach to a generalized fractional integral, *Appl. Math. Comput.*, 218(3), 860–865 (2011).

[9] Khalil, R., Horani, M. Al., Yousef, A. and Sababheh, M. A new definition of fractional derivative, *J. Comput. Appl. Math.*, 264, 65–70 (2014).

[10] Kilbas, A.A., Srivastava, H.M. and Trujillo, J.J. *Theory and applications of fractional differential equations*, (North-Holland Mathematics Studies, Elsevier Sci. B.V., Amsterdam), 204 (2006).

[11] Mititelu, S. Invex sets, *Stud. Cerc. Mat.*, 46(5), 529–532 (1994).

[12] Mohan, S.R. and Neogy, S.K. On invex sets and preinvex functions, *J. Math. Anal. Appl.*, 189(3), 901–908 (1995).

[13] Mubeen, S. and Habibullah, G.M. k–fractional integrals and application, *Int. J. Contemp. Math. Sci.*, 7(2), 89–94 (2012).

[14] Noor, M.A. Hermite–Hadamard integral inequalities for log–preinvex functions, *J. Math. Anal. Approx. Theory,* 2, 126–131 (2007).

[15] Sarikaya, M.Z. and Ertugral, F. On the generalized Hermite-Hadamard inequalities, *Annals of the University of Craiova-Mathematics and Computer Sci. Series,* 47(1), 193–213 (2020).

[16] Sarikaya, M.Z., Set, E., Yaldiz, H. and Basak, N. Hermite-Hadamard's inequalities for fractional integrals and related fractional inequalities, *Math. Comput. Model.,* 57(9–10), 2403–2407 (2013).

[17] Set, E. and Gozpinar, A. A study on Hermite–Hadamard type inequalities for s–convex functions via conformable fractional integrals, *Stud. Univ. Babes-Bolyai Math.*, 62(3), 309–323 (2017).

[18] Toplu, T., Kadakal, M. and Iscan, I. On n-polynomial convexity and some related inequalities, *AIMS Math.*, 5(2), 1304–1318 (2020).

[19] Weir, T. and Mond, B. Pre-invex functions in multiple objective optimization, *J. Math. Anal. Appl.*, 136(1), 29–38 (1988).

[20] Wu, S., Awan, M.U., Ullah, M.U., Talib, S. and Kashuri, A. Some integral inequalities for n-polynomial ζ-preinvex functions, *J. Funct. Spaces.* 2021(6697729), 9, (2021).

https://doi.org/10.1142/9789811293009_0004

Chapter 4

An Approach of Locally Convex and Barrelled Spaces Over an Arbitrary Field Equipped with a Valuation, Without Using "Convex" but Utilizing "Absolutely Convex"

V. Benekas

Department of Mathematics, University of Ioannina,
451 10 Ioannina, Greece
benekas@gmail.com

Locally convex spaces over an arbitrary field, which has a valuation and is complete with the metric induced by the valuation, are being studied. We do not define convex, but we proceed with absolutely convex. The notion of absolutely convex differs from the one in the case of a non-Archimedean field. Locally convex spaces are related to a collection of seminorms. Barreled spaces are defined and some properties are given. Finally, it is proved that a subspace of countable codimension of a barrelled space remains barrelled.

1. Introduction

Topological Vector Spaces (TVSs) have been developed over the set $\mathbb{R}$ of real numbers and the set $\mathbb{C}$ of complex numbers. N. Bourbaki in [5] and H.H. Schaefer in [11] gave the definition and some properties over an arbitrary valued field $\mathbb{K}$. It was shown that every Topological Vector Space (TVS) over $\mathbb{K}$ has an additive and multiplicative base for the neighborhoods of zero which consists of balanced and closed sets and also one which consists of balanced and open sets.

Locally convex spaces were defined primarily over $\mathbb{R}$ and $\mathbb{C}$. There were defined locally convex spaces over non-Archimedean valued fields, with the notion of convex set being different from the one in the classical case because $\mathbb{K}$ is not ordered. In non-Archimedean spaces, the notion of absolutely convex (convex and balanced) differs also from the one in the classical case. We do not give a definition of convex sets but use the concept "absolutely convex set" so that it has the same meaning as in the classical case. It is shown that there is a relation between the absolutely convex absorbing sets and the seminorms, as in the classical case. A locally convex space is a TVS whose topology has a base for the neighborhoods of zero consisting of absolutely convex sets.

TVSs over $\mathbb{R}$ or $\mathbb{C}$ have been developed in [7–10], and [13], with emphasis to locally convex spaces and a theory of non-convex Hausdorff TVSs was given in [1]. In [3] and [4], TVSs were studied over an arbitrary complete valued field.

2. Preliminaries

In the following, by $\mathbb{K}$, we denote a field endowed with a valuation, i.e., a function $|\cdot| : \mathbb{K} \to \mathbb{R}$, with the properties *(i)* $|x| \geq 0$, *(ii)* $|x| = 0$, iff $x = 0$, *(iii)* $|x + y| \leq |x| + |y|$, and *(iv)* $|xy| = |x||y|$, for all $x, y \in \mathbb{K}$. We assume that $\mathbb{K}$ is a complete metric space, with the metric generated by the valuation and that the valuation is non-trivial, i.e., there exists $x \in \mathbb{K}$ such that $|x| \neq 1$. The *value group* of $\mathbb{K}$ is the subgroup $\{|\lambda| : \lambda \in \mathbb{K}, \lambda \neq 0\}$ of the multiplicative group of positive real numbers and it is either dense or discrete, i.e., cyclic. We say, respectively, that the valuation of $\mathbb{K}$ is *dense* or *discrete*.

By *TVS* we mean a vector space X over $\mathbb{K}$, such that X is endowed with a topology under which addition and scalar multiplication are continuous functions, where $X \times X$ and $\mathbb{K} \times X$ are equipped with the respective product topologies. From the continuity, it follows that U

is a neighborhood of $x \in X$ if and only if $-x + U$ is a neighborhood of zero and if U is a neighborhood of zero and $\lambda \in \mathbb{K} \setminus \{0\}$, then λU is a neighborhood of zero.

A subset A of a vector space X over $\mathbb{K}$ is called *balanced* if $\lambda A \subset A$, for all $\lambda \in \mathbb{K}$ with $|\lambda| \leq 1$, and *absorbing* if for each $x \in X$ there exists $\varepsilon > 0$ such that $\lambda x \in A$ for all $\lambda \in \mathbb{K}$ with $|\lambda| < \varepsilon$. Finally, we say that a set A *absorbs* a set B (or B *is absorbed* by A) if there exists $\varepsilon > 0$ such that for all $\lambda \in \mathbb{K}$ with $|\lambda| < \varepsilon$, $\lambda B \subset A$, holds. If A is balanced, then it is absorbing iff for each $x \in X$ there exists $\lambda \in \mathbb{K} \setminus \{0\}$ such that $\lambda x \in A$ and it absorbs B iff there exists $\mu \in \mathbb{K} \setminus \{0\}$ such that $\mu B \subset A$.

A collection $\mathcal{A}$ of subsets of a TVS X is called *directed* if for each $A, B \in \mathcal{A}$ there exists $C \in \mathcal{A}$ such that $C \subset A \cap B$, *additive* if for each $A \in \mathcal{A}$ there exists $B \in \mathcal{A}$ such that $B + B \subset A$, and *multiplicative* if for all $A \in \mathcal{A}$ and $\lambda \in \mathcal{K} \setminus \{0\}$ there exists $B \in \mathcal{A}$ such that $\lambda B \subset A$.

The following theorems are known.

Theorem 1 ([5]). *Every TVS has an additive and multiplicative base for the neighborhoods of zero, which consists of balanced and closed sets and one which consists of balanced and open sets. Every neighborhood of zero is an absorbing set.*

Theorem 2 ([5]). *If $\mathcal{B}$ is a non-empty, additive, multiplicative, and directed collection of balanced and absorbing subsets of a vector space X, then there exists exactly one vector topology for X, for which $\mathcal{B}$ is a base for the neighborhoods of zero.*

The notion of string given in [1] was generalized in [3]:

A sequence $\mathcal{U} = (U_n)$ of subsets of a vector space is called a *string*, if for each $n \in \mathbb{N}$, (i) U_n is balanced and absorbing, (ii) $U_{n+1} + U_{n+1} \subset U_n$, and (iii) for each $\mu \in \mathbb{K}$, there exists $m \in \mathbb{N}$ such that $\mu U_m \subset U_n$.

Each U_n is called a *knot* of $\mathcal{U}$. A string $\mathcal{U} = (U_n)$ in a TVS is called *topological* if every knot is a neighborhood of zero.

If X and Y are two TVSs, Z a subspace of X, $\lambda \in \mathbb{K} \setminus \{0\}$, f, g two linear maps from X to Y, with g being onto, τ a vector topology on X, $\mathcal{U} = (U_n)$ and $\mathcal{V} = (V_n)$ two strings in X, and $\mathcal{W} = (W_n)$ a string in Y, then $(U_n \cap V_n)$, (λU_n), $(U_n + V_n)$, $(f^{-1}(W_n))$, and $(\overline{U_n}^{\tau})$ are strings in X, $(g(U_n))$ is a string in Y, and $(U_n \cap Z)$ is a string in Z. We denote the above strings with $\mathcal{U} \cap \mathcal{V}$, $\lambda\mathcal{U}$, $\mathcal{U} + \mathcal{V}$, $f^{-1}(\mathcal{W})$,

$\overline{\mathcal{U}}^{\tau}$, $g(\mathcal{U})$, and $\mathcal{U} \cap Z$, respectively. Also, $\mathcal{U} \subset \mathcal{V}$ means $U_n \subset V_n$, for all $n \in \mathbb{N}$. The kernel of $\mathcal{U}$, $\ker \mathcal{U}$, is defined as $\bigcap_{n\in\mathbb{N}} U_n$. By $\tau_{\mathcal{U}}$, the topology generated by $\mathcal{U}$, it is meant the vector topology which has as a base for the neighborhoods of zero all the knots of $\mathcal{U}$ and for a collection $\mathcal{F}$ of strings $\tau_{\mathcal{F}} = \sup_{\mathcal{U}\in\mathcal{F}} \tau_{\mathcal{U}}$. For $\mathcal{F}$ as above, it holds that $\tau_{\mathcal{F}}$ is Hausdorff iff $\bigcap_{\mathcal{U}\in\mathcal{F}} \ker \mathcal{U} = \{0\}$.

3. Locally Convex Spaces

A subset A of a vector space X over $\mathbb{K}$ is called *absolutely convex* as cited in [2], if $\sum_{i=1}^{n} \lambda_i A \subset A$, for all $\lambda_1, \ldots, \lambda_n \in \mathbb{K}$ with $\sum_{i=1}^{n} |\lambda_i| \leq 1$. An absolutely convex set is balanced. Note that in the classical case of TVSs over $\mathbb{R}$ or $\mathbb{C}$, a set A is called absolutely convex if it is balanced and convex, and this is equivalent with the definition given above, since, if for all $\mu_1, \mu_2 \in \mathbb{C}$ with $|\mu_1| + |\mu_2| \leq 1$, $\mu_1 A + \mu_2 A \subset A$ holds and if $\lambda_1, \lambda_2, \lambda_3 \in \mathbb{C} \setminus \{0\}$, with $|\lambda_1| + |\lambda_2| + |\lambda_3| \leq 1$, then $\sum_{i=1}^{3} \lambda_i A = (|\lambda_1| + |\lambda_2|) \left(\frac{\lambda_1}{|\lambda_1|+|\lambda_2|} A + \frac{\lambda_2}{|\lambda_1|+|\lambda_2|} A \right) + \lambda_3 A \subset (|\lambda_1| + |\lambda_2|) A + \lambda_3 A \subset A$.

A *seminorm* on X is a function $p : X \to \mathbb{R}$ such that for all $\lambda \in \mathbb{K}$ and $x, y \in X$, $p(\lambda x) = |\lambda| p(x)$ and $p(x + y) \leq p(x) + p(y)$ holds. It is called a *norm*, if $p(x) = 0$ implies $x = 0$.

If p is a seminorm on X and $r > 0$, then it is easy to see that the sets $\{x \in X : p(x) < r\}$ and $\{x \in X : p(x) \leq r\}$ are absolutely convex and absorbing.

If A is an absolutely convex absorbing set and for each $x \in X$,

$$p_A(x) = \inf \left\{ \sum_{k=1}^{n} |\lambda_k| : x \in \sum_{k=1}^{n} \lambda_k A,\ \lambda_1, \lambda_2, \ldots, \lambda_n \in \mathbb{K}, n \in \mathbb{N} \right\},$$

then p_A is a seminorm, [2]. This seminorm is called the *Minkowski functional* of A in X. The following proposition gives some properties of it.

Proposition 3. *Let X be a vector space over $\mathbb{K}$ and U an absolutely convex absorbing subset of X. Then*

(i) $\{x \in X : p_U(x) < 1\} \subset U \subset \{x \in X : p_U(x) \leq 1\}$.
(ii) *p_U is a norm if and only if U does not contains any one-dimensional subspace of X.*

(iii) *If the valuation of* $\mathbb{K}$ *is discrete,* $\rho \in (0,1)$ *is the generator of the value group and* p *is a seminorm on* X *such that* $\{x \in X : p(x) < 1\} \subset U \subset \{x \in X : p(x) \leq 1\}$*, then* $\rho p_U \leq p \leq p_U$ *holds.*

(iv) *If the valuation of* $\mathbb{K}$ *is dense, then for each* $x \in X$*,* $p_U(x) = \inf\{|\lambda| : x \in \lambda U,\ \lambda \in \mathbb{K}\}$*, and if* p *is a seminorm on* X *such that* $\{x \in X : p(x) < 1\} \subset U \subset \{x \in X : p(x) \leq 1\}$*, then* $p = p_U$*.*

Proof.

(i) The first part follows because U is absolutely convex and the second is trivial.

(ii) Assume p_U is a norm and F is an one-dimensional subspace of X with $F \subset U$. Choose $x \in F \setminus \{0\}$ and $\lambda \in \mathbb{K}$ with $0 < |\lambda| < 1$. Then, for all $n \in \mathbb{N}$, $\lambda^{-n}x \in U$, and so by (i), $p_U(x) \leq |\lambda|^n$. This gives the contradiction $p_U(x) = 0$. Conversely, assume that there exists $x \in X \setminus \{0\}$ such that $p_U(x) = 0$. Then, $F = \{\mu x : \mu \in \mathbb{K}\}$ is a one-dimensional subspace of X with $F \subset U$, since for all $\mu \in \mathbb{K}$, $p_U(\mu x) = |\mu| p_U(x) = 0 < 1$.

(iii) If $x \in X$, $\lambda_1, \ldots \lambda_n \in \mathbb{K}$, $u_1, \ldots, u_n \in U$, and $x = \sum_{k=1}^n \lambda_k u_k$, we get $p(u_k) \leq 1$, for all $k = 1, \ldots, n$ and thus $p(x) \leq \sum_{k=1}^n |\lambda_k| p(u_k) \leq \sum_{k=1}^n |\lambda_k|$. This gives $p \leq p_U$. Assume that $\rho = |\lambda| \in (0,1)$ is the generator of the the discrete valuation, where $\lambda \in \mathbb{K}$. Let $x \in X$. If $p(x) = 0$, then $p(\lambda^{-n}x) = 0 < 1$, for all $n \in \mathbb{N}$ and so $\lambda^{-n}x \in U$. This implies $p_U(x) \leq |\lambda|^n$ and thus $p_U(x) = 0$. If $p(x) \neq 0$, then there exists $k \in \mathbb{Z}$ such that $|\lambda|^{k+1} \leq p(x) < |\lambda|^k$, hence $\lambda^{-k}x \in U$, which implies that $p_U(x) \leq |\lambda|^k \leq |\lambda|^{-1} p(x)$.

(iv) Set for each $x \in X$, $q(x) = \inf\{|\lambda| : x \in \lambda U,\ \lambda \in \mathbb{K}\}$. Obviously, $p_U \leq q$, and on the other hand, given $x \in X$ and $\varepsilon > 0$, there exists $\mu \in \mathbb{K}$ such that $p_U(x) < |\mu| < p_U(x) + \varepsilon$, hence $p_U(\mu^{-1}x) < 1$ and so $x \in \mu U$. Thus, $q(x) \leq |\mu| < p_U(x) + \varepsilon$. Since the last is true for every $\varepsilon > 0$, we get that $q(x) \leq p_U(x)$. Let now p be a seminorm on X such that $\{x \in X : p(x) < 1\} \subset U \subset \{x \in X : p(x) \leq 1\}$. As in (iii), $p \leq p_U$. Suppose that there exists $x \in X$ such that $p(x) < p_U(x)$. Since the valuation is dense, choose $\mu \in \mathbb{K}$ such that $p(x) < |\mu| < p_U(x)$, then $p(\mu^{-1}x) < 1$, hence $\mu^{-1}x \in U$ and thus $p_U(x) \leq |\mu|$, which is a contradiction. □

Definition 4. A TVS is called *locally convex* if it has a base for the neighborhoods of zero, consisting of absolutely convex sets.

By a *Fréchet space*, we mean a locally convex space which is metrizable and complete.

As in the classical case, we can prove that if A, B, G, and F are subsets of a TVS, G is open, F is closed, and $\lambda \in \mathbb{K}$, then $\overline{A}+\overline{B} \subset \overline{A+B}$, $A+G$ is open and $\lambda\overline{A} \subset \overline{\lambda A}$. If $\lambda \neq 0$, then λG is open and λF is closed.

Lemma 5. *Let U be an absolutely convex set in a TVS. Then*

(i) *the closure $\overline{U}$ of U is absolutely convex,*
(ii) *if the interior U° of U is non-empty, then $0 \in U^\circ$ and U is a neighborhood of zero,*
(iii) *the interior U° of U is absolutely convex.*

Proof.

(i) If $\lambda_1, \lambda_2, \ldots, \lambda_n \in \mathbb{K}$ with $\sum_{k=1}^n |\lambda_k| \leq 1$, then $\sum_{k=1}^n \lambda_k\overline{U} \subset \overline{\sum_{k=1}^n \lambda_k U} \subset \overline{U}$.
(ii) If $x \in U^\circ$, there exists a neighborhood V of zero such that $x+V \subset U$. If $\lambda \in \mathbb{K}$ with $0 < |\lambda| \leq \frac{1}{2}$, then $\lambda V = -\lambda x + \lambda x + \lambda V \subset -\lambda x + \lambda U \subset -\lambda U + \lambda U \subset U$, since U is absolutely convex and $|\lambda| + |\lambda| \leq 1$. But $\lambda \neq 0$, so λV is a neighborhood of zero and the result follows.
(iii) If $U^\circ = \emptyset$, it is absolutely convex. Assume $U^\circ \neq \emptyset$. Then, $0 \in U^\circ$. Let now $\lambda_1, \ldots, \lambda_n \in \mathbb{K}$ with $\sum_{k=1}^n |\lambda_k| \leq 1$. If $\lambda_1 = \cdots = \lambda_n = 0$, then $\sum_{k=1}^n \lambda_k U^\circ = \{0\} \subset U^\circ$. If there exists $k_0 \in \{1, \ldots, n\}$ such that $\lambda_{k_0} \neq 0$, then $\lambda_{k_0} U^\circ$ is open, hence $\sum_{k=1}^n \lambda_k U^\circ$ is open. Thus, $\sum_{k=1}^n \lambda_k U^\circ \subset \sum_{k=1}^n \lambda_k U \subset U$ implies $\sum_{k=1}^n \lambda_k U^\circ \subset U^\circ$. □

Now we can give the correspondence to Theorems 1 and 2 for locally convex spaces.

Theorem 6. *Every locally convex space has an additive and multiplicative base for the neighborhoods of zero, consisting of absolutely convex and closed sets and one consisting of absolutely convex and open sets.*

Proof. Let $\mathcal{B}$ be the collection of all the absolutely convex closed neighborhoods of zero. $\mathcal{B}$ is an additive and multiplicative base for the neighborhoods of zero, consisting of absolutely convex closed sets. Indeed, if U is a neighborhood of zero, by Theorem 1, there exists a closed neighborhood V of zero such that $V \subset U$. Since the space is locally convex, there exists an absolutely convex neighborhood W of zero with $W \subset V$. It holds $\overline{W} \subset V \subset U$, and by Lemma 5 (i), $\overline{W} \in \mathcal{B}$. Thus, $\mathcal{B}$ is a base for the neighborhoods of zero. If $U \in \mathcal{B}$ and $\lambda \in \mathbb{K}$ with $0 < |\lambda| \leq \frac{1}{2}$, we get $\lambda U \in \mathcal{B}$, and since $\lambda U + \lambda U \subset U$, it follows that $\mathcal{B}$ is additive. Finally, if $U \in \mathcal{B}$ and $\lambda \in \mathbb{K} \setminus \{0\}$, then $V = \lambda^{-1} U \in \mathcal{B}$ and $\lambda V \subset U$, hence $\mathcal{B}$ is also multiplicative. The proof for open instead of closed is similar. □

Theorem 7. *Let $\mathcal{B}$ be a non-empty, additive, multiplicative, and directed collection of absolutely convex and absorbing sets of a vector space X. Then there exists exactly one locally convex vector topology on X, which has $\mathcal{B}$ as a base for the neighborhoods of zero.*

Corollary 8. *Let $\mathcal{B}$ be a non-empty collection of absolutely convex and absorbing sets of a vector space X. Then there exists exactly one locally convex vector topology on X, which has $\{\mu B : B \in \mathcal{B}, \mu \in \mathbb{K}, \mu \neq 0\}$ as a subbase for the neighborhoods of zero.*

Let p be a seminorm on a vector space X and $\varepsilon > 0$. Set $B_{p,\varepsilon} = \{x \in X : p(x) \leq \varepsilon\}$. The collection $\mathcal{B}_p = \{B_{p,\varepsilon} : \varepsilon > 0\}$ satisfies the hypothesis of Theorem 7. The vector topology which is generated by this collection is denoted by τ_p and is called the topology generated by the seminorm p. By Theorem 7, τ_p is locally convex.

Note that the supremum of a collection of vector topologies is a vector topology. If P is a collection of seminorms on X, define τ_P to be the supremum of the collection $\{\tau_p : p \in P\}$. It is called the vector topology generated by the set P of seminorms. A base for the neighborhoods for zero of τ_P is the collection $\mathcal{B} = \{\bigcap_{k=1}^{n} B_{p_k,\varepsilon_k} : p_1, \ldots, p_n \in P, \varepsilon_1 > 0, \ldots, \varepsilon_n > 0, n \in \mathbb{N}\}$ and τ_P is locally convex.

The next result characterizes fully, as in the classical case, the locally convex spaces.

Theorem 9. *A TVS (X, τ) is locally convex iff there exists a collection P of seminorms on X which generates the topology τ.*

Proof. Suppose that (X, τ) is locally convex and $\{U_i : i \in I\}$ is a base for the neighborhoods of zero, with each U_i being absolutely convex. Set $p_i = p_{U_i}$, the Minkowski functional of U_i, for all $i \in I$. It holds $\tau = \tau_P$, where $P = \{p_i : i \in I\}$. Indeed, if U is a τ_P-neighborhood of zero, there exist $i_1, \ldots, i_n \in I$, $\varepsilon_1 > 0, \ldots, \varepsilon_n > 0$ such that $\bigcap_{j=1}^n \{x \in X : p_{i_j}(x) \leq \varepsilon_j\} \subset U$. By choosing $\lambda \in \mathbb{K} \setminus \{0\}$ with $|\lambda| \leq \min\{\varepsilon_1, \ldots, \varepsilon_n\}$, we get $\lambda \bigcap_{j=1}^n U_{i_j} \subset U$, i.e., U is a τ-neighborhood of zero. On the other hand, if U is a τ-neighborhood of zero and $U_i \subset U$ for some $i \in I$, it follows that U is a τ_P-neighborhood of zero since $\{x \in X : p_i(x) < 1\} \subset U$. The converse follows from the fact that τ_P is locally convex. □

Proposition 10. *Let Y be a subspace of a locally convex space X. Then Y is locally convex and every continuous seminorm on Y can be extended to a continuous seminorm on X.*

Proof. The first part is trivial, since if $\{U_i : i \in I\}$ is a base for the zero neighborhoods in X, consisting of absolutely convex sets, then $\{U_i \cap Y : i \in I\}$ is a base for the zero neighborhoods in Y consisting of absolutely convex sets in Y.

Let now p be a continuous seminorm on Y. Since $V = \{y \in Y : p(y) \leq 1\}$ is a neighborhood of zero in Y, there exists an absolutely convex neighborhood U of zero in X with $U \cap Y \subset V$. Let $\lambda \in \mathbb{K}$ with $|\lambda| > 1$ and define for each $x \in X$,

$$q(x) = \inf_{y \in Y} [p(y) + |\lambda| p_U(x - y)].$$

Now, q is a seminorm on X. Indeed, let $x, y \in X$ and $\mu \in \mathbb{K}$. If $\mu = 0$, then $q(\mu x) = q(0) \leq p(0) + |\lambda| p_U(0) = 0$. If $\mu \neq 0$, then $q(\mu x) = \inf_{y \in Y} [p(y) + |\lambda| p_U(\mu x - y)] = \inf_{y \in Y} [p(\mu y) + |\lambda| p_U(\mu x - \mu y)] = |\mu| q(x)$. For every $z, w \in Y$, we have $q(x + y) \leq p(z + w) + |\lambda| p_U [x + y - (z + w)] \leq p(z) + |\lambda| p_U (x - z) + p(w) + |\lambda| p_U (y - w)$. Thus, $q(x + y) \leq q(x) + q(y)$.

It holds that $q \leq |\lambda| p_U$ and therefore q is continuous. It remains to show that $q = p$ on Y. So, let $y \in Y$. Clearly, $q(y) \leq p(y)$. For the converse, first we show that $|\lambda| p_U(y) \geq p(y)$. Indeed, if $p_U(y) = 0$, then $p_U(y) < |\lambda|^{-n}$, for all $n \in \mathbb{N}$ and so $p_U(\lambda^n y) < 1$, hence $\lambda^n y \in U \cap Y \subset V$, which implies that $p(y) = 0$. If $p_U(y) > 0$, there exists an integer k such that $|\lambda|^k \leq p_U(y) < |\lambda|^{k+1}$ and so $\lambda^{-k-1} y \in V$, which implies that $p(y) \leq |\lambda|^{k+1} \leq |\lambda| p_U(y)$. Hence, for each $z \in Y$, it holds $|\lambda| p_U(y - z) \geq p(y - z)$ and therefore

$p(z) + |\lambda| p_U(y - z) \geq p(z) + p(y - z) \geq p(y)$. By taking the infimum over $z \in Y$ on both sides, we get the result we wish. □

Proposition 11. *Let Y be a subspace of a locally convex space X and V an absolutely convex neighborhood of zero in Y. Then there exists an absolutely convex neighborhood U of zero in X such that $U \cap Y = V$. Moreover, if $z \in X \setminus \overline{Y}$, then U may be chosen so that $z \notin U$.*

Proof. Since V is a neighborhood of zero in Y, the Minkowski functional p_V in Y is continuous. By Proposition 10, there exists a continuous seminorm q on X which extends p_V. Now

$$U = \{x \in X : q(x) < 1\} \cup \{x \in V : q(x) = 1\}$$

is a neighborhood of zero in X, since $U \supset \{x \in X : q(x) < 1\}$ and q is continuous.

Moreover, U is absolutely convex. Indeed, let $\lambda_1, \ldots, \lambda_n$, $\mu_1, \ldots, \mu_m \in \mathbb{K}$, with $\sum_{i=1}^n |\lambda_i| + \sum_{j=1}^m |\mu_j| \leq 1$, $x_1, \ldots, x_n \in X$, with $q(x_i) < 1$, for all $i = 1, \ldots, n$, and $y_1, \ldots, y_m \in V$ with $q(y_j) = 1$ for all $j = 1, \ldots, m$. It suffices to show that $x = \sum_{i=1}^n \lambda_i x_i + \sum_{j=1}^m \mu_j y_j \in U$. Clearly, $q(x) \leq 1$. If $\sum_{i=1}^n |\lambda_i| > 0$, then $q(x) < 1$, which yields that $x \in U$. In the other case, $\lambda_i = 0$, for all $i = 1, \ldots, n$. Hence, $x = \sum_{j=1}^m \mu_j y_j$ and since V is absolutely convex, we get $x \in V$. Since $q(x) \leq 1$, we conclude $x \in U$.

Next, $U \cap Y = V$. Indeed, if $x \in U \cap Y$ and $x \notin V$, it follows that $q(x) < 1$. Since $x \in Y$, $p_V(x) < 1$ and so $x \in V$. Conversely, if $x \in V$ and $x \notin U$, we get $q(x) \neq 1$. But since $x \in Y$, $q(x) = p_V(x) \leq 1$, i.e., $q(x) < 1$, hence $x \in U$.

Finally, let $z \in X$, $z \notin \overline{Y}$. Then there exists an absolutely convex neighborhood W of zero in X such that $(z + W) \cap Y = \emptyset$ and $W \cap Y \subset V$. Let $\lambda \in \mathbb{K}$ with $|\lambda| > 1$ and p_W be the Minkowski functional of W in X. Define $p : X \to \mathbb{R}$, by $p(x) = \inf_{y \in Y} [p_V(y) + |\lambda| p_W(x - y)]$, for each $x \in X$. Then, as in Proposition 10 (by replacing p with p_V, U with W, and q with p), it can be proved that p is a continuous seminorm on X whose restriction to Y equals p_V. Set $U = \{x \in X : p(x) < 1\} \cup \{x \in V : p(x) = 1\}$. Then, as above (by using p instead of q), we get $U \cap Y = V$. Moreover, $z \notin U$. Indeed, assume that $z \in U$. Since $z \notin Y$, we have that $z \notin V$ and so $p(z) < 1$. Thus, there exists $y \in Y$ with $p_V(y) + |\lambda| p_W(z - y) < 1$. Hence, there are $\lambda_1, \ldots, \lambda_n \in \mathbb{K}$, such that $y \in \sum_{i=1}^{n} \lambda_i V$ and $\sum_{i=1}^{n} |\lambda_i| + |\lambda| p_W(z-y) < 1$ and there exist $\mu_1, \ldots, \mu_m \in \mathbb{K}$ such that $z-y \in \sum_{j=1}^{m} \mu_j W$ and $|\lambda| \sum_{j=1}^{m} |\mu_j| < 1 - \sum_{i=1}^{n} |\lambda_i|$. Therefore,

$$z = y + (z-y) \in \sum_{i=1}^{n} \lambda_i V + \sum_{j=1}^{m} \mu_j W \subset V + W$$

because $\sum_{i=1}^{n} |\lambda_i| < 1$ and $\sum_{j=1}^{m} |\mu_j| < |\lambda|^{-1} < 1$. So, we have shown that there are $v \in V$ and $w \in W$ such that $z = v + w$. Since W is balanced, $-w \in W$ and therefore

$$z - w = v \in (z + W) \cap V \subset (z + W) \cap Y = \emptyset,$$

a contradiction. □

4. Barrelled Locally Convex Spaces

A *barrel* in a locally convex space X over $\mathbb{K}$ is defined to be an absolutely convex absorbing closed set. This definition generalizes the corresponding in the classical case. X is called *barrelled* if every barrel is a neighborhood of zero. The next proposition gives a wide class of barrelled spaces.

Proposition 12. *Every locally convex space X of second category is barrelled.*

Proof. Let B be a barrel in X and $\lambda \in \mathbb{K}$ with $|\lambda| > 1$. Since B is absorbing, it follows that $X = \bigcup_{n \in \mathbb{N}} \lambda^n B$ and since it is of second category, the interior of $\lambda^{n_0} B$ is non-empty, for some $n_0 \in \mathbb{N}$. Hence, B has non-empty interior. By Lemma 5 (ii), B is a neighborhood of zero. □

Corollary 13. *Every Fréchet space is barrelled.*

Recall that if X is a topological space, a function $f : X \to \mathbb{R}$ is called *lower semicontinuous* at some $x \in X$ if for all real numbers a, $f(x) > a$ implies that there exists a neighborhood N of x such that $f(y) > a$, for all $y \in N$. It is called lower semicontinuous if it is lower semicontinuous at all $x \in X$. Equivalently, $f^{-1}\big((-\infty, a]\big)$ is closed in X for all real numbers a.

We shall see next that there is a characterization of barrelledness by using lower semicontinuity. We first need the following:

Lemma 14. *If B is a barrel in a TVS X, then there exists a lower semicontinuous seminorm p on X such that $\{x \in X : p(x) < 1\} \subset B \subset \{x \in X : p(x) \leq 1\}$. If the valuation of $\mathbb{K}$ is dense, then the Minkowski functional p_B of B is lower semicontinuous.*

Proof. Let $\{U_i : i \in I\}$ be an additive and multiplicative base for the neighborhoods of zero, consisting of balanced sets. Make I a partially ordered set by defining $i \leq j$ to mean $U_j \subset U_i$, for all $i, j \in I$, and define $p : X \to \mathbb{R}$ by $p(x) = \sup_{i\in I} \inf_{y\in U_i} p_B(x-y)$, for all $x \in X$.

Then, p is a seminorm on X. Indeed, let x in X, $\lambda \in \mathbb{K}$, and $i \in I$. Obviously, $p(0) = 0$. Assume $\lambda \neq 0$. Then there exists $j \in I$ such that $U_j \subset \frac{1}{\lambda}U_i$. We have $\inf_{y\in U_j} p_B(x-y) \geq \inf_{y\in \frac{1}{\lambda}U_i} p_B(x-y) = \inf_{y\in U_i} p_B(x-\frac{1}{\lambda}y)$. Since the last inequality holds for all $i \in I$, we get $p(x) \geq \frac{1}{|\lambda|} \sup_{i\in I} \inf_{y\in U_i} p_B(\lambda x - y) = \frac{1}{|\lambda|}p(\lambda x)$. Thus, $p(\lambda x) \leq |\lambda| p(x)$ and since this holds for all $\lambda \in \mathbb{K} \setminus \{0\}$, we have also $p(x) = p\left(\lambda \frac{1}{\lambda} x\right) \leq \frac{1}{|\lambda|}p(\lambda x)$, i.e., $|\lambda|p(x) < p(\lambda x)$. Let now $x, z \in X$ and $i \in I$. Then there exists $j \in I$ such that $U_j + U_j \subset U_i$, hence $\inf_{y\in U_i} p_B(x+z-y) \leq \inf_{y_1,y_2\in U_j} p_B\big(x+z-(y_1+y_2)\big) \leq \inf_{y_1,y_2\in U_j} [p_B(x-y_1) + p_B(z-y_2)] = \inf_{y_1\in U_j} p_B(x-y_1) + \inf_{y_2\in U_j} p_B(z-y_2) \leq p(x)+p(z)$ and since this holds for every $i \in I$, by taking the supremum over $i \in I$, we get $p(x+z) \leq p(x) + p(z)$. Note that p is well defined, since for all $x \in X$, $p(x) \leq p_B(x)$ and B is absorbing.

p is lower semicontinuous. Indeed, let $a \in \mathbb{R}$, $x \in X$, and $p(x) > a$. Then there exists $i \in I$ such that $\inf_{y\in U_i} p_B(x-y) > a$. Now, $x + U_i$ is a neighborhood of x, and since it is balanced, we have that for all $z \in x + U_i$, $x - z \in U_i$, so $p_B(z) = p_B\big(x - (x-z)\big) > a$. Note that p is the largest lower semicontinuous function f with $f \leq p_B$. This is true because if there exists $x \in X$ with $f(x) > p(x)$, then there exists $\varepsilon > 0$ such that $f(x) > p(x) + \varepsilon$. Hence, there is $i \in I$ such that for each $y \in x + U_i$, $f(y) > p(x) + \varepsilon$. But $p(x) \geq \inf_{z\in U_i} p_B(x-z)$. For each $z \in U_i$, we have $-z \in U_i$, so $x - z \in x + U_i$. Thus, $p_B(x-z) \geq f(x-z) > \inf_{w\in U_i} p_B(x-w) + \varepsilon$, and by taking infimum over $z \in U_i$, we get $\varepsilon \leq 0$, which is a contradiction.

Now we shall prove that $\{x \in X : p(x) < 1\} \subset B \subset \{x \in X : p(x) \leq 1\}$. Let $x \in X$ with $p(x) < 1$. Then for each $i \in I$ there exists $y_i \in U_i$ such that $p_B(x - y_i) < 1$, so $x - y_i \in B$. Now $(x - y_i)_{i \in I}$ is a net in B and $x - y_i \to x$, since $y_i \to 0$. This shows that $x \in B$ because B is closed. The other inclusion is obvious, since $p \leq p_B$.

The last assertion follows from Proposition 3 (iv). □

Proposition 15. *A locally convex space is barrelled iff every lower semicontinuous seminorm is continuous.*

Proof. If X is a locally convex barrelled space and p is a lower semicontinuous seminorm on X, then the set $\{x \in X : p(x) \leq 1\}$ is closed and since it is absorbing and absolutely convex, i.e., a barrel, it is a neighborhood of zero and this implies that p is continuous. The converse follows from the preceding lemma. □

By a closed map, we mean a function with closed graph.

Lemma 16. *Let (X_1, τ_1), (X_2, τ_2) be two locally convex spaces and $f : (X_1, \tau_1) \to (X_2, \tau_2)$ be a linear map. Then*

(i) *f is closed iff there exists a Hausdorff locally convex topology τ_3 on X_2 such that $\tau_3 \subset \tau_2$ and $f : (X_1, \tau_1) \to (X_2, \tau_3)$ is continuous,*
(ii) *if f is closed, then $\ker f$ is τ_1-closed,*
(iii) *if $\hat{f} : (X_1, \tau_1)/\ker f \to (X_2, \tau_2)$ is the linear map defined by $\hat{f}(x + \ker f) = f(x)$, for all $x \in X_1$, then f is closed iff $\hat{f}$ is closed.*

Proof. (i) Suppose f is closed and define $\tau_3 = \tau_{\mathcal{F}}$, where $\mathcal{F}$ is the collection of strings, of the form $f(\mathcal{U}) + \mathcal{V}$, where $\mathcal{U}$ and $\mathcal{V}$ are topological strings in (X_1, τ_1) and (X_2, τ_2), respectively. Then $\tau_3 = \tau_{\mathcal{F}} \subset \tau_2$ and $f : (X_1, \tau_1) \to (X_2, \tau_3)$ is continuous. If $\tau_{\mathcal{F}}$ is not Hausdorff, there exists $y \in \bigcap_{\mathcal{W} \in \mathcal{F}} \ker \mathcal{W}$ with $y \neq 0$. Let G be the graph of f. Since $(0, y)$ is not in the $\tau_1 \times \tau_2$-closure of G, there exist a τ_1-topological string $\mathcal{U} = (U_n)$ and a τ_2-topological string $\mathcal{V} = (V_n)$ such that $\big[(0, y) + (U_1 \times V_1)\big] \cap G = \emptyset$. But $y \in \ker \big[f(\mathcal{U}) + \mathcal{V}\big]$, so for some $u \in U_1$ and $v \in V_1$, we have $y = f(u) + v$ and we get the contradiction $(u, y - v) \in \big[(0, y) + (U_1 \times V_1)\big] \cap G$. Thus, $\tau_{\mathcal{F}}$ is Hausdorff and it is also locally convex since τ_1 and τ_2 are so.

The proof does not involve the valued field and is done as in the classical case. So, we skip the proof of the inverse and of (ii) and (iii). □

Lemma 17. *For every Fréchet space, there exists a string with all the knots being absolutely convex and closed sets, which generates the topology.*

Proof. If X is a Fréchet space, we can find a countable base $\{V_m : m \in \mathbb{N}\}$ of the neighborhoods of zero, consisting of absolutely convex closed sets. Choose $\lambda \in \mathbb{K}$ with $0 < |\lambda| < \frac{1}{2}$ and set $U_n^{(m)} = \lambda^{n-1} V_m$, for $m, n \in \mathbb{N}$ and $U_n = \bigcap_{m=1}^{n} U_n^{(m)}$. Then, $\mathcal{U} = (U_n)$ is a string and the linear topology which is generating from $\mathcal{F} = \{\mathcal{U}_m = (U_n^{(m)})_{n\in\mathbb{N}} : m \in \mathbb{N}\}$ coincides with $\tau_{\mathcal{U}}$. □

Proposition 18. *On a Fréchet space, there is no strictly coarser Hausdorff locally convex barrelled topology.*

Proof. Let (F, τ_1) be a Fréchet space and τ a Hausdorff locally convex barrelled topology on F, such that $\tau \subset \tau_1$. By the preceding lemma there exists a string $\mathcal{U} = (U_n)$ which generates τ_1 and each U_n is absolutely convex and τ_1-closed. Since τ is barrelled, each $\overline{U_n}$ is a τ-neighborhood of zero. The closure here and for the rest of the proof denotes the τ-closure. Let $n \in \mathbb{N}$ and $x \in \overline{U}$. Since for each $k \in \mathbb{N}$, $\overline{U_{n+k+1}}$ is a τ-neighborhood of zero, we can construct, by induction, a sequence $(x_k)_{k\in\mathbb{N}}$ in F, such that $x_k \in U_{n+k}$ and $x - \sum_{i=1}^{k} x_i \in \overline{U_{n+k+1}}$, for all $k \in \mathbb{N}$. Set $y_k = \sum_{i=1}^{k} x_i$, for each k. Then $(y_k)_{k\in\mathbb{N}}$ is a Cauchy sequence in (F, τ_1), and therefore it converges to some $y \in F$ in τ_1. Since for all $k \in \mathbb{N}$, $y_k = \sum_{i=1}^{k} x_i \in \sum_{i=1}^{k} U_{n+i} \subset U_n$ and U_n is τ_1-closed, it follows that $y \in U_n$. Let now $k \in \mathbb{N}$. Since $\overline{U_{k+1}}$ is a τ_1-neighborhood of zero, we can find $k_0 \in \mathbb{N}$, such that for all natural numbers $k_1 > k_0$ it holds $y - y_{k_1} \in \overline{U_{k+1}}$, and if we choose $k_1 > k_0$ and $k_1 > k$, we get $x - y = (x - y_{k_1}) + (y_{k_1} - y) \in \overline{U_{n+k+1}} + \overline{U_{k+1}} \subset \overline{U_{k+1}} + \overline{U_{k+1}} \subset \overline{U_k}$. Hence, $x - y \in \bigcap_{k=1}^{\infty} \overline{U_k}$. If $x \neq y$, since τ is Hausdorff, there exists a τ-neighborhood V of zero, such that $x - y \notin V + V$. But $\tau \subset \tau_1$ and so $U_k \subset V$, for some k. Then, $x - y \notin U_k + V$ and this gives the contradiction $x - y \notin \overline{U_k}$. Thus, $x = y \in U_n$. So, we have shown that $\overline{U_{n+1}} \subset U_n$. Since τ is barrelled, $\mathcal{U}$ is τ-topological. Thus, $\tau = \tau_1$. □

Definition 19. Let (X, τ) be a locally convex space and $\Phi = \{\tau' : \tau'$ barrelled locally convex topology on X, with $\tau' \supset \tau\}$. The topology $\tau^\beta = \sup \Phi$ is defined to be the *associated barrelled topology* for τ.

τ^β is well defined because the largest locally convex topology on X (the vector topology which is generated by all the absolutely convex absorbing sets) belongs to Φ. τ^β is the coarsest of all locally convex barrelled topologies on X which are finer than τ and is barrelled itself.

Proposition 20. *Let $f : (X_1, \tau_1) \to (X_2, \tau_2)$ be a continuous linear map between two locally convex spaces. Then, $f : (X_1, \tau_1^\beta) \to (X_2, \tau_2^\beta)$ is also continuous.*

Proof. Consider the collection Φ of all locally convex topologies τ' on X_2 for which $f : (X_1, \tau_1^\beta) \to (X_2, \tau')$ is continuous. Since $\tau_1 \subset \tau_1^\beta$, we get $\tau_2 \in \Phi$, i.e., $\Phi \neq \emptyset$. Set $\tau = \sup \Phi$. It follows that $f : (X_1, \tau_1^\beta) \to (X_2, \tau)$ is continuous. Let τ' be the topology which is generated by the set of all the barrels in (X_2, τ). If B is a barrel in (X_2, τ), then $f^{-1}(B)$ is a barrel in (X_1, τ_1^β), hence a neighborhood of zero. This shows that $\tau' \in \Phi$ and therefore $\tau' \subset \tau$. By the definition of τ', $\tau \subset \tau'$, so $\tau = \tau'$. Thus, τ is barrelled. Since $\tau_2 \subset \tau$, we get $\tau_2^\beta \subset \tau$ and the result follows. □

Now we are able to give another characteristic property for the locally convex barrelled spaces.

Theorem 21. *Let X be a locally convex space. The following are equivalent:*

(i) *X is barrelled.*
(ii) *For every Fréchet space F, every closed linear map from X to F is continuous.*
(iii) *For every Banach space F, every closed linear map from X to F is continuous.*

Proof. *(i)* $\Rightarrow$ *(ii)* : Assume that (X, τ) is barrelled, (F, τ_1) is a Fréchet space, and $f : (X, \tau) \to (F, \tau_1)$ is a closed linear map. By Lemma 16(i), there exists a Hausdorff locally convex topology τ_2 on F with $\tau_2 \subset \tau_1$ such that $f : (X, \tau) \to (F, \tau_2)$ is continuous. By Proposition 20, $f : (X, \tau^\beta) \to (F, \tau_2^\beta)$ is also continuous. Since τ is barrelled, it holds $\tau^\beta = \tau$, and since $\tau_2 \subset \tau_1$, we get, by using

Corollary 13, that $\tau_2^\beta \subset \tau_1$, so by Proposition 18, $\tau_2^\beta = \tau_1$. Thus, $f : (X, \tau) \to (F, \tau_1)$ is continuous.

$(ii) \Rightarrow (iii)$: This is obvious.

$(iii) \Rightarrow (i)$: Let p be a lower semicontinuous seminorm on (X, τ). By Proposition 15, it suffices to show that p is continuous. Both $W = \{x \in X : p(x) \leq 1\}$ and $G = \ker p = \{x \in X : p(x) \leq 0\}$ are τ-closed. Consider the quotient map $q : (X, \tau) \to (X, \tau)/G$ and let τ' be the quotient topology and $V = q(W)$. V is an absolutely convex absorbing set in X/G. Also, V is closed, since it is the complement of $q(X \setminus W)$ in X/G. The Minkowski functional p_V of V is a norm. Indeed, let $x \in X$ with $p_V\big(q(x)\big) = 0$ and $\varepsilon > 0$. There are $\lambda_1, \ldots, \lambda_n \in \mathbb{K}$ such that $q(x) \in \sum_{k=1}^n \lambda_k V$ and $\sum_{k=1}^n |\lambda_k| < \varepsilon$. There are also $x_1, \ldots, x_n \in W$ such that $q(x) = \sum_{k=1}^n \lambda_k q(x_k)$ and $\lambda_{n+1} \in \mathbb{K} \setminus \{0\}$ such that $\sum_{k=1}^{n+1} |\lambda_k| < \varepsilon$. Now $x - \sum_{k=1}^n \lambda_k x_k \in G$, and since $G \subset \lambda_{n+1} W$, there is $x_{n+1} \in W$ with $x - \sum_{k=1}^n \lambda_k x_k = \lambda_{n+1} x_{n+1}$, i.e., $x = \sum_{k=1}^{n+1} \lambda_k x_k$, hence $p(x) \leq \sum_{k=1}^{n+1} |\lambda_k| p(x_k) \leq \sum_{k=1}^{n+1} |\lambda_k| < \varepsilon$. This proves $p(x) = 0$, hence $q(x) = 0$, which means that p_V is a norm. Consider the collection $\mathcal{B} = \{q(U) + \mu V : U$ is an absolutely convex neighborhood of zero in (X, τ) and $\mu \in \mathbb{K}, \mu \neq 0\}$. By Theorem 7, $\mathcal{B}$ is a base for the neighborhoods of zero of a locally convex topology τ_0 on X/G. Obviously, $\tau_0 \subset \tau'$. We have $\cap \mathcal{B} = \bigcap_{\mu \in \mathbb{K} \setminus \{0\}} \overline{\mu V}^{\tau'} = \bigcap_{\mu \in \mathbb{K} \setminus \{0\}} \mu V = \{0\}$ because, if $x \in X$ is such that for every $\mu \in \mathbb{K}$, with $\mu \neq 0$, there is $w_\mu \in W$ such that $q(x) = \mu q(w_\mu)$ and it holds $p(x) = p(x - \mu w_\mu + \mu w_\mu) \leq p(x - \mu w_\mu) + |\mu| p(w_\mu) \leq |\mu|$. This proves $x \in G$ and therefore τ_0 is Hausdorff. Consider $\lambda \in \mathbb{K}$ with $0 < |\lambda| < 1$ and set $V_1 = \cap\{q(U) + V + \lambda^n V : U$ absolutely convex neighborhood of zero, $n \in \mathbb{N}\}$. Then, $V_1 = \overline{V}^{\tau_0}$. There exists $n_0 \in \mathbb{N}$ such that $|\lambda| + |\lambda|^{n_0+1} \leq 1$ and so $\lambda V + \lambda^{n_0+1} V \subset V$. Thus, $V_1 = \bigcap_{n \in \mathbb{N}} \overline{V + \lambda^n V}^{\tau'} \subset \overline{\lambda^{-1} V}^{\tau'} = \lambda^{-1} V$. It follows that $(X/G, \tau_{p_V})$ has a base for the neighborhoods of zero consisting of τ_0-closed sets. Since $\tau_0 \subset \tau_{p_V}$, we get, as in Problem D, Ch. 2, Sec. 7 in [8], that the continuous extension $\hat{I} : (\widehat{X/G}, \hat{\tau}_{p_V}) \to (\widetilde{X/G}, \hat{\tau}_0)$ of the identity map $I : (X/G, \tau_{p_V}) \to (X/G, \tau_0)$ is one-to-one, where $(\widehat{X/G}, \hat{\tau}_{p_V})$ and $(\widetilde{X/G}, \hat{\tau}_0)$ are the completions of $(X/G, \tau_{p_V})$ and $(X/G, \tau_0)$, respectively. Thus, $\widehat{X/G} \subset \widetilde{X/G}$ and $\hat{\tau}_0|_{\widehat{X/G}} \subset \hat{\tau}_{p_V}$. Since $\tau_0 \subset \tau'$, the identity map $i : (X/G, \tau') \to (X/G, \tau_0)$ is continuous, hence, $i : (X/G, \tau') \to (\widehat{X/G}, \hat{\tau}_0|_{\widehat{X/G}})$ is continuous. If $\hat{i} : (X, \tau) \to (\widehat{X/G}, \hat{\tau}_0|_{\widehat{X/G}})$ is defined by $\hat{i}(x) = i\big(q(x)\big)$, it follows easily that it is also continuous and since $\hat{\tau}_0|_{\hat{X/G}} \subset \hat{\tau}_{p_V}$, by Lemma 16 (i), $\hat{i} : (X, \tau) \to (\hat{X/G}, \hat{\tau}_{p_V})$ is closed. Since $(\hat{X/G}, \hat{\tau}_{p_V})$ is a Banach space, it follows by the hypothesis that $\hat{i} : (X, \tau) \to (\hat{X/G}, \hat{\tau}_{p_V})$ is continuous, hence $i : (X/G, \tau') \to (X/G, \tau_{p_V})$ is continuous. So, $p_V : (X, \tau)/G \to \mathbb{R}$ is continuous and therefore V is a neighborhood of zero in $(X, \tau)/G$. Since $q^{-1}(V) = q^{-1}(q(W)) = W + G = W$ and $q^{-1}(V)$ is a neighborhood of zero in (X, τ), it follows that $p : (X, \tau) \to \mathbb{R}$ is continuous. □

Proposition 22. *Let (F, τ_1) be a Hausdorff locally convex space. The following are equivalent:*

(i) *For every barrelled locally convex space (X, τ), every closed linear map $f : (X, \tau) \to (F, \tau_1)$ is continuous.*
(ii) *For every Hausdorff locally convex topology τ_0 in F, such that $\tau_0 \subset \tau_1$, it holds $\tau_1 \subset \tau_0^\beta$.*

Proof. $(i) \Rightarrow (ii)$: Let τ_0 be a Hausdorff locally convex topology on F and $\tau_0 \subset \tau_1$. Then, the identity map $i : (F, \tau_0) \to (F, \tau_1)$ is closed, thus $i : (F, \tau_0^\beta) \to (F, \tau_1)$ is closed, and by hypothesis, it is continuous. Hence, $\tau_1 \subset \tau_0^\beta$.

$(ii) \Rightarrow (i)$: Let (X, τ) be a barrelled locally convex space and $f : (X, \tau) \to (F, \tau_1)$ a closed linear map. By Lemma 16(i), there exists a Hausdorff locally convex topology τ_0 on F with $\tau_0 \subset \tau_1$ such that f is $\tau - \tau_0$ continuous. By Proposition 20, f is $\tau - \tau_0^\beta$ continuous. By hypothesis, $\tau_1 \subset \tau_0^\beta$ and so $f : (X, \tau) \to (F, \tau_1)$ is continuous. □

If X is a TVS over $\mathbb{R}$ or $\mathbb{C}$, or more generally, if the valued field $\mathbb{K}$ is locally compact and if $y \in X$ and V is a closed subset of X, the set $V + \mathrm{co}\{y\}$ is closed, where $\mathrm{co}\{y\} = \{\lambda y : \lambda \in \mathbb{K}, |\lambda| \leq 1\}$ is the absolutely convex hull of $\{y\}$ (the smallest set which is absolutely convex and contains y) because it is the sum of a closed and a compact set. The result is not necessarily true for a general valued field $\mathbb{K}$, but we have the following.

Lemma 23. *Let X be a TVS, $y \in X$, and V an absolutely convex closed subset of X, which does not absorbs y. Then, $V + co\{y\}$ is closed.*

Proof. Let $x \in \overline{V + \mathrm{co}\{y\}}$ and $(y_i)_{i \in I}$ a net in $V + \mathrm{co}\{y\}$ such that $y_i \xrightarrow[i \in I]{} x$. Then there are nets $(x_i)_{i \in I}$ in V and $(\lambda_i)_{i \in I}$ in $\mathbb{K}$, such that for each $i \in I$ it holds $|\lambda_i| \leq 1$ and $y_i = x_i + \lambda_i y$. Set $D = I \times I$ and define an order on it as follows. For $i_1, i_2, j_1, j_2 \in I$, $(i_1, j_1) \geq (i_2, j_2)$ iff $i_1 \geq i_2$ and $j_1 \geq j_2$. Then D becomes a directed set. For all $\delta = (i, j) \in D$, set $\lambda_\delta = \lambda_i - \lambda_j$ and $x_\delta = x_i - x_j$.

It holds $y_\delta = x_\delta + \lambda_\delta y \underset{\delta \in D}{\longrightarrow} 0$ in X. In fact, let U and W be neighborhoods of zero such that $W - W \subset U$. There exists $i_0 \in I$ such that for all $i, j \in I$ with $i, j \geq i_0$ we have $y_i, y_j \in x + W$. Then for all $i, j \in I$ with $(i, j) \geq (i_0, i_0)$, we have $y_{(i,j)} = y_i - y_j \in W - W \subset U$.

Now, $\lambda_\delta \underset{\delta \in D}{\longrightarrow} 0$ in $\mathbb{K}$. Indeed, let $\varepsilon > 0$ and $\lambda \in \mathbb{K}$ with $0 < |\lambda| < \varepsilon$. Set $J = \{\delta \in D : |\lambda_\delta| \geq |\lambda|\}$. Assume, to get a contradiction, that for each $\delta_0 \in D$ there exists $\delta \in J$ such that $\delta \geq \delta_0$. For the subnet $(y_\delta)_{\delta \in J}$ of $(y_\delta)_{\delta \in D}$, we have $\lambda\lambda_\delta^{-1} y_\delta = \lambda\lambda_\delta^{-1} x_\delta + \lambda y$. If W is a balanced neighborhood of zero in X, since $y_\delta \underset{\delta \in J}{\longrightarrow} 0$ in X, it holds $y_\delta \in W$, eventually for all $\delta \in J$ and since $|\lambda\lambda_\delta^{-1}| \leq 1$ for $\delta \in J$, we get $\lambda\lambda_\delta^{-1} y_\delta \in W$ eventually for all $\delta \in J$. Hence, $\lambda\lambda_\delta{}^{1} y_\delta \underset{\delta \in J}{\longrightarrow} 0$, thus $-\lambda\lambda_\delta^{-1} x_\delta \underset{\delta \in J}{\longrightarrow} \lambda y$ in X. Since $|-\lambda\lambda_\delta^{-1}| \leq 1$, it holds $-\lambda\lambda_\delta^{-1} x_\delta \in V$, for all $\delta \in J$, and thus, $\lambda y \in \overline{V} = V$. This is a contradiction since $\lambda \neq 0$ and V does not absorbs y. Hence, $|\lambda_\delta| < |\lambda| < \varepsilon$ eventually for all $\delta \in D$ and the result follows.

So, $(\lambda_i)_{i \in I}$ is a Cauchy net in $\mathbb{K}$ and there exists $\lambda \in \mathbb{K}$ with $|\lambda| \leq 1$ such that $x_i \underset{i \in I}{\longrightarrow} x - \lambda y$ in X. Thus, $x - \lambda y \in V$, i.e., $x \in V + \mathrm{co}\{y\}$. □

Now we are in a position to show that every subspace of countable codimension of a barrelled locally convex space is barrelled. First, we prove it for finite codimension. The analogous result in the classical case was proven for the finite case in [6] and for the infinite case in [12].

Theorem 24. *Let X be a locally convex space and F a subspace of finite codimension. Then for every barrel V in F, there exists a barrel W in E such that $W \cap F = V$.*

Proof. It suffices to prove the result for codimension 1. So, assume that F has codimension 1 in X.

Suppose first that the space generated by $\overline{V}$ is the whole space X. Then we take $W = \overline{V}$. W is clearly absolutely convex and closed in X. It is also absorbing, since, if $x \in X$, there are $\lambda_1, \ldots, \lambda_n \in \mathbb{K}$

such that $x \in \sum_{k=1}^{n} \lambda_k W$, and by choosing $\lambda \in \mathbb{K}$ with $|\lambda| \geq \max\{|\lambda_1|, \ldots, |\lambda_n|, n\}$, we get $x \in \lambda^2 W$. Hence, W is a barrel in X and $W \cap F = \overline{V} \cap F = V$.

Assume now that the space generated by $\overline{V}$ is not X. Then there exists $y \in X$ which is not absorbed by $\overline{V}$. Set $W = \overline{V} + \text{co}\{y\}$. W is absolutely convex and by the preceding Lemma it is closed. We shall show now that it is also absorbing. Since y is not absorbed by V, the space spanned by $\{y\}$ is complementary to F. Let $x \in X$. There are $z \in F$ and $\mu \in \mathbb{K}$ such that $x = z + \mu y$. Since V is absorbing in F and V is absolutely convex, there exists $\mu_1 \in \mathbb{K}$ with $|\mu_1| \geq |\mu|$ such that $z \in \mu_1 V$. But $\mu y \in \mu_1 \text{co}\{y\}$ because $\text{co}\{y\}$ is absolutely convex. So, it follows $x \in \mu_1 W$, which states that W is a barrel in X. Now it remains to show that $W \cap F = V$. Obviously, $V \subset W \cap F$. On the other hand, let $x \in W \cap F$. There is $v \in \overline{V}$ and $\lambda \in \mathbb{K}$ with $|\lambda| \leq 1$ such that $x = v + \lambda y$. Since $x \in F$, there is $\mu \in \mathbb{K} \setminus \{0\}$ such that $\mu x \in V$. Choose $\gamma \in \mathbb{K}$ with $|\gamma| \geq 1 + |\mu|^{-1}$. Then, $\lambda y = x - v \in \mu^{-1} V + \overline{V} \subset \gamma \overline{V}$. This shows that $\lambda = 0$ and completes the proof. □

Corollary 25. *Every subspace of finite codimension of a barrelled locally convex space is barrelled.*

For the countable codimension, we need the following.

Lemma 26. *Let X be a barrelled locally convex space and (V_n) an increasing sequence of absolutely convex subsets of X such that $V = \bigcup_{n=1}^{\infty} V_n$ is absorbing in X. If $\mu \in \mathbb{K}$ with $|\mu| \geq 2$ and $W = \bigcup_{n=1}^{\infty} \overline{V_n}$, then $W \subset \overline{V} \subset \mu W$.*

Proof. Obviously, $W \subset \overline{V}$. Let $x \in \overline{V}$ and assume that $x \notin \mu W$. Then for all $n \in \mathbb{N}$ we have $x \notin \mu \overline{V_n}$. Denote by $\mathcal{N}$ the set of all neighborhoods of zero in X. Since

$$\mu \overline{V_n} = \overline{\mu V_n} = \bigcap_{U \in \mathcal{N}} (\mu V_n + U) = \bigcap_{U, U_1 \in \mathcal{N}} (\mu V_n + U + U_1)$$

$$= \bigcap_{U \in \mathcal{N}} (\overline{\mu V_n + U}),$$

we get that for each $n \in \mathbb{N}$ there exists an absolutely convex neighborhood W_n of zero such that

$$x \notin \overline{\mu V_n + W_n}. \qquad (*)$$

Set $U = \bigcap_{n=1}^{\infty} \overline{V_n + W_n}$. Then U is a barrel in X. Indeed, clearly, it is absolutely convex and closed. It remains to show that it is absorbing. Let $x \in X$. Choose $n_0 \in \mathbb{N}$ and $\lambda, \lambda_1 \in \mathbb{K} \setminus \{0\}$ such that $\lambda x \in V_{n_0+1}$ and $\lambda_1 x \in \bigcap_{k=1}^{n_0} W_k$. Consider $\gamma \in \mathbb{K}$ with $|\gamma| = \min\{|\lambda|, |\lambda_1|\}$. Then $\gamma x \in W_k$, for $k \leq n_0$, and $\gamma x \in V_k$, for $k > n_0$. Thus, $\gamma x \in U$.

So, U is a neighborhood of zero, therefore $x \in V + U$ from which we get that there is $n \in \mathbb{N}$ such that $x \in V_n + U \subset V_n + \overline{V_n + W_n} \subset \overline{V_n + V_n + W_n} \subset \overline{\mu V_n + W_n}$. But this contradicts $(*)$. □

Proposition 27. *Let X be a barrelled locally convex space, F a subspace with countable codimension, V a barrel in F, and $\mu \in \mathbb{K}$ with $|\mu| \geq 2$. Then there exists an increasing sequence (V_n) of absolutely convex closed subsets of X, such that $W = \bigcup_{n=1}^{\infty} V_n$ is a neighborhood of zero in X, $W \cap F = V$, and $\overline{W} \cap F \subset \mu V$.*

Proof. Let $\{x_n : n \in \mathbb{N}\}$ be a Hamel basis of the complement of F and F_n the space spanned by $F \cup \{x_1, \ldots, x_n\}$, for $n \in \mathbb{N}$. By Theorem 24, for each $n \in \mathbb{N}$, there exists a barrel W_n in F_n such that $W_1 \cap F = V$ and $W_{n+1} \cap F_n = W_n$. Set for each n, $V_n = \overline{W}_n$ and $W = \bigcup_{n=1}^{\infty} V_n$. Since $W_n \subset W_{n+1}$, (V_n) is an increasing sequence of absolutely convex and closed subsets of X. If $x \subset X$, then $x \in F_{n_0}$ for some $n_0 \in \mathbb{N}$, and since W_{n_0} is absorbing, it follows that W is absorbing in X and so, by the preceding Lemma, $\overline{W} \subset \mu W$. $(*)$

Since $\overline{W}$ is a barrel in X, it is a neighborhood of zero, so by $(*)$, W is a neighborhood of zero. Let now $y \in W \cap F$. For some $n_0 > 1$, we have $y \in \overline{W_{n_0}} \cap F_{n_0} = W_{n_0}$, hence $y \in W_{n_0} \cap F \subset W_{n_0} \cap F_{n_0-1} = W_{n_0-1}$. It follows $y \in W_1 \cap F = V$. Since, $V \subset F$ and $V \subset W_1 \subset W$, we get $W \cap F = V$ and finally, by $(*)$, it follows $\overline{W} \cap F \subset \mu W \cap F = \mu(W \cap F) = \mu V$. □

Corollary 28. *Every subspace of countable codimension of a barrelled locally convex space is barrelled.*

References

[1] Adasch, N., Ernst, B. and Keim, D. *Topological vector spaces: The theory without convexity conditions*, Lecture Notes in Math. 639, Springer (1978).

[2] Eleizalde, J. B. *Convexidad y acotacion en espacios vectoriales topologicos sombre cuerpos cuasivalorados*, Doctoral Thesis, University of Zaragoza (1975).

[3] Benekas, V. and Katsaras, A. K. Topological vector spaces over valued fields, *Glasnik Matematici*, 28(48), 241–258 (1993).

[4] Benekas, V. and Katsaras, A. K. Topological vector spaces over valued fields II, *Glasnik Matematici*, 29(49), 311–332 (1994).

[5] Bourbaki, N. Éléments de mathématique, Livre V: Espaces vectoriels topologiques, Masson, Paris (1981).

[6] Dieudonné, J. Sur les propriétés de permanence de certains espaces vectoriels topologiques, *Ann. Soc. Pol. Math.* 25, 50–55 (1952).

[7] Husain, T. *The Open Mapping and Closed Graph Theorems in Topological Vector Spaces.* Friedr. Vieweg & Sohn, Braunschweig (1965).

[8] Kelley, J. and Namioka, I. *et. al.*, *Linear Topological Spaces.* Springer Berlin, Heidelberg (1963).

[9] Köthe, G. *Topologische Lineare Räume I.* Springer, Berlin (1966).

[10] A. P. Robertson and W. Robertson. *Topological Vector Spaces.* Cambridge, University Press (1964).

[11] H. H. Schaefer, *Topological vector spaces*, Springer-Verlag, Berlin, Heidelberg, New York (1971).

[12] Valdivia, M. Absolutely convex sets in barrelled spaces, *Ann. Inst. Fourier*, 21, 3–13 (1971).

[13] Wilansky, A. *Modern Methods in Topological Vector Spaces*, New York Mc Graw Hill (1978).

https://doi.org/10.1142/9789811293009_0005

Chapter 5

Random Stability of a System of Functional Equations

Abasalt Bodaghi

Department of Mathematics, West Tehran Branch, Islamic Azad University, Tehran, IRAN

abasalt.bodaghi@gmail.com

In this chapter, we introduce a system of additive-quadratic-cubic-quartic-quintic-sextic (AQCQQS-System) functional equations and describe it as a single equation. We also obtain some stability and hyperstability results concerning this equation in the setting of random normed spaces.

1. Introduction

Ulam [24] introduced the first stability problem concerning group homomorphisms in 1940. The famous Ulam stability problem was partially solved by Hyers [5] for the linear functional equation of Banach spaces as follows: Let X and Y be Banach spaces. Suppose that $f : X \longrightarrow Y$ satisfies

$$\|f(x+y) - f(x) - f(y)\| \leq \epsilon$$

for all $x, y \in X$ and for some $\epsilon \geq 0$. Then, there exists a unique additive mapping $T : X \longrightarrow Y$ such that

$$\|f(x) - T(x)\| \leq \epsilon$$

for all $x \in X$. Recall that a functional equation Γ is said to be *stable* if any function f satisfies the equation Γ approximately, then it must be near to an exact solution of Γ. Moreover, Γ is called *hyperstable* if any function ϕ fulfills Γ approximately (under some conditions), then it is an exact solution of Γ.

Hyers' theorem was generalized by Aoki [1] for additive mappings and by Th. M. Rassias [18] for linear mappings by considering an unbounded Cauchy difference. A generalization of the Rassias theorem was developed by Găvruţa [4] by replacing the unbounded Cauchy difference using a general control function in the spirit of the Rassias approach. Subsequently, many authors and mathematicians were attracted and motivated to investigate the stability problems of functional equations in various spaces; for more information and details, we refer to several articles and books such as [3, 6, 8–13, 15, 16, 19, 21, 25] and references therein.

In what follows, we recall some useful functional equations to achieve our purpose in this chapter:

$$\begin{aligned}
&A(2x+y) + A(2x-y) = 4A(x) \qquad \text{(additive equation)},\\
&Q(2x+y) + Q(2x-y)\\
&\quad = Q(x+y) + Q(x-y) + 6Q(x) \qquad \text{(quadratic equation)},\\
&C(2x+y) + C(2x-y)\\
&\quad = 2[C(x+y) + C(x-y)] + 12C(x) \qquad \text{(cubic equation)},\\
&\mathcal{Q}(2x+y) + \mathcal{Q}(2x-y)\\
&\quad = 4[\mathcal{Q}(x+y) + \mathcal{Q}(x-y)]\\
&\qquad + 24\mathcal{Q}(x) - 6\mathcal{Q}(y) \qquad \text{(quartic equation)},\\
&2\mathfrak{Q}(2x+y) + 2\mathfrak{Q}(2x-y) + \mathfrak{Q}(x+2y) + \mathfrak{Q}(x-2y)\\
&\quad = 20[\mathfrak{Q}(x+y) + \mathfrak{Q}(x-y)] + 90\mathfrak{Q}(x) \qquad \text{(quintic equation)},\\
&S(2x+y) + S(2x-y) + S(x+2y) + S(x-2y)\\
&\quad = 20[S(x+y) + S(x-y)]\\
&\qquad + 90[S(x) + S(y)] \qquad \text{(sextic equation)}.
\end{aligned}$$

The solutions and the Hyers–Ulam stability of the above mentioned have been studied in many articles, some of which are [3, 7, 14, 17] and [20].

For the set X, we denote $\overbrace{X \times X \times \cdots \times X}^{n-\text{times}}$ by X^n. In this chapter, we consider the 6-variable mapping $f : V^6 \longrightarrow W$ that satisfies the general system of additive-quadratic-cubic-quartic-quintic-sextic functional equations (AQCQQS-System) as mentioned earlier. Then, we unify this general system of functional equations as an equation. In continuation, we study some stability results concerning the mappings above in the setting of random normed spaces. Lastly, we show that such mappings under some conditions can be hyperstable.

2. Characterization of a System of Functional Equations

Throughout this chapter, $\mathbb{N}$ and $\mathbb{Q}$ are the set of all positive integers and rational numbers, respectively, and moreover $\mathbb{N}_0 := \mathbb{N} \cup \{0\}, \mathbb{R}_+ := [0, \infty)$. For any $l \in \mathbb{N}_0, n \in \mathbb{N}$, $t = (t_1, \ldots, t_n) \in \{-1, 1\}^n$, and $x = (x_1, \ldots, x_n) \in V^n$, we write $lx := (lx_1, \ldots, lx_n)$ and $tx := (t_1x_1, \ldots, t_nx_n)$, where lx stands, as usual, for the scaler product of l on x in the commutative group $(V, +)$.

Let V and W be vector spaces over $\mathbb{Q}$. Consider a mapping $f : V^6 \longrightarrow W$ that satisfies the general system of additive-quadratic-cubic-quartic-quintic-sextic functional equations (AQCQQS-System):

$$\begin{cases} f(2x_1 + x_2, y, z, u, v, w) + f(2x_1 - x_2, y, z, u, v, w) \\ \quad = 4f(x_1, y, z, u, v, w), \\ f(x, 2y_1 + y_2, z, u, v, w) + f(x, 2y_1 - y_2, z, u, v, w) \\ \quad = f(x, y_1 + y_2, z, u, v, w) + f(x, y_1 - y_2, z, u, v, w) \\ \qquad + 6f(x, y_1, z, u, v, w), \\ f(x, y, 2z_1 + z_2, u, v, w) + f(x, y, 2z_1 - z_2, u, v, w) \\ \quad = 2[f(x, y, z_1 + z_2, u, v, w) + f(x, y, z_1 - z_2, u, v, w) \\ \qquad + 12f(x, y, z_1, u, v, w), \\ f(x, y, z, 2u_1 + u_2, v, w) + f(x, y, z, 2u_1 - u_2, v, w) \\ \quad = 4[f(x, y, z, u_1 + u_2, v, w) + f(x, y, z, u_1 - u_2, v, w)] \\ \qquad + 24f(x, y, z, u_1, v, w) - 6f(x, y, z, u_2, v, w), \end{cases}$$

$$\begin{cases} 2[f(x,y,z,u,2v_1+v_2,w)+f(x,y,z,u,2v_1-v_2,w)] \\ \quad =-f(x,y,z,u,v_1+2v_2,w)-f(x,y,z,u,v_1-2v_2,w) \\ \qquad +20[f(x,y,z,u,v_1+v_2,w)+f(x,y,z,u,v_1-v_2,w)] \\ \qquad +90f(x,y,z,u,v_1,w), \\ f(x,y,z,u,v,2w_1+w_2)+f(x,y,z,u,v,2w_1-w_2) \\ \quad =-f(x,y,z,u,v,w_1+2w_2)-f(x,y,z,u,v,w_1-2w_2) \\ \qquad +20[f(x,y,z,u,v,w_1+w_2)+f(x,y,z,u,v,w_1-w_2)] \\ \qquad +90[f(x,y,z,u,v,w_1)+f(x,y,z,u,v,w_2)]. \end{cases} \tag{1}$$

It is clear that the function $f : \mathbb{R}^6 \longrightarrow \mathbb{R}$ defined by $f(x,y,z,u,v,w) = cxy^2z^3u^4v^5w^6$ is a solution of system (1), where $c \in \mathbb{R}$.

Let V be a vector space over $\mathbb{Q}$. Let also $n \in \mathbb{N}$ with $n \geq 2$ and $x_i^6 = (x_{i1}, \ldots, x_{i6}) \in V^6$, where $i \in \{1,2\}$. We shall denote x_i^6 by x_i when no confusion can arise. For $x_1, x_2 \in V^6$, set

$$\begin{aligned} \mathbb{X} = \{\mathfrak{A}_6 = (A_1,\ldots,A_6)|\ & A_1 = x_{11},\ A_2 \in \{x_{12}, x_{12} \pm x_{22}\}, \\ & A_3 \in \{x_{13}, x_{13} \pm x_{23}\},\ A_4 \in \{x_{14}, x_{24}, x_{14} \pm x_{24}\}, \\ & A_5 \in \{x_{15}, x_{15} \pm x_{25}, x_{15} \pm 2x_{25}\}, \\ & A_6 \in \{x_{16}, x_{26}, x_{16} \pm x_{26}, x_{16} \pm 2x_{26}\}\}. \end{aligned}$$

For $t_l \in \mathbb{N}_0$ with $0 \leq t_l \leq n$, where $l \in \{1,2\}$, consider the subset $\mathbb{X}_{(t_1,t_2,t_3)}^{([t_{4_1},t_{4_2}],[t_{5_1},t_{5_2}],[t_{6_1},t_{6_2},t_{6_3}])}$ of $\mathbb{X}$ as follows:

$$\begin{aligned} \mathbb{X}_{(t_1,t_2,t_3)}^{([t_{4_1},t_{4_2}],[t_{5_1},t_{5_2}],[t_{6_1},t_{6_2},t_{6_3}])} &:= \{\mathfrak{A}_6 \in \mathbb{X}|\ \mathrm{C}\{A_1 : A_1 = x_{11}\} = t_1, \\ & \mathrm{C}\{A_2 : A_2 = x_{12}\} = t_2,\ \mathrm{C}\{A_3 : A_3 = x_{13}\} = t_3, \\ & \mathrm{C}\{A_4 : A_4 = x_{14}\} = t_{4_1},\ \mathrm{C}\{A_4 : A_4 = x_{24}\} = t_{4_2}, \\ & \mathrm{C}\{A_5 : A_5 = x_{15}\} = t_{5_1},\ \mathrm{C}\{A_5 : A_5 = x_{15} \pm x_{25}\} = t_{5_2}, \\ & \mathrm{C}\{A_6 : A_6 = x_{16}\} = t_{6_1},\ \mathrm{C}\{A_6 : A_6 = x_{26}\} = t_{6_2}, \\ & \mathrm{C}\{A_6 : A_6 = x_{16} \pm x_{26}\} = t_{6_3}\}, \end{aligned}$$

where C means the cardinal of a set. Note that

$$t_2, t_3, t_{4_1}, t_{4_2}, t_{5_1}, t_{5_2}, t_{6_1}, t_{6_2}, t_{6_3} \in \{0,1\}$$

and t_1 is always equal to 1.

In the sequel, for the mappings satisfying system (1), we use the following notations:

$$f\left(\mathbb{X}_{(t_1,t_2,t_3)}^{([t_{4_1},t_{4_2}],[t_{5_1},t_{5_2}],[t_{6_1},t_{6_2},t_{6_3}])}\right) := \sum_{\mathfrak{A}_6 \in \mathbb{X}_{(t_1,t_2,t_3)}^{([t_{4_1},t_{4_2}],[t_{5_1},t_{5_2}],[t_{6_1},t_{6_2},t_{6_3}])}} f(\mathfrak{A}_6). \tag{2}$$

We remember that the binomial coefficient for all $n, r \in \mathbb{N}_0$ with $n \geq r$ is defined and denoted by $\binom{n}{r} := \frac{n!}{r!(n-r)!}$.

Definition 1. Let $j \in \{1, 2, \dots, 6\}$ be fixed. We say a mapping $f : V^6 \longrightarrow W$ satisfies the *j-power condition* if

$$f(z_1, \dots, 2z_j, \dots, z_6) = 2^j f(z_1, \dots, z_j, \dots, z_6)$$

for all $z_1, \dots, z_6 \in V^6$. Moreover, f has *zero condition* if $f(x) = 0$ for any $x \in V^6$ with at least one component which is equal to zero.

Remark 2. It is easily verified that if f satisfies system (1), then it has the zero condition. With the above notations, by a long and tedious computation, one can show that if a mapping $f : V^6 \longrightarrow W$ fulfilling system (1), then the equation

$$\begin{aligned} 2f(2x_1 + x_2) = \sum_{p_1=0} & \mathbb{B}_{(t_1,t_2,t_3)}^{([t_{4_1},t_{4_2}],[t_{5_1},t_{5_2}],[t_{6_1},t_{6_2},t_{6_3}])} \\ & \times f\left(\mathbb{X}_{(t_1,t_2,t_3)}^{([t_{4_1},t_{4_2}],[t_{5_1},t_{5_2}],[t_{6_1},t_{6_2},t_{6_3}])}\right) \end{aligned} \tag{3}$$

is valid for it for all $x_1, x_2 \in V^6$, where $f(\mathbb{X}_{(t_1,t_2,t_3)}^{([t_{4_1},t_{4_2}],[t_{5_1},t_{5_2}],[t_{6_1},t_{6_2},t_{6_3}])})$ is defined in (2) and

$$\begin{aligned} &\mathbb{B}_{(t_1,t_2,t_3)}^{([t_{4_1},t_{4_2}],[t_{5_1},t_{5_2}],[t_{6_1},t_{6_2},t_{6_3}])} \\ &\quad := (-1)^{2+t_{4_2}-t_{5_1}-t_{5_2}-t_{6_1}-t_{6_2}-t_{6_3}} \times 2^{5+t_2+t_3+t_{4_1}-t_{4_2}} \\ &\quad\quad \times 3^{t_2+t_3+t_{4_1}+t_{4_2}} \times 20^{t_{5_2}+t_{6_3}} \times 90^{t_{5_1}+t_{6_1}+t_{6_2}}. \end{aligned} \tag{4}$$

On the other hand, if f satisfies equation (3) and j-power condition for all $j \in \{1, 2, \dots, 6\}$, similar to the proof of [2, Proposition 2.3], one can show that system (1) is true for f.

3. Stability Results for (3)

In this section, we state the usual terminology, notations, and conventions of the theory of random normed spaces, as in [22] and [23]. The set of all probability distribution functions is denoted by

$$\begin{aligned}\Delta^+ := \{F : \mathbb{R}\cup\{-\infty,\infty\} \longrightarrow [0,1] |\ F \text{ is left} - \text{continuous}\\ \text{and nondecreasing on } \mathbb{R};\ \text{where } F(0)=0 \text{ and } F(+\infty)=1\}.\end{aligned}$$

Let us define $D^+ := \{F \in \Delta^+ | l^- F(+\infty) = 1\}$, where $l^- F(x)$ denotes the left limit of the function f at the point x. The set Δ^+ is partially ordered by the usual pointwise ordering of functions, that is, $F \leq G$ if and only if $F(t) \leq G(t)$ for all $t \in \mathbb{R}$. The maximal element for Δ^+ in this order is the distribution function ϵ_0 given by

$$\epsilon_0(t) = \begin{cases} 0, & \text{if } t \leq 0 \\ 1, & \text{if } t > 0. \end{cases}$$

Definition 3. ([22]) A mapping $\tau : [0,1] \times [0,1] \longrightarrow [0,1]$ is said to be a continuous triangular norm (briefly, a continuous t-norm) if τ satisfies the following conditions:

(i) τ is commutative and associative.
(ii) τ is continuous.
(iii) $\tau(a,1) = a$ for all $a \in [0,1]$.
(iv) $\tau(a,b) \leq \tau(c,d)$ whenever $a \leq c$ and $b \leq d$ for all $a,b,c,d \in [0,1]$.

Typical examples of continuous t-norms are $\tau_P(a,b) = ab$, $\tau_M(a,b) = \min\{a,b\}$, and $\tau_L(a,b)=\max\{a+b-1,0\}$.

Definition 4 ([23]). A random normed space (RN-space, in short) is a triple $(\mathcal{X},\mu,\tau)$, where $\mathcal{X}$ is a vector space, τ is a continuous t-norm, and μ is a mapping from $\mathcal{X}$ into D^+ such that the following conditions hold:

(RN1) $\mu_x(t) = \epsilon_0(t)$ for all $t > 0$ if and only if $x = 0$.
(RN2) $\mu_{\alpha x}(t) = \mu_x(t/|\alpha|)$ for all $x \in \mathcal{X}$, $\alpha \neq 0$ and all $t \geq 0$.
(RN3) $\mu_{x+y}(t+s) \geq \tau(\mu_x(t),\mu_y(s))$ for all $x,y \in \mathcal{X}$ and all $t,s \geq 0$.

Let $(\mathcal{X}, \|\cdot\|)$ be a normed space. Define the mapping $\mu : \mathcal{X} \longrightarrow D^+$ via $\mu_x(t) = \frac{t}{t+\|x\|}$ for all $x \in \mathcal{X}$ and all $t \geq 0$. Then $(\mathcal{X}, \mu, \tau_M)$ is a random normed space.

Definition 5. Let $(\mathcal{X}, \mu, \tau)$ be an RN-space:

(1) A sequence $\{x_n\}$ in $\mathcal{X}$ is said to be *convergent* to a point $x \in \mathcal{X}$ if, for every $t > 0$ and $\epsilon > 0$, there exists a positive integer N such that $\mu_{x_n - x}(t) > 1 - \epsilon$ whenever $n \geq N$.
(2) A sequence $\{x_n\}$ in $\mathcal{X}$ is called a *Cauchy sequence* if, for every $t > 0$ and $\epsilon > 0$, there exists a positive integer N such that $\mu_{x_n - x_m}(t) > 1 - \epsilon$ whenever $n \geq m \geq N$.
(3) An RN-space $(\mathcal{X}, \mu, \tau)$ is said to be *complete* if and only if every Cauchy sequence in $\mathcal{X}$ is convergent to a point in $\mathcal{X}$.

Theorem 6 ([22]). *If $(\mathcal{X}, \mu, \tau)$ is an RN-space and $\{x_n\}$ is a sequence such that $x_n \to x$, then $\lim_{n\to\infty} \mu_{x_n}(t) = \mu_x(t)$.*

For a t-norm τ and a given sequence $\{a_n\}$ in $[0, 1]$, we define $\tau_{j=1}^n a_j$ recursively by $\tau_{j=1}^1 a_j = a_1$ and $\tau_{j=1}^n a_j = \tau(\tau_{j=1}^{n-1} a_j, a_n)$ for all $n \geq 2$.

In this chapter, we prove the Hyers–Ulam stability of equation (3) in the setting of random normed spaces.

Here and subsequently, for a mapping $f : V^6 \longrightarrow W$, we consider the difference operator $\mathfrak{D}f : V^6 \times V^6 \longrightarrow W$ by

$$\begin{aligned}\mathfrak{D}f(x_1, x_2) &:= 2f(2x_1 + x_2)\\ &\quad - \sum_{p_1=0} \mathbb{B}_{(t_1,t_2,t_3)}^{([t_{4_1},t_{4_2}],[t_{5_1},t_{5_2}],[t_{6_1},t_{6_2},t_{6_3}])}\\ &\quad \times f\left(\mathbb{X}_{(t_1,t_2,t_3)}^{([t_{4_1},t_{4_2}],[t_{5_1},t_{5_2}],[t_{6_1},t_{6_2},t_{6_3}])}\right),\end{aligned}$$

where $f(\mathbb{X}_{(t_1,t_2,t_3)}^{([t_{4_1},t_{4_2}],[t_{5_1},t_{5_2}],[t_{6_1},t_{6_2},t_{6_3}])})$ and $\mathbb{B}_{(t_1,t_2,t_3)}^{([t_{4_1},t_{4_2}],[t_{5_1},t_{5_2}],[t_{6_1},t_{6_2},t_{6_3}])}$ are defined in (2) and (4), respectively. With this notation, we have the next stability result for functional equation (3).

Theorem 7. *Let V be a linear space, $(\mathcal{Z}, \Lambda, \tau_M)$ be an RN-space, and (W, μ, τ_M) be a complete RN-space. Suppose that*

$\psi : V^6 \times V^6 \longrightarrow \mathcal{Z}$ is a mapping such that for some $0 < \alpha < 2^{21}$,

$$\Lambda_{\psi(2x,0)}(t) \geq \Lambda_{\alpha\psi(x,0)}(t) \qquad (x \in V^6, t > 0) \tag{5}$$

and

$$\lim_{p\to\infty} \Lambda_{\psi(2^p x_1, 2^p x_2)}(2^{21p}t) = 1 \qquad (x, y \in V^6, t > 0). \tag{6}$$

If $f : V^6 \longrightarrow W$ is a mapping satisfying

$$\mu_{\mathfrak{D}f(x_1,x_2)}(t) \geq \Lambda_{\psi(x_1,x_2)}(t), \tag{7}$$

for all $x_1, x_2 \in V^6$ and all $t > 0$, then there exists a unique solution $\mathcal{F} : V^6 \longrightarrow W$ of (3) such that

$$\mu_{f(x)-\mathcal{F}(x)}(t) \geq \Lambda_{\psi(x,0)}\left(2^7(2^{21} - \alpha)t\right), \tag{8}$$

for all $x \in V^6$ and all $t > 0$. Moreover, if $\mathcal{F}$ has the j-power condition for any $j \in \{1, \ldots, 6\}$, then it fulfills system (1).

Proof. Putting $x_2 = 0$ in (7), we have

$$\mu_{(2^7 f(2x) - 2^{28} f(x))}(t) \geq \Lambda_{\psi(x,0)}\,(t), \tag{9}$$

$x : x_1 \in V^6$ and $t > 0$. Indeed, for the rest of the proof, we set x_1 by x unless otherwise stated explicitly. Thus, inequality (9) implies that

$$\mu_{\left(\frac{1}{2^{21}} f(2x) - f(x)\right)}(t) \geq \Lambda_{\psi(x,0)}\left(2^{28}t\right), \tag{10}$$

for all $x \in V^6$. Substituting x by $2^p x$ in (10) and applying (5), we get

$$\begin{aligned}
\mu_{\left(\frac{f(2^{p+1}x)}{2^{21(p+1)}} - \frac{f(2^p x)}{2^{21p}}\right)}(t) &\geq \Lambda_{\psi(2^p x,0)}(2^7 \times 2^{21(p+1)}t) \\
&\geq \Lambda_{\alpha^p\psi(x,0)}(2^7 \times 2^{21(p+1)}t) \\
&\geq \Lambda_{\psi(x,0)}\left(2^{28}\left(\frac{2^{21}}{\alpha}\right)^p t\right),
\end{aligned} \tag{11}$$

for all $x \in V^6$ and all non-negative integers p. Applying relations (10) and (11), we obtain

$$\begin{aligned}
&\mu_{\left(\frac{f(2^p x)}{2^{21p}} - f(x)\right)} \left(\frac{1}{2^{28}} \sum_{j=0}^{p-1} \left(\frac{\alpha}{2^{21}}\right)^j t \right) \\
&= \mu_{\left(\sum_{j=0}^{p-1} \left(\frac{f(2^{j+1}x)}{2^{21(j+1)}} - \frac{f(2^j x)}{2^{21j}}\right)\right)} \left(\frac{1}{2^{28}} \sum_{j=0}^{p-1} \left(\frac{\alpha}{2^{21}}\right)^j t \right) \\
&\geq (\tau_M)_{j=0}^{p-1} \left(\mu_{\left(\frac{f(2^{j+1}x)}{2^{21(j+1)}} - \frac{f(2^j x)}{2^{21j}}\right)} \left(\frac{1}{2^{28}} \left(\frac{\alpha}{2^{21}}\right)^j t \right) \right) \\
&= \mu_{\left(\frac{1}{2^{21}} f(2x) - f(x)\right)} \left(\frac{1}{2^{28}} t \right) \\
&\geq \Lambda_{\psi(x,0)}(t),
\end{aligned}$$

for all $x \in V^6$ and all non-negative integers p. The above relations necessitate that

$$\mu_{\left(\frac{f(2^p x)}{2^{21p}} - f(x)\right)}(t) \geq \Lambda_{\psi(x,0)} \left(\frac{t}{\frac{1}{2^{28}} \sum_{j=0}^{p-1} \left(\frac{\alpha}{2^{21}}\right)^j} \right). \tag{12}$$

Switching x into $2^l x$ in (12), we have

$$\mu_{\left(\frac{f(2^{p+l}x)}{2^{21(p+l)}} - \frac{f(2^l x)}{2^{21l}}\right)}(t) \geq \Lambda_{\psi(x,0)} \left(\frac{t}{\frac{1}{2^{28}} \sum_{j=l}^{l+p} \left(\frac{\alpha}{2^{21}}\right)^j} \right), \tag{13}$$

for all $x \in V^6$ and all integers $p \geq l \geq 0$. Due to the convergence of $\sum_{j=l}^{\infty} (\frac{\alpha}{2^{21}})^j$, we find that $\Lambda_{\psi(x,0)}(\frac{t}{\frac{1}{2^{28}} \sum_{j=l}^{l+p} (\frac{\alpha}{2^{21}})^j})$ goes to 1 as l and n tend to infinity, and so $\{\frac{f(2^p x)}{2^{21p}}\}$ is a Cauchy sequence in (W, μ, τ_M). The completeness of (W, μ, τ_M) as a RN-space implies that the mentioned sequence converges to some point $\mathcal{F}(x) \in W$. It follows from (12) that for each $\epsilon > 0$

$$\begin{aligned}
&\mu_{(\mathcal{F}(x) - f(x))}(t + \epsilon) \\
&\geq \tau_M \left(\mu_{\left(\mathcal{F}(x) - \frac{f(2^p x)}{2^{21p}}\right)}(\epsilon), \mu_{\left(\frac{f(2^p x)}{2^{21p}} - f(x)\right)}(t) \right)
\end{aligned}$$

$$\geq \tau_M \left(\mu_{\left(\mathcal{F}(x)-\frac{f(2^p x)}{2^{21p}}\right)}(\epsilon), \Lambda_{\psi(0,x)} \left(\frac{t}{\frac{1}{2^{28}} \sum_{j=0}^{p-1} \left(\frac{\alpha}{2^{21}}\right)^j} \right) \right),$$

for all $x \in V^6$. Taking p to infinity in the above inequality, we deduce that

$$\mu_{(\mathcal{F}(x)-f(x))}(t+\epsilon) \geq \Lambda_{\psi(x,0)} \left(2^7(2^{21}-\alpha)t\right). \tag{14}$$

Letting $\epsilon \to 0$ in (14), we get (8). Moreover, inequality (7) implies that

$$\mu_{\frac{1}{2^{21p}} \mathfrak{D} f(2^p x_1, 2^p x_2)}(t) \geq \Lambda_{\psi(2^p x_1, 2^p x_2)}(2^{21p} t), \tag{15}$$

for all $x_1, x_2 \in V^6$ and all $t > 0$. Once more, letting p to infinity in (15), by (6), we observe that the mapping $\mathcal{F}$ fulfills (3). If $\mathfrak{F} : V^6 \longrightarrow W$ is another mapping that satisfies (3) and (8), then

$$\begin{aligned}
&\mu_{\left(\frac{\mathfrak{F}(2^p x)}{2^{21p}} - \frac{\mathcal{F}(2^p x)}{2^{21p}}\right)}(t) \\
&\quad\geq \min\left\{ \mu_{\left(\frac{f(2^p x)}{2^{21p}} - \frac{\mathcal{F}(2^p x)}{2^{21p}}\right)}\left(\frac{t}{2}\right), \mu_{\left(\frac{\mathfrak{F}(2^p x)}{2^{21p}} - \frac{f(2^p x)}{2^{21p}}\right)}\left(\frac{t}{2}\right) \right\} \\
&\quad\geq \Lambda_{(\psi(2^p x,0))}\left(2^6 \times 2^{21p}(2^{21}-\alpha)t\right) \\
&\quad\geq \Lambda_{(\psi(x,0))}\left(\left(\frac{2^{21}}{\alpha}\right)^p 2^6 \times (2^{21}-\alpha)t \right),
\end{aligned}$$

for all $x \in V^6$. Therefore,

$$\begin{aligned}
\mu_{\mathfrak{F}(x)-\mathcal{F}(x)}(t) &= \lim_{p\to\infty} \mu_{\left(\frac{\mathfrak{F}(2^p x)}{2^{21p}} - \frac{\mathcal{F}(2^p x)}{2^{21p}}\right)}(t) \\
&\geq \lim_{p\to\infty} \Lambda_{(\psi(x,0))}\left(\left(\frac{2^{21}}{\alpha}\right)^p 2^6(2^{21}-\alpha)t \right) = 1.
\end{aligned}$$

The above relation implies that $\mathcal{F}(x) = \mathfrak{F}(x)$ for all $x \in V^6$. The last assertion follows from Remark 2 and now the proof is completed. □

The following corollary is a direct consequences of Theorem 7 concerning the stability of (3).

Corollary 8. *Let V be a linear space, $(\mathcal{Z}, \Lambda, \tau_M)$ be an RN-space, and (W, μ, τ_M) be a complete RN-space. Let also s be a real number* *such that* $s \in [0, 21)$ *and* $z_0 \in \mathcal{Z}$. *If* $f : V^6 \longrightarrow W$ *is a mapping such that*

$$\mu_{\mathfrak{D}f(x_1,x_2)}(t) \geq \Lambda_{(\sum_{i=1}^{2}\sum_{j=1}^{6}\|x_{ij}\|^s)z_0}(t), \tag{16}$$

for all $x_1, x_2 \in V^6$ *and all* $t > 0$, *then there exists a unique solution* $\mathcal{F} : V^6 \longrightarrow W$ *of* (3) *satisfying*

$$\mu_{f(x)-\mathcal{F}(x)}(t) \geq \Lambda_{\sum_{j=1}^{6}\|x_{1j}\|^s z_0}\left(2^7(2^{21} - 2^s)t\right),$$

for all $x \in V^n$ *and all* $t > 0$. *In particular, if* $\mathcal{F}$ *has the* j*-power condition for any* $j \in \{1, \ldots, 6\}$, *then it fulfills system* (1).

Proof. Putting $\psi(x_1, x_2) := (\sum_{i=1}^{2}\sum_{j=1}^{6}\|x_{ij}\|^s)z_0$ and applying Theorem 7 when $\alpha = 2^s$, we get the desired result. □

We have the upcoming stability result which is analogous to Theorem 7 with somewhat different method in the proof.

Theorem 9. *Let* V *be a linear space,* $(\mathcal{Z}, \Lambda, \tau_M)$ *be an RN-space, and* (W, μ, τ_M) *be a complete RN-space. Suppose that* $\psi : V^6 \times V^6 \longrightarrow \mathcal{Z}$ *is a mapping such that for some* $\alpha > 2^{21}$,

$$\Lambda_{\psi(\frac{x}{2}x,0)}(t) \geq \Lambda_{\psi(x,0)}(\alpha t) \qquad (x \in V^6, t > 0) \tag{17}$$

and

$$\lim_{n\to\infty} \Lambda_{2^{21p}\psi(2^p x_1, 2^p x_2)}(t) = 1 \qquad (x_1, x_2 \in V^6, t > 0). \tag{18}$$

If $f : V^6 \longrightarrow W$ *is a mapping satisfying*

$$\mu_{\mathfrak{D}f(x_1,x_2)}(t) \geq \Lambda_{\psi(x_1,x_2)}(t), \tag{19}$$

for all $x_1, x_2 \in V^6$ *and all* $t > 0$, *then there exists a unique solution* $\mathcal{F} : V^6 \longrightarrow W$ *of* (3) *such that*

$$\mu_{f(x)-\mathcal{F}(x)}(t) \geq \Lambda_{\psi(x,0)}\left(\frac{\alpha - 2^{28}}{\alpha}t\right), \tag{20}$$

for all $x \in V^6$ *and* $t > 0$. *In addition, if* $\mathcal{F}$ *has the* j*-power condition for any* $j \in \{1, \ldots, 6\}$, *then it fulfills system* (1).

Proof. Putting $x_2 = 0$ in (7), we have

$$\mu_{\left(f(x)-2^{21}f\left(\frac{x}{2}\right)\right)}(t) \geq \Lambda_{\psi\left(\frac{x}{2},0\right)}\left(2^7 t\right), \tag{21}$$

for all $x := x_1 \in V^6$ and $t > 0$. Indeed, for the rest of the proof, we set x_1 by x. It follows from (17) and (21) that

$$\mu_{\left(f(x)-2^{21}f\left(\frac{x}{2}\right)\right)}(t) \geq \Lambda_{\psi(x,0)}\left(2^7 \alpha t\right). \tag{22}$$

Replacing x by $\frac{x}{2^p}$ in (22), we get

$$\mu_{\left(2^{21p}f\left(\frac{x}{2^p}\right)-2^{21(p+1)}f\left(\frac{x}{2^{p+1}}\right)\right)}(t) \geq \Lambda_{\psi(x,0)}\left(2^7\left(\frac{\alpha}{2^{21}}\right)^p t\right),$$

for all $x \in V^6$ and $t > 0$ and all non-negative integers p. Since

$$f(x) - 2^{21p}f\left(\frac{x}{2^p}\right) = \sum_{j=0}^{p-1} 2^{21j}f\left(\frac{x}{2^j}\right) - 2^{21(j+1)}f\left(\frac{x}{2^{j+1}}\right),$$

we have

$$\mu_{\left(f(x)-2^{21p}f\left(\frac{x}{2^p}\right)\right)}\left(\frac{1}{2^7}\sum_{j=0}^{p-1}\left(\frac{2^{21}}{\alpha}\right)^j t\right) \geq \Lambda_{\psi(x,0)}(t), \tag{23}$$

for all $x \in V^6$ and $t > 0$. In other words,

$$\mu_{\left(f(x)-2^{21p}f\left(\frac{x}{2^p}\right)\right)}(t) \geq \Lambda_{\psi(x,0)}\left(\frac{t}{\frac{1}{2^7}\sum_{j=0}^{p-1}\left(\frac{2^{21}}{\alpha}\right)^j}\right), \tag{24}$$

for all $x \in V^6$ and $t > 0$. Substituting x by $\frac{x}{2^l}$ in (24), we have

$$\mu_{\left(2^{21l}f\left(\frac{x}{2^l}\right)-2^{21(p+l)}f\left(\frac{x}{2^{p+l}}\right)\right)}(t) \geq \Lambda_{\psi(x,0)}\left(\frac{t}{\frac{1}{2^7}\sum_{j=l}^{l+p}\left(\frac{2^{21}}{\alpha}\right)^j}\right), \tag{25}$$

for all $x \in V^6$ and all integers $p \geq l \geq 0$. Since the series $\sum_{j=l}^{\infty}\left(\frac{2^{21}}{\alpha}\right)^j$ is convergent, we see that $\Lambda_{\psi(x,0)}\left(\frac{t}{\frac{1}{2^7}\sum_{j=l}^{l+p}\left(\frac{2^{21}}{\alpha}\right)^j}\right)$ goes to 1 as l and n tend to infinity, and so $\left\{2^{21p}f\left(\frac{x}{2^p}\right)\right\}$ is a Cauchy sequence in (V, μ, τ_M). The completeness of (V, μ, τ_M) as a RN-space necessitates that the last sequence converges to some point $\mathcal{F}(x) \in W$. The remaining assertion goes through in a similar method to the corresponding part of Theorem 7. This finishes the proof. □

Corollary 10. *Let V be a linear space, $(\mathcal{Z}, \Lambda, \tau_M)$ be an RN-space, and (W, μ, τ_M) be a complete RN-space. Suppose that s is a real number such that $s \in [21, \infty)$ and $z_0 \in \mathcal{Z}$. If $f : V^6 \longrightarrow W$ is a mapping such that*

$$\mu_{\mathfrak{D}f(x_1,x_2)}(t) \geq \Lambda_{\left(\sum_{i=1}^{2}\sum_{j=1}^{6}\|x_{ij}\|^s\right)z_0}(t) \tag{26}$$

for all $x_1, x_2 \in V^6$ and all $t > 0$, then there exists a unique solution $\mathcal{F} : V^6 \longrightarrow W$ of (3) satisfying

$$\mu_{f(x)-\mathcal{Q}(x)}(t) \geq \Lambda_{\sum_{j=1}^{6}\|x_{1j}\|^{s_{1j}}z_0}\left(\frac{2^n(2^s - 2^{21})}{2^s}t\right)$$

for all $x \in V^6$ and all $t > 0$. In the case that $\mathcal{F}$ fulfills the j-power condition for any $j \in \{1, \ldots, 6\}$, it fulfills system (1).

Proof. Putting $\psi(x_1, x_2) := \left(\sum_{i=1}^{2}\sum_{j=1}^{6}\|x_{ij}\|^s\right)z_0$ and applying Theorem 7 when $\alpha = 2^s$, we get the desired result. □

Under some mild conditions, equation (3) can be hyperstable as follows.

Corollary 11. *Let V be a linear space, $(\mathcal{Z}, \Lambda, \tau_M)$ be an RN-space, and (W, μ, τ_M) be a complete RN-space. Let s_{ij} be non-negative real numbers such that $\sum_{i=1}^{2}\sum_{j=1}^{6} s_{ij} \neq 21$ and $z_0 \in \mathcal{Z}$. If $f : V^6 \longrightarrow W$ is a mapping such that*

$$\mu_{\mathcal{D}f(x,y)}(t) \geq \Lambda_{\prod_{i=1}^{2}\prod_{j=1}^{6}\|x_{ij}\|^{s_{ij}}z_0}(t)$$

for all $x_1, x_2 \in V^6$ and all $t > 0$, then f satisfies (3). Furthermore, if f satisfies the j-power condition for any $j \in \{1, \ldots, 6\}$, then system (1) is valid for it.

Proof. Putting $\psi(x_1, x_2) := \prod_{i=1}^{2}\prod_{j=1}^{6}\|x_{ij}\|^{s_{ij}}z_0$ in Theorem 7 and Theorem 9, we obtain the result. □

References

[1] Aoki, T. On the stability of the linear transformation in Banach spaces, *J. Math. Soc. Japan.* 2, 64–66 (1950).

[2] Bodaghi, A. and Shojaee, B. On an equation characterizing multi-cubic mappings and its stability and hyperstability, *Fixed Point Theory.* 22, 1, 83–92 (2021).

[3] Daras, N.J. and Rassias, M. Th. Approximation and computation in science and engineering, *Series Springer Optimization and Its Applications* (SOIA). 180, Springer (2022).

[4] Găvruţa, P. A generalization of the Hyers-Ulam-Rassias stability of approximately additive mappings, *J. Math. Anal. Appl.* 184, 431–436 (1994).

[5] Hyers, D.H. On the stability of the linear functional equation, *Proc. Natl. Acad. Sci., U.S.A.* 27, 222–224 (1941).

[6] Hyers, D.H., Isac, G. and Rassias, M. Th. Stability of Functional Equations in Several Variables, Birkhauser (1998).

[7] Jun, K.W. and Kim, H.M. On the Hyers-Ulam-Rassias stability of a general cubic functional equation, *Math. Inequal. Appl.* 6, 2, 289–302 (2003).

[8] Jung, S.-M., Lee, K.-S., Rassias M. Th. and Yang, S.-M. Approximation properties of solutions of a mean value-type functional inequality, II, *Mathematics.* 8, 1299 (2020) DOI: 10.3390/math8081299.

[9] Jung, S.-M., Mortici, C. and Rassias, M. Th. On a functional equation of trigonometric type, *Appl. Math. Comput.* 252, 294–303 (2015).

[10] Jung, S.-M., Popa, D. and Rassias, M. Th. On the stability of the linear functional equation in a single variable on complete metric groups, *J. Global Optimiz.* 59, 165–171 (2014).

[11] Jung, S.-M. and Rassias, M. Th. A linear functional equation of third order associated to the Fibonacci numbers, *Abstr. Appl. Anal.* 2014 (2014), Art. ID 137468, 7 pp. http://dx.doi.org/10.1155/2014/137468.

[12] Kannappan, P. *Functional Equations and Inequalities with Applications*, Springer, (2009).

[13] Kuczma, M. *An Introduction to the Theory of Functional Equations and Inequalities. Cauchy's Equation and Jensen's Inequality*, Birkhauser Verlag, Basel (2009).

[14] Lee, S., Im, S. and Hwang, I. Quartic functional equations, *J. Math. Anal. Appl.* 307, 387–394 (2005).

[15] Lee, Y-H., Jung, S-M. and Rassias, M. Th. Uniqueness theorems on functional inequalities concerning cubic-quadratic-additive equation, *J. Math. Inequal.* 12, 43–61 (2018).

[16] Lee, Y-H., Jung, S-M. and Rassias, M. Th. On an n-dimensional mixed type additive and quadratic functional equation, *Appl. Math. Comput.* 228, 13–16 (2014).

[17] Murali, R., Boobalan, M. and Raj, A.A. Stability of quintic functional equation in 2-Banach space, *Int. J. Math. Appl.* 4 (2016), 41–46.

[18] Rassias, M. Th. On the stability of the linear mapping in Banach Space, *Proc. Amer. Math. Soc.* 72(2), 297–300 (1978).

[19] Rassias, M. Th. Functional Equations and Inequalities, Kluwer Academic Publishers (2000).

[20] Ravi, K. and Sabarinathan, S. Generalized Hyers-Ulam stability of a sextic functional equation in paranormed spaces, *Int. J. Math. Trends Tech.,* 9, 60–69 (2014).

[21] Sahoo, P.K. and Kannappan, P. *Introduction to Functional Equations.* CRC Press, Boca Raton (2011)

[22] Schweizer, B. and Sklar, A. *Probabilistic metric spaces*, Elsevier, North Holand, New York (1983).

[23] Šerstnev, A.N. On the motion of a random normed space, *Dokl. Akad. Nauk SSSR,* 149, 280–283 (1963).

[24] Ulam, S.M. *Problems in Modern Mathematics*, Science Editions, Wiley, New York, (1964).

[25] Xu, T.Z., Wang, C. and Rassias, M. Th. On the stability of multi-additive mappings in non-Archimedean normed spaces, *J. Comput. Anal. Appl.* 18, 1102–1110 (2015).

https://doi.org/10.1142/9789811293009_0006

Chapter 6

Multi-quintic Mappings and Their Stability

Abasalt Bodaghi[*,§], Ali Bagheri Vakilabad[†,¶]
and Themistocles Michael Rassias[‡,||]

[*]*Department of Mathematics, West Tehran Branch, Islamic Azad University, Tehran, IRAN*
[†]*Department of Mathematics, Ardabil Branch, Islamic Azad University, Ardabil, IRAN*
[‡]*Department of Mathematics, National Technical University of Athens, Zografou Campus, 157 80, Athens, GREECE*
[§]*abasalt.bodaghi@gmail.com*
[¶]*alibagheri1385@yahoo.com*
[||]*trassias@math.ntua.gr*

In this chapter, we investigate multivariable mappings, which are quintic in each variable. We characterize the structure of such mappings and prove that such mappings can be unified as an equation. We also establish the stability of the quintic functional equation using a fixed point theorem in a normed space setting. Additionally, we present a few corollaries corresponding to some known stability outcomes. Lastly, using a characterization result, we supply an appropriate counterexample to invalidate the consequences for the case of singularity.

1. Introduction

We say a functional equation Γ is *stable* if any function f satisfies the equation Γ approximately, then it must be near to an exact solution of Γ. Moreover, Γ is *hyperstable* if any function ϕ fulfills Γ approximately (under some conditions), then it is an exact solution of Γ.

Over the last two decades, the stability problem for functional equations, initiated by Ulam [19] for group homomorphisms (answered by Hyers [11]) has been studied by Aoki [2], Th. M. Rassias [18], and Găvruţa [10] for multiple variables mappings including multi-additive, multi-quadratic, multi-cubic, and multi-quartic mappings [4, 5, 7, 8, 16], and [21]. Their definitions can be stated as follows:

Let V be a commutative group, W be a linear space over rational numbers, and n be an integer with $n \geq 2$. A mapping $f\colon V^n \longrightarrow W$ is called

(i) *multi-additive* if it satisfies Cauchy's functional equation $A(x+y) = A(x) + A(y)$ in each variable [8];
(ii) *multi-quadratic* if it fulfills the quadratic functional equation $Q(x+y)+Q(x-y) = 2Q(x)+2Q(y)$ in each variable [3,7,21];
(iii) *multi-cubic* if it satisfies the cubic equation $C(2x+y)+C(2x-y) = 2C(x+y)+2C(x-y)+12C(x)$ in each variable [5];
(iv) *multi-quartic* if it satisfies one of the following quartic equation in each variable [1,4,14,17]:

$$\begin{aligned}\mathfrak{Q}(x+2y)+\mathfrak{Q}(x-2y) &= 4\mathfrak{Q}(x+y)+4\mathfrak{Q}(x-y)\\ &\quad -6\mathfrak{Q}(x)+24\mathfrak{Q}(y),\\ \mathfrak{Q}(2x+y)+\mathfrak{Q}(2x-y) &= 4\mathfrak{Q}(x+y)+4\mathfrak{Q}(x-y)+24\mathfrak{Q}(x)\\ &\quad -6\mathfrak{Q}(y).\end{aligned}$$

In [20], Xu *et al.* obtained the general solution the quintic functional equation

$$\begin{aligned}&f(x+3y)-5f(x+2y)+10f(x+y)-10f(x)+5f(x-y)\\ &\quad -f(x-2y) = 120f(y)\end{aligned}$$

for the first time. They also investigated the Ulam stability problem for it in quasi-β-normed spaces via the fixed point method. Park *et al.* [15] presented a different form of the quintic functional equation as follows:

$$\begin{aligned} &f(3x+y) - 5f(2x+y) + f(2x-y) + 10f(x+y) - 5f(x-y) \\ &\quad = 10f(y) + f(3x) - 3f(2x) - 27f(x). \end{aligned} \tag{1}$$

Morali *et al.* [13] studied the Hyers–Ulam stability of the functional equation on

$$\begin{aligned} &2Q(2x+y) + 2Q(2x-y) + Q(x+2y) + Q(x-2y) \\ &\quad = 20[Q(x+y) + Q(x-y)] + 90Q(x) \end{aligned} \tag{2}$$

in 2-Banach spaces setting.

Motivated by the quintic functional equation (2), in this chapter, we introduce multi-quintic mappings that fulfill (2) in each component and include a characterization of such mappings. In fact, we prove that every multi-quintic mapping can be shown as a single functional equation and vice versa (under some additional conditions). Furthermore, we investigate the generalized Hyers–Ulam stability for multi-quintic mappings by applying the fixed point method in normed spaces as presented in [6]. Finally, by using a characterization result, we indicate an example for the non-stable multi-quintic functional equation.

2. Characterization of Multi-Quintic Mappings

Throughout this chapter, $\mathbb{N}$ and $\mathbb{Q}$ are the sets of all positive integers and rational numbers, respectively, and moreover, $\mathbb{N}_0 := \mathbb{N} \cup \{0\}, \mathbb{R}_+ := [0, \infty)$. For any $l \in \mathbb{N}_0, n \in \mathbb{N}$, $t = (t_1, \ldots, t_n) \in \{-1, 1\}^n$, and $x = (x_1, \ldots, x_n) \in V^n$, we write $lx := (lx_1, \ldots, lx_n)$ and $tx := (t_1x_1, \ldots, t_nx_n)$, where lx stands, as usual, for the scaler product of l on x in the commutative group $(V, +)$.

Definition 1. Let V and W be vector spaces over $\mathbb{Q}$, $n \in \mathbb{N}$. A several variables mapping $f : V^n \longrightarrow W$ is called *n-quintic* or *multi-quintic* if f satisfies (2) in each of its n arguments, that is,

$$\begin{aligned}2[&f(v_1,\ldots,v_{i-1},2v_i+v_i',v_{i+1},\ldots,v_n)\\&+f(v_1,\ldots,v_{i-1},2v_i-v_i',v_{i+1},\ldots,v_n)]\\&+f(v_1,\ldots,v_{i-1},v_i+2v_i',v_{i+1},\ldots,v_n)\\&+f(v_1,\ldots,v_{i-1},v_i-2v_i',v_{i+1},\ldots,v_n)\\=20[&f(v_1,\ldots,v_{i-1},v_i+v_i',v_{i+1},\ldots,v_n)\\&+f(v_1,\ldots,v_{i-1},v_i-v_i',v_{i+1},\ldots,v_n)]\\&+90f(v_1,\ldots,v_{i-1},v_i,v_{i+1},\ldots,v_n).\end{aligned}$$

Let $n \in \mathbb{N}$ with $n \geq 2$ and $x_i^n = (x_{i1},\ldots,x_{in}) \in V^n$, where $i \in \{1,2\}$. We shall denote x_i^n by x_i when no confusion can arise. For $x_1, x_2 \in V^n$, set

$$\mathbb{A}^n = \{\mathfrak{A}_n = (A_1,\ldots,A_n)|\ A_j \in \{x_{1j}, x_{1j} \pm x_{2j}, x_{1j} \pm 2x_{2j}\}\},$$

for all $j \in \{1,\ldots,n\}$. For $p_l \in \mathbb{N}_0$ with $0 \leq p_l \leq n$, where $l \in \{1,2\}$, consider the subset $\mathbb{A}^n_{(p_1,p_2)}$ of $\mathbb{A}^n$ as follows:

$$\begin{aligned}\mathbb{A}^n_{(p_1,p_2)} := \{&\mathfrak{A}_n \in \mathbb{A}^n|\ \mathrm{Card}\{A_j\colon A_j = x_{1j}\}\\&= p_1, \mathrm{Card}\{A_j\colon A_j = x_{1j} \pm x_{2j}\} = p_2\}.\end{aligned}$$

In the sequel, for the multi-quintic mappings, we use the following notations:

$$f(\mathbb{A}^n_{(p_1,p_2)}) := \sum_{\mathfrak{A}_n \in \mathbb{A}^n_{(p_1,p_2)}} f(\mathfrak{A}_n) \tag{3}$$

and

$$f(\mathbb{A}^n_{(p_1,p_2)}, z) := \sum_{\mathfrak{A}_n \in \mathbb{A}^n_{(p_1,p_2)}} f(\mathfrak{A}_n, z) \quad (z \in V).$$

We remember that the binomial coefficient for all $n, r \in \mathbb{N}_0$ with $n \geq r$ is defined and denoted by $\binom{n}{r} := \frac{n!}{r!(n-r)!}$.

Definition 2. We say a mapping $f : V^n \longrightarrow W$ satisfies the 5-*power condition* in the jth variable if

$$f(z_1, \ldots, z_{j-1}, 2z_j, z_{j+1}, \ldots, z_n) = 2^5 f(z_1, \ldots, z_{j-1}, z_j, z_{j+1}, \ldots, z_n)$$

for all $z_1, \ldots, z_n \in V^n$. The 5-power condition is also called the *quintic condition.*

Remark 1. It is easily verified that if f is a multi-quintic mapping, then it satisfies the quintic condition in each of variable. Note that the converse is not true. Here, by means of an example, we show that the quintic condition in all variables for a mapping f does not imply being multi-quintic. Let $(\mathcal{A}, \|\cdot\|)$ be a Banach algebra. Fix the vector a_0 in $\mathcal{A}$ (not necessarily unit). Define the mapping $f : \mathcal{A}^n \longrightarrow \mathcal{A}$ by $f(a_1, \ldots, a_n) = \prod_{j=1}^{n} \|a_j\|^5 a_0$ for $(a_1, \ldots, a_n) \in \mathcal{A}^n$. It is easy to check that the mapping h satisfies the quintic condition in all variables but not multi-quintic even for $n = 1$, which means that f does not satisfy equation (2).

In the following theorem, we characterize a multi-quintic mapping as an equation. In addition, this equation describe a multi-quintic mapping when we add the quintic condition to all variables.

Theorem 2. *If $f : V^n \longrightarrow W$ is a multi-quintic mapping, then it fulfills the equation*

$$2^n \sum_{q \in \{-1,1\}^n} f(2x_1 + qx_2) = \sum_{p_1=0}^{n} \sum_{p_2=0}^{n-p_1} (-1)^{n-p_1-p_2} 20^{p_2} 90^{p_1} f(\mathbb{A}^n_{(p_1,p_2)}), \tag{4}$$

for all $x_1, x_2 \in V^n$, where $f(\mathbb{A}^n_{(p_1,p_2)})$ is defined in (3). The converse holds provided that f satisfies the quintic condition in each variable.

Proof. Assume that f is multi-quintic. We prove that f satisfies equation (4) by induction on n. For $n = 1$, it is trivial that f satisfies equation (2). If (4) is valid for some positive integer $n > 1$, then

$$\begin{aligned}
2^{n+1} & \sum_{q\in\{-1,1\}^{n+1}} f\left(2x_1^{n+1} + qx_2^{n+1}\right) \\
&= -2^n \sum_{q\in\{-1,1\}^n} \sum_{t\in\{-1,1\}} f\left(x_1^n + qx_2^n, x_{1,n+1} + 2tx_{2,n+1}\right) \\
&\quad + 20 \times 2^n \sum_{q\in\{-1,1\}^n} \sum_{t\in\{-1,1\}} f\left(x_1^n + qx_2^n, x_{1,n+1} + tx_{2,n+1}\right) \\
&\quad + 90 \times 2^n \sum_{q\in\{-1,1\}^n} f\left(x_1^n + qx_2^n, x_{1,n+1}\right) \\
&= -\sum_{p_1=0}^{n} \sum_{p_2=0}^{n-p_1} \sum_{t\in\{-1,1\}} (-1)^{n-p_1-p_2} \times 20^{p_2} \times 90^{p_1} \\
&\quad \times f\left(\mathbb{A}^n_{(p_1,p_2)}, x_{1,n+1} + 2tx_{2,n+1}\right) \\
&\quad + 20 \sum_{p_1=0}^{n} \sum_{p_2=0}^{n-p_1} \sum_{t\in\{-1,1\}} (-1)^{n-p_1-p_2} \times 20^{p_2} \times 90^{p_1} f \\
&\quad \times \left(\mathbb{A}^n_{(p_1,p_2)}, x_{1,n+1} + tx_{2,n+1}\right) \\
&\quad + 90 \sum_{p_1=0}^{n} \sum_{p_2=0}^{n-p_1} (-1)^{n-p_1-p_2} \times 20^{p_2} \times 90^{p_1} f\left(\mathbb{A}^n_{(p_1,p_2)}, x_{1,n+1}\right) \\
&= \sum_{p_1=0}^{n+1} \sum_{p_2=0}^{n+1-p_1} (-1)^{n+1-p_1-p_2} \times 20^{p_2} \times 90^{p_1} f\left(\mathbb{A}^{n+1}_{(p_1,p_2)}\right).
\end{aligned}$$

This means that (4) holds for $n + 1$.

For the converse, fix $j \in \{1, \ldots, n\}$. Set

$$\begin{aligned}
f^*(2x_{1j}, x_{2j}) \colon &= f\left(x_{11}, \ldots, x_{1,j-1}, 2x_{1j} + x_{2j}, x_{1,j+1}, \ldots, x_{1n}\right) \\
&\quad + f\left(x_{11}, \ldots, x_{1,j-1}, 2x_{1j} - x_{2j}, x_{1,j+1}, \ldots, x_{1n}\right), \\
f^*(x_{1j}, 2x_{2j}) \colon &= f\left(x_{11}, \ldots, x_{1,j-1}, x_{1j} + 2x_{2j}, x_{1,j+1}, \ldots, x_{1n}\right) \\
&\quad + f\left(x_{11}, \ldots, x_{1,j-1}, x_{1j} - 2x_{2j}, x_{1,j+1}, \ldots, x_{1n}\right),
\end{aligned}$$

$$f^*(x_{1j},x_{2j})\colon = f\left(x_{11},\ldots,x_{1,j-1},x_{1j}+x_{2j},x_{1,j+1},\ldots,x_{1n}\right)\\ + f\left(x_{11},\ldots,x_{1,j-1},x_{1j}-x_{2j},x_{1,j+1},\ldots,x_{1n}\right),$$

and

$$f^*(x_{1j}) := f(x_1) = f(x_{11},\ldots,x_{1n}).$$

Putting $x_{2k}=0$ for all $k\in\{1,\ldots,n\}\backslash\{j\}$ in (4) and using the quintic condition, we get

$$\begin{aligned}
&2^n\times 2^{n-1}\times 2^{5(n-1)}f^*(2x_{1j},x_{2j})\\
&\quad = 2^n\times 2^{n-1}[f\left(2x_{11},\ldots,2x_{1,j-1},2x_{1j}+x_{2j},2x_{1,j+1},\ldots,2x_{1n}\right)\\
&\qquad + f\left(2x_{11},\ldots,2x_{1,j-1},2x_{1j}-x_{2j},2x_{1,j+1},\ldots,2x_{1n}\right)]\\
&\quad = \sum_{p_1=0}^{n-1}\sum_{p_2=0}^{n-1-p_1}\left[\binom{n-1}{p_1}\binom{n-1-p_1}{p_2}2^{n-1-p_1-p_2}\right.\\
&\qquad \left.\times 2^{p_2}\times(-1)^{n-p_1-p_2}\times 20^{p_2}\times 90^{p_1}\right]f^*(x_{1j},2x_{2j})\\
&\qquad +\sum_{p_1=0}^{n-1}\sum_{p_2=1}^{n-p_1}\left[\binom{n-1}{p_1}\binom{n-p_1}{p_2-1}2^{n-p_1-p_2}\right.\\
&\qquad \left.\times 2^{p_2-1}\times(-1)^{n-p_1-p_2}\times 20^{p_2}\times 90^{p_1}\right]f^*(x_{1j},x_{2j})\\
&\qquad +\sum_{p_1=1}^{n}\sum_{p_2=0}^{n-p_1}\left[\binom{n-1}{p_1-1}\binom{n-p_1}{p_2}2^{n-p_1-p_2}\right.\\
&\qquad \left.\times 2^{p_2}\times(-1)^{n-p_1-p_2}\times 20^{p_2}\times 90^{p_1}\right]f^*(x_{1j})\\
&\quad = -\sum_{p_1=0}^{n-1}\sum_{p_2=0}^{n-1-p_1}\left[\binom{n-1}{p_1}\binom{n-1-p_1}{p_2}(-2)^{n-1-p_1-p_2}\right.\\
&\qquad \left.\times 40^{p_2}\times 90^{p_1}\right]f^*(x_{1j},2x_{2j})\\
&\qquad +\sum_{p_1=0}^{n-1}\sum_{p_2=0}^{n-1-p_1}\left[\binom{n-1}{p_1}\binom{n-p_1}{p_2}2^{n-1-p_1-p_2}\times 2^{p_2}\right.
\end{aligned}$$

$$
\begin{aligned}
&\times (-1)^{n-1-p_1-p_2} \times 20^{p_2+1} \times 90^{p_1}\Bigg] f^*(x_{1j}, x_{2j}) \\
&+ \sum_{p_1=0}^{n-1} \sum_{p_2=0}^{n-1-p_1} \left[\binom{n-1}{p_1}\binom{n-1-p_1}{p_2} 2^{n-1-p_1-p_2} \times 2^{p_2}\right. \\
&\times (-1)^{n-1-p_1-p_2} \times 20^{p_2} \times 90^{p_1+1}\Big] f^*(x_{1j}) \\
= &- \sum_{p_1=0}^{n-1} \left[\binom{n-1}{p_1} 38^{n-1-p_1} \times 90^{p_1}\right] f^*(x_{1j}, 2x_{2j}) \\
&+ 20 \sum_{p_1=0}^{n-1} \left[\binom{n-1}{p_1} 38^{n-1-p_1} \times 90^{p_1}\right] f^*(x_{1j}, x_{2j}) \\
&+ 90 \sum_{p_1=0}^{n-1} \left[\binom{n-1}{p_1} 38^{n-1-p_1-p_2} \times 90^{p_1}\right] f^*(x_{1j}) \\
= &-128^{n-1} f^*(x_{1j}, 2x_{2j}) + 20 \times 128^{n-1} f^*(x_{1j}, x_{2j}) \\
&+ 90 \times 128^{n-1} f^*(x_{1j}). \qquad (5)
\end{aligned}
$$

Relation (5) implies that

$$2f^*(2x_{1j}, x_{2j}) = -f^*(x_{1j}, 2x_{2j}) + 20f^*(x_{1j}, x_{2j}) + 90f^*(x_{1j}).$$

The above equality means that f is quintic in the jth variable. □

By means of the above theorem, it is easily seen that the mapping $f : \mathbb{R} \to \mathbb{R}$ defined by $f(z_1, \ldots, z_n) = c\prod_{j=1}^{n} z_j^5$ satisfies (4) and so this equation is said to be *multi-quintic* functional equation.

3. Stability Results for (4)

Throughout, for two sets X and Y, the set of all mappings from X to Y is denoted by Y^X. Here, we highlight the following theorem which is a fundamental result in fixed point theory [6, Theorem 1]. This result plays a key tool to obtain our objective in this chapter.

Theorem 3. *Let hypotheses*

(A1) *Y is a Banach space, $\mathcal{S}$ is a non-empty set, $j \in \mathbb{N}$, $g_1, \ldots, g_j$: $\mathcal{S} \longrightarrow \mathcal{S}$, and $L_1, \ldots, L_j : \mathcal{S} \longrightarrow \mathbb{R}_+$,*

(A2) *$\mathcal{T} : Y^{\mathcal{S}} \longrightarrow Y^{\mathcal{S}}$ is an operator satisfying the inequality*

$$\|\mathcal{T}\lambda(x) - \mathcal{T}\mu(x)\| \leq \sum_{i=1}^{j} L_i(x) \, \|\lambda(g_i(x)) - \mu(g_i(x))\|,$$
$$\lambda, \mu \in Y^{\mathcal{S}}, x \in \mathcal{S},$$

(A3) *$\Lambda : \mathbb{R}_+^{\mathcal{S}} \longrightarrow \mathbb{R}_+^{\mathcal{S}}$ is an operator defined through*

$$\Lambda\delta(x) := \sum_{i=1}^{j} L_i(x)\delta(g_i(x))\delta \in \mathbb{R}_+^{\mathcal{S}}, x \in \mathcal{S}$$

hold and the function $\theta : \mathcal{S} \longrightarrow \mathbb{R}_+$ and the mapping ϕ: $\mathcal{S} \longrightarrow Y$ fulfill thc following two conditions:

$$\|\mathcal{T}\phi(x) - \phi(x)\| \leq \theta(x), \quad \theta^*(x) := \sum_{l=0}^{\infty} \Lambda^l \theta(x) < \infty \quad (x \in \mathcal{S}).$$

Then, there exists a unique fixed point ψ of $\mathcal{T}$ such that

$$\|\phi(x) - \psi(x)\| \leq \theta^*(x) \quad (x \in \mathcal{S}).$$

Moreover, $\psi(x) = \lim_{l\to\infty} \mathcal{T}^l \phi(x)$ for all $x \in \mathcal{S}$.

Here and subsequently, for the mapping $f : V^n \longrightarrow W$, we consider the difference operator $\mathfrak{D}f : V^n \times V^n \longrightarrow W$ by

$$\mathfrak{D}f(x_1, x_2) := 2^n \sum_{q\in\{-1,1\}^n} f(2x_1 + qx_2)$$
$$- \sum_{p_1=0}^{n} \sum_{p_2=0}^{n-p_1} (-1)^{n-p_1-p_2} 20^{p_2} 90^{p_1} f\left(\mathbb{A}_{(p_1,p_2)}^n\right),$$

where $f\left(\mathbb{A}_{(p_1,p_2)}^n\right)$ is defined in (3). With this notation, we have the following Găvruţa stability result for equation (4).

Theorem 4. *Let $\beta \in \{-1, 1\}$, V be a linear space, and W be a Banach space. Suppose that $\phi : V^n \times V^n \longrightarrow \mathbb{R}_+$ is a mapping satisfying*

$$\lim_{l\to\infty}\left(\frac{1}{2^{5n\beta}}\right)^l \phi(2^{\beta l}x_1, 2^{\beta l}x_2) = 0, \tag{6}$$

for all $x_1, x_2 \in V^n$ and

$$\Phi(x) = \frac{1}{2^{5n\frac{\beta+1}{2}+2n}}\sum_{l=0}^{\infty}\left(\frac{1}{2^{5n\beta}}\right)^l \phi\left(2^{\beta l+\frac{\beta-1}{2}}x, 0\right) < \infty, \tag{7}$$

for all $x \in V^n$. Assume also $f : V^n \longrightarrow W$ satisfying the inequality

$$\|\mathfrak{D}f(x_1, x_2)\| \leq \phi(x_1, x_2), \tag{8}$$

for all $x_1, x_2 \in V^n$. Then, there exists a solution $\mathcal{Q} : V^n \longrightarrow W$ of (4) such that

$$\|f(x) - \mathcal{Q}(x)\| \leq \Phi(x), \tag{9}$$

for all $x \in V^n$. Moreover, if $\mathcal{Q}$ satisfies the quintic condition in each variable, then it is a unique multi-quintic mapping.

Proof. Putting $x = x_1$ and $x_2 = 0$ in (8), we have

$$\left\|2^{2n}f(2x) - Tf(x)\right\| \leq \phi(x, 0), \tag{10}$$

for all $x \in V^n$, where

$$T = \sum_{p_1=0}^{n}\sum_{p_2=1}^{n-p_1}\binom{n}{p_1}\binom{n-p_1}{p_2}2^{n-p_1-p_2} \times 2^{p_2} \times (-1)^{n-p_1-p_2} \times 20^{p_2} \times 90^{p_1}.$$

On the other hand, we have

$$\begin{aligned} T &= \sum_{p_1=0}^{n}\sum_{p_2=1}^{n-p_1}\binom{n}{p_1}\binom{n-p_1}{p_2}(-2)^{n-p_1-p_2} \times 40^{p_2} \times 90^{p_1} \\ &= \sum_{p_1=0}^{n}\binom{n}{p_1}38^{n-p_1}90^{p_1} = (38+90)^n = 2^{7n}. \end{aligned} \tag{11}$$

It follows from (10) and (11) that

$$\|f(2x) - 2^{5n}f(x)\| \leq \frac{1}{2^{2n}}\phi(x,0), \tag{12}$$

for all $x \in V^n$. Set

$$\xi(x) := \frac{1}{2^{5n\frac{\beta+1}{2}+2n}}\phi(2^{\frac{\beta-1}{2}}x,0), \quad \text{and} \quad \mathcal{T}$$

$$\xi(x) := \frac{1}{2^{5n\beta}}\xi(2^{\beta}x) \quad (\xi \in W^{V^n}).$$

Then, relation (12) can be rewritten as

$$\|f(x) - \mathcal{T}f(x)\| \leq \xi(x) \quad (x \in V^n).$$

Define $\Lambda\eta(x) := \frac{1}{2^{5n\beta}}\eta(2^{\beta}x)$ for all $\eta \in \mathbb{R}_+^{V^n}, x \in V^n$. We now see that Λ has the form given in (A3) with $\mathcal{S} = V^n$, $g_1(x) = 2^{\beta}x$, and $L_1(x) = \frac{1}{2^{5n\beta}}$ for all $x \in V^n$. Furthermore, for each $\lambda, \mu \in W^{V^n}$ and $x \in V^n$, we obtain

$$\|\mathcal{T}\lambda(x) - \mathcal{T}\mu(x)\| = \left\| \frac{1}{2^{5n\beta}}\left[\lambda(2^{\beta}x) - \mu(2^{\beta}x)\right] \right\| \leq L_1(x)$$
$$\|\lambda(g_1(x)) - \mu(g_1(x))\|.$$

The above relation shows that the hypothesis (A2) holds. By induction on l, one can check that for any $l \in \mathbb{N}_0$ and $x \in V^n$, we have

$$\Lambda^l\xi(x) := \left(\frac{1}{2^{5n\beta}}\right)^l \xi(2^{\beta l}x) = \frac{1}{2^{5n\frac{\beta+1}{2}+2n}}\left(\frac{1}{2^{5n\beta}}\right)^l \phi\left(2^{\beta l+\frac{\beta-1}{2}}x,0\right), \tag{13}$$

for all $x \in V^n$. It now follows from (7) and (13) that all assumptions of Theorem 3 are fulfilled. Hence, there exists a mapping $\mathcal{Q} : V^n \longrightarrow W$ such that

$$\mathcal{Q}(x) = \lim_{l\to\infty}(\mathcal{T}^l f)(x) = \frac{1}{2^{5n\beta}}\mathcal{Q}(2^{\beta}x) \qquad (x \in V^n),$$

and moreover (9) holds. We claim that

$$\|\mathfrak{D}(\mathcal{T}^l f)(x_1,x_2)\| \leq \left(\frac{1}{2^{5n\beta}}\right)^l \phi(2^{\beta l}x_1, 2^{\beta l}x_2) \tag{14}$$

for all $x_1, x_2 \in V^n$ and $l \in \mathbb{N}_0$. We argue by induction on l. It is clear that (14) is true for $l = 0$ by (8). Assume that (14) is valid for an

$l \in \mathbb{N}_0$. Then

$$\begin{aligned}
&\left\|\mathfrak{D}(\mathcal{T}^{l+1}f)(x_1,x_2)\right\| \\
&= \Bigg\|2^n \sum_{q\in\{-1,1\}^n} (\mathcal{T}^{l+1}f)(2x_1+qx_2) \\
&\qquad - \sum_{p_1=0}^{n}\sum_{p_2=0}^{n-p_1}(-1)^{n-p_1-p_2}\times 20^{p_2}\times 90^{p_1}(\mathcal{T}^{l+1}f)\left(\mathbb{A}^n_{(p_1,p_2)}\right)\Bigg\| \\
&= \frac{1}{2^{5n\beta}}\Bigg\| \sum_{q\in\{-1,1\}^n} (\mathcal{T}^{l}f)(2^\beta(2x_1+qx_2)) \\
&\qquad - \sum_{p_1=0}^{n}\sum_{p_2=0}^{n-p_1}(-1)^{n-p_1-p_2}\times 20^{p_2}\times 90^{p_1}(\mathcal{T}^{l+1}f)\left(2^\beta\mathbb{A}^n_{(p_1,p_2)}\right)\Bigg\| \\
&= \frac{1}{2^{5n\beta}}\left\|\mathfrak{D}(\mathcal{T}^{l}f)(2^\beta x_1,2^\beta x_2)\right\| \\
&\le \left(\frac{1}{2^{5n\beta}}\right)^{l+1}\phi(2^{\beta(l+1)}x_1,2^{\beta(l+1)}x_2),
\end{aligned} \tag{15}$$

for all $x_1,x_2 \in V^n$. Letting $l \to \infty$ in (14) and using (6), we reach $\mathfrak{D}\mathcal{Q}(x_1,x_2)=0$ for all $x_1,x_2 \in V^n$. This means that the mapping $\mathcal{Q}$ satisfies (4). If $\mathcal{Q}$ has the quintic condition in each component, then it is multi-quintic by Theorem 2. Finally, assume that $\mathcal{Q}' : V^n \longrightarrow W$ is another multi-quintic mapping satisfying equation (4) and inequality (9), and fix $x \in V^n$, $j \in \mathbb{N}$. Then

$$\begin{aligned}
&\|\mathcal{Q}(x)-\mathcal{Q}'(x)\| \\
&= \left\|\frac{1}{2^{5n\beta j}}\mathcal{Q}\left(2^{\beta j}x\right)-\frac{1}{2^{5n\beta j}}\mathcal{Q}'\left(2^{\beta j}x\right)\right\| \\
&\le \frac{1}{2^{5n\beta j}}\left(\left\|\mathcal{Q}\left(2^{\beta j}x\right)-f\left(2^{\beta j}x\right)\right\|+\left\|\mathcal{Q}'\left(2^{\beta j}x\right)-f\left(2^{\beta j}x\right)\right\|\right) \\
&\le \frac{1}{2^{5n\beta j}}2\Phi\left(2^{\beta j}x\right) \\
&\le 2\frac{1}{2^{5n\frac{\beta+1}{2}+2n}}\sum_{l=j}^{\infty}\left(\frac{1}{2^{5n\beta}}\right)^l\phi\left(2^{\beta l+\frac{\beta-1}{2}}x,0\right).
\end{aligned}$$

Consequently, letting $j \to \infty$ and using the fact that series (7) is convergent for all $x \in V^n$, we obtain $\mathcal{Q}(x) = \mathcal{Q}'(x)$ for all $x \in V^n$, which finishes the proof. □

Equation (4) can be stable when the norm $\mathfrak{D}f(x_1, x_2)$ is controlled by a positive number δ. This is shown in the following corollary.

Corollary 5. *Let δ be a non-negative real number. Suppose that V is a normed space and W is a Banach space. If $f : V^n \longrightarrow W$ is a mapping satisfying the inequality*

$$\|\mathfrak{D}f(x_1, x_2)\| \leq \delta$$

for all $x_1, x_2 \in V^n$. Then, there exists a solution $\mathcal{Q} : V^n \longrightarrow W$ of (4) such that

$$\|f(x) - \mathcal{Q}(x)\| \leq \frac{\delta}{2^{2n}(2^{5n} - 1)}.$$

for all $x \in V^n$. In particular, moreover, if $\mathcal{Q}$ satisfies the quintic condition, then it is a unique multi-quintic mapping.

In the following example, we show that the conditions of Theorem 4 can happen when ϕ has been chosen suitable.

Example 1. Let δ and ε be non-negative real numbers. Suppose that $\alpha, p_{ij} > 0$ for $i \in \{1, 2\}$ and $j \in \{1, \ldots, n\}$ fulfills $\sum_{i=1}^{2} \sum_{j=1}^{n} p_{ij} \neq 5n$ and $\alpha \neq 5n$. Let V be a normed space and W be a Banach space. If $f : V^n \longrightarrow W$ is a mapping satisfying the inequality

$$\|\mathfrak{D}f(x_1, x_2)\| \leq \sum_{i=1}^{2} \sum_{j=1}^{n} \|x_{ij}\|^{\alpha} \varepsilon + \prod_{i-1}^{2} \prod_{j-1}^{n} \|x_{ij}\|^{p_{ij}} \delta$$

for all $x_1, x_2 \in V^n$. Then, there exists a solution $\mathcal{Q} : V^n \longrightarrow W$ of (4) such that

$$\|f(x) - \mathcal{Q}(x)\| \leq \frac{\delta}{|2^{5n} - 2^{\alpha}|} \sum_{j=1}^{n} \|x_{1j}\|^{\alpha},$$

for all $x = x_1 \in V^n$. Further, if $\varepsilon = 0$, then f is a solution of (4) multi-quintic. In particular, if f satisfies the quintic condition, then it is a unique multi-quintic mapping.

We bring an elementary lemma without proof as follows.

Lemma 2. *If a function $g : \mathbb{R} \longrightarrow \mathbb{R}$ is a continuous and satisfies (2), then it has the form $g(x) = cx^5$, for all $x \in \mathbb{R}$, where $c = f(1)$.*

In the following result, we extend Lemma 2 for multivariable functions. For doing this, we use an idea taken from the proof of [12, Theorem 13.4.3].

Proposition 3. *Let $f : \mathbb{R}^n \longrightarrow \mathbb{R}$ be a continuous n-quintic function. Then, there exists a constant $c \in \mathbb{R}$ such that*

$$f(x_1, \dots, x_n) = c \prod_{j=1}^{n} x_j^5 \tag{16}$$

for all $x_1, \dots, x_n \in \mathbb{R}$.

Proof. We argue the proof by induction on n. For $n = 1$, (16) holds in view of Lemma 2. Let (16) hold for an $n \in \mathbb{N}$. Assume that $f : \mathbb{R}^{n+1} \longrightarrow \mathbb{R}$ is a continuous $(n+1)$-quintic function. Fix the n variables $x_1, \dots, x_n$. Then, the function $y \mapsto f(x_1, \dots, x_n, y)$ as a function of y is quintic and continuous, and so by Lemma 2, there exists a constant $c \in \mathbb{R}$ such that

$$f(x_1, \dots, x_n, y) = cy^5, \quad (y \in \mathbb{R}). \tag{17}$$

Note that c depends on $x_1, \dots, x_n$, and indeed

$$c = c(x_1, \dots, x_n). \tag{18}$$

Letting $y = 1$ in (17) and applying (18), we get

$$c = c(x_1, \dots, x_n) = f(x_1, \dots, x_n, 1).$$

Since f is $(n+1)$-quintic, it follows that c is an n-quintic function and hence by the induction hypothesis, there exists a real number c' such that

$$c = c(x_1, \dots, x_n) = c' \prod_{j=1}^{n} x_j^5. \tag{19}$$

Now, the result follows from (17) and (19). □

Remark 4. Note that in the proof of Proposition 3, only the continuity of f with respect to each variable separately was used.

Therefore, the result is again true if and only if f is supposed separately continuous with respect to each variable. On the other hand, by virtue of the proof of Proposition 3, if the continuity condition of f is removed, then the theorem remains valid for a function $f : \mathbb{Q}^p \longrightarrow \mathbb{Q}$. We use this fact to make a non-stable example.

Here, we present a non-stable example for the multi-quintic mappings on $\mathbb{Q}^n$. Indeed, we show the hypothesis $\alpha \neq 5n$ cannot be removed in Example 1. Remember that the method of the proof is taken from [9].

Example 5. Let $\delta \in \mathbb{Q}$, $\delta > 0$ and $n \in \mathbb{N}$ and consider

$$S \geq 2^{2n} + \sum_{p_1=0}^{n} \sum_{p_2=0}^{n-p_1} 20^{p_2} \times 90^{p_1},$$

where $S \in \mathbb{Q}$. Put $\mu = \frac{2^{5n}-1}{2^{10n}S}\delta$. Define the function $\psi : \mathbb{Q}^n \longrightarrow \mathbb{Q}$ through

$$\psi(r_1, \ldots, r_n) = \begin{cases} \mu \prod_{j=1}^{n} r_j^5 & \text{for all } r_j \text{ with } |r_j| < 1, \\ \mu & \text{otherwise.} \end{cases}$$

Moreover, consider the function $f : \mathbb{Q}^n \longrightarrow \mathbb{Q}$ defined by

$$f(r_1, \ldots, r_n) = \sum_{l=0}^{\infty} \frac{\psi(2^l r_1, \ldots, 2^l r_n)}{2^{5nl}}, \qquad (r_j \in \mathbb{Q}).$$

It is obvious that ψ is bounded by μ. Indeed, for each $(r_1, \ldots, r_n) \in \mathbb{Q}^n$, we have $|f(r_1, \ldots, r_n)| \leq \frac{2^{5n}}{2^{5n}-1}\mu$. It follows from the last inequality that

$$|\mathfrak{D} f(x_1, x_2)| \leq \frac{2^{5n}}{2^{5n}-1}\mu S. \tag{20}$$

We claim that

$$|\mathfrak{D} f(x_1, x_2)| \leq \delta \sum_{i=1}^{2} \sum_{j=1}^{n} |x_{ij}|^{5n}, \tag{21}$$

for all $x_1, x_2 \in \mathbb{Q}^n$. We have three cases as follows:

(i) If $x_1 = x_2 = 0$, then it is clear that (21) holds.
(ii) Let $x_1, x_2 \in \mathbb{Q}^n$ with

$$\sum_{i=1}^{2}\sum_{j=1}^{n}|x_{ij}|^{5n} < \frac{1}{2^{5n}}. \tag{22}$$

Thus, there exists a positive integer N such that

$$\frac{1}{2^{5n(N+1)}} < \sum_{i=1}^{2}\sum_{j=1}^{n}|x_{ij}|^{5n} < \frac{1}{2^{5nN}} \tag{23}$$

and hence

$$|x_{ij}|^{5n} < \sum_{i=1}^{2}\sum_{j=1}^{n}|x_{ij}|^{5n} < \frac{1}{2^{5nN}}. \tag{24}$$

Relation (24) implies that $2^N|x_{ij}| < 1$ for all $i \in \{1,2\}$ and $j \in \{1,\ldots,n\}$. Therefore, $2^{N-1}|x_{ij}| < 1$. If $y_1, y_2 \in \{x_{ij}|\, i \in \{1,2\},\ j \in \{1,\ldots,n\}\}$, then

$$2^{N-1}|\, y_1 \pm y_2| < 1,\ 2^{N-1}|\, 2y_1 \pm y_2| < 1,\ 2^{N-1}|\, y_1 \pm 2y_2| < 1.$$

Since, ψ is multi-quintic function on $(-1,1)^n$, we have $\mathcal{D}\psi\left(2^l x_1, 2^l x_2\right) = 0$ for all $l \in \{0,1,2,\ldots,N-1\}$. We conclude from the last equality and (23) that

$$\begin{aligned}
\frac{\left|\mathcal{D}f\left(2^l x_1, 2^l x_2\right)\right|}{\sum_{i=1}^{2}\sum_{j=1}^{n}|x_{ij}|^{5n}} &\le \sum_{l=N}^{\infty}\frac{\left|\mathcal{D}\psi\left(2^l x_1, 2^l x_2\right)\right|}{2^{5nl}\sum_{i=1}^{2}\sum_{j=1}^{n}|x_{ij}|^{5n}}\\
&\le \sum_{l=0}^{\infty}\frac{\mu S}{2^{5n(l+N)}\sum_{i=1}^{2}\sum_{j=1}^{n}|x_{ij}|^{5n}}\\
&\le \mu 2^{5n} S\sum_{l=0}^{\infty}\frac{1}{2^{5nl}}\\
&= \mu S\frac{2^{10n}}{2^{5n}-1} = \delta,
\end{aligned}$$

for all $x_1, x_2 \in \mathbb{Q}^n$, and thus (21) is true in this case.
(iii) Assume that $\sum_{i=1}^{2}\sum_{j=1}^{n}|x_{ij}|^{5n} \ge \frac{1}{2^{5n}}$. Using (20), we have

$$\frac{\left|\mathcal{D}f\left(2^l x_1, 2^l x_2\right)\right|}{\sum_{i=1}^{2}\sum_{j=1}^{n}|x_{ij}|^{n}} \le 2^{5n}\frac{2^{5n}}{2^{5n}-1}\mu S = \delta.$$

Therefore, f satisfies in (21) for all $x_1, x_2 \in \mathbb{Q}^n$. Suppose the assertion is false, that is, there exist a number $\eta > 0$ and a multi-quintic function $\mathcal{Q} : \mathbb{Q}^n \longrightarrow \mathbb{Q}$ such that

$$|f(r_1, \ldots, r_n) - \mathcal{Q}(r_1, \ldots, r_n)| < \eta \sum_{j=1}^{n} |r_j|^5.$$

Without loss of generality, one can take a number $b \in [0, \infty)$ such that $\eta \sum_{j=1}^{n} |r_j|^5 \leq b \prod_{j=1}^{n} |r_j|^5$ and so

$$|f(r_1, \ldots, r_n) - \mathcal{Q}(r_1, \ldots, r_n)| < b \prod_{j=1}^{n} |r_j|^5,$$

for all $(r_1, \ldots, r_n) \in \mathbb{Q}^n$. It follows now from Proposition 3 that there is a constant $c \in \mathbb{R}$ such that $\mathcal{Q}(r_1, \ldots, r_n) = c \prod_{j=1}^{n} r_j^5$ for all $(r_1, \ldots, r_n) \in \mathbb{Q}^n$ and therefore

$$|f(r_1, \ldots, r_n)| \leq (|c| + b) \prod_{j=1}^{n} |r_j|^5, \tag{25}$$

for all $(r_1, \ldots, r_n) \in \mathbb{Q}^n$. On the other hand, one can choose $N \in \mathbb{N}$ such that $N\mu > |c| + b$. If $r = (r_1, \ldots, r_n) \in \mathbb{Q}^n$ such that $r_j \in \left(0, \frac{1}{2^{N-1}}\right)$ for all $j \in \{1, \ldots, n\}$, then $2^l r_j \in (0, 1)$ for all $l = 0, 1, \ldots, N - 1$. Hence,

$$\begin{aligned} |f(r_1, \ldots, r_n)| &= \left| \sum_{l=0}^{\infty} \frac{\psi\left(2^l r_1, \ldots; 2^l r_2\right)}{2^{5nl}} \right| = \left| \sum_{l=0}^{N} \frac{\mu 2^{5nl} \prod_{j=1}^{n} r_j^5}{2^{5nl}} \right| \\ &= N\mu \prod_{j=1}^{n} |r_j|^5 > (|c| + b) \prod_{j=1}^{n} |r_j|^5, \end{aligned}$$

which leads us to a contradiction with (25).

References

[1] Abbasbeygi, Z., Bodaghi, A. and Gharibkhajeh, A. On an equation characterizing multi-quartic mappings and its stability, *Int. J. Nonlinear Anal. Appl.* 13(1), 991–1002 (2022).

[2] Aoki, T. On the stability of the linear transformation in Banach spaces, *J. Math. Soc. Japan.* 2, 64–66 (1950).

[3] Bodaghi, A. Functional inequalities for generalized multi-quadratic mappings, *J. Inequ. Appl.* 2021(145) (2021). https://doi.org/10.1186/s13660-021-02682-z.

[4] Bodaghi, A., Park, C. and Mewomo, O.T. Multiquartic functional equations, *Adv. Difference Equa.* 2019(312) 2019. https://doi.org/10.1186/s13662-019-2255-5.

[5] Bodaghi, A. and Shojaee, B. On an equation characterizing multi-cubic mappings and its stability and hyperstability, *Fixed Point Theory.* 22(1), 83–92 (2021).

[6] Brzdęk, J., Chudziak, J. and Zs. Páles, A fixed point approach to stability of functional equations, *Nonlinear Anal.* 74, 6728–6732 (2011).

[7] Ciepliński, K. On the generalized Hyers-Ulam stability of multi-quadratic mappings, *Comput. Math. Appl.* 62, 3418–3426 (2011).

[8] Ciepliński, K. Generalized stability of multi-additive mappings, *Appl. Math. Lett.* 23, 1291–1294 (2010).

[9] Gajda, Z. On stability of additive mappings, *Internat. J. Math. Math, Sci.* 14, 431–434 (1991).

[10] Găvruţa, P. A generalization of the Hyers-Ulam-Rassias stability of approximately additive mappings, *J. Math. Anal. Appl.* 184, 431–436 (1994).

[11] Hyers, D.H. On the stability of the linear functional equation, *Proc. Natl. Acad. Sci., U.S.A.* 27, 222–224 (1941).

[12] Kuczma, M. *An Introduction to the Theory of Functional Equations and Inequalities. Cauchy's Equation and Jensen's Inequality*, Birkhauser Verlag, Basel (2009).

[13] Murali, R., Boobalan, M. and Raj, A.A. Stability of quintic functional equation in 2-Banach space, *Int. J. Math. Appl.* 4, 41–46 (2016).

[14] Park, C., Bodaghi, A. and Xu, T.-Z. On an equation characterizing multi-Jensen-quartic mappings and its stability, *J. Math. Inequa.* 15(1), 333–347 (2021).

[15] Park, C., Cui, J. and Eshaghi Gordji, M. Orthogonality and quintic functional equations, *Acta Math. Sinica, English Series.* 29, 1381–1390 (2013).

[16] Park, C.-G. Multi-quadratic mappings in Banach spaces, *Proc. Amer. Math. Soc.* 131, 2501–2504 (2002).

[17] Rassias, J.M. Solution of the Ulam stability problem for quartic mappings, *Glasnik Matematicki Series III.* 34(2), 243–252 (1999).

[18] Rassias, M.Th. On the stability of the linear mapping in Banach Space, *Proc. Amer. Math. Soc.* 72(2), 297–300 (1978).

[19] Ulam, S.M. *Problems in Modern Mathematics*, Science Editions, Wiley, New York (1964).

[20] Xu, T.Z., Rassias, J.M., Rassias, M.J. and Xu, W.X. A fixed point approach to the stability of quintic and sextic functional equations in quasi-β-normed spaces, *J. Inequal. Appl.* 2010(423231), 23, (2011) DOI: 10.1155/2010/423231.

[21] Zhao, X., Yang, X. and Pang, C.-T. Solution and stability of the multi-quadratic functional equation, *Abstr. Appl. Anal.* 2013 (2013), Art. ID 415053, 8 pp. http://dx.doi.org/10.1155/2013/415053.

https://doi.org/10.1142/9789811293009_0007

Chapter 7

On Classical Orthogonal Polynomials on Lattices and Some Characterization Theorems

K. Castillo[*,§], D. Mbouna[†,¶], and J. Petronilho[*,‖]

[*]*University of Coimbra, CMUC, Dep. Mathematics, 3001-501 Coimbra, Portugal*

[†]*Institute of Mathematics of the University of Seville (IMUS), Edificio Celestino Mutis-CITUS II, Av. de la Reina Mercedes, s/n, 41012, Seville, Spain*

[§]*kenier@mat.uc.pt*

[¶]*dmbouna@univ-lr.fr*

[‖]*josep@mat.uc.pt*

In this chapter, are given necessary and sufficient conditions for the regularity of solutions of the functional equation appearing in the theory of classical orthogonal polynomials. In addition, we also present the functional Rodrigues formula and a closed formula for the recurrence coefficients. We finally used these results to solve some interesting research problems concerning characterization theorems.

1. Introduction

Let $\mathcal{P}$ be the vector space of all polynomials with complex coefficients and let $\mathcal{P}^*$ be its algebraic dual. $\mathcal{P}_n$ denotes the space of all polynomials with degree less than or equal to n. Define $\mathcal{P}_{-1} := \{0\}$. A simple set in $\mathcal{P}$ is a sequence $(P_n)_{n\geq 0}$ such that $P_n \in \mathcal{P}_n \setminus \mathcal{P}_{n-1}$ for each n. A simple set $(P_n)_{n\geq 0}$ is called an orthogonal polynomial sequence (OPS) with respect to $\mathbf{u} \in \mathcal{P}^*$ if

$$\langle \mathbf{u}, P_n P_k \rangle = h_n \delta_{n,k} \quad (n, k = 0, 1, \ldots;\ h_n \in \mathbb{C} \setminus \{0\}),$$

where $\langle \mathbf{u}, f \rangle$ is the action of $\mathbf{u}$ on $f \in \mathcal{P}$. $\mathbf{u}$ is called regular, or quasi-definite, if there exists an OPS with respect to it. The left multiplication of a functional $\mathbf{u}$ by a polynomial ϕ is defined by

$$\langle \phi\mathbf{u}, f \rangle = \langle \mathbf{u}, \phi f \rangle \quad (f \in \mathcal{P}).$$

A sequence $(\mathbf{a}_n)_{n\geq 0}$ is called dual basis for the sequence of simple set polynomials $(P_n)_{n\geq 0}$ if $\langle \mathbf{a}_n, P_m \rangle = \delta_{n,m}$, for each $n, m = 0, 1, \ldots$. Consequently, if $(P_n)_{n\geq 0}$ is a monic OPS with respect to $\mathbf{u} \in \mathcal{P}^*$, then the corresponding dual basis is explicitly given by

$$\mathbf{a}_n = \left\langle \mathbf{u}, P_n^2 \right\rangle^{-1} P_n \mathbf{u}. \tag{1}$$

Any functional $\mathbf{u} \in \mathcal{P}^*$ ($\mathcal{P}$ being endowed with an appropriate strict inductive limit topology, see, e.g., [9, 20]) can be written as

$$\mathbf{u} = \sum_{n=0}^{\infty} \langle \mathbf{u}, P_n \rangle \, \mathbf{a}_n,$$

in the sense of the weak dual topology in $\mathcal{P}^*$, where $(\mathbf{a}_n)_{n\geq 0}$ is the dual basis for the sequence of polynomials $(P_n)_{n\geq 0}$. These are some basic facts of the algebraic theory on orthogonal polynomials due to Maroni (see [18–21]). For a recent survey on the subject, we refer the reader to [9].

1.1. *Lattices*

Definition 1. A lattice in this framework is a mapping $x(s)$ given by

$$x(s) := \begin{cases} \mathfrak{c}_1 q^{-s} + \mathfrak{c}_2 q^s + \mathfrak{c}_3 & \text{if} \quad q \neq 1, \\ \mathfrak{c}_4 s^2 + \mathfrak{c}_5 s + \mathfrak{c}_6 & \text{if} \quad q = 1, \end{cases} \tag{2}$$

$s \in \mathbb{C}$, where $q > 0$ (fixed) and $\mathfrak{c}_j$ $(1 \leq j \leq 6)$ are (complex) constants, that may depend on q, such that $(\mathfrak{c}_1, \mathfrak{c}_2) \neq (0,0)$ if $q \neq 1$, and $(\mathfrak{c}_4, \mathfrak{c}_5, \mathfrak{c}_6) \neq (0,0,0)$ if $q = 1$.

In the case $q = 1$, the lattice is called quadratic if $\mathfrak{c}_4 \neq 0$, and linear if $\mathfrak{c}_4 = 0$, and in the case $q \neq 1$, it is called q-quadratic if $\mathfrak{c}_1\mathfrak{c}_2 \neq 0$, and q-linear if $\mathfrak{c}_1\mathfrak{c}_2 = 0$ (cf. [3]). Note that

$$\frac{x\big(s+\frac{1}{2}\big)+x\big(s-\frac{1}{2}\big)}{2} = \alpha x(s) + \beta,$$

where α and β are given by

$$\alpha := \frac{q^{1/2}+q^{-1/2}}{2}, \qquad \beta := \begin{cases} (1-\alpha)\mathfrak{c}_3 & \text{if } q \neq 1, \\ \mathfrak{c}_4/4 & \text{if } q = 1. \end{cases} \tag{3}$$

The lattice $x(s)$ fulfills (cf. [3]):

$$\begin{aligned} \frac{x(s+n)+x(s)}{2} &= \alpha_n x_n(s) + \beta_n, \\ x(s+n) - x(s) &= \gamma_n \nabla x_{n+1}(s), \end{aligned}$$

for each $n = 0, 1, \ldots$, where $x_\mu(s) := x\big(s+\frac{\mu}{2}\big)$, $\nabla f(s) := f(s) - f(s-1)$, and $(\alpha_n)_{n\geq 0}$, $(\beta_n)_{n\geq 0}$, and $(\gamma_n)_{n\geq 0}$ are sequences of numbers generated by the following system of difference equations:

$$\alpha_0 = 1, \quad \alpha_1 = \alpha, \quad \alpha_{n+1} - 2\alpha\alpha_n + \alpha_{n-1} = 0, \tag{4}$$

$$\beta_0 = 0, \quad \beta_1 = \beta, \quad \beta_{n+1} - 2\beta_n + \beta_{n-1} = 2\beta\alpha_n, \tag{5}$$

$$\gamma_0 = 0, \quad \gamma_1 = 1, \quad \gamma_{n+1} - \gamma_{n-1} = 2\alpha_n, \tag{6}$$

for each $n = 1, 2, \ldots$. The explicit solutions of these difference equations are

$$\alpha_n = \frac{q^{n/2}+q^{-n/2}}{2}, \tag{7}$$

$$\beta_n = \begin{cases} \beta \left(\dfrac{q^{n/4}-q^{-n/4}}{q^{1/4}-q^{-1/4}} \right)^2 & \text{if} \quad q \neq 1 \\ \beta\, n^2 & \text{if} \quad q = 1, \end{cases} \tag{8}$$

$$\gamma_n = \begin{cases} \dfrac{q^{n/2} - q^{-n/2}}{q^{1/2} - q^{-1/2}} & \text{if} \quad q \neq 1 \\ n & \text{if} \quad q = 1. \end{cases} \tag{9}$$

We assume that $\alpha_{-1} := \alpha$ and $\gamma_{-1} := -1$, consistently with (7) and (9).

1.2. *Operators on lattices*

Definition 2. Let $x(s)$ be a lattice given by (2). The x-derivative operator on $\mathcal{P}$, D_x, and the x-average operator on $\mathcal{P}$, S_x, are the operators on $\mathcal{P}$ defined for each $f \in \mathcal{P}$ so that $\deg(\mathrm{D}_x f) = \deg f - 1$, $\deg(\mathrm{S}_x f) = \deg f$, and

$$\mathrm{D}_x f(x(s)) = \frac{f\big(x(s+\frac{1}{2})\big) - f\big(x(s-\frac{1}{2})\big)}{x(s+\frac{1}{2}) - x(s-\frac{1}{2})}, \tag{10}$$

$$\mathrm{S}_x f(x(s)) = \frac{f\big(x(s+\frac{1}{2})\big) + f\big(x(s-\frac{1}{2})\big)}{2}, \tag{11}$$

where it is understood that $\mathrm{D}_x f = f'$ and $\mathrm{S}_x f = f$ whenever $x(s) = \mathfrak{c}_6$.

The relations (10) and (11) appear in [22, (3.2.4)–(3.2.5)] up to a shift in the variable s. The operators D_x and S_x on $\mathcal{P}$ induce two operators on the dual space $\mathcal{P}^*$, namely, $\mathbf{D}_x : \mathcal{P}^* \to \mathcal{P}^*$ and $\mathbf{S}_x : \mathcal{P}^* \to \mathcal{P}^*$, via the following definition (cf. [11]):

Definition 3. Let $x(s)$ be a lattice given by (2). For each $\mathbf{u} \in \mathcal{P}^*$, the functionals $\mathbf{D}_x\mathbf{u} \in \mathcal{P}^*$ and $\mathbf{S}_x\mathbf{u} \in \mathcal{P}^*$ are defined by

$$\langle \mathbf{D}_x\mathbf{u}, f\rangle := -\langle \mathbf{u}, \mathrm{D}_x f\rangle, \quad \langle \mathbf{S}_x\mathbf{u}, f\rangle := \langle \mathbf{u}, \mathrm{S}_x f\rangle \quad (f \in \mathcal{P}). \tag{12}$$

We call $\mathbf{D}_x\mathbf{u}$ the x-derivative of $\mathbf{u}$ and $\mathbf{S}_x\mathbf{u}$ the x-average of $\mathbf{u}$.

Hereafter, $z := x(s)$ being a lattice given by (2). We consider two fundamental polynomials, U_1 and U_2, introduced in [11], defined by

$$\mathrm{U}_1(z) = \begin{cases} (\alpha^2 - 1)(z - \mathfrak{c}_3) & \text{if} \quad q \neq 1, \\ \dfrac{1}{2}\mathfrak{c}_4 & \text{if} \quad q = 1 \end{cases} \tag{13}$$

and

$$\mathrm{U}_2(z) = \begin{cases} (\alpha^2 - 1)\big((z - \mathfrak{c}_3)^2 - 4\mathfrak{c}_1\mathfrak{c}_2\big) & \text{if} \quad q \neq 1, \\ \mathfrak{c}_4(z - \mathfrak{c}_6) + \dfrac{1}{4}\mathfrak{c}_5^2 & \text{if} \quad q = 1. \end{cases} \tag{14}$$

We point out some useful properties (see [8, 16, 23] and references therein).

Proposition 4. *Let $f, g \in \mathcal{P}$ and $\mathbf{u} \in \mathcal{P}^*$. Then the following properties hold:*

$$\mathrm{D}_x(fg) = (\mathrm{D}_x f)(\mathrm{S}_x g) + (\mathrm{S}_x f)(\mathrm{D}_x g), \tag{15}$$

$$\mathrm{S}_x(fg) = (\mathrm{D}_x f)(\mathrm{D}_x g)\mathrm{U}_2 + (\mathrm{S}_x f)(\mathrm{S}_x g), \tag{16}$$

$$f\mathrm{S}_x g = \mathrm{S}_x\big((\mathrm{S}_x f - \alpha^{-1}\mathrm{U}_1\,\mathrm{D}_x f)g\big) - \alpha^{-1}\mathrm{U}_2\mathrm{D}_x\big(g\mathrm{D}_x f\big), \tag{17}$$

$$f\mathrm{D}_x g = \mathrm{D}_x\big((\mathrm{S}_x f - \alpha^{-1}\mathrm{U}_1\,\mathrm{D}_x f)g\big) - \alpha^{-1}\mathrm{S}_x\big(g\mathrm{D}_x f\big), \tag{18}$$

$$\mathbf{D}_x(f\mathbf{u}) = \big(\mathrm{S}_x f - \alpha^{-1}\mathrm{U}_1\mathrm{D}_x f\big)\,\mathbf{D}_x\mathbf{u} + \alpha^{-1}(\mathrm{D}_x f)\mathbf{S}_x\mathbf{u}, \tag{19}$$

$$\mathbf{S}_x(f\mathbf{u}) = \big(\alpha\mathrm{U}_2 - \alpha^{-1}\mathrm{U}_1^2\big)(\mathrm{D}_x f)\mathbf{D}_x\mathbf{u} + \big(\mathrm{S}_x f + \alpha^{-1}\mathrm{U}_1\mathrm{D}_x f\big)\mathbf{S}_x\mathbf{u}, \tag{20}$$

$$\mathrm{D}_x^n\mathrm{S}_x f = \alpha_n\mathrm{S}_x\mathrm{D}_x^n f + \gamma_n\mathrm{U}_1\mathrm{D}_x^{n+1} f \quad (n = 0, 1, \ldots), \tag{21}$$

$$\alpha\mathbf{D}_x^n\mathbf{S}_x\mathbf{u} = \alpha_{n+1}\mathbf{S}_x\mathbf{D}_x^n\mathbf{u} + \gamma_n\mathrm{U}_1\mathbf{D}_x^{n+1}\mathbf{u} \quad (n = 0, 1, 2, \ldots). \tag{22}$$

The following results can be proved by mathematical induction on n (see alternatively [8]).

Proposition 5. *For the lattice $x(s) = \mathfrak{c}_1 q^{-s} + \mathfrak{c}_2 q^s + \mathfrak{c}_3$, the following holds:*

$$\mathrm{D}_x z^n = \gamma_n z^{n-1} + u_n z^{n-2} + v_n z^{n-3} + \cdots, \tag{23}$$

$$\mathrm{S}_x z^n = \alpha_n z^n + \widehat{u}_n z^{n-1} + \widehat{v}_n z^{n-2} + \cdots, \tag{24}$$

for each $n = 0, 1, \ldots$, *where* α_n *and* γ_n *are given by* (7) *and* (9), *and*

$$u_n := \big(n\gamma_{n-1} - (n-1)\gamma_n\big)\mathfrak{c}_3, \tag{25}$$

$$v_n := \big(n\gamma_{n-2} - (n-2)\gamma_n\big)\mathfrak{c}_1\mathfrak{c}_2 \tag{26}$$

$$+\tfrac{1}{2}\big(n(n-1)\gamma_{n-2} - 2n(n-2)\gamma_{n-1} + (n-1)(n-2)\gamma_n\big)\mathfrak{c}_3^2,$$

$$\widehat{u}_n := n(\alpha_{n-1} - \alpha_n)\mathfrak{c}_3, \tag{27}$$

$$\widehat{v}_n := n(\alpha_{n-2} - \alpha_n)\mathfrak{c}_1\mathfrak{c}_2 + n(n-1)(\alpha-1)\alpha_{n-1}\mathfrak{c}_3^2. \tag{28}$$

We introduce an operator $\mathrm{T}_{n,k} : \mathcal{P} \to \mathcal{P}$ ($n = 0, 1, \ldots; k = 0, 1, \ldots, n$), defined for each $f \in \mathcal{P}$ as follows: if $n = k = 0$, set

$$\mathrm{T}_{0,0} f := f, \tag{29}$$

and if $n \geq 1$ and $0 \leq k \leq n$, define recurrently

$$\mathrm{T}_{n,k} f := \mathrm{S}_x \mathrm{T}_{n-1,k} f - \frac{\gamma_{n-k}}{\alpha_{n-k}} \mathrm{U}_1 \mathrm{D}_x \mathrm{T}_{n-1,k} f + \frac{1}{\alpha_{n+1-k}} \mathrm{D}_x \mathrm{T}_{n-1,k-1} f, \tag{30}$$

with the conventions $\mathrm{T}_{n,k} f := 0$ whenever $k > n$ or $k < 0$. Note that

$$\deg \mathrm{T}_{n,k} f \leq \deg f - k.$$

We are ready to state the following.

Theorem 6 (Leibniz's formula). Let $\mathbf{u} \in \mathcal{P}^*$ and $f \in \mathcal{P}$. Then

$$\mathbf{D}_x^n(f\mathbf{u}) = \sum_{k=0}^{n} \mathrm{T}_{n,k} f \, \mathbf{D}_x^{n-k} \mathbf{S}_x^k \mathbf{u} \quad (n = 0, 1, \ldots), \tag{31}$$

where $\mathrm{T}_{n,k} f$ is a polynomial defined by (29)–(30).

Corollary 7. *Consider the lattice* $x(s) := \mathfrak{c}_1 q^{-s} + \mathfrak{c}_2 q^s + \mathfrak{c}_3$. *Let* $\mathbf{u} \in \mathcal{P}^*$ *and* $f \in \mathcal{P}_2$. *Write* $f(z) = az^2 + bz + c$, *with* $a, b, c \in \mathbb{C}$. *Then*

$$\mathbf{D}_x^n(f\mathbf{u}) = \left(\frac{a\alpha}{\alpha_n\alpha_{n-1}}(z - \mathfrak{c}_3)^2 + \frac{f'(\mathfrak{c}_3)}{\alpha_n}(z - \mathfrak{c}_3) + f(\mathfrak{c}_3) \right.$$

$$\left. + \frac{4a(1-\alpha^2)\gamma_n\mathfrak{c}_1\mathfrak{c}_2}{\alpha_{n-1}}\right)\mathbf{D}_x^n\mathbf{u}$$

$$+\frac{\gamma_n}{\alpha_n}\left(\frac{a(\alpha_n+\alpha\alpha_{n-1})}{\alpha_{n-1}^2}(z-\mathfrak{c}_3)+f'(c_3)\right)\mathbf{D}_x^{n-1}\mathbf{S}_x\mathbf{u}$$

$$+\frac{a\gamma_n\gamma_{n-1}}{\alpha_{n-1}^2}\mathbf{D}_x^{n-2}\mathbf{S}_x^2\mathbf{u}. \tag{32}$$

2. Classical OPS on Lattices

This section concerns classical OPS on lattices and their associated regular functionals. We start by reviewing some basic definitions and needed properties.

Definition 8. Let $x(s)$ be a lattice given by (2). $\mathbf{u} \in \mathcal{P}^*$ is called x-classical if it is regular and there exist nonzero polynomials $\phi \in \mathcal{P}_2$ and $\psi \in \mathcal{P}_1$ such that

$$\mathbf{D}_x(\phi\mathbf{u}) = \mathbf{S}_x(\psi\mathbf{u}). \tag{33}$$

An OPS with respect to a x-classical functional is called an x-classical OPS (or a classical OPS on the lattice x).

Definition 8 appears in [11]. We refer to (33) as *x-Geronimus–Pearson functional equation on the lattice x*, or, simply, *x-GP functional equation*. Our principal goal is to state necessary and sufficient conditions, involving only ϕ and ψ, such that a given functional $\mathbf{u} \in \mathcal{P}^*$ satisfying the x-GP functional equation (33) becomes regular. In order to move on, we need to introduce notations and to state some preliminary needed properties.

We denote by $P_n^{[k]}$ the monic polynomial of degree n defined by

$$P_n^{[k]}(z) := \frac{D_x^k P_{n+k}(z)}{\prod_{j=1}^k \gamma_{n+j}} = \frac{\gamma_n!}{\gamma_{n+k}!}D_x^k P_{n+k}(z) \quad (k,n=0,1,\ldots). \tag{34}$$

Here, as usual, it is understood that $\mathrm{D}_x^0 f = f$, empty product equals one, and

$$\gamma_0! := 1, \quad \gamma_{n+1}! := \gamma_1 \ldots \gamma_n\gamma_{n+1} \quad (n=0,1,\ldots).$$

Definition 9. Let $\phi \in \mathcal{P}_2$ and $\psi \in \mathcal{P}_1$. (ϕ,ψ) is called an x-admissible pair if

$$d_n \equiv d_n(\phi,\psi,x) := \tfrac{1}{2}\gamma_n\,\phi'' + \alpha_n\psi' \neq 0 \quad (n=0,1,\ldots).$$

This is an analogous for lattices of the corresponding definitions for the D-classical and (q,ω)-classical cases (cf. [2,17,20]).

Following [11], given $\mathbf{u} \in \mathcal{P}^*$, $\phi \in \mathcal{P}_2$, and $\psi \in \mathcal{P}_1$, we define recursively polynomials $\phi^{[k]} \in \mathcal{P}_2$ and $\psi^{[k]} \in \mathcal{P}_1$ by

$$\phi^{[0]} := \phi, \quad \psi^{[0]} := \psi, \tag{35}$$

$$\phi^{[k+1]} := \mathrm{S}_x\phi^{[k]} + \mathrm{U}_1\mathrm{S}_x\psi^{[k]} + \alpha\mathrm{U}_2\mathrm{D}_x\psi^{[k]}, \tag{36}$$

$$\psi^{[k+1]} := \mathrm{D}_x\phi^{[k]} + \alpha\mathrm{S}_x\psi^{[k]} + \mathrm{U}_1\mathrm{D}_x\psi^{[k]}, \tag{37}$$

and functionals $\mathbf{u}^{[k]} \in \mathcal{P}^*$ by

$$\mathbf{u}^{[0]} := \mathbf{u}, \quad \mathbf{u}^{[k+1]} := \mathbf{D}_x\big(\mathrm{U}_2\psi^{[k]}\mathbf{u}^{[k]}\big) - \mathbf{S}_x\big(\phi^{[k]}\mathbf{u}^{[k]}\big) \tag{38}$$

$(k = 0, 1, \ldots)$. $\mathbf{u}^{[k]}$ may be seen as the higher-order x-derivative of $\mathbf{u}$. Next, we provide explicit representations for the polynomials $\phi^{[k]}$ and $\psi^{[k]}$.

Proposition 10. *Consider the lattice $x(s) := \mathfrak{c}_1q^{-s} + \mathfrak{c}_2q^s + \mathfrak{c}_3$. Let $\phi \in \mathcal{P}_2$ and $\psi \in \mathcal{P}_1$, so there are $a, b, c, d, e \in \mathbb{C}$ such that*

$$\phi(z) = az^2 + bz + c, \quad \psi(z) = dz + e.$$

Then the polynomials $\phi^{[k]}$ and $\psi^{[k]}$ defined by (35)–(37) are given by

$$\psi^{[k]}(z) = \big(a\gamma_{2k} + d\alpha_{2k}\big)(z - \mathfrak{c}_3) + \phi'(\mathfrak{c}_3)\gamma_k + \psi(\mathfrak{c}_3)\alpha_k, \tag{39}$$

$$\begin{aligned}\phi^{[k]}(z) &= \big(d(\alpha^2 - 1)\gamma_{2k} + a\alpha_{2k}\big)\big((z - \mathfrak{c}_3)^2 - 2\mathfrak{c}_1\mathfrak{c}_2\big)\\ &+ \big(\phi'(\mathfrak{c}_3)\alpha_k + \psi(\mathfrak{c}_3)(\alpha^2 - 1)\gamma_k\big)(z - \mathfrak{c}_3) + \phi(\mathfrak{c}_3) + 2a\mathfrak{c}_1\mathfrak{c}_2,\end{aligned} \tag{40}$$

for each $k = 0, 1, 2 \ldots$.

Proof. By mathematical induction on n. □

Theorem 11. [8] *Let $\mathbf{u} \in \mathcal{P}^*$. Suppose that $\mathbf{u}$ is regular and satisfies (33), where $(\phi, \psi) \in \mathcal{P}_2 \times \mathcal{P}_1 \setminus \{(0,0)\}$. Then (ϕ, ψ) is a x-admissible pair and $\mathbf{u}^{[k]}$ is regular for each $k = 1, 2, \ldots$. Moreover, if $(P_n)_{n\geq 0}$ is the monic OPS with respect to $\mathbf{u}$, then $\big(P_n^{[k]}\big)_{n\geq 0}$ is the monic OPS with respect to $\mathbf{u}^{[k]}$.*

The converse of this theorem is known in the following form (see [5]).

Theorem 12. *If* $(P_n)_{n\geq 0}$ *and* $(P_n^{[k]})_{n\geq 0}$*, for some* k*,* $k=1,2,\ldots$*, are both orthogonal polynomials sequences, then* $(P_n)_{n\geq 0}$ *is* x*-classical OPS.*

We can now state the main result of this paragraph (see [8] for its proof).

Theorem 13. *Consider the lattice*

$$x(s)=\mathfrak{c}_1 q^{-s}+\mathfrak{c}_2 q^{s}+\mathfrak{c}_3.$$

Let $\mathbf{u}\in\mathcal{P}^*\setminus\{\mathbf{0}\}$ *and suppose that there exist* $(\phi,\psi)\in\mathcal{P}_2\times\mathcal{P}_1\setminus\{(0,0)\}$ *such that*

$$\mathbf{D}_x(\phi\mathbf{u})=\mathbf{S}_x(\psi\mathbf{u}). \tag{41}$$

Set $\phi(z)=az^2+bz+c$*,* $\psi(z)=dz+e$ $(a,b,c,d,e\in\mathbb{C})$*,* $d_n=a\gamma_n+d\alpha_n$*,* $e_n=\phi'(\mathfrak{c}_3)\gamma_n+\psi(\mathfrak{c}_3)\alpha_n$*, and* $\phi^{[n]}$ *and* $\psi^{[n]}$ *be given by* (39)–(10)*. Then,* $\mathbf{u}$ *is regular if and only if* (ϕ,ψ) *is an* x*-admissible pair and* $\psi^{[n]}\nmid\phi^{[n]}$ *for each* $n=0,1,\ldots$*; that is to say,* $\mathbf{u}$ *is regular if and only if the following conditions hold:*

$$d_n\neq 0,\quad \phi^{[n]}\left(\mathfrak{c}_3-\frac{e_n}{d_{2n}}\right)\neq 0\quad (n=0,1,\ldots). \tag{42}$$

Under such conditions, the monic OPS $(P_n)_{n\geq 0}$ *with respect to* $\mathbf{u}$ *satisfies*

$$P_{n+1}(z)=(z-B_n)P_n(z)-C_nP_{n-1}(z)\quad (n=0,1,\ldots), \tag{43}$$

with $P_{-1}(z)=0$*, where the recurrence coefficients are given by*

$$B_n=\mathfrak{c}_3+\frac{\gamma_n e_{n-1}}{d_{2n-2}}-\frac{\gamma_{n+1}e_n}{d_{2n}}, \tag{44}$$

$$C_{n+1}=-\frac{\gamma_{n+1}d_{n-1}}{d_{2n-1}d_{2n+1}}\phi^{[n]}\left(\mathfrak{c}_3-\frac{e_n}{d_{2n}}\right) \tag{45}$$

$(n=0,1,\ldots)$*. Moreover, the following functional Rodrigues formula holds:*

$$P_n\mathbf{u}=k_n\mathbf{D}_x^n\mathbf{u}^{[n]},\quad k_n:=(-\alpha)^{-n}\prod_{j=1}^{n}d_{n+j-2}^{-1}\quad (n=0,1,\ldots). \tag{46}$$

The following asymptotic behavior of the above sequence $(B_n)_{n\geq 0}$ is useful (see [15]).

Corollary 14. *Assume the hypothesis (and hence the conclusions) of Theorem* 13. *Then*

$$S_n := \sum_{j=0}^{n-1}(B_j - \mathfrak{c}_3) = -\frac{\gamma_n e_{n-1}}{d_{2n-2}}, \quad n = 0, 1, 2, \ldots. \tag{47}$$

In addition, setting $u := (q^{1/2} - q^{-1/2})^{-1}$, the following holds:

(a) *For $0 < q < 1$ and $d - 2au \neq 0$, we have*

$$\lim_{n\to\infty}(B_n - \mathfrak{c}_3) = \lim_{n\to\infty} q^{\pm n/2}(B_n - \mathfrak{c}_3) = 0, \tag{48}$$

$$\lim_{n\to\infty} q^{-n}(B_n - \mathfrak{c}_3) = -\frac{q^{-1/2}(\psi(\mathfrak{c}_3) - 4\alpha u^2\phi'(\mathfrak{c}_3))}{u(d - 2au)}, \tag{49}$$

$$S := \sum_{j=0}^{\infty}(B_j - \mathfrak{c}_3) = \frac{\psi(\mathfrak{c}_3) - 2u\phi'(\mathfrak{c}_3)}{(q-1)(d-2au)}. \tag{50}$$

(b) *For $1 < q < \infty$ and $d + 2au \neq 0$, we have*

$$\lim_{n\to\infty}(B_n - \mathfrak{c}_3) = \lim_{n\to\infty} q^{\pm n/2}(B_n - \mathfrak{c}_3) = 0, \tag{51}$$

$$\lim_{n\to\infty} q^{n}(B_n - \mathfrak{c}_3) = \frac{q^{1/2}(\psi(\mathfrak{c}_3) - 4\alpha u^2\phi'(\mathfrak{c}_3))}{u(d + 2au)} \tag{52}$$

$$S = \sum_{j=0}^{\infty}(B_j - \mathfrak{c}_3) = \frac{\psi(\mathfrak{c}_3) + 2u\phi'(\mathfrak{c}_3)}{(q^{-1}-1)(d+2au)}. \tag{53}$$

Applying a limiting process (as $q \to 1$) on Theorem 13, we obtain the following result for quadratic lattices.

Theorem 15. *Consider the lattice*

$$x(s) = 4\beta s^2 + \mathfrak{c}_5 s + \mathfrak{c}_6.$$

Let $\mathbf{u} \in \mathcal{P}^* \setminus \{\mathbf{0}\}$ and suppose that there exist $(\phi, \psi) \in \mathcal{P}_2 \times \mathcal{P}_1 \setminus \{(0,0)\}$ such that

$$\mathbf{D}_x(\phi\mathbf{u}) = \mathbf{S}_x(\psi\mathbf{u}). \tag{54}$$

Set $\phi(z) := az^2 + bz + c$ and $\psi(z) := dz + e$. Then $\mathbf{u}$ *is regular if and only if*

$$d_n \neq 0, \quad \phi^{[n]}\left(-\beta n^2 - \frac{e_n}{d_{2n}}\right) \neq 0 \quad (n = 0, 1, \ldots), \tag{55}$$

where $d_n := an + d$, $e_n := bn + e + 2\beta dn^2$, *and*

$$\phi^{[n]}(z) = az^2 + (b + 6\beta nd_n)z + \phi(\beta n^2) + 2\beta n\psi(\beta n^2) \\ - \frac{n}{4}(16\beta\mathfrak{c}_6 - \mathfrak{c}_5^2)d_n.$$

Under these conditions, the monic OPS $(P_n)_{n\geq 0}$ *with respect to* $\mathbf{u}$ *satisfies* (43) *with*

$$B_n = \frac{ne_{n-1}}{d_{2n-2}} - \frac{(n+1)e_n}{d_{2n}} - 2\beta n(n-1), \tag{56}$$

$$C_{n+1} = -\frac{(n+1)d_{n-1}}{d_{2n-1}d_{2n+1}}\phi^{[n]}\left(-\beta n^2 - \frac{e_n}{d_{2n}}\right) \tag{57}$$

$(n = 0, 1, \ldots)$. *Moreover, the following functional Rodrigues formula holds:*

$$P_n\mathbf{u} = k_n\mathbf{D}_x^n\mathbf{u}^{[n]}, \quad k_n := (-1)^n\prod_{j=1}^{n} d_{n+j-2}^{-1} \quad (n = 0, 1, \ldots). \tag{58}$$

The following corollary can be easily proved.

Corollary 16. *Under the assumptions of Theorem* 15 *(and hence the conclusions), the following holds:*

$$\lim_{n\to\infty}\frac{C_{n+1}}{n^4} = \begin{cases} \beta^2, & a \neq 0, \\ 16\beta^2, & a = 0. \end{cases} \qquad \lim_{n\to\infty}\frac{B_n}{n^2} = \begin{cases} -2\beta, & a \neq 0, \\ -8\beta, & a = 0. \end{cases} \tag{59}$$

For the case of a q-linear lattice in Theorem 13 (respectively linear lattice in Theorem 15), we refer the reader to [2] (respectively [9,17]).

3. Applications: Characterization Theorems

The objective of this paragraph is to show how previous results can be used to solve some interesting problems. For this purpose, let us recall some monic OPS needed here. We consider a monic $(P_n)_{n\geq 0}$ satisfying the following three-term recurrence relation (TTRR):

$$P_{-1}(z)=0, \quad P_{n+1}(z)=(z-B_n)P_n(z)-C_nP_{n-1}(z) \quad (C_n\neq 0), \tag{60}$$

with

$$B_n=\frac{\langle \mathbf{u}, zP_n^2\rangle}{\langle \mathbf{u}, P_n^2\rangle}, \quad C_{n+1}=\frac{\langle \mathbf{u}, P_{n+1}^2\rangle}{\langle \mathbf{u}, P_n^2\rangle}.$$

Recall that the monic Askey–Wilson polynomial, $(Q_n(\cdot;a_1,a_2, a_3,a_4|q))_{n\geq 0}$, satisfies (60) (see [13, (14.1.5)]) with

$$B_n=a_1+\frac{1}{a_1}-\frac{\begin{matrix}(1-a_1a_2q^n)(1-a_1a_3q^n)(1-a_1a_4q^n)\\(1-a_1a_2a_3a_4q^{n-1})\end{matrix}}{a_1(1-a_1a_2a_3a_4q^{2n-1})(1-a_1a_2a_3a_4q^{2n})}$$
$$-\frac{a_1(1-q^n)(1-a_2a_3q^{n-1})(1-a_2a_4q^{n-1})(1-a_3a_4q^{n-1})}{(1-a_1a_2a_3a_4q^{2n-1})(1-a_1a_2a_3a_4q^{2n-2})},$$

$$C_{n+1}=(1-q^{n+1})(1-a_1a_2a_3a_4q^{n-1})$$
$$\times\frac{\begin{matrix}(1-a_1a_2q^n)(1-a_1a_3q^n)(1-a_1a_4q^n)(1-a_2a_3q^n)\\(1-a_2a_4q^n)(1-a_3a_4q^n)\end{matrix}}{\begin{matrix}4(1-a_1a_2a_3a_4q^{2n-1})(1-a_1a_2a_3a_4q^{2n})^2\\(1-a_1a_2a_3a_4q^{2n+1})\end{matrix}}$$

and subject to the following restrictions (see [8]):

$$(1-a_1a_2a_3a_4q^n)(1-a_1a_2q^n)(1-a_1a_3q^n)$$
$$\times(1-a_1a_4q^n)(1-a_2a_3q^n)(1-a_2a_4q^n)(1-a_3a_4q^n)\neq 0.$$

The monic Meixner polynomials of the second kind, $(M_n(\cdot;b_1,b_2))_{n\geq 0}$, are defined by (see [10, p. 179, (3.17)])

$$\begin{aligned} zM_n(z;b_1,b_2) &= M_{n+1}(z;b_1,b_2)-b_1(2n+b_2)M_n(z;b_1,b_2)\\ &\quad +(b_1^2+1)n(n+b_2-1)M_{n-1}(z;b_1,b_2),\\ &\quad \times M_{-1}(z;b_1,b_2)=0, \end{aligned} \tag{61}$$

where b_1 and b_2 are parameters so that $b_1^2\neq -1$ and $b_2\neq 0, -1,-2,\ldots$, while the monic Al-Salam polynomials $(A_n(\cdot;a,b|q))_{n\geq 0}$

satisfy (60) with $B_n = (a+b)q^n/2$ and $C_{n+1} = (1-abq^n)(1-q^{n+1})/4$ for each $n = 0, 1, \ldots$ (see [12]). In addition, the continuous monic dual q-Hahn polynomial, $(R_n(x; a, b|q))_{n\geq 0}$, satisfies (60) with $a_n = (a + a^{-1} - a(1-q^n)(1-bcq^{n-1}) - (1-abq^n)(1-acq^n)/a)/2$ and $b_n = (1-abq^n)(1-acq^n)(1-bcq^n)(1-q^{n+1})/4$.

3.1. *A first case*

Let $(P_n)_{n\geq 0}$ be a monic OPS such that

$$\mathrm{D}_x P_{n+1}(z) = k_n \mathrm{S}_x P_n(z) \quad (n = 0, 1, \ldots;\ k_n \in \mathbb{C}). \tag{62}$$

Our goal in this section is to characterize all such OPS for quadratic and q-quadratic lattices. These results are presented in [7].

Lemma 17. *Let $(P_n)_{n\geq 0}$ be a monic OPS with respect to $\mathbf{u} \in \mathcal{P}^*$. Assume that (62) holds. Then $\mathbf{u}$ is x-classical. Moreover,*

$$\mathbf{D}_x(\phi \mathbf{u}) = \mathbf{S}_x(\psi \mathbf{u}), \tag{63}$$

where

$$\psi(z) = B_0 - z,$$

$$\phi(z) = \begin{cases} (\alpha - \alpha^{-1})(z - c_3)(z - B_0) + \alpha^{-1} C_1, & q \neq 1, \\ 2\beta(z - B_0) + C_1, & q = 1. \end{cases}$$

Proof. By identifying the leading coefficient on each member of (62) using (23)–(24), we obtain $k_n = \alpha_n^{-1}\gamma_{n+1}$. Let $(\mathbf{a}_n)_{n\geq 0}$ and $(\mathbf{a}_n^{[1]})_{n\geq 0}$ be the dual basis associated with the sequences $(P_n)_{n\geq 0}$ and $(P_n^{[1]})_{n\geq 0}$, respectively. We claim that

$$\mathbf{S}_x \mathbf{a}_n^{[1]} = \alpha_n \mathbf{a}_n. \tag{64}$$

Indeed, by (62), we have

$$\left\langle \mathbf{S}_x \mathbf{a}_n^{[1]}, P_j \right\rangle = \left\langle \mathbf{a}_n^{[1]}, \mathrm{S}_x P_j \right\rangle = k_j^{-1}\gamma_{j+1} \left\langle \mathbf{a}_n^{[1]}, P_j^{[1]} \right\rangle = \alpha_j \delta_{n,j},$$

$$(j = 0, 1, \ldots).$$

Hence,

$$\mathbf{S}_x \mathbf{a}_n^{[1]} = \sum_{j=0}^{+\infty} \left\langle \mathbf{S}_x \mathbf{a}_n^{[1]}, P_j \right\rangle \mathbf{a}_j = \alpha_n \mathbf{a}_n.$$

We now apply the operator $\mathbf{D}_x$ to (64), using (22) (for $n=1$ and replacing $\mathbf{u}$ by $\mathbf{a}_n^{[1]}$) and the fact that $\mathbf{D}_x\mathbf{a}_n^{[1]} = -\gamma_{n+1}\mathbf{a}_{n+1}$, to obtain

$$-\alpha\alpha_n\mathbf{D}_x\mathbf{a}_n = (2\alpha^2-1)\gamma_{n+1}\mathbf{S}_x\mathbf{a}_{n+1} + \gamma_{n+1}\mathrm{U}_1\mathbf{D}_x\mathbf{a}_{n+1}.$$

Taking into account that $\mathrm{U}_1\mathbf{D}_x\mathbf{a}_{n+1} = \alpha\mathbf{D}_x(\mathrm{U}_1\mathbf{a}_{n+1}) - (\alpha^2-1)\mathbf{S}_x\mathbf{a}_{n+1}$ holds, the previous equality becomes

$$\mathbf{D}_x(\alpha_n\mathbf{a}_n + \gamma_{n+1}\mathrm{U}_1\mathbf{a}_{n+1}) = -\alpha\gamma_{n+1}\mathbf{S}_x\mathbf{a}_{n+1}.$$

The desired result follows from the previous equation by taking $n=0$ and by using successively (1) and (60). □

Theorem 18. *Consider the lattice $x(s) = \mathfrak{c}_1q^{-s} + \mathfrak{c}_2q^s + \mathfrak{c}_3$ with $\mathfrak{c}_1\mathfrak{c}_2 \neq 0$. Then, up to an affine transformation of the variable, the only monic OPS, $(P_n)_{n\geq 0}$, satisfying*

$$\mathrm{D}_xP_{n+1}(z) = \alpha_n^{-1}\gamma_{n+1}\mathrm{S}_xP_n(z) \tag{65}$$

are those of the Askey–Wilson polynomials

$$P_n(z) = 2^n(\mathfrak{c}_1\mathfrak{c}_2)^{n/2}Q_n\left(\frac{z-\mathfrak{c}_3}{2\sqrt{\mathfrak{c}_1\mathfrak{c}_2}}; a, -a, iq^{-1/2}/a, -iq^{-1/2}/a\middle|q\right),$$

with $a \notin \{\pm q^{(n-1)/2}, \pm iq^{-n/2} \,|\, n=0,1,\dots\}$.

Proof. Let $\mathbf{u}\in\mathcal{P}^*$ be the regular functional with respect to which $(P_n)_{n\geq 0}$ is an OPS. We claim that B_n, in (60), is given by

$$B_n = \mathfrak{c}_3 \quad (n=0,1,\dots). \tag{66}$$

Indeed, recall that $P_n(z) = z^n + f_nz^{n-1} + \cdots$, where $f_0 = 0$ and $f_n = -\sum_{j=0}^{n-1}B_j$ $(n=1,2,\dots)$. We then use (23)–(24) and identify the second coefficient of higher degree in both sides of (65). This gives

$$\frac{\alpha_n}{\gamma_{n+1}}f_{n+1} = \frac{\alpha_{n-1}}{\gamma_n}f_n + \frac{n\alpha - \alpha_n\gamma_n}{\gamma_n\gamma_{n+1}}\mathfrak{c}_3.$$

By the telescopic sum method, we have

$$f_n = -\frac{\gamma_n}{\alpha_{n-1}}\left(B_0 + \mathfrak{c}_3\sum_{j=1}^{n-1}\frac{\alpha_j\gamma_j - \alpha j}{\gamma_j\gamma_{j+1}}\right) = -\frac{\gamma_n}{\alpha_{n-1}}(B_0 - \mathfrak{c}_3) - n\mathfrak{c}_3.$$

Therefore, we obtain

$$B_n = f_n - f_{n+1} = \mathfrak{c}_3 + \frac{\alpha}{\alpha_{n-1}\alpha_n}(B_0 - \mathfrak{c}_3). \tag{67}$$

By Lemma 17, **u** satisfies (63), where $\phi(z) = -(\alpha-\alpha^{-1})(z-B_0)(z-\mathfrak{c}_3) - \alpha^{-1}C_1$ and $\psi(z) = z - B_0$. From (44), we obtain

$$B_n = \mathfrak{c}_3 + (1+q)(B_0 - \mathfrak{c}_3)q^{n-2}\frac{(q-1)(1-q^{2n-2}) + (1+q)q^{n-1}}{(1+q^{2n-3})(1+q^{2n-1})}. \tag{68}$$

It is not hard to see that the equality between (67) and (68) requires $B_0 = \mathfrak{c}_3$ and so (66) follows. Therefore, the above expressions for ϕ and ψ reduce to

$$\phi(z) = -(\alpha-\alpha^{-1})(z-\mathfrak{c}_3)^2 - \alpha^{-1}C_1, \quad \psi(z) = z - \mathfrak{c}_3.$$

It follows that C_1 can be regarded as the only free parameter. From (45), we obtain $d_n = \alpha^{-1}\alpha_{n-1}$ and

$$\phi^{[n]}\left(\mathfrak{c}_3 - \frac{e_n}{d_{2n}}\right) = \frac{1}{2\alpha}(\mathfrak{c}_1\mathfrak{c}_2(1-q^{-1})(1-q^n)(1+q^{-n+1}) - 2C_1). \tag{69}$$

Choose a parameter r as a solution of the quadratic equation

$$(q-1)\mathfrak{c}_1\mathfrak{c}_2Z^2 + 2(C_1 + 2(\alpha^2-1)\mathfrak{c}_1\mathfrak{c}_2)Z - (1-q^{-1})\mathfrak{c}_1\mathfrak{c}_2 = 0,$$

that is,

$$r = \frac{C_1 + 2(\alpha^2-1)\mathfrak{c}_1\mathfrak{c}_2}{(1-q)\mathfrak{c}_1\mathfrak{c}_2} \pm \sqrt{q^{-1} + \left(\frac{C_1 + 2(\alpha^2-1)\mathfrak{c}_1\mathfrak{c}_2}{(1-q)\mathfrak{c}_1\mathfrak{c}_2}\right)^2}.$$

Instead of C_1, we may consider r as the free parameter and to express C_1 in terms of r as follows:

$$C_1 = \frac{1}{2}(1-q^{-1})(1+r^{-1})(1-rq)\mathfrak{c}_1\mathfrak{c}_2.$$

Therefore, (69) can be rewritten as

$$\phi^{[n]}\left(\mathfrak{c}_3 - \frac{e_n}{d_{2n}}\right) = \mathfrak{c}_1\mathfrak{c}_2\frac{1-q}{2\alpha}(1+rq^n)(1-r^{-1}q^{n-1})q^{-n}.$$

It follows from the regularity conditions that the free parameter r should satisfy the condition $r \notin \{q^{n-1}, -q^{-n} \,|\, n = 0, 1, \ldots\}$. Moreover, (45) yields

$$C_{n+1} = \mathfrak{c}_1\mathfrak{c}_2\frac{(1+q^{n-2})(1-q^{n+1})(1+rq^n)(1-r^{-1}q^{n-1})}{(1+q^{2n-2})(1+q^{2n})}. \tag{70}$$

Thus,

$$P_n(x) = 2^n(\mathfrak{c}_1\mathfrak{c}_2)^{n/2}Q_n\left(\frac{z-\mathfrak{c}_3}{2\sqrt{\mathfrak{c}_1\mathfrak{c}_2}}; \sqrt{r}, -\sqrt{r}, i/\sqrt{rq}, -i/\sqrt{rq}\,|\, q\right),$$

and setting $r = a^2$, the theorem follows. □

Following the same ideas, we may easily prove rigorously the following one.

Corollary 19. *Consider the lattice* $x(s) = 4\beta s^2 + \mathfrak{c}_5 s + \mathfrak{c}_6$ *with* $(\beta, \mathfrak{c}_5) \neq (0,0)$. *Then there exist monic OPS,* $(P_n)_{n\geq 0}$, *satisfying*

$$\mathrm{D}_x P_{n+1}(z) = (n+1)\mathrm{S}_x P_n(z) \tag{71}$$

if and only if $\beta = 0$. *In this case, up to an affine transformation of the variable, these polynomials are those of Meixner of the second kind*

$$P_n(z) = \left(\frac{i\mathfrak{c}_5}{2}\right)^n M_n\left(\frac{2i(B_0 - z)}{\mathfrak{c}_5}; 0, -\frac{4C_1}{\mathfrak{c}_5^2}\right),$$

with $B_0, C_1 \in \mathbb{C}$ *and* $4C_1/\mathfrak{c}_5^2 \notin \mathbb{N}_0$.

3.2. A second case

In this section, instead of (62), we characterize monic OPS satisfying the following structure relation

$$\mathrm{D}_x P_n(z) = c_n P_{n-1}(z) \quad (n = 0, 1, \ldots) \tag{72}$$

for quadratic and q-quadratic lattices. We start by showing that all monic OPS satisfying (72) are x-classical and then we prove that the coefficients of the associated TTRR satisfy a system of nonlinear equations. These results appear in [15].

Lemma 20. *We consider the q-quadratic lattice* $x(s) = \mathfrak{c}_1 q^{-s} + \mathfrak{c}_2 q^s + \mathfrak{c}_3$. *Let* $\mathbf{u} \in \mathcal{P}^*$ *be a regular functional such that its corresponding monic OPS* $(P_n)_{n\geq 0}$ *satisfies* (72) *subject to the condition* $c_n \neq 0$, *for each* $n = 0, 1, \ldots$. *Then* $\mathbf{u}$ *is x-classical. Moreover,* $D_x(\phi\mathbf{u}) = S_x(\psi\mathbf{u})$, *with* ψ *and* ϕ *polynomials given by*

$$\psi(z) = z - B_0, \quad \phi(z) = (\mathfrak{a}z - \mathfrak{b})(z - B_0) - (\mathfrak{a} + \alpha)C_1, \tag{73}$$

where

$$\mathfrak{a} := \alpha\frac{2C_1 - C_2}{C_2}, \quad \mathfrak{b} := \beta - B_0 + 2\alpha\frac{B_1 C_1}{C_2}. \tag{74}$$

(Here, B_0, B_1, C_1, *and* C_2 *are coefficients of the TTRR* (60) *satisfied by* $(P_n)_{n\geq 0}$.*)*

Proof. By identifying the leading coefficients of each member of (72), we obtain $c_n = \gamma_n$. Let $(\mathbf{a}_n)_{n\geq 0}$ be the dual basis associated with the monic OPS $(P_n)_{n\geq 0}$ satisfying (72). We claim that

$$\mathbf{D}_x\mathbf{u} = R_1\mathbf{u}, \quad R_1(z) := \frac{B_0 - z}{C_1}. \tag{75}$$

Indeed, fix $j \in \mathbb{N}_0$. Using (72), we deduce $\langle \mathbf{D}_x\mathbf{a}_0, P_j\rangle = -\langle \mathbf{a}_0, D_x P_j\rangle = -\gamma_j\delta_{0,j-1}$. This implies

$$\mathbf{D}_x\mathbf{a}_0 = \sum_{j=0}^{\infty}\langle \mathbf{D}_x\mathbf{a}_0, P_j\rangle\,\mathbf{a}_j = -\mathbf{a}_1.$$

Hence, (75) holds using (1) and (60). Applying D_x to both sides of (60), and using (15), yields

$$\begin{aligned} S_x P_n(z) = &-(\alpha z + \beta)D_x P_n(z) + D_x P_{n+1}(z) + B_n D_x P_n(z) \\ &+ C_n D_x P_{n-1}(z). \end{aligned}$$

Using (72) and (60), we obtain

$$S_x P_n(z) = r_n^{[1]} P_n(z) + r_n^{[2]} P_{n-1}(z) + r_n^{[3]} P_{n-2}(z) \tag{76}$$

for each $n = 0, 1, 2, \ldots$, where

$$r_n^{[1]} := c_{n+1} - \alpha c_n, \quad r_n^{[2]} := c_n(B_n - \alpha B_{n-1} - \beta),$$
$$r_n^{[3]} := C_n c_{n-1} - \alpha C_{n-1} c_n.$$

For a fixed $j \in \mathbb{N}_0$, using (76), we obtain

$$\langle \mathbf{S}_x \mathbf{a}_0, P_j \rangle = \langle \mathbf{a}_0, S_x P_j \rangle = r_j^{[1]} \delta_{0,j} + r_j^{[2]} \delta_{0,j-1} + r_j^{[3]} \delta_{0,j-2}.$$

Therefore,

$$\mathbf{S}_x \mathbf{a}_0 = \sum_{j=0}^{\infty} \langle \mathbf{S}_x \mathbf{a}_0, P_j \rangle \, \mathbf{a}_j = r_0^{[1]} \mathbf{a}_0 + r_1^{[2]} \mathbf{a}_1 + r_2^{[3]} \mathbf{a}_2,$$

and so

$$\mathbf{S}_x \mathbf{u} = R_2 \mathbf{u}, \quad R_2(z) := r_0^{[1]} + \frac{r_1^{[2]}}{C_1} P_1(z) + \frac{r_2^{[3]}}{C_2 C_1} P_2(z). \tag{77}$$

Next, on the one hand, applying successively (77), (22), and (75), we obtain

$$\mathbf{D}_x(R_2 \mathbf{u}) = \mathbf{D}_x \mathbf{S}_x \mathbf{u} = \frac{2\alpha^2 - 1}{\alpha} \mathbf{S}_x(R_1 \mathbf{u}) + \frac{\mathrm{U}_1}{\alpha} \mathbf{D}_x(R_1 \mathbf{u}). \tag{78}$$

On the other hand, using (19) with $f = \mathrm{U}_1$, we obtain

$$\frac{\mathrm{U}_1}{\alpha} \mathbf{D}_x(R_1 \mathbf{u}) = \mathbf{D}_x(\mathrm{U}_1 R_1 \mathbf{u}) - \frac{\alpha^2 - 1}{\alpha} \mathbf{S}_x(R_1 \mathbf{u}). \tag{79}$$

Thus, combining (78) and (79), we obtain

$$\mathbf{D}_x((R_2 - \mathrm{U}_1 R_1)\mathbf{u}) = \mathbf{S}_x(\alpha R_1 \mathbf{u}).$$

This leads us to define $\psi(x) := z - B_0$ and $\phi(z) := -\alpha^{-1} C_1 \big(R_2(z) - \mathrm{U}_1(z) R_1(z)\big)$. Clearly, $\deg \psi = 1$, $\deg \phi \leq 2$ and $\mathbf{D}_x(\phi \mathbf{u}) = \mathbf{S}_x(\psi \mathbf{u})$. Finally, setting without lost of generality $c_0 := 0$ and $C_0 := 0$, we obtain

$$\mathfrak{a} := \frac{1}{2}\phi''(0) = -\frac{1}{\alpha}\left(\frac{r_2^{[5]}}{C_2} + \alpha^2 - 1\right) = 2\alpha \frac{C_1}{C_2} - \alpha.$$

Similarly, we also have

$$\phi'(0) = -\mathfrak{a} B_0 - \beta + B_0 - (\mathfrak{a} + \alpha) B_1,$$
$$\phi(0) = -(\mathfrak{a} + \alpha) C_1 - B_0 \left(-\beta + B_0 - (\mathfrak{a} + \alpha) B_1\right).$$

Hence, the desired result is proved. □

Lemma 21. *For the q-quadratic lattice $x(s)=\mathfrak{c}_1q^{-s}+\mathfrak{c}_2q^s+\mathfrak{c}_3$, let $(P_n)_{n\geq 0}$ be a monic OPS satisfying (72). Then the coefficients B_n and C_n of the TTRR (60) satisfied by $(P_n)_{n\geq 0}$ fulfill the following system of difference equations:*

$$c_{n+2}-2\alpha c_{n+1}+c_n=0, \tag{80}$$

$$t_{n+2}-2\alpha t_{n+1}+t_n=0, \quad t_n:=c_n/C_n=k_1q^{n/2}+k_2q^{-n/2}, \tag{81}$$

$$t_{n+3}(B_{n+2}-\mathfrak{c}_3)-(t_{n+2}+t_{n+1})(B_{n+1}-\mathfrak{c}_3)+t_n(B_n-\mathfrak{c}_3)=0, \tag{82}$$

$$\begin{aligned}(t_{n+1}+t_{n+2})(C_{n+1}-\mathfrak{c}_1\mathfrak{c}_2)-2(1+\alpha)t_n(C_n-\mathfrak{c}_1\mathfrak{c}_2)&\\ +(t_{n-1}+t_{n-2})(C_{n-1}-\mathfrak{c}_1\mathfrak{c}_2)&\\ =t_n[(B_n-\mathfrak{c}_3)^2-2\alpha(B_n-\mathfrak{c}_3)(B_{n-1}-\mathfrak{c}_3)+(B_{n-1}-\mathfrak{c}_3)^2]&,\end{aligned} \tag{83}$$

$$c_{n+1}(B_{n+1}-\mathfrak{c}_3)+(1-2\alpha)(B_n-\mathfrak{c}_3)+c_n(B_{n-1}-\mathfrak{c}_3)=0. \tag{84}$$

Proof. Applying the operator S_x to both sides of (60) and using (16), we deduce

$$\begin{aligned}\mathrm{U}_2(z)D_xP_n(z)+(\alpha z+\beta)S_xP_n(z)&=S_xP_{n+1}(z)+B_nS_xP_n(z)\\ &\quad+C_nS_xP_{n-1}(z).\end{aligned}$$

Using successively (14), (72), (76), and (60), we obtain a vanishing linear combination of the polynomials $P_{n+3}, P_{n+2}, \ldots, P_{n-3}$. Thus, setting

$$t_n:=c_n/C_n, \quad n=1,2,3,\ldots, \tag{85}$$

after straightforward computations, we obtain (80)–(81) together with the following equations:

$$-t_{n+2}B_{n+1}+(t_{n+1}+t_n)B_n-t_{n-1}B_{n-1}=2\beta(t_n+t_{n+1}), \tag{86}$$

$$\begin{aligned}t_nB_n^2+t_nB_{n-1}^2-(t_{n-1}+t_{n+1})B_nB_{n-1}-2\beta t_n(B_n+B_{n-1})&\\ -(t_{n+2}+t_{n+1})C_{n+1}+2(1+\alpha)t_nC_n-(t_{n-2}+t_{n-1})C_{n-1}&\\ =(\delta-\beta^2)t_n&,\end{aligned} \tag{87}$$

$$c_{n+1}B_{n+1}+(1-2\alpha)(c_n+c_{n+1})B_n+c_nB_{n-1}=2\beta(c_n+c_{n+1}), \tag{88}$$

where $\delta = (\alpha^2 - 1)(\mathfrak{c}_3^2 - 4\mathfrak{c}_2\mathfrak{c}_2)$. (82) (respectively (83) and (84)) is obtained from (86) (respectively (87) and (88)). This completes the proof. □

Remark 22. According to Lemma 21, the coefficients B_n and C_n of the TTRR (60) of any monic OPS $(P_n)_{n\geq 0}$ fulfilling (72) must fulfill (80)–(84). However, we need to take into account some initial conditions which are specified. Indeed, for instance, it is clear that

$$B_n = \mathfrak{c}_3, \quad C_{n+1} = \mathfrak{c}_1\mathfrak{c}_2 \quad (n = 0, 1, 2, \ldots)$$

provide a solution of system (80)–(84). The corresponding monic OPS is

$$P_n(x) = 2^n(\mathfrak{c}_1\mathfrak{c}_2)^{n/2}\widehat{U}_n\left(\frac{z - \mathfrak{c}_3}{2\sqrt{\mathfrak{c}_1\mathfrak{c}_2}}\right) \quad (n = 0, 1, 2, \ldots),$$

where $(\widehat{U}_n)_{n\geq 0}$ is the monic Chebyshev polynomials of the second kind. However, this sequence $(P_n)_{n\geq 0}$ does not provide a solution of (72) (see (91)).

The system of equations (80)–(84) is nonlinear and so, in general, it is not easy to solve it. Nevertheless, in view of Lemma 20, an OPS satisfying (72) is x-classical and so the results presented in previous paragraph are useful to find the explicit expressions for the coefficients of the TTRR satisfied by the OPS under analysis (see Theorem 13 and Corollary 14). We see that some patterns appear associated with the system of equations (80)–(84) which will allow us to solve the system.

Recall that from (81), we have

$$t_n = \frac{c_n}{C_n} = k_1 q^{n/2} + k_2 q^{-n/2} \quad (n = 1, 2, 3, \ldots), \tag{89}$$

where k_1 and k_2 are two complex numbers. Since $c_n \neq 0$, for $n = 1, 2, 3, \ldots$, then k_1 and k_2 cannot vanish simultaneously. Recall also that we defined $c_0 = C_0 = 0$, and so we define

$$t_0 := k_1 + k_2,$$

by compatibility with (89).

Theorem 23. *Up to an affine transformation of the variable, the only monic OPS $(P_n)_{n\geq 0}$ satisfying (72) are the monic q-Hermite polynomials of Rogers.*

Proof. Let $(P_n)_{n\geq 0}$ be a monic OPS satisfying (72). We know that $c_n = \gamma_n$ and therefore by identifying the coefficient of the second term with the higher degree in (72), we also obtain

$$B_n = \mathfrak{c}_3 + (\gamma_{n+1} - \gamma_n)(B_0 - \mathfrak{c}_3).$$

Since by assumption $t_n \neq 0$ for each $n = 1, 2, \ldots$, it is not hard to see that the above expression of B_n satisfies (82) if and only if $B_0 = \mathfrak{c}_3$. Hence,

$$B_n = \mathfrak{c}_3 \quad (n = 0, 1, 2, \ldots). \tag{90}$$

In addition, from (44), we obtain

$$\frac{\gamma_{n+1} e_n}{d_{2n}} = \frac{\gamma_n e_{n-1}}{d_{2n-2}} \quad (n = 0, 1, 2, \ldots).$$

Since $e_n = \phi'(\mathfrak{c}_3)\gamma_n + \psi(\mathfrak{c}_3)\alpha_n$, we find $e_0 = 0$ (because $\gamma_0 = 0$, and from (73), $\psi(\mathfrak{c}_3) = \mathfrak{c}_3 - B_0 = 0$), and so $e_n = 0$, for $n = 0, 1, 2, \ldots$, and consequently $\phi'(\mathfrak{c}_3) = 0$. Then, from (73), we obtain $\mathfrak{b} = \mathfrak{a}\mathfrak{c}_3$. Taking $n = 3$ in (72) and using (23)–(26), we obtain

$$C_2 = 2(2\alpha^2 - 1)(C_1 - \mathfrak{c}_1\mathfrak{c}_2) + 2\mathfrak{c}_1\mathfrak{c}_2. \tag{91}$$

Therefore, using (73), we have $\phi(z) = \mathfrak{a}(z - \mathfrak{c}_3)^2 - (\mathfrak{a} + \alpha)C_1$ and $\psi(z) = z - \mathfrak{c}_3$, where

$$\mathfrak{a} = \frac{2\alpha C_1}{C_2} - \alpha = \frac{2\alpha(1 - \alpha^2)(C_1 - \mathfrak{c}_1\mathfrak{c}_2)}{(2\alpha^2 - 1)(C_1 - \mathfrak{c}_1\mathfrak{c}_2) + \mathfrak{c}_1\mathfrak{c}_2}. \tag{92}$$

Recall that

$$t_n = \frac{\gamma_n}{C_n} = k_1 q^{n/2} + k_2 q^{-n/2}, \quad k_1 = \frac{(1+q)C_1 - C_2}{q^{1/2}(q-1)C_1C_2},$$
$$k_2 = \frac{(q^{-1}+1)C_1 - C_2}{q^{-1/2}(q^{-1}-1)C_1C_2}. \tag{93}$$

Equation (83) reduces to

$$(t_{n+1} + t_{n+2})(C_{n+1} - \mathfrak{c}_1\mathfrak{c}_2) - 2(1+\alpha)t_n(C_n - \mathfrak{c}_1\mathfrak{c}_2)$$
$$+ (t_{n-1} + t_{n-2})(C_{n-1} - \mathfrak{c}_1\mathfrak{c}_2) = 0. \tag{94}$$

Next, define $r_n := t_n + t_{n+1} = aq^{n/2} + bq^{-n/2}$, where $a := k_1(1 + q^{1/2})$ and $b := k_2(1 + q^{-1/2})$. By setting $K_c := (r_2(C_2 - \mathfrak{c}_1\mathfrak{c}_2) - r_0(C_1 - \mathfrak{c}_1\mathfrak{c}_2))/(1 - q^{-1/2})$, we see that (94) reads as

$$\begin{aligned} &r_{n+1}(C_{n+1} - \mathfrak{c}_1\mathfrak{c}_2) - r_{n-1}(C_n - \mathfrak{c}_1\mathfrak{c}_2) \\ &\quad = r_n(C_n - \mathfrak{c}_1\mathfrak{c}_2) - r_{n-2}(C_{n-1} - \mathfrak{c}_1\mathfrak{c}_2). \end{aligned}$$

By applying this relation successively, we obtain

$$r_{n+1}(C_{n+1} - \mathfrak{c}_1\mathfrak{c}_2) - r_{n-1}(C_n - \mathfrak{c}_1\mathfrak{c}_2) = K_c(1 - q^{-1/2}).$$

Multiplying this equation by r_n and applying a telescoping process to the resulting equation, we obtain

$$C_{n+1} = \mathfrak{c}_1\mathfrak{c}_2 + \frac{\begin{array}{c} r_0r_1(C_1 - \mathfrak{c}_1\mathfrak{c}_2)q^{n/2} \\ +K_c(aq^n + (b - aq^{-1/2})q^{n/2} - bq^{-1/2}) \end{array}}{(aq^n + b)(aq^{n+1} + b)} q^{(n+1)/2} \tag{95}$$

for each $n = 0, 1, 2, \ldots$.

We claim that

$$(C_2 - (1 + q)C_1)(C_2 - (1 + q^{-1})C_1) = 0. \tag{96}$$

Indeed, suppose that (96) does not hold. Then, by (93), we would have $k_1k_2 \neq 0$, and we may write

$$C_{n+1} = \frac{\gamma_{n+1}}{t_{n+1}} = \frac{u(q^{n+1} - 1)}{k_1q^{n+1} + k_2}, \quad u^{-1} = q^{1/2} - q^{-1/2} \quad (n = 0, 1, 2, \ldots). \tag{97}$$

Assume without lost of generality that $0 < q < 1$. Then taking successively limits as $n \to \infty$ in the expressions for C_{n+1} and $q^{-(n+1)/2}(C_{n+1} - \mathfrak{c}_1\mathfrak{c}_2)$ given by (95) and (97), we obtain $u + k_2\mathfrak{c}_1\mathfrak{c}_2 = 0$ and $K_c = 0$. Now, the equality between (95) and (97) implies

$$\begin{aligned} &2(1 + \alpha)(u - k_1\mathfrak{c}_1\mathfrak{c}_2)k_1^2q^{2n+2} \\ &\quad + k_1(4(1 + \alpha)^2(u - k_1\mathfrak{c}_1\mathfrak{c}_2)k_2 - r_0r_1(C_1 - \mathfrak{c}_1\mathfrak{c}_2))q^{n+1} \\ &\quad + k_2(2(1 + \alpha)(u - k_1\mathfrak{c}_1\mathfrak{c}_2)k_2 - r_0r_1(C_1 - \mathfrak{c}_1\mathfrak{c}_2)) = 0, \end{aligned}$$

for each $n = 0, 1, 2, \ldots$. This implies $u - k_1\mathfrak{c}_1\mathfrak{c}_2 = 0$ and $r_0r_1(C_1 - \mathfrak{c}_1\mathfrak{c}_2) = 0$. Consequently, $C_{n+1} = \mathfrak{c}_1\mathfrak{c}_2$ for each $n = 0, 1, 2, \ldots$, which contradicts (91). Hence, (96) holds and so $k_1k_2 = 0$.

Suppose that $k_1 = 0$, i.e., $C_2 = (1+q)C_1$. Then, from (91), we find $C_1 = (1-q)\mathfrak{c}_1\mathfrak{c}_2$ and so, from (92), we obtain $\mathfrak{a} = -1/(2u)$. Since $\mathfrak{b} = \mathfrak{a}\mathfrak{c}_3$ and $B_0 = \mathfrak{c}_3$, using (45) in Theorem 13, we obtain

$$C_{n+1} = (1-q^{n+1})\mathfrak{c}_1\mathfrak{c}_2 \quad (n = 0, 1, 2, \ldots). \tag{98}$$

One easily sees that the expressions for B_n and C_{n+1} given by (90) and (98) satisfy (84) and so the system of equations (80)–(84) is satisfied.

Similarly, if $k_2 = 0$, i.e., $C_2 = (1+q^{-1})\mathfrak{c}_1\mathfrak{c}_2$, we obtain $\mathfrak{a} = 1/(2u)$ and

$$C_{n+1} = (1-q^{-n-1})\mathfrak{c}_1\mathfrak{c}_2 \quad (n = 0, 1, 2, \ldots), \tag{99}$$

which together with (90) also fulfill the system of equations (80)–(84). Thus,

$$\phi(z) = \pm\frac{1}{2u}\left((z-\mathfrak{c}_3)^2 - \mathfrak{c}_1\mathfrak{c}_2\right) \quad \textit{and} \quad \psi(z) = z - \mathfrak{c}_3.$$

Hence, we conclude that

$$P_n(z) = 2^n(\mathfrak{c}_1\mathfrak{c}_2)^{n/2} A_n\left(\frac{z-\mathfrak{c}_3}{2\sqrt{\mathfrak{c}_1\mathfrak{c}_2}}; 0, 0|q\right)$$

or

$$P_n(z) = 2^n(\mathfrak{c}_1\mathfrak{c}_2)^{n/2} A_n\left(\frac{z-\mathfrak{c}_3}{2\sqrt{\mathfrak{c}_1\mathfrak{c}_2}}; 0, 0|q^{-1}\right),$$

where $(A_n(;a,b|q))_{n\geq 0}$ is the monic Al-Salam polynomial. As a matter of fact, this $(P_n)_{n\geq 0}$ is the sequence of monic q-Hermite polynomials of Rogers. Thus, the proof is complete. □

Following the ideas of this proof, one can easily prove the following result.

Corollary 24. *Consider the quadratic lattice* $x(s) = 4\beta s^2 + \mathfrak{c}_5 s + \mathfrak{c}_6$ *with* $(\beta, \mathfrak{c}_5) \neq (0, 0)$. *Then* (72) *has no OPS solutions.*

We denote by $\mathcal{D}_q$ the Askey–Wilson operator, obtained from the D_x operator by restricting the q-quadratic lattice to $x(s) = (q^{-s} + q^s)/2$. It is important to note that the previous theorem was solved in [1] using a different method and the advantage of this method developed here using Theorem 13 lies on the fact that we can use it to solve equations of type (72) in a more general form. Indeed, the following problem [12, Conjecture 24.7.8] is a conjecture posed by Ismail.

Conjecture 25. *Let $(P_n)_{n\geq 0}$ be a monic OPS and π be a polynomial of degree at most 2 which does not depend on n. If $(P_n)_{n\geq 0}$ satisfies*

$$\pi(x)\mathcal{D}_q P_n(x) = (a_n x + b_n)P_n(x) + c_n P_{n-1}(x), \tag{100}$$

then $(P_n)_{n\geq 0}$ are continuous q-Jacobi polynomials, Al-Salam–Chihara polynomials, or special or limiting cases of them. The same conclusion holds if π has degree $s+1$ and the condition (100) is replaced by

$$\pi(x)\mathcal{D}_q P_n(x) = \sum_{k=-r}^{s} c_{n,k} P_{n+k}(x), \tag{101}$$

for positive integers r, s, and a polynomial π which does not depend on n.

Although (100) is a simple relation, this problem was only solved recently due to the complexity of the Askey–Wilson operator and his properties. Using the method developed here, it is proved in [6] that the only solutions of (100) are some particular cases of the Al-Salam–Chihara polynomials, the Chebyshev polynomials of the first kind, and the continuous q-Jacobi polynomials. The following relation is proved in [4, 14]:

$$\begin{aligned}
(\alpha^2-1)(x^2-1)\mathcal{D}_q R_n(x;1,-1,q^{1/4}|q^{1/2}) &= (\alpha^2-1)\gamma_n \\
&\times R_{n+1}(x;1,-1,q^{1/4}|q^{1/2}) + (c_{n+1} - \alpha c_n + (1-\alpha)\alpha_n B_n) \\
&\times R_n(x;1,-1,q^{1/4}|q^{1/2}) + ((B_n - \alpha B_{n-1})c_n + (1-\alpha^2)\gamma_n C_n) \\
&\times R_{n-1}(x;1,-1,q^{1/4}|q^{1/2}) + (c_{n-1}C_n - \alpha c_n C_{n-1}) \\
&\times R_{n-2}(x;1,-1,q^{1/4}|q^{1/2}),
\end{aligned}$$

where B_n, c_n, and C_n are given by

$$B_n = \frac{1}{2}((1+q^{-1/2})q^{n/2} + 1 - q^{-1/2})q^{(2n+1)/4},$$

$$C_n = \frac{1}{4}(1+q^{(n-1)/2})(1-q^{n/2})(1-q^{n-1/2}), \; c_n = C_n q^{-(2n-1)/4},$$

and $(R_n(\cdot;a,b,c|q))_{n\geq 0}$ the continuous monic dual q-Hahn polynomial. This equation is of type (101) with $\deg \pi = 2$, $r = 1$, and $s = 2$ giving therefore a counterexample to (101).

In view of this section, it is clear that Theorem 13 presented here is useful to derive some characterization theorems on OPS and to solve some challenging problem in the subject.

Acknowledgments

This work is partially supported by the Centre for Mathematics of the University of Coimbra (funded by the Portuguese Government through FCT/MCTES, DOI 10.54499/UIDB/00324/2020). The first author is supported by FCT through DOI 10.54499/2022.00143. CEECIND/CP1714/CT0002. The second author is supported by the Grant QUALIFICA (PROGRAMA: AYUDAS A ACCIONES COMPLEMENTARIAS DE I+D+i) by Junta de Andalucía, grant number QUAL21 005 USE.

References

[1] Al-Salam, W. A characterization of the Rogers q-Hermite polynomials, *Internat. J. Math. & Math. Sci.* 18, 641–648 (1995).

[2] Álvarez-Nodarse, R., Castillo, K., Mbouna, D. and Petronilho, J. On classical orthogonal polynomials related to Hahn's operator, *Integral Transforms Spec. Funct.* 31, 487–505 (2020).

[3] Atakishiev, N.M., Rahman, M. and Suslov, S.K. On classical orthogonal polynomials, *Constr. Approx.* 11, 181–226 (1995).

[4] Castillo, K. and Mbouna, D. A counterexample to a conjecture of M. Ismail, *J. Math. Phys.* 65, 033507 (2024).

[5] Castillo, K. and Mbouna, D. Proof of two conjectures on Askey-Wilson polynomials, *Proc. Amer. Math. Soc.*, 151, 1655–1661 (2023).

[6] Castillo, K., Mbouna, D. and Petronilho, J. A characterization of continuous q-Jacobi, Chebyshev of the first kind and Al-Salam Chihara polynomials, *J. Math. Anal. Appl.* 514, 126358 (2022).

[7] Castillo, K., Mbouna, D. and Petronilho, J. Remarks on Askey-Wilson polynomials and Meixner polynomials of the second kind, *Ramanujan J.*, 58, 1159–1170 (2022).

[8] Castillo, K., Mbouna, D. and Petronilho, J. On the functional equation for classical orthogonal polynomials on lattices, *J. Math. Anal. Appl.* 515, 126390 (2022).

[9] Castillo, K. and Petronilho, J. Classical orthogonal polynomials revisited, *Results Math.* 78 (2023) Article number: 155.

[10] Chihara, T.S. An introduction to orthogonal polynomials. Gordon and Breach, New York (1978).

[11] Foupouagnigni, M., Kenfack-Nangho, M. and Mboutngam, S. Characterization theorem of classical orthogonal polynomials on nonuniform lattices: the functional approach, *Integral Transforms Spec. Funct.* 22, 739–758 (2011).

[12] Ismail, M.E.H. Classical and quantum orthogonal polynomials in one variable. With two chapters by W. Van Assche. With a foreword by R. Askey., Encyclopedia of *Mathematics and its Applications* Cambridge University Press, Cambridge, 98 (2005).

[13] Koekoek, R., Lesky, P.A. and Swarttouw, R.F. Hypergeometric orthogonal polynomials and their $q-$Analogues, *Springer Monographs in Mathematics*, Springer-Verlag, Berlin, 2010.

[14] Mbouna, D. A structure relation for some specific orthogonal polynomials, *Mediterr. J. Math.* 20, 237 (2023).

[15] Mbouna, D. On some problems in the theory of orthogonal polynomials, Doctoral dissertation, University of Coimbra (2021).

[16] Mboutngam, S., Foupouagnigni, M. and Njionou Sadjang, P. On the modifications of semi-classical orthogonal polynomials on nonuniform lattices, *J. Math. Anal. Appl.* 445, 819–836 (2017).

[17] Marcellán, F. and Petronilho, J. On the solution of some distributional differential equations: existence and characterizations of the classical moment functionals, *Integral Transforms Spec. Funct.* 2, 185–218 (1994).

[18] Maroni, P. Sur quelques espaces de distributions qui sont des formes linéaires sur l'espace vectoriel des polynômes, In C. Brezinski *et al.* (eds.), *Simposium Laguerre, Bar-le-Duc, Lecture Notes in Math.* Springer-Verlag, 1171, 184–194 (1985).

[19] Maroni, P., Le calcul des formes linéaires et les polynômes orthogonaux semiclassiques, In M. Alfaro *et al.* (eds.), *Orthogonal Polynomials and Their Applications, Lecture Notes in Math.* Springer-Verlag, 1329, 279–290 (1988).

[20] Maroni, P. Une théorie algébrique des polynômes orthogonaux. Applications aux polynômes orthogonaux semiclassiques, In C. Brezinski *et al.* (eds.), *Orthogonal Polynomials and Their Applications*, Proc. Erice 1990, *IMACS, Ann. Comp. App. Math.* 9, 95–130 (1991).

[21] Maroni, P. Fonctions eulériennes. Polynômes orthogonaux classiques, Téchniques de l'Ingénieur, traité Généralités (*Sciences Fondamentales*), A 154, 1–30 (1994).

[22] Nikiforov, A.F., Suslov, S. and Uvarov, V.B. Classical orthogonal polynomials of a discrete variable. Translated from the Russian. *Springer Series in Computational Physics.* Springer-Verlag, Berlin (1991).

[23] Njionou Sadjang, P., Koepf, W. and Foupouagnigni, M. On moments of classical orthogonal polynomials, *J. Math. Anal. Appl.* 424, 122–151, (2015).

https://doi.org/10.1142/9789811293009_0008

Chapter 8

A Closed-Loop Supply Chain Network Model for the Provision of 5G Services with UAVs and Trustworthiness Investments

G. Colajanni*, P. Daniele†, and D. Sciacca‡

Department of Mathematics and Computer Science, University of Catania, Viale Andrea Doria, 6, 95125 Catania, Italy

**gabriella.colajanni@unict.it*

†*patrizia.daniele@unict.it*

‡*daniele.sciacca@unict.it*

In this chapter, we provide a closed-loop supply chain network model for the provision of 5G services. It is based on a three-tiered closed loop network in which, in the forward chain, the users or devices on the ground require services, executed by a fleet of UAVs organized as a FANET, controlled by a fleet of manager UAVs, whose role is to receive service requests and send them to executor UAVs. In the reverse chain, the fleet of executor UAVs, once the requests are performed, provide the services to devices and users on the ground. Through a system optimization perspective, service providers seek to maximize their profit, given by the difference between the revenue associated with the provision of services and the costs due to the transmission and the execution of services. Taking into account the trustworthiness levels of each UAV belonging to the fleet at the highest level of the network, service providers also seek to minimize the investment costs to increase such levels and the penalty associated with not executed services due to unreliability of the executor UAVs. For the proposed optimization problem, we provide a variational formulation and some numerical simulations are performed.

1. Introduction

In an increasingly connected world, the speed of provision of technological services, their safety, and their reliability in terms of performance have become the primary objectives for service providers. Over the decades, we have witnessed an increasingly rapid improvement in the supply of technological services and new technologies, which now belong to any field of application, from telecommunications to agriculture, from education to transport, from emergency management to safety of cities. They are adapting more and more to innovations and to the different requests of services from users and devices. If the first communication infrastructure (1G mobile network), introduced in the 1980s/1990s, presented a largely analogical standard with the limit that each frequency band could communicate only one user at a time, the contemporary 5G infrastructure, due to the innovative technology used, allows you to connect a large number of devices at the same time, ensuring very high reliability, a higher connection speed, lower latency, and better performance than the current 4 LTE. 5G technology aims to offer unprecedented performance and efficiency in supporting not only mobile phone services but also Internet of Things (IoT) applications (see [11]). This means that everything and everyone will be more virtually connected, opening the doors to services that have only recently been seen as ahead of their time. 5G technology is able to provide faster connectivity speed, increased capacity, very low latency, lower power consumption, and greater network reliability. All these improvements make companies perfectly connected and optimize the user experience. Furthermore, 5G will represent the safest network ever developed, capable of creating a safer space for data transmission: this aspect is of fundamental importance in sectors, such as health and finance (see [19] and [20]).

The virtuality of the new 5G infrastructure and the possibility of integrating many different technologies without the need for physical connections have provided for the integration of UAVs (Unmanned Aerial Vehicles), such as drones and balloons, in the supply chain of

5G services. The development of UAV technology has led to possible synergies with 5G in consumer, commercial, and civil applications (see [1, 13], and [18]), as well as humanitarian ones. Having fully connected UAVs means seeing images with high resolution in real time, issuing commands that are received by the UAVs with very low latency and, precisely for this reason, giving a guarantee of safety and reliability in managing a UAV, often even heavy, that needs of a high reactivity to the command. However, there are currently significant limitations that restrict how effectively UAVs can perform, and the issue of permits to carry out flights is one of the main problems. The issue of the permit to carry out the flights, in fact, is based on a careful assessment of the risks of each particular situation, risks often linked to the loss of the signal, which can make it difficult to control the drone causing accidents. 5G, thanks to its complexity and solidity, can in this sense give guarantees in the continuity of communication with the network in a constant manner, drastically reducing risk factors.

The connectivity, on the other hand, is, actually, the most challenging issue to solve. Most UAVs require continuous communication with their managers in order to ensure safety and effective operations. Traditionally, the operational range of UAVs has been limited by the range of the radio controller (RC) resulting in most of them operating at low altitudes. However, despite significant efforts to develop proprietary radio control solutions, UAV designers have been able to improve the range by only a few kilometers and only within the vision line of sight (VLOS). The advent of 4G and its subsequent progression to 5G have proven to be revolutionary. The safety assurance provided by the 5G network makes possible important operations, such as surveillance, the urgent transport of life-saving products that would be feasible by flight beyond the line of sight (BVLOS). The drone connected to the 5G network is, in fact, double-stranded to the BVLOS flight, which is currently not feasible except for more complex and strictly regulated operations and which is not envisaged for the civilian world except on an experimental basis. In conclusion, through superior connectivity, UAVs are expected to autonomously perform complex missions, transmit and upload large amounts of high-definition data and video to the cloud, and travel much greater distances (see [12]).

The actual and great potential of the involvement of UAVs in the provision of 5G services has led us, therefore, to develop the optimization model that we will describe in detail in the following section. The optimization problem is based on a closed-loop supply chain network consisting of three main levels, that of users or devices on the ground, that of the fleet of UAVs that execute the commands and services requested by users, controlled and orchestrated by some manager UAVs, that constitute the middle level of the network. Once the services from the fleet of UAVs are performed, they are sent to devices and users on the ground. This network structure brings us to a closed-loop supply chain, whose forward chain is constituted by the fleet of UAVs, the manager UAVs, and the devices and users, while the revers chain consists of the fleet of UAVs and the devices and users on the ground. The aim of service providers is to maximize their profit, maximizing the revenues associated with the provision of services and, simultaneously, minimizing the costs associated with the transmission of services and the execution of them. Moreover, taking into account the trustworthiness level of the entire network, service providers aim to minimize also the investment costs associated with the increase of such level and the penalty related to not executed items due to unreliability of an UAV of the fleet.

The rest of this chapter is organized as follows. In Section 2, we develop and describe the mathematical model. Section 3 provides different equivalent variational formulations of the proposed optimization model, and, finally, Section 4 is devoted to numerical simulations.

2. The Mathematical Model

The supply chain network for the 5G applications and services provisioning consists of three levels as depicted in Figure 1 (see [2,3,6], and [16] for other optimization models based on supply chain network). The network topology consists of the following:

- devices and end users $(1, \dots, d, \dots, D)$, at the first level, requiring A 5G applications and services $(1, \dots, a, \dots, A)$;
- manager UAVs $(1, \dots, m, \dots, M)$, at the intermediate level, which receive and manage the 5G application and service requests;
- a fleet of UAVs $(1, \dots, s, \dots, S)$ where 5G applications and services are actually executed.

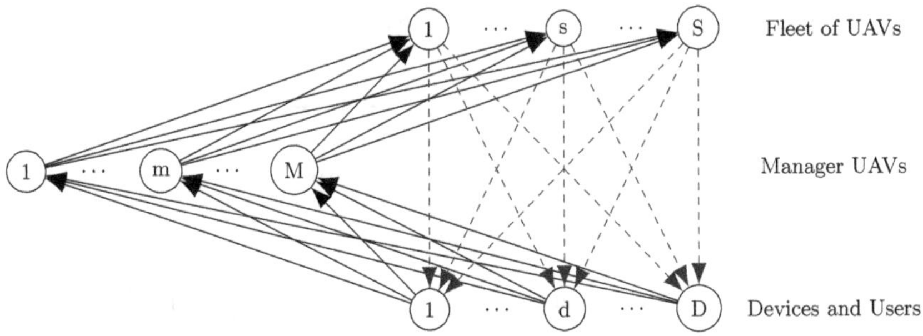

Fig. 1. Network topology.

We can assume that all the devices and users are located on the ground and each of them could require each type of 5G application or service a. The requests of 5G applications and services (from devices and users) are received by the manager UAVs on flight. In turn, each manager UAV m sends the received requests of 5G applications and services to the UAVs belonging to the fleet at the highest level of the network. Finally, each UAV s of the fleet executes the 5G applications and services received and sends them to the devices and users on the ground.

Therefore, the forward chain is constituted by the fleet of UAVs, the manager UAVs, and the devices and users, while the reverse chain is constituted by the fleet of UAVs and the devices and users. Note that we denoted the flows in the reverse chain by blue dashed arrows.

We highlight that not all the requested 5G applications and services could be executed because of the trustworthiness level of each UAV: some UAVs may be old or worn out, or their battery can run out, or they may be located in areas at risk (for natural disaster, for example). For this reason, we analyze the multi-level closed loop supply chain network.

The aim of this chapter is to propose a multi-objective optimization model which allows us to determine the optimal flows distribution and the optimal additional trustworthiness level of each UAV belonging to the fleet, in order to maximize the total profit (minimizing the costs) and minimize the penalty due to the unreliability of each UAV (that is, due to 5G applications and services that devices and users do not receive since they were not executed by UAVs).

Table 1. Variables notation and description.

Variable	Description
α_{dma}	The flow of application or service a requested by device or user d on the ground to the manager UAV m
β_{msa}	The flow of application or service a requests received by the UAV s belonging to the upper tier fleet from the manager UAV m
γ_{sda}	The flow of application or service a executed by any UAV $s \in \mathcal{F}$ in the upper tier fleet and sent to the user or device d on the ground
$x_s \in [0, (1 - r_s)]$	The additional trustworthiness level of the UAV s

The variables of the model and the corresponding descriptions are reported in Table 1.

Now, we present the parameters and cost functions. Let:

- p_a be the unit revenue obtained for the sale of application or service a executed and sent to the devices and users on the ground, $\forall a = 1, \ldots, A$;
- r_s be the initial trustworthiness level of the UAV s, $\forall s = 1, \ldots, S$; observe that the additional trustworthiness level of the UAV s (see x_s in Table 1) is between 0 and $1 - r_s$ so that the current trustworthiness level (given by the sum of the initial and additional trustworthiness levels, $r_s + x_s$) is between 0 and 1, where 1 means maximum trustworthiness, while 0 means untrustworthy;
- D_{da} be the total request of 5G application or service a from the device or user d, $\forall d = 1, \ldots, D$, $\forall a = 1, \ldots, A$;
- $\overline{C}_m$ be the maximum flow of application and service requests that the manager UAV m is able to control, $\forall m = 1, \ldots, M$;
- e_a be the running space requested to execute application or service a, $a = 1, \ldots, A$;
- $\overline{C}_s$ be the maximum capacity or running space that cannot be exceeded on the UAV s, $\forall s = 1, \ldots, S$;
- $\overline{B}$ be the maximum available budget to grow the trustworthiness level;
- c_{dm} be the cost to transmit requests of applications and services from device or user d to manager UAV m and let us assume c_{dm}

is a function of the transmitted flow:

$$c_{dm} = c_{dm}\left(\sum_{a=1}^{A} \alpha_{dma}\right) = c_{dm}(\alpha_{dm}), \quad \forall d = 1, \dots, D,$$
$$\forall m = 1, \dots, M,$$

where $\alpha_{dm} = (\alpha_{dma})_{a=1,\dots,A} \in \mathbb{R}^A_+$;

- c_{ms} be the cost function related to the transmission of the applications and service requests from manager UAV m to any UAV s belonging to the fleet at the highest level of the network and let us assume that such a cost is a function of the transmitted flow:

$$c_{ms} = c_{ms}\left(\sum_{a=1}^{A} \beta_{msa}\right) = c_{ms}(\beta_{ms}), \quad \forall m = 1, \dots, M,$$
$$\forall s = 1, \dots, S,$$

where $\beta_{ms} = (\beta_{msa})_{a=1,\dots,A} \in \mathbb{R}^A_+$;

- c_{sd} be the transmission cost of the applications and services from any UAV s to device or user d on the ground and let us assume c_{sd} is a function of the amount of transmitted services $\sum_{a=1}^{A} \gamma_{sda}$:

$$c_{sd} = c_{sd}\left(\sum_{a=1}^{A} \gamma_{sda}\right) = c_{sd}\left(\gamma_{sd}\right),$$
$$\forall s = 1, \dots, S, \ \forall d = 1, \dots, D,$$

where $\gamma_{sd} = (\gamma_{sda})_{a=1,\dots,A} \in \mathbb{R}^A_+$; furthermore, we assume that such transmission costs also depend on the distance $dist(d, m)$, $dist(m, s)$, and $dist(s, d)$, respectively, and such a dependence on distance will be taken into account in the analytical expression of costs;

- $c_s^{(E)}$ be the running costs of requested applications and services to the UAV $s = 1, \dots, S$, and let us assume they are functions of the total amount of performed applications and services:

$$c_s^{(E)} = c_s^{(E)}\left((r_s + x_s) \cdot \sum_{m=1}^{M} \sum_{a=1}^{A} \beta_{msa}\right) = c_s^{(E)}\left(x_s, \beta_s\right),$$
$$\forall s = 1, \dots, S,$$

where $\beta_s = (\beta_{msa})_{\substack{m=1,\dots,M\\a=1,\dots,A}} \in \mathbb{R}_+^{MA}$; note that such a cost depends from the current level of trustworthiness, given by the sum of the initial one (r_s) and the additional one (x_s);

- c_s be the investment cost function in order to increase the current trustworthiness level, by increasing the additional one x_s, of the UAV $s = 1, \dots, S$ and let us assume c_s is a function of both the additional trustworthiness level and the flow of application and service requests that s receives:

$$c_s\left(x_s, \sum_{m=1}^{M}\sum_{a=1}^{A}\beta_{msa}\right) = c_s\left(x_s, \beta_s\right), \quad \forall s = 1, \dots, S;$$

- P_a be the penalty to be paid for each service-a request not executed due to the unreliability of the UAV s, $\forall a = 1, \dots, A$; therefore, such a penalty expresses the consumer unsatisfaction and loss of demand and the total penalty cost to be paid for each UAV $s = 1, \dots, S$ is given by

$$(1 - (r_s + x_s)) \cdot \left(\sum_{a=1}^{A} P_a \cdot \left(\sum_{m=1}^{M}\beta_{msa}\right)\right).$$

The problem formulation is as follows:

$$\begin{aligned}
\max \sum_{s=1}^{S}\sum_{d=1}^{D}\sum_{a=1}^{A} p_a\gamma_{sda} &- \sum_{d=1}^{D}\sum_{m=1}^{M} c_{dm}\left(\sum_{a=1}^{A}\alpha_{dma}\right)\\
&- \sum_{m=1}^{M}\sum_{s=1} c_{ms}\left(\sum_{a=1}^{A}\beta_{msa}\right) - \sum_{s=1}^{S}\sum_{d=1}^{D} c_{sd}\left(\sum_{a=1}^{A}\gamma_{sda}\right)\\
&- \sum_{s=1}^{s} c_s^{(E)}\left((r_s + x_s) \cdot \sum_{m=1}^{M}\sum_{a=1}^{A}\beta_{msa}\right)\\
&- \sum_{s=1}^{S}(1 - (r_s + x_s)) \cdot \left(\sum_{a=1}^{A} P_a \cdot \left(\sum_{m=1}^{M}\beta_{msa}\right)\right)\\
&- \sum_{s=1}^{S} c_s\left(x_s, \sum_{m=1}^{M}\sum_{a=1}^{A}\beta_{msa}\right)
\end{aligned} \tag{1}$$

subject to

$$\sum_{m=1}^{M} \alpha_{dma} = D_{da} \quad \forall d = 1, \dots, D,\ \forall a = 1, \dots, A, \tag{2}$$

$$\sum_{a=1}^{A} \sum_{d=1}^{D} \alpha_{dma} \leq \bar{C}_m \quad \forall m = 1, \dots, M, \tag{3}$$

$$\sum_{s=1}^{S} \beta_{msa} \leq \sum_{d=1}^{D} \alpha_{dma} \quad \forall m = 1, \dots, M,\ \forall a = 1, \dots, A, \tag{4}$$

$$\sum_{m=1}^{M} \sum_{a=1}^{A} e_a \beta_{msa} \leq \bar{C}_s \quad \forall s = 1, \dots, S, \tag{5}$$

$$\sum_{d=1}^{D} \gamma_{sda} = (r_s + x_s) \cdot \left(\sum_{m=1}^{M} \beta_{msa} \right) \quad \forall s = 1, \dots, S,\ \forall a = 1, \dots, A, \tag{6}$$

$$\sum_{s=1}^{S} \gamma_{sda} \leq D_{da} \quad \forall d = 1, \dots, D,\ \forall a = 1, \dots, A, \tag{7}$$

$$\sum_{s=1}^{S} c_s \left(x_s, \sum_{m=1}^{M} \sum_{a=1}^{A} \beta_{msa} \right) \leq \overline{B}, \tag{8}$$

$$\alpha_{dma}, \beta_{msa}, \gamma_{sda} \in \mathbb{R}_+,\ \forall d = 1, \dots, D,\ \forall m = 1, \dots, M,$$
$$\forall s = 1, \dots, S,\ \forall a = 1, \dots, A, \tag{9}$$

$$x_s \in [0, (1 - r_s)] \quad \forall s = 1, \dots, S. \tag{10}$$

The objective function (1) represents the profit to maximize. It is given by the difference between the gain (that is the total revenue obtained by the sale of applications and services from UAVs to devices and users) and the total cost (that is the sum of all the transmission costs, the running costs, the costs to increase the trustworthiness level of each UAV, and the penalty to be paid for each not executed service request). The first constraint, (2), is the conservation law, which assures that the amount of service a requested by device or user d to all manager UAVs equals the request D_{da}. According to constraint (3), the maximum capacity $\bar{C}_m$ for manager UAV

m is not exceeded. Constraint (4) affirms that the flow of requests of application or service a that a manager UAV m could send to all the UAVs belonging to the fleet at the highest level of the network (for the execution) must be less than the flow of requests that m receives from all the devices and users on the ground. The requests of applications and services that each UAV s can receive must satisfy the capacity constraint (5) which states that the running space must not exceed the maximum allowed, $\overline{C}_s$. Constraint (2) determines the equality, for each application or service, between the flow of requests received by each UAV of the fleet (from all the manager UAVs), multiplied by the reliability level of the considered UAV s, and the flow of the applications and services sent to the devices and users. Constraint (7) establishes that each device or user d cannot receive more application or service a than requested. Constraint (8) guarantees that the costs to increase the trustworthiness level of the UAVs at the highest level of the network is less than or equal to the maximum available budget. The latest constraint families define the domain of the variables of the model.

3. The Variational Approach

In this section, we provide an alternative approach to problems (1)–(10) (see [5, 7, 8, 17], for various network optimization problems for which a variational approach is provided), advantageous from a computational point of view. Indeed, by determining the variational inequality associated with the above optimization problem, we can use the well-known computational procedure presented, for instance, in [14] as well as the theory of existence and uniqueness of the solutions.

3.1. *Theoretical preliminaries*

Given a continuous function $F : K \to \mathbb{R}^N$, where $K \subset \mathbb{R}^N$ is a given (non-empty) convex set, the finite-dimensional variational inequality problem consists of determining a vector $X^* \in K$ such that

$$\langle F(X^*), X - X^* \rangle = \sum_{i=1}^{N} F_i(X)(X_i - X_i^*) \geq 0, \quad \forall X \in K. \quad (11)$$

The advantage of using such a mathematical tool lies in the fact that it allows for a unified treatment of equilibrium problems and optimization problems. Indeed, many mathematical problems can be formulated as variational inequality problems, such as system of equations, complementarity problems, fixed points problems, Nash equilibrium problems, and the optimization ones (see [14] for a complete discussion of these equivalences). In our case, the equivalence between the solutions to an optimization problem and a variational inequality problem is fundamental. The following theorem holds.

Theorem 1. *Let X^* be a solution to the optimization problem:*

$$\min f(X)$$

subject to

$$X \in K, \tag{12}$$

where f is a continuously differentiable function and K is a closed convex set. Then, X^* is a solution to the variational inequality problem $VI(F, X)$:

$$\langle \nabla f(X^*), X - X^* \rangle \geq 0, \quad \forall X \in K, \tag{13}$$

where $F(X) \equiv \nabla f(X)$. Moreover, if $f(x)$ is a convex function and X^* is a solution to $VI(F, K)$, then X^* is a solution to the optimization problem (12).

Variational inequality theory is also a powerful tool in the qualitative analysis of equilibria. Indeed, a solid theory of existence and uniqueness of the solutions has been developed. The classic existence theorem states that a solution exists if the function F is continuous and the feasible set is compact. It is clear, therefore, that it is necessary to impose that the function f is continuously differentiable to guarantee an existence result.

Theorem 2. *If K is a compact convex set and $F(X)$ is a continuous function on K, then the variational inequality problem $VI(F, K)$ admits at least one solution X^*. Moreover, if $F(X)$ is strictly monotone on K, that is,*

$$\langle F(X^1) - F(X^2), X^1 - X^2 \rangle > 0, \quad \forall X^1,\ X^2 \in K,\ X^1 \neq X^2,$$

the solution is unique.

3.2. *The variational inequality formulation*

By adapting the notations and results of the previous section, in this section, we deduce the variational formulation of the optimization problem (1)–(10), also providing sufficient conditions for the existence and uniqueness of the solutions.

For this purpose, let $f(X) : K \to R^N$ be the objective function (1), more specifically,

$$\begin{aligned} f(X) = & \sum_{s=1}^{S}\sum_{d=1}^{D}\sum_{a=1}^{A} p_a \gamma_{sda} - \sum_{d=1}^{D}\sum_{m=1}^{M} c_{dm}\left(\sum_{a=1}^{A}\alpha_{dma}\right) \\ & - \sum_{m=1}^{M}\sum_{s=1}^{} c_{ms}\left(\sum_{a=1}^{A}\beta_{msa}\right) - \sum_{s=1}^{S}\sum_{d=1}^{D} c_{sd}\left(\sum_{a=1}^{A}\gamma_{sda}\right) \\ & - \sum_{s=1}^{s} c_s^{(E)}\left((r_s + x_s)\cdot\sum_{m=1}^{M}\sum_{a=1}^{A}\beta_{msa}\right) \\ & - \sum_{s=1}^{S}(1-(r_s+x_s))\cdot\left(\sum_{a=1}^{A}P_a\cdot\left(\sum_{m=1}^{M}\beta_{msa}\right)\right) \\ & - \sum_{s=1}^{S} c_s\left(x_s, \sum_{m=1}^{M}\sum_{a=1}^{A}\beta_{msa}\right), \end{aligned} \tag{14}$$

where $\alpha = (\alpha_{dma})_{\substack{d=1,\dots,D\\ m=1,\dots,M\\ a=1,\dots,A}} \in \mathbb{R}_+^{DMA}$, $\beta = (\beta_{msa})_{\substack{m=1,\dots,M\\ s=1,\dots,S\\ a=1,\dots,A}} \in \mathbb{R}_+^{MSA}$, $\gamma = (\gamma_{sda})_{\substack{s=1,\dots,S\\ d=1,\dots,D\\ a=1,\dots,A}} \in \mathbb{R}_+^{SDA}$, $x = (x_s)_{s=1,\dots,S} \in \mathbb{R}_+^S$ $X \equiv (\alpha,\beta,\gamma,x)$, K is the feasible set defined as follows:

$$K \equiv \{(\alpha,\beta,\gamma,x) : (2)-(10) \text{ hold}\} \subset \mathbb{R}_+^N, \tag{15}$$

and $N = DMA + MSA + SDA + S$.

Theorem 3. *Let us suppose that the cost functions c_{ms}, c_{sd}, $c_s^{(E)}$, c_s, $\forall m = 1,\dots,M$, $\forall s = 1,\dots,S$, $\forall d = 1,\dots,D$, are continuously differentiable and convex. Then, a vector X^* is a solution to the optimization problem (1)–(10) if and only if it is a* *solution to the variational inequality problem $VI(F, K)$:*

$$\langle \nabla f(X^*), X - X^* \rangle \geq 0, \quad \forall X \in K, \tag{16}$$

or equivalently to the variational inequality

$$\begin{aligned}
&\sum_{d=1}^{D}\sum_{m=1}^{M}\sum_{a=1}^{A}\left[\frac{\partial c_{dm}(\alpha^*_{dm})}{\partial \alpha_{dma}}\right] \times (\alpha_{dma} - \alpha^*_{dma})\\
&+\sum_{m=1}^{M}\sum_{s=1}^{S}\sum_{a=1}^{A}\left[\frac{\partial c_{ms}(\beta^*_{ms})}{\partial \beta_{msa}} + \frac{\partial c^{(E)}_s(x^*_s, \beta^*_s)}{\partial \beta_{msa}} + \frac{\partial c_s(x^*_s, \beta^*_s)}{\partial \beta_{msa}}\right.\\
&\left. + ((1 - (r_s + x^*_s))P_a\right] \times (\beta_{msa} - \beta^*_{msa})\\
&+\sum_{s=1}^{S}\sum_{d=1}^{D}\sum_{a=1}^{A}\left[\frac{\partial c_{sd}(\gamma^*_{sd})}{\partial \gamma_{sda}} - p_a\right] \times (\gamma_{sda} - \gamma^*_{sda})\\
&+\sum_{s=1}^{S}\left[\frac{\partial c^{(E)}_s(x^*_s, \beta^*_s)}{\partial x_s} - \sum_{a=1}^{A} P_a \sum_{m=1}^{M} \beta^*_{msa} + \frac{\partial c_s(x^*_s, \beta^*_s)}{\partial x_s}\right]\\
&\times (x_s - x^*_s) \geq 0, \quad \forall X \in K.
\end{aligned} \tag{17}$$

Moreover, variational inequality (17) *admits at least one solution.*

Proof. Since the cost functions c_{ms}, c_{sd}, $c^{(E)}_s$, c_s, for all $m = 1, \ldots, M$, $s = 1, \ldots, S$, $d = 1, \ldots, D$, are continuously differentiable and convex, the function f in (14) is continuously differentiable and convex. Therefore, the function $F \equiv \nabla f(X)$ is continuous and convex. Moreover, the feasible set K, for the assumption on the cost functions c_s, for all $s = 1, \ldots, S$, and the nature of the constraints (2)–(10) are clearly convex and closed. Compactness follows by observing that

- from constraints (2) and (3), we deduce that $\alpha_{dms} < +\infty$, for all d, m, and s;
- from constraints (4) and (5), we deduce that $\beta_{msa} < +\infty$, for all m, s, and a;
- from constraints (2) and (7), we deduce that $\gamma_{sda} < +\infty$, for all s, d, and a;
- from constraint (10), we have $x_s \leq (1 - r_s)$, for all s.

Hence, the hypotheses of Theorem 1 are satisfied, and in virtue of it, the thesis follows.

Variational inequality (17) is equivalent to variational inequality (16) since

$$F(X) = \nabla f(X) = \left(\frac{\partial f(X)}{\partial \alpha}, \frac{\partial f(X)}{\partial \beta}, \frac{\partial f(X)}{\partial \gamma}, \frac{\partial f(X)}{\partial x}\right)$$

$$= \begin{pmatrix} \left[\dfrac{\partial c_{dm}(\alpha_{dm})}{\partial \alpha_{dma}}\right]_{\substack{d=1,\dots,D \\ m=1,\dots,M,\ a=1,\dots,A}} \\ \left[\dfrac{\partial c_{ms}(\beta_{ms})}{\partial \beta_{msa}} + \dfrac{\partial c_s^{(E)}(x_s,\beta_s)}{\partial \beta_{msa}} \right. \\ \left. + \dfrac{\partial c_s(x_s,\beta_s)}{\partial \beta_{msa}} + (1-x_s)P_a\right]_{\substack{m=1,\dots,M \\ s=1,\dots,S,\ a=1,\dots,A}} \\ \left[\dfrac{\partial c_{sd}(\gamma_{sda})}{\partial \gamma_{sd}} - p_a\right]_{\substack{s=1,\dots,S \\ d=1,\dots,D,\ a=1,\dots,A}} \\ \left[\dfrac{\partial c_s^{(E)}(x_s,\beta_s)}{\partial x_s} + \sum_{a=1}^{A} P_a\beta_{sa} + \dfrac{\partial c_s(x_s,\beta_s)}{\partial x_s}\right]_{s=1,\dots,S} \end{pmatrix}^T.$$

Finally, in virtue of Theorem 2, variational inequality problem (17) admits at least one solution. □

To guarantee the uniqueness of the solution to variational inequality problem (16) (equiv. (17)), we provide the following sufficient condition for the strictly monotonocity of $F(X)$.

Theorem 4. *Let us suppose that cost functions c_{ms}, c_{sd}, $c_s^{(E)}$, c_s are strictly convex for all $m = 1,\dots,M;\ s = 1,\dots,S; d = 1,\dots,D$, and that cost functions $c^{(E)}$ and c_s are additive, that is, $c_s^{(E)}(x_s,\beta_s) = c_s^{(E),1}(x_s) + c_s^{(E),2}(\beta_s)$ and $c_s(x_s,\beta_s) = c_s^1(x_s) + c_s^2(\beta_s)$, for all $s = 1,\dots,S$. Then, variational inequality (16), equivalently variational inequality (17), admits a unique solution.*

Proof. The thesis follows by observing that the strict convexity of the cost functions and the additivity assumption on cost functions

$c^{(E)}$ and c_s implies the strict monotonocity of the function F. Indeed, let us suppose to consider X^1, $X^2 \in K$, with $X^1 \neq X^2$. We evaluate the following quantity:

$$
\begin{aligned}
&\langle F(X^1) - F(X^2), X^1 - X^2 \rangle \\
&= \sum_{d=1}^{D}\sum_{m=1}^{M}\sum_{a=1}^{A}\left[\frac{\partial c_{dm}(\alpha^1_{dm})}{\partial \alpha_{dma}} - \frac{\partial c_{dm}(\alpha^2_{dm})}{\partial \alpha_{dma}}\right] \times (\alpha^1_{dma} - \alpha^2_{dma}) \\
&\quad + \sum_{m=1}^{M}\sum_{s=1}^{S}\sum_{a=1}^{a}\left[\frac{\partial c_{ms}(\beta^1_{ms})}{\partial \beta_{msa}} - \frac{\partial c_{ms}(\beta^2_{ms})}{\partial \beta_{msa}}\right] \times (\beta^1_{msa} - \beta^2_{msa}) \\
&\quad + \sum_{m=1}^{M}\sum_{s=1}^{S}\sum_{a=1}^{a}\left[\frac{\partial c^{(E),2}_s(\beta^1_s)}{\partial \beta_{msa}} - \frac{\partial c^{(E),2}_s(\beta^2_s)}{\partial \beta_{msa}}\right] \times (\beta^1_{msa} - \beta^2_{msa}) \\
&\quad + \sum_{m=1}^{M}\sum_{s=1}^{S}\sum_{a=1}^{a}\left[\frac{\partial c^2_s(\beta^1_s)}{\partial \beta_{msa}} - \frac{\partial c^2_s(\beta^2_s)}{\partial \beta_{msa}}\right] \times (\beta^1_{msa} - \beta^2_{msa}) \\
&\quad + \sum_{s=1}^{S}\sum_{d=1}^{D}\sum_{a=1}^{A}\left[\frac{\partial c_{sd}(\gamma^1_{sd})}{\partial \gamma_{sda}} - \frac{\partial c_{sd}(\gamma^2_{sd})}{\partial \gamma_{sda}}\right] \times (\gamma^1_{sda} - \gamma^2_{sda}) \\
&\quad + \sum_{m=1}^{M}\sum_{s=1}^{S}\sum_{a=1}^{a}\left[\frac{\partial c^{(E),1}_s(x^1_s)}{\partial \beta_{msa}} - \frac{\partial c^{(E),1}_s(x^2_s)}{\partial \beta_{msa}}\right] \times (x^1_s - x^2_s) \\
&\quad + \sum_{m=1}^{M}\sum_{s=1}^{S}\sum_{a=1}^{a}\left[\frac{\partial c^1_s(x^1_s)}{\partial \beta_{msa}} - \frac{\partial c^1_s(x^2_s)}{\partial \beta_{msa}}\right] \times (x^1_s - x^2_s). \qquad (18)
\end{aligned}
$$

Assuming the cost functions c_{ms}, c_{sd}, $c^{(E)}_s$, c_s strictly convex for all $m = 1, \ldots, M$; $s = 1, \ldots, S; d = 1, \ldots, D$, and the cost functions $c^{(E)}$ and c_s additive, we can deduce that each addendum of the previous quantity is strictly positive. Therefore, the quantity (18) is strictly positive and, hence, $F(X)$ is strictly monotone. □

3.3. *Alternative variational inequality*

The particular structure of the feasible set K, where nonlinear constraints, equality, and inequality constraints are present, can create difficulties in computational terms. To avoid these difficulties, we provide an alternative variational inequality, equivalent to variational inequalities (16) and (17), where we relax the constraints on the objective function by the means of the Lagrange multipliers (see [4,5,8]).

Theorem 5. *A vector $X^* \in K$ is an optimal solution to problems (1)–(10) if and only if there exist the Lagrange multiplier vectors $\mu^{1*} \in \mathbb{R}^{DA}$, $\mu^{2*} \in \mathbb{R}^{SA}$, and $\lambda^{1*} \in \mathbb{R}_+$ such that the vector $(X^*, \mu^{1*}, \lambda^{2*}, \lambda^{3*}) \in K \times \mathbb{R}^{DA} \times \mathbb{R}^{SA} \times \mathbb{R}_+$ is a solution to the variational inequality:*

Find $(X^*, \mu^{1*}, \lambda^{2*}, \lambda^{3*}) \in K \times \mathbb{R}^{DA} \times \mathbb{R}^{SA} \times \mathbb{R}_+$ such that

$$
\begin{aligned}
&\sum_{d=1}^{D}\sum_{m=1}^{M}\sum_{a=1}^{A}\left[\frac{\partial c_{dm}(\alpha^*_{dm})}{\partial \alpha_{dma}} - \mu^{1*}_{da}\right] \times (\alpha_{dma} - \alpha^*_{dma})\\
&\quad + \sum_{m=1}^{M}\sum_{s=1}^{S}\sum_{a=1}^{A}\left[\frac{\partial c_{ms}(\beta^*_{ms})}{\partial \beta_{msa}} + \frac{\partial c^{(E)}_s(x^*_s, \beta^*_s)}{\partial \beta_{msa}}\right.\\
&\quad + (1-\mu^{2*}_{sa})\frac{\partial c_s(x^*_s, \beta^*_s)}{\partial \beta_{msa}} + ((1-(r_s + x^*_s))P_a\\
&\quad \left. + \mu^{1*}_{sa}(r_s + x^*_s)\right] \times (\beta_{msa} - \beta^*_{msa})\\
&\quad + \sum_{s=1}^{S}\sum_{d=1}^{D}\sum_{a=1}^{A}\left[\frac{\partial c_{sd}(\gamma^*_{sd})}{\partial \gamma_{sda}} - p_a - \mu^{1*}_{sa}\right] \times (\gamma_{sda} - \gamma^*_{sda})\\
&\quad + \sum_{s=1}^{S}\left[\frac{\partial c^{(E)}_s(x^*_s, \beta^*_s)}{\partial x_s} - \sum_{a=1}^{A} P_a \sum_{m=1}^{M}\beta^*_{msa}\right.\\
&\quad \left. + \left(1 - \sum_{a=1}^{A}\mu^{2*}_{sa}\right)\frac{\partial c_s(x^*_s, \beta^*_s)}{\partial x_s} - \sum_{a=1}^{A}\mu^{1*}_{sa} x_s \sum_{m=1}^{M}\beta_{msa}\right]\\
&\quad \times (x_s - x^*_s) \geq 0, \quad \forall X \in K. \qquad (19)
\end{aligned}
$$

4. Numerical Simulations

In this section, we provide some numerical simulations in order to more clearly illustrate the potential of the model proposed in this chapter. We first describe the setting (the network topology) and

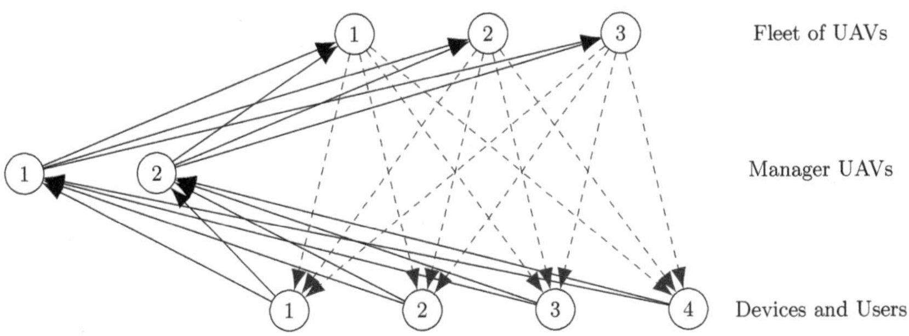

Fig. 2. Network topology for the numerical simulations.

illustrate the numerical data, and then we provide a standard solution and an optimized solution. Finally, we compare the standard and optimized solutions.

4.1. *Setting Description*

We consider $D = 4$ devices or users on the ground, $D = 2$ manager UAVs, $S = 3$ UAVs belonging to the fleet at the highest level of the network, and $S = 1$ type of applications and services.

The network topology is depicted in Figure 2.

The size of the network and the following numerical data have been chosen for illustrative purpose only:

$$p_1 = 20;\ D_{11} = 3, \quad D_{21} = 5,\ D_{31} = 7,\ D_{41} = 9;$$
$$\overline{C}_1 = 15,\ \overline{C}_2 = 10;\ P_1 = 1;\ e_1 = 1,\ \overline{C}_1 = 10,$$
$$\overline{C}_2 = 8,\ \overline{C}_3 = 6;\ r_1 = 0.4,\ r_2 = 0.3,\ r_3 = 0.2;\ \overline{B} = 100.$$

We assume that the cost to transmit requests of applications and services from device or user on the ground to any manager UAV has the following generic expression:

$$c_{dm}\left(\sum_{a=1}^{A}\alpha_{dma}\right) = \pi_{dm}\cdot\left(\sum_{a=1}^{A}\alpha_{dma}\right)^2 + \pi'_{dm}\cdot\left(\sum_{a=1}^{A}\alpha_{dma}\right),$$

where the values of π_{dm} and π'_{dm}, $\forall d = 1,\ldots,D$, $\forall m = 1,\ldots,M$, are reported in Table 2.

We assume that the cost function related to the transmission of the application and service requests from manager UAV to any UAV

Table 2. Values of the parameters for the transmission cost from d to m.

$\pi_{dm} = \pi'_{dm}$	$m = 1$	$m = 2$
$d = 1$	0.1	0.3
$d = 2$	0.1	0.2
$d = 3$	0.2	0.1
$d = 4$	0.3	0.1

Table 3. Values of the parameters for the transmission cost from m to s.

$\psi_{ms} = \psi'_{ms}$	$s = 1$	$s = 2$	$s = 3$
$m = 1$	0.1	0.3	0.4
$m = 2$	0.3	0.1	0.2

belonging to the fleet has the following generic expression:

$$c_{ms}\left(\sum_{a=1}^{A}\beta_{msa}\right) = \psi_{ms}\cdot\left(\sum_{a=1}^{A}\beta_{msa}\right)^2 + \psi'_{ms}\cdot\left(\sum_{a=1}^{A}\beta_{msa}\right),$$

where the values of ψ_{ms} and ψ'_{ms}, $\forall m = 1,\ldots,M$, $\forall s = 1,\ldots,S$, are reported in Table 3.

We assume that the transmission cost of the applications and services from any UAV s to each device or user on the ground has the following generic expression:

$$c_{sd}\left(\sum_{a=1}^{A}\gamma_{sda}\right) = \mu_{sd}\cdot\left(\sum_{a=1}^{A}\gamma_{sda}\right)^2 + \mu'_{sd}\cdot\left(\sum_{a=1}^{A}\gamma_{sda}\right),$$

where the values of μ_{sd} and μ'_{sd}, $\forall s = 1,\ldots,S$, $\forall d = 1,\ldots,D$, are reported in Table 4.

Table 4. Values of the parameters for the transmission cost from s to d.

$\mu_{sd} = \mu'_{sd}$	$d = 1$	$d = 2$	$d = 3$	$d = 4$
$s = 1$	0.2	0.1	0.3	0.4
$s = 2$	0.2	0.1	0.1	0.2
$s = 3$	0.4	0.3	0.1	0.2

Also the running costs of requested applications and services to the UAV s have a generic quadratic expression as shown by the following equation:

$$c_s^{(E)}\left((r_s + x_s) \cdot \sum_{m=1}^{M}\sum_{a=1}^{A}\beta_{msa}\right) = \phi_s \cdot \left((r_s + x_s) \cdot \sum_{m=1}^{M}\sum_{a=1}^{A}\beta_{msa}\right)^2 + \phi'_s \cdot \left((r_s + x_s) \cdot \sum_{m=1}^{M}\sum_{a=1}^{A}\beta_{msa}\right),$$

where $\phi_s = \phi'_s = 0.1, \ \forall s = 1, 2, 3$;

$$c_s(x_s) = \nu_s \cdot (x_s)^2 + \nu'_s \cdot (x_s),$$

where $\nu_s = \nu'_s = 0.1, \ \forall s = 1, 2, 3$.

4.2. *Standard approach*

The solution of the standard approach, which we use for the comparison with the optimal solution (that is, the results of the proposed optimization model), is obtained by a random assignment of flows, such that all the constraints are satisfied. We also assume that, analyzing the standard problem, it is not allowed to modify the trustworthiness levels of the UAVs belonging to the fleet at the highest layer of the network.

The choice of these assumptions is justified by reality, in which decisions are not made according to optimization criteria and in which, moreover, the possibility of increasing the levels of trustworthiness is often not analyzed.

Table 5. Standard solutions: α'_{dm1}.

α'_{dm1}	$m = 1$	$m = 2$
$d = 1$	0	3
$d = 2$	5	0
$d = 3$	7	0
$d = 4$	2	7

Table 6. Standard solutions: β'_{ms1}.

β'_{ms1}	$s = 1$	$s = 2$	$s = 3$
$m = 1$	0	8	6
$m = 2$	10	0	0

Table 7. Standard solutions: γ'_{sd1}.

γ'_{sd1}	$d = 1$	$d = 2$	$d = 3$	$d = 4$
$s = 1$	0	0	0	4
$s = 2$	1.2	0	0	1.2
$s = 3$	1.2	0	0	0

Therefore, in this first simulation, we assume that all the variables on the additional trustworthiness levels of the UAVs are equal to zero: $x'_s = 0$, $\forall s = 1, 2, 3$, and we randomly assign the value of α_{dma}, β_{msa}, γ_{sda}, $\forall d$, $\forall m$, $\forall s$, and $\forall a$ so that all the constraints (2)–(10) are satisfied. The standard solutions obtained are summarized in Tables 5, 6, and 7.

4.3. *Optimal approach*

The optimal approach is obtained using the optimal formulation proposed in the previous sections.

The optimal solution is computed through the Euler method (see [10] for a detailed description) using the Matlab program on an HP laptop with an AMD compute cores 2C + 3G processor, 8 GB RAM.

Table 8. Optimal solutions: α^*_{dm1}.

α^*_{dm1}	$m = 1$	$m = 2$
$d = 1$	3	0
$d = 2$	4.71	0.29
$d = 3$	3.38	3.62
$d = 4$	2.91	6.09

Table 9. Optimal solutions: β^*_{ms1}.

β^*_{ms1}	$s = 1$	$s = 2$	$s = 3$
$m = 1$	8.75	2.75	2.00
$m = 2$	1.25	5.25	3.50

Table 10. Optimal solutions: γ^*_{sd1}.

γ^*_{sd1}	$d = 1$	$d = 2$	$d = 3$	$d = 4$
$s = 1$	2.21	3.98	1.59	2.21
$s = 2$	0.73	1.02	2.81	3.45
$s = 3$	0.06	0	2.60	3.34

Table 11. Optimal solutions: x^*_s.

	$s = 1$	$s = 2$	$s = 3$
x^*_s	0.6	0.7	0.8

The standard solutions obtained are summarized in Tables 8, 9, 10, and 11.

4.4. *Comparison and analysis*

We now compare the results obtained with the standard approach and the optimized one. Table 12 shows the details of the results. Particularly, in Table 12, we reported the value of the objective function

Table 12. Comparison between the two approaches.

Approach	Objective function	Total profit	Transmission costs	Execution costs	Total penalty	Trustworthiness costs
Optimal	402.21	480.00	54.67	22.40	0.00	0.71
Standard	25.73	152.00	106.71	3.16	16.40	0.00
PC	1463.19%	215.79%	−48.77%	608.86%	−100%	-

obtained with the standard and the optimal solutions, respectively; the total profit, that is the only positive term of the objective function; the total transmission cost, given by the sum of the transmission costs between devices or users and manager UAVs, between manager UAVs and UAVs of the fleet and between UAVs at the highest layer and devices and users on the ground; the execution costs; the total penalty to be paid and the total cost to increase the trustworthiness levels of the UAVs.

We can easy observe that, despite the execution and trustworthiness costs being greater using the optimal approach (while using the standard approach, they are minor), the value of the objective function we aim to maximize is grater using our optimized approach than that using the standard one. The total profit of the optimal approach is also definitely grater than that of the standard approach. The total transmission cost of the optimal approach is less than that obtained using the standard approach. Finally, we have no penalties to pay, using our model, because the actual trustworthiness (given by the initial and the additional one) of each UAV is at the highest levels (they are equal to 1).

Furthermore, the last row of Table 12 reports the percentage change (PC), obtained as follows:

$$PC = \left(\frac{v_{opt} - v_{std}}{v_{std}} \cdot 100 \right) \%,$$

where v_{opt} is the value obtained by the optimal approach, while v_{std} is the value obtained by the standard one. It is evident that the obtained PC value for the objective function is very high, therefore, this result shows that using the strategic or optimization approach, we obtain an increase in profit of 1463% compared to that obtained with the standard approach. Note that the PC values for transmission costs and total penalty are negative, but since our problem aims to minimize them, this contributes to an improvement in objective function. The only component of the objective function that does not seem to improve it is the total execution cost because more services are performed.

Therefore, the results of this comparison indicate a huge advantage in carrying out optimization strategies.

5. Conclusions

In this chapter, we have developed a constrained closed-loop supply chain network model describing the provision of 5G services. In our framework, in the forward chain, users and devices on the ground, corresponding to the lowest level of the network, require 5G applications and services to services providers, each of which owns a fleet of executor UAVs organized as a FANET and orchestrated by some controller UAVs. In the reverse chain, instead, the fleet of UAVs execute applications and services and provide them to users and devices on the ground. Service providers, from a system optimization point of view, aim to better manage their UAVs and the flows related to them, in order to maximize their profit, given by the difference between the revenue associated with the provision of 5G applications and services and the costs due to transmission and execution of services. In our analysis, we have also taken into account one of the main characteristics of 5G networks, the trustworthiness level of the UAVs of the fleet. Indeed, it is unrealistic to assume that a UAV has maximum trustworthiness, since they may be old or worn, or they may be located in areas of high geological risk or simply, their battery is not fully charged (and does not have infinite life). For this reason, it is essential to take into account the trustworthiness levels of the UAVs in the mathematical formulation and, consequently, to analyze the closed-loop supply chain network, where the reverse chain allows us to describe the downward flows (from UAVs at the highest layer of the network to devices and users on the ground), which may be different from upward flows. Services providers, therefore, seek to increase levels of trustworthiness of each UAV of the fleet (because higher levels correspond to more provided services and, therefore, to greater profits) while minimizing the investment costs (to increase such levels) and, consequentially, minimizing the penalty due to the services requested by users but not provided because of the unreliability. The obtained optimization problem is clearly subject to a multitude of constraints, such as conservation laws, capacity constraints, and budget constraints. We have deduced a variational inequality problem formulation for which existence and uniqueness results we have provided. Finally, some numerical simulations, based on a $4 \times 2 \times 3$ network, have been performed. Particularly, two different approaches have been identified for simulations. The first one refers to the standard approach, which does not use any optimization technique and, moreover, does not take into account the possibility of increasing the levels of trustworthiness. On the contrary, the second approach is the optimal one, where the solutions are obtained by solving the optimization problem proposed in this chapter. Finally, we have conducted a comparison between the associated solutions obtained through the standard and the optimal approaches, highlighting how the use of our model leads to a considerably more advantageous result than the standard one.

In future works, we intend to analyze the contest of the quality of services, introducing also the possibility of adding new UAVs at the fleet for the execution of services.

Acknowledgements

The research was partially supported by the Spoke 1 "FutureHPC BigData" of the Italian Research Center on High-Performance Computing, Big Data and Quantum Computing (ICSC) funded by MUR Missione 4 Componente 2 Investimento 1.4: Potenziamento strutture di ricerca e creazione di "campioni nazionali di RS (M4C2-19)" – Next Generation EU (NGEU) and by the MUR research programs PRIN2022 founded by the European Union – Next Generation EU (Project "ACHILLES, eco-sustAinable effiCient tecHdrIven Last miLE logiStics" CUP: E53D23005640006). The research activity carried out in this paper was also partially supported by the EU under the Italian National Recovery and Resilience Plan (NRRP) of NextGenerationEU, partnership on "Telecommunications of the Future" (PE0000001 – program "RESTART") and by the European

Union (NextGeneration EU), through the MURR-PNNR project SAMOTHRACE (EC00000022). The research of Gabriella Colajanni was also partially supported by the Italian Ministry of University and Research (MUR) and the European Union for the PON project on Research and Innovation 2014-2020, D.M. 1062/2021. These supports are gratefully acknowledged.

References

[1] Alladi, T., Chamola, V., Sahu, N. and Guizani, M. *Applications of Blockchain in Unmanned Aerial Vehicles: A Review.* Vehicular Communications, 23, (2020). https://doi.org/10.1016/j.vehcom.2020.100249.

[2] Cappello, G.M., Colajanni, G., Daniele, P. and Sciacca, D. A constrained optimization model for the provision of services in a 5G network with multi-level cybersecurity investments. *Soft Comput.* (2022). https://doi.org/10.1007/s00500-022-07117-5.

[3] Colajanni, G. and Daniele, P. A mathematical network model and a solution algorithm for IaaS cloud computing. *Networks and Spatial Economics*, 22(2), 267–287 (2022).

[4] Colajanni, G., Daniele, P., Giuffrè, S. and Nagurney, A. Cybersecurity investments with nonlinear budget constraints and conservation laws: Variational equilibrium, marginal expected utilities, and Lagrange multipliers. *Int. Trans. Oper. Res.*, 25, 1443–1464 (2018).

[5] Colajanni, G., Daniele, P. and Sciacca, D. A projected dynamic system associated with a cybersecurity investment model with budget constraints and fixed demands. *J. Nonlinear Var. Anal.*, 4(1), 45–61 (2020). https://doi.org/10.23952/jnva.4.2020.1.05.

[6] Colajanni, G., Daniele, P. and Sciacca, D. On the provision of services with UAVs in disaster scenarios: A two-stage stochastic approach. *Oper. Res. Forum,* 3, 18 (2022). https://doi.org/10.1007/s43069-022-00127-x.

[7] Colajanni, G. and Sciacca, D. An Optimization model for services requests management in a 5G network architecture. In: Masone, A., Dal Sasso, V., Morandi, V. (eds.), *Optimization and Data Science: Trends and Applications.* AIRO Springer Series, Springer, Cham 6 (2021). https://doi.org/10.1007/978-3-030-86286-2_7.

[8] Daniele, P. and Sciacca, D. An optimization model for the management of green areas. *Int. Trans. Oper. Res.*, 28, 3094–3116 (2021).

[9] Davis, L. *Handbook of Genetic Algorithms.* Van Nostrand Reinhold, New York (1991).

[10] Dupuis, P. and Nagurney, A. Dynamical systems and variational inequalities. *Annals of Operations Research*, 44, 9–42 (1993).
[11] Li, S., Da Xu, L. and Zhao, S. 5G Internet of Things: A survey. *J. Ind. Inf. Integr*, 10, 1–9 (2018).
[12] Li, B., Fei, Z. and Zhang, Y. UAV Communications for 5G and beyond: Recent advances and future trends. In *IEEE Internet of Things Journal,* 6(2), 2241–2263 (2019).
[13] Mohammed, F., Idries, A., Mohamed, N., Al-Jaroodi, J. and Jawhar, I. UAVs for Smart Cities: Opportunities and Challenges. *2014 International Conference on Unmanned Aircraft Systems (ICUAS)*, 267–273, (2014) DOI: 10.1109/ICUAS.2014.6842265.
[14] Nagurney, A. *Network Economics: A Variational Inequality Approach*, Kluwer Academic Publishers, Boston, Massachusetts (1993).
[15] Nagurney, A. A Multiperiod Supply Chain Network Optimization Model with Investments in Labor Productivity Enhancements. *Oper. Res. Forum* 2, 68 (2021). https://doi.org/10.1007/s43069-021-00112-w.
[16] Nagurney, A. Supply chain networks, wages, and labor productivity: Insights from Lagrange analysis and computations. *J. Global Optim.* 83, 615–638 (2022).
[17] Nagurney, A., Daniele, P. and Shukla, S. A supply chain network game theory model of cybersecurity investments with nonlinear budget constraints, *Ann. Oper. Res.* 248(1), 405–427 (2017).
[18] Shakhatreh, H., Sawalmeh, A.H., AlL-Fuqaha, A., Dou, Z., Almaita, E., Khalil, I., Othman N.S., Khreishah, A. and Guizani, M. Unmanned Aerial Vehicles (UAVs): A survey on civil applications and key research challenges. In *IEEE Access*, vol. 7, 48572–48634 (2019). DOI: 10.1109/ACCESS.2019.2909530.
[19] Singh, R.P., Javaid, M., Haleem, A. and Suman, R. Internet of Things (IoT) applications to fight against COVID-19 pandemic. *Diabetes & Metabolic Syndrome: Clinical Research & Reviews*, 14(4), 521–524 (2020).
[20] Tian, M., Wang, L., Yan, S., Tian, X., Liu, Z. and Rodrigues, J.J.P.C. (2019). Research on financial technology innovation and application based on 5G network. In *IEEE Access*, 7, 138614–138623 (2019), DOI: 10.1109/ACCESS.2019.2936860.

https://doi.org/10.1142/9789811293009_0009

Chapter 9

Kuratowski's Topology in b-Metric Spaces

Stefan Czerwik

Silesian University of Technology,
Kaszubska 23, 44-100 Gliwice, Poland
steczerw@gmail.com

The problem of Kuratowski topology in b-metric spaces is considered. Also other problems in so-called b-metric spaces are investigated, using the construction of Paluszyński–Stempak.

Assume that (X, d) is a b-metric space. In the usual way, we can introduce on a b-metric space. We define a ball of a center $x_0 \in X$ and radius $r > 0$ as follows:

$$B(x_0, r) := \{z \in X : d(x_0, z) < r\}.$$

Clearly, a set $Y \subset X$ is called open, iff for every $z \in Y$ there exists $r_z > 0$ with $B(z, r_z) \subset Y$. Therefore, the family τ_d of all open subsets of X is a Kuratowski topology.

It can be shown (see [9] and [4]) that there is the metric which gives the topology. This metric is given by

$$\varrho(x, y) := \inf \left\{ \sum_{i=1}^{n} d^p(x_{i-1}, x_i) : x = x_0, \ldots, x_n = y \right\}, \qquad (1)$$

$n \in \mathbb{N}$, $x_i \in X$, $i = 0, 1, \ldots, n$, $0 < p \leqslant 1$.

One has (see [5] and [1]).

Theorem 1 ([9]). *Let d be a b-metric on a set $X \neq 0$, with $s \geqslant 1$ fixed. If*

$$(2s)^p = 2, \tag{2}$$

then the mapping $\varrho_p : X \times X \to [0, \infty)$, given by (1), is a metric on X, equivalent to d^p, such that

$$\varrho_p(x, y) \leqslant d^p(x, y) \leqslant 4\varrho_p(x, y), \tag{3}$$

for all $x, y \in X$.

The reader can find the detailed proof in [9]; see also interesting paper of prof. Stefan Cobzas, and others in [2].

We also have the following.

Theorem 2 ([2, 3]). *Let (X, d) be a b-metric space with $s \geqslant 1$, and let (2) be satisfied. Then the topologies τ_d and τ_ϱ are such that*

$$\tau_d = \tau_\varrho.$$

Moreover,

$$(x_n \to x) \Leftrightarrow d(x_n, x) \to 0, \tag{4}$$

for all sequences $(x_n) \subset X$ and $x \in X$.

Proof. We first show that

$$B_d(x, r^{\frac{1}{p}}) \subset B_\varrho(x, r), \tag{5}$$

where B_d and B_ϱ denote the bales with respect to b-metric d and "induced" metric ϱ.

In fact, $y \in B_d(x, r^{\frac{1}{p}})$ is equivalent to $d(x, y) < r^{\frac{1}{p}}$, and therefore

$$d^p(x, y) < r.$$

But since we have (3), then

$$\varrho_p(x, y) \leqslant d^p(x, y) < r,$$

i.e.,

$$y \in B_\varrho(x, r),$$

that is,

$$B_d(x, r^{\frac{1}{p}}) \subset B_\varrho(x, r).$$

Similarly,

$$B_\varrho(x, 4^{-1}r^p) \subset B_d(x, r). \tag{6}$$

By (5) and (6), one gets $\tau_d = \tau_\varrho$. Using these and the inequalities (3), we get easily (4), and the proof of Theorem 2 is finished (complete).

But the topology τ_d has some not very useful properties, concerning the openness of balls and the continuity of the function d.

Kirk and Shahzad in [8], 12.4 considered a special class of b-metric, which has not such problems. Clearly, they used a so-called strong b-metric, satisfying the following conditions:

$$d(x, y) = 0 \Leftrightarrow x = y,$$

$$d(x, y) = d(y, x)$$

and

$$d(x, y) \leqslant d(x, z) + sd(z, y),$$

for some fixed $s \geqslant 1$ for all $x, y, z \in X$.

By the symmetry, one has also

$$d(x, y) \leqslant sd(x, z) + d(z, y), \tag{7}$$

for all $x, y, z \in X$. You can see that (7) implies

$$d(x, y) \leqslant \min\left\{sd(x, z) + d(y, z); d(x, z) + sd(y, z)\right\}.$$

The Kuratowski topology, which is induced by a strong b-metric, has nice properties. We state and prove some of them now. □

Theorem 3. *The balls in (X, d) with a strong b-metric are open sets.*

Proof. Let $y \in B(x, r)$, $r > 0$ and

$$d(x, z) \leqslant d(x, y) + sd(y, z) < r, \quad z \in X$$

and $d(x, z) < r$ such that

$$sd(y, z) < r - d(y, x).$$

Then, if $z \in B(y, r')$, where $r' = r - d(x, y)$, then

$$d(x, z) \leqslant d(x, y) + sd(y, z) < d(x, y) + [r - d(x, y)] < r < r',$$

i.e.,

$$d(x,z) < r'.$$

So, if $r' = r - d(x,y)$ and $y \in B(x,r)$, then

$$B(y,r') \subset B(x,r).$$

Now we recall the ideas of generalized b-metric spaces and a completion of such spaces. A Cauchy sequence in a b-metric spaces (X,d) is any sequence of elements belonging to X such that

$$\lim_{m,n\to\infty} d(x_n, x_m) = 0.$$

It is clear that every convergent sequence is a Cauchy. The (X,d), d-b-metric in X is called complete b-metric iff every Cauchy sequence is convergent to some $x \in X$. In [8], the reader can find the proof that every strong b-metric space can admit a completion, but the similar answer for a generalized metric can be found in [5]. □

Remark 4. The existence of a completion of an arbitrary b-metric space is still open important problem. The author of this paper, in [5] (not published), has presented some ideas on this matter.

We have the following.

Theorem 5 ([3, 8]). *Let (X,d) be a complete b-metric space. If $f : X \to X$ satisfies*

$$d[f(x), f(y)] \leqslant \alpha d(x,y) \tag{8}$$

and

$$0 < \alpha < 1 \tag{9}$$

for all $x, y \in X$, then f has a unique fixed point $u \in X$, and $\{f^n(x)\}$ converges to u for every $x \in X$.

Proof. The functional ϱ_p defined by (1) is a metric on X and satisfies

$$\varrho_p \leqslant d^p \leqslant 4\varrho_p. \tag{10}$$

Therefore, for $y_i = f(x_i), i = 0, \ldots, n$, by (10),

$$\varrho_p[f(u), f(v)] \leqslant \sum_{k=0}^{n-1} d(y_k, y_{k+1})^p \leqslant \alpha^p \sum_{k=0}^{k-1} d(x_k, x_{k+1}),$$

but it is

$$q_p[f(u), f(v)] \leqslant \alpha^p \varrho_p(u, v),$$

for $u, v \in X$ with $0 < \alpha^p < 1$. Since (X, d) is complete, also (X, ϱ_p) is so. By Banach's fixed point principle, we have got the thesis. $\square$

For generalized b-metric, we get the following.

Theorem 6 ([5]). *Let (X, d) be a complete generalized b-metric space and let $f : X \to X$ satisfy*

$$d[f(x), f(y)] \leqslant \varphi[d(x, y)], d(x, y) < \infty, \tag{11}$$

where $\varphi : \mathbb{R}_+ \to \mathbb{R}_+$ is such that

(i) *φ is non-decreasing,*
(ii) $\lim_{n\to\infty} \varphi^n(t) = 0$,
(iii) *$\varphi(t) < \frac{t}{x}$, then for $x \in X$ either*
(iv) *for all $n \in \mathbb{N}$, $d[f^n(x), f^{n+1}(x)] = \infty$, or*
(v) *$\{f^n(x)\big|_{n=1}^{\infty}\}$, $f^n(x) \to u \in X$, and $u = f(u)$.*

Proof. Let $X = \bigcup_{i\in I} X_i$, where X_i are the equivalence classes with respect to "$\sim$": $x \sim y \Leftrightarrow d(x, y) < \infty$ (see Jang [6]). Case (iv) does not hold. Then

$$d[f^m(x), f^{m+1}(x)] < \infty$$

for some $m \in \mathbb{N}$. Now we have, by (11),

$$d[f^{m+1}(x), f^{m+2}(x)] < \infty,$$

and, by mathematical induction, $f^{m+1}(x) \in X_i$, $i \in I$. But for $z \in X_i$, $d[z, f^m(x)] < \infty$, and also

$$d[f(z), f^{m+1}(x)] \leqslant \varphi[d(z, f^m(x)] < \infty,$$

that is, $f_i = f\big|_{x_i} : x_i \to x_i$. One has

$$d[f_i(u), f_i(v)] \leqslant \varphi[d(u, v)], \quad u, v \in X_i.$$

By Theorem 3.1 and Theorem 4.2 of [xx], the sequence $\{f^{m+k}(x)\}_{k=0}^{\infty}$ is convergent to a fixed point z of f_i. But $f_i(z) = f(z) = z$, and the proof is complete. $\square$

Remark 7. We can get also other results on fixed points of the alternative (see [7] and [8]). For example, for $S = 1$, $\varphi(t) = t$, $t \geqslant 0$, $0 < \alpha < 1$, we get the Diaz and Margolis result on fixed point of the alternative, which was the first of this type in the literature.

References

[1] Rydi, H. and Czerwik, S. Fixed point theorems in generalized b-metric spaces. In: N.J. Daras, Th.M. Rassias (eds.), *Modern Discreate Mathematics and Analysis with Applications in Cryptography, Information systems and Modeling*, Springer, pp. 1–9 (2018).

[2] Bakhtin, I.A. The contraction mapping principle in quasimetric spaces, (Rassian) Funktionalnyi Analyz, *Ulianowskii Gos. Ped. Inst.*, 30, 26–37 (1989).

[3] Cobzas, S. and Czerwik, S. The completion of generalized b-metric spaces and fixed points, *Fixed Point Theory*, 21, 133–150 (2020).

[4] Czerwik, S. Fixed point theorems and special solutions of functional equations, *Univ, Sil., Katowice*, 7–83 (1980).

[5] Czerwik, S. Generalized Metric Spaces, *Muntes Touras Journal Pure and Applied Mathematics*, 4(3), 194–262 (2022).

[6] Jang, C.F.K. On generalized complete metric spaces, *Bull. Amer. Math. Soc.*, 75, 113–116 (1969).

[7] Khamsi, M.A. Generalized metric spaces, a survey, *J. Fixed Point Theory Appl.* 17, 455–475 (2015).

[8] Kirk, W. and Shahzad, N. *Fixed Point Theory in Distance Spaces*, Cham, Springer (2014).

[9] Paluszyński, M. and Stempak, K. On quasi-metric and metric spaces, *Proc. Amer. Math. Soc.*, 137, 4307–4312 (2009).

https://doi.org/10.1142/9789811293009_0010

Chapter 10

Cube-Free Values of $[n^c \tan^\theta(\log n)]$

S.I. Dimitrov

Faculty of Applied Mathematics and Informatics,
Technical University of Sofia, Bulgaria
Institute of Biophysics and Biomedical Engineering
Bulgarian Academy of Sciences, Bulgaria
sdimitrov@tu-sofia.bg
xyzstoyan@gmail.com

Let $[\cdot]$ be the floor function. In this chapter, we prove that when $1 < c < 2$ and $\theta > 1$ is fixed, then there exist infinitely many cube-free numbers of the form $[n^c \tan^\theta(\log n)]$.

1. Notations

Let x be a sufficiently large positive number. We denote by $\mu(n)$ the Möbius function. As usual, $[t]$ and $\{t\}$ denote the integer part, respectively, the fractional part of t. Moreover, $e(y) = e^{2\pi i y}$. Instead of $m \equiv n \pmod{k}$, we write for simplicity $m \equiv n(k)$. Throughout this chapter, we suppose that $1 < c < 2$ and $\gamma = \frac{1}{c}$. Assume that $\theta > 1$ is a fixed.

Denote

$$z = x^{\frac{4c-3}{10}}, \tag{1}$$

$$\psi(t) = \{t\} - 1/2, \tag{2}$$

$$\Delta_1 = e^{\pi\left[\frac{\log x}{\pi}\right] + \arctan 1}, \tag{3}$$

$$\Delta_2 = e^{\pi\left[\frac{\log x}{\pi}\right] + \arctan 2}, \tag{4}$$

$$S_c(x) = \sum_{\substack{\Delta_1 \le n < \Delta_2 \\ [n^c \tan^\theta(\log n)] \text{ is cube-free}}} 1, \tag{5}$$

$$\zeta(3) = \sum_{n=1}^{\infty} \frac{1}{n^3}. \tag{6}$$

2. Introduction and Statement of the Result

Zhang and Li [1] were the first to study the distribution of cube-free numbers of the form $[n^c]$. In 2017, they proved a very general result which implies that for any fixed $1 < c < \frac{11}{6}$ and $0 < \varepsilon < 10^{-10}$, one has

$$\#\{n \le x : [n^c] \text{ is cube-free}\} = \frac{x}{\zeta(3)} + \mathcal{O}(x^{1-\varepsilon}).$$

Subsequently, Deshouillers [2] improved the result of Zhang and Li by showing that for any fixed $1 < c < 2$, the asymptotic formula

$$\#\{n \le x : [n^c] \text{ is cube-free}\} = \frac{x}{\zeta(3)} + \mathcal{O}\left(x^{\frac{c+1}{3}} \log x\right)$$

holds. On the other hand, recently the author [3] proved that there exist infinitely many prime numbers of the form $[n^c \tan^\theta(\log n)]$. Motivated by these results, we study the existence of infinitely many cube-free numbers of the form $[n^c \tan^\theta(\log n)]$. More precisely, we establish the following theorem.

Theorem 1. *Let $1 < c < 2$ and $\theta > 1$ is a fixed. Then for the sum (5), the lower bound*

$$S_c(x) \gg x \tag{7}$$

holds.

3. Preliminary Lemmas

Lemma 2. *For every* $H \geq 1$, *we have*

$$\psi(t) = \sum_{1 \leq |h| \leq H} a(h)e(ht) + \mathcal{O}\left(\sum_{|h| \leq H} b(h)e(ht)\right),$$

where

$$a(h) \ll \frac{1}{|h|}, \quad b(h) \ll \frac{1}{H}. \tag{8}$$

Here $\psi(t)$ *is denoted by* (2).

Proof. See [4]. □

Lemma 3. *Let* $q \geq 0$ *be an integer. Suppose that* $f(t)$ *has* $q+2$ *continuous derivatives on* I *and that* $I \subseteq (N, 2N]$. *Assume also that there is some constant* F *such that*

$$|f^{(r)}(t)| \asymp FN^{-r} \tag{9}$$

for $r = 1, \ldots, q+2$. *Let* $Q = 2^q$. *Then*

$$\left|\sum_{n \in I} e(f(n))\right| \ll F^{\frac{1}{4Q-2}} N^{1-\frac{q+2}{4Q-2}} + F^{-1}N.$$

The implied constant depends only upon the implied constants in (9).

Proof. See ([5], Theorem 2.9). □

The following lemma is well known.

Lemma 4. *Let the function* $f(x)$ *be continuously differentiable, non-negative, and monotonic in the interval* $[a, b]$. *Then*

$$\left|\sum_{a < n \leq b} f(n) - \int_a^b f(x)\,dx\right| \leq \max\big(f(a), f(b)\big).$$

4. Beginning of the Proof

Using (1), (5), and the characteristic function of cube-free numbers

$$\sum_{d^3|n} \mu(d) = \begin{cases} 0, & \exists\, m \text{ s.t. } m^3|n, \\ 1, & \text{otherwise,} \end{cases}$$

we write

$$\begin{aligned} S_c(x) &= \sum_{\Delta_1 \le n < \Delta_2} \sum_{d^3|[n^c \tan^\theta(\log n)]} \mu(d) \\ &= \sum_{d \le \sqrt[3]{\Delta_2^c \tan^\theta(\log \Delta_2)}} \mu(d) \sum_{\substack{\Delta_1 \le n < \Delta_2 \\ [n^c \tan^\theta(\log n)] \equiv 0\,(d^3)}} 1 \\ &= S_c^{(1)}(x) + S_c^{(2)}(x), \end{aligned} \tag{10}$$

where

$$S_c^{(1)}(x) = \sum_{d \le z} \mu(d) \sum_{\substack{\Delta_1 \le n < \Delta_2 \\ [n^c \tan^\theta(\log n)] \equiv 0\,(d^3)}} 1, \tag{11}$$

$$S_c^{(2)}(x) = \sum_{z < d \le \sqrt[3]{\Delta_2^c \tan^\theta(\log \Delta_2)}} \mu(d) \sum_{\substack{\Delta_1 \le n < \Delta_2 \\ [n^c \tan^\theta(\log n)] \equiv 0\,(d^3)}} 1. \tag{12}$$

We shall estimate $S_c^{(1)}(x)$ and $S_c^{(2)}(x)$, respectively, in Sections 5 and 6. In Section 7, we shall finalize the proof of Theorem 1.

5. Estimation of $\mathbf{S_c^{(1)}(x)}$

From (11), we obtain

$$\begin{aligned} S_c^{(1)}(x) = \sum_{d \le z} \mu(d) & \sum_{(\Delta_1^c \tan^\theta(\log \Delta_1) - 1) d^{-3} < k < \Delta_2^c \tan^\theta(\log \Delta_2) d^{-3}} \\ & \times \sum_{\substack{\Delta_1 \le n < \Delta_2 \\ [n^c \tan^\theta(\log n)] = kd^3}} 1 \\ = \sum_{d \le z} \mu(d) & \sum_{(\Delta_1^c \tan^\theta(\log \Delta_1) - 1) d^{-3} < k < \Delta_2^c \tan^\theta(\log \Delta_2) d^{-3}} \end{aligned}$$

$$\times \sum_{\substack{\Delta_1 \le n < \Delta_2 \\ kd^3 \le n^c \tan^\theta(\log n) < kd^3+1}} 1$$

$$= \sum_{d \le z} \mu(d) \sum_{(\Delta_1^c \tan^\theta(\log \Delta_1)-1)d^{-3} < k < \Delta_2^c \tan^\theta(\log \Delta_2) d^{-3}}$$

$$\times \sum_{\Delta_1' \le n < \Delta_2'} 1 + \mathcal{O}(1)$$

$$= \sum_{d \le z} \mu(d) \sum_{(\Delta_1^c \tan^\theta(\log \Delta_1)-1)d^{-3} < k < \Delta_2^c \tan^\theta(\log \Delta_2) d^{-3}}$$

$$\times ([-\Delta_k'] - [-\Delta_k'']) + \mathcal{O}(1)$$

$$= S_c^{(3)}(x) + S_c^{(4)}(x) + \mathcal{O}(1), \tag{13}$$

where

$$[\Delta_k', \Delta_k'') \subset [\Delta_1, \Delta_2) \tag{14}$$

and the interval $[\Delta_k', \Delta_k'')$ is a solution of the system inequalities

$$\left| \begin{array}{l} n^c \tan^\theta(\log n) \ge kd^3 \\ n^c \tan^\theta(\log n) < kd^3 + 1 \end{array} \right. \tag{15}$$

where

$$S_c^{(3)}(x) = \sum_{d \le z} \mu(d) \sum_{(\Delta_1^c \tan^\theta(\log \Delta_1)-1)d^{-3} < k < \Delta_2^c \tan^\theta(\log \Delta_2) d^{-3}} (\Delta_k'' - \Delta_k'), \tag{16}$$

$$S_c^{(4)}(x) = \sum_{d \le z} \mu(d) \sum_{(\Delta_1^c \tan^\theta(\log \Delta_1)-1)d^{-3} < k < \Delta_2^c \tan^\theta(\log \Delta_2) d^{-3}}$$

$$\times \left(\psi(-\Delta_k'') - \psi(-\Delta_k')\right). \tag{17}$$

5.1. *Lower bound for* $\mathbf{S_c^{(3)}(x)}$

Consider the function $t(y)$ defined by

$$t = y^c \tan^\theta(\log y), \tag{18}$$

$$y \in [\Delta_1', \Delta_2']. \tag{19}$$

The first derivative of y as implicit function of t is

$$y' = \frac{y^{1-c}}{\big(c\tan(\log y) + \theta\sec^2(\log y)\big)\tan^{\theta-1}(\log y)}. \tag{20}$$

Using (15), (16), (18), Lemma 4, change of variables, and the mean-value theorem, we get

$$\begin{aligned}
S_c^{(3)}(x) &= \sum_{d\le z}\mu(d) \sum_{(\Delta_1^c\tan^\theta(\log\Delta_1)-1)d^{-3}<k<\Delta_2^c\tan^\theta(\log\Delta_2)d^{-3}} \\
&\quad\times\big(y(kd^3+1)-y(kd^3)\big) \\
&= \sum_{d\le z}\mu(d)\int_{(\Delta_1^c\tan^\theta(\log\Delta_1)-1)d^{-3}}^{\Delta_2^c\tan^\theta(\log\Delta_2)d^{-3}} \big(y(ud^3+1)-y(ud^3)\big)\,du \\
&\quad+\mathcal{O}\Big(z\Big(y(\Delta_1^c\tan^\theta(\log\Delta_1))-y(\Delta_1^c\tan^\theta(\log\Delta_1)-1)\Big)\Big) \\
&= \sum_{d\le z}\frac{\mu(d)}{d^3}\int_{\Delta_1^c\tan^\theta(\log\Delta_1)-1}^{\Delta_2^c\tan^\theta(\log\Delta_2)} \big(y(z+1)-y(z)\big)\,dz + \mathcal{O}(z|y'(\xi)|) \\
&= \sum_{d\le z}\frac{\mu(d)}{d^3}\int_{\Delta_1^c\tan^\theta(\log\Delta_1)-1}^{\Delta_2^c\tan^\theta(\log\Delta_2)} y'(\zeta_z)\,dz + \mathcal{O}(z|y'(\xi)|),
\end{aligned} \tag{21}$$

where

$$\left|\begin{array}{l}
y(\xi)\asymp\xi^\gamma \\
y(\zeta_z)\asymp\zeta_z^\gamma \\
z<\zeta_z<z+1 \\
\Delta_1^c\tan^\theta(\log\Delta_1)-1<\xi<\Delta_1^c\tan^\theta(\log\Delta_1).
\end{array}\right. \tag{22}$$

Now (1), (3), (4), (6), (14), (18)–(22), and the well-known asymptotic formula

$$\sum_{d\le z}\frac{\mu(d)}{d^3} = \frac{1}{\zeta(3)} + \mathcal{O}\left(\frac{1}{z^2}\right)$$

give us

$$S_c^{(3)}(x) \gg x. \tag{23}$$

5.2. *Upper bound for* $S_c^{(4)}(x)$

It is easy to see that (3) and (4) imply

$$\frac{2(\Delta_1^c \tan^\theta(\log \Delta_1) - 1)}{d^3} < \frac{\Delta_2^c \tan^\theta(\log \Delta_2)}{d^3}. \tag{24}$$

Taking into account (17) and (24), we split the range of k into dyadic subintervals to deduce

$$S_c^{(4)}(x) \ll (\log x) \sum_{d \le z} \left| S_c^{(5)}(x) \right|, \tag{25}$$

where

$$S_c^{(5)}(K) = \sum_{K < k \le 2K} \left(\psi(-\Delta_k'') - \psi(-\Delta_k') \right) \tag{26}$$

and where

$$\left| \begin{array}{l} K \asymp \dfrac{x^c}{d^3} \\ \dfrac{\Delta_1^c \tan^\theta(\log \Delta_1) - 1}{d^3} \le K \le \dfrac{\Delta_2^c \tan^\theta(\log \Delta_2)}{2d^3} \end{array} \right. . \tag{27}$$

Now (2), (26), and Lemma 2 yield

$$S_c^{(5)}(K) = S_c^{(6)}(K) + S_c^{(7)}(K) + S_c^{(8)}(K), \tag{28}$$

where

$$S_c^{(6)}(K) = \sum_{K < k \le 2K} \sum_{1 \le |h| \le H} a(h) \left(e(-h\Delta_k'') - e(-h\Delta_k') \right), \tag{29}$$

$$S_c^{(7)}(K) \ll \sum_{K < k \le 2K} \sum_{|h| \le H} b(h) e(-h\Delta_k'), \tag{30}$$

$$S_c^{(8)}(K) \ll \sum_{K < k \le 2K} \sum_{|h| \le H} b(h) e(-h\Delta_k''). \tag{31}$$

We first estimate $S_c^{(7)}(K)$. According to (14), (15), (27), and (30), the variable Δ_k' is an implicit function of k defined by

$$y^c \tan^\theta(\log y) = kd^3 \quad \text{for} \quad k \in (K, 2K) \tag{32}$$

and

$$y \subset [\Delta_1, \Delta_2). \tag{33}$$

Bearing in mind (3), (4), (18), (20), (27), (32), and (33), we find

$$y' \asymp K^{\gamma-1} d^{3\gamma}. \tag{34}$$

Proceeding in the same way, we derive

$$y'' \asymp K^{\gamma-2} d^{3\gamma}. \tag{35}$$

Now (8), (27), (30), (34), (35), and Lemma 3 with $q = 0$ lead to

$$\begin{aligned} S_c^{(7)}(K) &\ll \frac{K}{H} + \frac{1}{H} \sum_{1 \le h \le H} \left| \sum_{K < k \le 2K} e\big(- h\Delta'_k\big) \right| \\ &\ll \frac{K}{H} + \frac{1}{H} \sum_{1 \le h \le H} \left(h^{\frac{1}{2}} K^{\frac{\gamma}{2}} d^{\frac{3\gamma}{2}} + h^{-1} K^{1-\gamma} d^{-3\gamma} \right) \\ &\ll KH^{-1} + K^{\frac{\gamma}{2}} H^{\frac{1}{2}} d^{\frac{3\gamma}{2}} + K^{1-\gamma} H^{-1} d^{-3\gamma} \log H \\ &\ll x^c H^{-1} d^{-3} + x^{\frac{1}{2}} H^{\frac{1}{2}} + x^{c-1} H^{-1} d^{-3} \log H \\ &\ll x^c H^{-1} d^{-3} + x^{\frac{1}{2}} H^{\frac{1}{2}}. \end{aligned} \tag{36}$$

Arguing in the same way for the sum defined by (31), we get

$$S_c^{(8)}(K) \ll x^c H^{-1} d^{-3} + x^{\frac{1}{2}} H^{\frac{1}{2}}. \tag{37}$$

It remains to estimate the sum $S_c^{(6)}(K)$. By (8) and (29), we have

$$S_c^{(6)}(K) \ll \sum_{1 \le |h| \le H} \frac{1}{h} \left| \sum_{K < k \le 2K} \Phi_h(k) e\big(- h\Delta'_k\big) \right|, \tag{38}$$

where

$$\Phi_h(t) = e\big(h(\Delta'_t - \Delta''_t)\big) - 1. \tag{39}$$

From (3), (4), (18), (20), (27), (32), (33), (39), and the mean-value theorem, it follows

$$|\Phi_h(t)| \le 2\big|\sin\big(\pi h(\Delta'_t - \Delta''_t)\big)\big| \ll |h||\Delta'_t - \Delta''_t| \ll |h| x^{1-c}, \tag{40}$$

$$\Phi'_h(t) = 2\pi i h \left(\frac{\partial \Delta'_t}{\partial t} - \frac{\partial \Delta''_t}{\partial t} \right) e\big(h(\Delta'_t - \Delta''_t)\big) \ll |h| x^{1-2c} \tag{41}$$

for $t \in [K, 2K]$. Now (27), (32), (38), (40), (41), and Abel's summation formula give us

$$\begin{aligned} S_c^{(6)}(K) &\ll \sum_{1\le|h|\le H} \frac{1}{h}\left|\Phi_h(2K)\sum_{K<k\le 2K} e\big(-h\Delta_k'\big)\right| \\ &\quad + \sum_{1\le|h|\le H}\frac{1}{h}\int_K^{2K}\left|\Phi_h'(t)d^2\sum_{K<k\le t} e\big(-h\Delta_k'\big)\right| dt \\ &\ll x^{1-c}\big|S_c^{(9)}(K_1)\big|, \end{aligned} \tag{42}$$

where

$$S_c^{(9)}(K_1) = \sum_{1\le h\le H}\left|\sum_{K<k\le K_1} e\big(h\Delta_k'\big)\right| \tag{43}$$

for some number $K_1 \in (K, 2K]$. By (27), (34), (35), (43), and Lemma 3 with $q = 0$, we obtain

$$\begin{aligned} S_c^{(9)}(K_1) &\ll \sum_{1\le h\le H}\left(h^{\frac12}K^{\frac{\gamma}{2}}d^{\frac{3\gamma}{2}} + h^{-1}K^{1-\gamma}d^{-3\gamma}\right) \\ &\ll K^{\frac{\gamma}{2}}H^{\frac32}d^{\frac{3\gamma}{2}} + K^{1-\gamma}d^{-3\gamma}\log H \\ &\ll x^{\frac12}H^{\frac32} + x^{c-1}d^{-3}\log H. \end{aligned} \tag{44}$$

Put

$$H = \frac{x^{\frac{4c-3}{5}}}{d^2}. \tag{45}$$

Now (42), (44), and (45) imply

$$S_c^{(6)}(K) \ll x^{\frac{3-2c}{2}}H^{\frac32} + d^{-3}\log H \ll x^{\frac{3-2c}{2}}H^{\frac32}. \tag{46}$$

Taking into account (1), (25), (28), (36), (37), (45), and (46), we deduce

$$\begin{aligned} S_c^{(4)}(x) &\ll (\log x)\sum_{d\le z}\left(x^c H^{-1}d^{-3} + x^{\frac12}H^{\frac12} + x^{\frac{3-2c}{2}}H^{\frac32}\right) \\ &\ll x^{\frac{c+3}{5}}\log^2 x. \end{aligned} \tag{47}$$

5.3. *Lower bound for* $\mathbf{S_c^{(1)}(x)}$

Bearing in mind (13), (23), and (47), we derive

$$S_c^{(1)}(x) \gg x. \tag{48}$$

6. Upper Bound for $\mathbf{S_c^{(2)}(x)}$

By (1), (4), and (12), we obtain

$$\begin{aligned} S_c^{(2)}(x) &\ll \sum_{z<d\leq \sqrt[3]{\Delta_2^c \tan^\theta(\log \Delta_2)}} \sum_{\substack{\Delta_1\leq n<\Delta_2 \\ [n^c \tan^\theta(\log n)]\equiv 0\,(d^3)}} 1 \\ &\ll \sum_{z<d\leq \sqrt[3]{\Delta_2^c \tan^\theta(\log \Delta_2)}} \sum_{k\leq \frac{\Delta_2^c \tan^\theta(\log \Delta_2)}{d^3}} 1 \\ &\ll x^c \sum_{z<d\leq \sqrt[3]{\Delta_2^c \tan^\theta(\log \Delta_2)}} \frac{1}{d^3} \ll x^c z^{-2} \\ &= x^{\frac{c+3}{5}}. \end{aligned} \tag{49}$$

7. The End of the Proof

Summarizing (10), (48), and (49), we establish the lower bound (7). This completes the proof of Theorem 1.

References

[1] Zhang, J.L.M. Distribution of cube-free numbers with form $[n^c]$, *Front. Math. China.* 12, 1515–1525 (2017).

[2] Deshouillers, J.M. A remark on cube-free numbers in segal-piatetski-shapiro sequences, *Hardy-Ramanujan J.* 41, 127–132 (2018).

[3] Dimitrov. S.I. Prime numbers of the form $[n^c \tan^\theta(\log n)]$ *Indian J. Pure Appl. Math.* 55, 1198–1209 (2024).

[4] Vaaler, J.D. Some extremal functions in fourier analysis, *Bull. Amer. Math. Soc.* 12, 183–216 (1985).

[5] Graham, S.W. and Kolesnik, G. *Van der Corput's Method of Exponential Sums.* Cambridge University Press, New York (1991).

https://doi.org/10.1142/9789811293009_0011

Chapter 11

Some Inequalities of Trapezoid Type for Double Integral Mean of Absolutely Continuous Functions

Silvestru Sever Dragomir*

Applied Mathematics Research Group, ISILC,
Victoria University, Melbourne City,
MC 8001, Australia
Mathematical Sciences, School of Science, RMIT University,
GPO Box 2476V, Melbourne, Victoria 3001, Australia
**sever.dragomir@vu.edu.au*

In this chapter, we establish some Ostrowski-type inequalities for double integral mean of absolutely continuous functions.

1. Introduction

In 1999, Cerone and Dragomir [9] obtained the following *generalized trapezoid inequality* for absolutely continuous functions

*This chapter is dedicated to my granddaughters Audrey and Sienna.

$f:[a,b]\to\mathbb{C}$

$$\left|\left(\frac{x-a}{b-a}\right)f(a)+\left(\frac{b-x}{b-a}\right)f(b)-\frac{1}{b-a}\int_a^b f(t)\,dt\right|$$

$$\leq\begin{cases}\left[\frac{1}{4}+\left(\frac{x-\frac{a+b}{2}}{b-a}\right)^2\right](b-a)\|f'\|_{[a,b],\infty} & \text{if } f'\in L_\infty[a,b],\\ \frac{1}{(q+1)^{1/q}}\left[\left(\frac{x-a}{b-a}\right)^{q+1}\right. & \text{if } f'\in L_p[a,b],\\ \left.+\left(\frac{b-x}{b-a}\right)^{q+1}\right]^{1/q}(b-a)\|f'\|_{[a,b],p} & p>1,\ \frac{1}{p}+\frac{1}{q}=1,\\ \left[\frac{1}{2}+\left|\frac{x-\frac{a+b}{2}}{b-a}\right|\right](b-a)\|f'\|_{[a,b],1},\end{cases}\tag{1}$$

where the *Lebesgue norms* $\|\cdot\|_{[a,b],p}$ are defined by

$$\|g\|_{[a,b],p}:=\begin{cases}\operatorname*{essup}_{t\in[a,b]}|g(t)| & \text{if } f'\in L_\infty[a,b],\\ \left(\int_a^b|g(t)|^p\,dt\right)^{1/p} & \text{if } f'\in L_p[a,b],\ p\geq 1.\end{cases}$$

The best inequality one can get from (1) is for $x=\frac{a+b}{2}$, namely,

$$\left|\frac{f(a)+f(b)}{2}-\frac{1}{b-a}\int_a^b f(t)\,dt\right|$$

$$\leq\begin{cases}\frac{1}{4}(b-a)\|f'\|_{[a,b],\infty} & \text{if } f'\in L_\infty[a,b],\\ \frac{1}{2^{1/q}(q+1)^{1/q}}(b-a)^{1/q}\|f'\|_{[a,b],p} & \text{if } f'\in L_p[a,b],\\ p>1,\ \frac{1}{p}+\frac{1}{q}=1, & \\ \frac{1}{2}\|f'\|_{[a,b],1}, & \end{cases}\tag{2}$$

with $\frac{1}{4}$, $\frac{1}{2^{1/q}}$, and $\frac{1}{2}$ as the best possible constants.

For some related trapezoid-type inequalities, see [1–8], [10–14], and [16–23].

For the integrable function $f : [a, b] \to \mathbb{C}$, we consider the *double integral mean defined* by

$$\frac{1}{(b-a)^2} \int_a^b \int_a^b f\left(\frac{t+s}{2}\right) dtds.$$

Motivated by the trapezoid inequality above, it is thus natural to ask what is the distance between the double integral mean and the convex combination of two values, $f(x)$ and $f(y)$, with $x, y \in [a, b]$, in general, and $f(a)$ and $f(b)$, in particular.

Some answers for the absolutely continuous functions whose derivatives are essentially bounded or p-Lebesgue integrable are provided in the following.

2. Some Preliminary Facts

We need the following preliminary facts.

Lemma 1. *If $f : I \to \mathbb{C}$ is locally absolutely continuous on $\mathring{I}$, then for each distinct z, x, $y \in \mathring{I}$ and $\lambda \in \mathbb{R} \setminus \{0, 1\}$, we have*

$$f(z) = (1-\lambda) f(x) + \lambda f(y) + S_\lambda(z, x, y), \tag{3}$$

where the remainder $S_\lambda(z, x, y)$ is given by

$$\begin{aligned} S_\lambda(z, x, y) &:= (1-\lambda)(z-x) \int_0^1 f'((1-s)x + sz)\, ds \\ &\quad - \lambda(y-z) \int_0^1 f'((1-s)z + sy)\, ds. \end{aligned} \tag{4}$$

Proof. For any t, $s \in I$, $s \neq t$, one has, see [12],

$$\frac{f(s) - f(t)}{s-t} = \frac{1}{s-t} \int_t^s f'(u)\, du = \int_0^1 f'[(1-\lambda)s + \lambda t]\, d\lambda,$$

showing that

$$f(s) = f(t) + (s-t) \int_0^1 f'[(1-\lambda)s + \lambda t]\, d\lambda \tag{5}$$

for all t, $s \in I$.

Then, by (5), we have

$$f(z) = f(x) + (z - x) \int_0^1 f'[(1 - \lambda) z + \lambda x] \, d\lambda \tag{6}$$

and

$$f(z) = f(y) + (z - y) \int_0^1 f'[(1 - \lambda) z + \lambda y] \, d\lambda. \tag{7}$$

If we multiply (6) by $1 - \lambda$ and (7) by λ and add the obtained equalities, then we get equality (3), where the remainder is given by (4). □

The above identity (3) produces the following simple representations of interest.

We have for each distinct z, x, $y \in \mathring{I}$ that

$$f(z) = \frac{1}{y - x} [(y - z) f(x) + (z - x) f(y)] + L(z, x, y), \tag{8}$$

where

$$L(z, x, y) := \frac{(y - z)(z - x)}{y - x} \left[\int_0^1 f'((1 - s) x + sz) \, ds - \int_0^1 f'((1 - s) z + sy) \, ds \right] \tag{9}$$

and

$$f(z) = \frac{1}{y - x} [(z - x) f(x) + (y - z) f(y)] + P(z, x, y), \tag{10}$$

where

$$P(z, x, y) := \frac{1}{y - x} \left[(z - x)^2 \int_0^1 f'((1 - s) x + sz) \, ds - (y - z)^2 \int_0^1 f'((1 - s) z + sy) \, ds \right]. \tag{11}$$

We also have

$$f((1 - \lambda) x + \lambda y) = (1 - \lambda) f(x) + \lambda f(y) + S_\lambda(x, y), \tag{12}$$

where the remainder $S_\lambda(x,y)$ is given by

$$S_\lambda(x,y) := (1-\lambda)\lambda(y-x)\left[\int_0^1 f'((1-s\lambda)x+s\lambda y)\,ds - \int_0^1 f'((1-s-\lambda+s\lambda)x+(\lambda+s-s\lambda)y)\,ds\right] \quad (13)$$

and

$$f((1-\lambda)y+\lambda x) = (1-\lambda)f(x)+\lambda f(y)+P_\lambda(x,y), \quad (14)$$

where the remainder $P_\lambda(x,y)$ is given by

$$P_\lambda(x,y) := (y-x)\left[(1-\lambda)^2\int_0^1 f'((1-s+\lambda s)x+(1-\lambda)sy)\,ds - \lambda^2\int_0^1 f'((1-s)\lambda x+(1-\lambda+\lambda s)y)\,ds\right]. \quad (15)$$

Moreover, if we take in (3) $z=\frac{x+y}{2}$ for each distinct $x, y \in \mathring{I}$ and $\lambda\in\mathbb{R}\setminus\{0,1\}$, then we have

$$f\left(\frac{x+y}{2}\right) = (1-\lambda)f(x)+\lambda f(y)+S_\lambda(x,y), \quad (16)$$

where the remainder $S_\lambda(x,y)$ is given by

$$S_\lambda(x,y) := \frac{1}{2}(y-x)\left[(1-\lambda)\int_0^1 f'\left((1-s)x+s\frac{x+y}{2}\right)ds - \lambda\int_0^1 f'\left((1-s)\frac{x+y}{2}+sy\right)ds\right]. \quad (17)$$

In particular, for $\lambda - \frac{1}{2}$, we have

$$f\left(\frac{x+y}{2}\right) = \frac{f(x)+f(y)}{2}+S(x,y), \quad (18)$$

where

$$S(x,y) := \frac{1}{4}(y-x)\left[\int_0^1 f'\left((1-s)x+s\frac{x+y}{2}\right)ds - \int_0^1 f'\left((1-s)\frac{x+y}{2}+sy\right)ds\right]. \quad (19)$$

We also need some technical identities that are used to obtain the main results. They have been obtained in [15], however, for the sake of completeness, we provide here their proofs.

We recall the function *sign* defined by

$$\operatorname{sgn}(x) := \begin{cases} -1 & \text{if } x < 0, \\ 0 & \text{if } x = 0, \\ 1 & \text{if } x < 0. \end{cases}$$

The following simple lemma holds.

Lemma 2. *We have for any $a < b$, $d \in \mathbb{R}$, and $p > 0$ that*

$$\begin{aligned} &\int_a^b |x-d|^p \, dx \\ &\quad = \frac{1}{p+1}\left[\operatorname{sgn}(b-d)\,|b-d|^{p+1} + \operatorname{sgn}(d-a)\,|d-a|^{p+1}\right] \\ &\quad = \frac{1}{p+1}\left[(b-d)\,|b-d|^{p} + (d-a)\,|d-a|^{p}\right]. \end{aligned} \tag{20}$$

Proof. If $d \le a$, then

$$\begin{aligned} \int_a^b |x-d|^p \, dx = \int_a^b (x-d)^p \, dx &= \frac{1}{p+1}\left[(b-d)^{p+1} - (a-d)^{p+1}\right] \\ &= \left[\operatorname{sgn}(b-d)\,|b-d|^{p+1} + \operatorname{sgn}(d-a)\,|d-a|^{p+1}\right]. \end{aligned}$$

If $d \in [a,b]$, then

$$\begin{aligned} &\int_a^b |x-d|^p \, dx \\ &\quad = \int_a^d (d-x)^p \, dx + \int_d^b (x-d)^p \, dx \\ &\quad = \frac{1}{p+1}\left[(d-a)^{p+1} + (b-d)^{p+1}\right] \\ &\quad = \frac{1}{p+1}\left[\operatorname{sgn}(b-d)\,|b-d|^{p+1} + \operatorname{sgn}(d-a)\,|d-a|^{p+1}\right]. \end{aligned}$$

If $d \geq b$, then

$$\begin{aligned}
&\int_a^b |x-d|^p \, dx \\
&\quad = \int_a^b (d-x)^p \, dx = \frac{1}{p+1}\left[-(d-b)^{p+1} + (d-a)^{p+1}\right] \\
&\quad = \frac{1}{p+1}\left[\operatorname{sgn}(b-d)\,|b-d|^{p+1} + \operatorname{sgn}(d-a)\,|d-a|^{p+1}\right]
\end{aligned}$$

and the first equality in (20) is thus proved.

The second part follows by the fact that

$$x = \operatorname{sgn}(x)\,|x| \quad \text{for } x \in \mathbb{R}. \qquad \square$$

Further, we have the following representation as well.

Lemma 3. *We have for any $a < b$, $s \in [a,b]$, and $p > 0$ that*

$$\begin{aligned}
&\int_a^b \int_a^b \left|\frac{x+y}{2} - s\right|^p dxdy \\
&\quad = \frac{4}{(p+1)(p+2)}\left[(b-s)^{p+2} - 2\left|s - \frac{a+b}{2}\right|^{p+2} + (s-a)^{p+2}\right].
\end{aligned} \tag{21}$$

In particular, we have

$$\begin{aligned}
\int_a^b \int_a^b \left(\frac{x+y}{2} - a\right)^p dxdy &= \int_a^b \int_a^b \left(b - \frac{x+y}{2}\right)^p dxdy \\
&= \frac{2^{p+1}-1}{2^{p-1}(p+1)(p+2)}(b-a)^{p+2}
\end{aligned} \tag{22}$$

and

$$\int_a^b \int_a^b \left|\frac{x+y}{2} - \frac{a+b}{2}\right|^p dxdy = \frac{1}{2^{p-1}(p+1)(p+2)}(b-a)^{p+2}. \tag{23}$$

Proof. We denote

$$I_p(s) := \int_a^b \int_a^b \left|\frac{x+y}{2} - s\right|^p dxdy = \int_a^b \left(\int_a^b \left|\frac{x+y}{2} - s\right|^p dy\right) dx.$$

If we make the change of variable $z = \frac{1}{2}(x+y)$, where $y \in [a,b]$, then we have

$$dz = \frac{1}{2}dy, \ z \in \left[\frac{1}{2}(x+a), \frac{1}{2}(x+b)\right]$$

and

$$\begin{aligned} I_p(s) &= \int_a^b \left(\int_a^b \left|\frac{x+y}{2} - s\right|^p dy\right) dx \\ &= 2\int_a^b \left(\int_{\frac{1}{2}(x+a)}^{\frac{1}{2}(x+b)} |z-s|^p dz\right) dx. \end{aligned} \tag{24}$$

Using the representation (20), we have

$$\begin{aligned} \int_{\frac{1}{2}(x+a)}^{\frac{1}{2}(x+b)} |z-s|^p dz = \frac{1}{p+1}&\left[\left(\frac{x+b}{2} - s\right)\left|\frac{x+b}{2} - s\right|^p\right. \\ &\left.+ \left(s - \frac{x+a}{2}\right)\left|s - \frac{x+a}{2}\right|^p\right] \end{aligned} \tag{25}$$

for $s, x \in [a,b]$, and by (24), we get

$$\begin{aligned} I_p(s) = \frac{2}{p+1}\int_a^b &\left[\left(\frac{x+b}{2} - s\right)\left|\frac{x+b}{2} - s\right|^p\right. \\ &\left.+ \left(s - \frac{x+a}{2}\right)\left|s - \frac{x+a}{2}\right|^p\right] dx \end{aligned}$$

for $s \in [a,b]$.

We consider

$$I_{1,p}(s) := \int_a^b \left|\frac{x+b}{2} - s\right|^p \left(\frac{x+b}{2} - s\right) dx$$

and

$$I_{2,p}(s) := \int_a^b \left|s - \frac{x+a}{2}\right|^p \left(s - \frac{x+a}{2}\right) dx$$

for $s \in [a,b]$.

(a) For $s \in \left[a, \frac{a+b}{2}\right]$, we have

$$\frac{x+b}{2} - s \geq \frac{a+b}{2} - s \geq 0 \quad \text{for } x \in [a,b],$$

then

$$\begin{aligned} I_{1,p}(s) &= \int_a^b \left(\frac{x+b}{2} - s\right)^p \left(\frac{x+b}{2} - s\right) dx \\ &= \int_a^b \left(\frac{x+b}{2} - s\right)^{p+1} dx \\ &= \frac{2}{p+2}\left[(b-s)^{p+2} - \left(\frac{a+b}{2} - s\right)^{p+2}\right] \end{aligned}$$

for $s \in \left[a, \frac{a+b}{2}\right]$.

We have $s - \frac{x+a}{2} = 0$ for $x = 2s - a \in [a,b]$. Then

$$\begin{aligned} I_{2,p}(s) &= \int_a^b \left|s - \frac{x+a}{2}\right|^p \left(s - \frac{x+a}{2}\right) dx \\ &= \int_a^{2s-a} \left(s - \frac{x+a}{2}\right)^p \left(s - \frac{x+a}{2}\right) dx \\ &\quad + \int_{2s-a}^b \left(\frac{x+a}{2} - s\right)^p \left(s - \frac{x+a}{2}\right) dx \\ &= \int_a^{2s-a} \left(s - \frac{x+a}{2}\right)^{p+1} dx - \int_{2s-a}^b \left(\frac{x+a}{2} - s\right)^{p+1} dx \\ &= 2\frac{(s-a)^{p+2}}{p+2} - 2\frac{\left(\frac{b+a}{2} - s\right)^{p+2}}{p+2} \\ &= \frac{2}{p+2}\left[(s-a)^{p+2} - \left(\frac{b+a}{2} - s\right)^{p+2}\right] \end{aligned}$$

for $s \in \left[a, \frac{a+b}{2}\right]$.

In conclusion, for $s \in \left[a, \frac{a+b}{2}\right]$, we get

$$
\begin{aligned}
I_p(s) &= \frac{2}{p+1}\left[\frac{2}{p+2}\left[(b-s)^{p+2} - \left(\frac{a+b}{2} - s\right)^{p+2}\right]\right. \\
&\quad \left. + \frac{2}{p+2}\left[(s-a)^{p+2} - \left(\frac{b+a}{2} - s\right)^{p+2}\right]\right] \\
&= \frac{4}{(p+1)(p+2)}\left[(b-s)^{p+2} - 2\left(\frac{a+b}{2} - s\right)^{p+2}\right. \\
&\quad \left. + (s-a)^{p+2}\right]. \qquad (26)
\end{aligned}
$$

(b) Assume that $s \in \left[\frac{a+b}{2}, b\right]$. We have $\frac{x+b}{2} - s = 0$ for $x = 2s - b \in [a, b]$. Then

$$
\begin{aligned}
I_{1,p}(s) &= \int_a^b \left|\frac{x+b}{2} - s\right|^p \left(\frac{x+b}{2} - s\right) dx \\
&= \int_a^{2s-b} \left(s - \frac{x+b}{2}\right)^p \left(\frac{x+b}{2} - s\right) dx \\
&\quad + \int_{2s-b}^b \left(\frac{x+b}{2} - s\right)^p \left(\frac{x+b}{2} - s\right) dx \\
&= -\int_a^{2s-b} \left(s - \frac{x+b}{2}\right)^{p+1} dx + \int_{2s-b}^b \left(\frac{x+b}{2} - s\right)^{p+1} dx \\
&= -\frac{2}{p+2}\left(s - \frac{a+b}{2}\right)^{p+2} + \frac{2}{p+2}(b-s)^{p+2} \\
&= \frac{2}{p+2}\left[(b-s)^{p+2} - \left(s - \frac{a+b}{2}\right)^{p+2}\right]
\end{aligned}
$$

for $s \in \left[\frac{a+b}{2}, b\right]$.

If $s \in \left[\frac{a+b}{2}, b\right]$, then we have

$$
s - \frac{x+a}{2} \geq \frac{a+b}{2} - \frac{x+a}{2} = \frac{b-x}{2} \geq 0
$$

for $x \in [a,b]$ and then

$$\begin{aligned} I_{2,p}(s) &= \int_a^b \left| s - \frac{x+a}{2} \right|^p \left(s - \frac{x+a}{2} \right) dx \\ &= \int_a^b \left(s - \frac{x+a}{2} \right)^{p+1} dx \\ &= -2\frac{\left(s - \frac{b+a}{2}\right)^{p+2}}{p+2} + 2\frac{(s-a)^{p+2}}{p+2} \\ &= \frac{2}{p+2}\left[(s-a)^{p+2} - \left(s - \frac{b+a}{2} \right)^{p+2} \right] \end{aligned}$$

for $s \in \left[\frac{a+b}{2}, b\right]$.

Therefore,

$$\begin{aligned} I_p(s) &= \frac{2}{p+1}\left[\frac{2}{p+2}\left[(b-s)^{p+2} - \left(s - \frac{a+b}{2} \right)^{p+2} \right] \right. \\ &\quad \left. + \frac{2}{p+2}\left[(s-a)^{p+2} - \left(s - \frac{b+a}{2} \right)^{p+2} \right] \right] \\ &= \frac{4}{(p+1)(p+2)}\left[(b-s)^{p+2} - 2\left(s - \frac{a+b}{2} \right)^{p+2} \right. \\ &\quad \left. + (s-a)^{p+2} \right] \end{aligned} \tag{27}$$

for $s \in \left[\frac{a+b}{2}, b\right]$.

By utilizing (26) and (27), we get the desired result (21). □

Corollary 4. *With the assumptions of Lemma 3, we have*

$$\begin{aligned} &\int_a^b \int_a^b \int_a^b \left| \frac{x+y}{2} - s \right|^p dxdyds \\ &\quad = \frac{2^{p+2}-1}{2^{p-1}(p+1)(p+2)(p+3)}(b-a)^{p+3}. \end{aligned} \tag{28}$$

Proof. We observe that

$$\int_a^b (b-s)^{p+2}\,ds = \int_a^b (s-a)^{p+2}\,ds = \frac{(b-a)^{p+3}}{p+3}$$

and

$$\begin{aligned}\int_a^b \left|s-\frac{a+b}{2}\right|^{p+2} ds &= 2\int_{\frac{a+b}{2}}^b \left(s-\frac{a+b}{2}\right)^{p+2} ds \\ &= \frac{1}{2^{p+2}(p+3)}(b-a)^{p+3},\end{aligned}$$

therefore,

$$\begin{aligned}&\int_a^b \left[(b-s)^{p+2} - 2\left|s-\frac{a+b}{2}\right|^{p+2} + (s-a)^{p+2}\right] ds \\ &= \frac{2(b-a)^{p+3}}{p+3} - \frac{2}{2^{p+2}(p+3)}(b-a)^{p+3} \\ &= \frac{2^{p+2}-1}{2^{p+1}(p+3)}(b-a)^{p+3}.\end{aligned}$$

Now, by taking the integral over $s \in [a,b]$ in identity (21), we get (28). □

Remark 5. The case $p = 1$ is of interest in applications and produces the following equalities:

$$\int_a^b\int_a^b \left|\frac{x+y}{2}-s\right| dxdy = \frac{2}{3}\left[(b-s)^3 - 2\left|s-\frac{a+b}{2}\right|^3 + (s-a)^3\right]. \tag{29}$$

In particular, we have

$$\begin{aligned}\int_a^b\int_a^b \left(\frac{x+y}{2}-a\right) dxdy &= \int_a^b\int_a^b \left(b-\frac{x+y}{2}\right) dxdy \\ &= \frac{1}{2}(b-a)^3,\end{aligned} \tag{30}$$

$$\int_a^b\int_a^b \left|\frac{x+y}{2}-\frac{a+b}{2}\right| dxdy = \frac{1}{6}(b-a)^3, \tag{31}$$

and

$$\int_a^b \int_a^b \int_a^b \left| \frac{x+y}{2} - s \right| dxdyds = \frac{1}{8}(b-a)^4. \tag{32}$$

3. Two Point Inequalities for Integral Mean

We have the following.

Theorem 6. *Let $f : [a,b] \to \mathbb{C}$ be an absolutely continuous function on $[a,b]$. Then for any x, $y \in [a,b]$ and $\lambda \in [0,1]$, we have*

$$\begin{aligned} &\left| \frac{1}{b-a} \int_a^b f(t)\, dt - (1-\lambda) f(x) - \lambda f(y) \right| \\ &\quad \le (1-\lambda) \frac{1}{b-a} \int_a^b |t-x| \, \nu_{f'}(x,t)\, dt \\ &\quad\quad + \lambda \frac{1}{b-a} \int_a^b |y-t| \, \nu_{f'}(y,t)\, dt, \end{aligned} \tag{33}$$

where

$$0 \le \nu_{f'}(z,t) := \int_0^1 \left| f'((1-s)z + st) \right| ds, \quad and z,\ t \in [a,b].$$

In particular, we have the λ-weighted trapezoid inequality

$$\begin{aligned} &\left| \frac{1}{b-a} \int_a^b f(t)\, dt - (1-\lambda) f(a) - \lambda f(b) \right| \\ &\quad \le (1-\lambda) \frac{1}{b-a} \int_a^b |t-a| \, \nu_{f'}(a,t)\, dt \\ &\quad\quad + \lambda \frac{1}{b-a} \int_a^b |b-t| \, \nu_{f'}(b,t)\, dt \end{aligned} \tag{34}$$

for any $\lambda \in [0,1]$ and the Ostrowski-type inequality

$$\left| \frac{1}{b-a} \int_a^b f(t)\, dt - f(x) \right| \le \frac{1}{b-a} \int_a^b |t-x| \, \nu_{f'}(x,t)\, dt \tag{35}$$

for any $x \in [a,b]$.

Proof. From Lemma 1, we have

$$\begin{aligned} f(t) &= (1-\lambda) f(x) + \lambda f(y) \\ &\quad + (1-\lambda)(t-x) \int_0^1 f'((1-s)x + st)\, ds \\ &\quad - \lambda (y-t) \int_0^1 f'((1-s)t + sy)\, ds \end{aligned} \tag{36}$$

for any $x, y \in [a,b]$ and $\lambda \in [0,1]$.

Taking the integral mean in (36), we get the following identity of interest:

$$\begin{aligned} &\frac{1}{b-a}\int_a^b f(t)\,dt \\ &= (1-\lambda) f(x) + \lambda f(y) \\ &\quad + (1-\lambda)\frac{1}{b-a}\int_a^b (t-x)\left(\int_0^1 f'((1-s)x+st)\,ds\right)dt \\ &\quad - \lambda\frac{1}{b-a}\int_a^b (y-t)\left(\int_0^1 f'((1-s)t+sy)\,ds\right)dt \end{aligned} \tag{37}$$

for any $x, y \in [a,b]$ and $\lambda \in [0,1]$.

This implies that

$$\begin{aligned} &\left|\frac{1}{b-a}\int_a^b f(t)\,dt - (1-\lambda) f(x) - \lambda f(y)\right| \\ &\le (1-\lambda)\frac{1}{b-a}\left|\int_a^b (t-x)\left(\int_0^1 f'((1-s)x+st)\,ds\right)dt\right| \\ &\quad + \lambda\frac{1}{b-a}\left|\int_a^b (y-t)\left(\int_0^1 f'((1-s)t+sy)\,ds\right)dt\right| \\ &\le (1-\lambda)\frac{1}{b-a}\int_a^b \left|(t-x)\left(\int_0^1 f'((1-s)x+st)\,ds\right)\right|dt \\ &\quad + \lambda\frac{1}{b-a}\int_a^b \left|(y-t)\left(\int_0^1 f'((1-s)t+sy)\,ds\right)\right|dt \end{aligned}$$

$$\le (1-\lambda)\frac{1}{b-a}\int_a^b |t-x|\left(\int_0^1 \left|f'((1-s)x+st)\right| ds\right) dt$$

$$+\lambda\frac{1}{b-a}\int_a^b |y-t|\left(\int_0^1 \left|f'((1-s)t+sy)\right| ds\right) dt$$

$$= (1-\lambda)\frac{1}{b-a}\int_a^b |t-x|\left(\int_0^1 \left|f'((1-s)x+st)\right| ds\right) dt$$

$$+\lambda\frac{1}{b-a}\int_a^b |y-t|\left(\int_0^1 \left|f'((1-s)y+st)\right| ds\right) dt$$

$$= (1-\lambda)\frac{1}{b-a}\int_a^b |t-x|\,\nu_{f'}(x,t)\,dt$$

$$+\lambda\frac{1}{b-a}\int_a^b |y-t|\,\nu_{f'}(x,t)\,dt =: B_\lambda(x,y) \tag{38}$$

for any x, $y\in[a,b]$ and $\lambda\in[0,1]$.

This proves the desired inequality (33). □

Remark 7. We observe that for any bound $M(\cdot,\cdot)$ for the function $\nu_{f'}(\cdot,\cdot)$ on the square $[a,b]^2$, one can obtain the corresponding two-point Ostrowski inequality

$$\left|\frac{1}{b-a}\int_a^b f(t)\,dt-(1-\lambda)f(x)-\lambda f(y)\right|$$

$$\le (1-\lambda)\frac{1}{b-a}\int_a^b |t-x|\,M(x,t)\,dt$$

$$+\lambda\frac{1}{b-a}\int_a^b |y-t|\,M(y,t)\,dt \tag{39}$$

for any x, $y\in[a,b]$ and $\lambda\in[0,1]$.

In the following, we give only a few examples of such inequalities. The interested reader may obtain many more by choosing other bounds for the function $\nu_{f'}(\cdot,\cdot)$ on the square $[a,b]^2$.

For $g\in L_\infty[a,b]$ and any x, $y\in[a,b]$, we use the notation

$$\left|\|g\|_{[x,y],\infty}\right| := \begin{cases}\|g\|_{[x,y],\infty} & \text{if } x<y,\\ \|g\|_{[y,x],\infty} & \text{if } y<x.\end{cases}$$

Corollary 8. *Let $f:[a,b]\to\mathbb{C}$ be an absolutely continuous function on $[a,b]$. Assume that $f'\in L_\infty[a,b]$, then for any x, $y\in[a,b]$ and $\lambda\in[0,1]$, we have*

$$\begin{aligned}
&\left|\frac{1}{b-a}\int_a^b f(t)\,dt-(1-\lambda)f(x)-\lambda f(y)\right|\\
&\le(1-\lambda)\frac{1}{b-a}\int_a^b|t-x|\left|\|f'\|_{[x,t],\infty}\right|dt\\
&\quad+\lambda\frac{1}{b-a}\int_a^b|y-t|\left|\|f'\|_{[y,t],\infty}\right|dt\\
&\le\left[\frac{1}{4}+(1-\lambda)\left(\frac{x-\frac{a+b}{2}}{b-a}\right)^2+\lambda\left(\frac{y-\frac{a+b}{2}}{b-a}\right)^2\right]\\
&\quad\times\|f'\|_{[a,b],\infty}(b-a). \qquad (40)
\end{aligned}$$

In particular, we have the λ-weighted trapezoid inequality

$$\begin{aligned}
&\left|\frac{1}{b-a}\int_a^b f(t)\,dt-(1-\lambda)f(a)-\lambda f(b)\right|\\
&\le(1-\lambda)\frac{1}{b-a}\int_a^b(t-a)\|f'\|_{[t,a],\infty}\,dt\\
&\quad+\lambda\frac{1}{b-a}\int_a^b(b-t)\|f'\|_{[t,b],\infty}\,dt\\
&\le\frac{1}{2}\|f'\|_{[a,b],\infty}(b-a) \qquad (41)
\end{aligned}$$

for any $\lambda\in[0,1]$ and the refinement of Ostrowski inequality

$$\begin{aligned}
&\left|\frac{1}{b-a}\int_a^b f(t)\,dt-f(x)\right|\\
&\le\frac{1}{b-a}\int_a^b|t-x|\left|\|f'\|_{[x,t],\infty}\right|dt\\
&\le\left[\frac{1}{4}+\left(\frac{x-\frac{a+b}{2}}{b-a}\right)^2\right]\|f'\|_{[a,b],\infty}(b-a). \qquad (42)
\end{aligned}$$

Proof. Since $f' \in L_\infty[a,b]$, then by using the notation above, we have

$$\left|f'((1-s)z+st)\right| \le \left|\|f'\|_{[z,t],\infty}\right| \text{ for any } s \in [0,1],$$

which implies that

$$\nu_{f'}(x,t) \le \left|\|f'\|_{[x,t],\infty}\right| \text{ and } \nu_{f'}(y,t) \le \left|\|f'\|_{[y,t],\infty}\right|$$

for any t, x, $y \in [a,b]$.

On using (33), we have

$$\begin{aligned}
&\left|\frac{1}{b-a}\int_a^b f(t)\,dt - (1-\lambda)f(x) - \lambda f(y)\right| \\
&\quad\le (1-\lambda)\frac{1}{b-a}\int_a^b |t-x|\,\nu_{f'}(x,t)\,dt \\
&\qquad + \lambda\frac{1}{b-a}\int_a^b |y-t|\,\nu_{f'}(y,t)\,dt \\
&\quad\le (1-\lambda)\frac{1}{b-a}\int_a^b |t-x|\left|\|f'\|_{[x,t],\infty}\right|dt \\
&\qquad + \lambda\frac{1}{b-a}\int_a^b |y-t|\left|\|f'\|_{[y,t],\infty}\right|dt := C_\lambda(x,y) \qquad (43)
\end{aligned}$$

for any x, $y \in [a,b]$ and $\lambda \in [0,1]$.

Since

$$\left|\|f'\|_{[x,t],\infty}\right|, \quad \left|\|f'\|_{[y,t],\infty}\right| \le \|f'\|_{[a,b],\infty} \quad \text{for any } x,\ y \in [a,b],$$

then

$$\begin{aligned}
&\int_a^b |t-x|\left|\|f'\|_{[x,t],\infty}\right|dt \\
&\quad\le \|f'\|_{[a,b],\infty}\int_a^b |t-x|\,dt = \frac{(b-x)^2+(x-a)^2}{2}\|f'\|_{[a,b],\infty} \\
&\quad= \left[\frac{1}{4}(b-a)^2 + \left(x-\frac{a+b}{2}\right)^2\right]\|f'\|_{[a,b],\infty}
\end{aligned}$$

and

$$\int_a^b |y-t| \left| \|f'\|_{[y,t],\infty} \right| dt \le \left[\frac{1}{4}(b-a)^2 + \left(y - \frac{a+b}{2} \right)^2 \right] \|f'\|_{[a,b],\infty}$$

for any $x,\ y \in [a,b]$.

Therefore,

$$\begin{aligned} C(x,y) &\le (1-\lambda)\frac{1}{b-a}\left[\frac{1}{4}(b-a)^2 + \left(x-\frac{a+b}{2}\right)^2\right]\|f'\|_{[a,b],\infty} \\ &\quad + \lambda\frac{1}{b-a}\left[\frac{1}{4}(b-a)^2 + \left(y-\frac{a+b}{2}\right)^2\right]\|f'\|_{[a,b],\infty} \\ &= (1-\lambda)\left[\frac{1}{4} + \left(\frac{x-\frac{a+b}{2}}{b-a}\right)^2\right]\|f'\|_{[a,b],\infty}(b-a) \\ &\quad + \lambda\left[\frac{1}{4} + \left(\frac{y-\frac{a+b}{2}}{b-a}\right)^2\right]\|f'\|_{[a,b],\infty}(b-a) \\ &= \left[\frac{1}{4} + (1-\lambda)\left(\frac{x-\frac{a+b}{2}}{b-a}\right)^2 + \lambda\left(\frac{y-\frac{a+b}{2}}{b-a}\right)^2\right] \\ &\quad \times \|f'\|_{[a,b],\infty}(b-a), \end{aligned}$$

for any $x,\ y \in [a,b]$ and $\lambda \in [0,1]$, which proves the last part of (40). □

Remark 9. With the assumptions of Theorem 6, we have the midpoint inequality

$$\begin{aligned} \left| \frac{1}{b-a}\int_a^b f(t)\,dt - f\left(\frac{a+b}{2}\right) \right| &\le \frac{1}{b-a}\int_a^b \left| t - \frac{a+b}{2} \right| \left| \|f'\|_{[\frac{a+b}{2},t],\infty} \right| dt \\ &\le \frac{1}{4}\|f'\|_{[a,b],\infty}(b-a). \end{aligned} \tag{44}$$

We have the following.

Theorem 10. *Let $f : [a, b] \to \mathbb{C}$ be an absolutely continuous function on $[a, b]$. If $f' \in L_p [a, b]$, with $p > 1$ and $\frac{1}{p} + \frac{1}{q} = 1$, then for any x, $y \in [a, b]$ and $\lambda \in [0, 1]$, we have*

$$\begin{aligned}\left|\frac{1}{b-a}\int_a^b f(t)\,dt - (1-\lambda) f(x) - \lambda f(y)\right| &\le (1-\lambda)\frac{1}{b-a}\int_a^b |t-x|\,\nu_{f',p}(x,t)\,dt \\ &\quad + \lambda\frac{1}{b-a}\int_a^b |y-t|\,\nu_{f',p}(y,t)\,dt, \end{aligned} \tag{45}$$

where

$$0 \le \nu_{f',p}(z,t) := \left(\int_0^1 |f'((1-s)z+st)|^p\,ds\right)^{1/p}, \quad \text{and} \quad z,\ t \in [a, b].$$

In particular, we have the λ-weighted trapezoid inequality

$$\begin{aligned}\left|\frac{1}{b-a}\int_a^b f(t)\,dt - (1-\lambda) f(a) - \lambda f(b)\right| &\le (1-\lambda)\frac{1}{b-a}\int_a^b |t-a|\,\nu_{f',p}(a,t)\,dt \\ &\quad + \lambda\frac{1}{b-a}\int_a^b |b-t|\,\nu_{f',p}(b,t)\,dt \end{aligned} \tag{46}$$

for any $\lambda \in [0, 1]$ and the Ostrowski-type inequality

$$\left|\frac{1}{b-a}\int_a^b f(t)\,dt - f(x)\right| \le \frac{1}{b-a}\int_a^b |t-x|\,\nu_{f',p}(x,t)\,dt \tag{47}$$

for any $x \in [a, b]$.

Proof. For $p > 1$, we have

$$\int_0^1 |f'((1-s)z+st)|\,ds \le \left(\int_0^1 |f'((1-s)z+st)|^p\,ds\right)^{1/p}$$

for any z, $t \in [a, b]$.

From (38), we obtain

$$\begin{aligned}
&\left|\frac{1}{b-a}\int_a^b f(t)\,dt-(1-\lambda)f(x)-\lambda f(y)\right| \\
&\leq (1-\lambda)\frac{1}{b-a}\int_a^b |t-x|\left(\int_0^1 \left|f'((1-s)x+st)\right| ds\right)dt \\
&\quad +\lambda\frac{1}{b-a}\int_a^b |y-t|\left(\int_0^1 \left|f'((1-s)t+sy)\right| ds\right)dt \\
&\leq (1-\lambda)\frac{1}{b-a}\int_a^b |t-x|\left(\int_0^1 \left|f'((1-s)x+st)\right|^p ds\right)^{1/p}dt \\
&\quad +\lambda\frac{1}{b-a}\int_a^b |y-t|\left(\int_0^1 \left|f'((1-s)y+st)\right|^p ds\right)^{1/p}dt \\
&= (1-\lambda)\frac{1}{b-a}\int_a^b |t-x|\,\nu_{f',p}(x,t)\,dt \\
&\quad +\lambda\frac{1}{b-a}\int_a^b |y-t|\,\nu_{f',p}(y,t)\,dt \qquad (48)
\end{aligned}$$

for any $x,\ y\in[a,b]$ and $\lambda\in[0,1]$.

This proves the desired result (45). □

Corollary 11. *With the assumptions of Theorem* 10, *we have*

$$\begin{aligned}
&\left|\frac{1}{b-a}\int_a^b f(t)\,dt-(1-\lambda)f(x)-\lambda f(y)\right| \\
&\leq (1-\lambda)\frac{1}{b-a}\int_a^b |t-x|^{1/q}\left|\int_x^t \left|f'(\tau)\right|^p d\tau\right|^{1/p}dt \\
&\quad +\lambda\frac{1}{b-a}\int_a^b |t-y|^{1/q}\left|\int_y^t \left|f'(\tau)\right|^p d\tau\right|^{1/p}dt \\
&\leq \frac{\|f'\|_{[a,b],p}}{(1/q+1)(b-a)} \\
&\quad\times\left\{(1-\lambda)\left[(b-x)^{1/q+1}+(x-a)^{1/q+1}\right]\right. \\
&\quad\quad \left.+\lambda\left[(b-y)^{1/q+1}+(y-a)^{1/q+1}\right]\right\} \qquad (49)
\end{aligned}$$

for any $x,\ y\in[a,b]$ *and* $\lambda\in[0,1]$.

In particular, we have the λ-weighted trapezoid inequality

$$\begin{aligned}
&\left|\frac{1}{b-a}\int_a^b f(t)\,dt-(1-\lambda)f(a)-\lambda f(b)\right| \\
&\quad\leq(1-\lambda)\frac{1}{b-a}\int_a^b(t-a)^{1/q}\left(\int_a^t|f'(\tau)|^p\,d\tau\right)^{1/p}dt \\
&\qquad+\lambda\frac{1}{b-a}\int_a^b(b-t)^{1/q}\left(\int_t^b|f'(\tau)|^p\,d\tau\right)^{1/p}dt \\
&\quad\leq\frac{\|f'\|_{[a,b],p}}{(1/q+1)}(b-a) \qquad (50)
\end{aligned}$$

for any $\lambda\in[0,1]$ and the Ostrowski-type inequality

$$\begin{aligned}
&\left|\frac{1}{b-a}\int_a^b f(t)\,dt-f(x)\right| \\
&\quad\leq\frac{1}{b-a}\int_a^b|t-x|^{1/q}\left|\int_x^t|f'(\tau)|^p\,d\tau\right|^{1/p}dt \\
&\quad\leq\frac{\|f'\|_{[a,b],p}}{(1/q+1)(b-a)}\left[(b-x)^{1/q+1}+(x-a)^{1/q+1}\right] \qquad (51)
\end{aligned}$$

for any $x\in[a,b]$.

Proof. If $z\neq t$, we have

$$\int_0^1|f'((1-s)z+st)|^p\,ds=\frac{1}{t-z}\int_z^t|f'(\tau)|^p\,d\tau,$$

which implies that

$$\begin{aligned}
&|t-z|\left(\int_0^1|f'((1-s)z+st)|^p\,ds\right)^{1/p} \\
&\quad=|t-z|^{1-1/p}\left|\int_z^t|f'(\tau)|^p\,d\tau\right|^{1/p} \\
&\quad=|t-z|^{1/q}\left|\int_z^t|f'(\tau)|^p\,d\tau\right|^{1/p}.
\end{aligned}$$

Therefore,

$$\int_0^1 \left| f'((1-s)z+st) \right| ds \le |t-z|^{1/q} \left| \int_z^t |f'(\tau)|^p \, d\tau \right|^{1/p}$$

for any t, $z \in [a,b]$.

By utilizing (48), we get

$$\begin{aligned}
&\left| \frac{1}{b-a} \int_a^b f(t)\,dt - (1-\lambda) f(x) - \lambda f(y) \right| \\
&\le (1-\lambda) \frac{1}{b-a} \int_a^b |t-x| \left(\int_0^1 \left| f'((1-s)x+st) \right| ds \right) dt \\
&\quad + \lambda \frac{1}{b-a} \int_a^b |y-t| \left(\int_0^1 \left| f'((1-s)t+sy) \right| ds \right) dt \\
&\le (1-\lambda) \frac{1}{b-a} \int_a^b |t-x|^{1/q} \left| \int_x^t |f'(\tau)|^p \, d\tau \right|^{1/p} dt \\
&\quad + \lambda \frac{1}{b-a} \int_a^b |t-y|^{1/q} \left| \int_y^t |f'(\tau)|^p \, d\tau \right|^{1/p} dt =: D_\lambda(x,y)
\end{aligned}$$

for any x, $y \in [a,b]$ and $\lambda \in [0,1]$.

This proves the first inequality in (49).

Since

$$\left| \int_x^t |f'(\tau)|^p \, d\tau \right|^{1/p}, \quad \left| \int_y^t |f'(\tau)|^p \, d\tau \right|^{1/p} \le \|f'\|_{[a,b],p}$$

for any t, $x \in [a,b]$ and

$$\int_a^b |t-x|^{1/q} \, dt = \frac{(b-x)^{1/q+1} + (x-a)^{1/q+1}}{1/q+1},$$

$$\int_a^b |t-y|^{1/q} \, dt = \frac{(b-y)^{1/q+1} + (y-a)^{1/q+1}}{1/q+1},$$

then

$$
\begin{aligned}
D(x,y) &\leq (1-\lambda)\left\|f'\right\|_{[a,b],p}\left[\frac{(b-x)^{1/q+1}+(x-a)^{1/q+1}}{(1/q+1)(b-a)}\right] \\
&\quad +\lambda\left\|f'\right\|_{[a,b],p}\left[\frac{(b-y)^{1/q+1}+(y-a)^{1/q+1}}{(1/q+1)(b-a)}\right] \\
&= \frac{\left\|f'\right\|_{[a,b],p}}{(1/q+1)(b-a)}\left[(1-\lambda)\left[(b-x)^{1/q+1}+(x-a)^{1/q+1}\right]\right. \\
&\quad \left. +\lambda\left[(b-y)^{1/q+1}+(y-a)^{1/q+1}\right]\right]
\end{aligned}
$$

for any x, $y \in [a,b]$ and $\lambda \in [0,1]$.

This proves the last part of (49). □

Remark 12. With the assumptions of Theorem 10, we have the midpoint inequality

$$
\begin{aligned}
&\left|\frac{1}{b-a}\int_a^b f(t)\,dt - f\left(\frac{a+b}{2}\right)\right| \\
&\quad \leq \frac{1}{b-a}\int_a^b \left|t-\frac{a+b}{2}\right|^{1/q}\left|\int_{\frac{a+b}{2}}^t \left|f'(\tau)\right|^p d\tau\right|^{1/p} dt \\
&\quad \leq \frac{\left\|f'\right\|_{[a,b],p}}{2^{1/q}(1/q+1)}(b-a)^{1/q}. \qquad (52)
\end{aligned}
$$

4. Two Point Inequalities for Double Integral Mean

We also have the following.

Theorem 13. *Let $f : [a,b] \to \mathbb{C}$ be an absolutely continuous function on $[a,b]$. Then, for any x, $y \in [a,b]$ and $\lambda \in [0,1]$, we have*

$$
\left|\frac{1}{(b-a)^2}\int_a^b\int_a^b f\left(\frac{u+v}{2}\right)dudv - (1-\lambda)f(x) - \lambda f(y)\right|
$$

$$\leq (1-\lambda)\frac{1}{(b-a)^2}\int_a^b\int_a^b\left|\frac{u+v}{2}-x\right|\nu_{f'}\left(x,\frac{u+v}{2}\right)dudv$$
$$+\lambda\frac{1}{(b-a)^2}\int_a^b\int_a^b\left|y-\frac{u+v}{2}\right|\nu_{f'}\left(y,\frac{u+v}{2}\right)dudv \quad (53)$$

where, as above,

$$0\leq \nu_{f'}(z,t):=\int_0^1\left|f'((1-s)z+st)\right|ds, \quad \text{and} \quad z,\ t\in[a,b].$$

In particular, we have the λ-weighted trapezoid inequality

$$\left|\frac{1}{(b-a)^2}\int_a^b\int_a^b f\left(\frac{u+v}{2}\right)dudv-(1-\lambda)f(a)-\lambda f(b)\right|$$
$$\leq (1-\lambda)\frac{1}{(b-a)^2}\int_a^b\int_a^b\left(\frac{u+v}{2}-a\right)\nu_{f'}\left(a,\frac{u+v}{2}\right)dudv$$
$$+\lambda\frac{1}{(b-a)^2}\int_a^b\int_a^b\left(b-\frac{u+v}{2}\right)\nu_{f'}\left(b,\frac{u+v}{2}\right)dudv \quad (54)$$

for any $\lambda\in[0,1]$ and the Ostrowski-type inequality

$$\left|\frac{1}{(b-a)^2}\int_a^b\int_a^b f\left(\frac{u+v}{2}\right)dudv-f(x)\right|$$
$$\leq\frac{1}{(b-a)^2}\int_a^b\int_a^b\left|\frac{u+v}{2}-x\right|\nu_{f'}\left(x,\frac{u+v}{2}\right)dudv \quad (55)$$

for any $x\in[a,b]$.

Proof. From Lemma 1, we have

$$f\left(\frac{u+v}{2}\right)=(1-\lambda)f(x)+\lambda f(y)$$
$$+(1-\lambda)\left(\frac{u+v}{2}-x\right)\int_0^1 f'\left((1-s)x+s\frac{u+v}{2}\right)ds$$
$$-\lambda\left(y-\frac{u+v}{2}\right)\int_0^1 f'\left((1-s)\frac{u+v}{2}+sy\right)ds \quad (56)$$

for any $x,\ y,\ u,\ v\in[a,b]$ and $\lambda\in[0,1]$.

Taking the double integral mean in (56), we have

$$\begin{aligned}
&\frac{1}{(b-a)^2}\int_a^b\int_a^b f\left(\frac{u+v}{2}\right)dudv \\
&\quad = (1-\lambda) f(x) + \lambda f(y) \\
&\qquad + (1-\lambda)\frac{1}{(b-a)^2}\int_a^b\int_a^b \left(\frac{u+v}{2}-x\right) \\
&\qquad \times \left(\int_0^1 f'\left((1-s)x + s\frac{u+v}{2}\right)ds\right)dudv \\
&\qquad - \lambda\frac{1}{(b-a)^2}\int_a^b\int_a^b \left(y-\frac{u+v}{2}\right) \\
&\qquad \times \left(\int_0^1 f'\left((1-s)\frac{u+v}{2} + sy\right)ds\right)dudv
\end{aligned}$$

for any x, $y \in [a,b]$ and $\lambda \in [0,1]$.

This implies that

$$\begin{aligned}
&\left|\frac{1}{(b-a)^2}\int_a^b\int_a^b f\left(\frac{u+v}{2}\right)dudv - (1-\lambda) f(x) - \lambda f(y)\right| \\
&\quad \le (1-\lambda)\frac{1}{(b-a)^2}\int_a^b\int_a^b \left|\frac{u+v}{2}-x\right| \\
&\qquad \times \left|\int_0^1 f'\left((1-s)x + s\frac{u+v}{2}\right)ds\right|dudv \\
&\qquad + \lambda\frac{1}{(b-a)^2}\int_a^b\int_a^b \left|y-\frac{u+v}{2}\right| \\
&\qquad \times \left|\int_0^1 f'\left((1-s)\frac{u+v}{2} + sy\right)ds\right|dudv \\
&\quad \le (1-\lambda)\frac{1}{(b-a)^2}\int_a^b\int_a^b \left|\frac{u+v}{2}-x\right| \\
&\qquad \times \int_0^1 \left|f'\left((1-s)x + s\frac{u+v}{2}\right)\right|dsdudv
\end{aligned}$$

$$+\lambda\frac{1}{(b-a)^2}\int_a^b\int_a^b\left|y-\frac{u+v}{2}\right|$$
$$\times\int_0^1\left|f'\left((1-s)\frac{u+v}{2}+sy\right)\right|dsdudv$$
$$=(1-\lambda)\frac{1}{(b-a)^2}\int_a^b\int_a^b\left|\frac{u+v}{2}-x\right|\nu_{f'}\left(x,\frac{u+v}{2}\right)dsdudv$$
$$+\lambda\frac{1}{(b-a)^2}\int_a^b\int_a^b\left|y-\frac{u+v}{2}\right|\nu_{f'}\left(y,\frac{u+v}{2}\right)dudv \tag{57}$$

for any x, $y\in[a,b]$ and $\lambda\in[0,1]$. □

Remark 14. We observe that for any bound $M(\cdot,\cdot)$ for the function $\nu_{f'}(\cdot,\cdot)$ on the square $[a,b]^2$, one can obtain the corresponding two-point Ostrowski inequality

$$\left|\frac{1}{(b-a)^2}\int_a^b\int_a^b f\left(\frac{u+v}{2}\right)dudv-(1-\lambda)f(x)-\lambda f(y)\right|$$
$$\leq(1-\lambda)\frac{1}{(b-a)^2}\int_a^b\int_a^b\left|\frac{u+v}{2}-x\right|M\left(x,\frac{u+v}{2}\right)dudv$$
$$+\lambda\frac{1}{(b-a)^2}\int_a^b\int_a^b\left|y-\frac{u+v}{2}\right|M\left(y,\frac{u+v}{2}\right)dudv \tag{58}$$

for any x, $y\in[a,b]$ and $\lambda\in[0,1]$.

In the following, we give only a few examples of such inequalities. The interested reader may obtain many more by choosing other bounds for the function $\nu_{f'}(\cdot,\cdot)$ on the square $[a,b]^2$.

Corollary 15. *Let $f:[a,b]\to\mathbb{C}$ be an absolutely continuous function on $[a,b]$. Assume that $f'\in L_\infty[a,b]$, then for any x, $y\in[a,b]$ and $\lambda\in[0,1]$, we have*

$$\left|\frac{1}{(b-a)^2}\int_a^b\int_a^b f\left(\frac{u+v}{2}\right)dudv-(1-\lambda)f(x)-\lambda f(y)\right|$$
$$\leq(1-\lambda)\frac{1}{(b-a)^2}\int_a^b\int_a^b\left|\frac{u+v}{2}-x\right|\left|\|f'\|_{\left[x,\frac{u+v}{2}\right],\infty}\right|dudv$$

$$+\lambda\frac{1}{(b-a)^2}\int_a^b\int_a^b\left|y-\frac{u+v}{2}\right|\left|\|f'\|_{[y,\frac{u+v}{2}],\infty}\right|dudv$$

$$\leq\frac{2}{3(b-a)^2}(1-\lambda)\left[(b-x)^3-2\left|x-\frac{a+b}{2}\right|^3+(x-a)^3\right]$$

$$\times\|f'\|_{[a,b],\infty}$$

$$+\frac{2}{3(b-a)^2}\lambda\left[(b-y)^3-2\left|y-\frac{a+b}{2}\right|^3+(y-a)^3\right]$$

$$\times\|f'\|_{[a,b],\infty}. \tag{59}$$

In particular, we have the λ*-weighted trapezoid inequality*

$$\left|\frac{1}{(b-a)^2}\int_a^b\int_a^b f\left(\frac{u+v}{2}\right)dudv-(1-\lambda)f(a)-\lambda f(b)\right|$$

$$\leq(1-\lambda)\frac{1}{(b-a)^2}\int_a^b\int_a^b\left(\frac{u+v}{2}-a\right)\|f'\|_{[a,\frac{u+v}{2}],\infty}dudv$$

$$+\lambda\frac{1}{(b-a)^2}\int_a^b\int_a^b\left(b-\frac{u+v}{2}\right)\|f'\|_{[\frac{u+v}{2},b],\infty}dudv$$

$$\leq\frac{1}{2}\|f'\|_{[a,b],\infty}(b-a) \tag{60}$$

for any $\lambda\in[0,1]$ *and the refinement of Ostrowski inequality*

$$\left|\frac{1}{(b-a)^2}\int_a^b\int_a^b f\left(\frac{u+v}{2}\right)dudv-f(x)\right|$$

$$\leq\frac{1}{(b-a)^2}\int_a^b\int_a^b\left|\frac{u+v}{2}-x\right|\left|\|f'\|_{[x,\frac{u+v}{2}],\infty}\right|dudv$$

$$\leq\frac{2}{3(b-a)^2}\left[(b-x)^3-2\left|x-\frac{a+b}{2}\right|^3+(x-a)^3\right]\|f'\|_{[a,b],\infty} \tag{61}$$

for any $x\in[a,b]$.

Remark 16. With the assumption of Theorem 13, we have the midpoint inequality

$$\begin{aligned}&\left|\frac{1}{(b-a)^2}\int_a^b\int_a^b f\left(\frac{u+v}{2}\right)dudv - f\left(\frac{a+b}{2}\right)\right| \\ &\quad\leq \frac{1}{(b-a)^2}\int_a^b\int_a^b \left|\frac{u+v}{2}-\frac{a+b}{2}\right|\left|\left\|f'\right\|_{\left[\frac{a+b}{2},\frac{u+v}{2}\right],\infty}\right| dudv \\ &\quad\leq \frac{1}{6}(b-a)^3\left\|f'\right\|_{[a,b],\infty}. \end{aligned} \tag{62}$$

Further, we have the following.

Theorem 17. *Let $f:[a,b]\to\mathbb{C}$ be an absolutely continuous function on $[a,b]$. If $f'\in L_p[a,b]$, with $p>1$ and $\frac{1}{p}+\frac{1}{q}=1$, then for any x, $y\in[a,b]$ and $\lambda\in[0,1]$, we have*

$$\begin{aligned}&\left|\frac{1}{(b-a)^2}\int_a^b\int_a^b f\left(\frac{u+v}{2}\right)dudv-(1-\lambda)f(x)-\lambda f(y)\right| \\ &\quad\leq(1-\lambda)\frac{1}{(b-a)^2}\int_a^b\int_a^b\left|\frac{u+v}{2}-x\right|\nu_{f',p}\left(x,\frac{u+v}{2}\right)dudv \\ &\qquad+\lambda\frac{1}{(b-a)^2}\int_a^b\int_a^b\left|y-\frac{u+v}{2}\right|\nu_{f',p}\left(y,\frac{u+v}{2}\right)dudv \end{aligned} \tag{63}$$

where, as above,

$$\begin{aligned}0&\leq\nu_{f',p}(z,t) \\ &:=\left(\int_0^1\left|f'((1-s)z+st)\right|^p ds\right)^{1/p}, \quad \text{and} \quad z,\ t\in[a,b].\end{aligned}$$

In particular, we have the λ-weighted trapezoid inequality

$$\begin{aligned}&\left|\frac{1}{(b-a)^2}\int_a^b\int_a^b f\left(\frac{u+v}{2}\right)dudv-(1-\lambda)f(a)-\lambda f(b)\right| \\ &\quad\leq(1-\lambda)\frac{1}{(b-a)^2}\int_a^b\int_a^b\left(\frac{u+v}{2}-a\right)\nu_{f',p}\left(a,\frac{u+v}{2}\right)dudv \\ &\qquad+\lambda\frac{1}{(b-a)^2}\int_a^b\int_a^b\left(b-\frac{u+v}{2}\right)\nu_{f',p}\left(b,\frac{u+v}{2}\right)dudv \end{aligned} \tag{64}$$

for any $\lambda \in [0,1]$ *and the Ostrowski-type inequality*

$$\left|\frac{1}{(b-a)^2}\int_a^b\int_a^b f\left(\frac{u+v}{2}\right)dudv - f(x)\right|$$
$$\leq \frac{1}{(b-a)^2}\int_a^b\int_a^b \left|\frac{u+v}{2}-x\right| \nu_{f',p}\left(x,\frac{u+v}{2}\right)dudv \quad (65)$$

for any $x \in [a,b]$.

Proof. For $p > 1$, we have the inequality

$$\int_0^1 \left|f'\left[(1-s)z + s\left(\frac{u+v}{2}\right)\right]\right| ds$$
$$\leq \left(\int_0^1 \left|f'\left[(1-s)z + s\left(\frac{u+v}{2}\right)\right]\right|^p ds\right)^{1/p}$$

for any z, x, $y \in [a,b]$.

From (57), we have

$$\left|\frac{1}{(b-a)^2}\int_a^b\int_a^b f\left(\frac{u+v}{2}\right)dudv - (1-\lambda)f(x) - \lambda f(y)\right|$$
$$\leq (1-\lambda)\frac{1}{(b-a)^2}\int_a^b\int_a^b \left|\frac{u+v}{2}-x\right|$$
$$\times \int_0^1 \left|f'\left((1-s)x + s\frac{u+v}{2}\right)\right| dsdudv$$
$$+\lambda\frac{1}{(b-a)^2}\int_a^b\int_a^b \left|y-\frac{u+v}{2}\right|$$
$$\times \int_0^1 \left|f'\left((1-s)\frac{u+v}{2} + sy\right)\right| dsdudv$$
$$\leq (1-\lambda)\frac{1}{(b-a)^2}\int_a^b\int_a^b \left|\frac{u+v}{2}-x\right|$$
$$\times \left(\int_0^1 \left|f'\left[(1-s)x + s\left(\frac{u+v}{2}\right)\right]\right|^p ds\right)^{1/p} dsdudv$$

$$+\lambda \frac{1}{(b-a)^2} \int_a^b \int_a^b \left| y - \frac{u+v}{2} \right|$$

$$\times \left(\int_0^1 \left| f' \left[(1-s) y + s \left(\frac{u+v}{2} \right) \right] \right|^p ds \right)^{1/p} dsdudv$$

$$=: E_\lambda (x, y)$$

which proves (63). □

Corollary 18. *With the assumptions of Theorem 17, we have*

$$\left| \frac{1}{(b-a)^2} \int_a^b \int_a^b f\left(\frac{u+v}{2} \right) dudv - (1-\lambda) f(x) - \lambda f(y) \right|$$

$$\leq (1-\lambda) \frac{1}{(b-a)^2} \int_a^b \int_a^b \left| x - \frac{u+v}{2} \right|^{1/q}$$

$$\times \left| \int_{\frac{u+v}{2}}^x |f'(\tau)|^p d\tau \right|^{1/p} dudv$$

$$+\lambda \frac{1}{(b-a)^2} \int_a^b \int_a^b \left| y - \frac{u+v}{2} \right|^{1/q} \left| \int_{\frac{u+v}{2}}^y |f'(\tau)|^p d\tau \right|^{1/p} dudv$$

$$\leq \frac{4}{(1/q+1)(1/q+2)(b-a)^2} \|f'\|_{[a,b],p}$$

$$\times \left\{ (1-\lambda) \left[(b-x)^{1/q+2} - 2\left| x - \frac{a+b}{2} \right|^{1/q+2} + (x-a)^{1/q+2} \right] \right.$$

$$\left. +\lambda \left[(b-y)^{1/q+2} - 2\left| y - \frac{a+b}{2} \right|^{1/q+2} + (y-a)^{1/q+2} \right] \right. \tag{66}$$

for any x, $y \in [a,b]$ and $\lambda \in [0,1]$.

In particular, we have the λ-weighted trapezoid inequality

$$\left| \frac{1}{(b-a)^2} \int_a^b \int_a^b f\left(\frac{u+v}{2} \right) dudv - (1-\lambda) f(a) - \lambda f(b) \right|$$

$$\leq (1-\lambda) \frac{1}{(b-a)^2} \int_a^b \int_a^b \left(\frac{u+v}{2} - a \right)^{1/q}$$

$$\times \left| \int_{\frac{u+v}{2}}^{a} \left| f'(\tau) \right|^p d\tau \right|^{1/p} dudv$$

$$+ \lambda \frac{1}{(b-a)^2} \int_a^b \int_a^b \left(b - \frac{u+v}{2} \right)^{1/q} \left| \int_{\frac{u+v}{2}}^{y} \left| f'(\tau) \right|^p d\tau \right|^{1/p} dudv$$

$$\leq \frac{2^{1/q+1} - 1}{2^{1/q-1} (1/q+1)(1/q+2)} \left\| f' \right\|_{[a,b],p} (b-a)^{1/q} \tag{67}$$

for any $\lambda \in [0,1]$ *and the Ostrowski-type inequality*

$$\left| \frac{1}{(b-a)^2} \int_a^b \int_a^b f\left(\frac{u+v}{2} \right) dudv - f(x) \right|$$

$$\leq \frac{1}{(b-a)^2} \int_a^b \int_a^b \left| x - \frac{u+v}{2} \right|^{1/q} \left| \int_{\frac{u+v}{2}}^{x} \left| f'(\tau) \right|^p d\tau \right|^{1/p} dudv$$

$$\leq \frac{4}{(1/q+1)(1/q+2)(b-a)^2}$$

$$\times \left[(b-x)^{1/q+2} - 2\left| x - \frac{a+b}{2} \right|^{1/q+2} + (x-a)^{1/q+2} \right] \left\| f' \right\|_{[a,b],p} \tag{68}$$

for any $x \in [a,b]$.

Proof. For $p > 1$, we have

$$\int_0^1 \left| f'\left[(1-s) z + s \left(\frac{u+v}{2} \right) \right] \right| ds$$

$$\leq \left(\int_0^1 \left| f'\left[(1-s) z + s \left(\frac{u+v}{2} \right) \right] \right|^p ds \right)^{1/p}$$

for any z, u, $v \in [a,b]$.

Now, suppose that $z \neq \frac{u+v}{2}$. Then

$$\int_0^1 \left| f'\left[(1-s) z + s \left(\frac{u+v}{2} \right) \right] \right|^p ds$$

$$= \left(z - \frac{u+v}{2}\right)^{-1} \int_{\frac{u+v}{2}}^{z} |f'(\tau)|^p \, d\tau$$

$$= \left|z - \frac{u+v}{2}\right|^{-1} \left|\int_{\frac{u+v}{2}}^{z} |f'(\tau)|^p \, d\tau\right|,$$

namely,

$$\left(\int_0^1 \left|f'\left[(1-s)z + s\left(\frac{u+v}{2}\right)\right]\right|^p ds\right)^{1/p}$$

$$= \left|z - \frac{u+v}{2}\right|^{-1/p} \left|\int_{\frac{u+v}{2}}^{z} |f'(\tau)|^p \, d\tau\right|^{1/p}.$$

This implies that

$$\left|z - \frac{u+v}{2}\right| \left(\int_0^1 \left|f'\left[(1-s)z + s\left(\frac{u+v}{2}\right)\right]\right|^p ds\right)^{1/p}$$

$$= \left|z - \frac{u+v}{2}\right|^{1-1/p} \left|\int_{\frac{u+v}{2}}^{z} |f'(\tau)|^p \, d\tau\right|^{1/p}$$

$$= \left|z - \frac{u+v}{2}\right|^{1/q} \left|\int_{\frac{u+v}{2}}^{z} |f'(\tau)|^p \, d\tau\right|^{1/p}$$

for any z, u, $v \in [a,b]$.

Using the notation from the proof of Theorem 17, we get

$$E_\lambda(x,y) \le (1-\lambda)\frac{1}{(b-a)^2}\int_a^b\int_a^b \left|x - \frac{u+v}{2}\right|^{1/q}$$

$$\times \left|\int_{\frac{u+v}{2}}^{x} |f'(\tau)|^p \, d\tau\right|^{1/p} dudv$$

$$+ \lambda\frac{1}{(b-a)^2}\int_a^b\int_a^b \left|y - \frac{u+v}{2}\right|^{1/q}$$

$$\times \left|\int_{\frac{u+v}{2}}^{y} |f'(\tau)|^p \, d\tau\right|^{1/p} dudv,$$

which proves the first inequality in (66).

Since

$$\left|\int_{\frac{u+v}{2}}^{x}\left|f'(\tau)\right|^{p}d\tau\right|^{1/p},\quad\left|\int_{\frac{u+v}{2}}^{y}\left|f'(\tau)\right|^{p}d\tau\right|^{1/p}\leq\left\|f'\right\|_{[a,b],p}$$

for any x, y, u, $v\in[a,b]$, then

$$\begin{aligned}&\int_a^b\int_a^b\left|x-\frac{u+v}{2}\right|^{1/q}\left|\int_{\frac{u+v}{2}}^{x}\left|f'(\tau)\right|^{p}d\tau\right|^{1/p}dudv\\&\leq\left\|f'\right\|_{[a,b],p}\int_a^b\int_a^b\left|x-\frac{u+v}{2}\right|^{1/q}dudv\\&=\frac{4}{(1/q+1)(1/q+2)}\\&\quad\times\left\|f'\right\|_{[a,b],p}\left[(b-x)^{1/q+2}-2\left|x-\frac{a+b}{2}\right|^{1/q+2}+(x-a)^{1/q+2}\right]\end{aligned}$$

and

$$\begin{aligned}&\int_a^b\int_a^b\left|y-\frac{u+v}{2}\right|^{1/q}\left|\int_{\frac{u+v}{2}}^{y}\left|f'(\tau)\right|^{p}d\tau\right|^{1/p}dudv\\&\leq\frac{4}{(1/q+1)(1/q+2)}\left\|f'\right\|_{[a,b],p}\\&\quad\times\left[(b-y)^{1/q+2}-2\left|y-\frac{a+b}{2}\right|^{1/q+2}+(y-a)^{1/q+2}\right]\end{aligned}$$

for any x, $y\in[a,b]$.

This proves the last part of (66). □

Remark 19. With the assumption of Theorem 17, we have

$$\begin{aligned}&\left|\frac{1}{(b-a)^2}\int_a^b\int_a^b f\left(\frac{u+v}{2}\right)dudv-f\left(\frac{a+b}{2}\right)\right|\\&\leq\frac{1}{(b-a)^2}\int_a^b\int_a^b\left|\frac{a+b}{2}-\frac{u+v}{2}\right|^{1/q}\left|\int_{\frac{u+v}{2}}^{\frac{a+b}{2}}\left|f'(\tau)\right|^{p}d\tau\right|^{1/p}dudv\\&\leq\frac{1}{2^{1/q-1}(1/q+1)(1/q+2)}(b-a)^{1/q}\left\|f'\right\|_{[a,b],p}.\end{aligned}\tag{69}$$

References

[1] Alomari, M.W. A companion of the generalized trapezoid inequality and applications. *J. Math. Appl.* 36, 5–15 (2013).

[2] Barnett, N.S. and Dragomir, S.S. A trapezoid type inequality for double integrals. *Proceedings of the Third World Congress of Nonlinear Analysts*, Part 4 (Catania, 2000). *Nonlinear Anal.*, 47(4), 2321–2332 (2001).

[3] Barnett, N.S. and Dragomir, S.S. Perturbed version of a general trapezoidinequality. *Inequality Theory and Applications.* Nova Sci. Publ., Hauppauge, New York, 3, 1–12 (2003).

[4] Barnett, N.S. and Dragomir, S.S. A perturbed trapezoid inequality in terms of the third derivative and applications. *Inequality Theory and Applications.* Nova Sci. Publ., New York, 5, 1–11 (2007).

[5] Barnett, N.S., Dragomir, S.S. and Pearce, C.E.M. A quasi-trapezoid inequality for double integrals. *ANZIAM J.* 44(3), 355–364 (2003).

[6] Bica, A.M., Căuş, V.A. and Mureşan, S. Application of a trapezoid inequality to neutral Fredholm integro-differential equations in Banach spaces. *J. Inequal. Pure Appl. Math.* 7(5), 173, 11 (2006).

[7] Budak, H. and Sarikaya, M.Z. A companion of the generalised trapezoid inequality for functions of two variables with bounded variation and applications. *Int. J. Appl. Nonlinear Sci.* 2(4), 311–327 (2016).

[8] Cheng, X.-L. and Sun, J. A note on the perturbed trapezoid inequality. *J. Inequal. Pure Appl. Math.* 3(2), 29, 7 (2002).

[9] Cerone, P. and Dragomir, S.S. Trapezoidal-type rules from an inequalities point of view. *Handbook of Analytic-Computational Methods in Applied Mathematics*, 65–134, Chapman & Hall/CRC, Boca Raton, FL, 2000. Preprint *RGMIA Res. Rep. Coll.* 2(6), 8 (1999). http://rgmia.org/papers/v2n6/TrapRulesIneqRep.pdf.

[10] Cerone, P., Dragomir, S.S. and Pearce, C.E.M. A generalized trapezoid inequality for functions of bounded variation. *Turkish J. Math.* 24(2), 147–163 (2000).

[11] S.S. Dragomir, A Grüss' type integral inequality for mappings of r-Hölder's type and applications for trapezoid formula. *Tamkang J. Math.* 31(1), 43–47 (2000).

[12] Dragomir, S.S. Ostrowski type inequalities for isotonic linear functionals, *J. Ineq. Pure & Appl. Math.*, 3(4), 4 (2002).

[13] Dragomir, S.S. Improvements of Ostrowski and generalised trapezoidinequality in terms of the upper and lower bounds of the first derivative. *Tamkang J. Math.*, 34(3), 213–222 (2003).

[14] Dragomir, S.S. Refinements of the generalized trapezoid inequality in terms of the cumulative variation and applications. *Cubo* 17(2), 31–48 (2015).

[15] Dragomir, S.S. Some inequalities of Ostrowski type for double integral mean of absolutely continuous functions, Preprint *RGMIA Res. Rep. Coll.* 21 (2018).

[16] Dragomir, S.S., Cho, Y.J. and Kim, Y.H. On the trapezoid inequality for the Riemann-Stieltjes integral with Hölder continuous integrands and bounded variation integrators. *Inequality Theory and Applications.* Nova Sci. Publ., New York, 5, 71–79 (2007).

[17] Dragomir, S.S. and Mcandrew, A. On trapezoid inequality via a Grüss type result and applications. *Tamkang J. Math.* 31(3), 193–201 (2000).

[18] Dragomir, S.S., Pečarić, J. and Wang, S. The unified treatment of trapezoid, Simpson, and Ostrowski type inequality for monotonic mappings and applications. *Math. Comput. Modelling* 31(6–7), 61–70 (2000).

[19] Gunawan, H. A note on Dragomir-McAndrew's trapezoid inequalities. T*amkang J. Math.* 33(3), 241–244 (2002).

[20] Liu, W. and Park, J. Some perturbed versions of the generalized trapezoid inequality for functions of bounded variation. *J. Comput. Anal. Appl.* 22(1), 11–18 (2017).

[21] Mercer, P.R. Hadamard's inequality and trapezoid rules for the Riemann-Stieltjes integral. *J. Math. Anal. Appl.* 344(2), 921–926 (2008).

[22] Pachpatte, B.G. A note on a trapezoid type integral inequality. *Bull. Greek Math. Soc.* 49, 85–90 (2004).

[23] Ujević, N. Error inequalities for a generalized trapezoid rule. *Appl. Math. Lett.* 19(1), 32–37 (2006).

https://doi.org/10.1142/9789811293009_0012

Chapter 12

Mathematical Models and Advancements in Cardiac Hemodynamics

Anastasios C. Felias[*,#], Michail Th. Rassias[†], Michail A. Xenos[‡] and Minas E. Paschopoulos[*]

[*]*Faculty of Medicine, School of Health Sciences, University of Ioannina, 45110 Ioannina, Greece*
[†]*Department of Mathematics and Engineering Sciences, Hellenic Military Academy, 16673, Attiki, Greece*
[‡]*Department of Mathematics, University of Ioannina, Ioannina, 45110, Epirus, Greece*
[#]*tsfelias1995@gmail.com*

The emergence of advanced biomedical technologies has contributed to the development of biofluid dynamics. A major achievement of the latter is the prediction of hemodynamics, i.e., the flow patterns and wall stresses of the cardiovascular system in the human body. Notable exact results on hemodynamics are obtained by means of the Hagen–Poiseuille, Womersley, and Korteweg–de Vries–Burgers (KdV–B) mathematical models. The semi-exact homotopy analysis method is discussed and a numerical Fluid Structure Interaction (FSI) approach is followed. The semiexact and numerical results are compared with the exact ones. As of current analysis, the KdV–B mathematical model seems to be quite reliable as a cardiac hemodynamics model. It also compares very well with other advanced multi-dimensional numerical methods, such as the presented FSI numerical method.

1. The Hagen–Poiseuille and Womersley Models on Cardiac Hemodynamics

1.1. *Hagen–Poiseuille flow*

In fluid dynamics, the Hagen–Poiseuille model describes laminar flow through a pipe of length L and uniform (circular) cross-section of diameter D. Its applications range from hemodynamics, air flow in the pulmonary alveoli, to flows through hypodermic needles. The governing equations are derived from the Navier–Stokes momentum equations [8] in three-dimensional cylindrical coordinates, under the following considerations:

- The flow is steady, $\frac{\partial q}{\partial t} = 0$.
- The radial and azimuthal velocities are zero, with $q = (u, 0, 0)$.
- The flow is axisymmetric, $\frac{\partial u}{\partial \theta} = 0$.

The Hagen–Poiseuille equation, under the above assumptions, is

$$\frac{1}{r}\frac{d}{dr}\left(r\frac{du}{dr}\right) = \frac{1}{\mu}\frac{dp}{dx} \tag{1}$$

with boundary conditions $u(R) = \frac{du}{dr}(0) = 0$. In the above equation, the left-hand side is a function of r, while the right-hand side is a function of x, which means that both members should equal the same constant. Deriving the value of this constant is straightforward. Given the length L of the pipe, denoting as Δp, the pressure difference between the two ends of the pipe (high pressure minus low pressure),

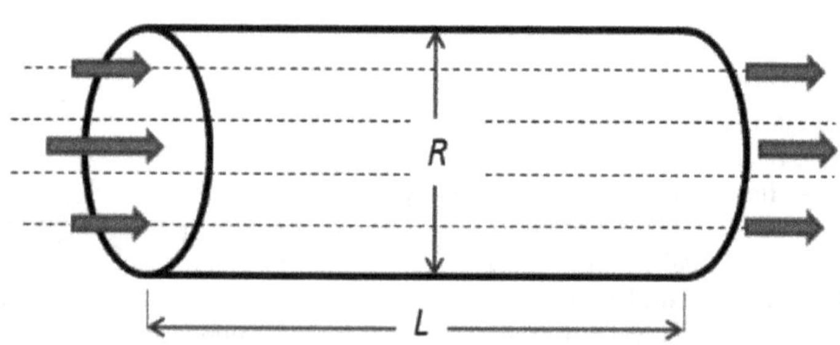

Fig. 1. A rectilinear cylindrical pipe filled with a fluid of density ρ and viscosity μ. The flow is laminar, axisymmetric, and parallel to the tube's axis.

then the constant will be given as

$$-\frac{dp}{dx} = \frac{\delta p}{L} = G \tag{2}$$

defined such that G is positive. The general solution of (1) will be given as

$$u(r) = -\frac{Gr^2}{4\mu} + C_1 \ln r + C_2 \tag{3}$$

which, due to the imposed boundary conditions, becomes

$$u(r) = \frac{G}{4\mu}\left(R^2 - r^2\right). \tag{4}$$

Considering the dimensionless distance,

$$y := \frac{r}{R} \in [0, 1], \tag{5}$$

(4) is transformed to

$$u(y) = \frac{GR^2}{4\mu}\left(1 - y^2\right). \tag{6}$$

We can easily see that $u(r)$ exhibits a maximum at $r = 0$, with the maximum velocity, U_{max}, given as

$$U_{max} = \frac{GR^2}{4\mu}. \tag{7}$$

The average velocity can be obtained by integration on the cross-section of the pipe:

$$\overline{u} = \frac{1}{\pi R^2}\int_0^R 2\pi r u dr = \frac{1}{2}U_{max}. \tag{8}$$

A quite significant quantity, which is at the same time easy to measure, is the flow rate $Q = \pi R^2 \overline{u}$. The rearrangement of this equation gives the Hagen–Poiseuille relation for the pressure drop in the pipe:

$$\Delta p = \frac{8\mu QL}{\pi R^4}. \tag{9}$$

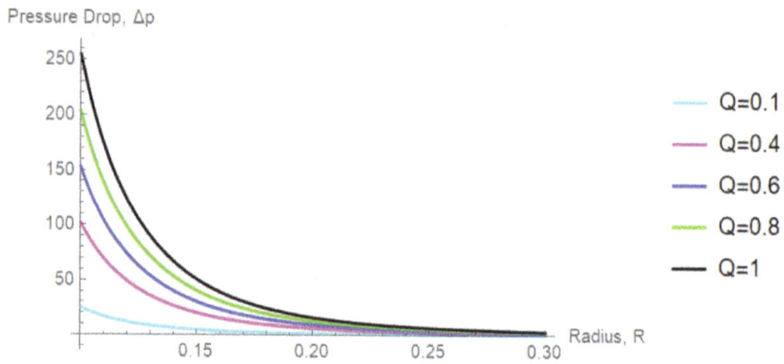

Fig. 2. Hagen–Poiseuille pressure drop, for different volume flow rates, Q, for $L = 1$ and $\mu = 0.01$.

Furthermore, relation (9) was obtained under the assumption of laminar flow of the fluid in the pipe. Hence, it is required that the *Reynolds* number

$$Re = \frac{\overline{u}D\rho}{\mu} \tag{10}$$

is kept low and less than the critical Reynolds number, Re_{cr}. According to experiments, in the case where the fluid in the tube is water, the critical number Reynolds is approximately equal to 2300 [8]. For $Re > Re_{cr}$, the flow ceases to be laminar and becomes turbulent [8].

1.1.1. *Hemodynamics and hypertension*

As seen before, the volume flow rate, Q, in the Hagen–Poiseuille flow is expressed as

$$Q = \frac{\pi R^4 \Delta p}{8\mu L}. \tag{11}$$

As the above relation pinpoints, the flow rate directly depends on the cross-sectional area of the pipe, in a quite sensitive manner. Even the smallest changes in the radius of the pipe, due to its elevation in the fourth power, bring about a noticeable change in the volumetric flow rate, provided that the other parameters do not change. This is often the reason why the HP flow is encountered in hemodynamics, explaining why small and narrow arteries lead to higher blood pressure. A slight narrowing of the veins reduces the flow rate, however, the heart tries to balance the latter, by working harder to raise blood pressure and maintain a constant blood flow. This also shows the link between arterial stenosis and high blood pressure.

1.2. *Womersley pulsatile flow*

In fluid dynamics, the Womersley pulsating flow is a flow governed by periodic fluctuations. Womersley was the first to derive such flow profiles in his study of blood flow in the cardiovascular system [10]. Pulse flow profiles are given by the Womersley equation, which describes laminar and non-permanent flow in a rectilinear cylindrical tube of length L and radius R, containing a fluid of density ρ, dynamic viscosity μ and kinematic viscosity $\nu = \frac{\mu}{\rho}$. The derivation of the exact solution of the velocity profile requires the following assumptions [11]:

- The fluid is homogeneous, incompressible, and Newtonian.
- The wall of the tube is rigid and cylindrical.
- The motion is axisymmetric and parallel to the axis of the tube.
- Axiosymmetry and the no-slip condition apply at the centre of the tube and at its wall, respectively.
- The pressure gradient is periodic in time.
- Gravity does not affect the fluid.

Based on the above assumptions, the Navier–Stokes [10] equations are simplified as follows:

$$\frac{\partial u}{\partial t} = -\frac{1}{\rho}\frac{\partial p}{\partial x} + \nu\left(\frac{\partial^2 u}{\partial r^2} + \frac{1}{r}\frac{\partial u}{\partial r}\right). \tag{12}$$

Considering the dimensionless, normalized variables,

$$y = \frac{r}{R}, \quad z = \frac{x}{R}, \quad \tau = \omega t, \quad u = \frac{v}{W} \tag{13}$$

via chain derivation, (12) is transformed into

$$\frac{R^2\omega}{\nu}\frac{\partial v}{\partial \tau} = -\frac{R}{\rho\nu W}\frac{\partial p}{\partial z} + \frac{\partial^2 v}{\partial y^2} + \frac{1}{y}\frac{\partial v}{\partial y}. \tag{14}$$

From (14), we obtain a natural normalization of the pressure:

$$p^* = \frac{Rp}{\rho\nu W} = \frac{Rp}{\mu W} \tag{15}$$

from which combined with (14), we obtain

$$\alpha^2 \frac{\partial v}{\partial \tau} = -\frac{\partial p^*}{\partial z} + \frac{\partial^2 v}{\partial y^2} + \frac{1}{y}\frac{\partial v}{\partial y} \tag{16}$$

which is the dimensionless, normalized Womersley equation.

Definition 1 ([10]). The dimensionless quantity,

$$\alpha := R\sqrt{\frac{\omega}{\nu}}, \tag{17}$$

expresses the frequency of the pulse flow in relation to its viscous effects and is defined as the Womersley number, symbolically α or Wo.

The flows where $\alpha \gg 1$ are usually characterized as dominated by inertia terms, since the viscous forces are negligible compared to the other terms in the equation (16) and the pressure gradient is used exclusively for the acceleration of the fluid particle. Such a situation usually occurs in large-diameter vessels (aorta), in high-frequency flows.

In the case where the flow has $\alpha \ll 1$ (small-diameter and/or low-frequency vessels), (16) is reduced to the normalized Hagen–Poiseuille equation, with the flow dominated by friction terms. Such a situation occurs in capillaries.

As (16) is linear, superposition of solutions for different harmonics is allowed. To proceed further, we assume that the pressure gradient is oscillatory and represent it by the relation

$$\frac{\partial p^*}{\partial z}(\tau) = \frac{\partial}{\partial z} e^{i}. \tag{18}$$

This consideration is justified by the fact that the heartbeat varies periodically in time and can therefore be expressed as a series of terms of the above form. Due to the harmonic form of the pressure gradient, we further assume that the velocity can be similarly represented as

$$v(y,\tau) = w(y)e^{i\tau}. \tag{19}$$

The transformation (19) converts (16) to

$$y^2\frac{d^2 w}{dy^2} + y\frac{dw}{dy} + \left(i^{3/2}\alpha y\right)^2 w = \frac{\partial \hat{p}}{\partial z} y^2. \tag{20}$$

The boundary conditions of the problem are $w(1) = \frac{dw}{dy}(0) = 0$. Through the transformation,

$$s = i^{3/2}\alpha y, \tag{21}$$

(20) is reduced to the non-homogeneous Bessel equation of order zero,

$$s^2\frac{d^2w}{ds^2} + s\frac{dw}{ds} + s^2 w = \frac{i}{\alpha^2}\frac{\partial \hat{p}}{\partial z}s^2, \tag{22}$$

with boundary conditions $w(i^{3/2}\alpha) = \frac{dw}{ds}(0) = 0$. Note that

$$w_p(s) = \frac{i}{\alpha^2}\frac{\partial \hat{p}}{\partial z} \tag{23}$$

is a partial solution of (22). Equation (22) is linear, so its general solution is obtained by summing the general solution of the homogeneous (22) with $\hat{w}_p$. The homogeneous (22) is solved analytically via Frobenius power series, around $s_0 = 0$, with its solution given via Bessel functions of first kind and zero order [17], as

$$w_H(s) = C_1 J_0(s) + C_2 Y_0(s). \tag{24}$$

Thus, the general solution of (22) is given as

$$w(s) = w_p(s) + w_H(s). \tag{25}$$

Respecting the boundary conditions, we arrive at the solution

$$w(s) = \frac{i}{\alpha^2}\frac{\partial \hat{p}}{\partial z}\left(1 - \frac{J_0(s)}{J_0(i^{3/2}\alpha)}\right). \tag{26}$$

Using in this phase, in reverse the aforementioned transformations, to return to our original variables, we finally get that

$$u(r,t) = \frac{i}{\rho\omega}\frac{\partial p}{\partial x}\left(1 - \frac{J_0(i^{3/2}\alpha r/R)}{J_0(i^{3/2}\alpha)}\right). \tag{27}$$

Finally, since (22) is linear, the real solution for the velocity profile is obtained by taking the real part of the complex function of (27):

$$\Re\left\{\frac{i}{\rho\omega}\frac{\partial p}{\partial x}\left(1 - \frac{J_0(i^{3/2}\alpha r/R)}{J_0(i^{3/2}\alpha)}\right)\right\}. \tag{28}$$

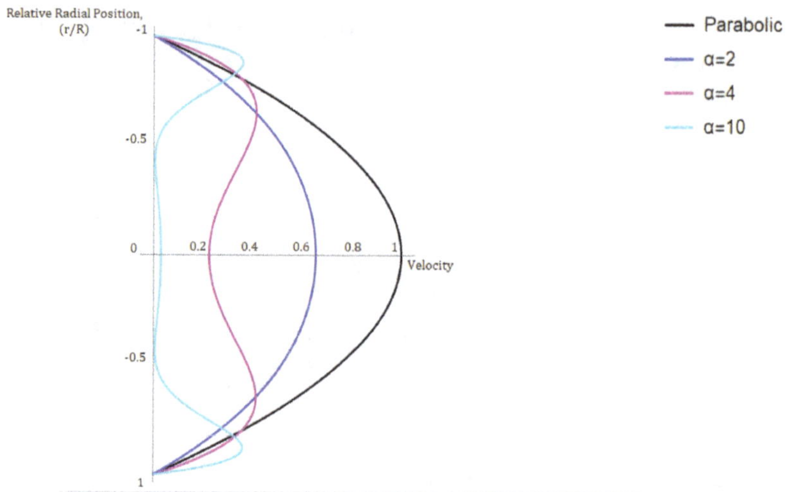

Fig. 3. Speed profiles for different values of the Womersley number. The parabolic profile comes from the Hagen–Poiseuille flow.

To be convinced of the validity of this claim, it suffices to see that if L is a linear (differential) operator and $u = \Re u + i/Imu$ such that $L[u] = 0$, then

$$0 = L[u] = L[\Re u + i\Im u] = L[\Re u] + iL[\Im u],$$

so $L[\Re u] = L[\Im u] = 0$, i.e., the real and imaginary parts of the solution are also solutions.

1.3. *Cardiovascular system and turbulence*

In the majority of the blood vessels, the flow is found to be laminar. Interestingly, under high flow conditions, especially in the ascending aorta, the laminar flow can be disturbed and become turbulent. Turbulence occurs when a critical Reynolds number, Re_{cr}, is exceeded. The Reynolds number can predict, under ideal conditions, the offset of turbulence.

The expression for the Reynolds number is

$$Re := \frac{U_{avg} D \rho}{\mu}, \tag{29}$$

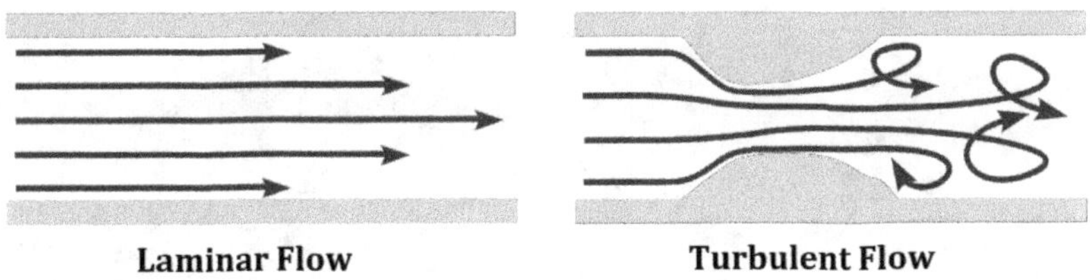

Fig. 4. Graphic illustration of laminar and turbulent flow in the cardiovascular system.

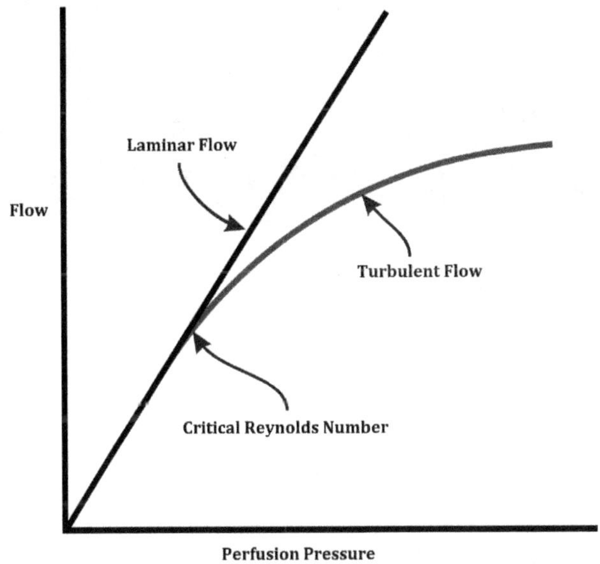

Fig. 5. Graphic illustration of the separation of laminar and turbulent flow, after exceeding the critical *Re*. Perfusion pressure is the pressure required to push blood through all blood vessels in a given area. As long as this pressure is high enough, blood will continue to flow through these vessels. This means that perfusion pressure is a critical part of how our body's circulatory system works.

where U_{avg} is the average velocity, D is the diameter of the vessel, ρ is the density of the blood, and μ is its viscosity [9].

In cardiac hemodynamics, the presence of turbulence is indicative of possible abnormal flows, thus contributing to further investigation. For example, in ultrasound, a turbulent flow will appear as a mosaic color, which is a mixture of the base color of the flow with shades of yellow added to it. All high velocity flows through weak valves appear

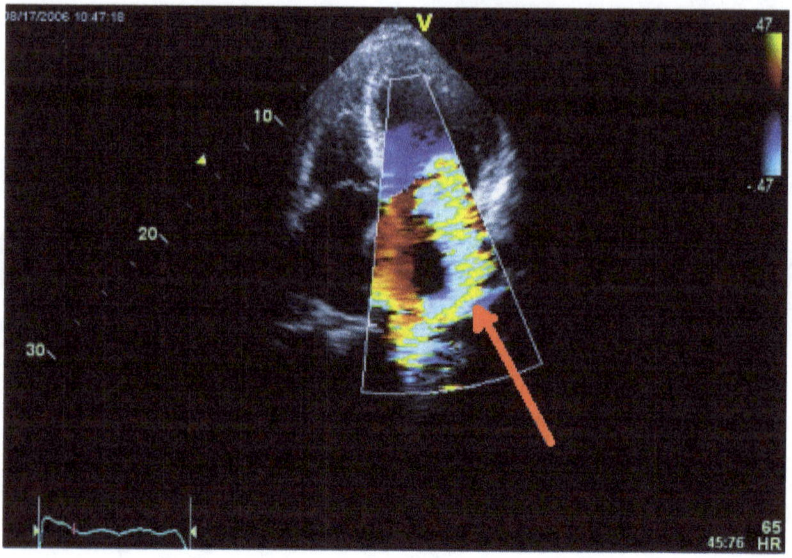

Fig. 6. Color Doppler of severe mitral regurgitation. We see that the jet of mitral regurgitation has a large extent in the Doppler color flow and extends to the posterior wall of the left atrium, a sign suggesting a high degree of regurgitant volume [12].

as jets of mosaic color [7, 9]. Color Doppler ultrasound echocardiography is extremely useful for rapid visual assessment of heart valve insufficiency and stenosis. It also indicates abnormal flows as in an atrial or ventricular septal abnormality. The increased cardiac output, even in the case of anatomically normal aortic valves, can cause physiological murmurs due to the turbulence. This sometimes occurs in pregnant women who have increased cardiac output (increased ejection velocities) and reduced blood viscosity due to anaemia. Both factors increase the Reynolds number, which increases the likelihood of turbulence [9]. Regarding cardiac output, a significant increase can be observed at as early as 5 weeks of pregnancy. Between the 16th and the 20th week, an increase of 50% above pre-pregnancy levels is expected. Typically, by the end of the 20th week, this increase stops and cardiac output levels remain stable until term. This significant hemodynamic change of increased cardiac output, coexisting with an underlying heart disease in pregnant women, may contribute toward decompensation and fetal death.

2. The Korteweg de Vries–Burgers Model on Cardiac Hemodynamics

2.1. *The KdV–B equation*

This section discusses the Korteweg–de Vries–Burgers (KdV–B) equation [16],

$$u_t + \gamma u u_x - \alpha u_{xx} + \beta u_{xxx} = 0, \quad u = u(x,t), \quad (x,t) \in \mathbb{R} \times (0,\infty), \tag{30}$$

serving as the simplest nonlinear and non-integrable wave model balancing nonlinearity, dissipation, and dispersion. Its derivation comes through considering the Womersley flow assumptions, along with a thin and viscoelastic tube [2, 4, 10]. Focus shall be set on its applications to cardiac hemodynamics [1, 4, 16], where the dicrotic notch stage is observed. The obtained analytical solutions are in qualitative agreement with the dicrotic notch stage.

2.1.1. *The hyperbolic tangent method*

Traveling wave solutions of the n-dimensional KdV–B and compound KdV–B equations ($n \in \mathbb{N}$) will be derived through the hyperbolic tangent, or Tanh, method [14, 16]. Conclusions will be drawn on the method's applicability to nonlinear evolution PDEs. Consider the

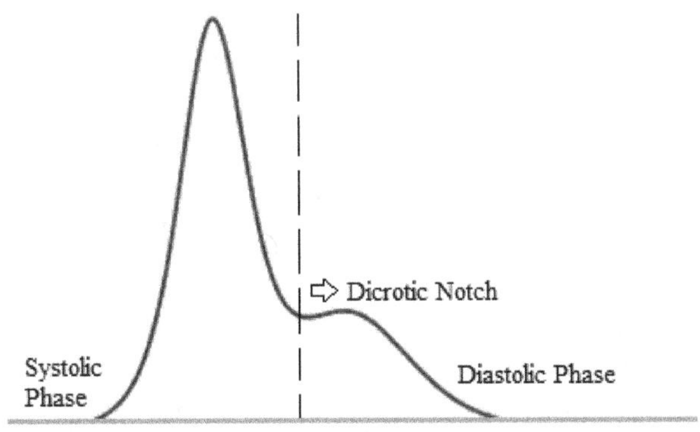

Fig. 7. The dicrotic notch, between the phases of systolic and diastolic flow, in the formation of the cardiac pulse.

traveling wave transform of u:

$$\begin{cases} u(x,t) = u(\zeta), \\ \zeta = \mu(x - \lambda t), \quad \mu > 0, \quad \lambda \neq 0. \end{cases} \tag{31}$$

Here, μ represents the wave number and λ the, perhaps unknown, velocity.

Now, (31) transforms the KdV–B equation, (30), into the ODE for $u(\zeta)$:

$$-\lambda u'(\zeta) + \gamma u(\zeta)u'(\zeta) - \alpha\mu u''(\zeta) + \beta\mu^2 u'''(\zeta) = 0. \tag{32}$$

Integrating (32) yields

$$-\lambda u(\zeta) + \frac{\gamma}{2}u^2(\zeta) - \alpha\mu u'(\zeta) + \beta\mu^2 u''(\zeta) = C. \tag{33}$$

The concept behind the tanh method lies on the key property of the tanh function's derivatives all being written in terms of the tanh function itself. The rather useful identity is used:

$$\operatorname{sech}^2\zeta = 1 - \tanh^2\zeta, \quad \zeta \in \mathbb{R}. \tag{34}$$

The point here is that the differential equation reduces to a polynomial equation for successive powers of tanh. Introducing the new variable,

$$y = \tanh\zeta, \tag{35}$$

the solution(s) sought shall take the form

$$u(y) = \sum_{n=0}^{N} a_n y^n, \tag{36}$$

limiting them to solitary and shock wave profiles. Chain differentiation provides with

$$\begin{cases} \dfrac{d}{d\zeta} = \dfrac{d}{dy}\dfrac{dy}{d\zeta} = (1-y^2)\dfrac{d}{dy}, \\ \dfrac{d^2}{d\zeta^2} = \dfrac{d}{d\zeta}\dfrac{d}{d\zeta} = (1-y^2)\left(-2y\dfrac{d}{dy} + (1-y^2)\dfrac{d^2}{dy^2}\right). \end{cases} \tag{37}$$

The positive integer value of N is determined after substituting (36), along with (37), into (33) and balancing the resulting highest order nonlinear terms with the highest order linear terms. As a result, it is definitely required that $a_{N+1} = 0$ and $a_N \neq 0$ for a particular N. It turns out that N equals 1 or 2 in most cases. As soon as N is determined, substitution of (36) into (33) produces an algebraic system for a_n, $n = 0, 1, \ldots, N$. Based on the problem under study, μ either remains fixed or undetermined, whereas the velocity λ is always a function of μ. In this chapter, $N = 2$, hence substituting (36) into (33) and setting the like powers of equal to zero, one obtains the following solutions:

$$\begin{cases} \lambda = \pm\dfrac{\sqrt{2}\sqrt{18\alpha^4 - 625C\beta^2\gamma}}{25\beta}, \\ \mu = \dfrac{\alpha}{10\beta}, \\ a_0 = \dfrac{3\alpha^2 + 25\beta\lambda}{25\beta\gamma}, \\ a_1 = -\dfrac{6\alpha^2}{25\beta\gamma}, \\ a_2 = -\dfrac{3\alpha^2}{25\beta\gamma}. \end{cases} \tag{38}$$

Combining (38) with (36), while using (35), provides with

$$u(\zeta) = \frac{\lambda}{\gamma} + \frac{3\alpha^2\left(\mathrm{sech}^2\zeta - 2\tanh\zeta\right)}{25\beta\gamma}. \tag{39}$$

Remark 2. The solution consists of both solitary and shock parts, a behavior also highlighted in the formational stages of the cardiac pulse [16].

2.1.2. *Homotopy analysis method*

The basic idea of this section comes from the topological concept of homotopy. For a deeper understanding behind the notion of the homotopy method, we refer to [5].

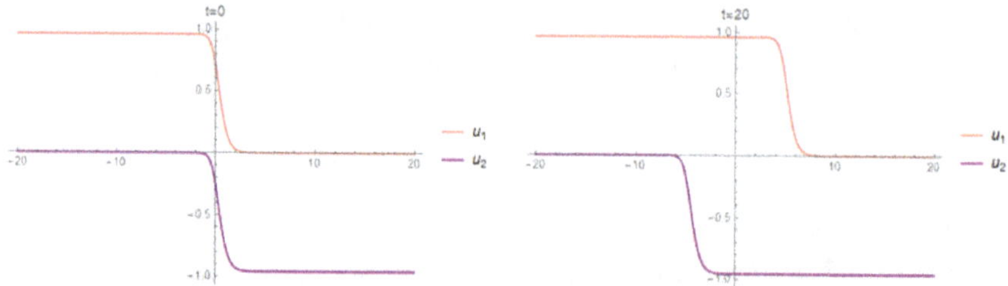

Fig. 8. A right- (u_1) and left-moving (u_2) wavefront solutions of KdV–B, for the values of $\alpha = 0.1$, $\beta = 0.01$, and $\gamma = 0.5$, with $x \in [-20, 20]$ and $t \in \{0, 20\}$.

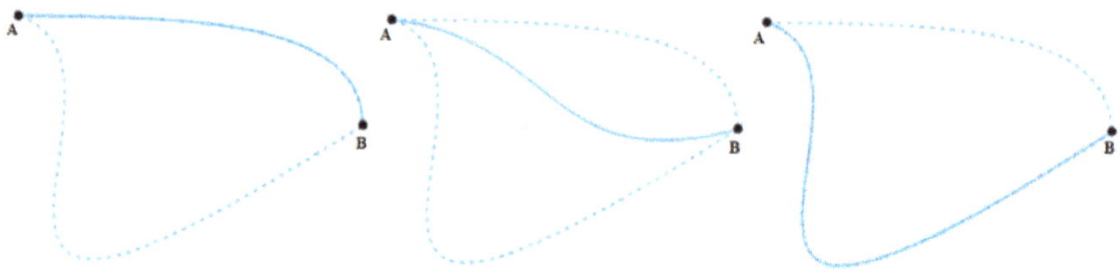

Fig. 9. A homotopy between two homotopic dotted paths.

The homotopy method will now be applied to KdV–B, where the initial guess is the wavefront:

$$u_0(x) = \frac{e^{-x}}{1 + e^{-x}}. \tag{40}$$

The linear operator is consider to be

$$L[u] := u_t, \tag{41}$$

with $L[u] = 0 \Leftrightarrow u = u(x)$. The $R_m(u_{m-1})$ term is computed as

$$R_m(u_{m-1}) = (u_{m-1})_t + \sum_{n=0}^{m-1} [u_{m-1-n}(u_n)_x] - \alpha(u_{m-1})_{xx} + \beta(u_{m-1})_{xxx}. \tag{42}$$

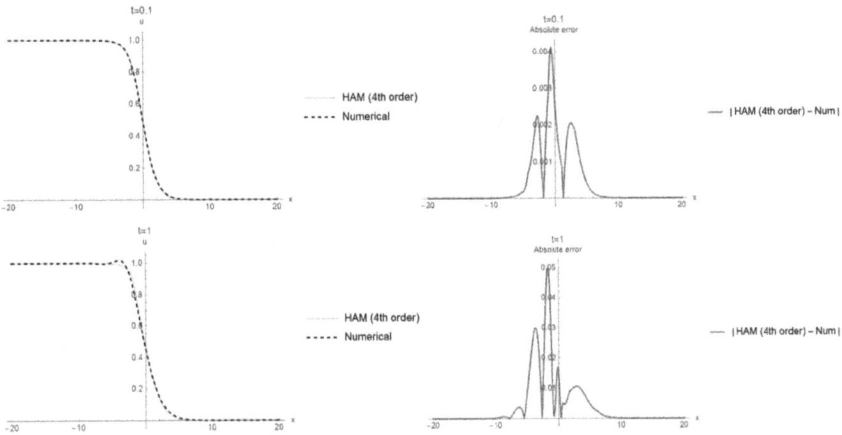

Fig. 10. Graphical representation, in two times, of the 4th-order homotopy solution of KdV–B, for $\hbar = -0.17$. The determination of the appropriate $\hbar$ comes by minimizing the sum of squared residuals (SSRs) between the approximate and numerical solutions [3]. It is further assumed that $\alpha = 0.2$, $\beta = 1$, and $\gamma = 0.1$. Regarding the black dashed line, it refers to the numerical solution of the equation.

2.2. *The 2-dimensional KdV–B equation*

Consider the 2-dimensional KdV–B equation [6]:

$$(u_t + \gamma u u_x - \alpha u_{xx} + \beta u_{xxx})_x + u_{yy} = 0, \quad u = u(x, y, t). \tag{43}$$

Application of the traveling wave transform,

$$\begin{cases} u = u(\zeta), \\ \zeta = \mu(x + y - \lambda t), \quad \mu > 0, \quad \lambda \neq 0, \end{cases} \tag{44}$$

transforms, through chain differentiation, the 2-dimensional KdV–B equation into

$$\left(-\lambda u' + \gamma u u' - \alpha\mu u'' + \beta\mu^2 u'''\right)' + u'' = 0, \quad u = u(\zeta). \tag{45}$$

Double integration with respect to ζ, taking integration constants to equal zero, provides with

$$\beta\mu^2 u'' - \alpha\mu u' + \frac{\gamma}{2}u^2 + (1 - \lambda)u = 0. \tag{46}$$

Under similar manipulations to those on the KdV–B case, one is led to the kink-form traveling wave solutions,

$$u(\zeta) = \frac{\lambda - 1}{\gamma} + \frac{3\alpha^2\left(\mathrm{sech}^2\zeta - 2\tanh\zeta\right)}{25\beta\gamma} \tag{47}$$

with

$$\begin{cases} \lambda = 1 \pm \dfrac{6\alpha^2}{25\beta}, \\ \mu = \dfrac{\alpha}{10\beta}. \end{cases} \tag{48}$$

2.2.1. *A generalization to the n-dimensional KdV–B equation*

Generalizing both (43) and the ideas followed in [6], one obtains the n-dimensional KdV–B equation:

$$\begin{aligned} &(u_t + \gamma u u_{x_1} - \alpha u_{x_1 x_1} + \beta u_{x_1 x_1 x_1})_{x_1} + \sum_{i=2}^{n} u_{x_i x_i} = 0, \\ &u = u(x_1, x_2, \ldots, x_n, t). \end{aligned} \tag{49}$$

Consider a generalization of (44),

$$\begin{cases} u = u(\zeta), \\ \zeta = \mu\left(\displaystyle\sum_{i=1}^{n} x_i - \lambda t\right), \quad \mu > 0, \quad \lambda \neq 0. \end{cases} \tag{50}$$

The tanh method, applied to (53), provides with the traveling wave solutions

$$u(\zeta) = \frac{\lambda - n + 1}{\gamma} + \frac{3\alpha^2\left(\mathrm{sech}^2\zeta - 2\tanh\zeta\right)}{25\beta\gamma}, \tag{51}$$

with

$$\begin{cases} \lambda = n - 1 \pm \dfrac{6\alpha^2}{25\beta}, \\ \mu = \dfrac{\alpha}{10\beta}, \end{cases} \qquad n \in \{2, \ldots\}. \tag{52}$$

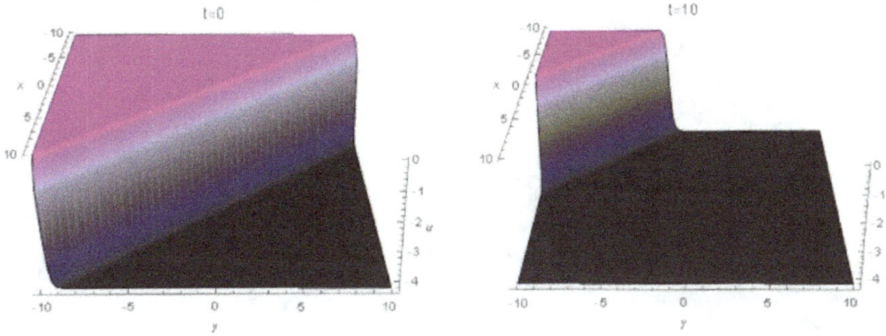

Fig. 11. A left-moving, kink-shaped, traveling wave of 2-dimensional KdV–B, for the values of $\alpha = 0.3$, $\beta = 0.01$, and $\gamma = 1$, with $(x, y) \in [-10, 10] \times [-10, 10]$ and $t \in \{0, 10\}$.

2.2.2. *The n-dimensional compound KdV–B equation*

Consider the n-dimensional compound KdV–B (cKdV–B)

$$\left(u_t + \gamma u u_{x_1} - \delta u^2 u_{x_1} - \alpha u_{x_1 x_1} + \beta u_{x_1 x_1 x_1}\right)_{x_1} + \sum_{i=2}^{n} u_{x_i x_i} = 0, \quad \delta > 0 \tag{53}$$

along with (50). The equation generalizes the one presented in [6], connecting the modified KdV and Burgers PDEs. More on these two are discussed in the process. Regarding cKdV–B, one obtains $N = 1$ and the tanh method provides with the traveling wave solution

$$u(\zeta) = \frac{3\sqrt{\beta}\gamma + \sqrt{6}\alpha\sqrt{\delta}}{6\sqrt{\beta}\delta} + \frac{\sqrt{6}\sqrt{\beta}\mu \tanh\zeta}{\sqrt{\delta}}, \tag{54}$$

with

$$\begin{cases} \lambda = n - 1 + \dfrac{\gamma^2}{6\delta} - \dfrac{\alpha\left(4\alpha + \dfrac{\sqrt{6}\sqrt{\beta}\gamma}{\sqrt{\delta}}\right)}{18\beta}, \\ \mu = \dfrac{\sqrt{\dfrac{-2\alpha^2\delta + 3\beta\left(\gamma^2 + 4\delta(n - 1 - \lambda)\right)}{\beta^2\delta}}}{2\sqrt{6}}, \end{cases} \quad n \in \mathbb{N}. \tag{55}$$

Remark 3. The tanh method allows for a straightforward algorithmic process to derive physically relevant kink and pulse form

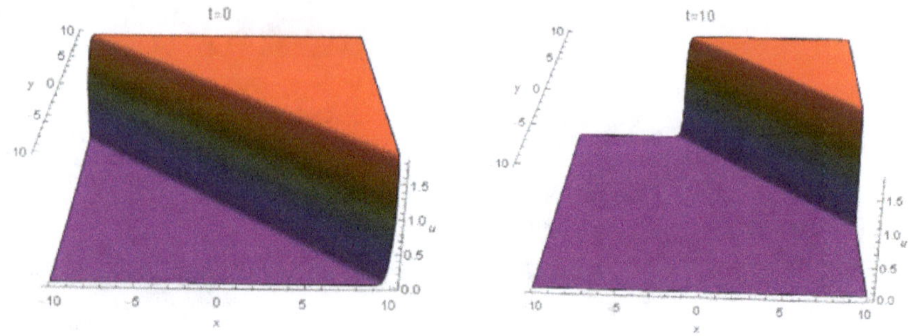

Fig. 12. A right-moving, kink-shaped, traveling wave of the 2-dimensional compound KdV–B, for the values of $\alpha = 0.1$, $\beta = 0.01$, $\gamma = \delta = 1$, with $(x, y) \in [-10, 10] \times [-10, 10]$ and $t \in \{0, 10\}$.

solutions for a wide variety of PDEs. Its novelty lies in circumventing integration and obtaining closed-form solutions at the cost of some algebraic manipulations.

2.3. *Comparison with numerical data from multi-dimensional numerical methods*

The emergence of advanced biomedical technologies has contributed to the development of biofluid dynamics. In hemodynamics, continuous effort is put on simulating the biomechanical behavior of arterial vessels, by means of advanced Fluid–Structure Interaction (FSI) techniques [15].

In what follows, cardiovascular data obtained from FSI methodology are compared with the time-evolved solution of the KdV–B equation. The FSI approach utilizes a mixed Euler–Lagrange formulation in studying aneurysmal blood flow during the cardiac cycle. The use of generalized curvilinear coordinates allows expressing the transport equations into a moving body-fitted reference frame. The physical velocities, u and v, are now transformed to U and V, being the contravariant velocity components, (56), considering the arbitrary motion of the domain. Moreover, the arbitrary grid motion velocities, $\dot{x}$, $\dot{y}$, are obtained numerically [13]:

$$U = (u - \dot{x})\, y_\eta - (v - \dot{y})\, x_\eta, \quad V = (v - \dot{y})\, x_\xi - (u - \dot{x})\, y_\xi, \tag{56}$$

$$q_1 = x_\eta^2 + y_\eta^2, \quad q_2 = x_\xi x_\eta + y_\xi y_\eta, \quad q_3 = x_\xi^2 + y_\xi^2, \tag{57}$$

In the above, x_ξ, x_η, y_ξ, and y_η denote the metrics of the transformation, computed locally on each cell volume [13]. Based on the above mixed formulation, the dimensionless transport equations are transformed to follow the exact geometrical vessel characteristics (58)–(60). Additionally, J is the determinant of the inverse Jacobian of the transformation, described in [15]:

$$\frac{\partial J}{\partial t} + \frac{\partial U}{\partial \xi} + \frac{\partial V}{\partial \eta} = 0, \tag{58}$$

$$\begin{aligned}\frac{\partial (Ju)}{\partial t} + \frac{\partial (Uu)}{\partial \xi} + \frac{\partial (Vu)}{\partial \eta} &= -\left(y_\eta \frac{\partial p}{\partial \xi} - y_\xi \frac{\partial p}{\partial \eta}\right) \\ &+ \frac{\partial}{\partial \xi}\left[\frac{1}{JRe}\left(q_1 \frac{\partial u}{\partial \xi} - q_2 \frac{\partial u}{\partial \eta}\right)\right] + \frac{\partial}{\partial \eta}\left[\frac{1}{JRe}\left(q_3 \frac{\partial u}{\partial \eta} - q_2 \frac{\partial u}{\partial \xi}\right)\right],\end{aligned} \tag{59}$$

$$\begin{aligned}\frac{\partial (Jv)}{\partial t} + \frac{\partial (Uv)}{\partial \xi} + \frac{\partial (Vv)}{\partial \eta} &= -\left(x_\xi \frac{\partial p}{\partial \eta} - x_\eta \frac{\partial p}{\partial \xi}\right) \\ &+ \frac{\partial}{\partial \xi}\left[\frac{1}{JRe}\left(q_1 \frac{\partial v}{\partial \xi} - q_2 \frac{\partial v}{\partial \eta}\right)\right] + \frac{\partial}{\partial \eta}\left[\frac{1}{JRe}\left(q_3 \frac{\partial v}{\partial \eta} - q_2 \frac{\partial v}{\partial \xi}\right)\right].\end{aligned} \tag{60}$$

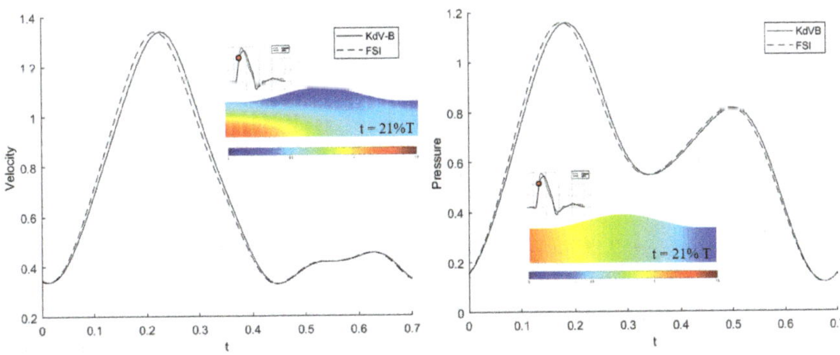

Fig. 13. Comparison of the FSI method with the KdV–B mathematical model, throughout the entire cardiac cycle. The values for the KdV–B equation, that provide similar results with the FSI equations, are $\alpha = 10^{-3}$, $\beta = 10^{-4}$, and $\gamma = 1$.

The finite volume method is used to numerically solve the aforementioned problem, under appropriate initial and boundary conditions. More information on FSI techniques can be found elsewhere [15]. Comparison of the FSI method with the KdV–B mathematical model, throughout the entire cardiac cycle, shows good agreement. The FSI equations are written in a dimensionalized form and the Reynolds number, Re, for the simulations was chosen to be, $Re = 10^3$. The values for the KdV–B equation, that provide similar results with the FSI equations, are $\alpha = 10^{-3}$, $\beta = 10^{-4}$, and $\gamma = 1$.

Remark 4. As of current analysis, the KdV–B mathematical model seems to be quite reliable as a cardiac hemodynamics model. It also compares very well with other advanced multi-dimensional numerical methods, such as the presented FSI numerical method.

References

[1] Antar, N. and Demiray, H. Weakly nonlinear waves in a prestressed thin elastic tube containing a viscous fluid. *International Journal of Engineering Science*, 37(14), 1859–1876 (1999).

[2] Demiray, H. Wave propagation through a viscous fluid contained in a prestressed thin elastic tube. *International Journal of Engineering Science*, 30(11), 1607–1620 (1992).

[3] Draper, N.R. and Smith, H. Applied Regression Analysis, John Wiley & Sons, 326 (1998).

[4] Erbay, H., Erbay, S. and Dost, S. Wave propagation in fluid filled nonlinear viscoelastic tubes. *Acta Mechanica*, 95(1–4), 87–102 (1992).

[5] Felias, A.C., Kyriakoudi, K.C., Mpiraki, K.N. and Xenos, M.A. Analytic and numerical solutions to nonlinear partial differential equations in biomechanics. *Analysis, Geometry, Nonlinear Optimization and Applications*, World Scientific Publishing, Singapore, pp. 331–403 (2023).

[6] Feng, Z. and Meng, Q. Exact solution for a two-dimensional kdv-burgers-type equation with nonlinear terms of any order. *Discrete & Continuous Dynamical Systems-B* 7(2), 285 (2007).

[7] Garfinkel, A., Shevtsov, J. and Guo, Y. Modeling life. Springer, Cham (2017).

[8] Gersten, K. Hermann schlichting and the boundary-layer theory. In *Hermann Schlichting–100 Years*, pp. 3–17. Springer (2009).

[9] Klabunde, R. *Cardiovascular Physiology Concepts.* Lippincott Williams & Wilkins (2011).

[10] Malatos, M., Raptis, A., Xenos, M.A., Kouvelos, G., Giannoukas, A. and Verhoeven, E. A multiscale model for hemodynamic properties prediction after fenestrated endovascular aneurysm repair. A pilot study. *Hell. Vasc. J,* 1(2), 73–79 (2019).

[11] Nield, D. and Kuznetsov, A. Forced convection with laminar pulsating flow in a channel or tube. *International Journal of Thermal Sciences,* 46(6), 551–560 (2007).

[12] Sawyer, D.B. and Vasan, R.S. *Encyclopedia of Cardiovascular Research and Medicine*, Elsevier (2017).

[13] Shyy, W., Udaykumar, H., Rao, M.M. and Smith, R.W. *Computational Fluid Dynamics with Moving Boundaries.* Courier Corporation (2012).

[14] Wazwaz, A.M. The tanh–coth and the sech methods for exact solutions of the jaulent–miodek equation. *Physics Letters A*, 366(1–2), 85–90 (2007).

[15] Xenos, M.A. An euler–lagrange approach for studying blood flow in an aneurysmal geometry. *Proceedings of the Royal Society A: Mathematical, Physical and Engineering Sciences*, 473(2199), 20160,774 (2017).

[16] Xenos, M.A. and Felias, A.C. Nonlinear dynamics of the kdv-b equation and its biomedical applications. In *Nonlinear Analysis, Differential Equations, and Applications*, pp. 765–793. Springer (2021).

[17] Abramowitz, M., Stegun, I.A. and Romer, R.H. *Handbook of Mathematical Functions with Formulas, Graphs, and Mathematical Tables*, American Association of Physics Teachers (1988).

https://doi.org/10.1142/9789811293009_0013

Chapter 13

Lyusternik–Schnirelmann Theory for Weighted Sobolev Spaces

Lucas Fresse[*,‡] and Viorica V. Motreanu[†,§]

[*] *Université de Lorraine, Institut Elie Cartan*
54506 Vandoeuvre lès Nancy, France,
[†] *Lycée Varoquaux, 10, rue Jean Moulin*
54510 Tomblaine, France,
[‡] *lucas.fresse@univ-lorraine.fr*
[§] *vmotreanu@gmail.com*

We consider eigenvalue problems for the weighted p-Laplacian in weighted Sobolev spaces. We use Lyusternik–Schnirelmann theory which allows us to show the existence of an infinite sequence of eigenvalues in this context. We also establish topological and variational properties of the spectrum. We rely on embedding assumptions for the weighted Sobolev spaces. This chapter ends with an example of domain and weights for which these embedding assumptions are fulfilled.

1. Introduction

There is a classical theory for the spectrum of the negative p-Laplacian on the Sobolev spaces $W^{1,p}(\Omega)$ (corresponding to an eigenvalue problem under Neumann boundary conditions) and $W_0^{1,p}(\Omega)$ (eigenvalue problem under Dirichlet boundary conditions).

More precisely, $\lambda \in \mathbb{R}$ is an eigenvalue of $(-\Delta_p, W^{1,p}(\Omega))$ if the equation

$$-\Delta_p u = \lambda |u|^{p-2} u \quad \text{in } W^{1,p}(\Omega)^*$$

has a non-trivial solution $u \in W^{1,p}(\Omega)$, called eigenfunction. In the quasi-linear case ($p \neq 2$), the negative p-Laplacian is a non-linear operator, hence the standard theory for linear operators cannot be used. The Lyusternik–Schnirelmann theory permits to show the existence of an infinite sequence of eigenvalues (λ_n) such that $\lim_{n\to+\infty} \lambda_n = +\infty$. Moreover, the spectrum enjoys topological properties: it is closed, the first eigenvalue is isolated, and the first two eigenvalues of the spectrum have variational characterizations. We refer to [1, 2, 5, 12]; see also, e.g., [15, §9.2] and references therein.

In this chapter, we consider an eigenvalue problem involving the weighted p-Laplacian $\Delta_{p,w} u = \operatorname{div}(w|\nabla u|^{p-2}\nabla u)$ in the context of weighted Sobolev spaces $W^{1,p}(\Omega, w_0, w)$ and $W_0^{1,p}(\Omega, w_0, w)$. Specifically, the eigenvalue problem for $W^{1,p}(\Omega, w_0, w)$ concerns the following equation:

$$-\Delta_{p,w} u = \lambda w_0 \xi |u|^{p-2} u \quad \text{in } W^{1,p}(\Omega, w_0, w)^*,$$

and it involves in fact three weights:

- $w, w_0 : \Omega \to (0, +\infty)$ which are (locally) integrable on the domain Ω and enter the definition of the weighted Sobolev space,
- $\xi : \Omega \to [0, +\infty)$, $\xi \not\equiv 0$, which belongs to $L^\infty(\Omega)$.

See Section 2 for a review on weighted Sobolev spaces and weighted p-Laplacian.

In this setting, we develop a Lyusternik–Schnirelmann approach for eigenvalue problems on weighted Sobolev spaces. We show the existence of a full sequence of eigenvalues with limit $+\infty$, and we establish topological properties of the spectrum and variational characterizations of the first eigenvalues. Our main results are stated in Section 3. We mainly focus on the eigenvalue problem for $W^{1,p}(\Omega, w_0, w)$, but the problem for $W_0^{1,p}(\Omega, w_0, w)$ is also considered.

Note that in the case where $w = w_0 = 1$ but $\xi \not\equiv 1$, the general theory for eigenvalues is quite similar to the classical theory (where $\xi \equiv 1$); see, e.g., [15, §9.2]. We also refer to [6, 7], and references therein, for other approaches to eigenvalue problems in weighted Sobolev spaces.

For making this chapter as self-contained as possible, we give a review of classical Lyusternik–Schnirelmann theory in Section 4. The proofs of the main theorems, which rely on this theory, are given in Section 5.

Beyond the presence of weights, the main differences with respect to the classical setting of $W^{1,p}(\Omega)$ and $W_0^{1,p}(\Omega)$ are the following:

- We make no assumption of regularity of the domain Ω nor of the weights w, w_0. As a result, we have no information on the regularity of solutions to the eigenvalue problem. This prevents us from relying on nonlinear regularity theory or maximum principles and from dealing with the notion of nodal domains of solutions. This also implies that for several of the results shown in Section 3, though the analogous results exist in the classical theory, the proofs that we give are quite different.
- In the classical setting, the theory heavily relies on the Sobolev embedding theorem. There is less knowledge on embedding properties for weighted Sobolev spaces. In this respect, we refer, for instance, to [11] and references therein. For showing our results, we rely on embedding assumptions formulated in Section 2.3. In Section 6, we give a general criterion and an example where these embedding assumptions are fulfilled.

2. Weighted Sobolev Spaces and Weighted p-Laplacian

2.1. *Doubly weighted Sobolev spaces*

Let Ω be a domain in $\mathbb{R}^N$ $(N \geq 1)$. We say that w is a weight on Ω if $w \in L^1_{\text{loc}}(\Omega)$ and $w(x) > 0$ for a.a. $x \in \Omega$. For $p \in (1, +\infty)$, we use the notation

$$\mathcal{W}_p(\Omega) = \{w \text{ weight on } \Omega : w^{\frac{1}{1-p}} \in L^1_{\text{loc}}(\Omega)\}.$$

Given a weight w on Ω, we consider the weighted Lebesgue space

$$L^p(\Omega, w) = \left\{u : \Omega \to \mathbb{R} \text{ measurable} : \int_\Omega |u(x)|^p w(x)\, dx < +\infty\right\},$$

which is a Banach space for the norm

$$\|u\|_{L^p(\Omega,w)} := \left(\int_\Omega |u(x)|^p w(x)\,dx\right)^{\frac{1}{p}}.$$

Since $p > 1$, the space $L^p(\Omega, w)$ is uniformly convex, and so it is reflexive (see, e.g., [7, §1.2]) and, by Hölder's inequality, $L^p(\Omega, w) \subset L^1_{\mathrm{loc}}(\Omega)$ whenever $w \in \mathcal{W}_p(\Omega)$.

Given $p \in (1, +\infty)$ and $w_0, w \in \mathcal{W}_p(\Omega)$, let the (w_0, w)-weighted Sobolev space

$$W^{1,p}(\Omega, w_0, w) = \left\{u \in L^p(\Omega, w_0) : \frac{\partial u}{\partial x_i} \in L^p(\Omega, w),\ \forall i = 1, \ldots, N\right\},$$

which is a Banach space for the norm

$$\|u\|_{W^{1,p}(\Omega,w_0,w)} := \|u\|_{L^p(\Omega,w_0)} + \|\,|\nabla u|\,\|_{L^p(\Omega,w)}$$

(where $\nabla u = (\frac{\partial u}{\partial x_1}, \ldots, \frac{\partial u}{\partial x_N})$ stands for the gradient of u in the distributional sense, which makes sense since $L^p(\Omega, w_0) \subset L^1_{\mathrm{loc}}(\Omega)$, and $|\cdot|$ stands for the Euclidean norm of $\mathbb{R}^N$). Moreover, the space $(W^{1,p}(\Omega, w_0, w), \|\cdot\|_{W^{1,p}(\Omega,w_0,w)})$ is uniformly convex (as shown in the proof of Theorem 4 in [10]), and thus the space $W^{1,p}(\Omega, w_0, w)$ is reflexive (see also [10, Proposition 1]).

Let $C_c^\infty(\Omega)$ denote the space of C^∞ functions with compact support in Ω, which is a linear subspace of $W^{1,p}(\Omega, w_0, w)$. We define $W_0^{1,p}(\Omega, w_0, w)$ as the closure of $C_c^\infty(\Omega)$ in $W^{1,p}(\Omega, w_0, w)$. In the sequel, we always consider it endowed with the topology induced by the seminorm

$$u \mapsto \|\,|\nabla u|\,\|_{L^p(\Omega,w)}. \tag{1}$$

In the case where $w_0, w \in \mathcal{W}_p(\mathbb{R}^N)$ and $|\mathbb{R}^N \setminus \Omega| > 0$ (here $|\cdot|$ denotes the Lebesgue measure of $\mathbb{R}^N$), it is known from [11, Proposition 2.4] that $W_0^{1,p}(\Omega, w_0, w)$ does not contain the nonzero constants and then the seminorm (1) is a norm on $W_0^{1,p}(\Omega, w_0, w)$.

If $w_0 = w = 1$, we obtain the usual Lebesgue space $L^p(\Omega)$ and the usual Sobolev spaces $W^{1,p}(\Omega)$ and $W_0^{1,p}(\Omega)$. We refer to [11] for proofs of the above properties and deeper considerations on the doubly weighted Sobolev spaces.

2.2. *Weighted p-Laplacian operators and eigenvalues*

Let $p \in (1, +\infty)$ and $w_0, w \in \mathcal{W}_p(\Omega)$. With the notation in Section 2.1, consider the w-weighted p-Laplacian operator

$$\Delta_{p,w} u = \operatorname{div}(w|\nabla u|^{p-2}\nabla u)$$

for all $u \in W^{1,p}(\Omega, w_0, w)$ (resp., for all $u \in W_0^{1,p}(\Omega, w_0, w)$). If $w_0 = w = 1$, it coincides with the usual p-Laplacian operator Δ_p.

Our goal is to develop the Lyusternik–Schnirelmann theory for the w-weighted p-Laplacian operator on $W^{1,p}(\Omega, w_0, w)$ (resp., on $W_0^{1,p}(\Omega, w_0, w)$). Precisely, let $\xi \in L^\infty(\Omega)$ be such that $\xi(x) \geq 0$ for a.a. $x \in \Omega$, $\xi \neq 0$, and consider the following eigenvalue problem, where $\lambda \in \mathbb{R}$ is a parameter:

Find $u \in W^{1,p}(\Omega, w_0, w)$ (resp., $u \in W_0^{1,p}(\Omega, w_0, w)$), $u \neq 0$, such that

$$-\Delta_{p,w} u = \lambda \xi w_0 |u|^{p-2} u.$$

Definition 1. (a) We say that $\lambda \in \mathbb{R}$ is an *eigenvalue of the negative w weighted p-Laplacian* $-\Delta_{p,w} : W^{1,p}(\Omega, w_0, w) \to W^{1,p}(\Omega, w_0, w)^*$ *with respect to* ξ if there exists $u \in W^{1,p}(\Omega, w_0, w)$, $u \neq 0$, such that

$$\int_\Omega w|\nabla u|^{p-2}\nabla u \cdot \nabla v \, dx = \lambda \int_\Omega \xi w_0 |u|^{p-2} uv \, dx \quad \text{for all } v \in W^{1,p}(\Omega, w_0, w). \tag{2}$$

(b) We say that $\lambda \in \mathbb{R}$ is an *eigenvalue of the negative w-weighted p-Laplacian* $-\Delta_{p,w} : W_0^{1,p}(\Omega, w_0, w) \to W_0^{1,p}(\Omega, w_0, w)^*$ *with respect to* ξ if there exists $u \in W_0^{1,p}(\Omega, w_0, w)$, $u \neq 0$, such that

$$\int_\Omega w|\nabla u|^{p-2}\nabla u \cdot \nabla v \, dx = \lambda \int_\Omega \xi w_0 |u|^{p-2} uv \, dx \quad \text{for all } v \in W_0^{1,p}(\Omega, w_0, w). \tag{3}$$

(c) For short, a number λ as in part (a) (resp., (b)) will be referred to as an eigenvalue of the triple $(-\Delta_{p,w}, W^{1,p}(\Omega, w_0, w), \xi)$ (resp., $(-\Delta_{p,w}, W_0^{1,p}(\Omega, w_0, w), \xi)$), and a function u as in part (a) (resp., (b)) will be called an *eigenfunction* of this triple corresponding to λ.

Remark 2. (a) If λ is an eigenvalue of $(-\Delta_{p,w}, W^{1,p}(\Omega, w_0, w), \xi)$ with eigenfunction u, then

$$\int_\Omega w|\nabla u|^p\,dx = \lambda \int_\Omega \xi w_0 |u|^p\,dx, \tag{4}$$

and therefore, $\lambda \geq 0$. Moreover, $\lambda = 0$ is an eigenvalue of $(-\Delta_{p,w}, W^{1,p}(\Omega, w_0, w), \xi)$ and the corresponding eigenfunctions are the constant functions (since $w(x) > 0$ for a.a. $x \in \Omega$).

(b) Let $w_0, w \in \mathcal{W}_p(\mathbb{R}^N)$ and $|\mathbb{R}^N \setminus \Omega| > 0$ so that $W_0^{1,p}(\Omega, w_0, w)$ does not contain the nonzero constants (see Section 2.1). If λ is an eigenvalue of $(-\Delta_{p,w}, W_0^{1,p}(\Omega, w_0, w), \xi)$ with eigenfunction u, then $u \in W_0^{1,p}(\Omega, w_0, w) \setminus \{0\}$ and satisfies (4). Thus $\nabla u \neq 0$ (since u is non-constant), which, combined with the fact that $w(x) > 0$ for a.a. $x \in \Omega$, yields $\lambda > 0$ and $\int_\Omega \xi w_0 |u|^p\,dx > 0$.

2.3. *Embedding assumptions on $W^{1,p}(\Omega, w_0, w)$ and equivalent norm*

Let $p \in (1, +\infty)$, $w_0, w \in \mathcal{W}_p(\Omega)$, and $\xi \in L^\infty(\Omega)$, $\xi \geq 0$ for a.a. $x \in \Omega$, $\xi \not\equiv 0$. We assume the following:

(E$_1$) (i) For each relatively compact set A of positive Lebesgue measure such that $\overline{A} \subset \Omega$, there exist $c = c(\Omega, p, w_0, w, A) > 0$ and $\rho \in L^\infty(A)$ such that

$$\|u - u_A\|_{L^p(\Omega, w_0)} \leq c\| \,|\nabla u|\, \|_{L^p(\Omega, w)} \quad \text{for all } u \in W^{1,p}(\Omega, w_0, w),$$

with $u_A = \int_A \rho(x) u(x)\,dx$.

(ii) There exists a compact embedding

$$W^{1,p}(\Omega, w_0, w) \subset L^p(\Omega, w_0).$$

Note that condition (E$_1$) (i) is satisfied in the classical case (when $w_0 = w = 1$ and Ω is bounded, or more generally if Ω is of finite width; see, e.g., [15, Theorem 1.41, Remark 1.42]), while condition (E$_1$) (ii) holds in the classical case provided that the boundary of Ω is Lipschitz continuous (see, e.g., [15, Theorem 1.49]). In Section 6, we give an example of a domain Ω and weights $w, w_0 \neq 1$ for which condition (E_1) holds.

Lemma 3. *Assume* (E_1) (i) *and* $w_0 \in L^1(\Omega)$. *Then, the map*

$$u \mapsto \|u\|_{W^{1,p}(\Omega,w_0,w),\xi} := \left(\int_\Omega w|\nabla u|^p\,dx + \int_\Omega \xi w_0|u|^p\,dx\right)^{\frac{1}{p}}$$

is a norm on $W^{1,p}(\Omega, w_0, w)$, *equivalent to the norm* $\|\cdot\|_{W^{1,p}(\Omega,w_0,w)}$.

Proof. Taking into account that $\xi \in L^\infty(\Omega)$, $\xi \geq 0$, and $\xi \not\equiv 0$, it is easy to see that $\|\cdot\|_{W^{1,p}(\Omega,w_0,w),\xi}$ is a norm and to find a constant $M_1 > 0$ such that

$$\|u\|_{W^{1,p}(\Omega,w_0,w),\xi} \leq M_1\|u\|_{W^{1,p}(\Omega,w_0,w)} \quad \text{for all } u \in W^{1,p}(\Omega, w_0, w).$$

We have

$$\{x \in \Omega : \xi(x)w_0(x) > 0\} = \bigcup_{n\geq 1}\left\{x \in \Omega : \xi(x)w_0(x) \geq \frac{1}{n}\right\}.$$

Since the set $\{x \in \Omega : \xi(x)w_0(x) > 0\}$ has positive Lebesgue measure, we can find $n_0 \geq 1$ such that $D := \{x \in \Omega : \xi(x)w_0(x) \geq \frac{1}{n_0}\}$ has positive Lebesgue measure. Let $K \subset \Omega$ be a compact subset such that $A := K \cap D$ has positive measure. By hypothesis (E_1)(i), there exist $c > 0$ and a map $\rho \in L^\infty(A)$ such that

$$\|u - u_A\|_{L^p(\Omega,w_0)} \leq c\|\,|\nabla u|\,\|_{L^p(\Omega,w)} \quad \text{for all } u \in W^{1,p}(\Omega, w_0, w), \tag{5}$$

with $u_A = \int_A \rho(x)u(x)\,dx$. Using Hölder's inequality, we see that

$$\begin{aligned} |u_A| &\leq n_0\|\rho\|_{L^\infty(A)} \int_A \xi w_0|u|\,dx \\ &\leq n_0\|\rho\|_{L^\infty(A)} \left(\int_\Omega \xi w_0|u|^p\,dx\right)^{\frac{1}{p}} \left(\int_\Omega \xi w_0\,dx\right)^{\frac{p-1}{p}}. \end{aligned} \tag{6}$$

Using (5) and (6), we find a constant $M_2 > 0$ such that

$$\begin{aligned} \|u\|_{L^p(\Omega,w_0)} &\leq \|u - u_A\|_{L^p(\Omega,w_0)} + \|w_0\|^{\frac{1}{p}}_{L^1(\Omega)}|u_A| \\ &\leq M_2\|u\|_{W^{1,p}(\Omega,w_0,w),\xi} \end{aligned}$$

for all $u \in W^{1,p}(\Omega, w_0, w)$, and thus

$$\|u\|_{W^{1,p}(\Omega,w_0,w)} \le (M_2+1)\|u\|_{W^{1,p}(\Omega,w_0,w),\xi}$$
$$\text{for all } u \in W^{1,p}(\Omega, w_0, w).$$

The proof is now complete. □

2.4. *Embedding assumptions on* $W_0^{1,p}(\Omega, w_0, w)$

In this section, we assume that $|\mathbb{R}^N \setminus \Omega| > 0$. Let $p \in (1,+\infty)$, $w_0, w \in \mathcal{W}_p(\mathbb{R}^N)$, and $\xi \in L^\infty(\Omega)$, $\xi(x) \ge 0$ for a.a. $x \in \Omega$, $\xi \ne 0$. The conditions ensure that $\| |\nabla\cdot| \|_{L^p(\Omega,w)}$ is a norm on $W_0^{1,p}(\Omega, w_0, w)$ (see Section 2.1), and for this normed space we assume the following:

(E$_2$) There exists a compact embedding

$$W_0^{1,p}(\Omega, w_0, w) \subset L^p(\Omega, w_0).$$

In particular, by the continuity of this embedding, there exists a constant $c = c(\Omega, p, w_0, w) > 0$ such that

$$\|u\|_{L^p(\Omega,w_0)} \le c\| |\nabla u| \|_{L^p(\Omega,w)} \quad \text{for all } u \in W_0^{1,p}(\Omega, w_0, w).$$

Then the norms $\| |\nabla\cdot| \|_{L^p(\Omega,w)}$ and $\|\cdot\|_{W^{1,p}(\Omega,w_0,w)}$ on $W_0^{1,p}(\Omega, w_0, w)$ are equivalent, so $W_0^{1,p}(\Omega, w_0, w)$ endowed with the norm $\| |\nabla\cdot| \|_{L^p(\Omega,w)}$ is a Banach space, which is in addition reflexive.

Note that condition (E$_2$) is satisfied in the classical case (when $w_0 = w = 1$ and Ω is a bounded domain). See Section 6 for an example where condition (E$_2$) is satisfied for weights $w_0, w \ne 1$.

2.5. *Maps of type* $(S)_+$

Recall that an operator $A : X \to X^*$, defined on a reflexive Banach space X, is said to be a map of type $(S)_+$ if every sequence $(u_n)_{n\ge1} \subset X$ such that

$$u_n \rightharpoonup u \text{ in } X \quad \text{and} \quad \limsup_{n\to+\infty}\langle A(u_n), u_n - u\rangle \le 0,$$

for some $u \in X$, satisfies $u_n \to u$ in X. (Hereafter, $\rightharpoonup$ stands for the weak convergence.)

Let $p \in (1, +\infty)$, $w_0, w \in \mathcal{W}_p(\Omega)$, and $\xi \in L^\infty(\Omega)$, $\xi \geq 0$ a.e. in Ω, $\xi \neq 0$.

Proposition 4. (a) *The operator* $A : W^{1,p}(\Omega, w_0, w) \to W^{1,p}(\Omega, w_0, w)^*$ *defined for all* $u, v \in W^{1,p}(\Omega, w_0, w)$ *by*

$$\langle A(u), v\rangle = \int_\Omega w|\nabla u|^{p-2}\nabla u \cdot \nabla v \, dx + \int_\Omega \xi w_0 |u|^{p-2} uv \, dx \tag{7}$$

is bounded, and, under condition (E$_1$) (ii), *it is a map of type* $(S)_+$.
(b) *The operator* $-\Delta_{p,w} : W^{1,p}(\Omega, w_0, w) \to W^{1,p}(\Omega, w_0, w)^*$ *given by*

$$\langle -\Delta_{p,w}(u), v\rangle = \int_\Omega w|\nabla u|^{p-2}\nabla u \cdot \nabla v \, dx \text{ for all } u, v \in W^{1,p}(\Omega, w_0, w) \tag{8}$$

is bounded, and, under condition (E$_1$) (ii), *it is a map of type* $(S)_+$.
(c) *Assuming* $w_0, w \in \mathcal{W}_p(\mathbb{R}^N)$ *and that* $|\mathbb{R}^N \setminus \Omega| > 0$, *the operator* $-\Delta_{p,w} : W_0^{1,p}(\Omega, w_0, w) \to W_0^{1,p}(\Omega, w_0, w)^*$ *given by*

$$\langle -\Delta_{p,w}(u), v\rangle = \int_\Omega w|\nabla u|^{p-2}\nabla u \cdot \nabla v \, dx \text{ for all } u, v \in W_0^{1,p}(\Omega, w_0, w)$$

is bounded, and, under the assumption that the embedding $W_0^{1,p}(\Omega, w_0, w) \subset L^p(\Omega, w_0)$ *is continuous, it is a map of type* $(S)_+$.

Proof. We prove part (a). The proof of parts (b) and (c) is similar and in fact simpler. Using Hölder's inequality, we have for all $v, z \in W^{1,p}(\Omega, w_0, w)$ that

$$\begin{aligned} |\langle A(v), z\rangle| &\leq \| \, |\nabla v| \, \|_{L^p(\Omega,w)}^{p-1} \| \, |\nabla z| \, \|_{L^p(\Omega,w)} \\ &\quad + \|\xi\|_{L^\infty(\Omega)} \|v\|_{L^p(\Omega,w_0)}^{p-1} \|z\|_{L^p(\Omega,w_0)}, \end{aligned}$$

from which we derive that

$$\|A(v)\|_{W^{1,p}(\Omega,w_0,w)^*} \leq \| \, |\nabla v| \, \|_{L^p(\Omega,w)}^{p-1} + \|\xi\|_{L^\infty(\Omega)} \|v\|_{L^p(\Omega,w_0)}^{p-1}.$$

This proves that A is bounded.

To show that A is a map of type $(S)_+$, let $(u_n)_{n\geq 1} \subset W^{1,p}(\Omega, w_0, w)$ be a sequence such that

$$u_n \rightharpoonup u \quad \text{in } W^{1,p}(\Omega, w_0, w) \quad \text{and} \quad \limsup_{n\to+\infty} \langle A(u_n), u_n - u\rangle \leq 0, \tag{9}$$

for some $u \in W^{1,p}(\Omega, w_0, w)$. Consider an arbitrary subsequence of $(u_n)_{n\geq 1}$, denoted again by $(u_n)_{n\geq 1}$, and we show that it has a subsequence which strongly converges to u in $W^{1,p}(\Omega, w_0, w)$.

Since A is bounded, we have that the boundedness of the sequence $(u_n)_{n\geq 1}$ in $W^{1,p}(\Omega, w_0, w)$ implies the boundedness of the sequence $(A(u_n))_{n\geq 1}$ in $W^{1,p}(\Omega, w_0, w)^*$, so along a relabeled subsequence, we have that

$$(A(u_n))_{n\geq 1} \text{ converges weakly in } W^{1,p}(\Omega, w_0, w)^*. \tag{10}$$

Since the map A is monotone (see (7)) and hemicontinuous (because it is continuous), we infer that A is maximal monotone (see, e.g., [15, Corollary 2.42]). Using the reflexivity of $W^{1,p}(\Omega, w_0, w)$, we deduce that A is pseudomonotone, and thus it is generalized pseudomonotone (see, e.g., [15, Propositions 2.60, 2.67]). Then, taking (9) and (10) into account, we obtain that

$$\lim_{n\to+\infty} \langle A(u_n), u_n\rangle = \langle A(u), u\rangle. \tag{11}$$

Since $u_n \rightharpoonup u$ in $W^{1,p}(\Omega, w_0, w)$, in view of hypothesis (E_1) (ii), along a relabeled subsequence, we have $u_n \to u$ in $L^p(\Omega, w_0)$. This implies that

$$\begin{aligned} &\lim_{n\to+\infty} \|u_n\|_{L^p(\Omega,w_0)} = \|u\|_{L^p(\Omega,w_0)}, \\ &\lim_{n\to+\infty} \int_\Omega \xi w_0 |u_n|^p \, dx = \int_\Omega \xi w_0 |u|^p \, dx. \end{aligned} \tag{12}$$

From (11) and (12), we get $\|u_n\|_{W^{1,p}(\Omega,w_0,w)} \to \|u\|_{W^{1,p}(\Omega,w_0,w)}$ as $n \to +\infty$. Since the space $(W^{1,p}(\Omega, w_0, w), \|\cdot\|_{W^{1,p}(\Omega,w_0,w)})$ is uniformly convex, it satisfies the Kadec–Klee property (see [14, p. 453]). We then conclude that $u_n \to u$ in $W^{1,p}(\Omega, w_0, w)$ as $n \to +\infty$. □

3. Spectrum of w-Weighted p-Laplacian Operators

3.1. *Lyusternik–Schnirelmann–type eigenvalues*

3.1.1. *Spectrum of* $(-\Delta_{p,w}, W^{1,p}(\Omega, w_0, w), \xi)$

Whenever A is a non-empty, closed, symmetric (i.e., $-A = A$) subset of a Banach space X, the Krasnosel'skiĭ genus of A is defined by

$$\operatorname{gen} A = \inf\{m \geq 1 : \text{there exists } \phi \in C(A, \mathbb{R}^m \setminus \{0\}) \text{ odd}\} \tag{13}$$

(see [4]). For instance, if A is a sphere in an m-dimensional subspace $V \subset X$, then $\operatorname{gen} A = m$.

Let $p \in (1, +\infty)$, $w_0, w \in \mathcal{W}_p(\Omega)$, and $\xi \in L^\infty(\Omega)$, $\xi(x) \geq 0$ for a.a. $x \in \Omega$, $\xi \neq 0$. With the norm $\|\cdot\|_{W^{1,p}(\Omega,w_0,w),\xi}$ defined in Lemma 3, we consider the set

$$M = \{u \in W^{1,p}(\Omega, w_0, w) : \|u\|^p_{W^{1,p}(\Omega,w_0,w),\xi} = p\}. \tag{14}$$

Let

$$\mathcal{K}(M) = \Big\{K \subset M : K \text{ is compact, symmetric,}$$
$$\int_\Omega \xi w_0 |u|^p \, dx > 0 \text{ for all } u \in K\Big\},$$

and, for every $n \geq 1$, let

$$\mathcal{K}_n = \{K \in \mathcal{K}(M) : \operatorname{gen} K \geq n\}. \tag{15}$$

Theorem 5. *Assume* (E_1) *and that* $w_0 \in L^1(\Omega)$. *Then there is a sequence of eigenvalues* $(\lambda_n)_{n\geq 0}$ *of* $(-\Delta_{p,w}, W^{1,p}(\Omega, w_0, w), \xi)$ *such that*

$$0 = \lambda_0 < \lambda_1 \leq \cdots \leq \lambda_n \leq \cdots, \qquad \lim_{n\to+\infty} \lambda_n = +\infty,$$

characterized by

$$\frac{1}{\lambda_n + 1} = \sup_{K\in\mathcal{K}_{n+1}} \min_{u\in K} \left(\frac{1}{p}\int_\Omega \xi w_0 |u|^p \, dx\right) \quad \textit{for all } n \geq 0. \tag{16}$$

3.1.2. *Spectrum of* $(-\Delta_{p,w}, W_0^{1,p}(\Omega, w_0, w), \xi)$

Assume that $|\mathbb{R}^N \setminus \Omega| > 0$. Let $p \in (1, +\infty)$, $w_0, w \in \mathcal{W}_p(\mathbb{R}^N)$, and $\xi \in L^\infty(\Omega)$, $\xi \geq 0$ a.e. in Ω, $\xi \neq 0$.

Let

$$M = \{u \in W_0^{1,p}(\Omega, w_0, w) : \| |\nabla u| \|_{L^p(\Omega,w)}^p = p\} \tag{17}$$

and, correspondingly, let the sets $\mathcal{K}_n$ (for $n \geq 1$) be as in (15).

Theorem 6. *Assume* (E$_2$). *There exists a sequence of eigenvalues* $(\lambda_n)_{n\geq 1}$ *of* $(-\Delta_{p,w}, W_0^{1,p}(\Omega, w_0, w), \xi)$ *such that*

$$0 < \lambda_1 \leq \lambda_2 \leq \cdots \leq \lambda_n \leq \cdots, \qquad \lim_{n\to+\infty} \lambda_n = +\infty,$$

characterized by

$$\frac{1}{\lambda_n} = \sup_{K\in\mathcal{K}_n} \min_{u\in K} \left(\frac{1}{p} \int_\Omega \xi w_0 |u|^p \, dx \right) \quad \textit{for all } n \geq 1. \tag{18}$$

In the case where $w_0 = w = 1$, from Theorems 5 and 6, we retrieve the sequence of the Lyusternik–Schnirelmann eigenvalues for the classical p-Laplacian operators under Neumann, respectively, Dirichlet, boundary conditions. The proofs of Theorems 5 and 6 are based on the Lyusternik–Schnirelmann principle, which is recalled in Section 4. The proof of Theorem 5 is done in Section 5.1 and the one of Theorem 6 is similar and is outlined in Section 5.2.

3.2. *Additional spectral properties*

3.2.1. *Spectrum of* $(-\Delta_{p,w}, W^{1,p}(\Omega, w_0, w), \xi)$

Let $p \in (1, +\infty)$, $w_0, w \in \mathcal{W}_p(\Omega)$, and $\xi \in L^\infty(\Omega)$, $\xi(x) \geq 0$ for a.a. $x \in \Omega$, $\xi \neq 0$. We denote by $\sigma(W^{1,p}(\Omega, w_0, w), \xi) \subset [0, +\infty)$ the set of all the eigenvalues of $(-\Delta_{p,w}, W^{1,p}(\Omega, w_0, w), \xi)$ (see Remark 2 (a)). In this section, we establish some topological properties of the spectrum and variational characterizations of the eigenvalues.

Proposition 7. *Assume that* $w_0 \in L^1(\Omega)$. *Let* λ *be an eigenvalue of* $(-\Delta_{p,w}, W^{1,p}(\Omega, w_0, w), \xi)$ *such that* $\lambda > 0$. *Then, every eigenfunction* u *corresponding to* λ *is nodal (i.e., sign changing).*

Proof. Arguing by contradiction, suppose that there is an eigenfunction u corresponding to $\lambda > 0$ such that u has constant sign, say $u \geq 0$ a.e. in Ω. Acting on (2) with the constant test function $v = 1 \in W^{1,p}(\Omega, w_0, w)$ (since $w_0 \in L^1(\Omega)$), we obtain $\lambda \int_\Omega \xi w_0 u^{p-1}\,dx = 0$, hence

$$\xi u^{p-1} = 0 \quad \text{a.e. in } \Omega. \tag{19}$$

Then (2) becomes

$$\int_\Omega w|\nabla u|^{p-2}\nabla u \cdot \nabla v\,dx = 0 \quad \text{for all } v \in W^{1,p}(\Omega, w_0, w).$$

Letting $v = u$ in the above equality, we obtain $\int_\Omega w|\nabla u|^p\,dx = 0$, hence $\nabla u = 0$ a.e. in Ω, and thus u is constant on Ω (cf. [8, p. 113]), say $u = c_0 \in \mathbb{R}$. This in conjunction with (19) and the fact that $\xi \neq 0$ yields $c_0 = 0$, which contradicts the property that u is an eigenfunction. Hence, u is nodal. □

Proposition 8. *Assume* (E_1) (ii). *Then the set* $\sigma(W^{1,p}(\Omega, w_0, w), \xi)$ *is closed.*

Proof. Let $(\mu_n)_{n\geq 1}$ be a sequence of eigenvalues of $(-\Delta_{p,w}, W^{1,p}(\Omega, w_0, w), \xi)$ such that $\mu_n \to \lambda$ as $n \to +\infty$. Let $(u_n)_{n\geq 1} \subset W^{1,p}(\Omega, w_0, w)$ be corresponding eigenfunctions normalized as $\|u_n\|_{W^{1,p}(\Omega,w_0,w)} = 1$ for all $n \geq 1$. Thus, along a relabeled subsequence, we have that $u_n \rightharpoonup u$ in $W^{1,p}(\Omega, w_0, w)$ as $n \to +\infty$. Since the embedding $W^{1,p}(\Omega, w_0, w) \subset L^p(\Omega, w_0)$ is compact (see (E_1) (ii)), we infer that $u_n \to u$ in $L^p(\Omega, w_0)$ as $n \to +\infty$. Using in (2) (written for the eigenvalue μ_n with eigenfunction u_n) the test function $v = u_n - u$, we see that

$$\int_\Omega w|\nabla u_n|^{p-2}\nabla u_n \cdot \nabla(u_n - u)\,dx - \mu_n \int_\Omega \xi w_0 |u_n|^{p-2}u_n(u_n - u)\,dx \tag{20}$$

for all $n \geq 1$. By Hölder's inequality, we can write

$$\left|\int_\Omega \xi w_0 |u_n|^{p-2}u_n(u_n - u)\,dx\right| \leq \|\xi\|_{L^\infty(\Omega)}\,\|u_n\|^{p-1}_{L^p(\Omega,w_0)}\,\|u_n - u\|_{L^p(\Omega,w_0)}.$$

Since $u_n \to u$ in $L^p(\Omega, w_0)$, the right-hand side of (20) converges to 0. Therefore, from (20) and (8), we obtain $\langle -\Delta_{p,w}(u_n), u_n - u\rangle \to 0$

as $n \to +\infty$. Recalling that $u_n \rightharpoonup u$ in $W^{1,p}(\Omega, w_0, w)$ as $n \to +\infty$ and taking into account that the operator $-\Delta_{p,w}$ on $W^{1,p}(\Omega, w_0, w)$ is a map of type $(S)_+$ (see Proposition 4 (b)), we infer that $u_n \to u$ in $W^{1,p}(\Omega, w_0, w)$ as $n \to +\infty$. It follows that $\|u\|_{W^{1,p}(\Omega,w_0,w)} = 1$ (hence $u \neq 0$). Moreover, passing to the limit as $n \to +\infty$ in (2) (written for the eigenvalue μ_n with eigenfunction u_n), we obtain that λ is an eigenvalue of $(-\Delta_{p,w}, W^{1,p}(\Omega, w_0, w), \xi)$ with corresponding eigenfunction u. This establishes that $\sigma(W^{1,p}(\Omega, w_0, w), \xi)$ is closed.

□

Under assumption (E_1) and that $w_0 \in L^1(\Omega)$, Theorem 5 provides a non-decreasing sequence $(\lambda_n)_{n\geq 0}$ of eigenvalues of $(-\Delta_{p,w}, W^{1,p}(\Omega, w_0, w), \xi)$. The following result studies the eigenvalue λ_1.

Theorem 9. *Assume* (E_1) *and that* $w_0 \in L^1(\Omega)$. *Then* λ_1 *is the second eigenvalue of* $(-\Delta_{p,w}, W^{1,p}(\Omega, w_0, w), \xi)$, *in other words, it satisfies*

$$\lambda_1 = \inf\{\lambda \in \sigma(W^{1,p}(\Omega, w_0, w), \xi) : \lambda > 0\} > 0.$$

Proof. We already know from Theorem 5 that $\lambda_1 > 0$. This implies that $\lambda_1 \geq \inf\{\lambda \in \sigma(W^{1,p}(\Omega, w_0, w), \xi) : \lambda > 0\}$. Then for completing the proof of the theorem, it remains to show that whenever $\lambda \in \sigma(W^{1,p}(\Omega, w_0, w), \xi)$ is an eigenvalue such that $\lambda > 0$, we have $\lambda \geq \lambda_1$.

Let $u \in W^{1,p}(\Omega, w_0, w)$ be an eigenfunction corresponding to λ. By Proposition 7, we know that u is nodal.

Let $u^+ = \max\{u, 0\}$ and $u^- = \max\{-u, 0\}$, which belong to $W^{1,p}(\Omega, w_0, w)$ and satisfy

$$\nabla(u^+) = \begin{cases} \nabla u & \text{a.e. in } \{x \in \Omega : u(x) > 0\}, \\ 0 & \text{a.e. in } \{x \in \Omega : u(x) \leq 0\}, \end{cases}$$

$$\nabla(u^-) = \begin{cases} 0 & \text{a.e. in } \{x \in \Omega : u(x) \geq 0\}, \\ -\nabla u & \text{a.e. in } \{x \in \Omega : u(x) < 0\} \end{cases}$$

(this holds from the definition of $W^{1,p}(\Omega, w_0, w)$, with $w_0, w \in \mathcal{W}_p(\Omega)$; see [9, Lemma 7.6] and also [10, Proposition 2]). Using in (2) successively the test functions $u^+, u^- \in W^{1,p}(\Omega, w_0, w)$, we have

$$\int_\Omega w|\nabla u^\pm|^p\,dx = \lambda \int_\Omega \xi w_0 (u^\pm)^p\,dx. \tag{21}$$

Since u is nodal, we have that $u^+ \neq 0$ and $u^- \neq 0$. We denote $V = \operatorname{span}\{u^+, u^-\}$. The linear independence of u^+ and u^- implies that the space V is of dimension 2. With the notation in Lemma 3, let

$$K := \{v \in V : \|v\|^p_{W^{1,p}(\Omega,w_0,w),\xi} = p\}.$$

Let an arbitrary $v \in K$. Then we can write $v = \alpha u^+ + \beta u^-$, with $\alpha, \beta \in \mathbb{R}$. Thus, we have

$$|v|^p = |\alpha|^p (u^+)^p + |\beta|^p (u^-)^p \text{ and } |\nabla v|^p = |\alpha|^p |\nabla u^+|^p + |\beta|^p |\nabla u^-|^p. \tag{22}$$

By means of (21) and (22), we note that

$$\begin{aligned} p = \|v\|^p_{W^{1,p}(\Omega,w_0,w),\xi} &= \int_\Omega w|\nabla v|^p\,dx + \int_\Omega \xi w_0 |v|^p\,dx \\ &= |\alpha|^p \left(\int_\Omega w|\nabla u^+|^p\,dx + \int_\Omega \xi w_0 (u^+)^p\,dx \right) \\ &\quad + |\beta|^p \left(\int_\Omega w|\nabla u^-|^p\,dx + \int_\Omega \xi w_0 (u^-)^p\,dx \right) \\ &= (\lambda+1) \left(|\alpha|^p \int_\Omega \xi w_0 (u^+)^p\,dx + |\beta|^p \int_\Omega \xi w_0 (u^-)^p\,dx \right) \\ &= (\lambda+1) \int_\Omega \xi w_0 |v|^p\,dx. \end{aligned} \tag{23}$$

Noting that K is a sphere in the 2-dimensional subspace V, we deduce that K is compact, symmetric, and $\operatorname{gen} K = 2$. Thus, from (23), we deduce that $K \in \mathcal{K}_2$ (see (15)), and so, using the variational characterization of λ_1 given by (16), we get

$$\frac{1}{\lambda_1 + 1} \geq \min_{v \in K} \left(\frac{1}{p} \int_\Omega \xi w_0 |v|^p\,dx \right) = \frac{1}{\lambda + 1}$$

(see (23) again). Hence, $\lambda_1 \leq \lambda$, and the proof of the theorem is complete. □

Corollary 10. *Assume* (E_1) *and that* $w_0 \in L^1(\Omega)$. *Then the eigenvalue* $\lambda_0 = 0$ *is isolated in* $\sigma(W^{1,p}(\Omega, w_0, w), \xi)$.

The following result gives another variational characterization of the second eigenvalue λ_1 of $(-\Delta_{p,w}, W^{1,p}(\Omega, w_0, w), \xi)$ (recall that Theorem 5 already provides a variational characterization of each eigenvalue in the sequence $(\lambda_n)_{n\geq 0}$).

Theorem 11. *Assume that* $w_0 \in L^1(\Omega)$. *Let*

$$Q = \left\{u \in W^{1,p}(\Omega, w_0, w) : \int_\Omega \xi w_0 |u|^p \, dx = 1, \int_\Omega \xi w_0 |u|^{p-2} u \, dx = 0\right\}.$$

Then, we have the following:

(a) $\lambda \geq \inf_{u \in Q} \| \, |\nabla u| \, \|^p_{L^p(\Omega,w)}$ *for each eigenvalue* $\lambda \in \sigma(W^{1,p}(\Omega, w_0, w), \xi)$ *with* $\lambda > 0$.
(b) *Assuming* (E_1), *if* $p \geq 2$, *then* $\lambda_1 = \inf_{u \in Q} \| \, |\nabla u| \, \|^p_{L^p(\Omega,w)}$.

Proof. (a) Let u be an eigenfunction corresponding to the eigenvalue λ. Since λ is positive, we have that the eigenfunction u is a non-constant function, so we see that $\| \, |\nabla u| \, \|_{L^p(\Omega,w)} \neq 0$. Thus, we can assume that the eigenfunction u satisfies $\| \, |\nabla u| \, \|^p_{L^p(\Omega,w)} = \lambda$.

Using (2) with the test functions $v = 1 \in W^{1,p}(\Omega, w_0, w)$ (since $w_0 \in L^1(\Omega)$), respectively, $v = u \in W^{1,p}(\Omega, w_0, w)$, we obtain

$$\int_\Omega \xi w_0 |u|^{p-2} u \, dx = 0 \quad \text{and} \quad \int_\Omega \xi w_0 |u|^p \, dx = \frac{1}{\lambda} \| \, |\nabla u| \, \|^p_{L^p(\Omega,w)} = 1.$$

It follows that $u \in Q$, which in conjunction with the fact that $\lambda = \| \, |\nabla u| \, \|^p_{L^p(\Omega,w)}$ yields the desired inequality.

(b) First, we see that

$$\text{the set } Q \text{ is sequentially weakly closed in } W^{1,p}(\Omega, w_0, w). \tag{24}$$

Indeed, let $(u_n)_{n\geq 1} \subset Q$ and $u \in W^{1,p}(\Omega, w_0, w)$ be such that

$$u_n \rightharpoonup u \quad \text{in } W^{1,p}(\Omega, w_0, w) \quad \text{as } n \to +\infty.$$

Then the compactness of the embedding $W^{1,p}(\Omega, w_0, w) \subset L^p(\Omega, w_0)$ (see (E_1) (ii)) ensures that $u_n \to u$ in $L^p(\Omega, w_0)$. This implies that

$w_0^{\frac{1}{p}} u_n \to w_0^{\frac{1}{p}} u$ in $L^p(\Omega)$, so, up to a subsequence, we may assume that $u_n \to u$ a.e. in Ω and there is $h \in L^p(\Omega)$ such that $w_0^{\frac{1}{p}} |u_n| \leq h$ a.e. in Ω (see, e.g., [3, p. 94]). By Lebesgue's dominated convergence theorem, it follows that

$$\int_\Omega \xi w_0 |u|^p \, dx = \lim_{n\to+\infty} \int_\Omega \xi w_0 |u_n|^p \, dx = 1$$

and, recalling that $w_0 \in L^1(\Omega)$,

$$\int_\Omega \xi w_0 |u|^{p-2} u \, dx = \lim_{n\to+\infty} \int_\Omega \xi w_0 |u_n|^{p-2} u_n \, dx = 0,$$

which shows that $u \in Q$, and this establishes (24).

Note that the map $u \mapsto \|u\|_{W^{1,p}(\Omega, w_0, w), \xi}$ is coercive on the reflexive Banach space $W^{1,p}(\Omega, w_0, w)$ (see Lemma 3, due to (E_1) (i)). Moreover, as this map is convex and continuous, it is weakly lower semicontinuous. Combining this with (24), we infer that there is $u \in Q$ such that

$$\|u\|_{W^{1,p}(\Omega, w_0, w), \xi} = \inf_{v \in Q} \|v\|_{W^{1,p}(\Omega, w_0, w), \xi},$$

which entails that

$$\| \, |\nabla u| \, \|_{L^p(\Omega, w)} = \inf_{v \in Q} \| \, |\nabla v| \, \|_{L^p(\Omega, w)}. \tag{25}$$

Since $p \geq 2$, the Lagrange multiplier rule implies that there exist $\lambda, \mu \in \mathbb{R}$ such that

$$\begin{aligned} p \int_\Omega & w |\nabla u|^{p-2} \nabla u \cdot \nabla v \, dx \\ &= \lambda p \int_\Omega \xi w_0 |u|^{p-2} u v \, dx + \mu (p-1) \int_\Omega \xi w_0 |u|^{p-2} v \, dx \end{aligned} \tag{26}$$

for all $v \in W^{1,p}(\Omega, w_0, w)$. Acting on (26) with the test function $v = \mu \in W^{1,p}(\Omega, w_0, w)$ (since $w_0 \in L^1(\Omega)$), and using that $u \in Q$, we obtain

$$\mu^2 (p-1) \int_\Omega \xi w_0 |u|^{p-2} \, dx = 0.$$

This ensures that $\mu = 0$, and thus we deduce from (26) that λ is an eigenvalue of $(-\Delta_{p,w}, W^{1,p}(\Omega, w_0, w), \xi)$ with corresponding eigenfunction u. This eigenvalue is positive since u is non-constant (because $u \in Q$), so $\lambda \geq \lambda_1$ (due to Theorem 9). From (26), with $v = u$, we derive

$$\| |\nabla u| \|_{L^p(\Omega,w)}^p = \lambda \int_\Omega \xi w_0 |u|^p \, dx = \lambda \geq \lambda_1.$$

This in conjunction with (25) yields $\inf_{v\in Q} \| |\nabla v| \|_{L^p(\Omega,w)}^p \geq \lambda_1$. Since the reversed inequality follows from part (a), the proof is complete. □

Lemma 12. *Let X be a Banach space and $\gamma \in C([-1,1], X)$ such that $\gamma(-1) = -\gamma(1)$. Let*

$$K_\gamma := \{\gamma(t) : t \in [-1,1]\} \cup \{-\gamma(t) : t \in [-1,1]\}.$$

Then, $\operatorname{gen} K_\gamma \geq 2$.

Proof. Let $S^1 = \{(\cos\theta, \sin\theta) : \theta \in [0, 2\pi]\} \subset \mathbb{R}^2$ and let $h : S^1 \to K_\gamma$ defined by

$$h(\cos\theta, \sin\theta) = \begin{cases} \gamma(\cos(\theta+\pi)) & \text{if } \theta \in [0,\pi], \\ -\gamma(\cos\theta) & \text{if } \theta \in [\pi, 2\pi]. \end{cases}$$

Then, $h \in C(S^1, K_\gamma)$ and it is odd. If $m \geq 1$ is such that there is an odd map $\phi \in C(K_\gamma, \mathbb{R}^m \setminus \{0\})$, then the composition $\phi \circ h \in C(S^1, \mathbb{R}^m \setminus \{0\})$ is odd, and this implies that $m \geq \operatorname{gen} S^1 = 2$. In view of (13), we deduce that $\operatorname{gen} K_\gamma \geq 2$. □

Coming back to the situation where $p \in (1, +\infty)$, we give another variational characterization of the second eigenvalue λ_1 of $(-\Delta_{p,w}, W^{1,p}(\Omega, w_0, w), \xi)$ (see Theorems 5 and 9).

Theorem 13. *Assume* (E_1) *and that* $w_0 \in L^1(\Omega)$, *and let*

$$S = \{u \in W^{1,p}(\Omega, w_0, w) : \int_\Omega \xi w_0 |u|^p \, dx = 1\},$$

$$u_0 := \left(\int_\Omega \xi w_0 \, dx\right)^{-\frac{1}{p}} \in S,$$

and $\Gamma = \{\gamma \in C([-1,1], S) : \gamma(-1) = -u_0, \ \gamma(1) = u_0\}.$

The second eigenvalue of $(-\Delta_{p,w}, W^{1,p}(\Omega, w_0, w), \xi)$ *is characterized by*

$$\lambda_1 = \min_{\gamma\in\Gamma} \max_{-1\le t\le 1} \| \, |\nabla\gamma(t)| \, \|^p_{L^p(\Omega,w)}.$$

Proof. We first show that

$$\lambda_1 \le \inf_{\gamma\in\Gamma} \max_{-1\le t\le 1} \| \, |\nabla\gamma(t)| \, \|^p_{L^p(\Omega,w)}. \tag{27}$$

Let $\gamma \in \Gamma$. We construct an element $\hat{\gamma} \in C([-1,1], M)$, with M as in (14), by letting

$$\hat{\gamma}(t) := \left(\frac{p}{\|\gamma(t)\|^p_{W^{1,p}(\Omega,w_0,w),\xi}}\right)^{\frac{1}{p}} \gamma(t) = \left(\frac{p}{1 + \| \, |\nabla\gamma(t)| \, \|^p_{L^p(\Omega,w)}}\right)^{\frac{1}{p}} \gamma(t)$$

for all $t \in [-1,1]$. The last equality is due to the fact that $\gamma(t) \in S$, and the same fact yields

$$\int_\Omega \xi w_0 |\hat{\gamma}(t)|^p \, dx = \frac{p}{1 + \| \, |\nabla\gamma(t)| \, \|^p_{L^p(\Omega,w)}} \quad \text{for all } t \in [-1,1]. \tag{28}$$

As $\hat{\gamma}(-1) = -p^{\frac{1}{p}} u_0 = -\hat{\gamma}(1)$, we can consider the set $K_{\hat{\gamma}}$ of Lemma 12 which is compact, symmetric, with $K_{\hat{\gamma}} \subset M$ and $\operatorname{gen} K_{\hat{\gamma}} \ge 2$. Taking also (28) into account, we get $K_{\hat{\gamma}} \in \mathcal{K}_2$ (see (15)). By Theorem 5 and (28), we deduce that

$$\frac{1}{1+\lambda_1} \ge \min_{u\in K_{\hat{\gamma}}} \frac{1}{p} \int_\Omega \xi w_0 |u|^p \, dx = \min_{-1\le t\le 1} \frac{1}{1 + \| \, |\nabla\gamma(t)| \, \|^p_{L^p(\Omega,w)}}.$$

It follows that

$$\lambda_1 \le \max_{-1\le t\le 1} \| \, |\nabla\gamma(t)| \, \|^p_{L^p(\Omega,w)} \quad \text{for all } \gamma \in \Gamma.$$

Since $\gamma \in \Gamma$ is arbitrary, this proves (27).

In view of (27), in order to complete the proof of the theorem, it suffices to construct $\gamma_1 \in \Gamma$ such that

$$\max_{-1\le t\le 1} \| \, |\nabla\gamma_1(t)| \, \|^p_{L^p(\Omega,w)} = \lambda_1. \tag{29}$$

To this end, let u be an eigenfunction corresponding to λ_1. Let $\eta : \mathbb{R} \to S$ be defined by

$$\eta(t) = \frac{u+t}{(\int_\Omega \xi w_0 |u+t|^p\,dx)^{\frac{1}{p}}} \quad \text{for all } t \in \mathbb{R}.$$

We have that

$$\eta(t) \to \pm \left(\int_\Omega \xi w_0\,dx \right)^{-\frac{1}{p}} = \pm u_0 \quad \text{in } W^{1,p}(\Omega, w_0, w) \text{ as } t \to \pm\infty.$$

Let $\gamma_1 : [-1,1] \to S$ be given by

$$\gamma_1(t) = \begin{cases} -u_0 & \text{if } t = -1, \\ \eta\left(\dfrac{t}{1-t^2}\right) & \text{if } t \in (-1,1), \\ u_0 & \text{if } t = 1 \end{cases}$$

so that $\gamma_1 \in \Gamma$. We show that γ_1 satisfies (29). We have

$$\|\,|\nabla \eta(t)|\,\|^p_{L^p(\Omega,w)} = \frac{\|\,|\nabla u|\,\|^p_{L^p(\Omega,w)}}{\int_\Omega \xi w_0 |u+t|^p\,dx},$$

hence

$$\frac{d}{dt}\|\,|\nabla \eta(t)|\,\|^p_{L^p(\Omega,w)} = \frac{-p\|\,|\nabla u|\,\|^p_{L^p(\Omega,w)}}{(\int_\Omega \xi w_0 |u+t|^p\,dx)^2} \int_\Omega \xi w_0 |u+t|^{p-2}(u+t)\,dx. \tag{30}$$

Since u is an eigenfunction corresponding to λ_1, we have $\int_\Omega \xi w_0 |u|^{p-2} u\,dx = 0$. Then, taking into account that $t \mapsto |t|^{p-2}t$ is increasing on $\mathbb{R}$, we obtain from (30) that the function $t \mapsto \|\,|\nabla \eta(t)|\,\|^p_{L^p(\Omega,w)}$ attains its maximum at $t = 0$. Using again that u is an eigenfunction corresponding to λ_1, we infer that

$$\max_{-1 \le t \le 1} \|\,|\nabla \gamma_1(t)|\,\|^p_{L^p(\Omega,w)} = \|\,|\nabla \eta(0)|\,\|^p_{L^p(\Omega,w)} = \frac{\|\,|\nabla u|\,\|^p_{L^p(\Omega,w)}}{\int_\Omega \xi w_0 |u|^p\,dx} = \lambda_1,$$

so γ_1 fulfills (29). □

Next, we study the monotonicity of the map $\xi \mapsto \lambda_1(\xi)$, where $\lambda_1(\xi)$ is the second eigenvalue of $(-\Delta_{p,w}, W^{1,p}(\Omega, w_0, w), \xi)$ (see Theorems 5 and 9).

Proposition 14. *Assume* (E_1) *and that* $w_0 \in L^1(\Omega)$. *If* $\xi, \xi' \in L^\infty(\Omega)$, $\xi \neq 0$, *and* $\xi'(x) > \xi(x) \geq 0$ *for a.a.* $x \in \Omega$, *then* $\lambda_1(\xi') < \lambda_1(\xi)$.

Proof. Let u be an eigenfunction corresponding to $\lambda_1(\xi)$. Since u is nodal (see Proposition 7), we have that u^+ and u^- are nonzero and linearly independent. Then, $V := \operatorname{span}\{u^+, u^-\}$ is a 2-dimensional subspace of $W^{1,p}(\Omega, w_0, w)$. By Lemma 3, the set

$$K := \left\{v \in V : \|v\|^p_{W^{1,p}(\Omega,w_0,w),\xi'} = p\right\}$$

is compact, symmetric, and we have $\operatorname{gen} K = 2$.

Let $v = \beta_1 u^+ + \beta_2 u^- \in K$, with $\beta_1, \beta_2 \in \mathbb{R}$. Using in (2) (written for the eigenvalue $\lambda_1(\xi)$ and eigenfunction u) the test functions $u^+, u^- \in W^{1,p}(\Omega, w_0, w)$, we see that

$$\begin{aligned} \int_\Omega w|\nabla v|^p\,dx &= |\beta_1|^p \int_\Omega w|\nabla u^+|^p\,dx + |\beta_2|^p \int_\Omega w|\nabla u^-|^p\,dx \\ &= \lambda_1(\xi)\left(|\beta_1|^p \int_\Omega \xi w_0|u^+|^p\,dx + |\beta_2|^p \int_\Omega \xi w_0 |u^-|^p\,dx\right) \\ &= \lambda_1(\xi) \int_\Omega \xi w_0 |v|^p\,dx. \end{aligned} \tag{31}$$

Moreover, since $0 \leq \xi < \xi'$, we have

$$\int_\Omega \xi w_0 |v|^p\,dx < \int_\Omega \xi' w_0 |v|^p\,dx. \tag{32}$$

Using the expression of $\lambda_1(\xi')$ in Theorem 5, the definition of K, (32) combined with the fact that $t \mapsto \frac{t}{t+a}$ is increasing on $[0, +\infty)$ for any

$a > 0$, and (31), we obtain

$$\begin{aligned}\frac{1}{\lambda_1(\xi')+1} &\geq \min_{v\in K}\frac{1}{p}\int_\Omega \xi' w_0|v|^p\,dx\\ &= \min_{v\in K}\frac{\int_\Omega \xi' w_0|v|^p\,dx}{\int_\Omega w|\nabla v|^p\,dx+\int_\Omega \xi' w_0|v|^p\,dx}\\ &> \min_{v\in K}\frac{\int_\Omega \xi w_0|v|^p\,dx}{\int_\Omega w|\nabla v|^p\,dx+\int_\Omega \xi w_0|v|^p\,dx} = \frac{1}{\lambda_1(\xi)+1}.\end{aligned}$$

This yields $\lambda_1(\xi') < \lambda_1(\xi)$, as asserted. □

3.2.2. *Spectrum of* $(-\Delta_{p,w}, W_0^{1,p}(\Omega, w_0, w), \xi)$

Assume $|\mathbb{R}^N \setminus \Omega| > 0$. Let $p \in (1, +\infty)$, $w_0, w \in \mathcal{W}_p(\mathbb{R}^N)$, and $\xi \in L^\infty(\Omega)$, $\xi(x) \geq 0$ for a.a. $x \in \Omega$, $\xi \neq 0$.

We denote by $\sigma(W_0^{1,p}(\Omega, w_0, w), \xi) \subset (0, +\infty)$ the set of all the eigenvalues of $(-\Delta_{p,w}, W_0^{1,p}(\Omega, w_0, w), \xi)$ (see Remark 2 (b)). Under condition (E_2), Theorem 6 provides a non-decreasing sequence $(\lambda_n)_{n\geq 1} \subset \sigma(W_0^{1,p}(\Omega, w_0, w), \xi)$ together with a variational characterization of each eigenvalue of the sequence.

The following result gives another variational characterization of λ_1 which turns out to be the first eigenvalue of $(-\Delta_{p,w}, W_0^{1,p}(\Omega, w_0, w), \xi)$.

Theorem 15. *Assume* (E_2). *Set*

$$Q = \left\{u \in W_0^{1,p}(\Omega, w_0, w) : \int_\Omega \xi w_0 |u|^p\,dx = 1\right\}.$$

Then,

$$\lambda_1 = \min \sigma(W_0^{1,p}(\Omega, w_0, w), \xi) = \inf_{u\in Q} \|\,|\nabla u|\,\|^p_{L^p(\Omega,w)}, \tag{33}$$

where the infimum is realized exactly on the set of eigenfunctions u corresponding to λ_1 with $u \in Q$.

Proof. Let $\lambda \in \sigma(W_0^{1,p}(\Omega, w_0, w), \xi)$ be arbitrary. Let u be an eigenfunction corresponding to λ such that $u \in Q$ (see Remark 2 (b)).

We then have

$$\inf_{v\in Q} \| \, |\nabla v| \, \|^p_{L^p(\Omega,w)} \le \| \, |\nabla u| \, \|^p_{L^p(\Omega,w)} = \lambda.$$

Since λ is arbitrarily chosen in the spectrum, we get

$$\inf_{v\in Q} \| \, |\nabla v| \, \|^p_{L^p(\Omega,w)} \le \inf \sigma(W_0^{1,p}(\Omega, w_0, w), \xi). \tag{34}$$

Now let an arbitrary $u \in Q$. Noting that $u \neq 0$, and so $\| \, |\nabla u| \, \|_{L^p(\Omega,w)} \neq 0$ (since $\| \, |\nabla \cdot| \, \|_{L^p(\Omega,w)}$ is a norm on $W_0^{1,p}(\Omega, w_0, w)$), we can consider the compact, symmetric set $K := \{-\frac{p^{\frac{1}{p}}}{\| |\nabla u| \|_{L^p(\Omega,w)}} u, \frac{p^{\frac{1}{p}}}{\| |\nabla u| \|_{L^p(\Omega,w)}} u\}$, which satisfies that $\operatorname{gen} K = 1$. Thus, we see that $K \in \mathcal{K}_1$ and then, from the variational characterization of λ_1 given by (18), we deduce that

$$\frac{1}{\lambda_1} \ge \frac{1}{\| \, |\nabla u| \, \|^p_{L^p(\Omega,w)}} \int_\Omega \xi w_0 |u|^p \, dx = \frac{1}{\| \, |\nabla u| \, \|^p_{L^p(\Omega,w)}}.$$

Since $u \in Q$ is arbitrary, we get $\lambda_1 \le \inf_{v\in Q} \| \, |\nabla v| \, \|^p_{L^p(\Omega,w)}$. Combining this and (34) leads to (33).

Now let $u \in Q$ be such that

$$\| \, |\nabla u| \, \|^p_{L^p(\Omega,w)} = \inf_{v\in Q} \| \, |\nabla v| \, \|^p_{L^p(\Omega,w)}.$$

Then the Lagrange multiplier rule yields $\lambda \in \mathbb{R}$ such that

$$p \int_\Omega w |\nabla u|^{p-2} \nabla u \cdot \nabla v \, dx = \lambda p \int_\Omega \xi w_0 |u|^{p-2} u v \, dx$$
$$\text{for all } v \in W_0^{1,p}(\Omega, w_0, w),$$

that is, λ is an eigenvalue of $(-\Delta_{p,w}, W_0^{1,p}(\Omega, w_0, w), \xi)$ with corresponding eigenfunction u. Acting on the previous formula with the test function $v = u$, we obtain $\lambda = \| \, |\nabla u| \, \|^p_{L^p(\Omega,w)}$. By (33), we conclude that u is an eigenfunction corresponding to $\lambda = \lambda_1$. Conversely, if $u \in Q$ is an eigenfunction corresponding to λ_1, then $\| \, |\nabla u| \, \|^p_{L^p(\Omega,w)} = \lambda_1 = \inf_{v\in Q} \| \, |\nabla v| \, \|^p_{L^p(\Omega,w)}$. This completes the proof. □

Proposition 16. *Assume* (E$_2$). *Then the set* $\sigma(W_0^{1,p}(\Omega, w_0, w), \xi)$ *is closed.*

Proof. Let $(\mu_n)_{n\geq 1}$ be a sequence of eigenvalues of $(-\Delta_{p,w}, W_0^{1,p}(\Omega, w_0, w), \xi)$ such that $\mu_n \to \lambda$ as $n \to +\infty$. Let $(u_n)_{n\geq 1} \subset W_0^{1,p}(\Omega, w_0, w)$ be a corresponding sequence of eigenfunctions. Acting on (3) (written for the eigenvalue μ_n and eigenfunction u_n) with the test function $v = u_n$, we see that

$$\| |\nabla u_n| \|^p_{L^p(\Omega,w)} = \mu_n \int_\Omega \xi w_0 |u_n|^p \, dx \quad \text{for all } n \geq 1. \tag{35}$$

Since $u_n \neq 0$ and $\| |\nabla \cdot | \|_{L^p(\Omega,w)}$ is a norm on $W_0^{1,p}(\Omega, w_0, w)$, we deduce that $\| |\nabla u_n| \|_{L^p(\Omega,w)} \neq 0$, and thus $\int_\Omega \xi w_0 |u_n|^p \, dx \neq 0$, so the sequence of eigenfunctions $(u_n)_{n\geq 1}$ can be assumed to satisfy $\int_\Omega \xi w_0 |u_n|^p \, dx = 1$ for all $n \geq 1$. Then (35) implies that the sequence $(u_n)_{n\geq 1}$ is bounded in $W_0^{1,p}(\Omega, w_0, w)$. Thus, along a relabeled subsequence, we have that $u_n \rightharpoonup u$ in $W_0^{1,p}(\Omega, w_0, w)$ as $n \to +\infty$. Owing to the compactness of the embedding $W_0^{1,p}(\Omega, w_0, w) \subset L^p(\Omega, w_0)$ (see (E$_2$)), we deduce that $u_n \to u$ in $L^p(\Omega, w_0)$ as $n \to +\infty$. Then, $\int_\Omega \xi w_0 |u_n|^p \, dx \to \int_\Omega \xi w_0 |u|^p \, dx$ as $n \to +\infty$. Therefore, $\int_\Omega \xi w_0 |u|^p \, dx = 1$, so $u \neq 0$.

Using now in (3) (written for the eigenvalue μ_n with eigenfunction u_n) the test function $v = u_n - u$, we see that

$$\int_\Omega w |\nabla u_n|^{p-2} \nabla u_n \cdot \nabla(u_n - u) \, dx = \mu_n \int_\Omega \xi w_0 |u_n|^{p-2} u_n (u_n - u) \, dx$$

for all $n \geq 1$. Arguing as for (20), we deduce that $\langle -\Delta_{p,w}(u_n), u_n - u\rangle \to 0$ as $n \to +\infty$. Since $u_n \rightharpoonup u$ in $W_0^{1,p}(\Omega, w_0, w)$ as $n \to +\infty$ and taking into account that the operator $-\Delta_{p,w}$ on $W_0^{1,p}(\Omega, w_0, w)$ is a map of type $(S)_+$ (see Proposition 4 (c)), we infer that $u_n \to u$ in $W_0^{1,p}(\Omega, w_0, w)$ as $n \to +\infty$. Passing to the limit as $n \to +\infty$ in (3) (written for the eigenvalue μ_n with eigenfunction u_n), we obtain that λ is an eigenvalue of $(-\Delta_{p,w}, W_0^{1,p}(\Omega, w_0, w), \xi)$ with corresponding eigenfunction u. Thus, $\lambda \in \sigma(W_0^{1,p}(\Omega, w_0, w), \xi)$, which proves that $\sigma(W_0^{1,p}(\Omega, w_0, w), \xi)$ is closed. □

Recall that λ_2 denotes the second eigenvalue of the sequence of eigenvalues given by Theorem 6.

Proposition 17. *Assume* (E_2). *If* λ *is an eigenvalue of* $(-\Delta_{p,w}, W_0^{1,p}(\Omega, w_0, w), \xi)$ *for which there exists an eigenfunction* u *which is nodal, then* $\lambda \geq \lambda_2$.

Proof. Since $u \in W_0^{1,p}(\Omega, w_0, w)$ is nodal, we have that $u^+ = \max\{u, 0\}$ and $u^- = \max\{-u, 0\}$ are both non–zero. Using in (3) the test functions $u^+, u^- \in W_0^{1,p}(\Omega, w_0, w)$ (see [10, Proposition 2]), we get

$$\int_\Omega w|\nabla u^\pm|^p\,dx = \lambda \int_\Omega \xi w_0 (u^\pm)^p\,dx. \tag{36}$$

Let $V = \operatorname{span}\{u^+, u^-\}$. Since u^+ and u^- are linearly independent, we deduce that $\dim V = 2$. Let

$$K := \{v \in V : \|\,|\nabla v|\,\|^p_{L^p(\Omega,w)} = p\}.$$

Let $v \in K$ be arbitrary. Then, $v = \alpha u^+ + \beta u^-$, with $\alpha, \beta \in \mathbb{R}$. By (36), we have that

$$\begin{aligned} p = \|\,|\nabla v|\,\|^p_{L^p(\Omega,w)} &= |\alpha|^p \int_\Omega w|\nabla u^+|^p\,dx + |\beta|^p \int_\Omega w|\nabla u^-|^p\,dx \\ &= \lambda\left(|\alpha|^p \int_\Omega \xi w_0 (u^+)^p\,dx + |\beta|^p \int_\Omega \xi w_0 (u^-)^p\,dx\right) = \lambda \int_\Omega \xi w_0 |v|^p\,dx. \end{aligned} \tag{37}$$

Since K is compact, symmetric, $\operatorname{gen} K = 2$, and due to (37), we see that $K \in \mathcal{K}_2$, hence, by the variational characterization of λ_2 in (18) and by (37), we obtain

$$\frac{1}{\lambda_2} \geq \min_{v \in K} \left(\frac{1}{p} \int_\Omega \xi w_0 |v|^p\,dx \right) = \frac{1}{\lambda},$$

hence $\lambda \geq \lambda_2$. □

The following result studies the monotonicity of the map $\xi \mapsto \lambda_1(\xi)$, where $\lambda_1(\xi)$ is the first eigenvalue of $(-\Delta_{p,w}, W_0^{1,p}(\Omega, w_0, w), \xi)$ given by Theorems 6 and 15.

Proposition 18. *Assume* (E_2). *For* $\xi, \xi' \in L^\infty(\Omega)$ *such that* $0 \leq \xi < \xi'$ *a.e. in* Ω *and* $\xi \neq 0$, *we have* $\lambda_1(\xi) > \lambda_1(\xi')$.

Proof. By Theorem 15, there is $u \in W_0^{1,p}(\Omega, w_0, w)$ such that $\int_\Omega \xi w_0 |u|^p \, dx = 1$ and $\| \, |\nabla u| \, \|^p_{L^p(\Omega,w)} = \lambda_1(\xi)$. Let $v = \left(\int_\Omega \xi' w_0 |u|^p \, dx\right)^{-\frac{1}{p}} u$. Using again Theorem 15, we have

$$\lambda_1(\xi') \leq \| \, |\nabla v| \, \|^p_{L^p(\Omega,w)} = \frac{\| \, |\nabla u| \, \|^p_{L^p(\Omega,w)}}{\int_\Omega \xi' w_0 |u|^p \, dx} < \frac{\| \, |\nabla u| \, \|^p_{L^p(\Omega,w)}}{\int_\Omega \xi w_0 |u|^p \, dx} = \lambda_1(\xi),$$

which is the desired conclusion. □

4. Lyusternik–Schnirelmann Principle

Let X be an infinite dimensional reflexive Banach space.

Let $\varphi : X \to \mathbb{R}$ be a map satisfying

H(φ) (i) $\varphi \in C^1(X, \mathbb{R})$ is even, $\varphi(0) = 0$;
(ii) φ' is completely continuous (i.e., $u_n \rightharpoonup u$ in X implies $\varphi'(u_n) \to \varphi'(u)$ in X^*),

and let $\psi : X \to \mathbb{R}$ be a map satisfying

H(ψ) (i) $\psi \in C^1(X, \mathbb{R})$ is even, $\psi(0) = 0$, and $M := \{u \in X : \psi(u) = 1\}$ is bounded;
(ii) ψ' is bounded, and if $u_n \rightharpoonup u$ in X, $\psi'(u_n) \rightharpoonup v$ in X^*, $\langle \psi'(u_n), u_n \rangle \to \langle v, u \rangle$, then $u_n \to u$ in X;
(iii) for all $u \in X \setminus \{0\}$, we have that $\langle \psi'(u), u \rangle > 0$ and $\lim_{t \to +\infty} \psi(tu) = +\infty$, moreover, $\inf_{v \in M} \langle \psi'(v), v \rangle > 0$.

Let $\overline{\text{conv}}\, M$ denote the closure of the convex hull of M. We also assume that the maps φ and ψ satisfy

H(φ, ψ) for $u \in \overline{\text{conv}}\, M$, we have

$$\langle \varphi'(u), u \rangle = 0 \Leftrightarrow \varphi(u) = 0.$$

Consider the following nonlinear eigenvalue problem:

$$\mu \psi'(u) = \varphi'(u), \quad (\mu, u) \in \mathbb{R} \times M. \tag{38}$$

Remark 19. If φ and ψ are positively p-homogeneous for some $p > 1$ (that is, $\varphi(tv) = t^p\varphi(v)$ and $\psi(tv) = t^p\psi(v)$ for all $v \in X$ and $t > 0$), then differentiating the above equalities at $t = 1$ and using the definition of M, we have

$$\varphi(u) = \frac{1}{p}\langle\varphi'(u), u\rangle = \frac{\mu}{p}\langle\psi'(u), u\rangle = \mu\psi(u) = \mu$$

whenever (μ, u) is a solution of (38).

With M in H(ψ) (i), for $n \geq 1$, let

$$\mathcal{K}_n = \{K \subset M : K \text{ is symmetric, compact, with } \varphi|_K > 0 \text{ and gen } K \geq n\}$$

and define the max-min values

$$c_n = \begin{cases} \sup\limits_{K\in\mathcal{K}_n} \min\limits_{u\in K} \varphi(u) & \text{if } \mathcal{K}_n \neq \emptyset, \\ 0 & \text{if } \mathcal{K}_n = \emptyset \end{cases}$$

so that $+\infty \geq c_1 \geq c_2 \geq \ldots \geq c_n \geq \ldots \geq 0$.

We now state the Lyusternik–Schnirelmann principle (see [16, p. 326–328]).

Theorem 20. *Assume that φ and ψ satisfy* H(φ), H(ψ), *and* H(φ, ψ). *Then, we have the following:*

(a) $c_1 < +\infty$ *and* $c_n \to 0$ *as* $n \to +\infty$.
(b) *If* $c_n > 0$, *then we can find a solution* (μ, u) *of* (38), *with* $\mu \neq 0$, $u \in M$, *and such that* $\varphi(u) = c_n$.
(c) *If* $c_n = c_{n+k} > 0$ *for some* $k \geq 0$, *then*

$$\text{gen}\,\{u \in M : (\mu, u) \text{ is a solution of } (38), \text{ for some } \mu, \text{ and } \varphi(u) = c_n\} \geq k+1.$$

(d) *If* $c_n > 0$ *for all* $n \geq 1$, *then there exists a sequence* $((\mu_n, u_n))_{n\geq 1}$ *of solutions of* (38) *with* $\varphi(u_n) = c_n$, $\mu_n \neq 0$ *for all* $n \geq 1$, *and* $\mu_n \to 0$ *as* $n \to +\infty$.
(e) *We strengthen* H(φ, ψ) *by assuming that, for* $u \in \overline{\text{conv}}\, M$, *we have*

$$\langle\varphi'(u), u\rangle = 0 \;\Leftrightarrow\; \varphi(u) = 0 \;\Leftrightarrow\; u = 0.$$

Then, $c_n > 0$ for all $n \geq 1$, and there exists a sequence $((\mu_n, u_n))_{n\geq 1}$ of solutions of (38) such that $\varphi(u_n) = c_n$, $\mu_n \neq 0$, $\mu_n \to 0$, and $u_n \rightharpoonup 0$ in X.

5. Proof of Theorems 5 and 6

5.1. *Proof of Theorem* 5

We aim to apply Theorem 20 to study the spectrum of $(-\Delta_{p,w}, W^{1,p}(\Omega, w_0, w), \xi)$. To this end, we consider the reflexive Banach space $X = W^{1,p}(\Omega, w_0, w)$, the map $\varphi : W^{1,p}(\Omega, w_0, w) \to \mathbb{R}$ defined by

$$\varphi(u) = \frac{1}{p}\int_\Omega \xi w_0|u|^p\, dx \quad \text{for all } u \in W^{1,p}(\Omega, w_0, w),$$

and the map $\psi : W^{1,p}(\Omega, w_0, w) \to \mathbb{R}$ defined by

$$\psi(u) = \frac{1}{p}\left(\int_\Omega w|\nabla u|^p\, dx + \int_\Omega \xi w_0|u|^p\, dx\right) \quad \text{for all } u \in W^{1,p}(\Omega, w_0, w).$$

Note from (14) that

$$M = \{u \in W^{1,p}(\Omega, w_0, w) : \psi(u) = 1\}. \tag{39}$$

Proceeding as in [11, Proposition 5.3], we have the following:

Lemma 21. *We have $\varphi, \psi \in C^1(W^{1,p}(\Omega, w_0, w), \mathbb{R})$ and, for $u, v \in W^{1,p}(\Omega, w_0, w)$,*

$$\langle \varphi'(u), v\rangle = \int_\Omega \xi w_0|u|^{p-2}uv\, dx, \tag{40}$$

$$\langle \psi'(u), v\rangle = \int_\Omega w|\nabla u|^{p-2}\nabla u\cdot\nabla v\, dx + \int_\Omega \xi w_0|u|^{p-2}uv\, dx. \tag{41}$$

We check that the above maps φ and ψ satisfy the hypotheses of Theorem 20.

Lemma 22. *Under hypotheses* (E_1) *and $w_0 \in L^1(\Omega)$, the maps φ, ψ satisfy conditions* $\mathrm{H}(\varphi)$, $\mathrm{H}(\psi)$, *and* $\mathrm{H}(\varphi, \psi)$.

Proof. First, note that the maps φ, ψ are even and that $\varphi(0) = \psi(0) = 0$. By Lemma 21, we have $\varphi, \psi \in C^1(W^{1,p}(\Omega, w_0, w), \mathbb{R})$. Let us show that φ' is completely continuous. Suppose that $u_n \rightharpoonup u$ in $W^{1,p}(\Omega, w_0, w)$ and let us check that from any relabeled subsequence of $(\varphi'(u_n))_{n\geq 1}$, we can extract a subsequence strongly converging to $\varphi'(u)$. From (40), denoting $p' = \frac{p}{p-1}$, we see that

$$|\langle \varphi'(u_n) - \varphi'(u), v\rangle| \\ \leq \|\xi\|_{L^\infty(\Omega)} \| |u_n|^{p-2}u_n - |u|^{p-2}u\|_{L^{p'}(\Omega,w_0)} \|v\|_{L^p(\Omega,w_0)}$$

for all $v \in W^{1,p}(\Omega, w_0, w)$, hence

$$\|\varphi'(u_n) - \varphi'(u)\|_{W^{1,p}(\Omega,w_0,w)^*} \\ \leq \|\xi\|_{L^\infty(\Omega)} \| |u_n|^{p-2}u_n - |u|^{p-2}u\|_{L^{p'}(\Omega,w_0)}. \tag{42}$$

Since $u_n \rightharpoonup u$ in $W^{1,p}(\Omega, w_0, w)$, in view of the compactness of the embedding $W^{1,p}(\Omega, w_0, w) \subset L^p(\Omega, w_0)$ (see (E_1) (ii)), along a relabeled subsequence, we have $u_n \to u$ in $L^p(\Omega, w_0)$. This implies that $w_0^{\frac{1}{p}} u_n \to w_0^{\frac{1}{p}} u$ in $L^p(\Omega)$, so, up to a subsequence, we may assume that $u_n \to u$ a.e. in Ω and there is $h \in L^p(\Omega)$ such that $w_0^{\frac{1}{p}}|u_n| < h$ a.e. in Ω. Applying Lebesgue's dominated convergence theorem, we obtain that $\| |u_n|^{p-2}u_n - |u|^{p-2}u\|_{L^{p'}(\Omega,w_0)} \to 0$ as $n \to +\infty$. This, combined with (42), yields $\varphi'(u_n) \to \varphi'(u)$ in $W^{1,p}(\Omega, w_0, w)^*$, so φ' is completely continuous. This shows that φ satisfies $\mathrm{H}(\varphi)$.

Owing to Lemma 3 (in view of (E_1) (i) and the fact that $w_0 \in L^1(\Omega)$), the set M in (14) (recall (39)) is bounded in $W^{1,p}(\Omega, w_0, w)$. This shows $\mathrm{H}(\psi)$ (i). Moreover, by Proposition 4 (a), we have that ψ' is bounded.

In order to show that ψ satisfies the second part of $\mathrm{H}(\psi)$ (ii), assume that $u_n \rightharpoonup u$ in $W^{1,p}(\Omega, w_0, w)$, $\psi'(u_n) \rightharpoonup v$ in $W^{1,p}(\Omega, w_0, w)^*$, and $\langle \psi'(u_n), u_n\rangle \to \langle v, u\rangle$ as $n \to +\infty$. Then we have

$$\lim_{n\to+\infty} \langle \psi'(u_n), u_n - u\rangle = \langle v, u\rangle - \langle v, u\rangle = 0.$$

Since ψ' is a map of type (S_+) (which is known from Proposition 4 (a), in view of hypothesis (E_1) (ii)), we deduce that $u_n \to u$ in $W^{1,p}(\Omega, w_0, w)$ as $n \to +\infty$. This proves that ψ satisfies $\mathrm{H}(\psi)$ (ii).

For all $u \in W^{1,p}(\Omega, w_0, w) \setminus \{0\}$, we have $\langle \psi'(u), u \rangle = \|u\|^p_{W^{1,p}(\Omega,w_0,w),\xi} > 0$ and $\psi(tu) = \frac{t^p}{p}\|u\|^p_{W^{1,p}(\Omega,w_0,w),\xi} \to +\infty$ as $t \to +\infty$. Moreover, $\langle \psi'(u), u \rangle = p$ if $u \in M$ (see (14)), and this ends the verification that ψ satisfies H(ψ).

Finally, we see that

$$\langle \varphi'(u), u \rangle = \int_\Omega \xi w_0 |u|^p \, dx = p\varphi(u) \quad \text{for all } u \in W^{1,p}(\Omega, w_0, w),$$

which shows that H(φ, ψ) is satisfied. This completes the proof of the lemma. □

The sets $\mathcal{K}_n$ defined in (15) satisfy the following.

Lemma 23. *For all $n \geq 1$, we have $\mathcal{K}_n \neq \emptyset$ and $c_n := \sup_{K \in \mathcal{K}_n} \min_{u \in K} \varphi(u) > 0$.*

Proof. Using that the set $\{x \in \Omega : \xi(x) > 0\}$ has positive measure, we can find functions $u_1, \ldots, u_n \in C^\infty_\mathrm{c}(\Omega)$ whose supports are pairwise disjoint such that

$$\int_\Omega \xi w_0 |u_k|^p \, dx > 0 \quad \text{for all } k \in \{1, \ldots, n\}. \tag{43}$$

Let $V = \operatorname{span}\{u_k\}_{k=1}^n$ and let $K = \{u \in V : \|u\|^p_{W^{1,p}(\Omega,w_0,w),\xi} = p\}$. Since K is a sphere in the n-dimensional space V, we obtain that K is compact, symmetric, and $\operatorname{gen} K = n$. Moreover, from (43), we have that $\varphi|_{V\setminus\{0\}} > 0$, so $\varphi|_K > 0$. Consequently, we have that $K \in \mathcal{K}_n$. Moreover, $c_n \geq \min_{u \in K} \varphi(u) > 0$. This proves the lemma. □

Proof of Theorem 5. We apply Theorem 20 (d) by virtue of Lemma 22. Taking the definition of c_n into account and since $c_n > 0$ for all $n \geq 1$ (see Lemma 23), we infer that there exists a sequence $\big((\mu_n, u_n)\big)_{n \geq 1}$ of solutions of (38) with $\mu_n \neq 0$ for all $n \geq 1$, $\mu_n \to 0$ as $n \to +\infty$, and

$$\mu_n = \varphi(u_n) = c_n = \sup_{K \in \mathcal{K}_n} \min_{u \in K} \varphi(u) > 0 \quad \text{for all } n \geq 1 \tag{44}$$

(where we use Remark 19 in view of the positive p-homogeneity of φ and ψ). Since $\mu_n = c_n$ for all $n \geq 1$, the sequence $(\mu_n)_{n \geq 1}$ is non-increasing.

Let u_0 be the constant function $u_0 \equiv \left(\frac{p}{\int_\Omega \xi w_0\, dx}\right)^{\frac{1}{p}} \in M$ (see (39) and since $w_0 \in L^1(\Omega)$) so that $K_0 := \{u_0, -u_0\} \in \mathcal{K}_1$. Observing that $\varphi(u) \le \psi(u) = 1$ for all $u \in M$, we see that

$$\mu_1 = \sup_{K \in \mathcal{K}_1} \min_{u \in K} \varphi(u) = \min_{u \in K_0} \varphi(u) = 1.$$

Using (38), (40), (41), and Definition 1 (a), we see that $\mu \neq 0$ is an eigenvalue for problem (38) if and only if $\frac{1}{\mu} - 1$ is an eigenvalue of $(-\Delta_{p,w}, W^{1,p}(\Omega, w_0, w), \xi)$. Let $\lambda_n = \frac{1}{\mu_{n+1}} - 1$ for all $n \ge 0$. Then, $\lambda_0 = 0$, $\lambda_n \to +\infty$ as $n \to +\infty$, $(\lambda_n)_{n \ge 0}$ is a non-decreasing sequence of eigenvalues of $(-\Delta_{p,w}, W^{1,p}(\Omega, w_0, w), \xi)$, and

$$\frac{1}{\lambda_n + 1} = \sup_{K \in \mathcal{K}_{n+1}} \min_{u \in K} \varphi(u) \quad \text{for all } n \ge 0,$$

which shows (16).

It remains to show that $\lambda_1 > \lambda_0 = 0$. Arguing by contradiction, we assume that $\lambda_1 = \lambda_0 = 0$, hence $\mu_2 = \mu_1 = 1$. This implies that $c_2 = c_1 = 1$ (see (44)). By Theorem 20 (c), we have that the set

$$\{u \in W^{1,p}(\Omega, w_0, w) : \|u\|^p_{W^{1,p}(\Omega, w_0, w), \xi} = p \text{ and } \int_\Omega \xi w_0 |u|^p\, dx = p\}$$

has genus ≥ 2. But this set only comprises the two constant functions $\pm \left(\frac{p}{\int_\Omega \xi w_0\, dx}\right)^{\frac{1}{p}}$, hence it has genus 1. This is a contradiction. The proof of the theorem is now complete. □

5.2. *Proof of Theorem* 6

We apply now the Lyusternik–Schnirelmann principle in Theorem 20 to the reflexive Banach space $X = W_0^{1,p}(\Omega, w_0, w)$ and to the maps $\varphi, \psi \in C^1(W_0^{1,p}(\Omega, w_0, w), \mathbb{R})$ defined by

$$\varphi(u) = \frac{1}{p} \int_\Omega \xi w_0 |u|^p\, dx \quad \text{and}$$

$$\psi(u) = \frac{1}{p} \int_\Omega w |\nabla u|^p\, dx \quad \text{for all } u \in W_0^{1,p}(\Omega, w_0, w).$$

For M given in (17), we have $M = \{u \in W_0^{1,p}(\Omega, w_0, w) : \psi(u) = 1\}$. With a similar proof as the one in Lemma 22, we see that under assumption (E_2), the maps φ, ψ defined above satisfy conditions $H(\varphi)$, $H(\psi)$, and $H(\varphi, \psi)$. As in the proof of Lemma 23, for all $n \geq 1$, the sets $\mathcal{K}_n$ defined in (15) (relatively to the above M) are non-empty and we have $c_n := \sup_{K \in \mathcal{K}_n} \min_{u \in K} \varphi(u) > 0$. To complete the proof, we apply Theorem 20 (d), Remark 2 (b) and argue similarly as in the proof of Theorem 5, noting that $\mu \neq 0$ is an eigenvalue for problem (38) if and only if $\frac{1}{\mu}$ is an eigenvalue of $(-\Delta_{p,w}, W_0^{1,p}(\Omega, w_0, w), \xi)$ (see Definition 1 (b)). □

6. Example of Domain and Weights Satisfying the Embedding Assumptions

In this section, we provide an example of a domain Ω and weights w and w_0 that satisfy the conditions (E_1) and (E_2). We rely on the following technical result. In the statement, q^* stands for the Sobolev critical exponent, given by

$$q^* = \begin{cases} \dfrac{Nq}{N-q} & \text{if } q < N, \\ +\infty & \text{if } q \geq N. \end{cases}$$

Proposition 24. *Let $p \in (1, +\infty)$ and $w, w_0 \in \mathcal{W}_p(\mathbb{R}^N)$. Assume that there are $q \in (1, p)$, $r \in (1, q^*)$, $r > p$, and a C^1-diffeomorphism*

$$\varphi : D \to E$$

between two domains $D, E \subset \mathbb{R}^N$, where D is a bounded domain with Lipschitz boundary, such that

$$K_1 := \int_D \left(\frac{\|\varphi'(x)\|^p}{|\det \varphi'(x)|\, w(\varphi(x))} \right)^{\frac{q}{p-q}} dx < +\infty$$

and

$$K_2 := \int_D (w_0(\varphi(x))\, |\det \varphi'(x)|)^{\frac{r}{r-p}}\, dx < +\infty.$$

Then, we have the following:

(a) *Condition* (E_1) *holds for the domain* $\Omega = E$.
(b) *Condition* (E_2) *holds for every subdomain* $\Omega \subset E$ *with* $|E \setminus \Omega| > 0$.

Proof. The proof contains several steps.

Step 1: Whenever $u \in W^{1,p}(E, w_0, w)$, we have $u \circ \varphi \in W^{1,q}(D)$ and

$$\left(\int_D |\nabla(u \circ \varphi)|^q \, dx\right)^{\frac{1}{q}} \le K_1^{\frac{p-q}{pq}} \left(\int_E |\nabla u|^p w(y) \, dy\right)^{\frac{1}{p}}. \tag{45}$$

Let $u \in W^{1,p}(E, w_0, w)$. First, we note that for every compact subset $K \subset D$, the image $\varphi(K) \subset E$ is compact, and the change of variable formula yields

$$\int_K |u \circ \varphi(x)| \, dx = \int_{\varphi(K)} |u(y)| \, |\det(\varphi^{-1})'(y)| \, dy < +\infty$$

(since $L^p(E, w_0) \subset L^1_{\text{loc}}(E)$), which implies that

$$u \circ \varphi \in L^1_{\text{loc}}(D). \tag{46}$$

Next, by applying Hölder's inequality and the change of variable formula, we have

$$\begin{aligned}\int_D |\nabla(u \circ \varphi)|^q \, dx &\le \int_D |\nabla u(\varphi(x))|^q \, \|\varphi'(x)\|^q \, dx \\ &\le K_1^{\frac{p-q}{p}} \left(\int_D |\nabla u(\varphi(x))|^p \, |\det \varphi'(x)| \, w(\varphi(x)) \, dx\right)^{\frac{q}{p}} \\ &= K_1^{\frac{p-q}{p}} \left(\int_E |\nabla u(y)|^p w(y) \, dy\right)^{\frac{q}{p}}.\end{aligned}$$

Since we assume $u \in W^{1,p}(E, w_0, w)$, this inequality ensures that $|\nabla(u \circ \varphi)| \in L^q(D)$, and we get (45). Moreover, by applying [13, §1.1.11] (and using (46)), it follows that $u \circ \varphi \in W^{1,q}(D)$.

Step 2: Whenever $v \in L^r(D)$, we have $v \circ \varphi^{-1} \in L^p(E, w_0)$ and

$$\left\|v \circ \varphi^{-1}\right\|_{L^p(E, w_0)} \le K_2^{\frac{r-p}{rp}} \|v\|_{L^r(D)}. \tag{47}$$

Let $v \in L^r(D)$. The change of variable formula and Hölder's inequality imply

$$\int_E |v \circ \varphi^{-1}(y)|^p w_0(y)\,dy = \int_D |v(x)|^p w_0(\varphi(x))\,|\det \varphi'(x)|\,dx$$
$$\leq K_2^{\frac{r-p}{r}} \left(\int_D |v(x)|^r\,dx \right)^{\frac{p}{r}} < +\infty.$$

Whence $v \circ \varphi^{-1} \in L^p(E, w_0)$, and relation (47) follows.

Step 3: Condition (E_1) (i) holds for $\Omega = E$.

Let $A \subset E$ be a subset of positive measure, such that $\overline{A}$ is compact and contained in E. Then, $B := \varphi^{-1}(A) \subset D$ is a subset of positive measure. In view of the Poincaré–Wirtinger inequality and the Sobolev embedding theorem (see, e.g., [15, Theorems 1.44, 1.49]), there is a constant $c_1 > 0$ such that

$$\|v - I_B(v)\|_{L^r(D)} \leq c_1 \|\,|\nabla v|\,\|_{L^q(D)} \quad \text{for all } v \in W^{1,q}(D), \tag{48}$$

where we denote $I_B(v) = \frac{1}{|B|} \int_B v(x)\,dx$ (recall that $r < q^*$).

Let $u \in W^{1,p}(E, w_0, w)$. Relations (45) and (48) give

$$\|u \circ \varphi - I_B(u \circ \varphi)\|_{L^r(D)} \leq c_1 K_1^{\frac{p-q}{pq}} \|\,|\nabla u|\,\|_{L^p(E,w)} \tag{49}$$

(in particular, $u \circ \varphi - I_B(u \circ \varphi) \in L^r(D)$). Note that

$$I_B(u \circ \varphi) = \frac{1}{|\varphi^{-1}(A)|} \int_{\varphi^{-1}(A)} u \circ \varphi(x)\,dx = \int_A \rho(y) u(y)\,dy$$

where we set $\rho := \frac{1}{|\varphi^{-1}(A)|} |\det(\varphi^{-1})'(\cdot)| \in L^\infty(A)$. Finally, (47) implies that

$$\left\| u - \int_A \rho(y) u(y)\,dy \right\|_{L^p(E,w_0)} \leq K_2^{\frac{r-p}{rp}} \|u \circ \varphi - I_B(u \circ \varphi)\|_{L^r(D)}.$$

Combining this with (49), we obtain that condition (E_1) (i) is verified.

Step 4: Condition (E_1) (ii) holds for $\Omega = E$.

Let $(u_n)_{n\geq 1}$ be a bounded sequence in $W^{1,p}(E,w_0,w)$. We can find $\alpha>0$ such that the set

$$A := \{x\in E : w_0(x)>\alpha\}$$

has positive measure. Moreover, there is a compact subset $K\subset E$ such that $A\cap K$ has positive measure. By Hölder's inequality, we have

$$\begin{aligned}&\int_{\varphi^{-1}(A\cap K)}|u_n\circ\varphi(x)|^q\,dx\\ &\leq\left(\int_{\varphi^{-1}(A\cap K)}|u_n\circ\varphi(x)|^p w_0(\varphi(x))\,|\det\varphi'(x)|\,dx\right)^{\frac{q}{p}}\\ &\quad\cdot\left(\int_{\varphi^{-1}(A\cap K)}(w_0(\varphi(x))\,|\det\varphi'(x)|)^{\frac{q}{q-p}}\,dx\right)^{\frac{p-q}{p}}\\ &\leq c_2\left(\int_{A\cap K}|u_n(y)|^p w_0(y)\,dy\right)^{\frac{q}{p}}\leq c_2\|u_n\|^q_{W^{1,p}(E,w_0,w)},\end{aligned}$$

for some constant $c_2>0$ (noting that the integral on the third line is finite because $\frac{q}{q-p}<0$ and $w_0(\varphi(x))>\alpha$ for a.a. $x\in\varphi^{-1}(A\cap K)$). We have that

$$v\mapsto\left(\int_{\varphi^{-1}(A\cap K)}|v(x)|^q\,dx+\|\,|\nabla v|\,\|^q_{L^q(D)}\right)^{\frac{1}{q}}$$

is a norm on $W^{1,q}(D)$ equivalent to the Sobolev norm: this follows from Lemma 3 applied with ξ given by $\xi|_{\varphi^{-1}(A\cap K)}\equiv 1$ and $\xi|_{D\setminus\varphi^{-1}(A\cap K)}\equiv 0$. By using also (45), we conclude that $(u_n\circ\varphi)_{n\geq 1}$ is a bounded sequence in $W^{1,q}(D)$. Now, since $r<q^*$, invoking the compact embedding $W^{1,q}(D)\subset L^r(D)$, there is a relabeled subsequence $(u_n\circ\varphi)_{n\geq 1}$ that converges to some u_0 in $L^r(D)$. By virtue of Step 2, it follows that $u_0\circ\varphi^{-1}\in L^p(E,w_0)$ and $(u_n)_{n\geq 1}$ converges to $u_0\circ\varphi^{-1}$ in $L^p(E,w_0)$. Altogether, this completes Step 4.

Steps 3–4 yield part (a) of the proposition. Note that for this it suffices to require $w,w_0\in\mathcal{W}_p(E)$. The assumption that $w,w_0\in\mathcal{W}_p(\mathbb{R}^N)$ is necessary only for Step 5.

Step 5: Condition (E_2) holds for any subdomain $\Omega \subset E$ such that $|E \setminus \Omega| > 0$.

Recall that whenever $u \in W_0^{1,p}(\Omega, w_0, w)$, the extension by zero $\overline{u}$ given by

$$\overline{u}(x) = \begin{cases} u(x) & \text{if } x \in \Omega, \\ 0 & \text{if } x \in E \setminus \Omega \end{cases}$$

belongs to $W^{1,p}(E, w_0, w)$; see [11, Proposition 2.3]. Taking a bounded subset $A \subset E \setminus \Omega$ of positive measure such that $\overline{A} \subset E$, Step 3 applied to $\overline{u}$ yields

$$\begin{aligned} \|u\|_{L^p(\Omega,w_0)} = \|\overline{u}\|_{L^p(E,w_0)} &= \left\| \overline{u} - \int_A \rho \overline{u}\, dx \right\|_{L^p(E,w_0)} \\ &\le c_3 \|\, |\nabla \overline{u}|\, \|_{L^p(E,w)} = c_3 \|\, |\nabla u|\, \|_{L^p(\Omega,w)}, \end{aligned}$$

for some $c_3 > 0$ independent of u. Thus the norm $\|\, |\nabla \cdot|\, \|_{L^p(\Omega,w)}$ is equivalent to the norm $\| \cdot \|_{W^{1,p}(\Omega,w_0,w)}$ in $W_0^{1,p}(\Omega, w_0, w)$.

Now, in order to show (E_2) for the subdomain Ω, let $(u_n)_{n\geq 1} \subset W_0^{1,p}(\Omega, w_0, w)$ be a bounded sequence. We have $\|\overline{u_n}\|_{W^{1,p}(E,w_0,w)} = \|u_n\|_{W^{1,p}(\Omega,w_0,w)}$, hence $(\overline{u_n})_{n\geq 1}$ is a bounded sequence in $W^{1,p}(E, w_0, w)$. Since the embedding $W^{1,p}(E, w_0, w) \subset L^p(E, w_0)$ is compact by virtue of Step 4, there is a relabeled subsequence $(\overline{u_n})_{n\geq 1}$ such that $\overline{u_n} \to u_0$ in $L^p(E, w_0)$ as $n \to +\infty$. It follows that $u_n = \overline{u_n}|_\Omega \to u_0|_\Omega$ in $L^p(\Omega, w_0)$. We have thus shown that the embedding $W_0^{1,p}(\Omega, w_0, w) \subset L^p(\Omega, w_0)$ is compact. The proof is complete. □

Example 25. For $N \geq 2$, any $\gamma_2, \ldots, \gamma_N \in [1, +\infty)$, and any $\ell \in (0, +\infty)$, we consider the domain

$$\Omega := \{(x_1, \ldots, x_N) \in (0, +\infty)^N : x_1 < \ell,\ \forall j = 2, \ldots, N,\ x_j < x_1^{\gamma_j}\}.$$

Let $p \in (1, N)$. Set $\gamma = 1 + \sum_{j=2}^N \gamma_j$. Let w, w_0 be defined by

$$w(x) = |x|^\alpha \quad \text{and} \quad w_0(x) = |x|^\beta \quad \text{for a.e. } x \in \mathbb{R}^N,$$

where

$$\alpha, \beta \in (-N, N(p-1)) \quad \text{are such that} \quad p - \gamma < \alpha < p + \beta. \tag{50}$$

The first part of (50) guarantees that

$$w, w_0 \in \mathcal{W}_p(\mathbb{R}^N).$$

We claim that conditions (E_1) and (E_2) hold.

To see this, we define a domain D and a diffeomorphism φ for which we will apply Proposition 24. Let

$$\mu := \min\left\{\frac{N}{\beta+\gamma}, \frac{N-p}{\alpha+\gamma-p}\right\} > \frac{N-p}{\beta+\gamma} > 0$$

(using the assumption that $0 < \alpha+\gamma-p < \beta+\gamma$). Hence, there is r such that

$$(p \leq)\frac{Np}{\mu(\beta+\gamma)} < r < \frac{Np}{N-p}.$$

Since the map $s \mapsto \frac{Ns}{N-s}$ is increasing and continuous on $(1, N)$, we can find $q \in (1, p)$ such that

$$r < \frac{Nq}{N-q}.$$

We consider the domain (with Lipschitz boundary)

$$D := \{(x_1, \ldots, x_N) \in (0, +\infty)^N : \forall j = 2, \ldots, N,\ x_j < x_1 < \ell^{\frac{1}{\mu}}\}$$

and the diffeomorphism

$$\varphi : D \to \Omega,\ (x_1, \ldots, x_N) \mapsto (x_1^{\mu}, x_2 x_1^{\mu\gamma_2 - 1}, \ldots, x_N x_1^{\mu\gamma_N - 1}).$$

We have

$$\varphi'(x) = \begin{pmatrix} \mu x_1^{\mu-1} & 0 & 0 & \cdots & 0 \\ (\mu\gamma_2 - 1)x_2 x_1^{\mu\gamma_2-2} & x_1^{\mu\gamma_2-1} & 0 & \cdots & 0 \\ \vdots & 0 & \ddots & \ddots & \vdots \\ \vdots & \vdots & \ddots & \ddots & 0 \\ (\mu\gamma_N - 1)x_N x_1^{\mu\gamma_N-2} & 0 & \cdots & 0 & x_1^{\mu\gamma_N-1} \end{pmatrix}.$$

Thus,

$$\det \varphi'(x) = \mu x_1^{\mu\gamma - N} \quad \text{for all } x \in D$$

and there are constants $c_1, c_2 > 0$ such that

$$x_1^\mu \leq |\varphi(x)| \leq c_1 x_1^\mu \quad \text{and} \quad \|\varphi'(x)\| \leq c_2 x_1^{\mu-1} \quad \text{for all } x \in D.$$

It follows that, for some $c_3 > 0$, we have

$$\begin{aligned} K_1 &= \int_D \left(\frac{\|\varphi'(x)\|^p}{|\det \varphi'(x)| w(\varphi(x))} \right)^{\frac{q}{p-q}} dx \\ &\leq c_3 \int_0^{\ell^{\frac{1}{\mu}}} \int_0^{x_1} \cdots \int_0^{x_1} \left(x_1^{p\mu - p - \mu\gamma + N - \mu\alpha} \right)^{\frac{q}{p-q}} dx_N \cdots dx_2 dx_1 \\ &= c_3 \int_0^{\ell^{\frac{1}{\mu}}} x_1^{N-1} \left(x_1^{p\mu - p - \mu\gamma + N - \mu\alpha} \right)^{\frac{q}{p-q}} dx_1 < +\infty \end{aligned}$$

since $p\mu - p - \mu\gamma + N - \mu\alpha \geq 0$ due to the choice of μ. In addition, for some $c_4 > 0$, we also have

$$\begin{aligned} K_2 &= \int_D (w_0(\varphi(x)) \, |\det \varphi'(x)|)^{\frac{r}{r-p}} dx \\ &\leq c_4 \int_0^{\ell^{\frac{1}{\mu}}} \int_0^{x_1} \cdots \int_0^{x_1} (x_1^{\mu\beta + \mu\gamma - N})^{\frac{r}{r-p}} dx_N \cdots dx_2 dx_1 \\ &= c_4 \int_0^{\ell^{\frac{1}{\mu}}} x_1^{N-1} (x_1^{\mu\beta + \mu\gamma - N})^{\frac{r}{r-p}} dx_1 < +\infty \end{aligned}$$

since $N - 1 + (\mu\beta + \mu\gamma - N)\frac{r}{r-p} > -1$, where the last inequality holds because we have chosen $r > \frac{Np}{\mu(\beta+\gamma)}$. Therefore, Proposition 24 applies, guaranteeing that (E$_1$) holds. Since ℓ is arbitrary, it also follows from the proposition that (E$_2$) holds.

References

[1] Anane, A. Simplicité et isolation de la première valeur propre du p-laplacien avec poids. *C. R. Acad. Sci. Paris Sér. I Math.*, 305, 725–728 (1987).

[2] Anane, A. and Tsouli, N. On the second eigenvalue of the p-Laplacian, In *Nonlinear Partial Differential Equations* (Fès, 1994), pp. 1–9, (Longman, Harlow, 1996).

[3] Brezis, H. *Functional Analysis, Sobolev Spaces and Partial Differential Equations.* Springer, New York (2011).

[4] Coffman, C.V. A minimum-maximum principle for a class of non-linear integral equations. *J. Analyse Math.*, 22, 391–419 (1969).

[5] Cuesta, M., de Figueiredo, D. and Gossez, J.-P. The beginning of the Fučik spectrum for the p-Laplacian. *J. Differential Equations*, 159, 212–238 (1999).

[6] Drábek, P., Ho, K. and Sarkar, A. On the eigenvalue problem involving the weighted p-Laplacian in radially symmetric domains. *J. Math. Anal. Appl.*, 468(2), 716–756 (2018).

[7] Drábek, P., Kufner, A. and Nicolosi, F. *Quasilinear Elliptic Equations with Degenerations and Singularities.* Walter de Gruyter & Co., Berlin (1997).

[8] Duistermaat, J.J. and Kolk, J.A.C. *Distributions: Theory and Applications.* Springer, Berlin–Heidelberg–New York (2006).

[9] Gilbarg, D. and Trudinger, N.S. *Elliptic Partial Differential Equations of Second Order.* Springer-Verlag, Berlin (2001).

[10] Gol'dshtein, V., Motreanu, D. and Motreanu, V.V. Non-homogeneous Dirichlet boundary value problems in weighted Sobolev spaces. *Complex Var. Elliptic Equ.*, 60(3), 372–391 (2015).

[11] Gol'dshtein, V., Motreanu, V.V. and Ukhlov, A. Embeddings of weighted Sobolev spaces and degenerate Dirichlet problems involving the weighted p-Laplacian. *Complex Var. Elliptic Equ.*, 56(10–11), 905–930 (2011).

[12] Lindqvist, P. On the equation $\operatorname{div}(|\nabla u|^{p-2}\nabla u) + \lambda|u|^{p-2}u = 0$. *Proc. Amer. Math. Soc.*, 109, 157–164 (1990).

[13] Maz'ja, V. *Sobolev Spaces.* Springer-Verlag, Berlin, Heidelberg (1985).

[14] Megginson, R.E. *An Introduction to Banach Space Theory.* Springer-Verlag, New York (1998).

[15] Motreanu, D., Motreanu, V.V. and Papageorgiou, N. *Topological and Variational Methods with Applications to Nonlinear Boundary Value Problems.* Springer, New York (2014).

[16] Zeidler, E. *Nonlinear Functional Analysis and Its Applications. III. Variational Methods and Optimization.* Springer-Verlag, New York (1985).

https://doi.org/10.1142/9789811293009_0014

Chapter 14

On the Stochastic Properties of Birth-Death Circuit Chains for a Generalized Sample Path Case

Chrysoula Ganatsiou

Department of Digital Systems, University of Thessaly, Gaiopolis Gampus, Larissa-Trikala Ring Road, 41500, Larissa, Greece
chganats@uth.gr, ganatsio@otenet.gr

By using the circuit representation theory of Markov processes, we study important stochastic properties of the corresponding Markov chains (circuit chains) describing uniquely the birth–death chains by directed weighted circuits in fixed environments for a generalized sample path case.

1. Introduction

It is known that the birth–death chain is a special case of homogeneous, aperiodic, irreducible, discrete, or continuous parameter Markov process with discrete state space where state changes can only happen between neighboring states which means that the state transitions are of only two types: *"births"* which increase the state variable by one and *"deaths"* which decrease the state variable by one. Here the states are represented as non-negative integer-valued without loss of generality ([17, 20]). It is also assumed that the birth and death events are independent of each other. Loosely speaking, this is a process which combines the property of a random walk with reflection at zero and the discrete or continuous parameter nature of the transitions times ([1, 2, 4, 6, 7, 9, 10]).

In general, the birth–death processes have many applications in epidemiology, biology, since they may be used to study the evolution of bacteria of the number of people with a disease within a population. Besides, the queuing model is an important application of the birth–death processes in computer networks and telecommunications since it can be used to optimize the size of the storage space, to determine the trade-off between throughput and inventory, as well as to exhibit the propagation of blockage ([18, 19, 21]). Usually, the birth–death chains are studied from the Markov chains' point of view, where the random mechanism of spatial motion is determined by the given transition probabilities (probabilities of jumps) at each state in a *non-random (fixed) environment.* Regarding the classification of states through the use of proper theorems, we have that the states are

(i) *positive recurrent* if and only if $\sum_{i=1}^{+\infty} r_i < +\infty$,
(ii) *transient* if and only if $\sum_{i=1}^{+\infty} r_i = +\infty$ and $\sum_{i=1}^{+\infty} \frac{1}{r_i \cdot p_i} < +\infty$,
(iii) *null recurrent* if and only if $\sum_{i=1}^{+\infty} r_i = \sum_{i=1}^{+\infty} \frac{1}{r_i \cdot p_i} = +\infty$ with $r_i = \frac{p_1 \cdots p_{i-1}}{q_1 \cdots q_i}, i \in \mathbb{N}^*$.

In parallel, in recent years, a systematic research has been developed (Kalpazidou [12], MacQueen [14], Qian Minping and Qian Min [15], Zemanian [22], and others) in order to investigate representations of the finite-dimensional distributions of Markov processes (discrete or continuous parameter) having an invariant measure, as decompositions in terms of the *circuit passage functions*

$$J_c(i,j) = \begin{cases} 1, & \text{if } i,j \text{ are } \textit{consecutive} \text{ states of } c, \\ 0, & \text{otherwise,} \end{cases}$$

for any directed sequence $c = (i_1, i_2, \ldots, i_v, i_1)$ of states called a *circuit*, $\nu > 1$ of the corresponding Markov process. This research has stimulated a motivation toward the representation of Markov processes through directed circuits and weights in terms of circuit passage functions in fixed or random environments, as well as the study of specific problems associated with Markov processes in a different way. The representations are called *circuit representations*, while the corresponding discrete or continuous parameter Markov chains generated by directed weighted circuits are called *circuit chains*.

Additionally, it is also known that the *entropy production* of Markov chains with transition probability matrix $P = (p_{ij})_{i,j}$ was defined at first as an entity arising in non-equilibrium statistical physics by the expression

$$E = \frac{1}{2}\sum_{i,j}(p_j p_{ji} - p_i p_{ij}) \log \frac{p_j p_{ji}}{p_i p_{ij}}, \quad p_{ij} > 0,$$

interpreted up to a constant factor (the Boltzmann constant multiplied with the temperature at which the reaction occurs) meaning the amount of entropy produced in any irreversible process and any irreversible thermodynamic cycle, for every non-equilibrium system of coupled chemical reactions in the case that some reactants are continuously introduced into the system and others are continuously withdrawn, where

$$K_{i,j} = p_j p_{ji} - p_i p_{ij}, \quad M_{ij} = \log \frac{p_j p_{ji}}{p_i p_{ij}}, \quad p_{ij} > 0,$$

indicate the reaction rates of the chemical components $i, j \in \{1, 2, \ldots, n\}$ involved in the reaction and the conjugated thermodynamic force of K_{ij}, respectively. Analogously for any irreducible, positive recurrent Markov chain with denumerable state space S and transition probability matrix $P = (p_{ij})_{i,j\in S}$, the *entropy production* of the chain is given by

$$E = \frac{1}{2}\sum_{i,j\in S}(\pi_i p_{ij} - \pi_j p_{ji}) \log \frac{\pi_i p_{ij}}{\pi_j p_{ji}}, \quad p_{ij} > 0, \quad i, j \in S,$$

where $\pi = (\pi_i / i \in S)$ is the invariant probability distribution (Kalpazidou [12]). Moreover, in the context of the circuit representation theory of Markov processes, the entropy production of any irreducible, positive recurrent Markov chain with denumerable state space S defined by directed circuits and weights with respect to some class $\{C, w_c\}$ through the circulation distribution $\{w_c/c \in C\}$ is defined by

$$E = \frac{1}{2} \sum_{c \in C} (w_c - w_{c_-}) \log \frac{w_c}{w_{c_-}} \tag{1}$$

expressing the total deviation from symmetry along the circuits occurring on the sample paths of the collection C of directed circuits, while c_- is the reversed circuit of the $c \in C$([12, 16]).

By using the circuit representation theory of Markov processes, this work arises as an attempt to study important stochastic properties of the corresponding Markov chains (circuit chains) describing uniquely the birth–death chains by directed weighted circuits in fixed environments for a generalized sample path case, such as the properties of transience and positive/null recurrence, as well as the reversibility and the entropy production giving a new perspective in the study of specific problems associated with birth–death chains in another way through directed weighted circuit representations ([3, 5]) (for the study of irreducible, denumerable circuit chains in general for a generalized sample path case, see [8]).

This chapter is organized as follows. In Section 2, we give useful information about the circuit representation theory of Markov processes in order to make the presentation of the work more comprehensible. In Section 3, the birth–death model in fixed environments for a generalized sample path case is considered and the unique representations by directed circuits and weights of the corresponding "*adjoint*" Markov chains (circuit chains) are investigated. These representations will give us the possibility to study suitable criteria regarding transience and recurrence as well as the reversibility and the entropy production of the abovementioned Markov chains, as it is described in Section 4 and Section 5, respectively. Throughout this chapter, we need the following notations: $\mathbb{N} = \{0, 1, 2, \ldots\}, \mathbb{N}^* = \{1, 2, 3, \ldots\}, \mathbb{Z} = \{\ldots, -1, 0, 1, \ldots\}$.

2. Preliminaries

Let S be a denumerable set. Let also C be a class of overlapping directed circuits in S such that for any $i \in S$ there exists a loop circuit (i,i). Recall that a directed circuit c in S is any periodic function $c : \mathbb{Z} \to S$, where $\mathbb{Z}$ is the set of integers. The smallest positive integer $p = p(c) \geq 1$ satisfying the equation $c(m+p) = c(m)$, for all $m \in \mathbb{Z}$, is called the *period of the circuit c*. A directed circuit c such that $p(c) = 1$ is called a *loop*. Then, the circuit c is completely determined by its *points* $c(m), m \in \mathbb{Z}$, that is, $c = (c(m), c(m+1), \ldots, c(m+p-1), c(m))$, while the directed pairs $(c(m+r-1), c(m+r)), r = 1, 2, \ldots, p-1$, are the *directed edges* of c. The corresponding *directed cycle* is defined by $\hat{c} = (c(m), c(m+1), \ldots, c(m+p-1))$. Furthermore, the equivalence class $c = \{c' : c'$ is a circuit in S such that $c(m+i) = c'(m)$, for some$i \in \mathbb{Z}$ and any $m \in Z\}$ can be characterized by a period $p(c)$ which is the common period of all circuits belonging to C. This allows us to call a circuit the corresponding class circuit. Next, we give the following definition of MacQueen [14] and Kalpazidou [12].

Definition 1. Let c be a directed circuit with period $p(c) \geq 1$. Then, we define by $J_c(i_1, i_2, \ldots, i_r)$ the number of distinct integers k, $0 \leq k \leq p(c) - 1$, such that $cot_k(m) = i_m, m = 1, 2, \ldots, r$, where $t_k(n) \equiv n + k$, $n \in \mathbb{Z}$. In this case, we say that *c passes through* $(i_1, i_2, \ldots, i_r)$ if and only if $J_c(i_1, i_2, \ldots, i_r) \neq 0$, where $J_c(i_1, i_2, \ldots, i_r)$ denotes the number of times c passes through $(i_1, i_2, \ldots, i_r)$. The functions $J_c : S^r \to \{0, 1\}$, $r \in \mathbb{N}^*$, are called the *rth-order passage functions associated with c*.

In particular, for $r = 2$, the collection $(J_c(i,j), i, j \in S)$ is called the *second-order passage matrix* associated with the directed circuit c, where

$$J_c(i,j) = \begin{cases} 1, & \text{if } (i,j) \text{ is an edge of } c \\ 0, & \text{otherwise} \end{cases}$$

such that

$$\sum_j J_c(i,j) = \sum_j J_c(j,i) \equiv J_c(i), \quad i \in S.$$

Here the value $J_c(i,j) = 1$, for some directed circuit c, depends upon the existence of (at least) two passage functions $J_{c_1}(\cdot,\cdot), J_{c_2}(\cdot,\cdot)$

associated with two directed circuits c_1 and c_2 such that $J_{c_1}(i,k) \times J_{c_2}(k,j) = 1$, for some $k \in S$. Hence, there is a communication between states for the sample paths of discrete skeletons associated with a corresponding Markov process generated by directed circuits and weight functions. This means that for a directed circuit $c = (i_1, i_2, \ldots, i_m, i_1)$, modulo the cyclic permutations, where $i_1, i_2, \ldots, i_m \in S$, $m > 1$, we may have the following definitions of MacQueen [14] and Kalpazidou [12].

Definition 2. Let c be a directed circuit with period $p(c) > 1$. Then we define by

$$J_c^{(n)}(i,j) = \begin{cases} 1, & \text{if there exists an } m \in \mathbb{Z} \\ & \text{such that } i = c(m), j = c(m+n) \\ 0, & \text{otherwise} \end{cases}$$

the *n-step passage function* associated with the directed circuit c, for any $i, j \in S, n \in \mathbb{N}^*$.

For $n = 1$, we have

$$J_c^{(1)}(i,j) = \begin{cases} 1, & \text{if there exists an } m \in \mathbb{Z} \\ & \text{such that } i = c(m), j = c(m+1) \\ 0, & \text{otherwise.} \end{cases}$$

Definition 3. Let c be a directed circuit with period $p(c) > 1$. Then, we define by

$$J_c(i) = \begin{cases} 1, & \text{if there exists an } m \in \mathbb{Z} \text{ such that } i = c(m) \\ 0, & \text{otherwise} \end{cases}$$

the *passage function* associated with the directed circuit c, for any $i \in S$.

Given a denumerable set S and an infinite denumerable class C of overlapping directed circuits with distinct points (except for the terminals) in S such that all the points of S can be reached from one another following paths of circuit-edges, that is, for each two distinct points i and j of S, there exists a finite sequence $c_1, c_2, \ldots, c_k, k \geq 1$, of circuits of C such that i lies on c_1 and j lies on c_k and any pair of consecutive circuits (c_n, c_{n+1}) has at least one point in common. We may assume also that the class C contains, among its elements, circuits with period greater or equal to 2. With each directed circuit, let us associate a *strictly positive weight* w_c which must be independent of the choice of the representative of c, that is, it must satisfy the consistency condition $w_{c \circ t_k} = w_c, k \in \mathbb{Z}$, where t_k is the translation of length k.

For a given class C of overlapping directed circuits and for a given sequence $(w_c)_{c \in C}$ of weights, we may define by

$$p_{ij} = \frac{\sum_{c \in C} w_c \cdot J_c^{(1)}(i,j)}{\sum_{c \in C} w_c \cdot J_c(i)} \tag{2}$$

the elements of a Markov transition matrix on S, if and only if $\sum_{c \in C} w_c \cdot J_c(i) < +\infty$, for any $i \in S$. This means that a given Markov transition matrix $P = (p_{ij}), i, j \in S$, can be represented by directed circuits and weights if and only if there exists a class of overlapping directed circuits C and a sequence of positive weights $(w_c)_{c \in C}$ such that formula (2) holds. In this case, the representation of the distribution of Markov process (with discrete or continuous parameter) having an invariant measure as decomposition in terms of the circuit passage functions is called *circuit representation* while the corresponding discrete parameter Markov chain generated by directed circuits is called *circuit chain* with Markov transition matrix P given by (2) and unique stationary distribution p (a solution of $p \cdot P = p$) defined by

$$p(i) = \sum_{c \in C} w_c \cdot J_c(i), i \in S.$$

It is known that the following classes of Markov chains may be represented uniquely by directed circuits and weights: (i) the recurrent Markov chains (15) and (ii) the reversible Markov chains.

We also assume that the following hypotheses are satisfied:

(a) the circuits determine an infinite oriented graph of bounded degree, that is, there is some integer $n_0 \geq 1$ such that the number of circuits that passes through any point of S is at most n_0,
(b) $\max_{c \in C} p(c) = R < +\infty$, and
(c) for every two points k and u of S there exists a finite sequence of circuits $c_1, \ldots, c_m$ and a finite path $k_0 = k, k_1, \ldots, k_m = u$ of points on $c_1, \ldots, c_m$ that connect k to u, that is, (k_n, k_{n+1}) is passed by $c_{n+1}, n = 0, \ldots, m-1$, in the sense of Definition 1 (Connectedness) (Kalpazidou [12], Section 2.2).

Let also $\{w_c(\cdot)/c \in C\}$ be a collection of non-negative functions defined on $T = [0, +\infty)$. Then, for any i, $j \in S$ and for any $t \in T$, we define

$$w(i,j,t) \equiv \sum_{c\in C} w_c(t) J_c(i,j)$$

such that

$$\sum_j w(i,j,t) = \sum_j w(j,i,t) \equiv w(i,t),\ t \in T,$$

satisfying the conditions $(w_1) - (w_4)$ (p. 66, Kalpazidou [12]). The non-negative functions $w_c(\cdot), c \in C$, which satisfy the conditions $(w_2) - (w_4)$ are called *weight functions* associated with the non-void class C of directed circuits in S, while the weight functions satisfying the conditions (w_1) and (w_2) are called *standard weight functions.* As it was proved regarding the existence of weight functions, there exists a non-void class C of directed circuits in S such that a collection of standard weight functions $\{w_c(\cdot)/c \in C\}$ verifying the conditions $(w_1) - (w_4)$. The matrix function $\{w(i,j,\cdot)/i,j \in S\}$ defines a denumerable class C of overlapping directed circuits in S and a collection $\{w_c(\cdot)/c \in C\}$ of non-negative functions defined on $[0,\infty)$. In the case that conditions $(w_2) - (w_4)$ are satisfied, any continuous parameter S-state Markov process whose transition matrix function $P(t) = (p_{ij}(t), i, j \in S)$ is defined by $p_{ij}(t) = w(i,j,t)/w(i,t), i,j \in S, t \in T$, is called a *circuit Markov process* or a *circuit process* associated with the collection $(C, w_c(\cdot))$. By associating with each $h > 0$ the discrete skeleton $\Xi_h = \{X_{nh}\}_{n\geq 0}$ with scale parameter $h > 0$, we may conclude that any Ξ_h is an aperiodic, irreducible Markov chain with transition probability matrix $P(h) = (p_{ij}(h), i, j \in S)$ in which we may apply a unique circuit and weight representation.

3. Auxiliary Results: Circuit and Weight Representations of Birth–Death Chains in Fixed Environments for a Generalized Sample Path Case

Let $0 \equiv t_s < t_{s+1} < \cdots < t_{s+\ell-1} < t_{s+\ell} < \cdots < +\infty$, $t_{s+\ell} \geq 0$, such that $(t_{s+\ell+1} - t_{s+\ell}) = (t_{s+\ell+2} - t_{s+\ell+1})$, for every $\ell = 0, 1, 2, \ldots, m-1, m, \ldots, s = r, r+1, \ldots, m+r-1, m+r, \ldots, r \in \mathbb{N}$, $m \in \mathbb{N}^*$. Let us also consider the Markov chain $\{X_t\}_{t \geq 0}$ on $\mathbb{N}$ (X_t expresses the current size of a *"population"* (|total births-total deaths|) at time $t \equiv t_{s+\ell}$ assuming that the system starts at state 0 at $t_s \equiv 0$, which describes a birth–death chain in fixed environments for a generalized sample path case given consideration of the directed circuits $c_\ell = (\ell, \ell+1, \ell)$ such that $\ell = c_\ell(t_{s+\ell}) = c_{\ell-1}(t_{s+\ell})$, where the state transitions are of only two types: $\ell \to (\ell+1)$ and $\ell \to (\ell-1)$, for every $\ell = 0, 1, 2, \ldots, m-1, m, \ldots, s = r, r+1, \ldots, m+r-1, m+r, \ldots, r \in \mathbb{N}$, $m \in \mathbb{N}^*$. Consequently, the elements of the corresponding Markov transition matrix (transition probabilities) are defined by

$$P(X_{t_{s+\ell+1}} = \ell+1 / \, X_{t_{s+\ell}} = \ell) = p_\ell(t_{s+\ell+1} - t_{s+\ell}),$$
$$P(X_{t_{s+\ell+1}} = \ell-1 / \, X_{t_{s+\ell}} = \ell) = q_\ell(t_{s+\ell+1} - t_{s+\ell}),$$

such that $p_\ell(t_{s+\ell+1} - t_{s+\ell}) + q_\ell(t_{s+\ell+1} - t_{s+\ell}) = 1$, $0 < p_\ell(t_{s+\ell+1} - t_{s+\ell}) \leq 1$, for every $\ell = 1, 2, \ldots, m-1, m, \ldots, s = r, r+1, \ldots, m+r-1, m+r, \ldots,$ $r \in \mathbb{N}, m \in \mathbb{N}^*$ with $p_0(t_{s+1} - t_s) = 1$.

Let us assume that $(p_\ell(t_{s+\ell+1} - t_{s+\ell}))_\ell$ is an arbitrary fixed sequence with $0 < p_\ell(t_{s+\ell+1} - t_{s+\ell}) \leq 1$, $p_0(t_{s+1} - t_s) = 1$, for every $\ell = 1, 2, \ldots, m-1, m, \ldots, s = r, r+1, \ldots, m+r-1, m+r, \ldots, r \in \mathbb{N}$, $m \in \mathbb{N}^*$. By considering the directed circuits $c_\ell = (\ell, \ell+1, \ell)$ such that $\ell = c_\ell(t_{s+\ell}) = c_{\ell-1}(t_{s+\ell})$ and the collection of weights $(w_{c_\ell}(t_{s+\ell+1} - t_{s+\ell}))_\ell$, for every $\ell = 0, 1, 2, \ldots, m-1, m, \ldots, s = r, r+1, \ldots, m+r-1, m+r, \ldots,$ $r \in \mathbb{N}$, $m \in \mathbb{N}^*$, we may obtain the corresponding transition probabilities

$$p_\ell(t_{s+\ell+1} - t_{s+\ell}) = \frac{w_{c_\ell}(t_{s+\ell+1} - t_{s+\ell})}{w_{c_{\ell-1}}(t_{s+\ell+1} - t_{s+\ell}) + w_{c_\ell}(t_{s+\ell+1} - t_{s+\ell})}$$

$$q_\ell(t_{s+\ell+1} - t_{s+\ell}) = \frac{w_{c_{\ell-1}}(t_{s+\ell+1} - t_{s+\ell})}{w_{c_{\ell-1}}(t_{s+\ell+1} - t_{s+\ell}) + w_{c_\ell}(t_{s+\ell+1} - t_{s+\ell})}$$
$$= 1 - p_\ell(t_{s+\ell+1} - t_{s+\ell}),$$

with

$$p_0(t_{s+1} - t_s) = 1,$$

for every $\ell = 1, 2, \ldots, m-1, m, \ldots, s = r, r+1, \ldots, m+r-1, m+r, \ldots,$ $r \in \mathbb{N}, m \in \mathbb{N}^*$. Since the class $C(\ell)$ contains the directed circuits $c_\ell = (\ell, \ell+1, \ell)$, $c_{\ell-1} = (\ell-1, \ell, \ell-1)$, we may take that the transition probability matrix with elements

$$p_{\ell j}(t_{s+\ell+1} - t_{s+\ell}) = \frac{\sum_{i=0}^{+\infty} w_{c_i}(t_{s+\ell+1} - t_{s+\ell}) \cdot J_{c_i}(\ell, j)}{\sum_{i=0}^{+\infty} w_{c_i}(t_{s+\ell+1} - t_{s+\ell}) \cdot J_{c_i}(\ell)} \tag{3}$$

(with $j = \ell - 1$ or $j = \ell + 1$)

$$p_{\ell\ell}(t_{s+\ell+1} - t_{s+\ell}) = 0, \tag{4}$$

where

$$J_{c_i}(\ell, j) = 1, \text{ if } (\ell, j) \text{ is an edge of the circuit } c_i,$$
$$J_{c_i}(\ell) = 1, \text{ if } \ell \text{ is a point of the circuit } c_i,$$

for every $\ell = 0, 1, 2, \ldots, m-1, m, \ldots, s = r, r+1, \ldots, m+r-1, m+r, \ldots, r \in \mathbb{N}, m \in \mathbb{N}^*, i \in \mathbb{N}$, expresses the representation of the abovementioned Markov chain $\{X_t\}_{t\geq 0}$ by directed circuits and weights.

Furthermore, let us consider the *"adjoint"* Markov chain $\{X'_t\}_{t\geq 0}$ on $\mathbb{N}$ whose the elements of the corresponding transition matrix are given by

$$P(X'_{t_{s+\ell+1}} = \ell + 1/X'_{t_{s+\ell}} = \ell) = q'_\ell(t_{s+\ell+1} - t_{s+\ell}),$$
$$P(X'_{t_{s+\ell+1}} = \ell - 1/X'_{t_{s+\ell}} = \ell) = p'_\ell(t_{s+\ell+1} - t_{s+\ell})$$

such that $p'_\ell(t_{s+\ell+1} - t_{s+\ell}) + q'_\ell(t_{s+\ell+1} - t_{s+\ell}) = 1$, $0 < q'_\ell(t_{s+\ell+1} - t_{s+\ell}) \leq 1$, with $q'_0(t_{s+1} - t_s) = 1$, for every $\ell = 1, 2, \ldots, m-1, m, \ldots, s = r, r+1, m+r-1, m+r, \ldots, r \in \mathbb{N}, m \in \mathbb{N}^*$.

Assuming that $(q'_\ell(t_{s+\ell+1} - t_{s+\ell}))_\ell$ is an arbitrary fixed sequence with $0 < q'_\ell(t_{s+\ell+1} - t_{s+\ell}) \leq 1$, $q'_0(t_{s+1} - t_s) = 0$ and considering the directed circuits $c'_\ell = (\ell+1, \ell, \ell+1)$ such that $\ell = c'_\ell(t_{s+\ell}) = c'_{\ell-1}(t_{s+\ell})$ and the collection of weights $(w_{c'_\ell}(t_{s+\ell+1} - t_{s+\ell}))_\ell$, for every $\ell = 0, 1, 2, \ldots, m-1, m, \ldots, s = r, r+1, \ldots, m+r-1, m+r, \ldots, r \in \mathbb{N}$, $m \in \mathbb{N}^*$, we may obtain that

$$q'_\ell(t_{s+\ell+1} - t_{s+\ell}) = \frac{w_{c'_\ell}(t_{s+\ell+1} - t_{s+\ell})}{w_{c'_{\ell-1}}(t_{s+\ell+1} - t_{s+\ell}) + w_{c'_\ell}(t_{s+\ell+1} - t_{s+\ell})}$$

$$p'_\ell(t_{s+\ell+1} - t_{s+\ell}) = \frac{w_{c'_{\ell-1}}(t_{s+\ell+1} - t_{s+\ell})}{w_{c'_{\ell-1}}(t_{s+\ell+1} - t_{s+\ell}) + w_{c'_\ell}(t_{s+\ell+1} - t_{s+\ell})}$$
$$= 1 - q'_\ell(t_{s+\ell+1} - t_{s+\ell}),$$

with

$$q'_0(t_{s+1} - t_s) = 0,$$

for every $\ell = 1, 2, \ldots, m-1, m, \ldots, s = r, r+1, \ldots, m+r-1, m+r, \ldots,\ r \in \mathbb{N}, m \in \mathbb{N}^*$. Here the class $C'(\ell)$ contains the directed circuits $c'_\ell = (\ell+1, \ell, \ell+1)$, $c'_{\ell-1} = (\ell, \ell-1, \ell)$. Hence, the corresponding Markov transition matrix with elements equivalent to that given by the abovementioned formulas (3) and (4) expresses the representation of the *"adjoint"* Markov chain $\{X'_t\}_{t\geq 0}$ by directed circuits and weights. So we have the following (11).

Proposition 4. *The "adjoint" Markov chains* $\{X_t\}_{t\geq 0}$, $\{X'_t\}_{t\geq 0}$ *defined as above have unique representations by directed circuits and weights.*

Proof.

(i) For *the Markov chain* $\{X_t\}_{t\geq 0}$. Since only the transitions from ℓ to $(\ell+1)$ and ℓ to $(\ell-1)$ are possible, we may consider the set of directed circuits $c_\ell = (\ell, \ell+1, \ell)$. It is obvious that there are two circuits through each point $\ell \in \mathbb{N}^* : c_\ell = (\ell, \ell+1, \ell)$, $c_{\ell-1} = (\ell-1, \ell, \ell-1)$ and only one through $0 : c_0 = (0, 1, 0)$. In order to solve the problem of uniqueness, we have to find the definition of the weights. To this direction by symbolizing by $w_\ell(t_{s+\ell+1} - t_{s+\ell})$ the weight of the circuit c_ℓ, for every $\ell = 0, 1, 2, \ldots, m-1, m, \ldots,$ $s = r, r+1, \ldots, m+r-1, m+r, \ldots, r \in \mathbb{N}, m \in \mathbb{N}^*$, we may obtain that the sequence $(w_\ell(t_{s+\ell+1} - t_{s+\ell}))_\ell$ must be a solution of

$$p_\ell(t_{s+\ell+1} - t_{s+\ell}) = \frac{w_\ell(t_{s+\ell+1} - t_{s+\ell})}{w_\ell(t_{s+\ell+1} - t_{s+\ell}) + w_{\ell-1}(t_{s+\ell+1} - t_{s+\ell})},$$
$$q_\ell(t_{s+\ell+1} - t_{s+\ell}) = 1 - p_\ell(t_{s+\ell+1} - t_{s+\ell}),$$

with

$$p_0(t_{s+1} - t_s) = 1$$

for every $\ell = 1, 2, \ldots, m-1, m, \ldots, s = r, r+1, \ldots, m+r-1, m+r, \ldots,$ $r \in \mathbb{N}$, $m \in \mathbb{N}^*$. Let us consider the sequence $(d_\ell(t_{s+\ell+1} - t_{s+\ell}))_\ell$

$$d_\ell(t_{s+\ell+1} - t_{s+\ell}) = \frac{w_\ell(t_{s+\ell+1} - t_{s+\ell})}{w_{\ell-1}(t_{s+\ell+1} - t_{s+\ell})},$$

for every $\ell = 1, 2, \ldots, m-1, m, \ldots, s = r, r+1, \ldots, m+r-1,$ $m+r, \ldots,$ $r \in \mathbb{N}$, $m \in \mathbb{N}^*$. Consequently, we may take that

$$d_\ell(t_{s+\ell+1} - t_{s+\ell}) = \frac{p_\ell(t_{s+\ell+1} - t_{s+\ell})}{q_\ell(t_{s+\ell+1} - t_{s+\ell})} = \frac{p_\ell(t_{s+\ell+1} - t_{s+\ell})}{1 - p_\ell(t_{s+\ell+1} - t_{s+\ell})},$$

for every $\ell = 1, 2, \ldots, m-1, m, \ldots, s = r, r+1, \ldots, m+r-1,$ $m+r, \ldots,$ $r \in \mathbb{N}$, $m \in \mathbb{N}^*$. It is obvious that for a given sequence $(p_\ell(t_{s+\ell+1} - t_{s+\ell}))_\ell$, the sequence $(d_\ell(t_{s+\ell+1} - t_{s+\ell}))_\ell$ exists and it is unique. This means that the sequence $(w_\ell(t_{s+\ell+1} - t_{s+\ell}))_\ell$ is defined uniquely by

$$\begin{aligned} w_\ell(t_{s+\ell+1} - t_{s+\ell}) &= w_0(t_{s+1} - t_s) \cdot d_1(t_{s+2} - t_{s+1}) \cdots \\ &\quad \times d_\ell(t_{s+\ell+1} - t_{s+\ell}), \end{aligned}$$

for every $\ell = 1, 2, \ldots, m-1, m, \ldots, s = r, r+1, \ldots, m+r-1,$ $m+r, \ldots,$ $r \in \mathbb{N}$, $m \in \mathbb{N}^*$ (the unicity is understood up to the factor $w_0(t_{s+1} - t_s)$).

(ii) *For the Markov chain* $\{X'_t\}_{t\geq 0}$. Following an analogous way, we may symbolize by $w'_\ell(t_{s+\ell+1} - t_{s+\ell})$ the weight of the circuit c'_ℓ, for every $\ell = 1, 2, \ldots, m-1, m, \ldots, s = r, r+1, \ldots, m+r-1,$ $m+r, \ldots,$ $r \in \mathbb{N}$, $m \in \mathbb{N}^*$. There are two circuits through each point $\ell \in \mathbb{N}^* : c'_\ell, c'_{\ell-1}$ and only one though $0 : c'_0$. So the sequence $(w'_\ell(t_{s+\ell+1} - t_{s+\ell}))_\ell$ must be a solution of

$$q'_\ell(t_{s+\ell+1} - t_{s+\ell}) = \frac{w'_\ell(t_{s+\ell+1} - t_{s+\ell})}{w'_{\ell-1}(t_{s+\ell+1} - t_{s+\ell}) + w'_\ell(t_{s+\ell+1} - t_{s+\ell})},$$

$$p'_\ell(t_{s+\ell+1} - t_{s+\ell}) = 1 - q'_\ell(t_{s+\ell+1} - t_{s+\ell}),$$

with

$$q'_0(t_{s+1} - t_s) = 1,$$

for every $\ell = 1, 2, \ldots, m-1, m, \ldots, s = r, r+1, \ldots, m+r-1, m+r, \ldots,\ r \in \mathbb{N},\ m \in \mathbb{N}^*$. By considering the sequence $(g_\ell(t_{s+\ell+1} - t_{s+\ell}))_\ell$, where

$$g_\ell(t_{s+\ell+1} - t_{s+\ell}) = \frac{w'_{\ell-1}(t_{s+\ell+1} - t_{s+\ell})}{w'_\ell(t_{s+\ell+1} - t_{s+\ell})},$$

we may take that

$$g_\ell(t_{s+\ell+1} - t_{s+\ell}) = \frac{p'_\ell(t_{s+\ell+1} - t_{s+\ell})}{q'_\ell(t_{s+\ell+1} - t_{s+\ell})} = \frac{1 - q'_\ell(t_{s+\ell+1} - t_{s+\ell})}{q'_\ell(t_{s+\ell+1} - t_{s+\ell})},$$

for every $\ell = 1, 2, \ldots, m-1, m, \ldots, s = r, r+1, \ldots, m+r-1, m+r, \ldots, r \in \mathbb{N},\ m \in \mathbb{N}^*$. It is clear that the sequence $(g_\ell(t_{s+\ell+1} - t_{s+\ell}))_\ell$ exists and it is unique for a given sequence $(q'_\ell(t_{s+\ell+1} - t_{s+\ell}))_\ell$. This means that the sequence of the weights $(w'_\ell(t_{s+\ell+1} - t_{s+\ell}))_\ell$ is defined uniquely by

$$w'_\ell(t_{s+\ell+1} - t_{s+\ell}) = \frac{w'_0(t_{s+1} - t_s)}{g_1(t_{s+2} - t_{s+1}) \cdot g_2(t_{s+3} - t_{s+2}) \cdots g_\ell(t_{s+\ell+1} - t_{s+\ell})},$$

for every $\ell = 1, 2, \ldots, m-1, m, \ldots, s = r, r+1, \ldots, m+r-1, m+r, \ldots, r \in \mathbb{N},\ m \in \mathbb{N}^*$ (the unicity is understood up to the factor $w'_0(t_{s+1} - t_s)$). □

4. Recurrence and Transience of the Markov Chains $\{X_t\}_{t\geq 0}$, $\{X'_t\}_{t\geq 0}$

Let us consider that for the Markov chains $\{X_t\}_{t\geq 0}$, $\{X'_t\}_{t\geq 0}$ there are unique invariant measures up to multiplicative factors $\mu_\ell(t_{s+\ell+1} - t_{s+\ell}) = w_{\ell-1}(t_{s+\ell+1} - t_{s+\ell}) + w_\ell(t_{s+\ell+1} - t_{s+\ell})$ with $\mu_0(t_{s+1} - t_s) = w_0(t_{s+1} - t_s)$ and $\mu'_\ell(t_{s+\ell+1} - t_{s+\ell}) = w'_{\ell-1}(t_{s+\ell+1} - t_{s+\ell}) + w'_\ell(t_{s+\ell+1} - t_{s+\ell})$ with $\mu'_0(t_{s+1} - t_s) = w'_0(t_{s+1} - t_s)$, respectively, for every $\ell = 1, 2, \ldots, m-1, m, \ldots, s = r, r+1, \ldots, m+r-1, m+r, \ldots, r \in \mathbb{N},\ m \in \mathbb{N}^*$. In the case of a recurrent irreducible chain, there is one and only one invariant measure (finite or not). In consequence, we may obtain the following.

Proposition 5.

(i) *The Markov chain* $\{X_t\}_{t\geq 0}$ *is positive recurrent if and only if*

$$\frac{\sum_{\ell=1}^{+\infty}[d_1(t_{s+2}-t_{s+1})\cdot d_2(t_{s+3}-t_{s+2})\cdots d_\ell(t_{s+\ell+1}-t_{s+\ell})]}{p_\ell(t_{s+\ell+1}-t_{s+\ell})}$$

$$< +\infty\left(or\ \frac{1}{w_0(t_{s+1}-t_s)}\cdot\frac{\sum_{\ell=1}^{+\infty} w_\ell(t_{s+\ell+1}-t_{s+\ell})}{p_\ell(t_{s+\ell+1}-t_{s+\ell})} < +\infty\right)$$

for every $\ell = 1, 2, \ldots, m-1, m, \ldots, s = r, r+1, \ldots, m+r-1, m+r, \ldots, r \in \mathbb{N}$, $m \in \mathbb{N}^*$.

(ii) *The Markov chain* $\{X'_t\}t\geq 0$ *is positive recurrent if and only if*

$$\sum_{\ell=1}^{+\infty}\frac{1}{g_1(t_{s+2}-t_{s+1})\cdot g_2(t_{s+3}-t_{s+2})\cdots g_\ell(t_{s+\ell+1}-t_{s+\ell})}$$

$$< +\infty\left(or\ \frac{1}{w'_0(t_{s+1}-t_s)}\cdot\sum_{\ell=1}^{+\infty} w'_\ell(t_{s+\ell+1}-t_{s+\ell}) < +\infty\right)$$

and

$$\frac{\sum_{\ell=1}^{+\infty}[g_1(t_{s+2}-t_{s+1})\cdot g_2(t_{s+3}-t_{s+2})\cdots g_\ell(t_{s+\ell+1}-t_{s+\ell})]}{p'_\ell(t_{s+\ell+1}-t_{s+\ell})}$$

$$= +\infty\left(or\ w'_0(t_{s+1}-t_s)\cdot\sum_{\ell=1}^{+\infty}\frac{1}{p'_\ell(t_{s+\ell+1}-t_{s+\ell})\cdot w'_\ell(t_{s+\ell+1}-t_{s+\ell})} = +\infty\right)$$

for every $\ell = 1, 2, \ldots, m-1, m, \ldots, s = r, r+1, \ldots, m+r-1, m+r, \ldots, r \in \mathbb{N}$, $m \in \mathbb{N}^*$.

We shall use the following ([13]).

Lemma 6. *Let there be an irreducible Markov chain on* $\mathbb{N}$. *Then if there exists a strictly increasing function that is harmonic on the complement of a finite interval and that is bounded, the chain is transient. In the case that such a function is unbounded, the chain is recurrent.*

By considering the harmonic function ($h_\ell(t_{s+\ell+1}-t_{s+\ell}), \ell = 1, 2, \ldots, m-1, m, \ldots, s = r, r+1, \ldots, m+r-1, m+r, \ldots, r \in \mathbb{N}$, $m \in \mathbb{N}^*$, on $\mathbb{N}^*$) and by using a method theorem regarding Foster's

Markov chain theorem ([13]), the study of the criteria of recurrence and transience for the Markov chain $\{X_t\}_{t\geq 0}$ is based on the solution of

$$p_\ell(t_{s+\ell+1} - t_{s+\ell}) \cdot h_{\ell+1}(t_{s+\ell+2} - t_{s+\ell+1}) + q_\ell(t_{s+\ell+1} - t_{s+\ell}) \cdot h_{\ell-1}(t_{s+\ell} - t_{s+\ell-1}) = h_\ell(t_{s+\ell+1} - t_{s+\ell})$$

or

$$p_\ell(t_{s+\ell+1} - t_{s+\ell}) \cdot h_{\ell+1}(t_{s+\ell+2} - t_{s+\ell+1}) + q_\ell(t_{s+\ell+1} - t_{s+\ell}) \cdot h_{\ell-1}(t_{s+\ell} - t_{s+\ell-1}) + q_\ell(t_{s+\ell+1} - t_{s+\ell}) \cdot h_\ell(t_{s+\ell+1} - t_{s+\ell}) - q_\ell(t_{s+\ell+1} - t_{s+\ell}) \cdot h_\ell(t_{s+\ell+1} - t_{s+\ell}) = h_\ell(t_{s+\ell+1} - t_{s+\ell})$$

or

$$\frac{p_\ell(t_{s+\ell+1} - t_{s+\ell})}{q_\ell(t_{s+\ell+1} - t_{s+\ell})} = \frac{h_\ell(t_{s+\ell+1} - t_{s+\ell}) - h_{\ell-1}(t_{s+\ell} - t_{s+\ell-1})}{h_{\ell+1}(t_{s+\ell+2} - t_{s+\ell+1}) - h_\ell(t_{s+\ell+1} - t_{s+\ell})},$$

for every $\ell = 1, 2, \ldots, m-1, m, \ldots, s = r, r+1, \ldots, m+r-1, m+r, \ldots, r \in \mathbb{N}$, $m \in \mathbb{N}^*$. By putting

$$b_\ell(t_{s+\ell+1} - t_{s+\ell}) = \frac{h_\ell(t_{s+\ell+1} - t_{s+\ell}) - h_{\ell-1}(t_{s+\ell} - t_{s+\ell-1})}{h_{\ell+1}(t_{s+\ell+2} - t_{s+\ell+1}) - h_\ell(t_{s+\ell+1} - t_{s+\ell})},$$

we take that

$$b_\ell(t_{s+\ell+1} - t_{s+\ell}) = \frac{p_\ell(t_{s+\ell+1} - t_{s+\ell})}{q_\ell(t_{s+\ell+1} - t_{s+\ell})} = \frac{1 - q_\ell(t_{s+\ell+1} - t_{s+\ell})}{q_\ell(t_{s+\ell+1} - t_{s+\ell})},$$

that is, the equation of the definition of the sequence $(g_\ell(t_{s+\ell+1} - t_{s+1}))_\ell$ for the Markov chain $\{X'_t\}_{t\geq 0}$ such that $q_\ell(t_{s+\ell+1} - t_{s+\ell}) = q'_\ell(t_{s+\ell+1} - t_{s+\ell})$, for every $\ell = 1, 2, \ldots, m-1, m, \ldots, s = r, r+1, \ldots, m+r-1, m+r, \ldots, r \in \mathbb{N}$, $m \in \mathbb{N}^*$. This means that the strictly increasing harmonic functions of the Markov chains $\{X_t\}_{t\geq 0}$ are in correspondence with the weight representations of the Markov chain $\{X'_t\}_{t\geq 0}$ such that

$$\begin{aligned} q'_\ell(t_{s+\ell+1} - t_{s+\ell}) &= q_\ell(t_{s+\ell+1} - t_{s+\ell}), \\ p'_\ell(t_{s+\ell+1} - t_{s+\ell}) &= 1 - q'_\ell(t_{s+\ell+1} - t_{s+\ell}), \\ 1 - q_\ell(t_{s+\ell+1} - t_{s+\ell}) &= p_\ell(t_{s+\ell+1} - t_{s+\ell}), \end{aligned} \tag{5}$$

for every $\ell = 1, 2, \ldots, m-1, m, \ldots, s = r, r+1, \ldots, m+r-1, m+r, \ldots, r \in \mathbb{N}$, $m \in \mathbb{N}^*$. In order to express this kind of duality, we call the Markov chain $\{X'_t\}_{t\geq 0}$ the *"adjoint"* of the Markov chain $\{X_t\}_{t\geq 0}$ and reciprocally in the case that relation (5) is satisfied.

Equivalently, for the Markov chain $\{X'_t\}_{t\geq 0}$, we consider the harmonic function $(h'_\ell(t_{s+\ell+1} - t_{s+\ell}), \ell = 1, 2, \ldots, m-1, m, \ldots, s = r, r+1, \ldots, m+r-1, m+r, \ldots, r \in \mathbb{N}, m \in \mathbb{N}^*$, on $\mathbb{N}^*)$ which satisfies the equation

$$\begin{aligned} &q'_\ell(t_{s+\ell+1} - t_{s+\ell}) \cdot h'_{\ell+1}(t_{s+\ell+2} - t_{s+\ell+1}) \\ &\quad + p'_\ell(t_{s+\ell+1} - t_{s+\ell}) \cdot h'_{\ell-1}(t_{s+\ell} - t_{s+\ell-1}) = h'_\ell(t_{s+\ell+1} - t_{s+\ell}) \end{aligned}$$

or

$$\begin{aligned} &q'_\ell(t_{s+\ell+1} - t_{s+\ell}) \cdot h'_{\ell+1}(t_{s+\ell+2} - t_{s+\ell+1}) + p'_\ell(t_{s+\ell+1} - t_{s+\ell}) \\ &\quad \cdot h'_{\ell-1}(t_{s+\ell} - t_{s+\ell-1}) + p'_\ell(t_{s+\ell+1} - t_{s+\ell}) \cdot h'_\ell(t_{s+\ell+1} - t_{s+\ell}) \\ &\quad - p'_\ell(t_{s+\ell+1} - t_{s+\ell}) \cdot h'_\ell(t_{s+\ell+1} - t_{s+\ell}) = h'_\ell(t_{s+\ell+1} - t_{s+\ell}) \end{aligned}$$

or

$$\frac{p'_\ell(t_{s+\ell+1} - t_{s+\ell})}{q'_\ell(t_{s+\ell+1} - t_{s+\ell})} = \frac{h'_{\ell+1}(t_{s+\ell+2} - t_{s+\ell+1}) - h'_\ell(t_{s+\ell+1} - t_{s+\ell})}{h'_\ell(t_{s+\ell+1} - t_{s+\ell}) - h'_{\ell-1}(t_{s+\ell} - t_{s+\ell-1})}$$

for every $\ell = 1, 2, \ldots, m-1, m, \ldots, s = r, r+1, \ldots, m+r-1, m+r, \ldots, r \in \mathbb{N}$, $m \in \mathbb{N}^*$. If we put

$$u_\ell(t_{s+\ell+1} - t_{s+\ell}) = \frac{h'_{\ell+1}(t_{s+\ell+2} - t_{s+\ell+1}) - h'_\ell(t_{s+\ell+1} - t_{s+\ell})}{h'_\ell(t_{s+\ell+1} - t_{s+\ell}) - h'_{\ell-1}(t_{s+\ell} - t_{s+\ell-1})},$$

we obtain that

$$u_\ell(t_{s+\ell+1} - t_{s+\ell}) = \frac{p'_\ell(t_{s+\ell+1} - t_{s+\ell})}{q'_\ell(t_{s+\ell+1} - t_{s+\ell})} = \frac{p'_\ell(t_{s+\ell+1} - t_{s+\ell})}{1 - p'_\ell(t_{s+\ell+1} - t_{s+\ell})},$$

for every $\ell = 1, 2, \ldots, m-1, m, \ldots, s = r, r+1, \ldots, m+r-1, m+r, \ldots, r \in \mathbb{N}$, $m \in \mathbb{N}^*$, which is the equation of the definition of the sequence $(d_\ell(t_{s+\ell+1} - t_{s+\ell}))_\ell$ for the Markov chain $\{X_t\}_{t\geq 0}$ such that

$$p'_\ell(t_{s+\ell+1} - t_{s+\ell}) = p_\ell(t_{s+\ell+1} - t_{s+\ell}),$$

$$q'_\ell(t_{s+\ell+1} - t_{s+\ell}) = 1 - p'_\ell(t_{s+\ell+1} - t_{s+\ell}),$$
$$1 - p_\ell(t_{s+\ell+1} - t_{s+\ell}) = q_\ell(t_{s+\ell+1} - t_{s+\ell}),$$

for every $\ell = 1, 2, \ldots, m-1, m, \ldots, s = r, r+1, \ldots, m+r-1, m+r, \ldots, r \in \mathbb{N}, m \in \mathbb{N}^*$. By taking a similar approximation of that given for the Markov chain $\{X_t\}_{t\geq 0}$, we may say that the strictly increasing harmonic functions of the Markov chain $\{X'_t\}_{t\geq 0}$ are in correspondence with the weight representations of the Markov chain $\{X\}_{t\geq 0}$ such that equivalent equations of (5) are satisfied. Consequently, we obtain the following.

Proposition 7. *The Markov chain $\{X_t\}_{t\geq 0}$ is transient if and only if the "adjoint" Markov chain $\{X'_t\}_{t\geq 0}$ is positive recurrent and reciprocally. The "adjoint" $\{X_t\}_{t\geq 0}, \{X'_t\}_{t\geq 0}$ are null recurrent simultaneously. In particular, we have the following:*

(i) *The Markov chain $\{X_t\}_{t\geq 0}$ is transient if and only if*

$$\frac{1}{w'_0(t_{s+1} - t_s)} \cdot \sum_{\ell=1}^{+\infty} w'_\ell(t_{s+\ell+1} - t_{s+\ell}) < +\infty$$

and

$$w'_0(t_{s+1} - t_s) \cdot \sum_{\ell=1}^{+\infty} \frac{1}{p'_\ell(t_{s+\ell+1} - t_{s+\ell}) \cdot w'_\ell(t_{s+\ell+1} - t_{s+\ell})} = +\infty$$

for every $\ell = 1, 2, \ldots, m-1, m, \ldots, s = r, r+1, \ldots, m+r-1, m+r, \ldots, r \in \mathbb{N}$, $m \in \mathbb{N}^*$.

(ii) *The Markov chain $\{X'_t\}_{t\geq 0}$ is transient if and only if*

$$\frac{1}{w_0(t_{s+1} - t_s)} \cdot \sum_{\ell=1}^{+\infty} \frac{w_\ell(t_{s+\ell+1} - t_{s+\ell})}{p_\ell(t_{s+\ell+1} - t_{s+\ell})} < +\infty,$$

for every $\ell = 1, 2, \ldots, m-1, m, \ldots, s = r, r+1, \ldots, m+r-1, m+r, \ldots, r \in \mathbb{N}$, $m \in \mathbb{N}^*$.

(iii) *The "adjoint" Markov chains $\{X_t\}_{t\geq 0}$, $\{X'_t\}_{t\geq 0}$ are null recurrent in the case that*

$$\frac{1}{w_0(t_{s+1} - t_s)} \cdot \sum_{\ell=1}^{+\infty} \frac{w_\ell(t_{s+\ell+1} - t_{s+\ell})}{p_\ell(t_{s+\ell+1} - t_{s+\ell})} = +\infty$$

and

$$\frac{1}{w'_0(t_{s+1}-t_s)} \cdot \sum_{\ell=1}^{+\infty} w'_\ell(t_{s+\ell+1}-t_{s+\ell}) = +\infty,$$

for every $\ell = 1, 2, \ldots, m-1, m, \ldots, s = r, r+1, \ldots, m+r-1, m+r, \ldots, r \in \mathbb{N}$, $m \in \mathbb{N}^*$.

Proof. The proof is an application of Lemma 6 as well as of Proposition 5. □

5. Reversibility: Entropy production of the chain $\{X_t\}_{t\geq 0}$

Let us consider the birth–death circuit chain $\{X_t\}_{t\geq 0}$ defined as above which is positive recurrent under the condition of Proposition 7(ii) (or Proposition 5(i)) and let us suppose that such a chain has an invariant probability distribution $\pi = (\pi_\ell)_\ell$, for every $\ell = 0, 1, 2, \ldots, m-1, m, \ldots, s = r, r+1, \ldots, m+r-1, m+r, \ldots, r \in \mathbb{N}$, $m \in \mathbb{N}^*$. Then since it satisfies the detailed balance equation

$$\pi_\ell p_{\ell j}(t_{s+\ell+1}-t_{s+\ell}) = \pi_j p_{j\ell}(t_{s+\ell+1}-t_{s+\ell}),$$

for every $\ell = 0, 1, 2, \ldots, m-1, m, \ldots, s = r, r+1, \ldots, m+r-1$, $m+r, \ldots, r \in \mathbb{N}$, $m \in \mathbb{N}^*$ (with $j = \ell - 1$ or $j = \ell + 1$), we may say that the chain is *reversible* following a necessary and sufficient criterion for a chain to be reversible in terms of its probability matrix defined by the abovementioned formulas (3) and (4).

We may also conclude that such a chain is reversible following a necessary and sufficient criterion in terms of circuits known as the Kolmogorov's criterion (Kalpazidou [12]). This means that the irreducible, positive recurrent Markov chain $\{X_t\}_{t\geq 0}$ under the condition of Proposition 7(ii) defined uniquely by directed circuits and weights (see Proposition 4(i)) is reversible given consideration of the directed circuits $c_\ell = (\ell, \ell+1, \ell)$ such that $\ell = c_\ell(t_{s+\ell}) = c_{\ell-1}(t_{s+\ell})$, where the state transitions are of only two types: $\ell \to (\ell+1)$ and $\ell \to (\ell-1)$, for every $\ell = 0, 1, 2, \ldots, m-1, m, \ldots, s = r, r+1, \ldots,$ $m+r-1, m+r, \ldots, r \in \mathbb{N}$, $m \in \mathbb{N}^*$.

Another criterion (Kalpazidou [12]) regarding definition (1) of the entropy production of a Markov chains in terms of the circuits, where

$$(w_{c_l}(t_{s+l+1}-t_{s+l})-w_{c_{l\_}}(t_{s+l+1}-t_{s+l}))\cdot\log\frac{w_{c_l}(t_{s+l+1}-t_{s+l})}{w_{c_{l\_}}(t_{s+l+1}-t_{s+l})}$$

expresses the deviation from symmetry along the circuit c_l, leads to the conclusion that the Markov chain $\{X_t\}_{t\geq 0}$ defined as above, which is positive recurrent under the condition of Proposition 7(ii), is reversible since the weights $(w_{c_l}(t_{s+l+1}-t_{s+l}))_l$, $c_l\in C(l)$, satisfy the relation $w_{c_l}(t_{s+l+1}-t_{s+l})=w_{c_{l\_}}(t_{s+l+1}-t_{s+l})$, where $c_{l\_}$ is the reversed circuit of c_l for every $\ell=0,1,2,\ldots,m-1,m,\ldots,s=r$, $r+1,\ldots,m+r-1,m+r,\ldots,r\in\mathbb{N}$, $m\in\mathbb{N}^*$.

Consequently, we may obtain the following.

Proposition 8. *For the irreducible, positive recurrent Markov chain $\{X_t\}_{t\geq 0}$ under the condition of Proposition 7(ii) defined as above, the following results are pairwise equivalent:*

(i) *The Markov chain $\{X_t\}_{t\geq 0}$ is reversible.*
(ii) *The Markov chain $\{X_t\}_{t\geq 0}$ satisfies the detailed balance equation.*
(iii) *The weights $(w_{c_l}(t_{s+l+1}-t_{s+l}))_l$ of $\{X_t\}_{t\geq 0}$ satisfy the condition*

$$w_{c_l}(t_{s+l+1}-t_{s+l})=w_{c_{l\_}}(t_{s+l+1}-t_{s+l}),$$

for every $\ell=0,1,2,\ldots,m-1,m,\ldots,s=r,r+1,\ldots,m+r-1,m+r,\ldots,r\in\mathbb{N}$, $m\in\mathbb{N}^*$.

(iv) *The entropy production of the Markov chain $\{X_t\}_{t\geq 0}$ is equal to 0, that is,*

$$E=\frac{1}{2}\cdot\sum_{c_l\in C(l)}(w_{c_l}(t_{s+l+1}-t_{s+l})-w_{c_{l\_}}(t_{s+l+1}-t_{s+l}))\cdot\log$$
$$\times\frac{w_{c_l}(t_{s+l+1}-t_{s+l})}{w_{c_{l\_}}(t_{s+l+1}-t_{s+l})}=0$$

for every $\ell=0,1,2,\ldots,m-1,m,\ldots,s=r,r+1,\ldots,m+r-1$, $m+r,\ldots,r\in\mathbb{N}$, $m\in\mathbb{N}^*$.

References

[1] Ganatsiou, Ch. On the application of Markov chains' cycle representation to a random walk with jumps in a fixed, random environment, in *Proceedings of the 33th International Conference on Stochastic Processes and Their Applications*, Berlin (2009).
[2] Ganatsiou, Ch. On cycle representations of random walks in fixed, random environments, in *Proceedings of the 58th World Congress of the International Statistical Institute,* Dublin, pp. 4628–4631 (2011).
[3] Ganatsiou, Ch. On cycle representations of discrete-time birth-death processes, in *Proceedings of the 29th European Meeting of Statisticians*, Budapest (2013).
[4] Ganatsiou, Ch. On the study of transience and recurrence of the Markov chain defined by directed weighted circuits associated with a random walk in fixed environment, *J. Prob. Stat.*, 2013, 424601 (2013), https://doi.org/10.1155/2013/424601.
[5] Ganatsiou, Ch. *et al.*, On discrete-time birth death processes defined by directed weighted circuits, *JP. J. Biostatist.* 11(2), 103–116 (2014).
[6] Ganatsiou, Ch. On the study of circuit chains associated with a random walk with jumps in fixed, random environments, in *Proceedings of the 7th European Congress of Mathematics*, Berlin (2016).
[7] Ganatsiou, Ch. On the study of circuit chains associated with a random walk with jumps in fixed, random environments: criteria of recurrence and transience, in *Mathematical Analysis and Applications*, ed. by T. Rassias, P. Pardalos. Springer Optimization and Its Applications, Springer, Berlin, 154, pp. 185–203 (2019).
[8] Ganatsiou, Ch. On irreducible denumerable continuous parameter circuit chains: Analysis of a generalized sample path case, *Quaestiones Mathematicae*, 45(1), 41–53 (2022).
[9] Ganatsiou, Ch. On the approximation of extinction time for the discrete-time birth-death circuit chains in random environments, In: Rassias T., Daras N. J. (eds.), *Approximation and Computation in Science and Engineering*, Springer, 333–348 (2022).
[10] Ganatsiou, Ch. On the expected extinction time of the adjoint circuit chains associated with a random walk with jumps in random environments, In: A. Nikeghbali, P. Pardalos, A. Raigorodskii, M. Th. Rassias (eds.), *High Dimensional Optimization and Probability: With a view towards Data Sciences*, Springer, pp. 219–239 (2022).
[11] Ganatsiou, Ch. On birth-death circuit chains in fixed environments: analysis of a generalized sample path case, in *Proceedings of the 34th European Meeting of Statisticians (EMS–2023)*, Warsaw (3–7 July, 2023).

[12] Kalpazidou, *S. Cycle Representations of Markov Processes,* New York (1995).

[13] Karlin, S., and Taylor, H. *A First Course in Stochastic Processes,* 2nd ed., Academic Press (1975).

[14] MacQueen, J. Circuit processes, *Ann. Probability,* 9(4), 604–610 (1981).

[15] Qian Minping, Qian Min, Circulation for recurrent Markov chains, *Z. Wahrsch. Verw. Gebiete,* 59(2), 203–210 1982.

[16] Qian Minping, Qian Min, The reversibility and the entropy production of Markov processes, *Contemp. Math.*, 118(2), 255–261 (1991).

[17] Nowak, M.A. *Evolutionary Dynamics: Exploring the Equations of Life,* Harvard University Press (2006).

[18] Parthasarathy, P.R. and Lenin, R.B. Birth and death process (BDP) models with applications - queueing, communication systems, chemical models, biological models: The state-of-the-art with a time–dependent perspective, *American Journal of Mathematical and Management Sciences,* 24(1), 51 (2004).

[19] Steele, J.M. *Stochastic Calculus and Financial Applications, Springer,* New York (2001).

[20] Wang, Z.K. and Yang, X.Q. *The Birth and Death Processes and Markov Chains,* Springer (1992).

[21] Wilkinson, D.J. *Stochastic Modeling for Systems Biology,* 2nd edition, CRC Press (2011).

[22] Zemanian, A.H. *Infinite Electrical Networks,* Cambridge University Press (1991).

https://doi.org/10.1142/9789811293009_0015

Chapter 15

Recurrent Neural Networks for Solving Time-Varying Linear Optimization Problems

D. Gerontitis[*,‡] and P. Tzekis[†,§]

[*]Department of Information and Electronic Engineering,
[†]International Hellenic University, Thessaloniki, Greece
[‡]dimitrios_gerontitis@yahoo.gr, dimger@iee.ihu.gr
[§]ptzekis@gmail.com, ptzekis@ihu.gr

Activation functions (AFs) are continuously enhancing the fast convergent neural models. It has been pushed to the forefront in recent years to study the issues of linear systems and optimization problems. Following AFs, gradient neural networks (GNNs) have been constructed for static situations, while Zhang neural networks (ZNNs) are applicable for time-varying systems. In the context of this chapter, different categories of recurrent neural networks are constructed and applied for solving time-varying matrix equations. At the same time, simulations in Matlab's environment show the effectiveness of the proposed models.

1. Introduction

Construction of fast convergent neural models based on different activation functions (AFs) is an interesting part of research in fields which are related with artificial intelligence, especially in the area of recurrent neural networks (RNNs). The first category of RNN models were the gradient neural networks (GNNs) [1–6] which were used for constant matrix equations solving. On the other side, for the study of time-varying systems, Zhang neural networks (ZNNs) [7–12] have been proposed and investigated for the solution of different time-variant matrix equations. It must be mentioned that the first ZNN models were based on the linear activation function [7–9], while in [10, 11], the authors proposed power-sigmoid AFs for acceleration of linear ZNN's convergence speed property. From this fact, it is observable how important challenge can be for researchers in the construction of different activation functions which will accelerate the convergence property of each RNN formula. With the passage of time, different activation functions have been created, such as Li activation function [13–16], tunable [18, 19], and piecewise sign function [17] while other scientists for the improvement of speed ability constructed different dynamical systems, such as finite zeroing neural networks (FTZNN) [23–26] and zeroing neural networks based on hyperpower iterative methods (ZNNHM) [22]. Another way for acceleration of the convergence speed is the application of time-varying parameter $\exp(t)$ [27–30] in front of each RNN formula.

In view of previous comments, it will be more appropriate to construct a novel nonlinear activation function whose main intention is to accelerate the convergence property of each nonlinear ZNN dynamical system based on different parameters. Main contributions of this chapter are summarized as follows:

(1) An effective activation function is applied for constant linear systems solving and in the study of quadratic optimization/minimization problem.
(2) Compared to ZNN based on Li, tunable, and VAF activation functions, the proposed model based on nonlinear activation function named extended exponential sign activation function (EESAF) demonstrates accelerated convergence property.
(3) Through theoretical analysis, we prove that the new model converges to solution of linear systems $\mathrm{Av} = \mathrm{b}$.
(4) Numerical experiments are presented with the aim of demonstrating superiority of the EESAF over already existed activation functions.

2. Preliminaries: Linear Systems

2.1. *Mathematical preliminaries*

In the sector of artificial intelligence, recurrent neural networks (RNNs) are represented as a dynamical system [34] in the form

$$\dot{\mathrm{v}}(t) = \theta(\mathrm{v}(t), t), \quad t \geq 0, \tag{1}$$

where $\mathrm{v}(t)$ is an appropriately sized system state, $\mathrm{v}(0) = \mathrm{v}_0$ constitutes an appropriately sized initial state, and $\mathrm{v}(t) = 0$ represents the equilibrium state of system (1). The first category of recurrent neural networks were the gradient neural networks (GNNs) which have been used for the solution of constant matrix equations $\mathrm{Av} = \mathrm{b}$ and were defined by the formula

$$\dot{\mathrm{v}}(t) = -k\mathrm{A}^{\mathrm{T}}(\mathrm{Av(t)} - \mathrm{b}),$$

where A^{T} is the transpose of the non-singular square matrix A and k is a positive parameter whose increasing values accelerate the convergence speed of the proposed GNN model. With the passage of time, the challenge for scientists was the construction of a dynamical system similar to (1) for the research and solution of time-varying problems. The answer to this was given by the use of Zhang neural networks (ZNNs) for the study of time-varying Sylvester equations (TVSE):

$$\mathrm{A}(t)\mathrm{v}(t) - \mathrm{v}(t)\mathrm{B}(t) = -\mathrm{C}(t), \tag{2}$$

where $\mathrm{A}(t) \in \mathbb{R}^{\nu\times\nu}$, $\mathrm{B}(t) \in \mathbb{R}^{\mu\times\mu}$, $\mathrm{C}(t) \in \mathbb{R}^{\nu\times\mu}$, and $\mathrm{v}(t)$ is the unknown matrix which must be computed. The first step for solving (TVSE) (2) is the definition of error matrix function $\mathrm{E}(t) = \mathrm{A}(t)\mathrm{v}(t) - \mathrm{v}(t)\mathrm{B}(t) + \mathrm{C}(t)$ and then the application of linear ZNN formula [7 9]:

$$\dot{\mathrm{E}}(t) = -k\mathrm{E}(t), \tag{3}$$

where k is a positive parameter whose increasing values accelerate the convergence speed of the linear ZNN formula (3) (as in GNN models). Dynamical system which arises from formula (3) based on (TVSE) (2) is given by following relationship:

$$\begin{aligned}\dot{\mathrm{v}}(t) = {} & [\mathrm{I} - \mathrm{A}(t)]\dot{\mathrm{v}}(t) - \dot{\mathrm{A}}(t)\mathrm{v}(t) + \dot{\mathrm{v}}(t)\mathrm{B}(t) + \mathrm{v}(t)\dot{\mathrm{B}}(t) - \dot{\mathrm{C}}(t) \\ & - k\left(\mathrm{A}(t)\mathrm{v}(t) - \mathrm{v}(t)\mathrm{B}(t) + \mathrm{C}(t)\right).\end{aligned} \tag{4}$$

The authors in [7] proved that the state matrix of (4) converges to theoretical solution of (2) with exponential convergence rate. In [10], for the acceleration of the convergence speed in linear ZNN (3) formula, the following nonlinear pattern is considered:

$$\dot{\mathrm{E}}(t) = -k\mathrm{F}(\mathrm{E}(t)), \tag{5}$$

for matrix inversion problem, where F(.): $\mathbb{R}^{\nu\times\nu} \rightarrow \mathbb{R}^{\nu\times\nu}$ denotes a matrix mapping of neural networks which element includes any monotonically increasing odd activation function of one variable. Dynamical system (5) gives the impulse for new scientists to focus their research on RNN models, especially in the study of nonlinear ZNN models which implies the following lemma from [44–46].

Lemma 1. *Given time-varying matrices* $\mathrm{A}(t)$, $\mathrm{B}(t)$, *and* $\mathrm{C}(t)$, *if any monotonically increasing and odd activation function array* F(.) *is used, then the state matrix* v(t) *of Zhang neural network* (5) *starting from any initial state* $\mathrm{v}_0 = \mathrm{v}(0)$ *always converges to time-varying theoretical solution* $\mathrm{v}^*(t)$ *of Sylvester equation* (2).

Previous lemma is very important because it strengthens the fact that the application of any odd and increasing activation function $[\mathrm{f}(u)]_{ij}$ as an element in matrix F(.) of (5) can be used for the solution of (2). Following notes about the study of (TVSE) (2) must be mentioned:

- If $\mathrm{B}(t) = 0$ and $\mathrm{C}(t) = -\mathrm{I}$, where 0 and I are zero and identity matrices, respectively, equation (2) represents the time-dependent matrix inversion problem [43].
- If $\mathrm{A}(t) = \mathrm{D}^{\mathrm{T}}(t)$ and $\mathrm{B}(t) = -\mathrm{D}(t)$, formula (2) represents the known Lyapunov matrix equation $\mathrm{D}^{\mathrm{T}}(t)\mathrm{v}(t) + \mathrm{v}(t)\mathrm{D}(t) + \mathrm{C}(t) = 0$ [40–42].
- If $\mathrm{A}(t) = \mathrm{A}$, $\mathrm{B}(t) = 0$, and $\mathrm{C}(t) = \mathrm{b}$, relationship (2) is simplified in the solution of linear systems which will be occupied in this survey.

Before theoretical analysis, the following definitions and proposition from [31–35, 37–39] must be referred.

Definition 2. The origin of (1) is globally finite-time stable if it is globally and asymptotically stable and there exists a locally bounded setting-time function $T : \mathbb{R}^{\nu} \mapsto \mathbb{R}_{+} \cup \{0\}$, such that $\mathrm{v}(t, \mathrm{v}_0) = 0$ for all $t \geq T(\mathrm{v}_0)$.

Definition 3. The origin of (1) is globally predefined-time stable if is globally finite-stable and the settling-time function T is globally bounded, i.e., there exists a constant $t_x \in \mathbb{R}_{+}$ satisfying $t_x \geq T(\mathrm{v}_0)$ for all $\mathrm{v}_0 \in R^{\nu}$.

Proposition 4. *Given inequality*

$$\dot{V}(t) \leq -g_1 V^{\mathrm{p}}(t) - g_2 V^{\mathrm{w}}(t), \tag{6}$$

where $g_1, g_2 > 0$, $0 < \mathrm{p} < 1$, *and* $\mathrm{w} > 1$ *are constant parameters, then inequality* (6) *is fixed-time stable and the predefined convergence time* t_p *of dynamical system* (6) *satisfies*

$$t_p \leq \frac{1}{g_1(1-\mathrm{p})} + \frac{1}{g_2(\mathrm{w}-1)}. \tag{7}$$

An RNN formula which satisfies previous properties based on specific nonlinear activation function is applied and analyzed in the following subsection for the study of constant systems.

2.2. *Linear systems*

The solution of linear algebraic equations plays an important role in applied mathematics' domain because of its appearance in different scientific fields like data analysis [47, 48], robotics [50–54], and scattering problems [55–59], so for this reason, the following equation (under study) is considered.

Linear equation under study:

$$\mathrm{Av} = \mathrm{b}, \tag{8}$$

where A is a $\nu \times \nu$ non-singular square matrix, v is the solution of (8), and b is a known $\nu \times 1$ vector. By defining the error matrix function $\mathrm{e}(t) = \mathrm{Av}(t) - \mathrm{b}$, the goal of this research manuscript is to construct and apply an odd-increasing AF in the linear ZNN formula $\dot{\mathrm{e}}(t) = -k\mathrm{e}(t)$ [7–9] which will vanish each element of the error function $\mathrm{e}(t)$ to zero, in finite time period. Previous nonlinear AFs for finite estimation of the convergence time are referred as follows:

(I) The **sign** activation function [17]

$$\operatorname{sign}(u) = \begin{cases} 1, & u > 0 \\ 0, & u = 0 \\ -1, & u < 0. \end{cases} \tag{9}$$

(II) The **Li** or sign-bi-power (**sbp**) activation function [13–16]

$$\operatorname{sbp}(u) = \operatorname{sgn}^{\mathrm{q}}(\mathrm{u}) + \operatorname{sgn}^{\frac{1}{\mathrm{q}}}(\mathrm{u}), \tag{10}$$

where

$$\operatorname{sgn}^{\mathrm{q}}(u) = |\mathrm{u}|^{\mathrm{q}}\operatorname{sign}(u) = \begin{cases} |\mathrm{u}|^{\mathrm{q}}, & u > 0 \\ 0, & u = 0 \\ -|\mathrm{u}|^{\mathrm{q}}, & u < 0 \end{cases} \tag{11}$$

and $\mathrm{q} \in (0, 1)$.

(III) The **tunable** activation function [18–21]

$$\mathrm{G(u)} = (|\mathrm{u}|^{\mathrm{q}} + |\mathrm{u}| + |\mathrm{u}|^{\frac{1}{\mathrm{q}}})\operatorname{sign}(\mathrm{u}). \tag{12}$$

(IV) The versatile activation function (**VAF**) [34–36]

$$\mathrm{G(u)} = (a_1|\mathrm{u}|^{\mathrm{q}} + a_2|\mathrm{u}|^{\mathrm{w}})\operatorname{sign}(u) + a_3\mathrm{u} + a_4\operatorname{sign}(\mathrm{u}), \tag{13}$$

where $a_1, a_2 > 0$, $a_3, a_4 \geq 0$, and $\mathrm{w} > 1$.

In this chapter, the following odd-increasing extended exponential sign activation function activation (**EESAF**) order m is considered:

$$\begin{aligned} \Phi_m(u) &= \mathrm{EESAF}_m(u) \\ &= \left(a_1 \left(\prod_{\mathrm{p}=1}^{\mathrm{m}} \exp(|\mathrm{u}|^{\mathrm{q_p}}) - 1 \right) + a_2|\mathrm{u}|^{\mathrm{w}} \right) \operatorname{sign}(\mathrm{u}), \end{aligned} \tag{14}$$

where $\mathrm{q_1, q_2, \ldots, q_m} \in (0, 1)$ with $\mathrm{q_1} \leq \mathrm{q_2} \cdots \leq \mathrm{q_m}$ and $\mathrm{w} \in (1, +\infty)$. The ZNN model based on (14) which is referred to in the solution of (8) is given by

$$\dot{\mathrm{v}}(t) = [\mathrm{I} - \mathrm{A}]\dot{\mathrm{v}}(t) - \Phi_m(\mathrm{A}\mathrm{v}(t) - \mathrm{b}). \tag{15}$$

Evolution formula (15) for the solution of linear equation (8) is called extended exponential Zhang neural network order m and is symbolized as EEZNN. Simulink implementation of (14) for $m = 2$ is represented in Fig. 1.

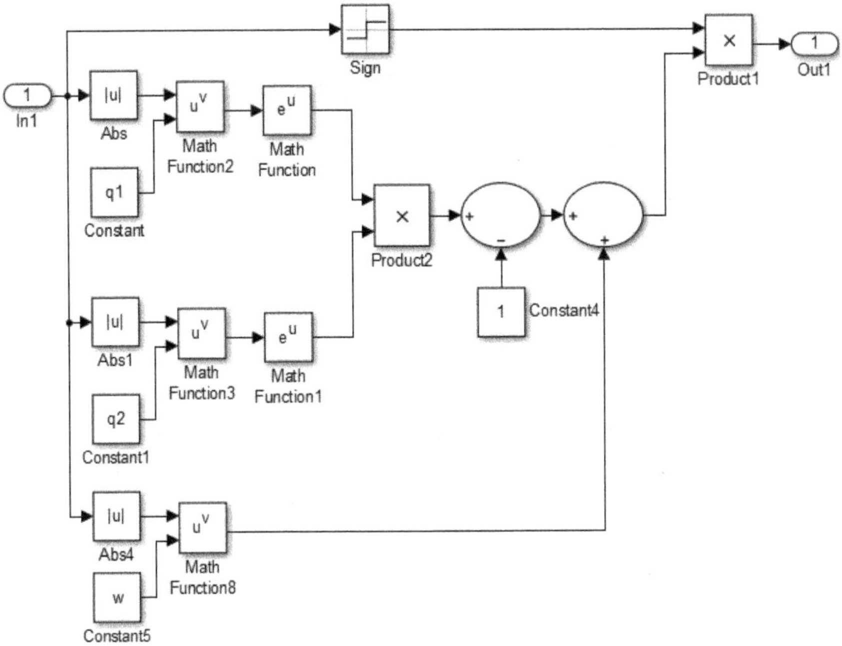

Fig. 1. Simulink model of $\Phi_2(u)$.

Lemma 5. *Given non-singular constant matrix* A *and vector* b, *if odd and monotonically increasing activation function* $\Phi_m(u)$ *in* (14) *is used, then the state vector* $\mathrm{v}(t)$ *of* EEZNN *model* (15) *starting from initial state* $\mathrm{v}(0)$ *always converges to theoretical solution of* (8).

Proof. Due to AF (14) being an odd and increasing function, Lemma 1 proves the above claim. □

For predefined estimation of the convergence time period in (15), the following theorem is introduced.

Theorem 6. *With known non-singular matrix* A *and vector* b, *if* $\Phi_m(u)$ *order* m *in* (14) *is used, then the neural state* $\mathrm{v}(t)$ *of* (15) *starting from the initial condition* $\mathrm{v}(0) \in \mathbb{R}^\nu$ *converges to theoretical solution of* $\mathrm{Av} = \mathrm{b}$ *in predefined time* t_x*:*

$$t_x \leq \frac{1}{a_1 k(1 - \mathrm{q_m})} + \frac{1}{a_2 k(\mathrm{w} - 1)}. \tag{16}$$

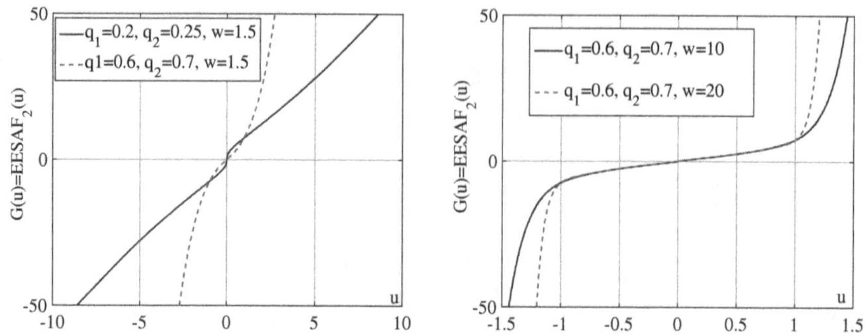

Fig. 2. $\Phi_2(u)$ AF for different values of $\mathrm{q}_1, \mathrm{q}_2, \mathrm{w}$.

Proof. Lyapunov candidate $V_i(t) = |e_i(t)|^2$ is considered for system (15) where $e_i(t)$ is the *ith* element of error function $e(t) = \mathrm{Av}(t) - \mathrm{b}$ and $i = 1, 2, \cdots, \nu$, then

$$\dot{V}_i(t) = 2e_i(t)\dot{e}_i(t),$$

therefore

$$\dot{V}_i(t) = 2e_i(t)\dot{e}_i(t) = -2ke_i(t)\Phi_m(e_i(t)),$$

which implies

$$\dot{V}_i(t) = -2ke_i(t)\left(a_1\left(\prod_{\mathrm{p}=1}^{m}\exp(|e_i(t)|^{\mathrm{q_p}}) - 1\right) + a_2|e_i(t)|^{\mathrm{w}}\right) \times \mathrm{sign}(e_i(t)).$$

In this step of the proof, the following cases about sign of $e_i(t)$ have been concluded:

- If $e_i(t) = 0$, then $\dot{V}_i(t) = 0$.
- If $e_i(t) > 0$, then $e_i(t) = |e_i(t)|$ and $\mathrm{sign}(e_i(t)) = 1$, which implies that

$$\dot{V}_i(t) = -2k|e_i(t)|\left(a_1\left(\prod_{\mathrm{p}=1}^{m}\exp(|e_i(t)|^{\mathrm{q_p}}) - 1\right) + a_2|e_i(t)|^{\mathrm{w}}\right), \tag{17}$$

where $\dot{V}_i(t) < 0$ and the following inequality

$$\dot{V}_i(t) = -2k|e_i(t)| \left(a_1 \left(\exp\left(\sum_{\mathrm{p}=1}^{m} |e_i(t)|^{\mathrm{q_p}} \right) - 1 \right) + a_2|e_i(t)|^{\mathrm{w}} \right),$$

which is similar to

$$\begin{aligned}\dot{V}_i(t) &\leq -2k|e_i(t)| \left(a_1 \sum_{\mathrm{p}=1}^{m} |e_i(t)|^{\mathrm{q_p}} + a_2|e_i(t)|^{\mathrm{w}} \right) \\ &\leq -2ka_1|e_i(t)|^{\mathrm{q_m}+1} - 2ka_2|e_i(t)|^{\mathrm{w}+1}. \end{aligned} \tag{18}$$

- If $e_i(t) < 0$, then $e_i(t) = -|e_i(t)|$ and $\operatorname{sign}(e_i(t)) = -1$, the same relationship (17) is obtained and the same results as in positive $e_i(t)$ case have been extracted which guarantees the negative definiteness of $\dot{V}_i$ for $e_i \neq 0$ and $\dot{V}_i = 0$ for $e_i = 0$. By Lyapunov stability theory [10, 49, 52], $e_i(t)$ globally converges to zero for any $i = 1, 2, \ldots, \nu$.

Summarizing in vector form,

$$\dot{V}(t) \leq -2a_1kV^{\frac{\mathrm{q_m}+1}{2}}(t) - 2a_2kV^{\frac{\mathrm{w}+1}{2}}(t). \tag{19}$$

From Proposition 4, the predefined convergence time of (19) is obtained by

$$t_x \leq \frac{1}{a_1k(1-\mathrm{q_m})} + \frac{1}{a_2k(\mathrm{w}-1)}, \tag{20}$$

and the proof is completed. □

For the special case that all the elements of error matrix satisfies $|e_i(t)| \leq 1$, the following corollary arises.

Corollary 7. *With known non-singular matrix* A *and vector* b, *if* $\Phi_m(u)$ *order* m (14) *is used and especially all elements of the error function satisfy* $|e_i(t)| \leq 1$, *where* $i = 1, 2, \ldots, \nu$, *then the neural state* $\mathrm{v}(t)$ *of* (15) *starting from initial condition* $\mathrm{v}(0) \in \mathbb{R}^\nu$ *converges to theoretical solution* v^* *of* (8) *in predefined time period* t_{x_a}*:*

$$t_{x_a} \leq \frac{1}{a_1km(1-\mathrm{q_m})} + \frac{1}{a_2k(\mathrm{w}-1)}. \tag{21}$$

Proof. From relationship (18),

$$\dot{V}_i(t) \leq -2k|e_i(t)| \left(a_1 \sum_{\mathrm{p}=1}^{m} |e_i(t)|^{\mathrm{q_p}} + a_2|e_i(t)|^{\mathrm{w}} \right)$$

because of $|e_i(t)| \leq 1$, then $|e_i(t)|^{\mathrm{q_1}} \geq |e_i(t)|^{\mathrm{q_2}} \geq |e_i(t)|^{\mathrm{q_3}} \geq \cdots \geq |e_i(t)|^{\mathrm{q_{m-1}}} \geq |e_i(t)|^{\mathrm{q_m}}$, and the following differential inequality (in vector form) is concluded:

$$\dot{V}(t) \leq -2kma_1V^{\frac{\mathrm{q_m}+1}{2}}(t) - 2ka_2V^{\frac{\mathrm{w}+1}{2}}(t), \tag{22}$$

with the same way as in Theorem 6 and from Proposition 4, predefined estimation about the convergence time is obtained:

$$t_{x_a} \leq \frac{1}{a_1km(1-\mathrm{q_m})} + \frac{1}{a_2k(\mathrm{w}-1)}, \tag{23}$$

so the proof is finished. □

Following theoretical result from [34–37, 39] represents the predefined convergence time produced by versatile activation function which is mentioned as follows.

Proposition 8. *The predefined convergence time t_l of* ZNN *model activated by versatile activation function* (13) *is estimated by*

$$t_l \leq \frac{1}{a_1k(1-\mathrm{q})} + \frac{1}{a_2k(\mathrm{w}-1)} = t_l'. \tag{24}$$

It is observable that if $|e_i(t)| \leq 1$ and $\mathrm{q_1} = \mathrm{q_2} = \mathrm{q_3} = \cdots = \mathrm{q_{m-1}} = \mathrm{q_m} = \mathrm{q}$, then we can write the following:

- The convergence time of EEZNN is obtained by

$$t_x = t_{x_a} \leq \frac{1}{a_1km(1-\mathrm{q})} + \frac{1}{a_2k(\mathrm{w}-1)} = t_x'. \tag{25}$$

- The predefined convergence time of the nonlinear ZNN dynamical system activated by $\Phi_m(u)$ (14) is shorter than VAF (13) and are

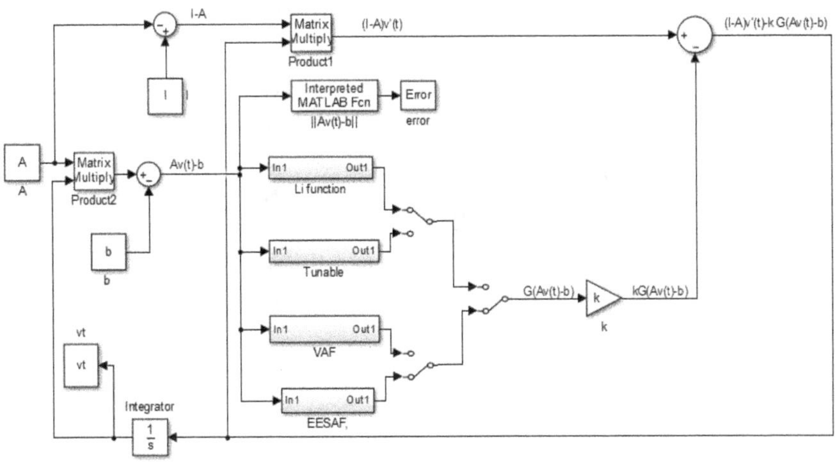

Fig. 3. Simulink implementation of (15).

related to

$$t_x' - t_l' = \left(\frac{1}{a_1 k(1-\mathrm{q})}\right)\left(\frac{1}{m} - 1\right) \leq 0 \Longrightarrow t_x' \leq t_l'. \tag{26}$$

- As number m of the set $\mathrm{q}_1, \mathrm{q}_2, \ldots, \mathrm{q}_{\mathrm{m}}$ is increasing, the left side of predefined convergence time t_x vanishes to 0 with fast pace. Observation strengthens the fact that the number m of parameters $\mathrm{q}_1, \mathrm{q}_2, \ldots, \mathrm{q}_m$ which were involved in the proposed AF can be applied for the minimization of convergence time which strengthens the use of (14) against previous AFs.
- It is observable from previous theoretical analysis that the predefined convergence time of EEZNN dynamical system depends on $k, a_1, a_2, \mathrm{q}_1, \mathrm{q}_m, \mathrm{w}, m$ and not on the values of initial condition $e(0)$ such as in the finite-time convergence analysis of sbp and tunable activation functions [13–16, 18–21]. Simulink implementation of (15) is represented in Fig. 3.

Following numerical examples have been considered for the application of the proposed EEZNN formula (15) where without loss of generality, $a_1 = a_2 = a_3 = a_4 = 1$.

2.3. *Numerical examples*

Example 2.3.1

For the validation of the proposed EEZNN dynamical system, the following matrix A from [43] and vector b have been defined:

$$\mathrm{A} = \begin{pmatrix} 1 & -2 & 3 \\ 0 & -3 & -2 \\ 1 & 2 & 3 \end{pmatrix}, \quad \mathrm{b} = \begin{pmatrix} 1 \\ 2 \\ 3 \end{pmatrix}. \tag{27}$$

The theoretical solution of $\mathrm{Av} = \mathrm{b}$ is equal to

$$\mathrm{v}^* = \mathrm{A}^{-1}\mathrm{b} = \begin{pmatrix} 7.25 \\ 0.50 \\ -1.75 \end{pmatrix}. \tag{28}$$

Simulink implementation, under the parameters $\mathrm{q}_1 = 0.3$, $\mathrm{q}_2 = 0.4, \mathrm{q}_3 = 0.6$, $\mathrm{w} = 1.5$, $k = 1, 10$ and for zero initial condition $\mathrm{v}(0) = 0$, produces the graphical results presented in Figs. 4, 5, and 6. From Fig. 4, all the elements of error function for $t = 0$ satisfy $|e_i(0)| \geq 1$ and vanish to zero with the passage of time.

It is observable from Fig. 5 that $\Phi_m(u)$ includes faster convergence speed property against previous Li, tunable, and versatile AFs because $||e(t)||_F = ||\mathrm{Av(t)} - \mathrm{b}||_F$ vanish to zero in shorter time. Furthermore, from Fig. 6, the red dashed lines which correspond to the solution produced by the EEZNN model (15) which converges to the theoretical solution (blue lines) in the predefined time

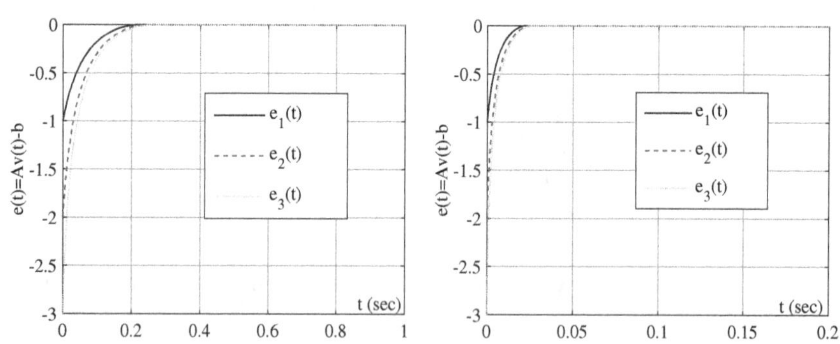

Fig. 4. $e_i(t)$ behavior under $k = 1$ and 10 for Example 2.3.1.

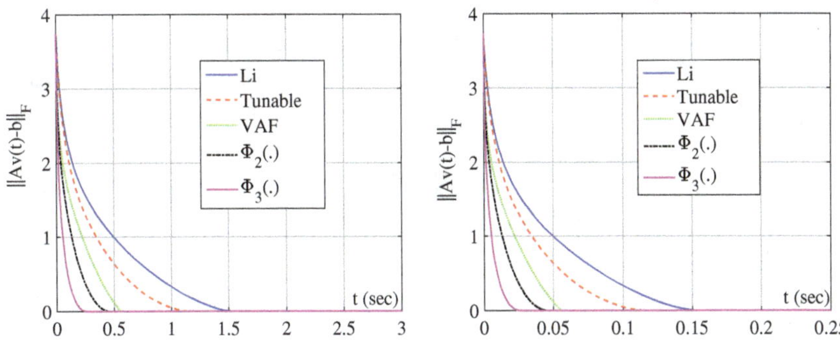

Fig. 5. Error comparison of $\Phi_m(u)$ with previous AFs when $m = 2, 3$ for Example 2.3.1.

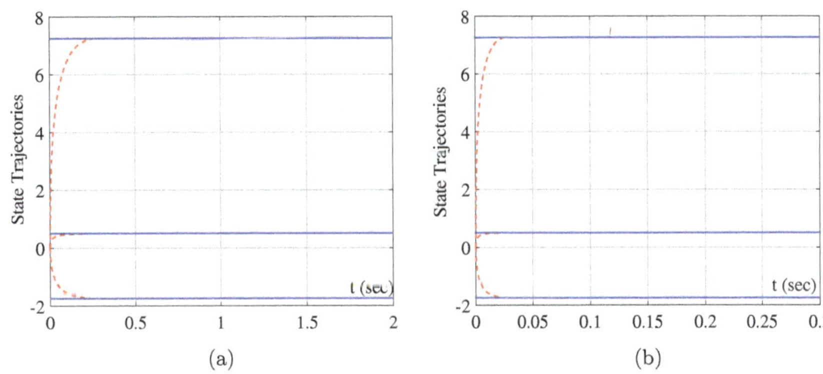

Fig. 6. State trajectories by $\Phi_3(u)$ under (a) $k = 1$ and (b) $k = 10$ for Example 2.3.1.

period. Figure 7 proves that the EESAF can work for different initial conditions v(0) and under different parameters $\mathrm{q_p}$, w, $\mathrm{p} = 1, 2, \ldots, m$. Figure 8 represents the numerical result by Simulink implementation for $m = 3, k = 1$ at time instant $t = 0.3$ sec which is in agreement with theoretical solution v*. In the following example, the behavior of the proposed EEZNN formula for the solution of ill-conditioned systems is examined.

Example 2.3.2

It is really interesting to contemplate the behavior of (14) in the special case that system's component A includes big condition

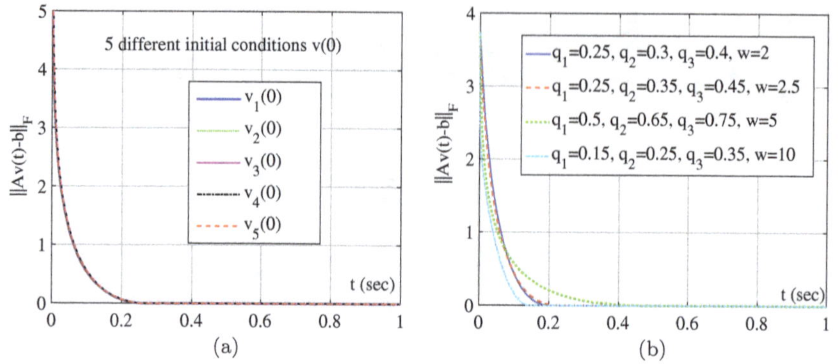

Fig. 7. $\Phi_3(u)$ error behavior (a) for different values of v(0) (b) with different combinations of q_k, w for Example 2.3.1.

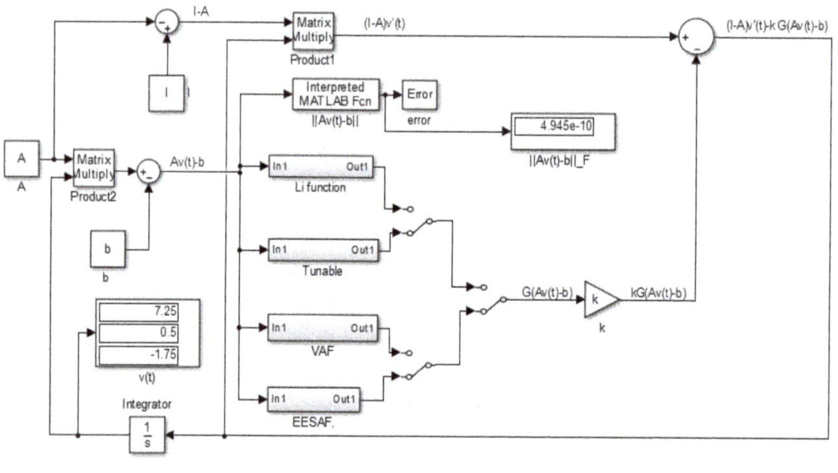

Fig. 8. Numerical result by Simulink model from Fig. 3 for Example 2.3.1.

number (ill-conditioned). So, for this reason, the following matrix with condition number cond(A) = 5.7689e + 04 from [4, 22] has been considered:

$$\mathrm{A} = \begin{pmatrix} 1 & 0 & 0 & 0 & 0 & 0 \\ 1 & 1 & 1 & 1 & 1 & 1 \\ 1 & 2 & 4 & 8 & 16 & 32 \\ 1 & 3 & 9 & 27 & 81 & 243 \\ 1 & 4 & 16 & 64 & 256 & 1024 \\ 1 & 5 & 25 & 125 & 625 & 3125 \end{pmatrix}, \quad \mathrm{b} = \begin{pmatrix} 10 \\ 20 \\ 30 \\ 40 \\ 50 \\ 60 \end{pmatrix}, \tag{29}$$

with theoretical solution

$$\mathrm{v}^* = \begin{pmatrix} 0 \\ 0 \\ 0 \\ 0 \\ 10 \\ 10 \end{pmatrix}. \tag{30}$$

Simulink implementation from Fig. 3, under the parameters $\mathrm{q}_1 = 0.35, \mathrm{q}_2 = 0.4, \mathrm{q}_3 = 0.45$, $\mathrm{w} = 2$ and the zero initial condition $\mathrm{v}(0) = 0$, is extracted in Figs. 9 and 10.

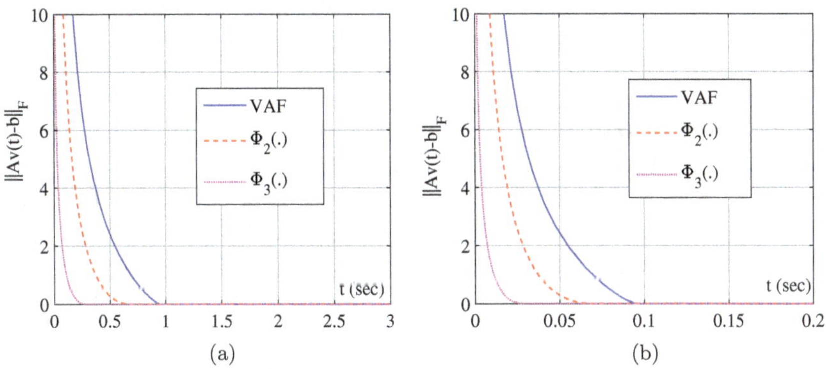

Fig. 9. Error comparison between VAF and $\Phi_m(u)$ with (a) $k = 1$ and (b) $k = 10$ for Example 2.3.2.

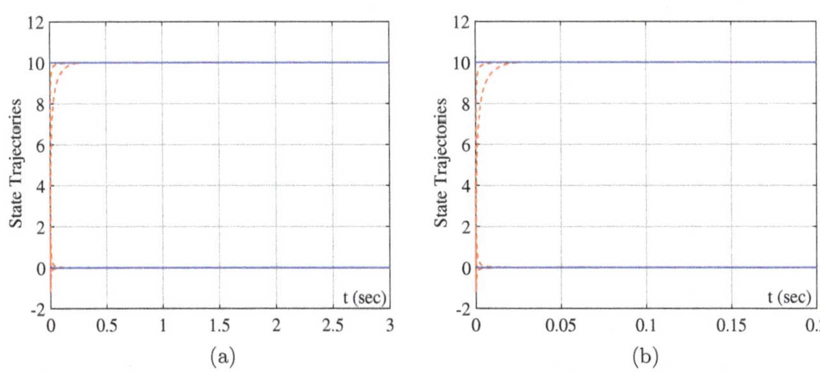

Fig. 10. State trajectories by EEZNN dynamical system under (a) $k = 1$ and (b) $k = 10$ for Example 2.3.2.

It is observable from Fig. 9 that the extended exponential sign AF (14) includes faster convergence property against VAF because Frobenius norm $||e(t)||_F = ||\mathrm{Av(t)} - \mathrm{b}||_F$ vanishes to zero in shorter time, while Fig. 10 strengthens the fact that the proposed activation function works and for ill-conditioned matrices since red dashed lines produced by (15) are in agreement with blue lines corresponding to theoretical solution v*. It is a really important result that novel formula EEZNN can work and for ill-conditioned systems with accelerated convergence property. In the following example, a high-order system is considered for the application of the EEZNN model (15).

Example 2.3.3

In the numerical experiment from [60] for the application of evaluation-interpolation technique concerning matrix manipulations, the following 11×11 system needed to be solved:

$$\begin{pmatrix}
0.11 & 0.03 & 0.01 & 0.04 & 0.01 & 0.01 & 0.00 & 0.00 & 0.01 & 0.00 & 0.00\\
0.10 & 0.10 & 0.10 & 0.03 & 0.03 & 0.01 & 0.01 & 0.01 & 0.00 & 0.00 & 0.00\\
0.30 & 0.23 & 0.17 & 0.16 & 0.12 & 0.09 & 0.08 & 0.06 & 0.06 & 0.04 & 0.03\\
0.71 & 0.12 & 0.02 & 0.60 & 0.10 & 0.50 & 0.14 & 0.02 & 0.72 & 0.12 & 0.61\\
0.82 & 0.09 & 0.01 & 0.74 & 0.08 & 0.67 & 0.13 & 0.01 & 1.09 & 0.11 & 0.98\\
0.56 & 0.40 & 0.30 & 0.41 & 0.30 & 0.31 & 0.26 & 0.19 & 0.27 & 0.19 & 0.20\\
0.51 & 0.07 & 0.01 & 0.37 & 0.05 & 0.26 & 0.06 & 0.01 & 0.34 & 0.04 & 0.24\\
0.20 & 0.10 & 0.05 & 0.09 & 0.05 & 0.04 & 0.03 & 0.01 & 0.02 & 0.01 & 0.01\\
0.28 & 0.24 & 0.21 & 0.15 & 0.13 & 0.08 & 0.07 & 0.06 & 0.05 & 0.04 & 0.02\\
0.46 & 0.37 & 0.30 & 0.31 & 0.25 & 0.21 & 0.19 & 0.15 & 0.16 & 0.13 & 0.11\\
0.28 & 0.27 & 0.26 & 0.15 & 0.14 & 0.08 & 0.08 & 0.07 & 0.04 & 0.04 & 0.02
\end{pmatrix}
\begin{pmatrix} \mathrm{v}_1\\ \mathrm{v}_2\\ \mathrm{v}_3\\ \mathrm{v}_4\\ \mathrm{v}_5\\ \mathrm{v}_6\\ \mathrm{v}_7\\ \mathrm{v}_8\\ \mathrm{v}_9\\ \mathrm{v}_{10}\\ \mathrm{v}_{11} \end{pmatrix}
=
\begin{pmatrix} -0.02\\ -0.01\\ -0.10\\ -0.86\\ -1.20\\ -0.36\\ -0.47\\ -0.05\\ -0.09\\ -0.23\\ -0.08 \end{pmatrix} \tag{31}$$

with theoretical solution

$$v^* = \begin{pmatrix} +0.0085 \\ -0.0252 \\ +0.0158 \\ +0.0381 \\ +0.0843 \\ -2.5223 \\ +1.3324 \\ -0.1684 \\ +0.2517 \\ -0.0099 \\ +0.0055 \end{pmatrix},$$

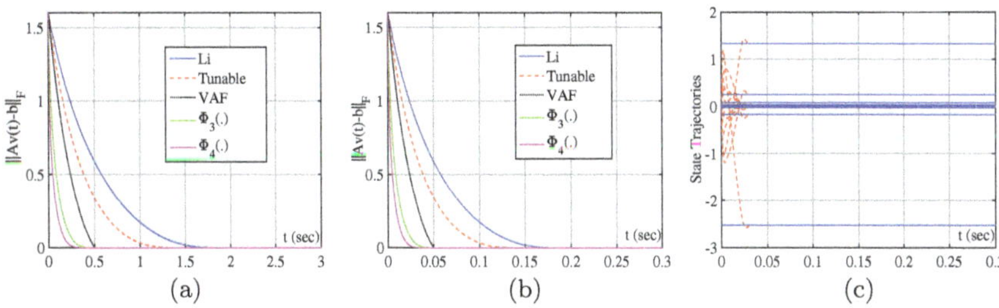

Fig. 11. $||\mathrm{Av}(t) - \mathrm{b}||_F$ for (a) $k = 1$, (b) $k = 10$, and (c) $\mathrm{v}(t)$ trajectories for $k = 10$.

where parameters v_i, $i = 1, \dots, 11$, represent unknown coefficients of a two-variable interpolation polynomial which are used for computation of matrix inverse and must be calculated. The system's determinant is equal with $\det(\mathrm{A}) = 6.0369 \times 10^{-16}$ and the condition number with $\mathrm{cond}(\mathrm{A}) = 2.4495 \times 10^3$. For the choice of $\mathrm{q} = 0.5$, $\mathrm{w} = 2$, $\mathrm{q}_1 = 0.5, \mathrm{q}_2 = 0.6, \mathrm{q}_3 = 0.7, \mathrm{q}_4 = 0.8$, $k = 1, 10$, and under zero initial condition, EEZNN's Simulink implementation produces graphical results, as shown in Figs. 11 and 12, respectively.

From Fig. 11, researchers can note that the proposed AF (14) against sbp (10), tunable (12), and VAF (13) represents faster convergence speed, while for increasing values of parameter m, the

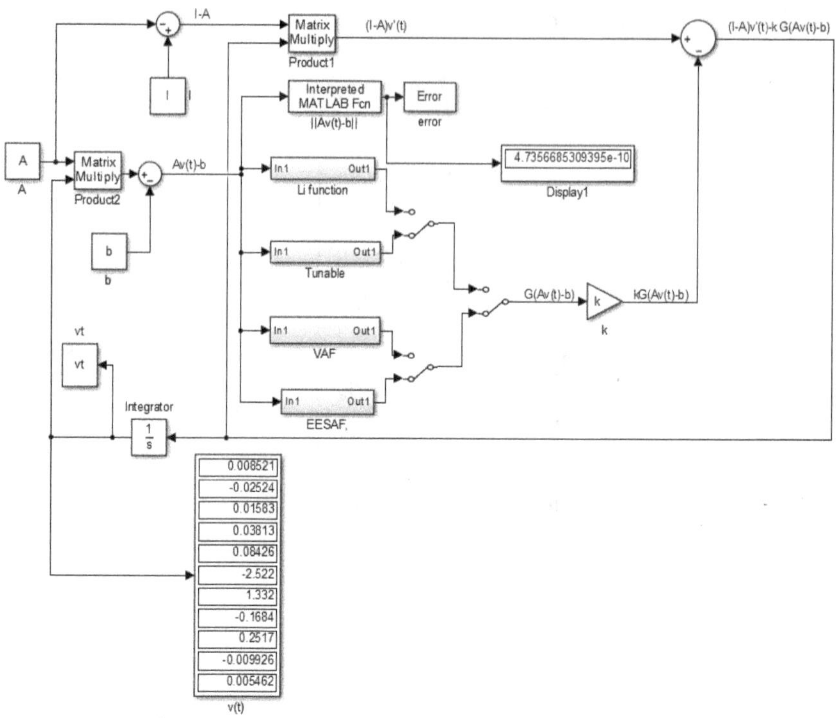

Fig. 12. Numerical result by Simulink model from Fig. 3 for Example 2.3.3.

residual error $||\mathrm{Av}(t) - \mathrm{b}||_F$ vanishes to zero in a shorter time than other AFs. In Fig. 11(c), the red dashed lines of trajectories produced by Simulink implementation in Fig. 3 are in agreement with the blue lines of theoretical solution v^* for $k = 1, 10$. Figure 12 represents numerical results produced by dynamical system (15) which approximates perfect theoretical solution v^* and strengthens the conclusion that the proposed AF (14) can work for high-order matrices with big condition number and for determinant near to zero.

From previous numerical examples, there is an important advantage of new activation function's (14) application against considered previous AFs. In the frames of this research manuscript, really interesting due to the implementation of evolution formula $\dot{\mathrm{E}}(t) = -k\Phi_m(\mathrm{E}(t))$ for the solution of time-varying quadratic problems which are encountered in many scientific fields, such as A/D converter signal circuits [63] and redundant manipulators [54, 64].

3. Time-Varying Quadratic Minimization Problem

In this survey, the following time-varying convex quadratic program subject to time-varying linear equality constraints from [61, 62] has been considered

$$min : \mathrm{v}^{\mathrm{T}}(t)\mathrm{P}(t)\mathrm{v}(t)/2 + \mathrm{q}^{\mathrm{T}}(t)\mathrm{v}(t), \tag{32}$$

subject to

$$\mathrm{A}(t)\mathrm{v}(t) = \mathrm{b}(t),$$

where matrices $\mathrm{P}(t) \in \mathbb{R}^{\mu\times\mu}$, $\mathrm{A}(t) \in \mathbb{R}^{\nu\times\mu}$ are positive definite and full rank respectively at any time instant $t \geq 0$, $\mathrm{q}(t) \in \mathbb{R}^{\mu}$, $\mathrm{b}(t) \in \mathbb{R}^{\nu}$ are time-variant vector coefficients, and $\mathrm{v}(t) \in \mathbb{R}^{\mu}$ is the unknown vector to be **optimized**. To monitor solving process of (32) under constraint $\mathrm{A}(t)\mathrm{v}(t) = \mathrm{b}(t)$, the following error function has been considered from [61]:

$$e(t) = W(t)y(t) + u(t), \tag{33}$$

where

$$W(t) = \begin{pmatrix} \mathrm{P}(t) & \mathrm{A}^{\mathrm{T}}(t) \\ \mathrm{A}(t) & O \end{pmatrix}, \quad y(t) = \begin{pmatrix} \mathrm{v}(t) \\ \lambda(t) \end{pmatrix},$$

$$u(t) = \begin{pmatrix} \mathrm{q}(t) \\ -\mathrm{b}(t) \end{pmatrix} \in \mathbb{R}^{\mu+\nu}, \tag{34}$$

and $\lambda(t) \in \mathbb{R}^{\nu}$ denotes the Lagrange multiplier vector. Aim of this chapter for minimization problem (32) solving is the construction of an RNN dynamical system, such as error function $e(t) \in \mathbb{R}^{\mu+\nu}$ in (33) must converge to zero matrix. For this reason, it is really interesting the application of EEZNN formula which analyzed in previous section for the solution of matrix equation $W(t)y(t) + u(t) = 0$, such as

$$\dot{e}(t) = -k\Phi_m(e(t)), \tag{35}$$

which implies

$$W(t)\dot{y}(t) = -\dot{W}(t)y(t) - k\Phi_m(W(t)y(t) + u(t)) + \dot{u}(t). \tag{36}$$

Predefined convergence time of dynamical system (36) satisfies the following propositions.

Proposition 9. *With known time-variant matrix $W(t)$ and vector $u(t)$, if the $\Phi_m(u)$ order m by (14) is used, then the neural state $y(t)$ of (36) starting from the initial condition $y(0)$ converges to theoretical solution of $W(t)y(t) = -u(t)$ in predefined time t_x*

$$t_x \le \frac{1}{a_1 k(1-\mathrm{q_m})} + \frac{1}{a_2 k(\mathrm{w}-1)}. \tag{37}$$

Proposition 10. *With known time-variant matrix $W(t)$ and vector $u(t)$, if the $\Phi_m(u)$ order m by (14) is used and especially all elements of the error matrix function (33) satisfy $|e(t)| \le 1$, then the neural state $y(t)$ of (36) starting from the initial condition $y(0)$ converges to theoretical solution of $W(t)y(t) = -u(t)$ in predefined time t_{x_a}:*

$$t_{x_a} \le \frac{1}{a_1 km(1-\mathrm{q_m})} + \frac{1}{a_2 k(\mathrm{w}-1)}. \tag{38}$$

Proofs here are exactly the same as in Theorem 6 and Corollary 7 of previous section and for this reason have been omitted. It must be mentioned that minimization problem (32) can be transformed into optimization problem:

$$max : -\mathrm{v}^{\mathrm{T}}(t)\mathrm{P}(t)\mathrm{v}(t)/2 - \mathrm{q}^{\mathrm{T}}(t)\mathrm{v}(t), \tag{39}$$

and the same solution process of (39) as previously can be applied. For validation of (36), the following experiment has been considered.

3.1. *Numerical experiment*

The following time-variant quadratic problem subject to linear inequalities $\mathrm{A}(t)\mathrm{v}(t) = \mathrm{b}(t)$ from [61] is defined:

$$\begin{aligned} min : &\frac{1}{4}(\cos(6t)+2)\mathrm{v}_1^2(t) + \frac{1}{4}(\sin(6t)+2)\mathrm{v}_2^2(t) + \cos(6t)\mathrm{v}_1(t)\mathrm{v}_2(t) \\ &+ \sin(8t)\mathrm{v}_1(t) + \cos(8t)\mathrm{v}_2(t), \end{aligned} \tag{40}$$

under the subject

$$\begin{cases} \sin(9t)\mathrm{v}_1(t) - \cos(9t)\mathrm{v}_2(t) = \cos(8t), \\ \cos(9t)\mathrm{v}_1(t) + \sin(9t)\mathrm{v}_2(t) = \sin(8t). \end{cases} \tag{41}$$

Applying the previous process for settlement of (40), the following matrices have been denoted:

$$\mathrm{P}(t)=\begin{pmatrix}\frac{1}{2}(\cos(6t)+2) & \cos(6t)\\ \cos(6t) & \frac{1}{2}(\sin(6t)+2)\end{pmatrix},\quad \mathrm{q}(t)=\begin{pmatrix}\sin(8t)\\ \cos(8t)\end{pmatrix}, \tag{42}$$

$$\mathrm{A}(t)=\begin{pmatrix}\sin(9t) & -\cos(9t)\\ \cos(9t) & \sin(9t)\end{pmatrix},\quad \mathrm{b}(t)=\begin{pmatrix}\cos(8t)\\ \sin(8t)\end{pmatrix},\quad O=\begin{pmatrix}0 & 0\\ 0 & 0\end{pmatrix}, \tag{43}$$

which are used for the construction of $W(t), u(t)$, respectively. In this step, for the solution of (40) under the subject (41), vector $y(t)$ from matrix equation $W(t)y(t)+u(t)=0$ must be computed where

$$y(t)=\begin{pmatrix}\mathrm{v}_1(t)\\ \mathrm{v}_2(t)\\ \lambda_1(t)\\ \lambda_2(t)\end{pmatrix}. \tag{44}$$

For the parameters $\mathrm{q}_1=0.1, \mathrm{q}_2=0.2, \mathrm{q}_3=0.3, \mathrm{q}_4=0.4, \mathrm{q}_5=0.5, \mathrm{q}_6=0.6, \mathrm{w}=2.5$, $k=1,5,10$, and zero initial condition, Simulink implementation from Fig. 13 extracts Figs. 14–17.

Figure 14 represents the residual error $||W(t)y(t)+u(t)||_F$ which vanishes to zero with a faster rate for increasing values of parameter m, while k is equal to 1, 5, and 10. Figure 15 strengthens the fact that solutions $\mathrm{v}_1(t), \mathrm{v}_2(t)$ produced by EEZNN (red line) converge to theoretical solutions $\mathrm{v}_1^*(t), \mathrm{v}_2^*(t)$ (blue dashed lines) of minimization problem (40). Figure 16 represents the Lagrange multipliers $\lambda_1(t)$ and $\lambda_2(t)$ which are produced by EEZNN formula based on EESAF_6 for k =1 and 10. It is important to mention that EEZNN formula can solve the quadratic minimization problem for a different initial condition than zero as depicts in Fig. 17. So one more important advantage of the proposed EEZNN dynamical system is that it can be used for time-varying quadratic problems solving.

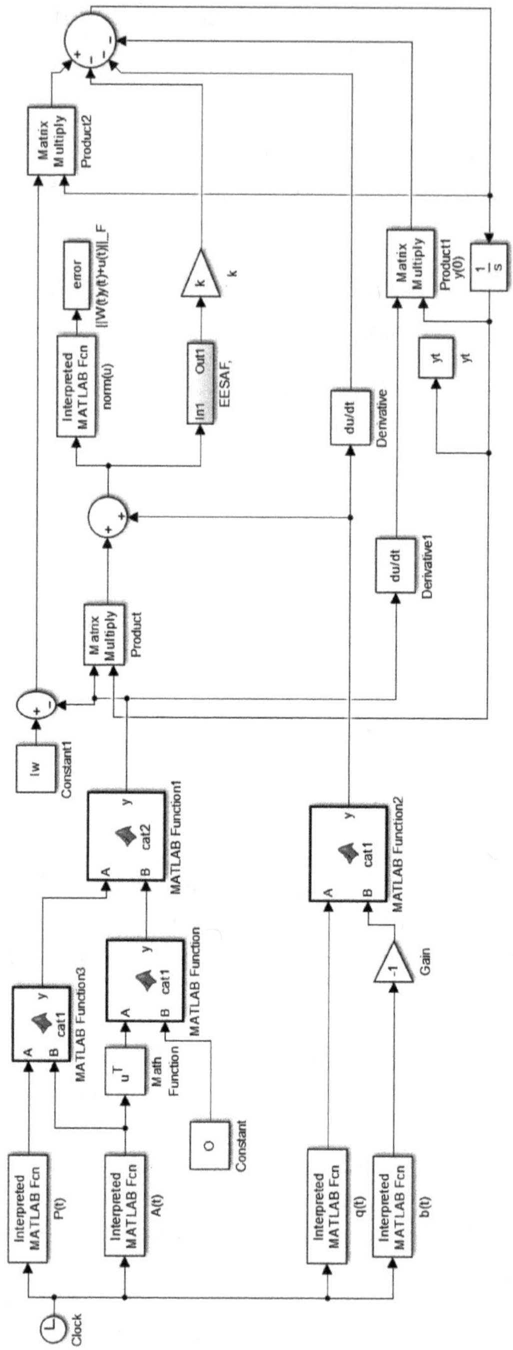

Fig. 13. Simulink model of dynamical system (36).

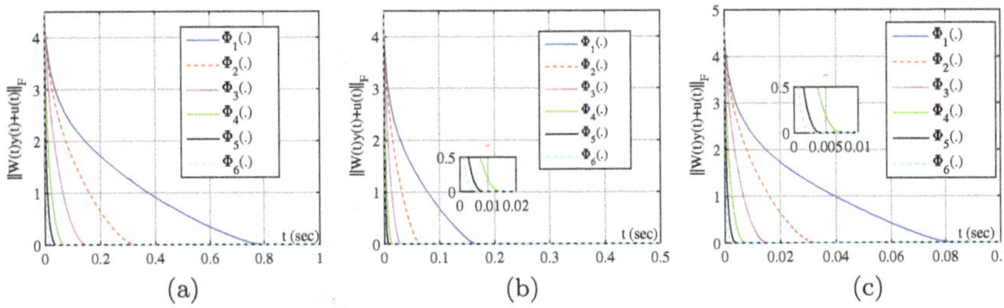

Fig. 14. Error behavior by $\Phi_m(u)$ under (a) $k = 1$, (b) $k = 5$, and (c) $k = 10$ for minimization problem (40).

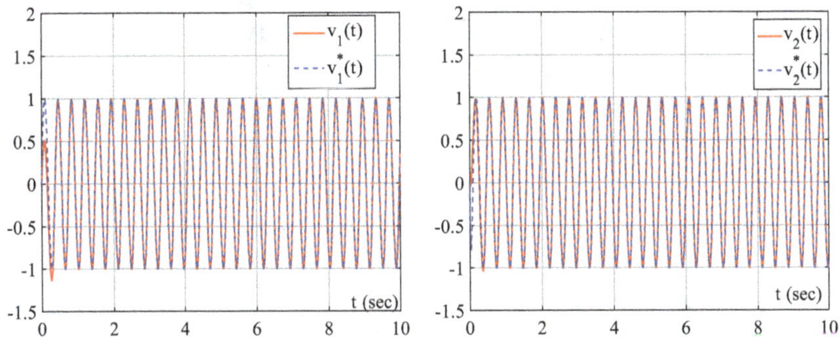

Fig. 15. State trajectories of EEZNN (36).

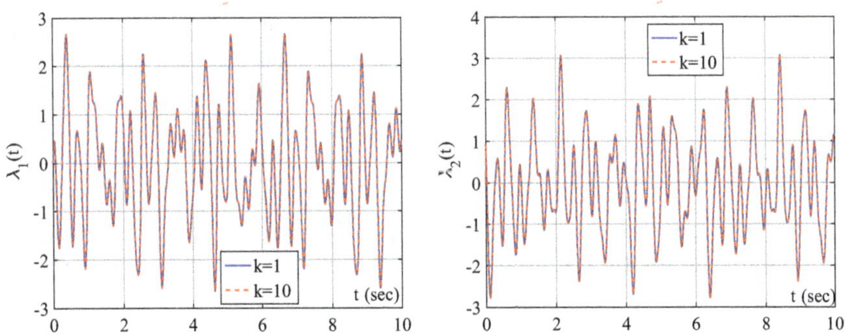

Fig. 16. Lagrange multiplier solution by $\Phi_6(u)$ for $k = 1$ and $k = 10$.

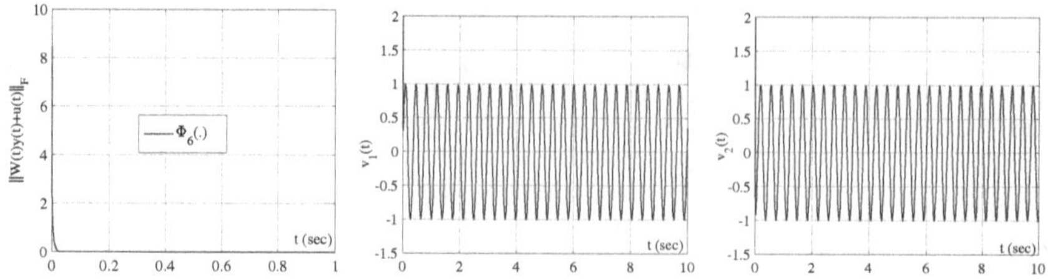

(a) (b) (c)

Fig. 17. For $k = 1$ and $y(0) = 2$, (a) error behavior by $\Phi_6(u)$, (b) $\mathrm{v}_1(t)$, and (c) $\mathrm{v}_2(t)$.

Previous results indicate that the proposed family of EEZNN model can be efficiently applied for the study of important numerical linear algebra's problems and mathematical modeling of these in science and engineering which involves systems of linear equation solving.

4. Conclusion

An extended exponential zeroing neural network, initiated by new activation function, named extended exponential sign activation function (EESAF) is proposed for the solution of linear algebraic systems $\mathrm{Av} = \mathrm{b}$ and time-varying quadratic problems which play important role in numerical linear algebra sector. It is verified theoretically and numerically that the neural state $\mathrm{v}(t)$ converges to theoretical solution of (8) in predefined time period. Convergence rate can be changed effectively by increasing the number of parameters m which are used in novel AF (14). Numerical experiments strengthen the conclusion that EEZNN evolution dynamics gives better results in a finite time estimation and the extended exponential sign activation function order m implies faster convergence property against previous Li, tunable, and VAF activation functions. One more advantage of the proposed EEZNN model is that it can be applied for quadratic minimization problem-solving. Future directions of this research survey will include the following:

- extension of the proposed EEZNN formula in computation of generalized inverses,

- application of EEZNN formula in solving algebraic Riccati equations which play important role in optimal control theory problems,
- application of EEZNN dynamical system for linear time-variant matrix Diophantine equation solving,
- extension of the proposed EEZNN model in finding solutions of tensor equations with Einstein product,
- discretization of EEZNN formula for greater accuracy in already existing results,
- application of EEZNN model in the study of Stewart's platform which plays an important role in communication systems.

References

[1] Chen, K. Robustness analysis of Wang neural network for online linear equation solving, *Electronics Letters*, 48(22), 1391–1392 (2012).

[2] Zhang, Y., Chen, K. and Tan, H.Z. Performance analysis of gradient neural network exploited for online time-varying matrix inversion, *IEEE Transactions on Automatic Control*, 54(8), 1940–1945 (2009).

[3] Wang, J. Electronic realisation of recurrent neural network for solving simultaneous linear equations, *Electronics Letters*, 28(5), 493–495 (1992).

[4] Osowski, S. Neural networks in interpolation problems, *Neurocomputing*, 5(2–3), 105–118 (1993).

[5] Wang, J. A recurrent neural network for real-time matrix inversion, *Applied Mathematics and Computation*, 55(1), 89–100 (1993).

[6] Raida, Z. Improvement of convergence properties of Wang neural network, *Electronics Letters*, 30(22), 1865–1866 (1994).

[7] Zhang, Y., Jiang, D. and Wang, J. A recurrent neural network for solving Sylvester equation with time-varying coefficients, *IEEE Transactions on Neural Networks*, 13(5), 1053–1063 (2002).

[8] Zhang, Y., Qiu, B., Jin, L., Guo, D. and Yang, Z. Infinitely many Zhang functions resulting in various ZNN models for time-varying matrix inversion with link to Drazin inverse, *Information Processing Letters*, 115(9), 703–706 (2015).

[9] Zhang, Y., Yi, C. and Ma, W. Simulation and verification of Zhang neural network for online time-varying matrix inversion, *Simulation Modelling Practice and Theory*, 17(10), 1603–1617 (2009).

[10] Zhang, Y. and Ge, S.S. Design and analysis of a general recurrent neural network model for time-varying matrix inversion, *IEEE Transactions on Neural Networks*, 16(6), 1477–1490 (2005).

[11] Zhang, Y., Guo, X., Ma, W., Chen, K. and Cai, B. MATLAB simulink modeling and simulation of Zhang neural network for online time-varying matrix inversion, *IEEE International Conference on Networking, Sensing and Control (ICNSC)*, 1480–1485 (2008).

[12] Zhang, Y., Li, Z. and Li, K. Complex-valued Zhang neural network for online complex-valued time-varying matrix inversion, *Applied Mathematics and Computation*, 217(24), 10066–10073 (2011).

[13] Guo, D. and Zhang, Y. Li-function activated ZNN with finite-time convergence applied to redundant-manipulator kinematic control via time-varying Jacobian matrix pseudoinversion, *Applied Soft Computing*, 24, 158–168 (2014).

[14] Lv, X., Xiao, L. and Tan, Z. Improved Zhang neural network with finite-time convergence for time-varying linear system of equations solving, *Information Processing Letters*, 147, 88–93 (2019).

[15] Xiao, L., Li, K., Tan, Z., Zhang, Z., Liao, B., Chen, K., Jin, L. and Li, S. Nonlinear gradient neural network for solving system of linear equations, *Information Processing Letters*, 142, 35–40 (2019).

[16] Xiao, L. A finite-time convergent Zhang neural network and its application to real-time matrix square root finding, *Neural Computing and Applications*, 31(2), 793–800 (2019).

[17] Zhang, Y., Ding, Y., Qiu, B., Zhang, Y. and Li, X. Signum-function array activated ZNN with easier circuit implementation and finite-time convergence for linear systems solving, *Information Processing Letters*, 124, 30–34, (2017).

[18] Ding, L., Xiao, L., Zhou, K., Lan, Y. and Zhang, Y. Two novel finite time convergent recurrent neural networks for tackling complex-valued systems of linear equation, *Filomat*, 34(15), 5009–5018 (2020).

[19] Miao, P., Shen, Y., Huang, Y. and Wang, Y.W. Solving time-varying quadratic programs based on finite-time Zhang neural networks and their application to robot tracking, *Neural Computing and Applications*, 26(3), 693–703 (2015).

[20] Jian, Z., Xiao, L., Dai, J., Tang, Z. and Liu, C. Design and analysis of new zeroing neural network models with improved finite-time convergence for time-varying reciprocal of complex matrix, *IEEE Transactions on Industrial Informatics*, 16(6), 3838–3848 (2019).

[21] Xiao, L., Zhang, Y., Dai, J., Zuo, Q. and Wang, S. Comprehensive analysis of a new varying parameter zeroing neural network for time

varying matrix inversion, *IEEE Transactions on Industrial Informatics*, 17(3), 1604–1613 (2020).

[22] Stojanović, I., Stanimirović, P.S., Zivković, I.S., Gerontitis, D. and Wang, X. Z. ZNN models for computing matrix inverse based on hyperpower iterative methods, *Filomat*, 31(10), 2999–3014 (2017).

[23] Xiao, L. A new design formula exploited for accelerating Zhang neural network and its application to time-varying matrix inversion, *Theoretical Computer Science*, 647, 50–58 (2016).

[24] Kong, Y., Lu, H.J., Xue, Y. and Xia, X.H. Terminal neural computing: Finite time convergence and its applications, *Neurocomputing*, 217, 133–144 (2016).

[25] Xiao, L., Zhang, Y., Li, K., Liao, B. and Tan, Z. A novel recurrent neural network and its finite-time solution to time-varying complex matrix inversion, *Neurocomputing*, 331, 483–492 (2019).

[26] Xiao, L. Accelerating a recurrent neural network to finite-time convergence using a new design formula and its application to time-varying matrix square root, *Journal of the Franklin Institute*, 354(13), 5667–5677 (2017).

[27] Zhang, Z., Deng, X., Qu, X., Liao, B., Kong, L.D. and Li, L. A varying-gain recurrent neural network and its application to solving online time-varying matrix equation, *IEEE Access*, 6, 77940–77952 (2018).

[28] Stanimirović, P.S., Gerontitis, D., Tzekis, P., Behera, R. and Sahoo, J.K. Simulation of varying parameter recurrent neural network with application to matrix inversion, *Mathematics and Computers in Simulation*, 185, 614–628 (2021).

[29] Gerontitis, D., Behera, R., Sahoo, J.K. and Stanimirović, P.S. Improved finite-time zeroing neural network for time-varying division, *Studies in Applied Mathematics*, 146(2), 526–549 (2021).

[30] Mo, C., Gerontitis, D. and Stanimirović, P.S. Solving the time-varying tensor square root equation by varying-parameters finite-time Zhang neural network, *Neurocomputing*, 445, 309–325 (2021).

[31] Mishra, J.P., Li, C., Yu, X. and Jalili, M. Fixed-time converging terminal surface with non-singular control design for second-order systems, *IFAC-PapersOnLine*, 50(1), 5139–5143 (2017).

[32] Polyakov, A. Nonlinear feedback design for fixed-time stabilization of linear control systems, *IEEE Transactions on Automatic Control*, 57(8), 2106–2110 (2011).

[33] Aouiti, C. and Miaadi, F. A new fixed-time stabilization approach for neural networks with time-varying delays, *Neural Computing and Applications*, 32(8), 3295–3309 (2020).

[34] Xiao, L., Zhang, Y., Dai, J., Chen, K., Yang, S., Li, W., Liao, B., Ding, L. and Li, J. A new noise-tolerant and predefined-time ZNN model

for time-dependent matrix inversion, *Neural Networks*, 117, 124–134 (2019).

[35] Dai, J., Jia, L. and Xiao, L. Design and analysis of two prescribed-time and robust ZNN models with application to time-variant Stein matrix equation, *IEEE Transactions on Neural Networks and Learning Systems*, 32(4), 1668–1677 (2020).

[36] Tan, Z., Hu, Y. and Chen, K. On the investigation of activation functions in gradient neural network for online solving linear matrix equation, *Neurocomputing*, 413, 185–192 (2020).

[37] Jin, J. and Gong, J. A noise-tolerant fast convergence ZNN for dynamic matrix inversion, *International Journal of Computer Mathematics*, 98(11), 2202–2219 (2021).

[38] Jin, J. A robust zeroing neural network for solving dynamic nonlinear equations and its application to kinematic control of mobile manipulator, *Complex & Intelligent Systems*, 7(1), 87–99 (2021).

[39] Xiao, L., Cao, Y., Dai, J., Jia, L. and Tan, H. Finite-time and predefined-time convergence design for zeroing neural network: Theorem, method, and verification, *IEEE Transactions on Industrial Informatics*, 17(7), 4724–4732 (2020).

[40] Ding, F. and Chen, T. Gradient based iterative algorithms for solving a class of matrix equations, *IEEE Transactions on Automatic Control*, 50(8), 1216–1221 (2005).

[41] Ding, F. and Zhang, H. Gradient-based iterative algorithm for a class of the coupled matrix equations related to control systems, *IET Control Theory & Applications*, 8(15), 1588–1595 (2014).

[42] Li, Z.Y., Zhou, B., Lam, J. and Wang, Y. Positive operator based iterative algorithms for solving Lyapunov equations for Ito stochastic systems with Markovian jumps, *Applied Mathematics and Computation*, 217(21), 8179–8195 (2011).

[43] Dai, J., Chen, Y., Xiao, L., Jia, L. and He, Y. Design and analysis of a hybrid GNN-ZNN model with a fuzzy adaptive factor for matrix inversion, *IEEE Transactions on Industrial Informatics*, 18(4), 2434–2442 (2021).

[44] Zhang, Y., Fan, Z. and Li, Z. Zhang neural network for online solution of time-varying Sylvester equation, In: *Advances in Computation and Intelligence: Second International Symposium. Berlin Springer* (ISICA), 4683, 276–285, (2007).

[45] Li, S., Chen, S. and Liu, B. Accelerating a recurrent neural network to finite-time convergence for solving time-varying Sylvester equation by using a sign-bi-power activation function, *Neural Processing Letters*, 37(2), 189–205 (2013).

[46] Khalil, H.K. *Nonlinear Systems*, 3rd edition. Prentice Hall (2002).

[47] Greenberg, B.G. and Sarhan, A.E. Matrix inversion, its interest and application in analysis of data, *Journal of the American Statistical Association*, 54(288), 755–766 (1959).

[48] Wu, M., Yin, B., Vosoughi, A., Studer, C., Cavallaro, J.R. and Dick, C. Approximate matrix inversion for high-throughput data detection in the large-scale MIMO uplink, *IEEE International Symposium on Circuits and Systems (ISCAS)*, 2155–2158 (2013).

[49] Bernstein, D.S., *Matrix mathematics*, Princeton University Press (2009).

[50] Sturges, R.H. Analog matrix inversion (robot kinematics), *IEEE Journal of Robotics and Automation*, 4(2), 157–162 (1988).

[51] Zhang, Y. Towards piecewise-linear primal neural networks for optimization and redundant robotics, *IEEE International Conference on Networking, Sensing and Control*, 374–379 (2006).

[52] Ge, S.S., Lee, T.H. and Harris, C.J. *Adaptive Neural Network Control of Robotic Manipulators*, London, U.K., World Scientific, 19 (1998).

[53] Kong, Y., Hu, T., Lei, J. and Han, R. A Finite-time convergent neural network for solving time-varying linear equations with inequality constraints applied to redundant manipulator, *Neural Processing Letters*, 54(1), 1–20 (2021).

[54] Li, Z and Li, S. Kinematic control of manipulator with remote center of motion constraints synthesised by a simplified recurrent neural network, *Neural Processing Letters*, 54(2), 1035–1054 (2021).

[55] Tsitsas, N.L., Kaklamani, D.I. and Uzunoglu, N.K. Rigorous integral equation analysis of nonsymmetric coupled grating slab waveguides, *Journal of the Optical Society of America A*, 23(11), 2888–2905 (2006).

[56] Tsitsas, N.L., Kaklamani, D.I. and Uzunoglu, N.K. Integral equation analysis of coupling in symmetric grating-assisted optical waveguides, *Journal of the Optical Society of America A*, 23(6), 1488–1502 (2006).

[57] Tsitsas, N.L. and Uzunoglu, N.K. Scattering by a grating slab waveguide with regular plane regions grooves: Integral equation modeling, *IEEE Transactions on Magnetics*, 45(3), 1080–1083 (2009).

[58] Gerontitis, D.K. and Tsitsas, N.L. Scattering by an all-dielectric metasurface including a periodic arrangement of arbitrary scatterers, *IEEE International Conference on Microwaves, Antennas, Communications and Electronic Systems (COMCAS)*, 1–4 (2019).

[59] Tsitsas, N.L. Second-kind Fredholm integral-equation analysis of scattering by layered dielectric gratings, *IET Microwaves, Antennas & Propagation*, 15(10), 1194–1205 (2021).

[60] Hadjifotinou, K.G. and Karampetakis, N.P. An algorithm for two-variable rational interpolation suitable for matrix manipulations with the evaluation-interpolation method, *Numerical Linear Algebra with Applications*, 28(5), e2378 (2021).

[61] Li, Z. and Zhang, Y. Time-varying quadratic programming by Zhang neural network equipped with a time-varying design parameter $\gamma(t)$, *International Symposium on Neural Networks (ISNN) Springer*, Berlin, Heidelberg, 101–108 (2011).

[62] Jin, L., Zhang, Y., Li, S. and Zhang, Y. Modified ZNN for time-varying quadratic programming with inherent tolerance to noises and its application to kinematic redundancy resolution of robot manipulators, *IEEE Transactions on Industrial Electronics*, 63(11), 6978–6988 (2016).

[63] Tank, D. and Hopfield, J. Simple 'neural' optimization network: An A/D converter signal decision circuit and a linear programming circuit, *IEEE Transactions on Circuits and Systems*, 33(5), 533–541 (1986).

[64] Wang, J. and Zhang, Y. Recurrent neural networks for real-time computation of inverse kinematics of redundant manipulators, *Machine Intelligence: Quo Vadis?*, 299–319 (2004).

https://doi.org/10.1142/9789811293009_0016

Chapter 16

Solution Estimates and Stability Conditions for Linear Differential Equations with Unbounded Operators in a Banach Space

Michael Gil'

Department of Mathematics
Ben Gurion University of the Negev
P.0. Box 653, Beer-Sheva 84105, Israel
gilmi@bezeqint.net

This chapter is a brief survey of the results of the author on solution estimates for linear differential equations with constant and variable unbounded operators on Banach and Hilbert spaces and their applications to the stability theory. In particular, we consider a semigroup generated by the sum of operators A and B. Besides, the solution estimates and stability conditions are derived in terms of the commutator $AB-BA$. We also consider a differential equation in a Hilbert space with an unbounded variable operator coefficient having a bounded derivative.

An essential part of this chapter is devoted to the equation $du(t)/dt = (A+B(t))u(t)$ $(t \geq 0)$ with an unbounded constant operator A and a bounded variable operator $B(t)$. In particular, we suggest stability conditions in terms of the commutator $AB(t) - B(t)A$. In addition, we investigate the above-pointed equation, assuming that $B(t)$ satisfies the Lipschitz condition and the generalized Lipschitz condition

$\|B(t) - B(\tau)\| \le q(|t - \tau|)$ $(t, \tau \ge 0)$, where $q(.)$ is a non-negative continuous function of a non-negative argument.

Illustrative examples are also given. Some of the results presented in this chapter are new.

1. Introduction

This chapter is a brief survey of the results of the author on solution estimates for linear differential equations with constant and variable unbounded operators in a Banach and Hilbert spaces. The obtained estimates give us stability conditions under various assumptions.

The literature on the asymptotic behavior of solutions and stability of abstract linear differential equations is very rich, cf. [1,2,6,7,9, 12,13,15,31,33–35,38]. In addition, the works [3,4,8,10,27,29,36] and references given therein deal with the stability of strongly continuous operator semigroups (C_0-semigroups), which are deeply connected with the equations with constant operators. The problem of stability analysis of differential equations continues to attract the attention of many specialists despite its long history. It is still one of the most burning problems of the theory of differential equations because of the absence of its complete solution.

Let $\mathcal{X}$ be a Banach space with a norm $\|.\|$. For a linear operator C, $\mathrm{Dom}(C)$ is the domain, $\sigma(C)$ is the spectrum, and $\alpha(C) = \sup \mathrm{Re}\, \sigma(C)$. By $\mathcal{L}(\mathcal{X})$ we denote the set of all bounded operators in $\mathcal{X}$. If $C \in \mathcal{L}(\mathcal{X})$, then $\|C\|$ is its operator norm.

The survey consists of 15 sections. Sections 2–7 deal with constant operators, and in Sections 8–15, we consider equations with variable operators.

In Sections 2–6, we investigate a semigroup generated by the sum of two operators A and B. Besides, the solution estimates and stability conditions are derived in terms of the commutator $AB - BA$. In Section 2, the commutator is assumed to be unbounded, and in Sections 3–6, it is assumed to be bounded. It should be noted that, to the best of our knowledge, stability conditions for semigroups in terms of the commutator have been considered only in [21, 22, 24]. They are discussed below.

As is well known [10], if A generates a semigroup, in many cases, the condition $\alpha(A) < 0$ provides the stability of that semigroup. In Section 7, we consider the following problem: let an operator A with a Hilbert–Schmidt resolvent be stable, i.e., $\alpha(A) < 0$. What conditions provide the stability of the perturbed operator?

Furthermore, one of the important classes of ordinary differential equations (ODEs) is the class of linear equations with slowly time-varying coefficients, cf. [5]. One of the results on the stability of ODE with slowly time-varying coefficients has been published in [16] (see also [19]). In Section 8, the main result from [16] is extended to a differential equation in a Hilbert space with an unbounded variable operator coefficient having a bounded derivative.

In Section 9, we consider the equation

$$du(t)/dt = (A + B(t))u(t) \quad (t \geq 0) \tag{1}$$

with a constant unbounded operator A and a bounded variable operator $B(t)$. For that equation, we suggest stability conditions in terms of the commutator $AB(t)-B(t)A$. Besides, we particularly generalize some results from Sections 2–6.

In Sections 10 and 11, we extend the well-known Dyson–Phillips theorem [11, Theorem 1.10] for semigroups to equations with variable operators.

Section 11 deals with equation (1) with operator $B(t)$ having the Lipschitz property $\|B(t) - B(s)\| \leq \text{const}|t - s|$ $(t, s \geq 0)$.

In Sections 12–15, we consider equation (1), assuming that $B(t)$ satisfies the generalized Lipschitz condition $\|B(t)-B(\tau)\| \leq q(|t-\tau|)$ $(t, \tau \geq 0)$, where $q(.)$ is a non-negative continuous function of a non-negative argument.

2. A Semigroup Generated by the Sum of Two Operators with an Unbounded Commutator

Throughout this section, A is the generator of an analytic semigroup e^{At} $(t \geq 0)$ on $\mathcal{X}$, and B is a bounded linear operator in $\mathcal{X}$ mapping $\text{Dom}(A)$ into itself.

Due to [11, Corollary 1.3.5], $A + B$ generates a C_0 semigroup $e^{(A+B)t}$.

Put $K := AB - BA$ and assume that there is a linear operator S having a bounded left-inverse operator S_l^{-1} (i.e., $S_l^{-1}Sx = x, x \in \mathrm{Dom}(A)$) such that

$$m(K,S) := \|KS_l^{-1}\| < \infty \tag{2}$$

and

$$J(S) := \int_0^\infty \|e^{Bt}\| \, \|Se^{At}\| dt < \infty. \tag{3}$$

From (3), the integral $J_0 := \int_0^\infty \|e^{Bt}\|\|e^{At}\|dt$ converges, since

$$J_0 = \int_0^\infty \|e^{Bt}\|\|S_l^{-1}Se^{At}\|dt \leq \|S_l^{-1}\|J(S).$$

Put $X(t) = e^{(A+B)t}$, $Y(t) = e^{At}e^{Bt}$ $(t \geq 0)$,

$$\|X\|_{C(0,\infty)} := \sup_{s\geq 0} \|X(s)\| \text{ and } \|X\|_{L^1(0,\infty)} := \int_0^\infty \|X(s)\|ds.$$

Clearly, $\|Y\|_{L^1(0,\infty)} \leq J_0$ and $\|Y\|_{C(0,\infty)} < \infty$, provided condition (3) holds.

The following result has been proved in [24].

Theorem 1. *Let the conditions* (2), (3), *and*

$$m(K,S)J_0J(S) < 1 \tag{4}$$

hold. Then

$$\|Y - X\|_{C(0,\infty)} \leq \frac{m(K,S)J_0J(S)\|Y\|_{C(0,\infty)}}{1 - m(K,S)J_0J(S)} \tag{5}$$

and

$$\|Y - X\|_{L^1(0,\infty)} \leq \frac{m(K,S)J_0^2J(S)}{1 - m(K,S)J_0J(S)}. \tag{6}$$

This theorem is sharp: if A and B commute, then Theorem 1 gives us the equality $X(t) = Y(t)$. Furthermore, since $\|Y\|_{L^1(0,\infty)} \leq J_0$, from (5) and (6), it directly follows that

$$\|X\|_{C(0,\infty)} \leq \frac{\|Y\|_{C(0,\infty)}}{1 - m(K,S)J_0 J(S)}$$

and

$$\|X\|_{L^1(0,\infty)} \leq \frac{J_0}{1 - m(K,S)J_0 J(S)}.$$

For example, if $-A$ is sectorial and $\alpha(A) < \alpha_A < 0$, then as is well known [28, Th. 1.4.3, p. 26], $\|e^{At}\| \leq c_A e^{\alpha_A t}$ ($c_A = const \geq 1$; $t \geq 0$) and for any $\nu \in (0,1)$,

$$\|(-A)^\nu e^{At}\| \leq c_\nu t^{-\nu} exp\,(-\delta_\nu t) \ \ (0 < \delta_\nu \leq |\alpha_A|; c_\nu = const \geq 1; \ \ t \geq 0).$$

Since B is bounded, there are constants $c_B = const \geq 1$ and $\alpha_B \in \mathbf{R}$, such that $\|e^{Bt}\| \leq c_B e^{\alpha_B t}$ ($t \geq 0$). So if $|\alpha_A| \geq \delta_\nu$ and $-\delta_\nu + \alpha_B < 0$, in the considered case,

$$J_0 = \int_0^\infty \|e^{Bt}\|\|e^{At}\|dt \leq c_A c_B \int_0^\infty e^{(\alpha_A+\alpha_B)t}dt = \frac{c_A c_B}{|\alpha_A + \alpha_B|}$$

and

$$J(S) = J_\nu := \int_0^\infty \|e^{Bt}\|\|(-A)^\nu e^{At}\|dt \leq c_\nu c_B \int_0^\infty t^{-\nu} e^{(\alpha_B - \delta_\nu)t}dt = \frac{c_A c_B \Gamma(1-\nu)}{|\delta_\nu - \alpha_B|^{1-\nu}},$$

where $\Gamma(1+x) = \int_0^\infty s^x e^{-s} ds$ ($x \in (-1,\infty)$) is the Euler gamma function.

Note that for any selfadjoint negative definite operator A in a Hilbert space, one has $\|f(A)\| = \sup_{s \leq \alpha(A)} |f(s)|$, for a function f bounded on $\sigma(A)$. Consequently, $\|e^{At}\| = e^{\alpha(A)t}$ ($t \geq 0$) and

$$\|(-A)^\nu e^{At}\| = \sup_{s \leq \alpha(A)} (-s)^\nu e^{st} = \phi_\nu(A,t) \ \ (0 < \nu < 1),$$

where

$$\phi_\nu(A,t) = \begin{cases} (\nu/t)^\nu e^{-\nu} & \text{if } t \le \frac{\nu}{|\alpha(A)|}, \\ |\alpha(A)|^\nu e^{\alpha(A)t} & \text{if } t \ge \frac{\nu}{|\alpha(A)|}. \end{cases}$$

So if A is negative definite and $S = (-A)^\nu$, then

$$J_0 \le \int_0^\infty \|e^{(B+\alpha(A))t}\| dt \text{ and } J(S) \le \int_0^\infty \|e^{Bt}\| \phi_\nu(A,t) dt,$$

provided that the integrals converge.

3. Equations with Constant Operators and Bounded Commutators

In this section, we refine Theorem 1 in the case when the commutator is bounded. That assumption enables us to remove the requirement that the semigroup generated by A is analytic.

Recall that $\mathcal{L}(\mathcal{X})$ is the set of all bounded operators in $\mathcal{X}$. If $C \in \mathcal{L}(\mathcal{X})$, then $\|C\|$ is the operator norm of C.

Throughout this section, A is the generator of a C_0-semigroup $(e^{At})_{t\ge0}$ on $\mathcal{X}$, and B is in $\mathcal{L}(\mathcal{X})$ and maps $\mathrm{Dom}(A)$ into itself. Due to [11, Corollary 1.3.5], $A+B$ generates a C_0 semigroup $e^{(A+B)t}$.

Assume that

$$J_0 = \int_0^\infty \|e^{At}\| \|e^{Bt}\| dt < \infty \tag{7}$$

and the commutator

$$AB - BA \text{ is bounded,} \tag{8}$$

i.e., the norms of $(AB-BA)x$ ($x \in \mathrm{Dom}(A)$, $\|x\| = 1$) are uniformly bounded. So $AB - BA$ can be extended to $\mathcal{X}$ as a bounded operator. That extension is denoted by K.

Again put $X(t) = e^{(A+B)t}$, $Y(t) = e^{At}e^{Bt}$ ($t \ge 0$), and define $\|X\|_{C(0,\infty)}$ and $\|X\|_{L^1(0,\infty)}$ as in the previous section.

Clearly, $\|Y\|_{C(0,\infty)} < \infty$ and $\|Y\|_{L^1(0,\infty)} < J_0$, provided condition (7) holds.

Theorem 2. *Let the conditions* (7), (8), *and*

$$\|K\| J_0^2 < 1$$

hold. Then

$$\|Y - X\|_{C(0,\infty)} \le \frac{\|K\| J_0^2 \|Y\|_{C(0,\infty)}}{1 - \|K\| J_0^2} \tag{9}$$

and

$$\|Y - X\|_{L^1(0,\infty)} \le \frac{J_0^3 \|K\|}{1 - \|K\| J_0^2}. \tag{10}$$

This theorem is proved in [21]. It is sharp: if A and B commute, then Theorem 2 gives us the equality $X(t) = Y(t)$. Furthermore, since $\|Y\|_{L^1(0,\infty)} < J_0$, from (9) and (10), it directly follows that

$$\|X\|_{C(0,\infty)} \le \frac{\|Y\|_{C(0,\infty)}}{1 - \|K\| J_0^2}$$

and

$$\|X\|_{L^1(0,\infty)} \le \frac{J_0}{1 - \|K\| J_0^2}.$$

For instance, let

$$\|e^{At}\| \le c_A e^{\alpha_A t}, \|e^{Bt}\| \le c_B e^{\alpha_B t} \ \ (c_A, c_B = const \ge 1;$$
$$\alpha_A, \alpha_B \text{ are real constants}; t \ge 0),$$

and $\alpha_A + \alpha_B < 0$. Then

$$J_0 \le \frac{c_A c_B}{|\alpha_A + \alpha_B|}.$$

4. Exponential Stability Conditions via Commutators for Semigroups on a Hilbert Space

Let $\mathcal{H}$ be a complex Hilbert space $\mathcal{H}$ with a scalar product $\langle .,. \rangle$ and the norm $\|.\| = \sqrt{\langle .,. \rangle}$. Throughout this section, A and B are linear operators on $\mathcal{H}$ with $\mathrm{Dom}(B) \supseteq \mathrm{Dom}(A)$. In addition, it is supposed that A and $A+B$ generate C_0-semigroups e^{At} and $e^{t(A+B)}$, respectively.

We consider the following problem: let e^{At} be exponentially stable, i.e.,

$$\|e^{At}\| \leq ce^{-\nu t} \quad (t \geq 0; c = \text{const} \geq 1, \nu = \text{const} > 0).$$

What are the conditions that provide the exponential stability of $e^{t(A+B)}$? In this section, we suggest exponential stability conditions for $e^{t(A+B)}$, assuming that

$$B\mathrm{Dom}(A^2) \subseteq \mathrm{Dom}(A), \tag{11}$$

and the operator

$$[A, B] = AB - BA \text{ is bounded.} \tag{12}$$

Besides, B can be unbounded. Since A generates a C_0-semigroup, $\mathrm{Dom}(A^2)$ is dense in $\mathcal{H}$, cf. [30, Theorem I.2.3]. So the operators AB and BA are defined on $\mathrm{Dom}(A^2)$. Thus, (12) means that $[A, B]$ is defined and uniformly bounded on $\mathrm{Dom}(A^2)$ and therefore admits the bounded extension, which again is denoted by K.

Introduce the operator

$$W := \int_0^\infty e^{A^*t} e^{At} dt.$$

This integral converges in the operator norm, since e^{At} is exponentially stable, and

$$\|W\| \leq \int_0^\infty \|e^{At}\|^2 dt \leq c^2 \int_0^\infty e^{-2\nu t} dt = \frac{c^2}{2\nu}.$$

The integral

$$\zeta(A) := 2 \int_0^\infty \|e^{At}\| \int_0^t \|e^{sA}\| \|e^{(t-s)A}\| ds\, dt$$

also converges, and

$$\zeta(A) \leq 2c^3 \int_0^\infty e^{-\nu t} \int_0^t e^{-\nu s} e^{-\nu(t-s)} ds\, dt = 2c^3 \int_0^\infty e^{-2\nu t} t dt = \frac{c^3}{2\nu^2}. \tag{13}$$

In addition, assume that

$$\Lambda(B) := \sup_{h\in \mathrm{Dom}(B);\|h\|=1} \mathrm{Re}\,\langle Bh, h\rangle < \infty \tag{14}$$

and put

$$\psi(W,B) := \begin{cases} 2\Lambda(B)\|W\| & \text{if } \Lambda(B) > 0,\\ 0 & \text{if } \Lambda(B) \le 0. \end{cases}$$

Theorem 3. *Let conditions* (11), (12), *and* (14) *hold, and let* e^{At} *be exponentially stable. If, in addition,*

$$\psi(W,B) + \|K\|\zeta(A) < 1,$$

then $e^{t(A+B)}$ *is also exponentially stable.*

This theorem is proved in [22, Theorem 1.1]. It is sharp. Indeed, let A and B be commuting normal operators, with $\alpha(A) := \sup \mathrm{Re}\,\sigma(A) < 0$. $\Lambda(B) = \alpha(B) > 0$. Then $\|e^{At}\| = e^{\alpha(A)t}$ $(t \ge 0)$, $\|W\| \le \frac{1}{2|\alpha(A)|}$. Consequently,

$$\psi(W,B) = \frac{\alpha(B)}{|\alpha(A)|}.$$

By Theorem 3, $e^{t(A+B)}$ is stable if $\alpha(B) < |\alpha(A)|$. But $\|e^{t(A+B)}\| = e^{(\alpha(A)+\alpha(B))t}$ $(t \ge 0)$. Therefore, in the considered case. $e^{t(A+B)}$ is stable, provided $\alpha(A) + \alpha(B) < 0$. So Theorem 3 is really sharp.

5. Example to Theorem 3

Let $\mathcal{H} = L^2(0,1)$, where $L^2(0,1)$ is the space of square-integrable functions defined on $[0,1]$ with the traditional scalar product. Let $a(x)$ be a complex-valued function having a bounded measurable derivative, b be a complex constant,

$$(Af)(x) = \frac{d^2 f(x)}{dx^2} + a(x)f(x) \text{ and } (Bf)(x) = bf'(x)$$
$$(0 < x < 1, f \in \mathrm{Dom}\,(A))$$

with

$$\mathrm{Dom}\,(A) = \{h \in L^2(0,1) : h'' \in L^2(0,1), h(0) = h(1) = 0\}.$$

Then the commutator is defined by $(Kf)(x) = -ba'(x)f(x)$ and $\|K\| = |b|\sup_x |a'(x)|$. Assume that $q := \max_x \operatorname{Re} a(x) < \pi^2$. Since the largest eigenvalue of the operator defined on Dom(A) by d^2/dx^2 is $-\pi^2$, we easily obtain

$$\sup_{h \in \mathrm{Dom}(A); \|h\|=1} \operatorname{Re} \langle Ah, h\rangle \leq q - \pi^2 < 0.$$

So A is dissipative. Besides, $A + B$, A^*, and $A^* + B^*$ are right-bounded. So $A + B$ and A generate C_0-semigroups, cf. [30, Section I.4]. In addition, we have

$$\|e^{At}\| \leq \exp\left[-t(\pi^2 - q)\right] \quad (t \geq 0).$$

Hence, by (13),

$$\zeta(A) \leq \frac{1}{2(\pi^2 - q)^2}.$$

Since $\langle f', f\rangle = -\langle f, f'\rangle$ $(f \in \mathrm{Dom}\ (A))$, we have $\Lambda(B) = 0$ and consequently, $\psi(W, B) = 0$. Thus, due to Theorem 3, the semigroup generated by the operator $\tilde{A} = A + B$ defined by

$$(\tilde{A}f)(x) = \frac{d^2 f(x)}{dx^2} + bf'(x) + a(x)f(x) \quad (0 < x < 1, f \in \mathrm{Dom}\ (A))$$

is exponentially stable, provided

$$|b| \sup_x |a'(x)| < 2(\pi^2 - q)^2.$$

6. A Particular case of Theorem 3

In this section, we refine Theorem 3, assuming that

$$\lambda(A) = \inf_{h \in \mathrm{Dom}(A); \|h\|=1} \operatorname{Re} \langle Ah, h\rangle > -\infty. \tag{15}$$

For example, let

$$(Af)(x) = \frac{df(x)}{dx} + a(x)f(x) \quad (0 < x < 1; f \in L^2(0,1))$$

with

$$\mathrm{Dom}\ (A) = \{h \in L^2(0,1) : h' \in L^2(0,1), h(0) = h(1)\}$$

and a complex bounded measurable function $a(x)$. Simple calculations show that in this case $\lambda(A) = \inf_x \operatorname{Re}\ a(x) > -\infty$.

Furthermore, put

$$\hat{\psi}(W,B) := \begin{cases} 2\Lambda(B)\|W\|, & \text{if } \Lambda(B) \geq 0, \\ \frac{\Lambda(B)}{|\lambda(A)|} & \text{if } \Lambda(B) < 0. \end{cases}$$

Theorem 4. *Let conditions* (11), (12), (14), *and* (15) *hold, and let* e^{At} *be exponentially stable. If, in addition,*

$$\hat{\psi}(W,B) + \|K\|\zeta(A) < 1,$$

then $e^{t(A+B)}$ *is also exponentially stable.*

The proof of this theorem can be found in [22, Theorem 4.1].

Since $\hat{\psi}(W,B) < \psi(W,B)$, provided $\Lambda(B) < 0$, under condition (15), Theorem 4 refines Theorem 3.

7. Perturbations of Operators with Hilbert–Schmidt Resolvents

Let A and $\tilde{A}$ be closed linear operators on a complex separable Hilbert space $\mathcal{H}$, and

$$\alpha(A) := \sup \operatorname{Re} \sigma(A) < 0. \tag{16}$$

Assuming that

$$\operatorname{Dom}(\tilde{A}) = \operatorname{Dom}(A) \text{ and } q := \|A - \tilde{A}\| < \infty, \tag{17}$$

we suggest conditions that provide the inequality

$$\alpha(\tilde{A}) < 0. \tag{18}$$

Condition (18) plays an important role in various applications. In particular, if (18) holds and $\tilde{A}$ generates a semigroup, then that semigroup is asymptotically stable for a wide class of operators, cf. [10]. To the best of our knowledge, the conditions that provide inequality (18) for non-selfadjoint operators were almost not investigated in the available literature.

Let $A = A_R + iA_I$, where A_R and A_I are self-adjoint operators (the Hermitian components of A). It is supposed that

$$A_I \in \mathcal{L}(\mathcal{H}) \text{ and } A^{-1} \text{ is a Hilbert-Schmidt operator.} \tag{19}$$

Denote by λ_k the eigenvalues of A with their multiplicities taken into account and enumerated in the non-decreasing order of their absolute values: $|\lambda_k| \leq |\lambda_{k+1}|$ $(k = 1, 2, \ldots)$. In [26], it is checked that conditions (16) and (19) imply that

$$\gamma(A) := \sum_{k=1}^{\infty} \frac{1}{(\text{Re } \lambda_k)^2} < \infty.$$

Put

$$\hat{\zeta}(A) := \frac{1}{|\alpha(A)|} \left(\frac{2\|A_I\|}{\pi|\alpha(A)|} + 1 \right) \exp\, [32\gamma(A)\|A_I\|^2].$$

Theorem 5. *Let A_R be upper-bounded, and conditions* (16), (17), *and* (19) *hold. In addition, let*

$$2q\hat{\zeta}(A) < 1.$$

Then (18) *is valid.*

The proof of this theorem can be found in [26].

8. Equations with Differentiable Operators on a Hilbert Space

Throughout this section, $A(t)$ $(t \geq 0)$ is a closed operator on a separable Hilbert space $\mathcal{H}$ with a constant dense domain D_0 having a strong derivative $A'(t)$ uniformly bounded on $[0, \infty)$; namely, $A(t)h$ is strongly differentiable for all $h \in D_0$ and

$$\sup_{t \geq 0} \|A'(t)\| := \sup_{t \geq 0, h \in D_0} \|A'(t)h\|/\|h\| < \infty.$$

Our main object in this section is the equation

$$du/dt = A(t)u \quad (t > 0, u = u(t)). \tag{20}$$

A solution to (20) for a given $u_0 \in D_0$ is a function $u : [0, \infty) \to D_0$ having at each point $t > 0$, a strong derivative, at zero the right strong derivative, and satisfying (20) for all $t > 0$ and $u(0) = u_0$. The existence, uniqueness, and continuous dependence on initial vectors of solutions under considerations are assumed.

For more about various solvability results, see, for instance, [30]. For example, if $A(t) = A_0 + B(t)$, where A_0 is a constant operator on $\mathcal{H}$ generating a C_0-semigroup, and $B(t)$ is uniformly bounded on $[0,\infty)$ operator mapping $\mathrm{Dom}(A_0)$ into itself, then the existence, uniqueness, and continuous dependence on initial vectors of solutions is due to the well-known Theorem II.3.4 from the book by S. Krein [30].

It is also assumed that for each $t \geq 0$, $A(t)$ generates an exponentially stable semigroup $e^{A(t)s}$ $(s \geq 0)$, i.e., $\|e^{A(t)s}\| \leq m_0 exp\,[-\nu_0 s]$, where m_0, ν_0 are positive constants which can depend on t. So

$$q_0(t) := 2\int_0^\infty \|e^{A(t)s}\|^2 ds < \infty \quad (t \geq 0).$$

Equation (20) is said to be L^2-*stable*, if there is a constant $m_1 > 0$, such that

$$\int_0^\infty \|u(t)\|^2 dt \leq m_1 \|u(0)\|^2$$

for any solution $u(t)$ of (20). It is exponentially stable, if there are constants $m_2, m_3 > 0$, such that $\|u(t)\| \leq m_2 e^{-m_3 t}\|u(0)\|$ $(t \geq 0)$; it is uniformly exponentially stable, if $\|u(t)\| \leq m_2 e^{-m_3(t-\tau)}\|u(\tau)\|$ $(t \geq \tau \geq 0)$ with $m_2, m_3 > 0$ independent of τ.

Theorem 6. *Let $A(t)$ have a constant dense domain D_0, a strong derivative uniformly bounded on $[0,\infty)$, and generate an exponentially stable C_0-semigroup for each $t \geq 0$. In addition, assume that*

$$\sup_{t\geq 0} q_0^2(t)\|A'(t)\| < 2.$$

Then any solution $u(t)$ of (20) satisfies the inequality

$$\int_\tau^\infty \|u(s)\|^2 ds \leq \eta(\tau)\|u(\tau)\|^2 \quad (\tau \geq 0), \tag{21}$$

where

$$\eta(\tau) = \frac{q_0(\tau)}{\inf_{t\geq 0}(2 - q_0^2(t)\|A'(t)\|)},$$

and therefore (20) is L^2-stable.

This theorem is proved in [23].

Let $U(s, \tau)$ $(s \geq \tau \geq 0)$ be the evolution operator of (20): $U(s, \tau)u(\tau) = u(s)$. For given $x \in D_0$, take $u(\tau) = x$. Then (21) can be rewritten as

$$\int_{\tau}^{\infty} \|U(s, \tau)x\|^2 ds \leq \eta(\tau)\|x\|^2 \quad (\tau \geq 0).$$

Now the Datko theorem (see Corollary 3.2.4 from [6]) implies.

Corollary 7. *Under the hypothesis of Theorem* 6, *let* $\sup_{t \geq 0} q_0(t) < \infty$. *Then* (20) *is uniformly exponentially stable.*

This corollary means that under consideration the L^2-stability and uniform exponential stability are equivalent, if $\sup_{t \geq 0} q_0(t) < \infty$.

9. Stability Conditions for Equations with Variable Operators via Commutators

Throughout this section, A is a generator of a C_0-semigroup e^{At} on a Banach space $\mathcal{X}$ and $B(t)$ $(t \geq 0)$ is a variable bounded piecewise strongly continuous operator mapping Dom(A) into itself for each $t \geq 0$. We investigate exponential stability conditions for the equation

$$\frac{dy(t)}{dt} = (A + B(t))y(t) \quad (t \geq 0). \tag{22}$$

It is assumed that the commutator $K(t) = AB(t) - B(t)A$ is uniformly bounded on $[0, \infty)$, i.e., the norms of $(AB(t) - BA(t))x$ $(x \in \text{Dom } (A), \|x\| = 1)$ are uniformly bounded. So $AB(t) - B(t)A$ can be extended to $\mathcal{X}$ as a bounded operator. That extension is also denoted by $K(t)$. Moreover,

$$\kappa := \sup_{t \geq 0} \|K(t)\| < \infty. \tag{23}$$

A solution $y(t)$ to equation (22) with a given $y(0) = y_0 \in \text{Dom } (A)$ and the uniform exponential stability are defined as above.

Since $B(t)$ is bounded and maps the domain of A into itself, and A is a generator of a C_0-semigroup, the existence, uniqueness, and continuous dependence on initial vectors of solutions are due to the well-known Theorem II.3.4 [30].

Equation (22) can be considered as the equation

$$\frac{dy(t)}{dt} = C(t)y(t) \tag{24}$$

with a variable linear operator $C(t)$. Observe that $C(t)$ in the considered case has a special form: it is the sum of operators A and $B(t)$. This fact allows us to use the information about the coefficients more completely than the theory of differential equations (24) containing an arbitrary operator $C(t)$.

Let $U_B(t,s)$ $(t \geq s \geq 0)$ be the evolution operator of the equation

$$\dot{u}(t) = B(t)u(t) \quad (t \geq 0).$$

Suppose that there are constants $c_0 \geq 1$ and $b_0 \in \mathbf{R}$, such that

$$\|U_B(t,s)\| \leq c_0 \exp\left[b_0(t-s)\right] \;\; (t \geq s \geq 0). \tag{25}$$

Theorem 8. *Let conditions* (23) *and* (25) *hold and the operator* $A+b_0I$ *generate an exponentially stable semigroup* $e^{(A+b_0I)t}$. *In addition, let*

$$\zeta := \kappa c_0 \left(\int_0^\infty \|e^{(A+b_0I)t}\|dt\right)^2 < 1.$$

Then equation (22) *is exponentially stable.*

This theorem is proved in [18]. It is sharp: if $B(t)f = b_0 f$ $(f \in \mathcal{X})$, then $\zeta = 0$ and Theorem 8 gives us necessary and sufficient exponential stability conditions for (22): $e^{(A+b_0I)t}$ should be exponentially stable.

10. The Dyson–Phillips-Type Approach to Equations with Variable Operators

Recall that $C([a,b],\mathcal{X})$ is the Banach space of $\mathcal{X}$-valued continuous functions defined on a real segment $[a,b]$ and equipped with the sup-norm $\|.\|_{C(a,b)}$. The aim of this section is to generalize the well-known Dyson–Phillips theorem for strongly continuous semigroups [11, Theorem 1.10] to the evolution operators of differential equations with variable operators in a Banach space.

Recall the Dyson–Phillips theorem. Let $T(t)$ be a strongly continuous semigroup generated on $\mathcal{X}$ by an operator A_0 and let $B_0 \in \mathcal{L}(\mathcal{X})$. For a finite $t_0 > 0$, introduce the Volterra operator V_0 by

$$(V_0 f)(t) = \int_0^t T(t-s)B_0 f(s)ds \ \ (0 \le t \le t_0;\ f \in C([0,t_0],\mathcal{X})).$$

Due to Lemma 1.9 from [11], V_0 is a quasi-nilpotent operator.

Theorem 9. *(The Dyson–Phillips theorem). The strongly continuous semigroup $S(t)$ generated on $\mathcal{X}$ by $A_0 + B_0$, where A_0 is a generator of a strongly continuous semigroup $T(t)$, can be obtained as*

$$S(t) = \sum_{k=0}^{\infty} S_n(t), \tag{26}$$

where $S_0(t) = T(t)$ and

$$S_{n+1}(t) = V_0 S_n(t) = \int_0^t T(t-s)B_0 S_n(s)ds.$$

The series in (26) converges in the operator norm. From this theorem, we have

$$\|T(t) - S(t)\| \le m_1 t \ \ (m_1 = const; 0 < t \le t_0),$$

[11, Corollary 1.11].

To formulate the main result of this section, consider the equation

$$dx(t)/dt = A(t)x(t) \ \ (0 \le t \le t_0) \tag{27}$$

with an unbounded variable operator $A(t)$, having a constant dense in $\mathcal{X}$ domain $\text{Dom}(A(t)) \equiv D_0$ and closed for each $t \in [0, t_0]$.

By the Cauchy problem in $\{0 \le s \le t \le t_0\}$, we mean the problem of finding for each fixed $s \in [0, t_0)$ a solution $x(t, s)$ of (27) on $[s, t_0]$, satisfying a given initial condition

$$x(s,s) = x_0 \in D_0. \tag{28}$$

The Cauchy problems (27)–(28) are said to be well-posed if the following conditions hold:

(1) For each $s \in [0, t_0]$ and any $x_0 \in D_0$, there exists a unique solution $x(t, s)$ of (27) on the segment $[s, t_0]$ satisfying condition (28).

(2) The function $x(t,s)$ and its derivative $x'_t(t,s)$ are continuous for s,t in $\{0 \le s \le t \le t_0\}$.

(3) The solution depends continuously on the initial data in the sense that if the sequence $\{x_{0n} \in D_0\}$ converges to zero, then the corresponding solutions $x_n(t,s)$ converge to zero uniformly relative to t and s in $\{0 \le s \le t \le t_0\}$.

If the Cauchy problem is well-posed, then it is possible to introduce the evolution operator $U(t,s)$ $(0 \le s \le t \le t_0)$, which assigns to each element $x_0 \in D_0$ the value at the point t of the solution of problems (27)–(28) on the segment $[s,t_0]$: $x(t,s) = U(t,s)x_0$. $U(t,s)$ is defined on D_0 and admits a continuous extension to space $\mathcal{X}$, which we shall denote also by $U(t,s)$. About various properties of $U(t,s)$, see, for instance, [6, Section 3.1] and [30, Section II.3.1].

Together with equation (27), consider the equation

$$\frac{d}{dt}y(t) = (A(t) + B(t))y(t) \quad (0 \le t \le t_0), \tag{29}$$

where $B(t)$ is a bounded variable operator defined and strongly continuous on $[0,t_0]$. It is assumed that the Cauchy problem to equation (29) is also well-posed and denote its evolution operator by $\tilde{U}(t,s)$.

Furthermore, for a fixed $s \in [0,t_0)$, define the operator V by

$$(Vf)(t) = \int_s^t U(t,\tau)B(\tau)f(\tau)d\tau \quad (f \in C([s,t_0],\mathcal{X})).$$

In [14, Section 3], it is shown that there is a constant $\gamma(V,[s,t_0])$ such that

$$\|V^k\|_{C(s,t_0)} \le \frac{\gamma^k(V,[s,t_0])}{k!} \quad (k = 1,2,\dots). \tag{30}$$

Consequently,

$$\sqrt[k]{\|V^k\|_{C(s,t_0)}} \to 0 \text{ as } k \to \infty.$$

So V is quasi-nilpotent and

$$(I - V)^{-1} = \sum_{k=0}^{\infty} V^k.$$

Hence,

$$\|(I - V)^{-1}\|_{C(s,t_0)} \le \sum_{k=0}^{\infty} \frac{\gamma^k(V,[s,t_0])}{k!} = e^{\gamma(V,[s,t_0])}. \tag{31}$$

Theorem 10. *Let $A(t)$ be strongly continuous on D_0 and $B(t)$ be bounded strongly continuous on $\mathcal{X}$. Let the Cauchy problems to equations (27) and (29) be well-posed. Then the evolution operator $\tilde{U}(t,s)$ of (29) can be obtained as*

$$\tilde{U}(t,s) = \sum_{k=0}^{\infty} \tilde{U}_n(t,s), \tag{32}$$

where $\tilde{U}_0(t,s) = U(t,s)$ and

$$\tilde{U}_{n+1}(t,s) \\ = [V\tilde{U}_n(.,s)](t) := \int_s^t U(t,\tau)B(\tau)\tilde{U}_n(\tau,s)d\tau \quad (n = 0, 1, \ldots).$$

The series in (32) converges in the operator norm for all $0 \le s \le t \le t_0$.

This theorem has been proved in [25]. From (32), we have

$$\tilde{U}(t,s) - U(t,s) = \left[\sum_{k=1}^{\infty} V^k U(.,s)\right](t) = [V(I-V)^{-1}U(.,s)](t)$$

and

$$\tilde{U}(t,s) = \left[\sum_{k=0}^{\infty} V^k U(.,s)\right](t) = [(I-V)^{-1}U(.,s)](t).$$

Now inequalities (30) and (31) imply the following.

Corollary 11. *Under the hypothesis of Theorem 10, one has*

$$\sup_{s\le t\le t_0} \|\tilde{U}(t,s) - U(t,s)\| \le \gamma(V,[s,t_0])e^{\gamma(V,[s,t_0])} \sup_{s\le t\le t_0} \|U(t,s)\|$$

and

$$\sup_{s\le t\le t_0} \|\tilde{U}(t,s)\| \le e^{\gamma(V,[s,t_0])} \sup_{s\le t\le t_0} \|U(t,s)\|.$$

11. Application of Theorem 10

In this section, by virtue of Theorem 10, we obtain a stability test for equation (29).

Lemma 12. *Let the Cauchy problems for* (27) *and* (29) *be well-posed and*

$$\|U(t,s)\| \leq c\exp\left[\int_s^t \alpha(t_1)dt_1\right] \quad (c = \text{const} \geq 1;\ t \geq s \geq 0),$$

where $\alpha(t)$ $(t \geq 0)$ *is a piece-wise continuous function. Then*

$$\|\tilde{U}(t,s)\| \leq c\exp\left[\int_s^t (\alpha(\tau) + c\|B(\tau)\|)d\tau\right].$$

Proof. Due to (32), we can write

$$\tilde{U}(t,0) = U(t,0) + \int_0^t U(t,\tau)B(\tau)\tilde{U}(\tau,0)\|d\tau,$$

and therefore,

$$\begin{aligned}\|\tilde{U}(t,0)\| &\leq \|U(t,0)\| + \int_0^t \|U(t,\tau)\|\|B(\tau)\|\|\tilde{U}(\tau,0)\|d\tau \\ &\leq c\exp\left[\int_0^t \alpha(t_1)dt_1\right] + c\int_0^t \exp\left[\int_\tau^t \alpha(t_1)dt_1\right] \\ &\quad \times \|B(\tau)\|\|\tilde{U}(\tau,0)\|d\tau.\end{aligned}$$

Hence, with

$$w(t) = \exp\left[-\int_0^t \alpha(t_1)dt_1\right]\|\tilde{U}(t,0)\|,$$

we have

$$w(t) \leq c + c\int_0^t \|B(\tau)\|d\tau.$$

Now the Gronwall lemma yields

$$w(t) \leq c\exp\left[c\int_0^t \|B(\tau)\|d\tau\right].$$

This proves the lemma for $s = 0$. Replacing zero by $s > 0$, we finish the proof. Q.E.D.

Corollary 13. *Under the hypotheses of Lemma* 12, *let*

$$\nu := \sup_{t\geq s\geq 0} \frac{1}{t-s}\int_s^t (\alpha(\tau)+c\|B(\tau)\|)d\tau < 0.$$

Then

$$\|\tilde{U}(t,s)\| \leq c\exp[\nu(t-s)] \ \ (t\geq s\geq 0),$$

and therefore (29) *is uniformly exponentially stable.*

12. Equations with the Lipschitz Property

Again consider the equation

$$du(t)/dt = (A_0 + B(t))u(t) \quad (t\geq 0), \tag{33}$$

where A_0 is a constant operator on $\mathcal{X}$ with a dense domain $\mathrm{Dom}(A_0)$, generating a C_0-semigroup $e^{A_0 t}$, and $B(t)$ is a bounded operator in $\mathcal{X}$ with the Lipschitz property

$$\|B(t)-B(\tau)\| \leq q_0|t-\tau| \quad (q_0 = const \geq 0;\ t,\tau\geq 0). \tag{34}$$

Moreover, $B(t)$ *maps* $\mathrm{Dom}(A_0)$ *into itself.* The existence, uniqueness, and continuous dependence on initial vectors of solutions under consideration are due to the well-known Theorem II.3.4 [30].

In this section, we suggest an estimate for the evolution operator in a Banach space, which gives us stability conditions when (34) holds and q_0 is sufficiently small.

Due to [11, Corollary 1.3.5], the operator $A(\tau) = A_0 + B(\tau)$ generates a C_0-semigroup $e^{A(\tau)t}$. It is assumed that there are a real constant ν and a positive one a_ν, independent of τ, such that

$$\|\exp[A(\tau)t]\| \leq a_\nu e^{\nu t} \quad (\tau, t\geq 0). \tag{35}$$

Recall that $\cosh(x) = (e^x + e^{-x})/2$ $(x\in\mathbf{R})$.

Theorem 14. *Let conditions* (34) *and* (35) *hold. Then for all* $t \geq t_0 \geq 0$, *the evolution operator of* (33) *satisfies the inequality*

$$\|U(t,t_0)\| \leq a_\nu e^{\nu(t-t_0)} \cosh(\sqrt{a_\nu q_0}(t-t_0)) \\ \leq a_\nu \exp\left[(\nu + \sqrt{a_\nu q_0})(t-t_0)\right].$$

This theorem is proved in [20]. It implies the following.

Corollary 15. *Let the conditions* (34), (35), *and*

$$\nu + \sqrt{a_\nu q_0} < 0 \tag{36}$$

hold. Then equation (33) *is uniformly exponentially stable.*

Corollary 15 and Theorem 14 are sharp in the following sense: if $B(t)$ is constant, then $q_0 = 0$ and (36) becomes the necessary condition for the exponential stability.

13. Equations with the Generalized Lipschitz Property

Again consider the equation

$$du(t)/dt = (A_0 + B(t))u(t) \quad (t \geq 0), \tag{37}$$

where A_0 is a constant operator on $\mathcal{X}$ with a dense domain, generating a C_0-semigroup $e^{A_0 t}$, and $B(t)$ is a bounded operator in $\mathcal{X}$ with the generalized Lipschitz property:

$$\|B(t) - B(\tau)\| \leq q(|t-\tau|) \quad (t,\tau \geq 0), \tag{38}$$

where $q(t)$ is a non-negative continuous function of a non-negative argument with $q(0) = 0$. In addition, $B(t)$ $(t \geq 0)$ maps $\mathrm{Dom}(A_0)$ into itself. Due to [11, Corollary 1.3.5], the operator $A(\tau) = A_0 + B(\tau)$ generates a C_0-semigroup $e^{A(\tau)t}$.

It is also assumed that

$$\|\exp[A(\tau)t]\| \leq \psi(t) \quad (\tau, t \geq 0), \tag{39}$$

where $\psi(t)$ is a non-negative continuous function independent of τ and

$$\hat{\psi}_0 := \sup_{t \geq 0} \psi(t) < \infty.$$

The existence, uniqueness, and continuous dependence on initial vectors of solutions under considerations are due to the well-known Theorem II.3.4 [30].

Theorem 16. *Let the conditions* (38), (39), *and*

$$\zeta_0 := \int_0^\infty q(s)\psi(s)ds < 1 \tag{40}$$

hold. Then any solution $u(t)$ *of* (37) *satisfies the inequality*

$$\sup_{t\geq 0} \|u(t)\| \leq \frac{\hat{\psi}_0\|u(0)\|}{1-\zeta_0}. \tag{41}$$

The proof of this theorem is present in the following section. Inequality (41) means that (37) is Lyapunov stable.

14. Proof of Theorem 16

Consider the equation

$$u(t) = \exp[A(\tau)t]u(0) + \int_0^t \exp[A(\tau)(t-s_1)][B(s) - B(\tau)]u(s_1)ds \tag{42}$$

with ($u(0) \in \mathrm{Dom}(A_0)$), $A(t) = A_0 + B(t)$ and an arbitrary fixed $\tau \geq 0$. Due to Corollary VI.7.8 [11], it is equivalent to

$$\begin{aligned} du(t)/dt &= A(\tau)u(t) + [B(t) - B(\tau)]u(t) \\ &= A(\tau)u(t) + [A(t) - A(\tau)]u(t) = A(t)u(t). \end{aligned}$$

So (42) and (37) are equivalent, and

$$\begin{aligned} \|u(t)\| \leq{}& \|\exp[A(\tau)t]\|\|u(0)\| \\ &+ \int_0^t \|\exp[A(\tau)(t-t_1)]\|\|B(t_1) - B(\tau)\|\|u(t_1)\|dt_1. \end{aligned}$$

According to (38) and (39),

$$\|u(t)\| \leq \hat{\psi}_0\|u(0)\| + \int_0^t \psi(t-t_1)q(|t_1-\tau|)\|u(t_1)\|dt_1.$$

Taking $\tau = t$, we obtain

$$\|u(t)\| \le \hat{\psi}_0 \|u(0)\| + \int_0^t \psi(t-t_1)q(t-t_1)\|u(t_1)\|dt_1,$$

and therefore, for any positive finite T,

$$\begin{aligned}\sup_{t\le T}\|u(t)\| &\le \hat{\psi}_0\|u(0)\| + \sup_{t\le T}\|u(t)\| \sup_{t\le T}\int_0^t \psi(t-t_1)q(t-t_1)dt_1 \\ &= \hat{\psi}_0\|u(0)\| + \sup_{t\le T}\|u(t)\| \int_0^T \psi(s)q(s)ds \le \hat{\psi}_0\|u(0)\| \\ &\quad + \sup_{t\le T}\|u(t)\|\zeta_0.\end{aligned}$$

Now (40) implies

$$\sup_{t\le T}\|u(t)\| \le \frac{\hat{\psi}_0\|u(0)\|}{1-\zeta_0}.$$

Hence, letting $T \to \infty$ we get the required result. Q.E.D.

15. Example to Theorem 16

Consider the equation

$$\frac{\partial u(t,x)}{\partial t} = \frac{\partial}{\partial x} c(x) \frac{\partial u(t,x)}{\partial x} + b(t)u(t,x) \quad (0 < x < 1), \tag{43}$$

with the boundary conditions

$$u(t,0) = u(t,1) = 0 \quad (t > 0), \tag{44}$$

where $c(x)$ is a scalar differentiable function with $c_0 = \inf_x c(x) > 0$, and $b(t) = (b_{jk}(t))_{j,k=1}^2$ is the matrix

$$b(t) = \begin{pmatrix} -1 & a\sin(\omega t) \\ -a\sin(\omega t) & -1 \end{pmatrix} \quad (a, \omega = const > 0).$$

Since

$$\sin x - \sin y = 2\cos\frac{1}{2}(x+y)\sin\frac{1}{2}(x-y) \quad (x, y \in \mathbf{R}),$$

we obtain

$$\|b(t) - b(s)\|_2 \le 2a|\sin \frac{\omega(t-s)}{2}|.$$

Here $\|.\|_2$ is the spectral matrix norm (the operator norm in the complex 2-dimensional Euclidean space $\mathbf{C}^2$).

Take $\mathcal{X} = L^2([0,1], \mathbf{C}^2)$ — the space of square integrable 2-vector functions defined on $[0,1]$ with the scalar product

$$(v, w) = \int_0^1 \sum_{k=1}^2 v_k(x)\overline{w}_k(x)dx, \ (v(x) = (v_k(x))_{k=1}^2,$$
$$w(x) = (w_k(x))_{k=1}^2 \in L^2([0,1], \mathbf{C}^2)).$$

In the considered case, $A_0 = \frac{d}{dx}c(x)\frac{d}{dx}$ with the domain

$$\mathrm{Dom}(A_0) = \{h \in L^2([0,1], \mathbf{C}^2) : h'' \in L^2([0,1], \mathbf{C}^2); h(0) = h(1) = 0\}$$

and $B(t)u(t) = b(t)u(t)$. Since A_0 and $B(t)$ commute, we can write $e^{A(t)s} = e^{sA_0}e^{sB(t)}$ $(s \ge 0)$. Simple calculations show that

$$(A_0h, h) = -(c(x)h', h') \le -c_0(h', h') = c_0(h'', h)$$
$$\le -c_0\pi^2(h, h) \ \ (h \in \mathrm{Dom}(A_0)).$$

Hence, $\|e^{sA_0}\| \le e^{-c_0\pi^2 s}$ $(s \ge 0)$. In addition, $\|B(t)\| \le \|b(t)\|_2$ and $\|e^{sB(t)}\| \le \|e^{sb(t)}\|_2$. So condition (38) holds with $q(t) = 2a|\sin(\frac{\omega t}{2})|$.

Due to Example 2.7.3 from [14], we get

$$\|e^{sb(t)}\|_2 \le e^{\alpha(b(t))s}(1 + g(b(t))s),$$

where $g(b(t)) = |b_{12}(t) - b_{21}(t)|$ and

$$\alpha(b(t)) = \max\{\mathrm{Re}\, \lambda_1(b(t)), \mathrm{Re}\, \lambda_2(b(t))\},$$

$\lambda_1(b(t)), \lambda_2(b(t))$ denote the eigenvalues of $b(t)$. Since $b(t)$ is a 2×2 matrix, its eigenvalues are simply calculated. We have

$$\|e^{t(A_0+B(\tau))}\| \le e^{-t(c_0\pi^2 - \alpha(b(\tau)))}(1 + g(b(\tau))t).$$

Obviously, $\alpha_0 := \sup_{t\geq 0} \alpha(b(t)) < \infty$ and $g_0 := \sup_{t\geq 0} g(b(t)) \leq 2a$. So condition (39) holds with

$$\psi(t) = e^{-t(c_0\pi^2-\alpha_0)}(1+2at).$$

Thus,

$$\zeta_0 \leq \hat{\zeta} := 2a\int_0^\infty e^{-t(c_0\pi^2-\alpha_0)}(1+2at)\left|\sin\left(\frac{\omega t}{2}\right)\right| dt.$$

If $\hat{\zeta} < 1$, then due to Theorem 16, equation (43) with the boundary conditions (15.2) is Lyapunov stable.

References

[1] Alabau, F., Cannarsa, P. and Komornik, V. Indirect internal stabilization of weakly coupled evolution equations. *J. Evol. Equ.* 2(2), 127–150 (2002).

[2] Andrica, D. and Rassias, Th. M. (eds.), *Differential and Integral Inequalities*, Springer Optimization and Its Applications, Springer Nature, Switzerland, 151 (2019).

[3] Buse, C., Khan, A., Rahmat, G. and Saierli, O. Weak real integral characterizations for exponential stability of semigroups in reflexive spaces, *Semigroup Forum*, 88, 195–204 (2014).

[4] Buse, C. and Niculescu, C. A condition of uniform exponential stability for semigroups. *Math. Inequal. Appl.* 11(3), 529–536 (2008).

[5] Bylov, B.F., Grobman, B.M., Nemyckii, V.V. and Vinograd, R.E. *The Theory of Lyapunov Exponents*, Nauka, Moscow (1966). In Russian.

[6] Chicone, C. and Latushkin, Yu. *Evolution Semigrous in Dynamical Systems and Differential Equations*, Amer. Math. Soc., Math. Surv. and Monographs, 70 (1999).

[7] Cialdea, A. and Lanzara, F. Stability of solutions of evolution equations, *Rend. Lincei Mat. Appl.* 24, 451–469 (2013).

[8] Curtain, R. and Zwart, H. *Introduction to Infinite-Dimensional Systems Theory*, Springer, NY (1995).

[9] Dragan, V. and Morozan, T. Criteria for exponential stability of linear differential equations with positive evolution on ordered Banach spaces, *IMA Journal of Mathematical Control and Information*, 27, 267–307 (2010).

[10] Eisner, T. *Stability of Operators and Operator Semigroups*, Operator Theory: Advances and Applications, Birkhäuser Verlag, Basel, 209 (2010).

[11] Engel, K.-J. and Nagel, R. *One-Parameter Semigroups for Linear Evolution Equations*, Springer, New York (2000).

[12] Fourrier, N. and Lasiecka, I. Regularity and stability of a wave equation with a strong damping and dynamic boundary conditions. *Evol. Equ. Control Theory*, 2(4), 631–667 (2013).

[13] Gil', M.I. Stability of linear evolution equations in lattice normed spaces, *Zeitschrift fur Analysis* 15(4), 949–959 (1996).

[14] Gil', M.I. *Operator Functions and Localization of Spectra*, Lecture Notes In Mathematics, Springer-Verlag, Berlin, 1830 (2003).

[15] Gil', M.I. Integrally small perturbations of semigroups and stability of partial differential equations, *International Journal of Partial Differential Equations*, 2013, 207581, 5 (2013).

[16] Gil', M.I. Stability of linear nonautonomous multivariable systems with differentiable matrices, *Systems & Control Letters*, 81, 31–33 (2015).

[17] Gil', M.I. *Operator Functions and Operator Equations*, World Scientific, New Jersey (2018).

[18] Gil', M.I. Stability of evolution equations with small commutators in a Banach space, *Rendiconti Lincei Math. Applications*, 29(4), 589–596 (2018).

[19] Gil', M.I. Inequalities for solutions of linear differential equations in a Banach space and integro–differential equations. In: Andrica D., Rassias T. (eds.), *Differential and Integral Inequalities*. Springer Optimization and Its Applications, Springer, Berlin, 151 (2019).

[20] Gil', M.I. A norm estimate for evolution operators of equations with the Lipschitz property in a Banach space, *Rendiconti Lincei-Matematica e Applicazioni*, 30(1), 169–174 (2019).

[21] Gil', M.I. Semigroups of sums of two operators with small commutators, *Semigroup Forum*, 98(1), 22–30 (2019).

[22] Gil', M.I. Stability conditions for perturbed semigroups on a Hilbert Space via commutators, *Communications in Advanced Mathematical Sciences*, 2(2), 129–134 (2019).

[23] Gil', M.I. Stability of linear equations with differentiable operators in a Hilbert space *IMA Journal of Mathematical Control and Information*, 37(1), 19–26 (2020).

[24] Gil', M.I. Norm estimates for a semigroup generated by the sum of two operators with an unbounded commutator, *Results in Mathematics*, 75, 1–10 (2020).

[25] Gil', M.I. On Dyson–Phillips type approach to differential equations with variable operators in a Banach space, *Annali di Matematica Pura ed Applicata*, 1–11, Published online July 7 (2021).

[26] Gil', M.I. On location of the spectrum of an operator with a Hilbert-Schmidt resolvent in the left half-plane, *Methods of Functional Analysis and Topology* (accepted for publication), (27)(4), 340–347 (2021).

[27] Guo, B. and Zwart, H. On the relation between stability of continuous- and discrete-time evolution equations via the Cayley transform, *Integral Equations Operator Theory*, 54, 349–383 (2006).

[28] Henry, D. *Geometric Theory of Semilinear Parabolic Equations.* Lectures Notes in Mathematics, No. 840. Springer-Verlag, New York (1981).

[29] Heymann, R. Eigenvalues and stability properties of multiplication operators and multiplication semigroups, *Math. Nachr.*, 287(5–6), 574–584 (2014).

[30] Krein, S.G. *Linear Equations in a Banach Space, Amer. Math. Soc.*, Providence, R.I. (1971).

[31] Laasri, H. and El-Mennaoui, O. Stability for non-autonomous linear evolution equations with Lp-maximal regularity, *Czechoslovak Mathe. J.*, 63(138), 887–908 (2013).

[32] Matrai, T. On perturbations preserving the immediate norm continuity of semigroups, *J. Math. Anal. Appl.*, 341, 961–974 (2008).

[33] Nicaise, S. Convergence and stability analyses of hierarchic models of dissipative second order evolution equations. *Collect. Math.*, 68(3), 433–462 (2017).

[34] Oostveen, J. *Strongly Stabilizable Distributed Parameter Systems*, SIAM, Frontiers in Applied Mathematics, Philadelphia (2000).

[35] Pardalos, P.M. and Rassias, T.M. (eds.), *Mathematics Without Boundaries*, Surveys in Interdisciplinary Research, VIII, Springer, NY (2014).

[36] L. Paunonen, H. Zwart, A Lyapunov approach to strong stability of semigroups, *Systems & Control Letters* 62, 673–678 (2013).

[37] Preda, C. and Preda, P. Lyapunov operator inequalities for exponential stability of Banach space semigroups of operators, *Applied Mathematics Letters* 25, 401–403 (2012).

[38] Pucci, P. and Serrin, J.P. Asymptotic stability for nonautonomous wave equation, *Comm. Pure Appl. Math., XLXX,* 177–216 (1996).

[39] Seifert, C. and Wingert, D. On the perturbation of positive semigroups, *Semigroup Forum* 91, 495–501 (2015).

https://doi.org/10.1142/9789811293009_0017

Chapter 17

Numerical Solution of a Moving Boundary Problem Related to Phase Change Material

D. Goeleven*,‡ and R. Oujja*,§

*University of La Réunion, Saint-Denis, La Réunion, France

‡daniel.goeleven@univ-reunion.fr

§rachid.oujja@univ-reunion.fr

The numerical solution of a moving boundary problem issued from the thermal behavior in phase change materials is given. The applied numerical method is based on a backward Euler scheme combined with a duality iterative algorithm. Some numerical results are given to show efficiency of this numerical method.

1. Introduction and State Problem

Thermal energy storage systems containing phase change materials (PCM) are advanced technical means of improving the energy efficiency and environmental impact of buildings. Heat transfers in PCM can be described using partial differential equations posed in domains whose boundaries are to be determined as part of the solution. Such problems are usually referred to as moving boundary problems. The base of Stefan problem was started with reference to the early work of J. Stefan [11, 12] who was interested in melting of the polar ice cap. Analytical solutions for Stefan problems are discussed in [3–8, 10]. These problems are highly nonlinear and thus it is usually difficult to find analytical solutions. Due to this, numerical methods have been used more commonly.

There are two numerical techniques to obtain the solution of the Stefan problems. The first one is the front-tracking method, in which the position of the moving boundary is continuously tracked by some iterative procedure. The other approach is to use a front-fixing method by fixing the moving boundary.

In this chapter, we opt for a front fixing method. We first give a partial differential equation formulation using subdifferential monotone operators to describe heat transfer problems. Then, we apply a backward Euler scheme for time discretization of the equation, combined with an iterative method to approximate the solution.

We consider the case of a building roof equipped with PCM represented by the two-dimensional domain $\Omega = [l_1, l_2] \times [0, 1]$ in the $x = (x_1, x_2)$ coordinates. The transient local two-dimensional equation of the energy balance under enthalpy reads

$$\begin{aligned}\frac{\partial \chi(x,t)}{\partial t} &+ \alpha \nabla.(\chi(x,t)\mathbf{u}) \\ &= \nabla.(\lambda(x,t)\nabla T(x,t)),\ t \in [0, \mathcal{T}],\ x \in \Omega, \qquad (1)\end{aligned}$$

where $\mathcal{T} > 0$ is given. The function T represents the temperature, λ is the thermal conductivity of the roof, and χ is the enthalpy defined by

$$\chi \in H(T) = \begin{cases} \rho_s c_s (T - T_m) & \text{if } T < T_m, \\ [0, \rho_l L_m] & \text{if } T = T_m, \\ \rho_l c_l (T - T_m) + \rho_l L_m & \text{if } T > T_m. \end{cases} \qquad (2)$$

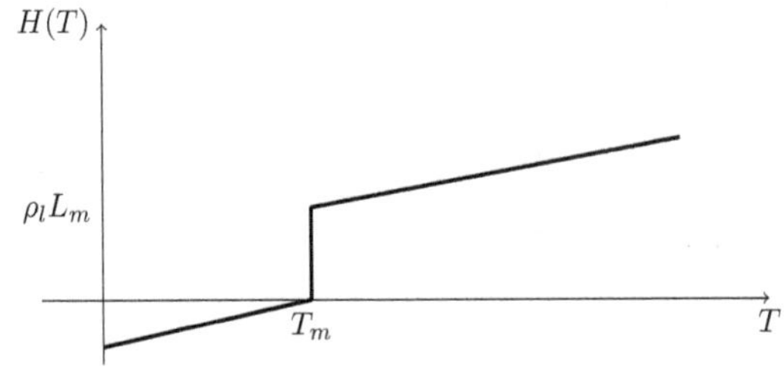

Here c_i $(i = s, l)$ represents the heat capacity, ρ_i $(i = s, l)$ the volume density, T_m the melt temperature, and L_m the latent heat of the roof.

Equation (1) is completed with the initial condition

$$\chi(x, 0) = \chi_0(x) \quad \text{a.e. } x \in \Omega \tag{3}$$

and the boundary conditions

$$(\forall t \in [0, \mathcal{T}], \forall x_2 \in [0, 1]) : T(l_1, x_2, t) = T_G, \;\; T(l_2, x_2, t) = T_D, \tag{4}$$

$$(\forall t \in [0, \mathcal{T}], \forall x_1 \in [l_1, l_2]) : \frac{\partial T}{\partial x_2}(x_1, 0, t) = \frac{\partial T}{\partial x_2}(x_1, 1, t) = 0, \tag{5}$$

where T_G and T_D are given values of boundary temperature and χ_0 is the initial enthalpy function in Ω.

The set-valued function defined in (2) is the subdifferential of the function $\Phi : \mathbb{R} \to \mathbb{R} \cup \{+\infty\}$ defined by

$$\Phi(T) = \begin{cases} \frac{1}{2}\rho_s c_s (T - T_m)^2, & \text{if } T \leq T_m, \\ \frac{1}{2}\rho_l c_l (T - T_F)^2 + \rho_l L_F (T - T_m), & \text{if } T \geq T_m. \end{cases}$$

This function is proper, convex, and lower semi-continuous, thus H is a maximal monotone operator. Our mathematical approach of problems (1)–(2) is based on classical results for monotone maps ([1, 2, 9]).

Let us first introduce the perturbed operator $L = H - \omega I$ for some $0 \leq \omega \ll 1$. And for a parameter $\beta > 0$ such that $\beta\omega < 1$, we consider the map $I + \beta L$ defined by

$$(I+\beta L)(T) = \begin{cases} (1-\beta\omega)T + \beta\rho_s c_s(T - T_m) & \text{for } T < T_m, \\ [(1-\beta\omega)T_m, (1-\beta\omega)T_m + \beta\rho_l L_m] & \text{for } T = T_m, \\ (1-\beta\omega)T + \beta\rho_l c_l(T - T_m) + \beta\rho_l L_m & \text{for } T > T_m. \end{cases}$$

It is easy to see that for all $f \in \mathbb{R}$ there exists a unique $y \in \mathbb{R}$ such that $f \in (I+\beta L)(y)$. The resolvent $J_\beta^L = (I+\beta L)^{-1}$ of L is defined by

$$J_\beta^L(y) = \begin{cases} \dfrac{y + \beta\rho_s c_s T_m}{1 - \beta\omega + \beta\rho_s c_s} & \text{for } y < (1-\beta\omega)T_m, \\ T_m \ \ \text{for } (1-\beta\omega)T_m < y < 1-\beta\omega)T_m + \beta\rho_l L_m, \\ \dfrac{y + \beta\rho_l c_l T_m - \beta\rho_l L_m}{1 - \beta\omega + \beta\rho_l c_l} & \text{for } y > (1-\beta\omega)T_m + \beta\rho_l L_m. \end{cases}$$

It is a single-valued and contraction map on $L^2(\Omega)$ ([2]). The map $L_\beta = \frac{I - J_\beta^L}{\beta}$ is the Moreau–Yosida approximation of L. This map is maximal monotone, single-valued, and $\frac{1}{\beta}$-Lipschitz continuous.

Our approach is based on the following property:

Proposition 1.

$$\forall u, y \in L^2(\Omega) : \ u \in L(y) \iff u = L_\beta(y + \beta u). \tag{6}$$

Proof. Let be $u = L_\beta(x)$. Then

$$\begin{aligned} u = \frac{I - J_\beta^L}{\beta}(x) &\iff \beta u = x - J_\beta^L(x) \\ &\iff J_\beta^L(x) = x - \beta u \\ &\iff x \in (I + \beta L)(x - \beta u) = x - \beta u + \beta L(x - \beta u) \\ &\iff \beta u \in \beta L(x - \beta u) \\ &\iff u \in L(x - \beta u) \end{aligned}$$

and by taking $x - \beta u = y$, we get $u \in L(y) \iff u = L_\beta(y + \beta u)$. □

2. Discrete Formulation

We shall consider here the one-dimensional equation. So the computational domain is assumed to be $[L_1, L_2]$ and the enthalpy equation is

$$\frac{\partial \chi(x,t)}{\partial t} + \alpha \nabla.(\chi(x,t)\mathbf{u}) = \frac{\partial}{\partial x}\left(\lambda(x,t)\frac{\partial T(x,t)}{\partial x}\right), \quad t \in [0, T_f], \; x \in [L_1, L_2], \tag{7}$$

$$\chi \in H(T). \tag{8}$$

We define on $[L_1, L_2]$ a uniform cartesian mesh of M nodes, with spacing $\Delta x = (L_2 - L_1)/M$. Subscript i at a variable refers to its value at the discrete location $x_i = i\Delta x$. The flux going from node $i-1$ to node i is approximated as

$$\alpha\chi(x,t) - \lambda(x,t)\frac{\partial T(x,t)}{\partial x} \simeq \alpha\chi_{i-1} - \frac{\lambda_{i-1} + \lambda_i}{2}\frac{T_i - T_{i-1}}{\Delta x}. \tag{9}$$

The sliding velocity $\mathbf{u}$ is assumed from left to right. The time variable is discretized into uniform time steps of length $\Delta t = T_f/M$ and a superscript n on a variable refers to its value at time $t_n = n\Delta t$. Combining the spatial and temporal discretization above in an implicit scheme, the following discrete equation at node i can be obtained as follows:

$$\Delta x^2 \frac{\chi_i^n - \chi_i^{n-1}}{\Delta t} + \alpha\Delta x(\chi_i^n - \chi_{i-1}^n) = s_i^n T_{i+1}^n - (s_i^n + s_{i-1}^n)T_i^n + s_{i-1}^n T_{i-1}^n, \tag{10}$$

where $s_i = \dfrac{\lambda_i + \lambda_{i+1}}{2}$.

Conditions (2) for the discrete variables are, for all i and n:

$$\chi_i^n \in H(T_i^n). \tag{11}$$

Thus, we obtain the following discretization of problems (7)–(8):

$$-s_{i-1}^n T_{i-1}^n + (s_i^n + s_{i-1}^n)T_i^n - s_i^n T_{i+1}^n = -\left(\frac{\Delta x^2}{\Delta t} + \alpha\Delta x\right)\chi_i^n + \alpha\Delta x\chi_{i-1}^n + \frac{\Delta x^2}{\Delta t}\chi_i^{n-1}, \tag{12}$$

$$\chi_i^n \in H(T_i^n). \tag{13}$$

Now, we define for all i and n $\gamma_i^n = \chi_i^n - \omega T_i^n$ and the operator $L = H - \omega I$ for some $0 < \omega < 1$. We can write the discrete equation by using the new variable γ_i^n as follows:

$$\begin{aligned} &-s_{i-1}^n T_{i-1}^n + (s_i^n + s_{i-1}^n) T_i^n \\ &\quad -s_i^n T_{i+1}^n + \omega \left(\tfrac{\Delta x^2}{\Delta t} + \alpha \Delta x\right) T_i^n - \alpha\omega\Delta x T_{i-1}^n \qquad (14) \\ &= -\left(\frac{\Delta x^2}{\Delta t} + \alpha\Delta x\right)\gamma_i^n + \alpha\Delta x \gamma_{i-1}^n + \frac{\Delta x^2}{\Delta t}\gamma_i^{n-1} + \omega\frac{\Delta x^2}{\Delta t} T_i^{n-1}, \qquad (15) \\ &\qquad \gamma_i^n \in L(T_i^n) = H(T_i^n) - \omega T_i^n. \qquad (16) \end{aligned}$$

From equivalence (6), we have

$$\gamma_i^n \in L(T_i^n) \iff \gamma_i^n = L_\beta(T_i^n + \beta\gamma_i^n)$$

for some parameter β satisfying $0 < \beta\omega < 1$. Equivalently, we get the discrete problem:

Find (T_i^n, γ_i^n) for $i = 2, \ldots, M-1$ and $n = 2, \ldots, N$ such that

$$\begin{aligned} &-s_{i-1}^n T_{i-1}^n + (s_i^n + s_{i-1}^n) T_i^n \\ &\quad - s_i^n T_{i+1}^n + \omega \left(\frac{\Delta x^2}{\Delta t} + \alpha \Delta x\right) T_i^n - \alpha\omega\Delta x T_{i-1}^n \qquad (17) \\ &= -\left(\frac{\Delta x^2}{\Delta t} + \alpha\Delta x\right)\gamma_i^n + \alpha\Delta x \gamma_{i-1}^n + \frac{\Delta x^2}{\Delta t}\gamma_i^{n-1} + \omega\frac{\Delta x^2}{\Delta t} T_i^{n-1}, \\ &\qquad \gamma_i^n = L^\lambda(T_i^n + \lambda\gamma_i^n). \qquad (18) \end{aligned}$$

3. Iterative Approximation

The nodal variables are obtained as limits of an iterative process. So by adding a second superscript to denote the iteration number, the algorithm reads the following:

Find T_i^n and γ_i^n, for $i = 2, \ldots, M-1$ and $n = 1, \ldots, N$ as

$$(T_i^n, \gamma_i^n) = \lim_{k\to\infty} (T_i^{n,k}, \gamma_i^{n,k}), \quad (T_i^{n,0}, \gamma_i^{n,0}) = (T_i^{n-1}, \gamma_i^{n-1}).$$

Then we follow the following method:

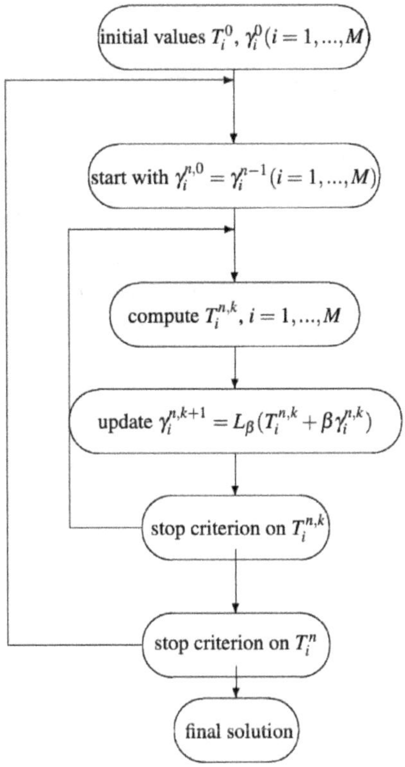

Flow diagram of numerical method

(1) Start with the initial value $\gamma_i^{n,0} = \gamma_i^{n-1}(i = 1, \ldots, M)$.

(2) For $\gamma_i^{n,k}(i = 1, \ldots, M)$ known, compute $T_i^{n,k}(i = 2, \ldots, M-1)$ solution to

$$
\begin{aligned}
-s_{i-1}^{n,k-1}T_{i-1}^{n,k} &+ (s_i^{n,k-1} + s_{i-1}^{n,k-1})T_i^{n,k} \\
&- s_i^{n,k-1}T_{i+1}^{n,k} + \omega\left(\frac{\Delta x^2}{\Delta t} + \alpha\Delta x\right)T_i^{n,k} - \alpha\omega\Delta x T_{i-1}^{n,k} \\
= -\left(\frac{\Delta x^2}{\Delta t} + \alpha\Delta x\right)\gamma_i^{n,k} &+ \alpha\Delta x\gamma_{i-1}^{n,k} + \frac{\Delta x^2}{\Delta t}\gamma_i^{n-1} + \omega\frac{\Delta x^2}{\Delta t}T_i^{n-1}.
\end{aligned}
$$

(3) Update $\gamma_i^n(i = 2, \ldots, M-1)$ as

$$\gamma_i^{n,k+1} = L_\beta(T_i^{n,k} + \beta\gamma_i^{n,k}).$$

(4) Go to (1) until stop criterion is reached.

4. Numerical Example

We present here the numerical results obtained by applying methods (17)–(18) to approximate the solution of problems (7)–(8) on the spacial interval $[L_1, L_2] = [0, 1]$ and for time interval $[0, T_f] = [0, 0.1]$. We take for volume density $\rho_l = \rho_s = 1$, for heat capacity $c_l = c_s = 1$, and for thermal conductivity $\lambda_s = 3$ and $\lambda_l = 0.5$. We fixed the melt temperature at $T_m = 0$ and the latent heat of the roof at $L_m = 1$. For the initial temperature, we have considered the function $T(x, 0) = 10 - 20x$, and we fixed the spacial step discretization $\Delta x = 0.01$ and time discretization $\Delta t = 0.01$.

We present in Figure 1 the distribution of the temperature T^n in the interval $[0, 1]$. In Figure 2, we give the discrete norm $\|T^{n,k+1} - T^{n,k}\| = (\sum_{i=1}^{M} |T_i^{n,k+1} - T_i^{n,k}|^2)^{1/2}$ along the iterative method for $n = 1, \ldots, 6$.

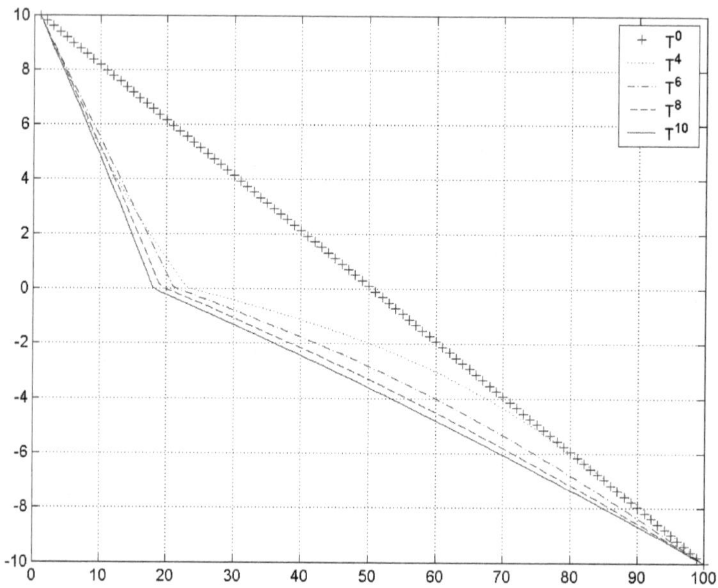

Fig. 1. Distribution of temperature T^n in space interval $[0, 1]$.

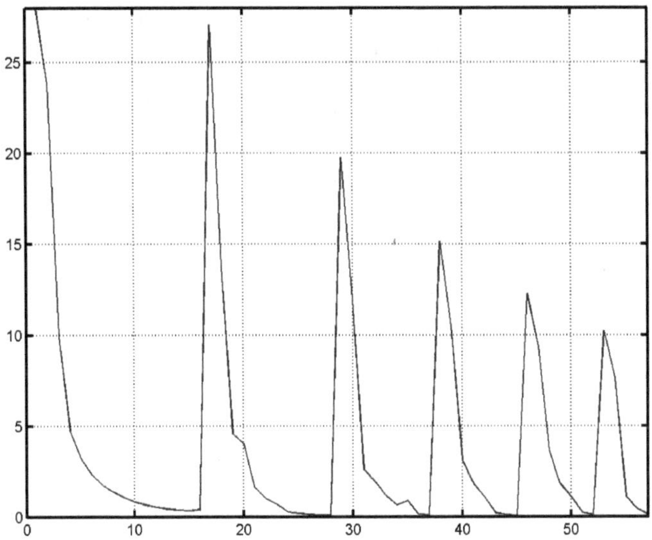

Fig. 2. Discrete norm $\|T^{n,k+1} - T^{n,k}\|$ for $n = 1, \ldots, 6$.

References

[1] Bermudez, A. and Moreno, C. Duality methods for solving variational inequalities, *Comput. Math. Appl.*, 7, 43–58 (1981).

[2] Brezis, H. *Opérateurs maximaux monotones et semi-groupes de contractions dans les espaces de Hilbert.* North-Holland, Amsterdam (1973).

[3] Carslaw, H.S. and Jaeger, J.C. *Conduction of Heat in Solids.* Clarendon Press, Oxford (1959).

[4] Crank, J. *Free and Moving Boundary Problems.* Clarendon Press, Oxford (1984).

[5] Douglas, J. A uniqueness theorem for the solution of a Stefan problem *Proc. Am. Math. Soc.* 8, 402–408 (1957).

[6] Evans, G.W. A note on the existence of a solution to a problem of Stefan, *Q. Appl. Math*, 9, 185–193 (1951).

[7] Friedman, A. *Partial Differential Equations of Parabolic Type.* Prentice-Hall, New Jersey (1964).

[8] Hill, J.M. *One-dimensional Stefan Problems: An introduction*, Longman Scientific Technical, Harlow (1987).

[9] Pazy, A. *Semigroups of non-linear contractions in Hilbert Spaces*, Problems in Nonlinear Analysis (C.I.M.E. Ed. Cremonese, Roma, 1971).

[10] Rubinstein, L.I. *The Stefan Problem*, Amer. Math. Soc. Translated from Russian (1971).

[11] Stefan, J. Uber die theorie der eisbildung inbesondee uber die eisbindung im polarmeere, *Ann. Phys. U. Chem.* 42, 269–286 (1891).

[12] Stefan, J. Uber Einige Probleme der Theorie der Warmeleitung, *Sitzungsber Wiener Akad Math Naturwiss Abt* 98, 473–484 (1889).

https://doi.org/10.1142/9789811293009_0018

Chapter 18

On Approximation of Certain Modified Szász–Mirakyan Operators

Vijay Gupta[*,‡], Gunjan Agrawal[**,§], and Michael Th. Rassias[†,¶]

[*]*Department of Mathematics,*
Netaji Subhas University of Technology,
Sector 3 Dwarka, New Delhi 110078, India
[†]*Department of Mathematics and Engineering Sciences*
Hellenic Military Academy Vari Attikis, Greece
[**]*School of Information Technology,*
Vivekananda Institute of Professional Studies-Technical Campus,
AU-Block, Outer Ring Road, Pitampura, Delhi 110034, India
[‡]*vijaygupta2001@hotmail.com*
[§]*gunjan.guptaa88@gmail.com*
[¶]*mthrassias@yahoo.com*

This chapter deals with the study of certain modification of semiexponential Szász–Mirakyan operators. Here, we derive the rate of point-wise convergence and obtain the basic convergence theorem in simultaneous approximation.

1. Introduction

In the year 1978, Ismail and May [1] considered the exponential-type operators and studied their approximation properties. These operators, say $L_\lambda(f,t)$, are given by

$$L_\lambda(f,t) = \int_{-\infty}^{\infty} W(\lambda,t,v)f(v)dv.$$

The kernel $W(\lambda,t,v)$ satisfies the following partial differential equation:

$$\frac{\partial}{\partial t}W(\lambda,t,v) = \frac{\lambda}{p(t)}W(\lambda,t,v)(v-t), \tag{1}$$

where $p(t)$ is an analytic and positive function and

$$\int_{-\infty}^{\infty} W(\lambda,t,v)dv = 1 \tag{2}$$

is the normalization condition. In [1], it is proved that for each $p(t)$, the approximation operators $L_\lambda(f,t)$ are determined uniquely by the differential equation (1) and the normalization condition (2). In the past few decades, many operators of exponential type viz. Bernstein, Baskakov, Ismail–May, Szász–Mirakyan, Post–Widder, Gauss–Weierstrass, and Cismaşiu have been intensively studied by researchers. The basic results essential to determine convergence are related to the moments, which have been recently collected by Gupta and Rassias in [2] for many operators. For a detailed study of these operators, one may refer to [3, 4]. It has been observed that these operators preserve both constant as well as linear functions. In 2005, the work of Ismail and May was extended by Tyliba and Wachnicki [5] by defining a new class of operators which extends the class of exponential-type operators. These operators do not preserve linear functions and are defined as

$$(L_\lambda^\beta f)(t) = \int_{-\infty}^{\infty} W^\beta(\lambda,t,v)f(v)dv,$$

with $\int_{-\infty}^{\infty} W^\beta(\lambda,t,v)dv = 1$ (normalization condition), where $W^\beta(\lambda,t,v)$–the kernel of L_λ^β is a positive function satisfying

$$\frac{\partial}{\partial t}W^\beta(\lambda,t,v) = \frac{\lambda}{p(t)}W^\beta(\lambda,t,v)(v-t) - \beta W^\beta(\lambda,t,v),$$

and β is a non-negative real number. Very recently, Herzog [6] studied some important properties of such operators and termed them as semiexponential-type operators. Here, we analyze the approximation properties of semiexponential Szász–Mirakyan operators by replacing the function with the weights of Baskakov basis functions under integration. Different types of modifications of the Szász–Mirakyan operators have been studied by several researchers in the past. In the year 1983, the modification of Szász–Mirakyan operators was introduced by Prasad *et al.* [7] by considering the weights of the Baskakov basis functions. However, there were certain mistakes in [7], which were later corrected by Gupta [8]. Later, Gupta and Srivastava [9] studied the approximation properties of Szász–Mirakyan–Baskakov-type operators. The asymptotic formula and error estimation for such operators were investigated by Gupta–Gupta in [10,11]. Many other direct results for the modified Szász–Mirakyan operators were investigated in [8,12–14]. Also, the post-quantum variant of such operators has been discussed in [15]. The simultaneous approximation properties of different operators have been studied by various researchers, see, for instance, [8,16–19]. Our main focus is to derive the rate of convergence in simultaneous approximation.

We consider the class $C_\gamma[0,\infty)$ to be the collection of all such functions $f \in C[0,\infty)$ satisfying $|f(t)| \leq K(1+t)^\gamma$ for some $K, \gamma > 0$. For $f \in C_\gamma[0,\infty)$, the semiexponential extension of Szász–Mirakyan operators is defined as follows:

$$(S_n^\beta f)(x) = \sum_{k=0}^{\infty} s_{nk}^\beta(x) \frac{< v_{nk}, f >}{< v_{nk}, 1 >}, \quad \beta > 0, \tag{3}$$

where the semiexponential Szász–Mirakyan basis function is given by

$$s_{nk}^\beta(x) = e^{-(n+\beta)x} \frac{(n+\beta)^k x^k}{k!}$$

and the Baskakov basis function with the rising factorial $(n)_k = n(n+1)\cdots(n+k-1)$, $(n)_0 = 1$, is given as

$$v_{nk}(t) = \frac{(n)_k}{k!} \frac{t^k}{(1+t)^{n+k}}.$$

Also, the operators defined by (3) can be expressed as

$$(S_n^\beta f)(x) = \int_0^\infty J_n^\beta(t,x) f(t)dt,$$

where

$$J_n^\beta(t,x) = e^{-(n+\beta)x} \sum_{k=0}^{\infty} \frac{(n+\beta)^k x^k}{k!} \frac{(n-1)_{k+1}}{k!} \frac{t^k}{(1+t)^{n+k}}.$$

This chapter consists of the estimation of the moments and central moments of the operators defined by (3), along with some other basic lemmas and remarks. We provide the convergence estimates for such operators in simultaneous approximation.

2. Auxiliary Results

In this section, we present certain results that are required in the sequel.

Lemma 1. *For $m \in N \cup \{0\}$, let the function $\mu_{n,m}^\beta(x)$ be defined as*

$$\mu_{n,m}^\beta(x) = \sum_{k=0}^{\infty} s_{nk}^\beta(x) \left(\frac{k}{n} - x\right)^m,$$

then

$$\mu_{n,0}^\beta(x) = 1, \quad \mu_{n,1}^\beta(x) = \frac{\beta x}{n},$$

$$\mu_{n,2}^\beta(x) = \frac{x\left(\beta + n + \beta^2 x\right)}{n^2},$$

$$\mu_{n,3}^\beta(x) = \frac{\beta + n + 3\beta^2 x + 3\beta n x + \beta^3 x^2}{n^3}.$$

Consequently, $\mu_{n,m}^\beta(x) = O_x(n^{-[(m+1)/2]})$, where $[s]$ stands for the integral part of s.

Lemma 2. *Let us denote* $e_i = t^i, \quad i = 0, 1, 2, \ldots,$ *then the following result holds*:

$$(S_n^\beta e_0)(x) = 1, \quad (S_n^\beta e_1)(x) = \frac{1 + (n+\beta)x}{n-2},$$

$$(S_n^\beta e_2)(x) = \frac{(n+\beta)^2 x^2 + 4(n+\beta)x + 2}{(n-2)(n-3)},$$

$$(S_n^\beta e_3)(x) = \frac{1}{(n-2)(n-3)(n-4)} \times \Big\{(n+\beta)^3 x^3 + 9(n+\beta)^2 x^2 + 18(n+\beta)x + 6\Big\},$$

$$(S_n^\beta e_4)(x) = \frac{1}{(n-2)(n-3)(n-4)(n-5)} \times \Big\{(n+\beta)^4 x^4 + 16(n+\beta)^3 x^3 + 72(n+\beta)^2 x^2 + 96(n+\beta)x + 24\Big\},$$

$$(S_n^\beta e_5)(x) = \frac{1}{(n-2)(n-3)(n-4)(n-5)(n-6)} \times \Big\{(n+\beta)^5 x^5 + 25(n+\beta)^4 x^4 + 200(n+\beta)^3 x^3 + 600(n+\beta)^2 x^2 + 600(n+\beta)x + 120\Big\},$$

$$(S_n^\beta e_6)(x) = \frac{1}{(n-2)(n-3)(n-4)(n-5)(n-6)(n-7)} \times \Big\{(n+\beta)^6 x^6 + 36(n+\beta)^5 x^5 + 450(n+\beta)^4 x^4 + 2400(n+\beta)^3 x^3 + 5400(n+\beta)^2 x^2 + 4320(n+\beta)x + 720\Big\}.$$

Proof. By definition of the operators (3), we have

$$\begin{aligned}(S_n^\beta e_r)(x) &= e^{-(n+\beta)x} \sum_{k=0}^{\infty} \frac{(n+\beta)^k x^k}{k!} \int_0^\infty \frac{(n-1)_{k+1}}{k!} \frac{t^{k+r}}{(1+t)^{n+k}} dt \\ &= \frac{(n-r-2)!}{(n-2)!} e^{-(n+\beta)x} \sum_{k=0}^{\infty} \frac{(n+\beta)^k x^k}{k!} \frac{(k+r)!}{k!}.\end{aligned}$$

The values of different moments can easily be determined for different values of r. □

Remark 3. From Lemma 2, we observe that the rth-order moment can be expressed as follows:

$$(S_n^\beta e_r)(x) = \frac{1}{(n-2)(n-3)...(n-r-1)}\{(n+\beta)^r x^r + r^2(n+\beta)^{r-1}x^{r-1} + \frac{r^2(r-1)^2}{2}(n+\beta)^{r-2}x^{r-2} + \text{ lower degree terms of } x\}.$$

Lemma 4. *For $m \in N \cup \{0\}$, let us denote by $U_{n,m}^\beta(x)$ the mth-order central moment of the operators (3), i.e.,*

$$U_{n,m}^\beta(x) = (n-1)\sum_{k=1}^{\infty} s_{nk}^\beta(x)\int_0^\infty (t-x)^m v_{nk}(t)dt,$$

then few moments are given as

$$U_{n,0}^\beta(x) = 1,$$
$$U_{n,1}^\beta(x) = \frac{1+(\beta+2)x}{n-2},$$
$$U_{n,2}^\beta(x) = \frac{x^2\left(\beta^2+6\beta+n+6\right)+2x(2\beta+n+3)+2}{(n-2)(n-3)}.$$

In general, we have

$$U_{n,m}^\beta(x) = O_x(n^{-[(m+1)/2]}).$$

The proof follows by using linearity property and applying Lemma 2, we omit the details.

Lemma 5. *There exist the polynomials $\xi_{i,j,r}(x)$ independent of n, β, and k such that*

$$x^r\frac{d^r}{dx^r}[s_{nk}^\beta(x)] = \sum_{\substack{2i+j\le r\\ i,j\ge 0}} (n+\beta)^i(k-(n+\beta)x)^j\xi_{i,j,r}(x)[s_{nk}^\beta(x)].$$

Proof. To derive the result, it is sufficient to show that

$$\frac{d^r}{dx^r}[e^{-(n+\beta)x}(n+\beta)^k x^k]$$
$$= \sum_{\substack{2i+j\le r\\ i,j\ge 0}} (n+\beta)^i (k-(n+\beta)x)^j \xi_{i,j,r}(x) e^{-(n+\beta)x}(n+\beta)^k x^{k-r}.$$

We make use of the principle of mathematical induction. First, for $r=1$, we have

$$\frac{d}{dx}[e^{-(n+\beta)x}(n+\beta)^k x^k]$$
$$= \frac{(k-nx)}{x} e^{-(n+\beta)x}(n+\beta)^k x^k - \beta e^{-(n+\beta)x}(n+\beta)^k x^k.$$

Hence, the result holds for $r=1$, as in this case, $i=0$, $j=1$, $\xi_{i,j,r}(x)=1$.

Next, we assume that the result holds for r, then we can write

$$\frac{\partial^{r+1}}{\partial x^{r+1}}\left[e^{-(n+\beta)x}(n+\beta)^k x^k\right]$$
$$= \frac{\partial}{\partial x}\left[\sum_{\substack{2i+j\le r\\ i,j\ge 0}} (n+\beta)^i (k-(n+\beta)x)^j \xi_{i,j,r}(x) e^{-(n+\beta)x}(n+\beta)^k x^{k-r}\right]$$
$$= \sum_{\substack{2i+j\le r\\ i,j\ge 0}} (n+\beta)^{i+1} (k-(n+\beta)x)^{j-1} (-j\xi_{i,j,r}(x)) e^{-(n+\beta)x}(n+\beta)^k x^{k-r}$$
$$+ \sum_{\substack{2i+j\le r\\ i,j\ge 0}} (n+\beta)^i (k-(n+\beta)x)^j \xi'_{i,j,r}(x) e^{-(n+\beta)x}(n+\beta)^k x^{k-r}$$
$$+ \sum_{\substack{2i+j\le r\\ i,j\ge 0}} (n+\beta)^{i+1} (k-(n+\beta)x)^j (-\xi_{i,j,r}(x)) e^{-(n+\beta)x}(n+\beta)^k x^{k-r}$$
$$+ \sum_{\substack{2i+j\le r\\ i,j\ge 0}} (n+\beta)^i (k-(n+\beta)x)^{j+1} \xi_{i,j,r}(x) e^{-(n+\beta)x}(n+\beta)^k x^{k-r-1}$$
$$+ \sum_{\substack{2i+j\le r\\ i,j\ge 0}} (n+\beta)^{i+1} (k-(n+\beta)x)^j \xi_{i,j,r}(x) e^{-(n+\beta)x}(n+\beta)^k x^{k-r}$$
$$+ \sum_{\substack{2i+j\le r\\ i,j\ge 0}} (n+\beta)^i (k-(n+\beta)x)^j (-r\xi_{i,j,r}(x)) e^{-(n+\beta)x}(n+\beta)^k x^{k-r}.$$

Hence, we get

$$
\begin{aligned}
&\frac{\partial^{r+1}}{\partial x^{r+1}}\left[e^{-(n+\beta)x}(n+\beta)^k x^k\right] \\
&= \sum_{\substack{2(i-1)+(j+1)\le r\\ i\ge 1, j\ge 0}} (n+\beta)^i (k-(n+\beta)x)^j (-(j+1) \\
&\qquad\qquad \xi_{i-1,j+1,r}(x)) e^{-(n+\beta)x}(n+\beta)^k x^{k-r} \\
&\quad + \sum_{\substack{2i+j\le r\\ i,j\ge 0}} (n+\beta)^i (k-(n+\beta)x)^j \xi'_{i,j,r}(x) e^{-(n+\beta)x}(n+\beta)^k x^{k-r} \\
&\quad + \sum_{\substack{2(i-1)+j\le r\\ i\ge 1, j\ge 0}} (n+\beta)^i (k-(n+\beta)x)^j (-\xi_{i-1,j,r}(x)) e^{-(n+\beta)x}(n+\beta)^k x^{k-r} \\
&\quad + \sum_{\substack{2i+(j-1)\le r\\ i\ge 0, j\ge 1}} (n+\beta)^i (k-(n+\beta)x)^j \xi_{i,j-1,r}(x) e^{-(n+\beta)x}(n+\beta)^k x^{k-r-1} \\
&\quad + \sum_{\substack{2(i-1)+j\le r\\ i\ge 1, j\ge 0}} (n+\beta)^i (k-(n+\beta)x)^j \xi_{i-1,j,r}(x) e^{-(n+\beta)x}(n+\beta)^k x^{k-r} \\
&\quad + \sum_{\substack{2i+j\le r\\ i,j\ge 0}} (n+\beta)^i (k-(n+\beta)x)^j (-r\xi_{i,j,r}(x)) e^{-(n+\beta)x}(n+\beta)^k x^{k-r}.
\end{aligned}
$$

The above expression has the desired form expressed as

$$
\begin{aligned}
\xi_{i,j,r+1}(x) &= -x(j+1)\xi_{i-1,j+1,r}(x) + x\xi'_{i,j,r}(x) - x\xi_{i-1,j,r}(x) + \xi_{i,j-1,r}(x) \\
&\quad + x\xi_{i-1,j,r}(x) - rx\xi_{i,j,r}(x) \\
&= -x(j+1)\xi_{i-1,j+1,r}(x) + x\xi'_{i,j,r}(x) + \xi_{i,j-1,r}(x) - rx\xi_{i,j,r}(x),
\end{aligned}
$$

where $2i+j\le r+1$ and $i,j\ge 0$ with the condition that $q_{i,j,r}(x)=0$ when any one of the constraints $2i+j\le r;\quad i,j\ge 0$ is not satisfied. Hence, the result is true for $r+1$.

This completes the proof of the lemma. □

3. Approximation

This section consists of certain direct results, which provide the convergence in ordinary and simultaneous approximations.

Theorem 6. *For $f \in C[0,\infty)$, we have*

$$\begin{aligned}
&\|S_n^\beta f(x) - f(x+t)\|_{C[0,a]} \\
&\quad \leq \left\{1 + \frac{a^2\left(\beta^2+6\beta+n+6\right)+2a(2\beta+n+3)+2}{(n-2)(n-3)}\right\}\omega\left(f,\frac{1}{\sqrt{n}}\right),
\end{aligned}$$

where $t \in [0,a]$ for $a > 0$ and ω is the usual first-order modulus of continuity and $||.||$ is the sup-norm on $[0,a]$.

The proof of the above result is obvious; we use second-order central moment given in Lemma 4.

Theorem 7. *For $f \in C_\gamma[0,\infty)$, if rth derivative of f exists in $(0,\infty)$, then*

$$\lim_{n\to\infty}[S_n^\beta f(x)]^{(r)} = f^{(r)}(x).$$

Proof. From Taylor's expansion, we have

$$f(t) = \sum_{k=0}^{r}\frac{f^{(k)}(x)}{k!}(t-x)^k + \phi(t,x)(t-x)^r,$$

where $\phi(t,x) \to 0$ as $t \to x$.

Thus,

$$\begin{aligned}
[S_n^\beta f(x)]^{(r)} &= \int_0^\infty J_n^{\beta(r)}(t,x)f(t)dt \\
&= \sum_{k=0}^{r}\frac{f^{(k)}(x)}{k!}\int_0^\infty J_n^{\beta(r)}(t,x)(t-x)^k dt \\
&\quad + \int_0^\infty J_n^{\beta(r)}(t,x)\phi(t,x)(t-x)^r dt \\
&:= I_1 + I_2.
\end{aligned}$$

First consider

$$\begin{aligned}I_1 &= \sum_{k=0}^{r} \frac{f^{(k)}(x)}{k!} \sum_{u=0}^{k} \binom{k}{u} (-x)^{k-u} \frac{d^r}{dx^r} \int_0^\infty J_n^\beta(t,x) t^u dt \\ &= \frac{d^r}{dx^r} \left(\frac{(n+\beta)^r}{(n-2)(n-3)\cdots(n-(r+1))} x^r + lower\ degree\ terms\ of\ x \right) \\ &\quad \cdot \frac{f^{(r)}(x)}{r!} \\ &= \frac{(n+\beta)^r}{(n-2)(n-3)\cdots(n-(r+1))} f^{(r)}(x),\end{aligned}$$

which converges to $f^{(r)}(x)$ as $n \to \infty$. Next, applying Lemma 5, we get

$$\begin{aligned}|I_2| \leq \sum_{\substack{2i+j\leq r \\ i,j\geq 0}} (n+\beta)^i \frac{|\xi_{i,j,r}|}{x^r} \sum_{k=1}^{\infty} |k-(n+\beta)x|^j s_{nk}^\beta(x) \\ \cdot \int_0^\infty v_{nk}(t) |\phi(t,x)| (t-x)^r dt := I_3 + I_4.\end{aligned}$$

Since $\phi(t,x) \to 0$ as $t \to x$, for a given $\epsilon > 0$, there exists a $\delta > 0$ such that $|\phi(t,x)| < \epsilon$ whenever $|t-x| < \delta$. Further, if s is any positive integer satisfying $s \geq \{\gamma, r\}$, then we can find an absolute constant M_1 such that $|(t-x)^r \phi(t,x)| \leq M_1 |t-x|^s$ for $|t-x| \geq \delta$. With $M_2 = \sup_{2i+j\leq r} x^{-r} |\xi_{i,j,r}|$, we have

$$\begin{aligned}I_3 \leq \sum_{\substack{2i+j\leq r \\ i,j\geq 0}} (n+\beta)^i \sum_{k=1}^{\infty} |k-(n+\beta)x|^j s_{nk}^\beta(x). \left\{ \epsilon \int_{|t-x|<\delta} v_{nk}(t) |t-x|^r dt \right. \\ \left. + \int_{|t-x|\geq\delta} v_{nk}(t) M_1 |t-x|^s dt \right\} := I_5 + I_6.\end{aligned}$$

Applying Lemmas 1 and 4 and using Cauchy–Schwarz inequality, we obtain

$$I_5 \leq \epsilon \sum_{\substack{2i+j\leq r\\ i,j\geq 0}} (n+\beta)^i \sum_{k=0}^{\infty} s_{nk}^{\beta}(x)|k-(n+\beta)x|^j \left\{ \int_0^{\infty} v_{nk}(t)dt \right\}^{\frac{1}{2}}$$

$$\cdot \left\{ \int_0^{\infty} v_{nk}(t)(t-x)^{2r}dt \right\}^{\frac{1}{2}}$$

$$\leq \epsilon M_2 \sum_{\substack{2i+j\leq r\\ i,j\geq 0}} 0(n^i)O(n^{\frac{j}{2}})O(n^{\frac{-r}{2}}) = \epsilon O(1).$$

Again using Lemmas 1 and 4 and applying Cauchy–Schwarz inequality, we obtain

$$I_6 \leq M_3 \sum_{\substack{2i+j\leq r\\ i,j\geq 0}} (n+\beta)^i \sum_{k=1}^{\infty} s_{nk}^{\beta}(x)|k-(n+\beta)x|^j$$

$$\left\{ \int_{|t-x|\geq\delta} v_{nk}(t)|t-x|^s dt \right\}^{\frac{1}{2}} \left\{ \int_0^{\infty} v_{nk}(t)(t-x)^{2r}dt \right\}$$

$$\leq \epsilon M_3 \sum_{\substack{2i+j\leq r\\ i,j\geq 0}} (n+\beta)^i \left\{ s_{nk}^{\beta}(x)(k-(n+\beta)x)^{2j} \right\}^{\frac{1}{2}}$$

$$\cdot \left\{ \sum_{k=0}^{\infty} s_{nk}^{\beta}(x) \int_0^{\infty} v_{nk}(t)(t-x)^{2s}dt \right\}^{\frac{1}{2}}$$

$$= \sum_{\substack{2i+j\leq r\\ i,j\geq 0}} O(n^i)O(n^{\frac{j}{2}})O(n^{\frac{-r}{2}}) = o(1).$$

As ϵ is arbitrary, hence $I_3 = o(1)$. Also, $I_2 = o(1)$ as $I_4 \to 0$ as $n \to \infty$. Finally, from the estimates of I_1 and I_2, we are led to the desired result. □

Theorem 8. *For $f \in C_\gamma[0,\infty)$, if $(r+2)$th derivative of f exists at a fixed point in $(0,\infty)$, then*

$$\lim_{n\to\infty} n\left[[S_n^\beta f(x)]^{(r)} - f^{(r)}(x)\right] = \frac{r(r+3+2\beta)}{2} f^{(r)}(x) + f^{(r+1)}(x)$$
$$[(\beta+2)x + r(x+1) + 1] f^{(r+1)}(x)$$
$$+\frac{x(x+2)}{2} f^{(r+2)}(x).$$

Proof. From Taylor's expansion, we can write

$$f(t) = \sum_{k=0}^{r+2} \frac{f^{(k)}(x)}{k!}(t-x)^k + \phi(t,x)(t-x)^{r+2},$$

where $\phi(t,x) \to 0$ as $t \to x$.

On applying Lemma 4, we have

$$n\left[[S_n^\beta f(x)]^{(r)} - f^{(r)}(x)\right]$$
$$= n\left[\sum_{k=0}^{r+2} \frac{f^{(k)}(x)}{k!} \int_0^\infty J_n^{\beta(r)}(t,x)(t-x)^k dt - f^{(r)}(x)\right]$$
$$+n\int_0^\infty J_n^{\beta(r)}(t,x)\phi(t,x)(t-x)^{r+2}dt$$
$$:= I_1 + I_2.$$

First consider

$$I_1 = n\left[\sum_{k=0}^{r+2} \frac{f^{(k)}(x)}{k!} \sum_{u=0}^{k} \binom{k}{u}(-x)^{k-u} \frac{d^r}{dx^r} \int_0^\infty J_n^\beta(t,x) t^u dt - f^{(r)}(x)\right]$$
$$= n\frac{f^{(r)}(x)}{r!}\left[S_n^{\beta(r)} e_r(x) - r!\right] + n\frac{f^{(r+1)}(x)}{(r+1)!}$$
$$\times \left\{(-x)(r+1)S_n^{\beta(r)} e_r(x) + S_n^{\beta(r)} e_{r+1}(x)\right\}$$
$$+n\frac{f^{(r+2)}(x)}{(r+2)!}\left\{x^2\frac{(r+1)(r+2)}{2} S_n^{\beta(r)}\right.$$
$$\left.\times e_r(x) + (-x)(r+2)S_n^{\beta(r)} e_{r+1}(x) + S_n^{\beta(r)} e_{r+2}(x)\right\}.$$

Using Remark 3, we obtain

$$\begin{aligned}
I_1 &= nf^{(r)}(x)\left\{\frac{(n+\beta)^r}{(n-2)(n-3)\cdots(n-(r+1))}-1\right\}\\
&\quad+n\frac{f^{(r+1)}(x)}{(r+1)!}\left\{(-x)(r+1)r!\frac{(n+\beta)^r}{(n-2)(n-3)\cdots(n-(r+1))}\right.\\
&\quad+(r+1)!x\frac{(n+\beta)^{r+1}}{(n-2)(n-3)\cdots(n-(r+2))}\\
&\quad\left.+r!(r+1)^2\frac{(n+\beta)^r}{(n-2)(n-3)\cdots(n-(r+2))}\right\}\\
&\quad+n\frac{f^{(r+2)}(x)}{(r+2)!}\left\{x^2\frac{(r+1)(r+2)}{2}r!\frac{(n+\beta)^r}{(n-2)(n-3)\cdots(n-(r+1))}\right.\\
&\quad+(-x)(r+2)\left((r+1)!x\frac{(n+\beta)^{r+1}}{(n-2)(n-3)\cdots(n-(r+2))}\right.\\
&\quad\left.+r!(r+1)^2\frac{(n+\beta)^r}{(n-2)(n-3)\cdots(n-(r+2))}\right)\\
&\quad+\left(\frac{(r+2)!}{2}x^2\frac{(n+\beta)^{r+2}}{(n-2)(n-3)\cdots(n-(r+3))}\right.\\
&\quad+(r+1)!x(r+2)^2\frac{(n+\beta)^{r+1}}{(n-2)(n-3)\cdots(n-(r+3))}\\
&\quad\left.\left.+r!\frac{(r+2)^2(r+1)^2}{2}\frac{(n+\beta)^r}{(n-2)(n-3)\cdots(n-(r+3))}\right)+O(n^{-(r+3)})\right\}\\
&= nf^{(r)}(x)\left\{\frac{(n+\beta)^r}{(n-2)(n-3)\cdots(n-(r+1))}-1\right\}\\
&\quad+nf^{(r+1)}(x)\left\{\frac{(n+\beta)^r\left[x(n+\beta)+(r+1)-x(n-r-2)\right]}{(n-2)(n-3)\cdots(n-(r+2))}\right\}\\
&\quad+nf^{(r+2)}(x)\left\{\frac{(n+\beta)^r}{2(n-2)(n-3)\cdots(n-(r+3))}\right.\\
&\quad\cdot\left\{\beta^2x^2+r(x+1)((2\beta+5)x+3)\right.\\
&\quad\left.+6\beta x^2+4\beta x+r^2(x+1)^2+6x^2+6x+2+n(x^2+2x)\right\}\\
&\quad\left.+O(n^{-(r+3)})\right\}.
\end{aligned}$$

Taking limit $n \to \infty$, we have the desired result provided $I_2 \to 0$ as $n \to \infty$ which follows along the lines of Theorem 7. □

As a special case of the above theorem, we have the following Voronovskaja-type asymptotic formula in ordinary approximation.

Corollary 9. *For $f \in C_\gamma[0,\infty)$, if f' and f'' exist at a fixed point in $(0,\infty)$, then*

$$\lim_{n\to\infty} n\left[S_n^\beta f(x) - f(x)\right] = [(\beta+2)x+1]f'(x) + \frac{x(x+2)}{2}f''(x).$$

References

[1] Ismail, M. and May, C.P. On a family of approximation operators, *J. Math. Anal. Appl.* 63, 446–462 (1978).

[2] Gupta, V. and Rassias, M.T. *Moments of Linear Positive Operators and Approximation.* Springer (2020).

[3] Gupta, V. and Rassias, M.T. *Computation and Approximation.* Springer (2021).

[4] Gupta, V. and Agarwal, R.P. *Convergence Estimates in Approximation Theory.* Springer (2014).

[5] Tyliba, A. and Wachnicki, E. On some class of exponential type operators, *Comment. Math.* 45, 59–73 (2005).

[6] Herzog, M. Semi-Exponential Operators, *Symmetry.* 13, 637 (2021).

[7] Prasad, G., Agrawal, P.N. and Kasana, H.S. Approximation of functions on $[0,\infty)$ by a new sequence of modified Szász operators, *Math. Forum.* 6(2), 1–11 (1983).

[8] Gupta, V. A note on modified Szász operators, *Bull. Inst. Math. Academia Sinica.* 21(3), 275–278 (1993).

[9] Gupta, V. and Srivastava, G.S. On the convergence of derivative by Szász-Mirakyan Baskakov type operators, *The Math. Student.* 64(1–4), 195–205 (1995).

[10] Gupta, V. and Gupta, P. Rate of convergence by Szász-Mirakyan Baskakov type operators, *Istanbul Univ. Fen Fakult. Mat. Dergisi.* 57–58, 71–78 (1998–1999).

[11] Gupta, V. and Gupta, P. Direct theorem in simultaneous approximation for Szász-Mirakyan Baskakov type operators, *Kyungpook Math. J.* 41(2), 243–249 (2001).

[12] Kasana, H.S. On approximation of unbounded functions by linear combinations of modified Szász-Mirakyan operators, *Acta Math. Hungar.* 61(3–4), 283–290 (1993).

[13] Kasana, H.S., Prasad, G., Agrawal, P.N. and Sahai. A. Modified Szász operators. In *Conference on Mathematical Analysis and its Applications*, Kuwait, Pergamon Press, Oxford, pp. 29–41 (1985).

[14] Mazhar, S.M. and Totik, V. Approximation by modified Szász operators, *Acta. Sci. Math.* 49, 257–269 (1985).

[15] Gupta, V. (p, q)-Szász–Mirakyan–Baskakov Operators, *Complex Anal. Oper. Theory.* 12, 17–25 (2018).

[16] Gupta, V. and Agrawal, G. Approximation for modification of exponential type operators connected with $x(x+1)^2$, *Revista de la Real Academia de Ciencias Exactas, Físicas y Naturales. Serie A. Matemáticas.* 114(3) (2020). https://doi.org/10.1007/s13398-020-00889-1.

[17] Gupta, V. and Srivastava, G.S. Simultaneous approximation by Baskakov–Szász type operators, *Bull. Math. Soc. Sci. Math. Roumanie (N. S.).* 37(3–4), 73–85 (1993).

[18] Gupta, V. and Agrawal, G. Approximation for link Ismail-May operators, *Ann. Funct. Anal.* 11, 728–747 (2020).

[19] Govil, N.K. and Gupta, V. Simultaneous approximation for the Bézier variant of Baskakov-Beta operators, *Math. Comput. Modelling.* 44 (11–12), 1153–1159 (2006).

https://doi.org/10.1142/9789811293009_0019

Chapter 19

Certain Extension of Riemann–Liouville Fractional Derivative Operator by Using Wiman's Function

Shilpi Jain[*,††], Rahul Goyal[†,‡‡], R.P Agarwal[‡,§§],
Praveen Agarwal[§,¶,‖,¶¶,‖‖], and Shaher Momani[#,**]

[*] *Department of Mathematics, Poornima University, Jaipur-303905, India*
[†] *Department of Mathematics, Anand International College of Engineering, Jaipur-303012, India*
[‡] *Department of Mathematics and Systems Engineering Florida Institute of Technology Melbourne, FL 32901 USA*
[§] *Department of Mathematics, Anand International College of Engineering, Jaipur-303012, India*
[¶] *Nonlinear Dynamics Research Center (NDRC), Ajman University, UAE*
[‖] *International Center for Basic and Applied Sciences, Jaipur 302029, India*
[#] *Nonlinear Dynamics Research Center (NDRC), Ajman University, UAE*
[**] *Department of Mathematics, School of Science, The University of Jordan, Amman 11942, Jordan*
[††] *shilpijain1310@gmail.com*
[‡‡] *rahul.goyal01@anandice.ac.in*
[§§] *agarwalr@fit.edu*
[¶¶] *praveen.agarwal@anandice.ac.in*
[‖‖] *goyal.praveen2011@gmail.com*

The objective of this chapter is to study the extension of the Riemann–Liouville fractional derivative operator by using 2-parameter Mittag-Leffler function given by Wiman. We have to study some generating relations, Mellin transforms, and relations to extended hypergeometric functions for this extended Riemann–Liouville fractional derivative operator.

1. Introduction and Preliminaries

The theory of fractional calculus (FC) has successfully been utilized to describe the fractal problems and real-life problems in engineering mathematics. It is a branch of mathematics that studies the many various changes of defining real number powers or complex number powers of the differentiation operator D and beginning a calculus for such operators generalizing the classical one. It has become an interesting area of research since it has various applications in different subjects of science and engineering, such as fluid flow, electrical networks, and probability theory. Riemann–Liouville fractional derivative operator plays an important role in fractional calculus because of its applications in different branches of science. In recent years, the concept of extensions of fractional operators, special functions, and applications of fractional differential equations attracts the attention of many researchers [2–13].

The classical Riemann–Liouville fractional derivative operator is defined as follows [14]:

$$D_z^u[f(z)] = \begin{cases} \frac{1}{\Gamma(-u)} \int_0^z (z-t)^{-u-1} f(t)dt, & (\Re(u) < 0) \\ \\ \frac{d^m}{dz^m}\{D_z^{u-m} f(z)\}, & (m-1 \le \Re(u) < m,\ m \in \mathbb{N}), \end{cases} \tag{1}$$

and by the definition, we have

$$\frac{d^m}{dz^m}\{D_z^{u-m} f(z)\} = \frac{1}{\Gamma(-u+m)} \int_0^z (z-t)^{-u+m-1} f(t)dt. \tag{2}$$

The theory of extension and modification of classical Riemann–Liouville fractional derivative operator have been started since 2000.

In 2010, Özarslan *et al.* [15] have extended classical Riemann–Liouville fractional derivative operator as follows:

$$D_z^{u,r}[f(z)] = \begin{cases} \frac{1}{\Gamma(-u)} \int_0^z (z-t)^{-u-1} e^{\left(\frac{-rz^2}{t(z-t)}\right)} f(t)dt, & (\Re(u) < 0) \\ \frac{d^m}{dz^m}\{D_z^{u-m,r} f(z)\}, & (m-1 \leq \Re(u) < m,\ m \in \mathbb{N}), \end{cases} \tag{3}$$

where $\Re(r) > 0$.

We easily seen that

$$D_{z,}^{u,r}[z^k] = \frac{B(k+1,-u;r)}{\Gamma(-u)} z^{k-u}. \tag{4}$$

Remark 1. If we set $r = 0$ in (3), we get classical Riemann–Liouville fractional derivative operator (1):

$$D_z^{u,0}[f(z)] = D_z^u[f(z)]. \tag{5}$$

In the above extension, Özarslan *et al.* have used exponential function to extend classical Riemann–Liouville fractional derivative operator, and they also discussed various generating relations, Mellin transforms, and relations with other special functions for this extended operator. They have introduced exponential function as kernel in the integral part of classical Riemann–Liouville fractional derivative operator and then the integral part reduces to extended beta function defined in [16].

Very recently, Goyal *et al.* [17] have extended the classical beta function by using 2-parameter Mittag-Leffler function $E_{r_1,r_2}[z]$ given by Wiman [1] and studied various properties of this extended beta function. They have introduced Wiman's function as kernel in the integral of classical Euler beta function:

$$B_{(u_1,u_2)}^{(u)}(y_1,y_2) = \int_0^1 t^{y_1-1}(1-t)^{y_2-1} E_{u_1,u_2}\left(\frac{-u}{t(1-t)}\right) dt, \tag{6}$$

where $\min\{\Re(y_1),\ \Re(y_2)\} > 0$, $\Re(u_1) > 0, \Re(u_2) > 0$, $u \geq 0$ and $E_{u_1,u_2}(z)$ is 2-parameter Mittag-Leffler function.

Later, Jain *et al.* [18] have extended Gauss hypergeometric function and confluent hypergeometric function by using extended beta function (6), given by Goyal *et al.* Particularly, they studied many basic properties like integral representations and Mellin transforms of these extended hypergeometric functions.

The extended Gauss hypergeometric function is defined as follows [18]:

$$F^{(s)}_{(s_1,s_2)}(q_0, q_1, q_2; z) = \sum_{n=0}^{\infty} \frac{B^{(s)}_{(s_1,s_2)}(q_1 + n, q_2 - p_1)}{B(q_1, q_2 - q_1)}(q_0)_n \frac{z^n}{n!}, \tag{7}$$

where $\Re(q_2) > \Re(q_1) > 0$, $\min\{\Re(s_1), \Re(s_2)\} > 0$, $s \geq 0$, $|z| < 1$, and $B^{(s)}_{(s_1,s_2)}(w_1, w_2)$ is extended beta function.

The integral representation of the extended Gauss hypergeometric function is defined as follows [18]:

$$\begin{aligned} F^{(s)}_{(s_1,s_2)}&(q_0, q_1, q_2; z) \\ &= \frac{1}{B(q_1, q_2 - q_1)} \int_0^1 t^{q_1 - 1}(1-t)^{q_2 - q_1 - 1}(1 - zt)^{-q_0} E_{s_1,s_2}\left(\frac{-s}{t(1-t)}\right) dt, \end{aligned} \tag{8}$$

where ($\Re(q_2) > \Re(q_1) > 0$, $\min\{\Re(s_1), \Re(s_2)\} > 0$, $s \geq 0$ and $|z| < 1$).

$$\Phi^{(s)}_{(s_1,s_2)}(q_1, q_2; z) = \sum_{n=0}^{\infty} \frac{B^{(s)}_{(s_1,s_2)}(q_1 + n, q_2 - q_1)}{B(q_1, q_2 - q_1)} \frac{z^n}{n!}, \tag{9}$$

where ($\Re(q_2) > \Re(q_1) > 0$, $\min\{\Re(s_1), \Re(s_2)\} > 0$ and $s \geq 0$).

The integral representation of the extended confluent hypergeometric function is defined as follows [18]:

$$\begin{aligned} \Phi^{(s)}_{(s_1,s_2)}&(q_1, q_2; z) \\ &= \frac{1}{B(q_1, q_2 - q_1)} \int_0^1 t^{q_1 - 1}(1-t)^{q_2 - q_1 - 1} e^{zt} E_{s_1,s_2}\left(\frac{-s}{t(1-t)}\right) dt, \end{aligned} \tag{10}$$

where ($\Re(q_2) > \Re(q_1) > 0$, $\min\{\Re(s_1), \Re(s_2)\} > 0$ and $s \geq 0$).

2. Main Results

In this section, we define a new extension of the classical Riemann–Liouville fractional derivative operator by using 2-parameter Mittag-Leffler function. We have introduced Wiman function as kernel in the integral part of classical Riemann–Liouville fractional derivative operator and then the integral part reduces to extended beta function defined in [17]. We have also established some good results for this extended Riemann–Liouville fractional derivative operator.

Definition 2.

$$D^{u,(r)}_{z,(r_1,r_2)}[f(z)] = \begin{cases} \frac{1}{\Gamma(-u)}\int_0^z (z-t)^{-u-1}E_{r_1,r_2}\left(\frac{-rz^2}{t(z-t)}\right)f(t)dt, & (\Re(u)<0) \\ \frac{d^m}{dz^m}\{D^{u-m,(r)}_{z,(r_1,r_2)}f(z)\}, & (m-1\leq \Re(u)<m,\ m\in\mathbb{N}), \end{cases} \tag{11}$$

where $(\min\{\Re(r_1), \Re(r_2)\} > 0, \Re(r) > 0$ and $E_{r_1,r_2}(z)$ is 2-parameter Mittag-Leffler function).

It is easily seen that

$$D^{u-m,(r)}_{z,(r_1,r_2)}[f(z)] = \frac{1}{\Gamma(-u+m)}\int_0^z (z-t)^{-u+m-1}E_{r_1,r_2}\left(\frac{-rz^2}{t(z-t)}\right)f(t)dt. \tag{12}$$

Remark 3. (i) If we set $r_1 = r_2 = 1$ in (11), we get extended Riemann–Liouville fractional derivative operator (3):

$$D^{u,(r)}_{z,(1,1)}[f(z)] = D^{u,r}_z[f(z)]. \tag{13}$$

(ii) If we take $r_1 = r_2 = 1$ and $r = 0$ in (11), we get classical Riemann–Liouville fractional derivative operator (1):

$$D^{u,(0)}_{z,(1,1)}[f(z)] = D^{u,0}_z[f(z)] = D^u_z[f(z)]. \tag{14}$$

Theorem 4. *Assume* $\Re(u) < 0$, $\Re(k) < -1$, *then*

$$D^{u,(r)}_{z,(r_1,r_2)}[z^k] = \frac{B^{(r)}_{(r_1,r_2)}(k+1,-u)}{\Gamma(-u)} z^{k-u}. \quad (15)$$

Proof. From the definition of extended Riemann–Liouville fractional derivative operator (11), we get

$$D^{u,(r)}_{z,(r_1,r_2)}[z^k] = \frac{1}{\Gamma(-u)} \int_0^z (z-t)^{-u-1} E_{r_1,r_2}\left(\frac{-rz^2}{t(z-t)}\right) t^k dt. \quad (16)$$

Now put $t = xz$ in the above equation, then changing the limit from $t = 0, t = z$ to $x = 0, x = 1$, $dt = zdx$ and rearranging the terms, we get

$$D^{u,(r)}_{z,(r_1,r_2)}[z^k] = \frac{z^{k-u}}{\Gamma(-u)} \int_0^1 x^k (1-x)^{-u-1} E_{r_1,r_2}\left(\frac{-r}{x(1-x)}\right) dx. \quad (17)$$

Then from the definition of extended beta function (6), we get our desired result:

$$D^{u,(r)}_{z,(r_1,r_2)}[z^k] = \frac{B^{(r)}_{(r_1,r_2)}(k+1,-u)}{\Gamma(-u)} z^{k-u}. \quad (18)$$

□

Theorem 5. *Consider* $\Re(u) < 0$, $\Re(k) > 0$, $\Re(l) > 0$ *and* $|z| < 1$, *then*

$$D^{k-u,(r)}_{z,(r_1,r_2)}[z^{k-1}(1-z)^{-l}] = \frac{\Gamma(k)}{\Gamma(u)} z^{u-1} F^{(r)}_{(r_1,r_2)}(l,k,u;z). \quad (19)$$

Proof. From the definition of extended Riemann–Liouville fractional derivative operator (11), we get

$$D^{k-u,(r)}_{z,(r_1,r_2)}[z^{k-1}(1-z)^{-l}] = \frac{1}{\Gamma(u-k)} \int_0^z (z-t)^{u-k-1} E_{r_1,r_2}\left(\frac{-rz^2}{t(z-t)}\right) t^{k-1}(1-t)^{-l} dt. \quad (20)$$

Taking out z from integral, we have

$$D_{z,(r_1,r_2)}^{k-u,(r)}[z^{k-1}(1-z)^{-l}]$$
$$=\frac{z^{u-k-1}}{\Gamma(u-k)}\int_0^z(1-\frac{t}{z})^{u-k-1}E_{r_1,r_2}\left(\frac{-rz^2}{t(z-t)}\right)t^{k-1}(1-t)^{-l}dt. \tag{21}$$

Then set $t=xz$ in the above equation, then changing the limit from $t=0, t=z$ to $x=0, x=1$, $dt=zdx$ and rearranging the terms, we get

$$D_{z,(r_1,r_2)}^{k-u,(r)}[z^{k-1}(1-z)^{-l}]$$
$$=\frac{z^{u-1}}{\Gamma(u-k)}\int_0^1(1-x)^{u-k-1}E_{r_1,r_2}\left(\frac{-r}{x(1-x)}\right)x^{k-1}(1-xz)^{-l}dt. \tag{22}$$

By the integral formula of extended Gauss hypergeometric function (7), we get

$$D_{z,(r_1,r_2)}^{k-u,(r)}[z^{k-1}(1-z)^{-l}]=\frac{z^{u-1}}{\Gamma(u-k)}B(k,u-k)F_{(r_1,r_2)}^{(r)}(l,k,u;z). \tag{23}$$

Then by using the relationship between gamma function and classical beta function $B(x',y')=\frac{\Gamma(x')\Gamma(y')}{\Gamma(x'+y')}$ in the above equation, we get our desired result:

$$D_{z,(r_1,r_2)}^{k-u,(r)}[z^{k-1}(1-z)^{-l}]=\frac{\Gamma(k)}{\Gamma(u)}z^{u-1}F_{(r_1,r_2)}^{(r)}(l,k,u;z). \tag{24}$$

□

Theorem 6. *Assume f(z) to be the holomorphic function in the disc* $|z|<\delta$ *and has the taylor series expansion* $f(z)=\sum_{k=0}^{\infty}b_kz^k$. *Then,*

$$D_{z,(r_1,r_2)}^{u,(r)}[z^{\beta-1}f(z)]=\frac{z^{\beta-1}}{\Gamma(-u)}\sum_{k=0}^{\infty}b_kB_{(r_1,r_2)}^{(r)}(k+\beta,-u)z^{k-u}, \tag{25}$$

provided that $\Re(u)<0$, $k\in N$, $\Re(\beta)>0$, *and* $|z|<\delta$.

Proof. Then from the definition of the extended Riemann–Liouville fractional derivative operator (11), we get

$$D^{u,(r)}_{z,(r_1,r_2)}[z^{\beta-1}f(z)] = \frac{1}{\Gamma(-u)}\int_0^z (z-t)^{-u-1}E_{r_1,r_2}\left(\frac{-rz^2}{t(z-t)}\right)z^{\beta-1}f(t)dt. \tag{26}$$

Then, by using $f(t) = \sum_{k=0}^{\infty} b_k t^k$ in the above equation, we get

$$D^{u,(r)}_{z,(r_1,r_2)}[z^{\beta-1}f(z)]$$
$$= \frac{1}{\Gamma(-u)}\int_0^z (z-t)^{-u-1}E_{r_1,r_2}\left(\frac{-rz^2}{t(z-t)}\right)\left(z^{\beta-1}\sum_{k=0}^{\infty} b_k t^k\right)dt. \tag{27}$$

Since the given series $\sum_{k=0}^{\infty} b_k t^k$ is uniformly convergent in the disc $| z |< \delta$, we can change the order of integration and summation as follows:

$$D^{u,(r)}_{z,(r_1,r_2)}[z^{\beta-1}f(z)]$$
$$= \sum_{k=0}^{\infty} b_k\left(\frac{1}{\Gamma(-u)}\int_0^z (z-t)^{-u-1}E_{r_1,r_2}\left(\frac{-rz^2}{t(z-t)}\right)t^{\beta+k-1}\right)dt. \tag{28}$$

Then from the definition of the extended Riemann–Liouville fractional derivative operator (11), we have

$$D^{u,(r)}_{z,(r_1,r_2)}[z^{\beta-1}f(z)] = \sum_{k=0}^{\infty} b_k\left(D^{u,(r)}_{z,(r_1,r_2)}[z^{\beta+k-1}]\right). \tag{29}$$

By using Theorem (2.3), in the above equation, we get

$$D^{u,(r)}_{z,(r_1,r_2)}[z^{\beta-1}f(z)] = \sum_{k=0}^{\infty} b_k \frac{B^{(r)}_{(r_1,r_2)}(k+\beta,-u)}{\Gamma(-u)} z^{\beta+k-u-1}. \tag{30}$$

On rearranging the terms, we get our desired result:

$$D^{u,(r)}_{z,(r_1,r_2)}[z^{\beta-1}f(z)] = \frac{z^{\beta-1}}{\Gamma(-u)}\sum_{k=0}^{\infty} b_k B^{(r)}_{(r_1,r_2)}(k+\beta,-u)z^{k-u}. \tag{31}$$

□

Corollary 7. *Assume f(z) to be the holomorphic function in the disc* $\mid z \mid < \delta$ *and has the taylor series expansion* $f(z) = \sum_{k=0}^{\infty} b_k z^k$. *Then*

$$D^{u,(r)}_{z,(r_1,r_2)}[f(z)] = \frac{1}{\Gamma(-u)}\sum_{k=0}^{\infty} b_k B^{(r)}_{(r_1,r_2)}(k+1,-u)z^{k-u}, \tag{32}$$

provided that $\Re(u) < 0$, $k \in N$, *and* $\mid z \mid < \delta$.

Proof. Taking $\beta = 1$ in (25), we get our desired result. □

Example 8. The following holds true:

$$D^{u,(r)}_{z,(r_1,r_2)}[e^z] = \frac{z^{-u}}{\Gamma(-u)}\sum_{k=0}^{\infty} B^{(r)}_{(r_1,r_2)}(k+1,-u)\frac{z^k}{k!}. \tag{33}$$

Taking $f(z) = e^z = \sum_{k=0}^{\infty} \frac{z^k}{k!}$ and $\beta = 1$ in Theorem (2.5), we obtained our desired result.

Theorem 9. *Consider* $\Re(u) < 0$, $\Re(\beta) > 0$, *and* $s > 0$, *then the Mellin transform for the extended Riemann–Liouville fractional derivative operator defined as* (11) *is given as follows:*

$$\boldsymbol{M}\left[D^{u,(r)}_{z,(r_1,r_2)}[z^{\beta}];s\right] = \frac{z^{\beta-u}}{\Gamma(-u)}\Gamma_0^{r_1,r_2}(s)B(\beta+s+1, -u+s). \tag{34}$$

Proof. From the definition of the Mellin transform, we have

$$\mathbf{M}\left[D^{u,(r)}_{z,(r_1,r_2)}[z^{\beta}];s\right] = \int_0^{\infty} r^{s-1}D^{u,(r)}_{z,(r_1,r_2)}[z^{\beta}]dr. \tag{35}$$

Then applying the definition of the classical Riemann–Liouville fractional derivative operator (1), to the right side of the above equation, we have

$$\mathbf{M}\big[D^{u,(r)}_{z,(r_1,r_2)}[z^{\beta}];s\big] = \int_0^{\infty} r^{s-1}\left(\frac{1}{\Gamma(-u)}\int_0^z (z-t)^{-u-1}E_{r_1,r_2}\left(\frac{-rz^2}{t(z-t)}\right)t^{\beta}dt\right)dr. \quad (36)$$

Taking out z from the integral, we have

$$\mathbf{M}\big[D^{u,(r)}_{z,(r_1,r_2)}[z^{\beta}];s\big] = \int_0^{\infty} r^{s-1}\left(\frac{z^{-u-1}}{\Gamma(-u)}\int_0^z (1-\frac{t}{z})^{-u-1}E_{r_1,r_2}\left(\frac{-rz^2}{t(z-t)}\right)t^{\beta}dt\right)dr. \quad (37)$$

Now take $t = xz$ in the above equation, then changing the limit from $t = 0, t = z$ to $x = 0, x = 1$, $dt = zdx$ and rearranging the terms, we have

$$\mathbf{M}\big[D^{u,(r)}_{z,(r_1,r_2)}[z^{\beta}];s\big] = \frac{z^{\beta-u}}{\Gamma(-u)}\int_0^{\infty} r^{s-1}\left(\int_0^1 (1-x)^{-u-1}E_{r_1,r_2}\left(\frac{-r}{x(1-x)}\right)x^{\beta}dx\right)dr. \quad (38)$$

Interchanging the order of integrations, we have

$$\mathbf{M}\big[D^{u,(r)}_{z,(r_1,r_2)}[z^{\beta}];s\big] = \frac{z^{\beta-u}}{\Gamma(-u)}\int_0^1 x^{\beta}(1-x)^{-u-1}\left(\int_0^{\infty} r^{s-1}E_{r_1,r_2}\left(\frac{-r}{x(1-x)}\right)dr\right)dx. \quad (39)$$

Substituting $v = \frac{r}{x(1-x)}$ and then changing the limits from $r = 0$, $r = \infty$ to $v = 0, v = \infty$ and $dr = x(1-x)dv$ in the above equation, we get

$$\mathbf{M}\big[D^{u,(r)}_{z,(r_1,r_2)}[z^{\beta}];s\big] = \frac{z^{\beta-u}}{\Gamma(-u)}\int_0^1 x^{\beta+s}(1-x)^{-u+s-1}\left(\int_0^{\infty} v^{s-1}E_{r_1,r_2}(-v)dv\right)dx. \tag{40}$$

From the known result in [19],

$$\Gamma_r^{(r_1,r_2)}(x_1) = \int_0^{\infty} t^{x_1-1}E_{r_1,r_2}(-t-rt^{-1})dt, \tag{41}$$

and using it with $r=0$ in the above equation, we get

$$\mathbf{M}\big[D^{u,(r)}_{z,(r_1,r_2)}[z^{\beta}];s\big] = \frac{z^{\beta-u}}{\Gamma(-u)}\int_0^1 x^{\beta+s}(1-x)^{-u+s-1}\left(\Gamma_0^{(r_1,r_2)}(s)\right)dx. \tag{42}$$

On rearranging the terms, we have

$$\mathbf{M}\big[D^{u,(r)}_{z,(r_1,r_2)}[z^{\beta}];s\big] = \frac{z^{\beta-u}\Gamma_0^{(r_1,r_2)}(s)}{\Gamma(-u)}\int_0^1 x^{\beta+s}(1-x)^{-u+s-1}dx. \tag{43}$$

Then from the definition of the classical Euler beta function defined in [20], we get our desired result:

$$\mathbf{M}\big[D^{u,(r)}_{z,(r_1,r_2)}[z^{\beta}];s\big] = \frac{z^{\beta-u}\Gamma_0^{r_1,r_2}(s)}{\Gamma(-u)}B(\beta+s+1,-u+s). \tag{44}$$

□

Theorem 10. *Assume $\Re(u)<0$, $\Re(k)>0$, $\mid z\mid<1$, and $s>0$, then another Mellin transform for the extended Riemann–Liouville fractional derivative operator is defined as follows:*

$$\boldsymbol{M}\big[D^{u,(r)}_{z,(r_1,r_2)}[(1-z)^{-k}];s\big] = \frac{z^{-u}}{\Gamma(-u)}\Gamma_0^{r_1,r_2}(s)\sum_{n=0}^{\infty}(k)_n B(n+s+1,-u+s)\frac{z^n}{n!}. \tag{45}$$

Proof. From the definition of Mellin transform, we have

$$\mathbf{M}\big[D^{u,(r)}_{z,(r_1,r_2)}[(1-z)^{-k}];s\big] = \int_0^\infty r^{s-1} D^{u,(r)}_{z,(r_1,r_2)}[(1-z)^{-k}]dr. \tag{46}$$

Then applying the definition of the classical Riemann–Liouville fractional derivative operator (1), to the right side of the above equation, we have

$$\mathbf{M}\big[D^{u,(r)}_{z,(r_1,r_2)}[(1-z)^{-k}];s\big] \\ = \int_0^\infty r^{s-1}\left(\frac{1}{\Gamma(-u)}\int_0^z (z-t)^{-u-1} E_{r_1,r_2}\left(\frac{-rz^2}{t(z-t)}\right)(1-t)^{-k}dt\right)dr. \tag{47}$$

The binomial expansion is defined as follows:

$$(1-t)^{-k} = \sum_{n=0}^{\infty} \frac{(k)_n}{n!} t^n, |t| < 1. \tag{48}$$

Using this result in the above equation (47), we have

$$\mathbf{M}\big[D^{u,(r)}_{z,(r_1,r_2)}[(1-z)^{-k}];s\big] \\ = \int_0^\infty r^{s-1}\left(\frac{1}{\Gamma(-u)}\int_0^z \left(\sum_{n=0}^{\infty}\frac{(k)_n}{n!}t^n\right)(z-t)^{-u-1} E_{r_1,r_2}\left(\frac{-rz^2}{t(z-t)}\right)dt\right)dr. \tag{49}$$

On rearranging the terms, we get

$$\mathbf{M}\big[D^{u,(r)}_{z,(r_1,r_2)}[(1-z)^{-k}];s\big] \\ = \int_0^\infty r^{s-1}\left(\sum_{n=0}^{\infty}\frac{(k)_n}{n!}\frac{1}{\Gamma(-u)}\int_0^z t^n(z-t)^{-u-1} E_{r_1,r_2}\left(\frac{-rz^2}{t(z-t)}\right)dt\right)dr. \tag{50}$$

On interchanging the order of integration and summation, we get

$$\mathbf{M}\big[D^{u,(r)}_{z,(r_1,r_2)}[(1-z)^{-k}];s\big] \\ = \sum_{n=0}^{\infty}\frac{(k)_n}{n!}\int_0^\infty r^{s-1}\left(\frac{1}{\Gamma(-u)}\int_0^z t^n(z-t)^{-u-1} E_{r_1,r_2}\left(\frac{-rz^2}{t(z-t)}\right)dt\right)dr. \tag{51}$$

Then from the definition of Mellin transform, we have

$$\mathbf{M}\big[D^{u,(r)}_{z,(r_1,r_2)}[(1-z)^{-k}];s\big]=\sum_{n=0}^{\infty}\frac{(k)_n}{n!}\mathbf{M}\big[D^{u,(r)}_{z,(r_1,r_2)}[z^n];s\big]. \tag{52}$$

Then using Theorem (2.8), and rearranging the terms, we get our desired result:

$$\begin{aligned}&\mathbf{M}\big[D^{u,(r)}_{z,(r_1,r_2)}[(1-z)^{-k}];s\big]\\&=\frac{z^{-u}}{\Gamma(-u)}\Gamma_0^{r_1,r_2}(s)\sum_{n=0}^{\infty}(k)_nB(n+s+1,-u+s)\frac{z^n}{n!}.\end{aligned} \tag{53}$$

□

Theorem 11. *For the extended hypergeometric function, the following generating relation holds true:*

$$\sum_{m=0}^{\infty}\frac{(\lambda)_m}{m!}F^{(r)}_{(r_1,r_2)}(\lambda+m,k,l;z)t^m=(1-t)^{-\lambda}F^{(r)}_{(r_1,r_2)}\left(\lambda,k,l;\frac{z}{(1-t)}\right), \tag{54}$$

where $|\frac{z}{(1-t)}|<1,\ \Re(\lambda)>0,$ *and* $\Re(k)>\Re(l)>0.$

Proof. Let the series identity

$$[(1-z)-t]^{-\lambda}=(1-t)^{-\lambda}\left(1-\frac{z}{(1-t)}\right)^{-\lambda}.$$

After rearranging the terms, we have

$$(1-z)^{-\lambda}\left(1-\frac{t}{1-z}\right)^{-\lambda}=(1-t)^{-\lambda}\left(1-\frac{z}{(1-t)}\right)^{-\lambda}.$$

Then using the binomial expansion of $(1-\frac{t}{1-z})^{-\lambda}$, we get

$$(1-z)^{-\lambda}\sum_{m=0}^{\infty}\frac{(\lambda)_m}{m!}\left(\frac{t}{1-z}\right)^m=(1-t)^{-\lambda}\left(1-\frac{z}{(1-t)}\right)^{-\lambda}. \tag{55}$$

Now multiplying both sides by z^{k-1}, we get

$$z^{k-1}(1-z)^{-\lambda}\sum_{m=0}^{\infty}\frac{(\lambda)_m}{m!}\left(\frac{t}{1-z}\right)^m = z^{k-1}(1-t)^{-\lambda}\left(1-\frac{z}{(1-t)}\right)^{-\lambda}. \tag{56}$$

Then using the Riemann–Liouville fractional derivative operator $D^{k-l,(r)}_{z,(r_1,r_2)}$ on both sides of the above equation, we have

$$\begin{aligned} &D^{k-l,(r)}_{z,(r_1,r_2)}\left[z^{k-1}(1-z)^{-\lambda}\sum_{m=0}^{\infty}\frac{(\lambda)_m}{m!}\left(\frac{t}{1-z}\right)^m\right] \\ &\quad = (1-t)^{-\lambda}D^{k-l,(r)}_{z,(r_1,r_2)}\left[z^{k-1}\left(1-\frac{z}{(1-t)}\right)^{-\lambda}\right]. \end{aligned} \tag{57}$$

After some calculations, we have

$$\begin{aligned} &\sum_{m=0}^{\infty}\frac{(\lambda)_m}{m!}D^{k-l,(r)}_{z,(r_1,r_2)}\left[z^{k-1}(1-z)^{-\lambda-m}\right]t^m \\ &\quad = (1-t)^{-\lambda}D^{k-l,(r)}_{z,(r_1,r_2)}\left[z^{k-1}\left(1-\frac{z}{(1-t)}\right)^{-\lambda}\right]. \end{aligned} \tag{58}$$

Then using Theorem (2.4), on both sides, we obtain our desired result:

$$\sum_{m=0}^{\infty}\frac{(\lambda)_m}{m!}F^{(r)}_{(r_1,r_2)}(\lambda+m,k,l;z)t^m = (1-t)^{-\lambda}F^{(r)}_{(r_1,r_2)}\left(\lambda,k,l;\frac{z}{(1-t)}\right). \tag{59}$$

□

Theorem 12. *For Prabhakar-type function, the following result holds true:*

$$\begin{aligned} &D^{\eta-u,(r)}_{z,(r_1,r_2)}\left[z^{\eta-1}E^{\gamma}_{\alpha,\beta}(z)\right] \\ &\quad = \frac{z^{u-1}}{\Gamma(u-\eta)}\sum_{k=0}^{\infty}\frac{(\gamma)_k}{\Gamma(\beta+k\alpha)}B^{(r)}_{(r_1,r_2)}(\eta+k,u-\eta)\frac{z^k}{k!}, \end{aligned} \tag{60}$$

where $E^{\gamma}_{\alpha,\beta}(z)$ is the Prabhakar-type function given in [21].

Proof. Using 3-parameter Mittag-Leffler function (Prabhakar-type function) defined as [21]

$$E^{\gamma}_{\alpha,\beta}(z) = \sum_{k=0}^{\infty} \frac{(\gamma)_k}{\Gamma(\beta + k\alpha)} \frac{z^k}{k!}.$$

Multiplying the above equation by $z^{\eta-1}$ and applying the Riemann–Liouville fractional derivative operator $D^{\eta-u,(r)}_{z,(r_1,r_2)}$, then we have

$$D^{\eta-u,(r)}_{z,(r_1,r_2)}\left[z^{\eta-1}E^{\gamma}_{\alpha,\beta}(z)\right] = D^{\eta-u,(r)}_{z,(r_1,r_2)}\left[z^{\eta-1}\sum_{k=0}^{\infty} \frac{(\gamma)_k}{\Gamma(\beta + k\alpha)} \frac{z^k}{k!}\right]. \tag{61}$$

By using Theorem (2.5), we have

$$D^{\eta-u,(r)}_{z,(r_1,r_2)}\left[z^{\eta-1}E^{\gamma}_{\alpha,\beta}(z)\right] = \sum_{k=0}^{\infty} \frac{(\gamma)_k}{\Gamma(\beta + k\alpha)k!} D^{\eta-u,(r)}_{z,(r_1,r_2)}\left[z^{\eta+k-1}\right]. \tag{62}$$

Then applying Theorem (2.3), we get our desired result:

$$\begin{aligned} &D^{\eta-u,(r)}_{z,(r_1,r_2)}\left[z^{\eta-1}E^{\gamma}_{\alpha,\beta}(z)\right] \\ &\quad = \frac{z^{u-1}}{\Gamma(u-\eta)} \sum_{k=0}^{\infty} \frac{(\gamma)_k}{\Gamma(\beta + k\alpha)} B^{(r)}_{(r_1,r_2)}(\eta + k, u - \eta) \frac{z^k}{k!}. \end{aligned} \tag{63}$$

□

Corollary 13. *The following result holds true:*

$$\begin{aligned} &D^{\eta-u,(r)}_{z,(r_1,r_2)}\left[z^{\eta-1}E_{\alpha,\beta}(z)\right] \\ &\quad = \frac{z^{u-1}}{\Gamma(u-\eta)} \sum_{k=0}^{\infty} \frac{1}{\Gamma(\beta + k\alpha)} B^{(r)}_{(r_1,r_2)}(\eta + k, u - \eta) z^k, \end{aligned} \tag{64}$$

where $E_{\alpha,\beta}(z)$ is Wiman's function given in [1].

Proof. Taking $\gamma = 1$ in Theorem (2.11), we get our desired result.

□

Corollary 14. *The following result holds true:*

$$D_{z,(r_1,r_2)}^{\eta-u,(r)}\left[z^{\eta-1}E_\alpha(z)\right]=\frac{z^{u-1}}{\Gamma(u-\eta)}\sum_{k=0}^{\infty}\frac{1}{\Gamma(1+k\alpha)}B_{(r_1,r_2)}^{(r)}(\eta+k,u-\eta)z^k, \tag{65}$$

where $E_\alpha(z)$ is the Mittag-Leffler function given in [22, 23].

Proof. On substituting $\gamma=1$ and $\beta=1$ in Theorem (2.11), we get our desired result. □

Corollary 15. *The following result holds true:*

$$D_{z,(r_1,r_2)}^{\eta-u,(r)}\left[z^{\eta-1}e^{(z)}\right]=\frac{z^{u-1}}{\Gamma(u-\eta)}\sum_{k=0}^{\infty}B_{(r_1,r_2)}^{(r)}(\eta+k,u-\eta)\frac{z^k}{k!}, \tag{66}$$

where $e^{(z)}$ is the exponential function.

Proof. On substituting $\gamma=1$, $\alpha=1$, and $\beta=1$ in Theorem (2.11), we get our desired result. □

Theorem 16. *For generalized hypergeometric function, the following result holds true:*

$$\begin{aligned}&D_{z,(r_1,r_2)}^{\eta-u,(r)}\left[z^{\eta-1}{}_pF_q(x_1,x_2..x_p;y_1,y_2..y_q;z)\right]\\&\quad=\frac{z^{u-1}}{\Gamma(u-\eta)}\frac{\prod_{j=1}^{q}\Gamma(y_j)}{\prod_{i=1}^{p}\Gamma(x_i)}\sum_{k=0}^{\infty}\frac{\prod_{i=1}^{p}\Gamma(x_i+k)}{\prod_{j=1}^{q}\Gamma(y_j+k)}B_{(r_1,r_2)}^{(r)}(\eta+k,u-\eta)\frac{z^k}{k!},\end{aligned} \tag{67}$$

where ${}_pF_q(x_1,x_2..x_p;y_1,y_2..y_q;z)$ is the generalized hypergeometric function given in [24].

Proof. On using the definition of the generalized hypergeometric function and following a similar rule as Theorem (2.11), we obtain our desired result. □

Theorem 17. *For the Wright–Fox function, the following result holds true:*

$$D^{\eta-u,(r)}_{z,(r_1,r_2)}\left[z^{\eta-1}{}_pW_q\left[z\left|\begin{matrix}(x_1,a_1)...(x_p,a_p)\\(y_1,b_1)...(x_q,b_q)\end{matrix}\right.\right]\cdot\right]$$
$$=\frac{z^{u-1}}{\Gamma(u-\eta)}\sum_{k=0}^{\infty}\frac{\prod_{i=1}^{p}\Gamma(x_i+a_ik)}{\prod_{j=1}^{q}\Gamma(y_j+a_jk)}B^{(r)}_{(r_1,r_2)}(\eta+k,u-\eta)\frac{z^k}{k!},\tag{68}$$

where ${}_pW_q\left[z\left|\begin{matrix}(x_1,a_1)...(x_p,a_p)\\(y_1,b_1)...(y_q,b_q)\end{matrix}\right.\right]$ *is the Wright–Fox function given in* [24].

Proof. On using the definition of the Wright–Fox function and following a similar rule as Theorem (2.11), we obtain our desired result. □

3. Conclusion

We conclude our investigation by remarking that the results presented in this chapter are easily converted by using the interesting known extension of the classical Riemann–Liouville fractional derivative operator and other special functions. The results presented here are new and very useful for the extension of other special functions. We are also trying to find certain possible applications of those results presented here for other research areas of science and engineering [25–32].

Acknowledgments

Praveen Agarwal and Rahul Goyal are very thankful to the NBHM (project 02011/12/2020NBHM(R.P)/R&D II/7867) for their necessary support and Shilpi Jain also thanks SERB (project number: MTR/2017/000194) for providing the necessary facility.

References

[1] Wiman, A. Über die Nullstellun der Funktionen $E_\alpha(x)$, *Acta Mathematica*, 29, 217–234 (1905).

[2] Agarwal, P., Al-Mdallal, Q., Cho, Y.J. and Jain, S. Fractional differential equations for the generalized Mittag-Leffler function. *Adv. Difference Equ.*, 1, 1–8, (2018).

[3] Jain, S., Agarwal, R.P., Agarwal, P. and Singh, P. Certain unified integrals involving a multivariate mittag–leffler function. *Axioms*, 10, 81 (2021).

[4] Choi, J. and Agarwal, P. Certain fractional integral inequalities involving hypergeometric operators. *East Asian J. Appl. Math.*, 30, 283–291 (2014).

[5] Çetinkaya, A., Agarwal, P., Kıymaz, I.O. and Agarwal, R.P. A comparative study on generating function relations for generalized hypergeometric functions via generalized fractional operators. *Adv. Diff. Equ.*, 1, 1–11 (2018).

[6] Srivastava, H.M., Agarwal, P. and Jain, S. Generating functions for the generalized Gauss hypergeometric functions. *Commun. Appl. Math. Comput.*, 247, 348–352 (2014).

[7] Agarwal, P., Choi, J. and Paris, R.B. Extended Riemann-Liouville fractional derivative operator and its applications. *Journal of Nonlinear Science and Applications (JNSA)*, 8(5), 451–466 (2015).

[8] Baleanu, D., Agarwal, P., Parmar, R.K., Alqurashi, M. and Salahshour, S. Extension of the fractional derivative operator of the Riemann-Liouville. *J. Nonlinear Sci. Appl.*, 10(6), 2914–2924 (2017).

[9] Salahshour, S., Ahmadian, A., Senu, N., Baleanu, D. and Agarwal, P. On analytical solutions of the fractional differential equation with uncertainty application to the basset problem. *Entropy*, 17(2), 885–902 (2015).

[10] Tariboon, J., Ntouyas, S.K. and Agarwal, P. New concepts of fractional quantum calculus and applications to impulsive fractional q-difference equations. *Adv Differ Equ.*, 18, 2015 (2015).

[11] Morales-Delgado, V.F., Gómez-Aguilar, J.F., Saad, K.M., Khan, M.A. and Agarwal, P. Analytic solution for oxygen diffusion from capillary to tissues involving external force effects: A fractional calculus approach, *Physica A: Statistical Mechanics and its Applications*, 523, 48–65 (2019).

[12] El-Sayed, A.A, and Agarwal, P. Numerical solution of multiterm variable-order fractional differential equations via shifted Legendre polynomials. *Numerical Solution of Multiterm Variable-Order*

Fractional Differential Equations via Shifted Legendre Polynomials, 42(11), 3978–3991 (2019).

[13] Agarwal, P., El-Sayed, A.A. and Tariboon, J. Vieta–Fibonacci operational matrices for spectral solutions of variable-order fractional integro-differential equations. *J. Comput. Appl. Math.*, 282, 113063, (2021).

[14] Srivastava, H.M. and Manocha, H.L. A treatise on generating functions, Ellis Horwood Limited, Chichester (1984).

[15] Özarslan, M.A. and Özergin, E. Some generating relations for extended hypergeometric functions via generalized fractional derivative operator. *Math. Comput. Model.*, 52, 1825–1833 (2010).

[16] Chaudhry, M.A., Qadir, A., Rafique, M. and Zubair, S.M. Extension of Euler's Beta function. *J. Comput. Appl. Math.*, 78, 19–32 (1997).

[17] Goyal, R., Agarwal, P., Momami, S. and Rassias, M.T. An Extension of Beta Function by Using wiman's function. *Axioms*, 10, 187 (2021).

[18] Jain, S., Goyal, R., Agarwal, P., Lupica, A. and Cesarano, Some results of extended beta function and hypergeometric functions by using Wiman's function. *Mathematics*, (submitted-under review) (2021).

[19] Atash, A.A., Barahmah, S.S. and Kulib, M.A On a new extensions of extended Gamma and Beta functions. *International Journal of Applied Mathematics and Statistics*, 3, 4–18 (2018).

[20] Rainville, E.D., Special Functions, Macmillan, New York, USA (1960)

[21] Prabhakar, T.R. A singular integral equation with a generalized Mittag-Leffler function in the kernel. *Yokohama Mathematical Journal*, 19, 7–15, (1971).

[22] Mittag-Leffler, G.M. Sur la nouvelle fonction $E_\alpha(x)$. *Comptes rendus de l'Académie des Sciences*, 137, 554–558 (1903).

[23] Gorenflo, R., Kilbas, A.A., Mainardi, F. and Rogosin, S.V. *Mittage Leffler Functions Related Topics and Application*, 2nd edition, Springer, New York, USA (2020).

[24] Olver, W.J.F., Lozier, W.D., Boisvert, F.R. and Clark, W.C. *NIST Handbook of Mathematical Functions*, Cambridge University Press, New York, USA (2010).

[25] Agarwal, P., Jleli, M., and Tomar, M. Certain Hermite-Hadamard type inequalities via generalized k-fractional integrals. *J. Inequal. Appl.*, 55, 2017 (2017).

[26] Agarwal, P., El-Sayed, A.A., Non-standard finite difference and Chebyshev collocation methods for solving fractional diffusion equation. *Physica A: Statistical Mechanics and its Applications*, 500, 40–49 (2018).

[27] Agarwal, P. Some inequalities involving Hadamard-type k-fractional integral operators. *Mathematical Methods in Applied Sciences*, 40, 3882–3891 (2017).

[28] Rekhviashvili, S., Pskhu, A., Agarwal, P. and Jain, S. Application of the fractional oscillator model to describe damped vibrations. *Turk J. Phys.*, 43, 236–242 (2019).

[29] Alderremy, A.A., Saad, K.M., Agarwal, P., Aly, S. and Jain, S. Certain new models of the multi space-fractional Gardner equation. *Physica A: Statistical Mechanics and its Applications*, 545, 123806 (2020).

[30] Sunarto, A., Agarwal, P., Sulaiman, J., Chew, J.V.L. and Momani, S. Quarter-Sweep Preconditioned Relaxation Method, Algorithm and Efficiency Analysis for Fractional Mathematical Equation. *Fractal and Fractional*, 5(3), 98 (2021).

[31] Günay, B., Agarwal, P., Guirao, J.L.G. and Momani, S. A Fractional Approach to a Computational Eco-Epidemiological Model with Holling Type-II Functional Response. *Symmetry*, 13(7), 1159 (2021).

[32] Serrano, E.Z., Bekiros, S., Garza, M.A.P., Castillo, C.P., Agarwal, P., Jahanshahi, H. and Aly, A.A, On chaos and projective synchronization of a fractional difference map with no equilibria using a fuzzy-based state feedback control. *Physica A: Statistical Mechanics and its Applications*, 578, 126100 (2021).

https://doi.org/10.1142/9789811293009_0020

Chapter 20

A Finite Element Scheme for an Initial Value Problem

Vassilios K. Kalpakides
National Technical University of Athens
School of Applied Mathematics and Physical Sciences,
Department of Mathematics, Zografou, GR-15780, Greece
kalpakides@math.ntua.gr

A new Hamilton principle of convolutional type, completely compatible with the initial conditions of an IVP, has been proposed in a recent publication [12]. In this chapter, the possible use of this principle for the formulation of an FE scheme adjusted to dynamical problems is investigated. To this end, an FE scheme based on a convolutional extremum principle for the harmonic oscillator — used as an exemplary initial value problem — is developed and presented in detail. Besides, from the local finite element analysis, a recurrent (one-step) algorithm arises which provides an approximate solution to the IVP, as well. The succeeded schemes are computationally tested for both free and forced vibration problems.

Keywords: FEM, initial value problem, convolution, Hamilton principle

1. Introduction

It is widely accepted that the initial value problems do not admit appropriate variational principles like the corresponding boundary value ones. This is due to the fact that the differential operator of dynamical problems, on account of the initial conditions, is not symmetric. If one considers solely the differential operator, independently of the initial conditions, it might be symmetric in some cases. Then it will be referred to as *formally symmetric* [18] and the formulation of a variational principle of Hamilton's type is possible. However, the initial conditions cannot emerge from the variational principle and must be artificially imposed on the differential equation.

As Tonti remarked in [18], the notion of symmetry is defined with respect to a bilinear form, often hidden in the background. Apparently, for differential operators, the symmetry is defined with respect to the L^2 inner product. Thus more precisely, one can state that *the operators of the initial value problems are not symmetric with respect to the symmetry rising from the* L^2 *inner product.* In other words, symmetry is not an inherent property of an operator; it should be regarded with respect to any specific bilinear form. An operator which is not symmetric with respect to the inner product might be symmetric with respect to another bilinear form.

From this viewpoint, Gurtn's attempt [9–11] to formulate variational principles for viscoelasticity and elastodynamics using the convolution was quite interesting. The same approach, a little improved, was followed by Reddy [16, 17] and later by a great number of authors [2–7,13–15] establishing what one can call *convolutional variational principles* [1]. Actually, in these works, the new variational principles have been obtained by the use of the convolution bilinear form instead of the standard one induced by L^2 inner product.

Once a convolutional variational principle for an initial value problem has been obtained, one may wonder whether or not it is possible to develop a finite element scheme analogous to the standard one for boundary value problems. Dargush [4] and his co-workers [5, 6, 8, 13] have presented quite an interesting work in this direction. They exploited their mixed convolutional variational principles to conclude in FE approximation schemes for initial boundary value problems. However, as concerns time evolution, these schemes are one-step marching algorithms and, to our best knowledge, the question on the possibility of a true FE scheme for the unified space–time continuum remains unanswered.

This chapter is a contribution toward answering this question. Here the possibility for the formulation of a true FE scheme for the time continuum is proved, in other words, a formulation of an FE scheme for an initial value problem is presented. Following up an earlier work [12], where a generalized Hamilton's principle in a convolutional form had been proposed, we present a fully developed finite element formulation which provides an approximate solution to the variational equation. A second approximate scheme is produced as a by-product of the FE formulation process. The local algebraic system that is formulated for the needs of the FE analysis provides a recurrent formula, like that one of [4], upon which a one-step algorithm is fully developed, as well. It is worth remarking that the adopted convolutional symmetry is "transferred" to the finite-dimensional version of the problem, where it appears as symmetry with respect to the second diagonal of the global stiffness matrix.

In Section 2, some basic definitions and propositions are recalled and wherever necessary are proved. A new definition for the convolution at a subdomain (element) as well as a corresponding variational principle at element level are provided. Also, the partition of the initial variational problem into a finite number of variational problems, one for each element, is presented. In Section 3, the FE scheme is analytically developed. First, a typical FE analysis is presented; starting from the element level, the local systems are formulated and appropriately assembled to the global algebraic system. Next, in Section 4, the inverse procedure is examined, by inserting a linear combination of the global shape functions into the variational equation, concluding in the same algebraic system. In Section 5, an alternative, one-step approximation scheme stemming from the local algebraic system is presented and its theoretical critical value for numerical stability is provided. Finally, in Section 6, some computational examples are presented supporting the theoretical considerations and the efficiency of the two numerical schemes to approximate the analytical solution. This chapter ends with Section 7, where the main conclusions are stated.

2. Prerequisites

2.1. *Definitions and propositions*

In this chapter, the simple problem of an harmonic oscillator is used as an example of an initial value problem. Thus, both theory and numerical examples concern the following problem.

Problem 1. *Find a u in $C^2[0,t]$ that satisfies the differential equation*

$$mu''(s) + ku(s) = f(s), \ s \in (0,t), \ t > 0 \tag{1}$$

and the initial conditions

$$u(0) = u_0, \quad u'(0) = v_0, \tag{2}$$

where $f \in C[0,\infty)$ is a periodic function, m and k are positive constants, and u_0 and v_0 are the initial data.

Since the notion of convolution is of central importance in our analysis, we recall its definition.

Definition 2. If f, g are L^2 functions over $(0,t)$, then their convolution is defined as

$$[g,h](t) = \int_0^t g(s)h(t-s)ds, \quad t > 0. \tag{3}$$

In what follows, if there is no danger of confusion, the argument in the above definition will be omitted.

Discretizing the interval $[0,t]$ in finite subintervals, a convolution over any finite interval of the form $[t_1,t_2]$, $t_1 > 0$ will be necessary. For that purpose, the following definition is proposed.

Definition 3. Let $[t_1,t_2]$ be any finite interval of $\mathbb{R}$ with $t_1 \geq 0$. The convolution of the functions $g,h \in L^2(t_1,t_2)$ is defined as

$$\begin{aligned}[g,h]_{t_1}^{t_2} &= \int_0^\tau g(t_1+s)h(t_2-s)ds \\ &= \int_{t_1}^{t_2} g(s)h(t_1+t_2-s)ds, \quad \tau = t_2 - t_1. \end{aligned} \tag{4}$$

It is easy to confirm that the above definition keeps all desired properties of the standard definition and it falls to it when $t_1 = 0$.

It has been proved [12] that the solution of the initial value problem (1)–(2) is related to the problem of finding a stationary value for the functional

$$I[u] = \frac{m}{2}\left[u', u'\right] + \frac{k}{2}[u, u] - [f, u] - [\tilde{f}, u], \quad u \in \mathcal{D}, \tag{5}$$

where $\tilde{f}(s) = f_0 \delta(s)$, $f_0 = m v_0$, δ is the Dirac function, and

$$\mathcal{D} = \{\phi \in C^2[0, t] \text{ and } \phi(0) = u_0\}. \tag{6}$$

More specifically, it holds as follows.

Proposition 4. *If the functional I attains a stationary value at some $u \in \mathcal{D}$, then that u will be a solution to the initial value problem* (1)–(2), *too.*

An analogous proposition holds for an interval of the form $[t_1, t_2]$. Consider the following.

Problem 5. *Find a function u in $C^2[t_1, t_2]$ which satisfies the following differential equation and initial conditions:*

$$mu''(s) + ku(s) = f(s), \; s \in (t_1, t_2), \; t_1 > 0, \tag{7}$$

$$u(t_1) = u_1, u'(t_1) = v_1. \tag{8}$$

This problem is related to the functional

$$I^t[u] = \frac{m}{2}\left[u', u'\right]_{t_1}^{t_2} + \frac{k}{2}[u, u]_{t_1}^{t_2} - [f, u]_{t_1}^{t_2} - [\tilde{f}^1, u]_{t_1}^{t_2}, \quad u \in \mathcal{D}^t, \tag{9}$$

where

$$\mathcal{D}^t = \{\phi \in C^2[t_1, t_2] \text{ and } \phi(t_1) = u_1\} \tag{10}$$

and

$$\tilde{f}^1(s) = f^1 \delta(s - t_1), \quad f^1 = m v_1.$$

The following statement determines the relation between Problem 5 and the functional I^t.

Proposition 6. *If the functional I^t attains a stationary value at some $u \in \mathcal{D}^t$, then that u will be a solution to problem* (7)–(8), *too.*

Proof The functional I^t is transformed by translating all functions within its argument to the left by an interval t_1:

$$h(s) = h^*(s - t_1).$$

For instance, the second term of the functional

$$[u, u]_{t_1}^{t_2} = \int_0^{t_2 - t_1} u(t_1 + s)u(t_2 - s)ds$$

becomes

$$\begin{aligned}
[u, u]_{t_1}^{t_2} &= \int_0^{t_2 - t_1} u^*(s)u^*(t_2 - t_1 - s)ds \\
&= \int_0^{\tau} u^*(s)u^*(\tau - s)ds = [u^*, u^*](\tau),
\end{aligned}$$

where $\tau = t_2 - t_1$.

Also, the last term becomes

$$\begin{aligned}
[\tilde{f^1}, u]_{t_1}^{t_2} &= \int_0^{t_2 - t_1} \tilde{f^1}(t_1 + s)u(t_2 - s)ds \\
&= \int_0^{\tau} \tilde{f^1}^*(s)u^*(\tau - s)ds = \left[\tilde{f^1}^*, u^*\right] \\
&= \int_0^{\tau} f^1\delta(s)u^*(\tau - s)ds = f^1 u^*(\tau).
\end{aligned}$$

Similarly, accounting for the rest terms of I^t, one concludes

$$\begin{aligned}
I^t[u] = I^*[u^*] &= \frac{m}{2}\left[u^{*\prime}, u^{*\prime}\right] + \frac{k}{2}[u^*, u^*] \\
&\quad - [f^*, u^*] - \left[\tilde{f^1}^*, u^*\right], \quad u \in \mathcal{D}^{t*},
\end{aligned} \tag{11}$$

where

$$\mathcal{D}^{t*} = \{\phi \in C^2[0, \tau] \text{ and } \phi(0) = u_1\}.$$

Note also that

$$\tilde{f^1}^* = f^1\delta(s), \ f^1 = mv_1.$$

Thus, if I^* attains a stationary value at u^*, according to Proposition 4, it will satisfy eq. (1) and the initial conditions (2), i.e.,

$$mu^{*\prime\prime}(z) + ku^*(z) = f^*(z), \ z \in (0, \tau), \ \tau > 0,$$
$$u^*(0) = u_1, u^{*\prime}(0) = v_1,$$

or, equivalently,

$$mu''(z + t_1) + ku(z + t_1) = f(z + t_1), \ z \in (0, \tau), \ \tau > 0,$$
$$u(t_1) = u_0, u'(t_1) = v_0.$$

Introducing the substitution

$$s = z + t_1,$$

the above equations can be written as follows:

$$mu''(s) + ku(s) = f(s), \ s \in (t_1, t_2),$$
$$u(t_1) = u_0, u'(t_1) = v_0$$

and the proof has been completed.

Finally, based on the above propositions, one can prove that the following statement holds.

Proposition 7. *Consider the functional*

$$I^\tau[u] = \frac{m}{2}\left[u', u'\right]_{t_1}^{t_2} + \frac{k}{2}\left[u, u\right]_{t_1}^{t_2} - [f, u]_{t_1}^{t_2} - [\tilde{f}^1 - \tilde{f}^2, u]_{t_1}^{t_2}, \quad u \in \mathcal{D}^\tau = C^2[t_1, t_2], \tag{12}$$

where $\tilde{f}^1 = f^1\delta(s - t_1)$, $\tilde{f}^2 = f^2\delta(s - t_2)$, *and* f^1, f^2 *are the given constants. If the functional* I^τ *attains a stationary value at some* $u \in \mathcal{D}^\tau$, *then* u *will be the solution of the following problem:*

$$mu''(s) + ku(s) = f(s), \ s \in (t_1, t_2), \tag{13}$$
$$u'(t_1) = f^1/m, \quad u'(t_2) = f^2/m. \tag{14}$$

Remark 8. Problems (13)–(14) are not well-posed. Note that the data in eq. (14) concern both ends of the time interval $[t_1, t_2]$, a fact that is not compatible with an initial value problem. In addition, the values of the velocity at t_1 and t_2 are generally unknown. Later on, it will become apparent that these values are not necessary for the numerical solution of the problem.

Remark 9. As concerns the smoothness of the solutions in the above propositions, note that it required more smoothness so as to establish equivalence between the variational and the classical problem. If one is interested only in the variational problem, then requiring a solution in $C^1[0,t]$ is enough. Moreover, if the weak form of the problem is of interest, $W_2^1(0,t)$ should be the appropriate space for a generalized solution.

2.2. *Partition of the problem*

To deal with an approximate solution of Problem 1, we apply a partition of the domain $[0,t]$ in n subintervals:

$$0 = s_1 < s_2 < \cdots < s_n < s_{n+1} = t.$$

Though elements of equal length are usually sufficient for a time discretization, one can take randomly the node locations provided the number of elements is even ($n = 2k,\ k \in \mathbb{Z}$) and the distribution of nodes in $[0,t]$ is symmetric with respect to the middle point of the interval. That means, the k node is necessarily located at the point $t/2$, the i-node is symmetric to $(n+2-i)$-node ($i = 1,\ldots,n+1$) and the e-element is symmetric and of equal length to $(n+1-e)$-element ($e = 1,\ldots,n$).

Let e be any element but the first one ($e > 1$) with end points s_e and s_{e+1}, respectively. Since there are no data for this element, to formulate a variational problem, we take the functional

$$\begin{aligned} I^e[u] &= \frac{m}{2}\left[u', u'\right]_{s_e}^{s_{e+1}} + \frac{k}{2}\left[u, u\right]_{s_e}^{s_{e+1}} - \left[f, u\right]_{s_e}^{s_{e+1}} \\ &\quad - \left([\tilde{f}_1^e, u]_{s_e}^{s_{e+1}} - [\tilde{f}_2^e, u]_{s_e}^{s_{e+1}}\right), \quad u \in \mathcal{D}^e, \end{aligned}$$

where

$$\mathcal{D}^e = C^2[s_e, s_{e+1}]$$

and

$$\tilde{f}_1^e = f_1^e \delta(s - s_e), \quad \tilde{f}_2^e = f_2^e \delta(s - s_{e+1})$$

with f_1^e and f_2^e as unknown constants.

If the functional I^e takes a stationary value at $u^e \in \mathcal{D}^e$, then that u^e will be a solution to the problem

$$\begin{aligned} mu''(s) + ku(s) &= f(s), \; s \in (s_e, s_{e+1}), \\ u'(s_e) &= f_1^e/m, u'(s_{e+1}) = f_2^e/m. \end{aligned} \tag{15}$$

Thus, the variational problems

$$I^e[u], \; u \in \mathcal{D}^e \to \text{stationary}$$

are linked with problems (15), for all e with $2 \leq e \leq n$.

The first element, $[s_1, s_2]$, satisfies the initial conditions of Problem 1, that is,

$$u(s_1) = u_0, \; u'(s_1) = v_0,$$

which, according to Proposition 2, leads to the variational problem

$$I^1[u], \; u \in \mathcal{D}^1 \to \text{stationary},$$

where

$$I^1[u] = \frac{m}{2}\left[u', u'\right]_{s_1}^{s_2} + \frac{k}{2}\left[u, u\right]_{s_1}^{s_2} - \left[f, u\right]_{s_1}^{s_2} - \left[\tilde{f_1^1}, u\right]_{s_1}^{s_2},$$

with

$$\tilde{f_1^1} = mv_1\delta(s - s_1)$$

and

$$\mathcal{D}^1 = \{\phi \in C^2[s_1, s_2] \text{ and } \phi(s_1) = u_1\}.$$

In conclusion, the variational form of Problem 1,

$$I[u], \; u \in \mathcal{D} \to \text{stationary}, \tag{16}$$

can be divided in n variational problems, one for each element:

$$\begin{aligned} &I^1[u], \; u \in \mathcal{D}^1 \to \text{stationary}, \\ &I^2[u], \; u \in \mathcal{D}^2 \to \text{stationary}, \\ &\qquad \vdots \\ &I^n[u], \; u \in \mathcal{D}^n \to \text{stationary}. \end{aligned} \tag{17}$$

Thus, the variational problem (16) is equivalent to the collection of the variational problems (17).

3. A Finite Element Scheme

Looking for an approximate solution of Problem 1, we relax the requirements for the smoothness of the solution to the variational problem, allowing for piecewise linear approximations. Thus, at element level, approximations in the form of a polynomial of first order are sought. Also, for the moment, we are not concerned with the initial conditions of Problem 1. These will be imposed at the last stage of the analysis, when the global algebraic system will be established.

Summing up, for approximating the solution, we consider the collection of variational problems:

$$I^e[u] \to \text{stationary in } X^e, \quad e = 1, \ldots, n, \tag{18}$$

where X^e is the space of first-order polynomials over $[s_e, s_{e+1}]$.

3.1. *Local analysis*

Let any element e of the partition be limited by the nodes e and $e+1$. Looking for a stationary point in X^e for the functional I^e, one may consider an approximation of the form

$$u^e(s) = u_1^e N_1^e(s) + u_2^e N_2^e(s), \tag{19}$$

where

$$N_1^e(s) = \frac{s_{e+1} - s}{s_{e+1} - s_e}, \quad N_2^e(s) = \frac{s - s_e}{s_{e+1} - s_e}, \quad s \in [s_e, s_{e+1}] \tag{20}$$

and u_1^e, u_2^e are constants to be determined.

Inserting the solution u^e into the functional I^e, one obtains

$$I^e[u^e] = \frac{m}{2}\left[u^{e\prime}, u^{e\prime}\right]_{s_e}^{s_{e+1}} + \frac{k}{2}\left[u^e, u^e\right]_{s_e}^{s_{e+1}} - \left[f, u^e\right]_{s_e}^{s_{e+1}} - \left[\tilde{f}_2^e, u^e\right]_{s_e}^{s_{e+1}} + \left[\tilde{f}_1^e, u^e\right]_{s_e}^{s_{e+1}}. \tag{21}$$

After a long calculation (see Appendix A), eq. (21) may be written in the form of a quadratic function of u_i^e:

$$\begin{aligned}\hat{I}^e(u_i^e) &= \frac{1}{2}\left(M_{ij}^e u_i^e u_j^e + K_{ij}^e u_i^e u_j^e\right) - F_i^e u_i^e - f_2^e u_1^e + f_1^e u_2^e, \\ &= \frac{1}{2}\mathcal{K}_{ij}^e u_i^e u_j^e - F_i^e u_i^e - f_2^e u_1^e + f_1^e u_2^e, \\ & \qquad\qquad\qquad\qquad e = 1, \ldots, n, \quad i, j = 1, 2\end{aligned}$$

or, in matrix denotation

$$\hat{I}^e(\mathbf{u}^e) = \frac{1}{2}\mathbf{u}^{eT}\mathcal{K}^e\mathbf{u}^e - \mathcal{F}^{eT}\mathbf{u}^e, \tag{22}$$

where

$$\mathcal{K}^e = \mathbf{M}^e + \mathbf{K}^e, \quad \mathcal{F}^e = \mathbf{F}^e + \mathbf{f}^e \tag{23}$$

are the *local stiffness matrix* and the *local force vector*, respectively.

The matrices $\mathbf{M}^e$, $\mathbf{K}^e$ and the vectors $\mathbf{F}^e$, $\mathbf{f}^e$ are given by the relations

$$M_{ij}^e = m\left[N_i^{e\prime}, N_j^{e\prime}\right]_{s_e}^{s_{e+1}}, \quad K_{ij}^e = k\left[N_i^e, N_j^e\right]_{s_e}^{s_{e+1}}, \tag{24}$$

$$F_i^e = \int_0^{\tau_e} f(s_e + s)N_i^e(s_{e+1} - s)ds, \quad \mathbf{f}^e = \begin{bmatrix} f_2^e \\ -f_1^e \end{bmatrix} = \begin{bmatrix} mu'(s_{e+1}) \\ -mu'(s_e) \end{bmatrix}, \tag{25}$$

for $i, j = 1, 2,\ e = 1, \ldots, n$.

A necessary condition for the function $\hat{I}^e$ to take a stationary value at $\mathbf{u}^e$ is the satisfaction of the following relations:

$$\frac{\partial \hat{I}^e}{\partial u_1^e} = 0, \quad \frac{\partial \hat{I}^e}{\partial u_2^e} = 0,$$

or, with the aid of eq. (22)

$$\mathcal{K}^e\mathbf{u}^e = \mathcal{F}^e, \tag{26}$$

for $e = 1, \ldots, n$.

Remark 10. Note that, due to the symmetry of the node locations that has been adopted, the local stiffness matrices enjoy the equalities

$$\mathcal{K}^e = \mathcal{K}^{n+1-e}, \quad e = 1, \ldots, n. \tag{27}$$

Remark 11. Also, with the aid of Definition 3 and eq. (20), one can easily verify the following useful relations:

$$\begin{aligned}
[f, N_1^e]_{s_e}^{s_{e+1}} &= \int_0^{T_e} f(s_e + s) N_1^e(s_{e+1} - s) ds \\
&= \int_{s_e}^{s_{e+1}} f(s) N_2^e(s) ds = (f, N_2^e)\,, \\
[f, N_2^e]_{s_e}^{s_{e+1}} &= \int_0^{T_e} f(s_e + s) N_2^e(s_{e+1} - s) ds \\
&= \int_{s_e}^{s_{e+1}} f(s) N_1^e(s) ds = (f, N_1^e)\,,
\end{aligned} \tag{28}$$

where the parentheses with comma (,) denotes the standard inner product of L^2.

3.2. *The assembly*

Now, one may unite the solutions at the different elements, accounting for the fact that the displacements at the common nodes of the neighboring elements should coincide, ensuring the continuity of the approximate solution.

Introducing global enumeration for the displacements

$$U_1 = u_1^1,\ U_{i+1} = u_2^i = u_1^{i+1},\ U_{n+1} = u_2^n, \quad i = 1, \dots, n-1, \tag{29}$$

the local systems provided by eqs. (26) are written in extensive form

$$\begin{aligned}
\mathcal{K}_{11}^1 U_1 + \mathcal{K}_{12}^1 U_2 &= F_1^1 - mu'(s_2), \\
\mathcal{K}_{21}^1 U_1 + \mathcal{K}_{22}^1 U_2 &= F_2^1 + mu'(s_1), \\
\\
\mathcal{K}_{11}^2 U_2 + \mathcal{K}_{12}^2 U_3 &= F_1^2 - mu'(s_3), \\
\mathcal{K}_{21}^2 U_2 + \mathcal{K}_{22}^2 U_3 &= F_2^2 + mu'(s_2), \\
&\cdots\cdots\cdots\cdots\cdots\cdots \\
&\cdots\cdots\cdots\cdots\cdots\cdots \\
\mathcal{K}_{11}^{n-1} U_{n-1} + \mathcal{K}_{12}^{n-1} U_n &= F_1^{n-1} - mu'(s_n),
\end{aligned}$$

$$\begin{aligned} \mathcal{K}_{21}^{n-1} U_{n-1} + \mathcal{K}_{22}^{n-1} U_n &= F_2^{n-1} + m u'(s_{n-1}), \\ \mathcal{K}_{11}^{n} U_n + \mathcal{K}_{12}^{n} U_{n+1} &= F_1^{n} - m u'(s_{n+1}), \\ \mathcal{K}_{21}^{n} U_n + \mathcal{K}_{22}^{n} U_{n+1} &= F_2^{n} + m u'(s_n). \end{aligned}$$

Adding those equations that have the last term in common, one obtains

$$\begin{aligned} \mathcal{K}_{21}^{1} U_1 + \mathcal{K}_{22}^{1} U_2 &= F_2^{1} + m u'(s_1), \\ \mathcal{K}_{11}^{1} U_1 + \left(\mathcal{K}_{12}^{1} + \mathcal{K}_{21}^{2}\right) U_2 + \mathcal{K}_{22}^{2} U_3 &= F_1^{1} + F_2^{2}, \\ &\cdots\cdots\cdots\cdots\cdots \\ &\cdots\cdots\cdots\cdots\cdots \\ \mathcal{K}_{11}^{n-1} U_{n-1} + \left(\mathcal{K}_{12}^{n-1} + \mathcal{K}_{22}^{n}\right) U_n + \mathcal{K}_{22}^{n} U_{n+1} &= F_1^{n-1} + F_2^{n}, \\ \mathcal{K}_{11}^{n} U_n + \mathcal{K}_{12}^{n} U_{n+1} &= F_1^{n} - m u'(s_{n+1}). \end{aligned}$$

Thus, a system of $n+1$ equations has arisen. The next step is to inverse (turn upside down) the sequence of the equations:

$$\begin{aligned} \mathcal{K}_{11}^{n} U_n + \mathcal{K}_{12}^{n} U_{n+1} &= F_1^{n} - m u'(s_{n+1}), \\ \mathcal{K}_{11}^{n-1} U_{n-1} + \left(\mathcal{K}_{12}^{n-1} + \mathcal{K}_{22}^{n}\right) U_n + \mathcal{K}_{22}^{n} U_{n+1} &= F_1^{n-1} + F_2^{n}, \\ &\cdots\cdots\cdots\cdots\cdots \\ &\cdots\cdots\cdots\cdots\cdots \\ \mathcal{K}_{11}^{1} U_1 + \left(\mathcal{K}_{12}^{1} + \mathcal{K}_{21}^{2}\right) U_2 + \mathcal{K}_{22}^{2} U_3 &= F_1^{1} + F_2^{2}, \\ \mathcal{K}_{21}^{1} U_1 + \mathcal{K}_{22}^{1} U_2 &= F_2^{1} + m u'(s_1). \end{aligned}$$

Recalling the element symmetry given by eq. (27), it is apparent that it holds

$$\mathcal{K}^{n} = \mathcal{K}^{1}, \quad \mathcal{K}^{n-1} = \mathcal{K}^{2}, \quad \ldots$$

and the system of algebraic equations takes its final form

$$\mathcal{K}_{11}^{1} U_n + \mathcal{K}_{12}^{1} U_{n+1} = F_1^{n} - m u'(s_{n+1}), \tag{30}$$

$$\mathcal{K}_{11}^{2} U_{n-1} + \left(\mathcal{K}_{12}^{2} + \mathcal{K}_{22}^{1}\right) U_n + \mathcal{K}_{22}^{1} U_{n+1} = F_1^{n-1} + F_2^{n}, \tag{31}$$

$$\cdots\cdots\cdots\cdots\cdots\cdots\cdots\cdots\cdots\cdots$$

$$\cdots\cdots\cdots\cdots\cdots\cdots\cdots\cdots\cdots\cdots$$

$$\mathcal{K}_{11}^{n} U_1 + \left(\mathcal{K}_{12}^{n} + \mathcal{K}_{21}^{n-1}\right) U_2 + \mathcal{K}_{22}^{n-1} U_3 = F_1^1 + F_2^2, \tag{32}$$

$$\mathcal{K}_{21}^{n} U_1 + \mathcal{K}_{22}^{n} U_2 = F_2^1 + m u'(s_1). \tag{33}$$

The system of eqs. (30)–(33) is written in matrix form

$$\begin{bmatrix} 0 & \cdots & \cdots & 0 & \mathcal{K}_{11}^1 & \mathcal{K}_{12}^1 \\ 0 & \cdots & 0 & \mathcal{K}_{11}^2 & \mathcal{K}_{12}^2 + \mathcal{K}_{21}^1 & \mathcal{K}_{22}^1 \\ \vdots & \vdots & \vdots & \vdots & \vdots & \vdots \\ \mathcal{K}_{11}^n & \mathcal{K}_{12}^n + \mathcal{K}_{21}^{n-1} & \mathcal{K}_{22}^{n-1} & 0 & \cdots & 0 \\ \mathcal{K}_{21}^n & \mathcal{K}_{22}^n & 0 & \cdots & \cdots & 0 \end{bmatrix} \begin{bmatrix} U_1 \\ U_2 \\ \vdots \\ U_n \\ U_{n+1} \end{bmatrix} = \begin{bmatrix} \mathcal{F}_1 \\ \mathcal{F}_2 \\ \vdots \\ \mathcal{F}_n \\ \mathcal{F}_{n+1} \end{bmatrix}$$

or,

$$\mathcal{K}\mathbf{U} = \mathcal{F}, \tag{34}$$

where

$$\mathcal{K} = \begin{bmatrix} \mathcal{K}_{11} & \mathcal{K}_{12} & \cdots & \mathcal{K}_{1n} & \mathcal{K}_{1(n+1)} \\ \mathcal{K}_{21} & \mathcal{K}_{22} & \cdots & \mathcal{K}_{2n} & \mathcal{K}_{2(n+1)} \\ \vdots & \vdots & \ddots & \vdots & \vdots \\ \mathcal{K}_{n1} & \mathcal{K}_{n2} & \cdots & \mathcal{K}_{nn} & \mathcal{K}_{n(n+1)} \\ \mathcal{K}_{(n+1)1} & \mathcal{K}_{(n+1)2} & \cdots & \mathcal{K}_{(n+1)n} & \mathcal{K}_{(n+1)(n+1)} \end{bmatrix}$$

$$= \begin{bmatrix} 0 & \cdots & \cdots & 0 & \mathcal{K}_{11}^1 & \mathcal{K}_{12}^1 \\ 0 & \cdots & 0 & \mathcal{K}_{11}^2 & \mathcal{K}_{12}^2 + \mathcal{K}_{21}^1 & \mathcal{K}_{22}^1 \\ \vdots & \vdots & \vdots & \vdots & \vdots & \vdots \\ \mathcal{K}_{11}^n & \mathcal{K}_{12}^n + \mathcal{K}_{21}^{n-1} & \mathcal{K}_{22}^{n-1} & 0 & \cdots & 0 \\ \mathcal{K}_{21}^n & \mathcal{K}_{22}^n & 0 & \cdots & \cdots & 0 \end{bmatrix} \tag{35}$$

and

$$\mathcal{F} = \begin{bmatrix} \mathcal{F}_1 \\ \mathcal{F}_2 \\ \vdots \\ \mathcal{F}_n \\ \mathcal{F}_{n+1} \end{bmatrix} = \begin{bmatrix} F_1^n \\ F_2^n + F_1^{n-1} \\ \vdots \\ F_1^1 + F_2^2 \\ F_2^1 \end{bmatrix} + \begin{bmatrix} -mu'(s_{n+1}) \\ 0 \\ \vdots \\ 0 \\ mu'(s_1) \end{bmatrix} \tag{36}$$

are the *global stiffness matrix* and the *global force vector*, respectively.

Remark 12. It is noted that the stiffness matrix, as it appears in eq. (35), does not take the typical form of the standard finite element method. Here, the non-vanishing entries lie along the second diagonal of the matrix. Actually, the notion of the matrix symmetry has been changed and now is regarded with respect to the second diagonal. The standard matrix symmetry comes out from the common scalar product of the finite dimensional Euclidean spaces. Here an alternative symmetry arises behind which is hidden the bilinear form

$$B(\mathbf{x}, \mathbf{y}) = x_1 y_{n+1} + x_2 y_n + \cdots + x_n y_2 + x_{n+1} y_1, \quad \mathbf{x}, \mathbf{y} \in \mathcal{E}^{n+1},$$

which is compatible with the convolution upon which this analysis is based.

Remark 13. The loading vector $\mathcal{F}$ is notable, too. The initial condition (2b) which applies at the first node ($s_1 = 0$) is pushed away from it and contributes to the last node ($s_{n+1} = t$). Certainly, the initial condition (2a) contributes to the first node through the set of admissible functions $\mathcal{D}$ (see eq. (6)). This redistribution of the two initial conditions to the initial and final time is essentially the mechanism that restores the inconsistency of the standard variational principles with the initial conditions.

3.3. *Imposition of the first initial condition*

At this stage, one may impose the first initial condition of the problem. The condition (2a) provides the position at the first node, i.e., $U_1 = u(0) = u_0$. One may deal with this condition as one does with the Dirichlet conditions in a boundary value problem. Thus, one may ignore the first equation of system (34) or, equivalently, may omit the first row and the first column of the global stiffness matrix and modify appropriately the global force vector.

As concerns the second condition (2b), it has been included in the functional I^1 (see eqs. (21) and (25b)), resulting in the appearance of the initial momentum $mu'(s_1)$ in the last row of the global force vector. In conclusion, one may establish an $n \times n$ stiffness matrix and a fully known n-dimensional force vector to determine the n unknown parameters $U_2, \ldots, U_{n+1}$.

4. Global Analysis

In the previous section, we started, as usual, from the local stiffness matrix and the corresponding linear system at each element and assembling we conclude in the global system. Since the presented approximation scheme is not the standard FEM, we reproduce this result starting from the global form of the sought solution.

The global finite element solution over the entire $[0, t]$ will be taken as a linear combination of the global shape functions which are composed from the local ones as

$$N_i(s) = \begin{cases} N_2^{i-1}(s), & s \in [s_{i-1}, s_i] \\ N_1^i(s), & s \in [s_i, s_{i+1}] \\ 0, & \text{elsewhere}, \end{cases} \quad i = 1, \ldots, n+1. \tag{37}$$

The functions N_i enjoy the relation

$$N_i(s_j) = \delta_{ij},$$

i.e, are linear independent and span the finite dimensional space of the piecewise linear functions over $[0, t]$:

$$X_{n+1} := \text{span}\{N_1, \ldots, N_{n+1}\} \subset W_2^1(0, t).$$

Thus, according to Proposition 1, to find an approximate solution of Problem 1, in the form of a linear combination of the global shape functions, is equivalent to the following:

Approximation problem *Find a solution to the variational problem*

$$I[u] \to \textit{stationary}, \tag{38}$$

in X_{n+1}.

Consider now such an approximate solution in the form of

$$u^{\text{fe}}(s) = \sum_{i=2}^{n+1} U_i N_i(s), \quad s \in [0, t], \tag{39}$$

where $U_i \in \mathcal{R}$, $i = 2, \dots n+1$, are the unknown displacements that has to be computed. Note that the first shape function N_1 has been omitted because the displacement U_1 is known by the first initial condition (see eq. (2a)).

It will be shown that the unknown parameters U_i can be determined by solving the linear system (34).

Indeed, inserting eq. (39) into the functional I (given by eq. (5)), one obtains

$$I[u^{fe}] = \hat{I}(U_i) = \frac{1}{2} M_{ij} U_i U_j + \frac{1}{2} K_{ij} U_i U_j - F_i U_i - \tilde{F}_i U_i \tag{40}$$

or

$$\begin{aligned} \hat{I}(\mathbf{U}) &= \frac{1}{2}\mathbf{U}^T \mathbf{M} \mathbf{U} + \frac{1}{2}\mathbf{U}^T \mathbf{K} \mathbf{U} - \mathbf{F}^T \mathbf{U} - \tilde{\mathbf{F}}^T \mathbf{U} \\ &= \frac{1}{2}\mathbf{U}^T \mathcal{K} \mathbf{U} - \mathcal{F}^T \mathbf{U}, \end{aligned}$$

where

$$\mathcal{K} = \mathbf{M} + \mathbf{K}, \quad \mathcal{F} = \mathbf{F} + \tilde{\mathbf{F}} \tag{41}$$

and

$$M_{ij} = m[N_i', N_j'], \; K_{ij} = k[N_i, N_j], \; F_i = [f, N_i], \; \tilde{F}_i = [\tilde{f}, N_i], \quad i, j = 2, \dots, n+1. \tag{42}$$

A necessary condition for $\hat{I}$ to attain a stationary value at U_i is to valid

$$\frac{\partial \hat{I}}{\partial U_i} = 0, \quad i = 2, \dots, n+1,$$

equivalently,

$$\mathcal{K}_{ij}U_j = \mathcal{F}_i, \quad i = 2, \ldots, n+1$$

or, in matrix form

$$\mathcal{K}\mathbf{U} = \mathcal{F}. \tag{43}$$

The next step is the determination of column vector $\mathcal{F}$ and the matrix $\mathcal{K}$ so as to confirm that they coincide with the ones given by eqs. (35) and (36), respectively.

4.1. *Determination of $\mathcal{F}$*

First, the column vector $\tilde{\mathbf{F}}$ is examined. Recalling the symmetry of the discretization (see Section 2.2), it is easy to verify that the global shape functions enjoy the symmetry

$$N_i(t-s) = N_m(s),$$

where m-node is the symmetric to the i-node, i.e.,

$$m = n + 2 - i.$$

Then one can easily calculate the n-dimensional vector $\tilde{\mathbf{F}}$

$$\tilde{F}_i = \left[\tilde{f}, N_i\right] = \int_0^t \tilde{f}(s)N_i(t-s)ds = \int_0^t f_0\delta(s-s_1)N_m(s)ds = f_0\delta_{1m},$$

for

$$i = 2, \ldots, n+1 \quad \text{corresponding to} \quad m = n, \ldots, 1$$

which implies

$$\tilde{\mathbf{F}} = \begin{bmatrix} 0 \\ \vdots \\ 0 \\ f_0 \end{bmatrix} = \begin{bmatrix} 0 \\ \vdots \\ 0 \\ mv_0 \end{bmatrix}. \tag{44}$$

Similarly, the vector $\mathbf{F}$ by virtue of eqs. (28) and (37) can be determined as follows:

$$[f, N_i] = \int_0^t f(s)N_i(t-s)ds = \int_0^t f(s)N_m(s)ds$$

$$
\begin{aligned}
&= \int_{s_{m-1}}^{s_{m+1}} f(s)N_m(s)ds = \int_{s_{m-1}}^{s_m} f(s)N_2^{m-1}(s)ds \\
&\quad + \int_{s_m}^{s_{m+1}} f(s)N_1^m(s)ds \\
&= \int_0^{\tau_m} f(s_{m-1}+s)N_1^{m-1}(s_m - s)ds \\
&\quad + \int_0^{\tau_m} f(s_{m-1}+s)N_2^m(s_{m+1} - s)ds \\
&\Rightarrow [f, N_i] = \left[f, N_1^{m-1}\right]_{s_{m-1}}^{s_m} + [f, N_2^m]_{s_m}^{s_{m+1}} \\
&\qquad\qquad = F_1^{m-1} + F_2^m, \quad \begin{array}{l} i = 2, \ldots, n+1 \\ m = n, \ldots, 1. \end{array}
\end{aligned}
$$

Thus, recalling the relation $m = n + 2 - i$, one concludes

$$
\begin{aligned}
F_i = [f, N_i] = F_1^{m-1} + F_2^m = F_1^{n+1-i} + F_2^{n+2-i}, \\
i = 2, \ldots, n+1, \quad m = n, \ldots, 1,
\end{aligned}
$$

or

$$
\mathbf{F} = \begin{bmatrix} F_1^{n-1} + F_2^n \\ \vdots \\ F_1^1 + F_2^2 \\ F_2^1 \end{bmatrix},
$$

and by turn, $\mathcal{F}$ takes its final form

$$
\mathcal{F} = \mathbf{F} + \tilde{\mathbf{F}} = \begin{bmatrix} F_1^{n-1} + F_2^n \\ \vdots \\ F_1^1 + F_2^2 \\ F_2^1 \end{bmatrix} + \begin{bmatrix} 0 \\ \vdots \\ 0 \\ mv_0 \end{bmatrix}, \tag{45}
$$

thus, apart from the first row, it coincides with eq. (36) of the previous section.

4.2. Determination of $\mathcal{K}$

If m is the symmetric to the j node, the following relation holds:

$$[N_i, N_j] = \int_0^t N_i(s)N_j(t-s)ds = \int_0^t N_i(s)N_m(s)ds = (N_i, N_m), \tag{46}$$

where $m = n + 2 - j$. Thus, for j running from 2 to $n+1$, m runs from n to 1, consequently, one can write

$$\mathbf{K} = k \begin{bmatrix} [N_2, N_2] & [N_2, N_3] & \dots & [N_2, N_{n+1}] \\ \vdots & \vdots & \dots & \vdots \\ [N_{n+1}, N_2] & [N_{n+1}, N_3] & \dots & [N_{n+1}, N_{n+1}] \end{bmatrix}$$
$$= k \begin{bmatrix} (N_2, N_n) & (N_2, N_{n-1}) & \dots & (N_2, N_1) \\ \vdots & \vdots & \dots & \vdots \\ (N_{n+1}, N_n) & (N_{n+1}, N_{n-1}) & \dots & (N_{n+1}, N_1) \end{bmatrix}. \tag{47}$$

Note that (N_i, N_m) may be nonzero only if the i and m nodes coincide or are neighboring, thus, for a certain i, there are only three cases for non-vanishing terms. Consider, for instance, the case $i = m$, that is, i is symmetric to j node:

$$\begin{aligned}
[N_i, N_j] &= (N_i, N_i) \quad (\text{for } i = n + 2 - j) \\
&= \int_0^t N_i^2(s)ds = \int_{s_{i-1}}^{s_{i+1}} N_i^2(s)ds \\
&= \int_{s_{i-1}}^{s_i} N_i^2(s)ds + \int_{s_i}^{s_{i+1}} N_i^2(s)ds \\
&= \int_{s_{i-1}}^{s_i} \left(N_2^{i-1}\right)^2(s)ds + \int_{s_i}^{s_{i+1}} \left(N_1^i\right)^2(s)ds \quad (\text{due to eq. (37)}) \\
&= \left(N_2^{i-1}, N_2^{i-1}\right) + \left(N_1^i, N_1^i\right) \\
&= \left[N_2^{i-1}, N_1^{i-1}\right]_{s_{i-1}}^{s_i} + \left[N_1^i, N_2^i\right]_{s_i}^{s_{i+1}} \quad (\text{due to eq. (28)}).
\end{aligned}$$

Thus, recalling eqs. (24) and (42), one concludes in

$$K_{ij} = K_{21}^{i-1} + K_{12}^i, \quad i \text{ symmetric to } j. \tag{48}$$

By the same manner, one can examine the cases $i = m - 1$ and $i = m + 1$ so as to prove the general relation:

$$K_{ij} = \begin{cases} K_{11}^{i-1}, & i = n + 1 - j \\ K_{21}^{i-1} + K_{12}^{i}, & i = n = 2 - j \\ K_{22}^{i}, & i = n + 3 - j \\ 0, & \text{otherwise} \end{cases} \tag{49}$$

and

$$M_{ij} = \begin{cases} M_{21}^{m-1} + M_{12}^{m}, & i = m \\ M_{11}^{m-1}, & i = m - 1 \\ M_{22}^{m}, & i = m + 1 \\ 0, & \text{otherwise.} \end{cases} \tag{50}$$

Thus, according to eq. (41), the matrix $\mathcal{K}$ is given by the relation

$$\mathcal{K}_{ij} = \begin{cases} \mathcal{K}_{21}^{m-1} + \mathcal{K}_{12}^{m}, & i = m \\ \mathcal{K}_{11}^{m-1}, & i = m - 1 \\ \mathcal{K}_{22}^{m}, & i = m + 1 \\ 0, & \text{otherwise} \end{cases} \tag{51}$$

or

$$\mathcal{K}_{ij} = \begin{bmatrix} 0 & \cdots & \cdots & \cdots & \mathcal{K}_{2(n-1)} & \mathcal{K}_{2n} & \mathcal{K}_{2(n+1)} \\ \vdots & \cdots & 0 & \mathcal{K}_{3(n-2)} & \mathcal{K}_{3(n-1)} & \mathcal{K}_{3n} & 0 \\ \vdots & \cdots & \cdots & \cdots & \cdots & \cdots & \vdots \\ 0 & \cdots & \cdots & \cdots & \cdots & \cdots & \vdots \\ \mathcal{K}_{(n-1)2} & \mathcal{K}_{(n-1)3} & \mathcal{K}_{(n-1)4} & 0 & \cdots & \cdots & \vdots \\ \mathcal{K}_{n2} & \mathcal{K}_{n3} & 0 & \cdots & \cdots & \cdots & \vdots \\ \mathcal{K}_{(n+1)2} & 0 & \cdots & \cdots & \cdots & \cdots & 0 \end{bmatrix}.$$

The latter, accounting for eq. (51), takes its final form

$$\mathcal{K}_{ij} = \begin{bmatrix} 0 & \cdots & \cdots & \cdots & \mathcal{K}_{11}^2 & \mathcal{K}_{21}^1 + \mathcal{K}_{12}^2 & \mathcal{K}_{22}^1 \\ \vdots & \cdots & 0 & \mathcal{K}_{11}^3 & \mathcal{K}_{21}^2 + \mathcal{K}_{12}^3 & \mathcal{K}_{22}^2 & 0 \\ \vdots & \cdots & \cdots & \cdots & \cdots & \cdots & \vdots \\ 0 & \cdots & \cdots & \cdots & \cdots & \cdots & \vdots \\ \mathcal{K}_{11}^{n-1} & \mathcal{K}_{21}^{n-2} + \mathcal{K}_{12}^{n-1} & \mathcal{K}_{22}^{n-2} & 0 & \cdots & \cdots & \vdots \\ \mathcal{K}_{21}^{n-1} + \mathcal{K}_{12}^{n} & \mathcal{K}_{22}^{n-1} & 0 & \cdots & \cdots & \cdots & \vdots \\ \mathcal{K}_{22}^{n} & 0 & \cdots & \cdots & \cdots & \cdots & 0 \end{bmatrix} \tag{52}$$

which is identical to the matrix given by eq. (35) if one omits the first row and the first column.

5. A One-step Marching Scheme

Returning to Section 3.1, note that the local linear system (26) may provide a recurrent algorithm for the computation of the field u and of its derivative u'. Indeed, consider the system corresponding to the e-element given by

$$\begin{aligned} \mathcal{K}_{11}^e U_e + \mathcal{K}_{12}^e U_{e+1} &= F_1^e - mV_{e+1}, \\ \mathcal{K}_{21}^e U_e + \mathcal{K}_{22}^e U_{e+1} &= F_2^e + mV_e, \end{aligned} \tag{53}$$

where V_e denotes the velocity at the time s_e, i.e., it holds

$$V_e = u'(s_e), \quad e = 1, \ldots, n. \tag{54}$$

The above system can be rewritten as

$$\begin{aligned} V_{e+1} &= \frac{1}{m}(-\mathcal{K}_{11}^e U_e - \mathcal{K}_{12}^e U_{e+1} + F_1^e), \\ U_{e+1} &= \frac{1}{\mathcal{K}_{22}^e}(\mathcal{K}_{21}^e U_e + mV_e + F_2^e). \end{aligned} \tag{55}$$

Thus, if the displacement and velocity are known at time s_e, one can compute U_{e+1} from eq. (55b) and, in turn, V_{e+1} from eq. (55a).

Besides, the approximation scheme can be written in the form of an implicit recurrent formula:

$$\begin{bmatrix} \mathcal{K}_{12}^e & m \\ \mathcal{K}_{22}^e & 0 \end{bmatrix} \begin{bmatrix} U_{e+1} \\ V_{e+1} \end{bmatrix} = \begin{bmatrix} -\mathcal{K}_{11}^e & 0 \\ -\mathcal{K}_{21}^e & m \end{bmatrix} \begin{bmatrix} U_e \\ V_e \end{bmatrix} + \begin{bmatrix} F_1^e \\ F_2^e \end{bmatrix}.$$

If one considers a uniform discretization, i.e., fixed time step, then the local stiffness matrix does not depend on the particular element and the approximation scheme takes the form

$$\mathbf{A}\mathbf{W}^{I+1} = \mathbf{B}\mathbf{W}^{I} + \mathbf{F}^{I}, \quad I \in \mathbb{Z}, \tag{56}$$

where

$$\mathbf{A} = \begin{bmatrix} \mathcal{K}_{12}^e & m \\ \mathcal{K}_{22}^e & 0 \end{bmatrix}, \quad \mathbf{B} = \begin{bmatrix} -\mathcal{K}_{11}^e & 0 \\ -\mathcal{K}_{21}^e & m \end{bmatrix}, \quad \mathbf{W}^I = \begin{bmatrix} U_I \\ V_I \end{bmatrix}, \quad \mathbf{F}^I = \begin{bmatrix} F_1^I \\ F_2^I \end{bmatrix}.$$

It is noted that the superscript e does not mean dependence of $\mathcal{K}^e$ on the element, i.e., on the time step. On the contrary, $\mathcal{K}^e$ is computed once and for all elements. The column vector $\mathbf{F}^I$, however, does depend on the time step and must be calculated at each time step from the given external force. Particularly, for a free vibration problem, where the external force $\mathbf{F}^I$ vanishes (or is constant), the recurrent formula (56) is completely independent of the element and the evolution of the system $\mathbf{W}$ will be given by

$$\mathbf{W}^{I+1} = \mathbf{C}^{(I+1)}\mathbf{W}^1, \tag{57}$$

where

$$\mathbf{C} = \mathbf{A}^{-1}\mathbf{B}$$

and $\mathbf{W}^1$ contains the initial data. To proceed further, one has to compute the matrix $\mathcal{K}^e$. Recalling eqs. (20), the local shape functions and their derivatives are written as

$$N_1^e = \frac{s_{e+1} - s}{\tau}, \quad N_2^e = \frac{s - s_e}{\tau}$$

and

$$N_1^{e\prime} = -\frac{1}{\tau}, \quad N_2^{e\prime} = \frac{1}{\tau},$$

where τ is the time step. Thus, the matrices $\mathbf{M}^e$ and $\mathbf{K}^e$ become

$$\mathbf{M}^e = \frac{m}{\tau}\begin{bmatrix} 1 & -1 \\ -1 & 1 \end{bmatrix}, \quad \mathbf{K}^e = \frac{k\tau}{3}\begin{bmatrix} 1/2 & 1 \\ 1 & 1/2 \end{bmatrix},$$

respectively, which in combination (eq. (23)) results in

$$\mathcal{K}^e = \begin{bmatrix} \frac{m}{\tau} + \frac{k\tau}{6} & -\frac{m}{\tau} + \frac{k\tau}{3} \\ -\frac{m}{\tau} + \frac{k\tau}{3} & \frac{m}{\tau} + \frac{k\tau}{6} \end{bmatrix}$$

and in turn $\mathbf{A}$ and $\mathbf{B}$ take the particular form

$$\mathbf{A} = \begin{bmatrix} -\frac{m}{\tau} + \frac{k\tau}{3} & m \\ \frac{m}{\tau} + \frac{k\tau}{6} & 0 \end{bmatrix}, \quad \mathbf{B} = \begin{bmatrix} -\frac{m}{\tau} - \frac{k\tau}{6} & 0 \\ \frac{m}{\tau} - \frac{k\tau}{3} & m \end{bmatrix}. \tag{58}$$

The next question concerns the stability of the approximation scheme (57)–(58). To secure the stability, one needs complex values of the generalized eigenvalue problem

$$\det(\mathbf{B} - \lambda\mathbf{A}) = 0$$

or

$$\lambda^2 - 2\lambda\frac{M-K}{M+K/2} + 1 = 0, \tag{59}$$

where

$$M = m/\tau \quad \text{and} \quad K = k\tau/3. \tag{60}$$

Requiring negative value for the determinant of eq. (65), one obtains

$$4\left[\left(\frac{M-K}{M+K/2}\right)^2 - 1\right] < 1 \Rightarrow K/M < 4,$$

or equivalently, on account of eqs. (60),

$$\tau^2 < 12m/k \quad \text{or} \quad (\tau\omega)^2 < 12, \tag{61}$$

where $\omega = \sqrt{k/m}$ is the frequency of the free oscillation. It can be rewritten in terms of the period $T = 2\pi/\omega$ as

$$\tau < \frac{\sqrt{12}}{2\pi}T \approx 0.551T. \tag{62}$$

If the time step, τ, fulfills the above inequality, then it is easy to confirm that for both conjugate eigenvalues, it holds

$$|\lambda_1| = |\lambda_2| = 1.$$

Thus, Eq. (61) (or eq. (62)) is the required condition for the numerical stability of the approximation scheme.

6. Computational Applications

In this section, some computational examples are presented to support the theoretical and numerical considerations of the previous sections. Uniform discretization (elements of equal length) is used so that the results of the two approximation schemes can be fully comparable. In that case, the local stiffness matrices are fixed for every element (see Appendix A):

$$\mathbf{K}^e = \frac{m\tau}{3}\begin{bmatrix} 1/2 & 1 \\ 1 & 1/2 \end{bmatrix}, \quad \mathbf{M}^e = \frac{m}{\tau}\begin{bmatrix} 1 & -1 \\ -1 & 1 \end{bmatrix}.$$

Taking into account eqs. (74) and (75) of Appendix B, the global matrices for a uniform discretization become

$$\mathbf{K} = \frac{k\tau}{3}\begin{bmatrix} 0 & \dots & \dots & 1/2 & 2 & 1/2 \\ 0 & \dots & 1/2 & 2 & 1/2 & 0 \\ \vdots & \vdots & \vdots & \vdots & \vdots & \vdots \\ 2 & 1/2 & 0 & 0 & \dots & 0 \\ 1/2 & 0 & 0 & \dots & \dots & 0 \end{bmatrix}, \quad \mathbf{M} = \frac{m}{\tau}\begin{bmatrix} 0 & \dots & \dots & -1 & 2 & -1 \\ 0 & \dots & -1 & 2 & -1 & 0 \\ \vdots & \vdots & \vdots & \vdots & \vdots & \vdots \\ 2 & -1 & 0 & 0 & \dots & 0 \\ -1 & 0 & 0 & \dots & \dots & 0 \end{bmatrix}.$$

All examples run with the constants $k = 9$ and $m = 1$, so as the natural frequency of the system to become $\omega = 3$.

First, both approximation schemes are tested for the free vibration problem. In that case, the external force vanishes, i.e., $f = 0$, and the system is excited by the initial velocity or the initial position. For the examples presented here, we have taken $u(0) = 0$ and $u'(0) = 2$. In Fig. 1, the values of the position versus time are shown for three distinct values of the element length (or time step). Simultaneously, comparison with the analytical solution is demonstrated in each case. Analogous results and comparisons are presented in Table 1.

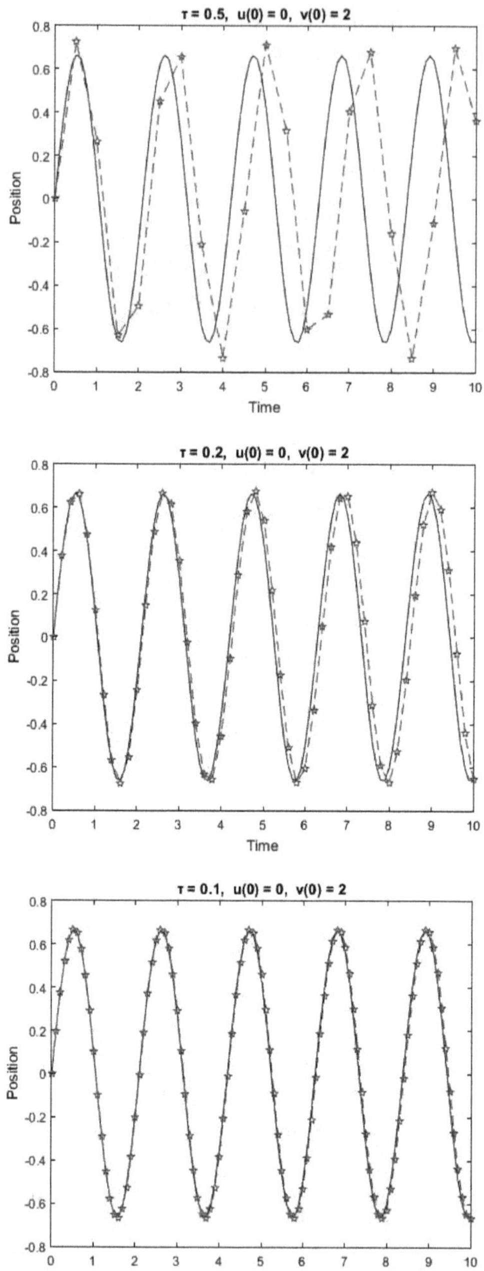

Fig. 1. FEM approximation for the free vibration for different values of the τ, the continuous line is the exact solution.

Table 1. Approximate solution of the free vibration problem for various time steps.

	FEM, One–step						
Time	$\tau = 0.1$	$\tau = 0.05$	$\tau = 0.025$	$\tau = 0.02$	$\tau = 0.0125$	$\tau = 0.01$	Exact
1	0.1018	0.0960	0.0946	0.0944	0.0942	0.0942	0.0941
2	−0.2012	−0.1900	−0.1872	−0.1869	−0.1865	−0.1864	−0.1863
3	0.2960	0.2801	0.2761	0.2756	0.2751	0.2750	0.2747
4	−0.3839	−0.3643	−0.3594	−0.3588	−0.3581	−0.3580	−0.3577
5	0.4628	0.4410	0.4354	0.4347	0.4340	0.4338	0.4335
6	−0.5309	−0.5085	−0.5026	−0.5019	−0.5012	−0.5010	−0.5007
7	0.5867	0.5654	0.5597	0.5590	0.5583	0.5581	0.5578
8	−0.6288	−0.6105	−0.6054	−0.6048	−0.6042	−0.6040	−0.6037
9	0.6563	0.6429	0.6390	0.6385	0.6379	0.6378	0.6376
10	−0.6685	−0.6619	−0.6595	−0.6592	−0.6589	−0.6588	−0.6587

The second example concerns a forced vibration problem. For this reason, an excitation term of the form

$$f(s) = f_0 \sin(\Omega t)$$

was taken, where Ω is the frequency and f_0 is the amplitude of the external force. The nodal forces, i.e., eqs. (25a), can be analytically calculated as (see Appendix C)

$$F_1^e = -(f_0/\Omega)\cos(\Omega s_{e+1}) + \left(f_0/(\tau_e\Omega^2)\right)\left[\sin(\Omega s_{e+1}) - \sin(\Omega s_e)\right],$$

$$F_2^e = (f_0/\Omega)\cos(\Omega s_e) - \left(f_0/(\tau_e\Omega^2)\right)\left[\sin(\Omega s_{e+1}) - \sin(\Omega s_e)\right].$$

In Fig. 2 and Table 2, the response of the system to the external excitation is demonstrated for various values of τ and is compared with the analytical solution. Also, in Fig. 3(a), the algorithms for the two approximation solutions were run for a longer time to reveal the characteristic pattern of the forced vibration and the particular case where the excitation frequency meets the natural velocity of the system is shown in Fig. 3(b).

It is remarkable that both approximation schemes, the FEM and the one-step, produce the same numbers, at least at the accuracy of four decimal digits which are used here. To find differences between the two schemes, one has to proceed further accounting for at least 12 additional decimal digits, as one can confirm by looking at the Table 3.

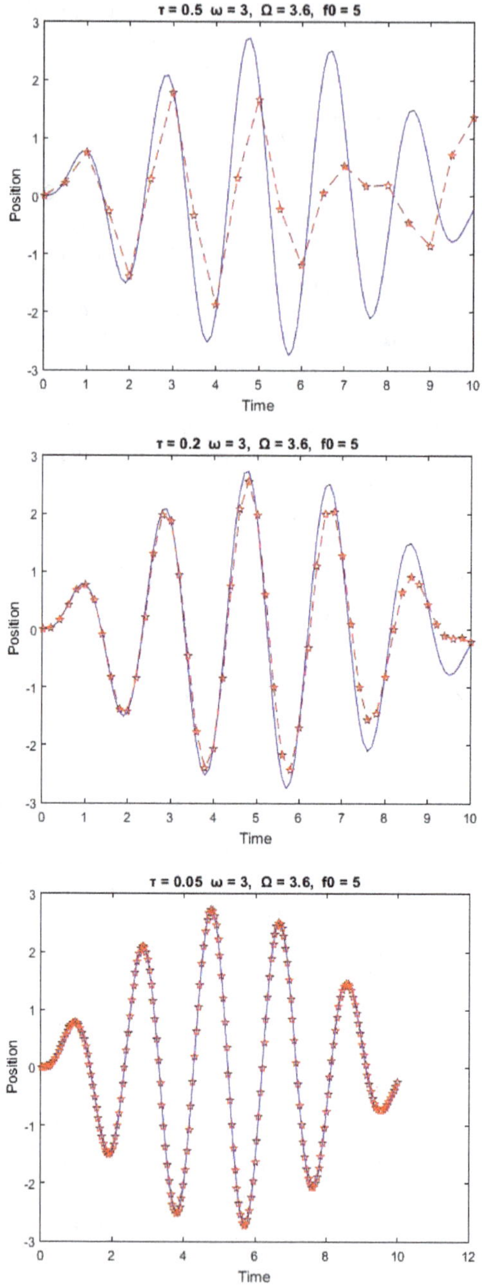

Fig. 2. FEM and one-step approximation schemes (asterisks) for the forced vibration, for different values of τ versus the exact solution (blue solid line).

Table 2. Approximate values of position for the forced vibration problem, $f_0 = 5$, $\Omega = 3.6$.

	FEM, One–step						
Time	$\tau = 0.1$	$\tau = 0.05$	$\tau = 0.025$	$\tau = 0.02$	$\tau = 0.0125$	$\tau = 0.01$	Exact
1	0.7713	0.7722	0.7725	0.7725	0.7725	0.7725	0.7726
2	−1.4247	−1.4253	−1.4254	−1.4254	−1.4255	−1.4255	−1.4255
3	1.8657	1.8638	1.8632	1.8631	1.8630	1.8630	1.8630
4	−2.0418	−2.0349	−2.0329	−2.0327	−2.0324	−2.0324	−2.0323
5	1.9523	1.9386	1.9348	1.9343	1.9338	1.9337	1.9335
6	−1.6478	−1.6270	−1.6212	−1.6205	−1.6197	−1.6196	−1.6192
7	1.2188	1.1927	1.1853	1.1844	1.1834	1.1832	1.1828
8	−0.7766	−0.7493	−0.7413	−0.7403	−0.7392	−0.7390	−0.7385
9	0.4296	0.4072	0.4001	0.3992	0.3982	0.3980	0.3976
10	−0.2607	−0.2506	−0.2464	−0.2458	−0.2452	−0.2450	−0.2448

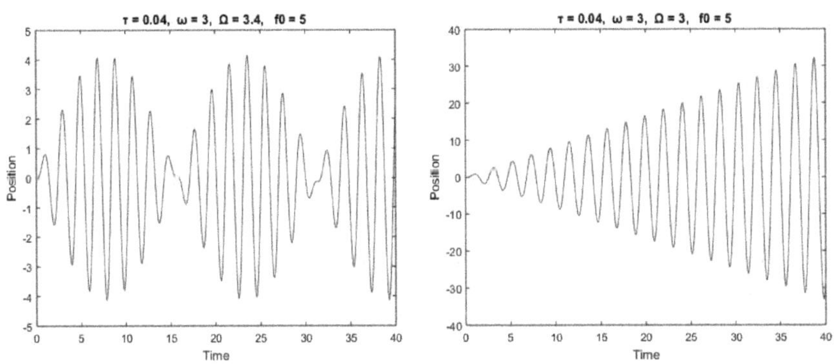

Fig. 3. (a) Characteristic pattern of a vibration imposed by external force. (b) Resonance.

7. Conclusions

In Ref. [12], a generalized Hamilton's principle has been developed which is free of the main inconsistency of the classical Hamilton's principle, i.e., its weakness to account properly for the initial conditions of a dynamical problem. The key role to this development is the use of the convolution instead of the L^2 inner product.

Table 3. Forced oscillation for $\Omega = 3.6$, $f_0 = 5$, F: FEM approximation, O: one-step approximation, $D = F - O$.

	$\tau = 0.5$			$\tau = 0.01$		
Time	F	O	$D \times 10^{14}$	F	O	$D \times 10^{12}$
1	0.7678	0.7678	0	0.7725	0.7725	−0.0356
2	−1.4215	−1.4215	0.0444	−1.4255	−1.4255	0.0429
3	1.8691	1.8691	0	1.8630	1.8630	−0.0551
4	−2.0581	−2.0581	0.2665	−2.0324	−2.0324	0.0226
5	1.9850	1.9850	−0.1110	1.9337	1.9337	−0.0628
6	−1.6952	−1.6952	0.2220	−1.6196	−1.6196	0.2720
7	1.2718	1.2718	−0.0888	1.1832	1.1832	−0.3257
8	−0.8185	−0.8185	0.2554	−0.7390	−0.7390	0.3303
9	0.4369	0.4369	−0.4441	0.3980	0.3980	−0.3147
10	−0.2060	−0.2060	0.5718	−0.2450	−0.2450	0.1495

In this chapter, this generalized principle is exploited to formulate new approximation schemes for a simple initial value problem. More specifically, a consistent finite element formulation is fully developed. The use of convolution introduces significant changes in the finite dimensional version of the problem. The stiffness matrix is now symmetric with respect to its second diagonal and the load vector is inversed. Besides, a recurrent formula stems from the consideration of the algebraic system at the element which provides an one-step approximation scheme, the numerical stability of which is examined. Both approximation schemes are tested by computational examples concerning the free and forced vibration of a one degree of freedom oscillator.

It seems that the analysis presented here may be applied to initial value problems in general. Actually, the essential conclusion of this work is the possibility to formulate FE schemes for initial value problems, provided that an appropriate convolutional variational formulation has been previously developed.

The next challenge is to further proceed to the formulation of FE scheme for a unified space–time continuum, i.e., to an undivided FE scheme for initial boundary value problems.

Appendices

A. Calculation of Local Matrices and Forces

The terms of eq. (21) are calculated one by one:

- $$[u^{e\prime}, u^{e\prime}]_{s_e}^{s_{e+1}}$$

$$
\begin{aligned}
&= \int_0^{\tau_e} u^{e\prime}(s_e + s)u^{e\prime}(s_{e+1} - s)ds \\
&= u_1^{e2} \int_0^{\tau_e} N_1^{e\prime}(s_e + s)N_1^{e\prime}(s_{e+1} - s)ds \\
&\quad + u_1^e u_2^e \int_0^{\tau_e} N_1^{e\prime}(s_e + s)N_2^{e\prime}(s_{e+1} - s)ds \\
&\quad + u_2^e u_1^e \int_0^{\tau_e} N_2^{e\prime}(s_e + s)N_1^{e\prime}(s_{e+1} - s)ds \\
&\quad + u_2^{e2} \int_0^{\tau_e} N_2^{e\prime}(s_e + s)N_2^{e\prime}(s_{e+1} - s)ds \\
&= u_1^{e2} \left[N_1^{e\prime}, N_1^{e\prime}\right]_{s_e}^{s_{e+1}} + 2u_1^e u_2^e \left[N_1^{e\prime}, N_2^{e\prime}\right]_{s_e}^{s_{e+1}} \\
&\quad + u_2^{e2} \left[N_2^{e\prime}, N_2^{e\prime}\right]_{s_e}^{s_{e+1}} \\
&\Rightarrow \frac{m}{2}[u^{e\prime}, u^{e\prime}]_{s_e}^{s_{e+1}} = \frac{1}{2}\left(u_1^{e2} M_{11}^e + 2u_1^e u_2^e M_{12}^e + u_2^{e2} M_{22}^e\right) \\
&= \frac{1}{2}\mathbf{u}^{eT}\mathbf{M}^e\mathbf{u}^e, \qquad (63)
\end{aligned}
$$

where

$$\mathbf{M}^e = M_{ij}^e = m\left[N_i^{e\prime}, N_j^{e\prime}\right]_{s_e}^{s_{e+1}}, \quad \begin{matrix} i = 1, 2, \\ e = 1, \ldots, n, \end{matrix} \qquad (64)$$

and

$$\mathbf{M}^e = \frac{m}{\tau^e}\begin{bmatrix} 1 & -1 \\ -1 & 1 \end{bmatrix}. \qquad (65)$$

- $$[u^e, u^e]_{s_e}^{s_{e+1}}$$

$$= \int_0^{\tau_e} u^e(s_e + s)u^e(s_{e+1} - s)ds$$

$$= u_1^{e2} \int_0^{\tau_e} N_1^e(s_e + s)N_1^e(s_{e+1} - s)ds$$

$$+ u_1^e u_2^e \int_0^{\tau_e} N_1^e(s_e + s)N_2^e(s_{e+1} - s)ds$$

$$+ u_2^e u_1^e \int_0^{\tau_e} N_2^e(s_e + s)N_1^e(s_{e+1} - s)ds$$

$$+ u_2^{e2} \int_0^{\tau_e} N_1^e(s_e + s)N_1^e(s_{e+1} - s)ds$$

$$= u_1^{e2} [N_1^e, N_1^e]_{s_e}^{s_{e+1}} + 2u_1^e u_2^e \left[N_1^e, N_2^{e\prime}\right]_{s_e}^{s_{e+1}} + u_2^{e2} [N_2^e, N_2^e]_{s_e}^{s_{e+1}}$$

$$\Rightarrow \frac{k}{2}[u^e, u^e]_{s_e}^{s_{e+1}} = \frac{1}{2}\left(u_1^{e2}K_{11}^e + 2u_1^e u_2^e K_{12}^e + u_2^{e2}K_{22}^e\right)$$

$$= \frac{1}{2}\mathbf{u}^{eT}\mathbf{K}^e\mathbf{u}^e, \tag{66}$$

where

$$\mathbf{K}^e = K_{ij}^e = k\left[N_i^e, N_j^e\right]_{s_e}^{s_{e+1}}, \qquad \begin{matrix} i, j = 1, 2, \\ e = 1, \ldots, n, \end{matrix} \tag{67}$$

and

$$\mathbf{K}^e = \frac{m\tau^e}{3}\begin{bmatrix} 1/2 & 1 \\ 1 & 1/2 \end{bmatrix}. \tag{68}$$

- $$[f, u^e]_{s_e}^{s_{e+1}} = \int_0^{\tau_e} f(s_e + s)u^e(s_{e+1} - s)ds$$

$$= \int_0^{\tau_e} f(s_e + s)\left(u_1^e N_1^e(s_{e+1} - s) + u_2^e N_2^e(s_{e+1} - s)\right)ds$$

$$= u_1^e \left(\int_0^{\tau_e} f(s_e + s) N_1^e(s_{e+1} - s) ds \right)$$

$$+ u_2^e \left(\int_0^{\tau_e} f(s_e + s) N_2^e(s_{e+1} - s) ds \right)$$

$$\Rightarrow [f, u^e]_{s_e}^{s_{e+1}} = u_1^e F_1^e + u_2^e F_2^e = \mathbf{u}^{eT} \mathbf{F}^e, \tag{69}$$

where

$$\mathbf{F}^e = F_i^e = \int_0^{\tau_e} f(s_e + s) N_i^e(s_{e+1} - s) ds, \ i = 1, 2, \ e = 1, \ldots, n.$$

- $$\left[\tilde{f}_2^e, u^e\right]_{s_e}^{s_{e+1}} = \int_0^{\tau_e} \tilde{f}_2^e(s_e + s) u^e(s_{e+1} - s) ds$$

$$= \int_0^{\tau_e} f_2^e \delta(s - s_{e+1}) u^e(s_{e+1} - s) ds$$

$$\Rightarrow \left[\tilde{f}_2^e, u^e\right]_{s_e}^{s_{e+1}} = f_2^e u_1^e. \tag{70}$$

- $$\left[\tilde{f}_1^e, u^e\right]_{s_e}^{s_{e+1}} = \int_0^{\tau_e} \tilde{f}_1^e(s_e + s) u^e(s_{e+1} - s) ds$$

$$= \int_0^{\tau_e} f_1^e \delta(s - s_e) u^e(s_{e+1} - s) ds$$

$$\Rightarrow \left[\tilde{f}_1^e, u^e\right]_{s_e}^{s_{e+1}} = f_1^e u_2^e. \tag{71}$$

B. Calculation of $[N_i, N_j]$ and $[N_i', N_j']$

The $n + 2 - j$ node is symmetric to the j node. The integral

$$[N_i, N_j] = (N_i, N_{n+2-j}) = \int_0^t N_i(s) N_{n+2-j}(s) ds$$

may be nonzero only if the i and $n + 2 - j$ nodes coincide or are neighboring, that is, for a particular i, there are only three cases for non-vanishing terms:

- *Case 1.* $n+2-j=i$

$$
\begin{aligned}
[N_i, N_j] &= \int_0^t N_i(s)N_j(t-s)ds \\
&= \int_0^t N_i(s)N_{n+2-j}(s)ds = \int_0^t N_i(s)N_i(s)ds \\
&= \int_{s_{i-1}}^{s_{i+1}} N_i(s)^2 ds = \int_{s_{i-1}}^{s_i} \left(\frac{s-s_{i-1}}{\tau_{i-1}}\right)^2 ds \\
&\quad + \int_{s_i}^{s_{i+1}} \left(\frac{s_{i+1}-s}{\tau_i}\right)^2 ds \\
&= \frac{\tau_{i-1}}{3} + \frac{\tau_i}{3} \quad \Rightarrow [N_i, N_j] = \frac{\tau_i + \tau_{i+1}}{3}.
\end{aligned}
$$

- *Case 2.* $n+2-j=i+1$

$$
\begin{aligned}
[N_i, N_j] &= \int_0^t N_i(s)N_{n+2-j}(s)ds \\
&= \int_0^t N_i(s)N_{i+1}(s)ds = \int_{s_i}^{s_{i+1}} N_i(s)N_{i+1}(s)ds \\
&= \int_{s_i}^{s_{i+1}} \left(\frac{s_{i+1}-s}{\tau_i}\right)\left(\frac{s-s_i}{\tau_i}\right) ds = \frac{\tau_i}{6} \Rightarrow [N_i, N_j] = \frac{\tau_i}{6}.
\end{aligned}
$$

- *Case 3.* $n+2-j=i-1$

$$
\begin{aligned}
[N_i, N_j] &= \int_0^t N_i(s)N_{n+2-j}(s)ds \\
&= \int_0^t N_i(s)N_{i-1}(s)ds = \int_{s_{i-1}}^{s_i} N_i(s)N_{i-1}(s)ds \\
&= \int_{s_{i-1}}^{s_i} \left(\frac{s-s_{i-1}}{\tau_{i-1}}\right)\left(\frac{\tau_i - s}{\tau_{i-1}}\right) ds = \frac{\tau_{i-1}}{6} \Rightarrow [N_i, N_j] \\
&= \frac{\tau_{i-1}}{6}.
\end{aligned}
$$

Thus, one concludes in the general formula

$$[N_i, N_j] = \begin{cases} \tau_i/6, & i = n+1-j \\ (\tau_{i-1}+\tau_i)/3, & i = n+2-j \\ \tau_i/6, & i = n+3-j \\ 0, & \text{otherwise} \end{cases}. \tag{72}$$

Analogously, one may prove

$$[N_i', N_j'] = \begin{cases} -1/\tau_i, & i = n+1-j \\ 1/\tau_{i-1}+1/\tau_i, & i = n+2-j \\ 1/\tau_{i-1}, & i = n+3-j \\ 0, & \text{otherwise.} \end{cases}. \tag{73}$$

After that, the matrices **K** and **M** are written as follows:

$$\mathbf{K} = \frac{k}{3}\begin{bmatrix} 0 & \dots & \dots & \tau_2/2 & \tau_1+\tau_2 & \tau_1/2 \\ 0 & \dots & \tau_3/2 & \tau_2+\tau_3 & \tau_2/2 & 0 \\ \vdots & \vdots & \vdots & \vdots & \vdots & \vdots \\ \tau_{n-1}+\tau_n & \tau_{n-1}/2 & 0 & 0 & \dots & 0 \\ \tau_n/2 & 0 & 0 & \dots & \dots & 0 \end{bmatrix} \tag{74}$$

and

$$\mathbf{M} = m\begin{bmatrix} 0 & \dots & \dots & -1/\tau_2 & 1/\tau_1+1/\tau_2 & -1/\tau_1 \\ 0 & \dots & -1/\tau_3 & 1/\tau_2+1/\tau_3 & -1/\tau_2 & 0 \\ \vdots & \vdots & \vdots & \vdots & \vdots & \vdots \\ 1/\tau_{n-1}+1/\tau_n & -1/\tau_{n-1} & 0 & 0 & \dots & 0 \\ -1\tau_n & 0 & 0 & \dots & \dots & 0 \end{bmatrix}, \tag{75}$$

respectively.

C. Calculation of Nodal Forces

From eqs. (20), (25a), and (63), one obtains

$$\begin{aligned} F_1^e &= \int_0^{\tau_e} f(s_e+s) N_1^e(s_{e+1}-s)ds \\ &= f_0 \int_0^{\tau_e} \cos\left(\Omega(s_e+s)\right) \frac{s_{e+1}-(s_{e+1}-s)}{\tau_e} ds \\ &= (f_0/\tau_e) \int_0^{\tau_e} \cos\left(\Omega(s_e+s)\right) s ds. \end{aligned} \tag{76}$$

Similarly,

$$\begin{aligned} F_2^e &= \int_0^{\tau_e} f(s_e+s) N_2^e(s_{e+1}-s)ds \\ &= f_0 \int_0^{\tau_e} \cos\left(\Omega(s_e+s)\right) \frac{(s_{e+1}-s)-s_e}{\tau_e} ds \\ &= (f_0/\tau_e) \int_0^{\tau_e} \cos\left(\Omega(s_e+s)\right) (\tau_e - s) ds. \end{aligned} \tag{77}$$

Also, it is easy to see that

$$\begin{aligned} &\int_0^{\tau_e} \cos\left(\Omega(s_e+s)\right) s ds \\ &\quad = (1/\Omega)\left[\sin(\Omega(s_e+\tau_e))\tau_e + (1/\Omega)\left(\cos(\Omega s_{e+1}) - \cos(\Omega s_e)\right)\right], \end{aligned}$$

with the aid of which eqs. (76) and (77) become

$$F_1^e = (f_0/\Omega)\sin(\Omega s_{e+1}) + (f_0/\tau_e\Omega^2)\left[\cos(\Omega s_{e+1}) - \cos(\Omega s_e)\right], \tag{78}$$

$$F_2^e = -(f_0/\Omega)\sin(\Omega s_e) - (f_0/\tau_e\Omega^2)\left[\cos(\Omega s_{e+1}) - \cos(\Omega s_e)\right]. \tag{79}$$

References

[1] Amiri-Hezaveh, P., Karimi, A.G. and Ostoja-Starzewski, M. IBVP for elctromagnetic–elastic materials: Variational approach. *Mathematics and Mechanics of Complex Systems*, 8, 47–67 (2020).

[2] Apostolakis, G. and Dargush. G.F. Mixed ariational principles for dynamic response of thermoelastic and poroelastic continua. *International Journal of Solids and Structures*, 50, 642–650 (2013).

[3] Apostolakis, G. and Dargush. G.F. Variational methods in irreversible thermooelasticity: Theoretical developments and minimum principles for the discrete form. *Acta Mechanica*, 224, 2065–2088 (2013).

[4] Dargush, G.F. Mixed convolved action for classical and fractional derivative dissipative dynamical systems. *Physical Review E*, 86, 066606 (2012).

[5] Dargush, G.F., Apostolakis, G., Darrall, B.T. and Kim. J. Mixed convolved action variational principles in heat diffusion. *Int. J. Heat Mass Transfer*, 100, 790–799 (2016).

[6] Dargush, G.F., Darrall, B.T., Kim, J., and Apostolakis, G. Mixed convolved action principles in linear continuum dynamics. *Acta Mechanica*, 226, 4111–4137 (2015).

[7] Dargush, G.F. and Kim, J.Mixed convolved action. *Physical Review E*, 85, 066606 (2012).

[8] Darrall, B.T. and Dargush, G.F. Variational principle and time-space finite element method for dynamic thermoelasticity based on mixed convolved action. *European Journal of Mechanics/A Solids*, 71, 351–364 (2018).

[9] Gurtin, M.E.Variational principles for linear theory of viscoelasticity. *Archives for Rational Mechanics and Analysis*, 13, 179–191 (1963).

[10] Gurtin, M.E. Variational principles for linear elastodynamics. *Archives for Rational Mechanics and Analysis*, 16, 34–50 (1964).

[11] Gurtin, M.E.Variational principles for linear initial-value problems. *Quarterly of Applied Mathematics*, 22, 252–256 (1964).

[12] Kalpakides, V.K. and Charalambopoulos, A. On Hamilton's principle for discrete and continuous systems. *Reports on Mathematical Physics*, 87, 225–248 (2021).

[13] Kim, J. Higher order temporal finite element methods through mixed formalisms. *SpringerPlus*, 3, 458 (2014).

[14] Peng, J.S., Lewis, R.W. and Zhang, J.Y. A semi-analutical approach for solving forced vibration problems based on convolution-type variational principle. *Computers and stractures*, 59, 167–177 (1996).

[15] Rafalski. A variational principle for the coupled thermoelastic problem. *International Journal of Enginerring Science*, 6, 465–471 (1968).
[16] Reddy, J.N. Modified Gurtin's variational principles in the linear dynamic theory of viscoelasticity. *International Journal of Solids and Structures*, 12, 227–235 (1976).
[17] Reddy. J.N. Variational principles for linear coupled dynamic theory of thermoviscoelasticity. *International Journal of Engineering Sciences*, 14, 605–616 (1976).
[18] E.T. Tonti. On the variational formulation for linear initial value problems. *Anal. Mat. Pura Appl.*, 95, 331–359 (1973).

https://doi.org/10.1142/9789811293009_0021

Chapter 21

Norm Inequalities for Fractional Integral Operators in Banach Spaces

Jichang Kuang
Department of Mathematics, Hunan Normal University, Changsha, Hunan, 410081, P. R. China
jckuang@163.com

In this chapter, we introduce some new fractional integral operators and fractional area balance operators in the Banach spaces. The corresponding norm inequalities are established. They are significant improvement and generalizations of many known and new classes of fractional integral operators.

1. Introduction

It is well known that fractional integral operator is one of the important operators in harmonic analysis with a background of partial differential equations. In fact, the solution of the Laplace equation $\triangle g = f$ for good functions on $\mathbb{R}^n$ can be represented by using the fractional integral operators acting on f. In the recent past, different versions of fractional integral operators have been developed which are useful in the study of different classes of differential and integral equations. These fractional integral operators act as ready tools to study the classes of differential and integral equations. Fractional integral operators were utilized extensively to study the classical inequalities. A generalization of classical inequalities by means of fractional integral operators is considered as an interesting subject area. Many mathematicians have established a variety of inequalities for those fractional integral and derivative operators, some of which have turned out to be useful in analyzing solutions of certain fractional integral and differential equations. Hence, fractional integral inequalities are very important in the theory and applications of differential equations. Such inequalities are also of great importance in the mathematical modeling of the fractional boundary value problems. See [1–9, 11–13, 16–18]. First, we recall the following definitions and some related results.

Definition 1 (cf. [1, 2], [6]). Let $f \in L[a,b]$, then Riemann–Liouville fractional integrals of f of order $\alpha > 0$ with $a \geq 0$ are defined by

$$T_1(f,x) = \frac{1}{\Gamma(\alpha)} \int_a^x (x-t)^{\alpha-1} f(t)dt \ x > a \tag{1}$$

and

$$T_2(f,x) = \frac{1}{\Gamma(\alpha)} \int_x^b (t-x)^{\alpha-1} f(t)dt, \ x < b, \tag{2}$$

respectively, where

$$\Gamma(\alpha) = \int_0^\infty t^{\alpha-1} e^{-t} dt \tag{3}$$

is the gamma function and when $\alpha = 0, T_1(f,x) = T_2(f,x) = f(x)$.

Definition 2 (cf. [3]). Let $f \in L[a,b]$, then Riemann–Liouville k-fractional integrals of f of order $\alpha > 0$ with $a \geq 0$ are defined by

$$T_3(f,x) = \frac{1}{k\Gamma_k(\alpha)} \int_a^x (x-t)^{(\alpha/k)-1} f(t)dt, \ x > a, \tag{4}$$

and

$$T_4(f,x) = \frac{1}{k\Gamma_k(\alpha)} \int_x^b (t-x)^{(\alpha/k)-1} f(t)dt, \ x < b, \tag{5}$$

respectively, where

$$\Gamma_k(\alpha) = \int_0^\infty t^{\alpha-1}e^{-(t^k/k)}dt, \ (\alpha > 0) \tag{6}$$

is the k-gamma function. Also, $\Gamma(x) = \lim_{k\to 1}\Gamma_k(x), \Gamma_k(\alpha) = k^{(\alpha/k)-1}\Gamma(\alpha/k)$, and $\Gamma_k(\alpha+k) = \alpha\Gamma_k(\alpha)$.

It is well known that the Mellin transform of f is defined by

$$M(f,\alpha) = \int_0^\infty t^{\alpha-1}f(t)dt, (\alpha > 0).$$

Therefore, the Mellin transform of the exponential function $\exp -t^k/k$ is the k-gamma function.

Definition 3 (cf. [4, 5]). Let $t^r f(t) \in L^1[a,b], a \geq 0$, then the generalized Riemann–Liouville fractional integrals of f of order (α, r) are defined by

$$T_5(f,x) = \frac{(r+1)^{1-\alpha}}{\Gamma(\alpha)}\int_a^x (x^{r+1} - t^{r+1})^{\alpha-1}t^r f(t)dt, \ x > a, \tag{7}$$

$$T_6(f,x) = \frac{(r+1)^{1-\alpha}}{\Gamma(\alpha)}\int_x^b (t^{r+1} - x^{r+1})^{\alpha-1}t^r f(t)dt, \ x < b, \tag{8}$$

and there are also (k,r) fractional of Riemann-type operators:

$$T_7(f,x) = \frac{(r+1)^{1-(\alpha/k)}}{k\Gamma_k(\alpha)}\int_a^x (x^{r+1} - t^{r+1})^{(\alpha/k)-1}t^r f(t)dt, \ x > a, \tag{9}$$

$$T_8(f,x) = \frac{(r+1)^{1-(\alpha/k)}}{k\Gamma_k(\alpha)}\int_x^b (t^{r+1} - x^{r+1})^{(\alpha/k)-1}t^r f(t)dt, \ x < b, \tag{10}$$

respectively, where $k, \alpha > 0, r \geq 0, x \in [a,b]$.

In particular, if $r = 0$, then Definition 3 reduces to Definition 1 and Definition 2.

Definition 4 (cf. [11, 12]). Let f be a conformable integrable function on $[a,b] \subset [0,\infty)$. The right-sided and left-sided generalized conformable fractional integrals T_9 and T_{10} of f of order $\alpha > 0$ are defined by

$$T_9(f,x) = \frac{1}{\Gamma(\alpha)} \int_a^x \left(\frac{x^{r+s} - t^{r+s}}{r+s} \right)^{\alpha-1} t^{r+s-1} f(t)dt, \ x > a, \quad (11)$$

and

$$T_{10}(f,x) = \frac{1}{\Gamma(\alpha)} \int_x^b \left(\frac{t^{r+s} - x^{r+s}}{r+s} \right)^{\alpha-1} t^{r+s-1} f(t)dt, \ x < b, \quad (12)$$

respectively, where $r \in \mathbb{R}^1, s \in (0,1], r+s \neq 0$.

In particular, if $s = 1$, then T_9, T_{10} reduce to T_5, T_6, respectively.

Definition 5 (cf. [8, 9]). Let $f \in L[a,b]$,$g : [a,b] \to (0,\infty)$ be an increasing function, and $g' \in C[a,b], \alpha > 0$. Then, g-Riemann–Liouville fractional integrals of f with respect to the function g on $[a,b]$ are defined by

$$T_{11}(f,x) = \frac{1}{\Gamma(\alpha)} \int_a^x g'(t)[g(x) - g(t)]^{\alpha-1} f(t)dt, \ x > a, \quad (13)$$

and

$$T_{12}(f,x) = \frac{1}{\Gamma(\alpha)} \int_x^b g'(t)[g(t) - g(x)]^{\alpha-1} f(t)dt, \ x < b, \quad (14)$$

respectively.

In 2018, S.S. Dragomir ([10]) introduced a new notion of the area balance function:

Definition 6 ([10]). Let $f \in L[a,b]$, then the area balance function of f is defined by

$$T_{13}(f,x) = \frac{1}{2} \left\{ \int_x^b f(t)dt - \int_a^x f(t)dt \right\}, x \in [a,b]. \quad (15)$$

In the following, let $AC[a,b]$ denote the class of absolutely continuous functions on $[a,b]$.

In 2020, Kuang ([16]) introduced the new notion of the generalized fractional integral operators and fractional area balance operators:

Definition 7 ([16]). Let $f \in L[a,b]$, $g : [a,b] \to (0,\infty)$ be an increasing function, and $g \in AC[a,b], k, c, \alpha > 0$, $a \geq 0$. Then, the generalized fractional integral operator T_{14} with respect to the function g on $[a,b]$ is defined by

$$T_{14}(f,x) = \frac{c}{k\Gamma_k(\alpha)} \int_a^b g'(t)|g(x) - g(t)|^{(\alpha/k)-1} f(t)dt, \qquad (16)$$

where $\Gamma_k(\alpha)$ is defined by (6).

Let

$$T_{15}(f,x) = \frac{c}{k\Gamma_k(\alpha)} \int_a^x g'(t)[g(x) - g(t)]^{(\alpha/k)-1} f(t)dt, \ x > a, \quad (17)$$

and

$$T_{16}(f,x) = \frac{c}{k\Gamma_k(\alpha)} \int_x^b g'(t)[g(t) - g(x)]^{(\alpha/k)-1} f(t)dt, \ x < b. \quad (18)$$

Then,

$$T_{14}(f,x) = T_{15}(f,x) + T_{16}(f,x). \qquad (19)$$

Just indicate that for a suitable and appropriate choice of the parameters and function, one can obtain various new and old results. For example, if $c = k = 1$ in (19), then (19) reduces to

$$T_{14}(f,x) = T_{11}(f,x) + T_{12}(f,x). \qquad (20)$$

If $c = (r+s)^{-\alpha}, g(t) = t^{r+s}, r \in \mathbb{R}^1, s \in (0,1], r+s \neq 0, k = 1$ in (19), then (19) reduces to

$$T_{14}(f,x) = T_9(f,x) + T_{10}(f,x). \qquad (21)$$

If $c = (r+1)^{-(\alpha/k)}, g(t) = t^{r+1}, r \geq 0$, in (19), then (19) reduces to

$$T_{14}(f,x) = T_7(f,x) + T_8(f,x). \qquad (22)$$

If $s = 1$ in (21), then (21) reduces to

$$T_{14}(f,x) = T_5(f,x) + T_6(f,x). \qquad (23)$$

If $r = 0$ in (22), then (22) reduces to

$$T_{14}(f, x) = T_3(f, x) + T_4(f, x). \tag{24}$$

If $k = 1$ in (24), then (24) reduces to

$$T_{14}(f, x) = T_1(f, x) + T_2(f, x). \tag{25}$$

We can also rewrite T_9 and T_{10} as

$$T_9(f, x) = \frac{(r+s)^{1-\alpha}}{\Gamma(\alpha)} \int_a^x (x^{r+s} - t^{r+s})^{\alpha-1} t^{r+s-1} f(t) dt, \ x > a,$$

and

$$T_{10}(f, x) = \frac{(r+s)^{1-\alpha}}{\Gamma(\alpha)} \int_x^b (t^{r+s} - x^{r+s})^{\alpha-1} t^{r+s-1} f(t) dt, \ x < b,$$

and then generalize them to

$$T_{17}(f, x) = \frac{(r+s)^{1-(\alpha/k)}}{k\Gamma_k(\alpha)} \int_a^x (x^{r+s} - t^{r+s})^{(\alpha/k)-1} t^{r+s-1} f(t) dt, \ x > a, \tag{26}$$

and

$$T_{18}(f, x) = \frac{(r+s)^{1-(\alpha/k)}}{k\Gamma_k(\alpha)} \int_x^b (t^{r+s} - x^{r+s})^{(\alpha/k)-1} t^{r+s-1} f(t) dt, \ x < b. \tag{27}$$

If $c = (r+s)^{-(\alpha/k)}, g(t) = t^{r+s}, r \in \mathbb{R}^1, s \in (0, 1], r + s \neq 0$ in (19), then (19) reduces to

$$T_{14}(f, x) = T_{17}(f, x) + T_{18}(f, x). \tag{28}$$

Remark 8. We can generalize the region from $[a, b]$ to $[0, \infty)$ by defining a function $f_1 : [0, \infty) \to \mathbb{R}^1$:

$$f_1(t) = \begin{cases} f(t), & t \in [a, b], \\ 0, & t \in [0, a) \bigcup (b, \infty), \end{cases}$$

and we can still f_1 as f, then the generalized conformable fractional integral operators T_{19}, T_{20} are defined by

$$T_{19}(f, x) = \frac{(r+s)^{1-(\alpha/k)}}{k\Gamma_k(\alpha)} \int_0^x (x^{r+s} - t^{r+s})^{(\alpha/k)-1} t^{r+s-1} f(t) dt$$

and

$$T_{20}(f,x) = \frac{(r+s)^{1-(\alpha/k)}}{k\Gamma_k(\alpha)} \int_x^{\infty} (t^{r+s} - x^{r+s})^{(\alpha/k)-1} t^{r+s-1} f(t)dt.$$

Let

$$T_{21}(f,x) = \frac{(r+s)^{1-(\alpha/k)}}{k\Gamma_k(\alpha)} \int_0^{\infty} |x^{r+s} - t^{r+s}|^{(\alpha/k)-1} t^{r+s-1} f(t)dt.$$

Then

$$T_{21}(f,x) = T_{19}(f,x) + T_{20}(f,x).$$

Definition 9 ([2]). Let $f \in L[a,b], a \geq 0$. The left-sided and right-sided Hadamard fractional integrals T_{22} and T_{23} of f of order $\alpha > 0$ are defined by

$$T_{22}(f,x) = \frac{1}{\Gamma(\alpha)} \int_a^x (\log x - \log t)^{\alpha-1} t^{-1} f(t)dt, \;\; x > a,$$

and

$$T_{23}(f,x) = \frac{1}{\Gamma(\alpha)} \int_x^b (\log t - \log x)^{\alpha-1} t^{-1} f(t)dt, \;\; x < b,$$

respectively.

We can generalize them to

$$T_{24}(f,x) = \frac{1}{k\Gamma_k(\alpha)} \int_a^x (\log x - \log t)^{(\alpha/k)-1} t^{-1} f(t)dt, \;\; x > a,$$

and

$$T_{25}(f,x) = \frac{1}{k\Gamma_k(\alpha)} \int_x^b (\log t - \log x)^{(\alpha/k)-1} t^{-1} f(t)dt, \;\; x < b.$$

If $c = 1, g(t) = \log t$ in (19), then (19) reduces to

$$T_{14}(f,x) = T_{24}(f,x) + T_{25}(f,x).$$

In particular, if $k = 1$, then

$$T_{14}(f,x) = T_{22}(f,x) + T_{23}(f,x).$$

Definition 10 ([16]). Under the assumptions of Definition 7, the fractional area balance operator T_{26} with respect to the function g on $[a,b]$ is defined by

$$T_{26}(f,x) = \frac{c}{k\Gamma_k(\alpha)}\left\{\int_x^b g'(t)[g(t)-g(x)]^{(\alpha/k)-1}f(t)dt - \int_a^x g'(t)[g(x)-g(t)]^{(\alpha/k)-1}f(t)dt\right\}, \tag{29}$$

where $\Gamma_k(\alpha)$ is defined by (6).

Using (17) and (18), we have

$$T_{26}(f,x) = T_{16}(f,x) - T_{15}(f,x). \tag{30}$$

Just indicate that for suitable and appropriate choices of the parameters and function, one can obtain various new and old results. For example, if $c = (r+s)^{-(\alpha/k)}, g(t) = t^{r+s}, r \in \mathbb{R}^1, s \in (0,1], r+s \neq 0$ in (30), then (30) reduces to

$$T_{26}(f,x) = T_{18}(f,x) - T_{17}(f,x). \tag{31}$$

If $g(t) = t, \alpha = k = 1, c = 1/2$ in (30), then T_{26} reduces to T_{13}. If $c = k = 1$ in (30), then (30) reduces to

$$T_{26}(f,x) = T_{12}(f,x) - T_{11}(f,x). \tag{32}$$

If $c = (r+s)^{-\alpha}, g(t) = t^{r+s}, r \in \mathbb{R}^1, s \in (0,1], r+s \neq 0, k = 1$ in (30), then (30) reduces to

$$T_{26}(f,x) = T_{10}(f,x) - T_9(f,x). \tag{33}$$

If $c = (r+1)^{-(\alpha/k)}, g(t) = t^{r+1}, r \geq 0$ in (30), then (30) reduces to

$$T_{26}(f,x) = T_8(f,x) - T_7(f,x). \tag{34}$$

If $k = 1$ in (34), then (34) reduces to

$$T_{26}(f,x) = T_6(f,x) - T_5(f,x). \tag{35}$$

If $r = 0$ in (34), then (34) reduces to

$$T_{26}(f,x) = T_4(f,x) - T_3(f,x). \tag{36}$$

If $k = 1$ in (36), then (36) reduces to

$$T_{26}(f, x) = T_2(f, x) - T_1(f, x). \tag{37}$$

If $c = 1, g(t) = \log t$ in (30), then (30) reduces to

$$T_{26}(f, x) = T_{22}(f, x) - T_{21}(f, x).$$

In particular, if $k = 1$, then

$$T_{26}(f, x) = T_{20}(f, x) - T_{19}(f, x).$$

Hence, definitions 7 and 9 unified and generalized many known and new classes of fractional integral operators. It is noted that all the fractional integral operators above are established for functions of one variable. We naturally ask how we generalize these results to functions of several variable? In 2014, Sarikaya ([17]) gives the definitions of Riemann–Liouville fractional integrals of two variable functions:

Definition 11 ([17–19]). Let $Q = [a, b] \times [c, d] = \bigcup_{k=1}^{4} Q_k$, where $Q_1 = [a, x] \times [c, y], Q_2 = [a, x] \times [y, d], Q_3 = [x, b] \times [c, y], Q_4 = [x, b] \times [y, d]$. The Riemann–Liouville fractional integrals $I_k (1 \leq k \leq 4)$ are defined by

$$I_1(f; x, y) = \frac{1}{\Gamma(\alpha)\Gamma(\beta)} \int_{Q_1} (x - t)^{\alpha-1}(y - s)^{\beta-1} f(t, s)dsdt,$$

$$I_2(f; x, y) = \frac{1}{\Gamma(\alpha)\Gamma(\beta)} \int_{Q_2} (x - t)^{\alpha-1}(s - y)^{\beta-1} f(t, s)dsdt,$$

$$I_3(f; x, y) = \frac{1}{\Gamma(\alpha)\Gamma(\beta)} \int_{Q_3} (t - x)^{\alpha-1}(y - s)^{\beta-1} f(t, s)dsdt,$$

and

$$I_4(f; x, y) = \frac{1}{\Gamma(\alpha)\Gamma(\beta)} \int_{Q_4} (t - x)^{\alpha-1}(s - y)^{\beta-1} f(t, s)dsdt.$$

If the Q in the above definition 11 is applied to the n-dimensional interval $Q = \bigcup_{k=1}^{n} Q_k (m \geq 2^n)$ in $\mathbb{R}_+^n$, where $Q_k = [a_k, b_k]$, then the problem becomes very complicated when n is large. The aim of this chapter is to introduce some new generalized fractional integral operators and fractional area balance operators on the Banach spaces, which includes $\mathbb{R}_+^n$ as the special case. In Section 2, we define generalized fractional integral operators and fractional area balance operators on the Banach space. In Section 3, some lemmas are derived. The corresponding integral operator inequalities are established in Section 4. They are significant improvements and generalizations of many known and new classes of fractional integral operators.

2. Generalized Fractional Integral Operators and Fractional Area Balance Operators

Throughout this chapter, we write

$$E_n = \Bigg\{ x = (x_1, x_2, \ldots, x_n) : x_k \geq 0, 1 \leq k \leq n,$$

$$\|x\| = \left(\sum_{k=1}^{n} |x_k|^\beta \right)^{1/\beta}, \beta > 0 \Bigg\}.$$

When $1 \leq \beta < \infty$, E_n is a Banach space. In particular, when $r = 2$, E_n is an n-dimensional Euclidean space $\mathbb{R}_+^n$. When $r = 1$, $\|x\| = \sum_{k=1}^{n} |x_k|$ is a Cartesian norm. Let

$$\begin{aligned} D_1 &= \{y = (y_1, y_2, \ldots, y_n) : y_k \geq 0, 1 \leq k \leq n, \\ &\quad 0 \leq \|y\| < \|x\|, x \in E_n\}, \\ D_2 &= \{y = (y_1, y_2, \ldots, y_n) : y_k \geq 0, 1 \leq k \leq n, \\ &\quad \|x\| < \|y\| < \infty, x \in E_n\}. \end{aligned}$$

Then, $E_n = D_1 \bigcup D_2$. The norm of operator $T : L^p(E_n) \to L_\omega^p(E_n) (1 \leq p < \infty)$ is defined by

$$\|T\| = \sup_{f \neq 0} \frac{\|Tf\|_{p,\omega}}{\|f\|_p}.$$

Definition 12. Let $f \in L(E_n)$, then the generalized conformable fractional integrals operator T_{27} is defined by

$$T_{27}(f,x) = \frac{(r+s)^{1-(\alpha/k)}}{k\Gamma_k(\alpha)} \times \int_{E_n} |\|x\|^{r+s} - \|y\|^{r+s}|^{(\alpha/k)-1} \|y\|^{r+s-1} f(y)dy, \tag{38}$$

where $\Gamma_k(\alpha)$ is defined by (6).

Let

$$T_{28}(f,x) = \frac{(r+s)^{1-(\alpha/k)}}{k\Gamma_k(\alpha)} \times \int_{D_1} [\|x\|^{r+s} - \|y\|^{r+s}]^{(\alpha/k)-1} \|y\|^{r+s-1} f(y)dy \tag{39}$$

and

$$T_{29}(f,x) = \frac{(r+s)^{1-(\alpha/k)}}{k\Gamma_k(\alpha)} \times \int_{D_2} [\|y\|^{r+s} - \|x\|^{r+s}]^{(\alpha/k)-1} \|y\|^{r+s-1} f(y)dy. \tag{40}$$

Thus,

$$T_{27}(f,x) = T_{28}(f,x) + T_{29}(f,x). \tag{41}$$

The generalized conformable fractional area balance operator is defined by

$$T_{30}(f,x) = T_{29}(f,x) - T_{28}(f,x). \tag{42}$$

In particular, if $n = \beta = 1$ in Definition 12, then T_{27}, T_{28}, and T_{29} reduce to T_{21}, T_{19}, and T_{20}, respectively.

Definition 13. Let $f \in L(E_n)$, $g : (0,\infty) \to (0,\infty)$ be an increasing function, and $g \in AC(0,\infty), k, c, \alpha > 0, a \geq 0$. Then, the generalized fractional integrals operator T_{31} with respect to the function g on E_n is defined by

$$T_{31}(f,x) = \frac{c}{k\Gamma_k(\alpha)} \int_{E_n} g'(\|y\|)|g(\|x\|) - g(\|y\|)|^{(\alpha/k)-1} f(y)dy, \tag{43}$$

where $\Gamma_k(\alpha)$ is defined by (6).

Let

$$T_{32}(f,x) = \frac{c}{k\Gamma_k(\alpha)} \int_{D_1} g'(\|y\|)[g(\|x\|) - g(\|y\|)]^{(\alpha/k)-1} f(y)dy \tag{44}$$

and

$$T_{33}(f,x) = \frac{c}{k\Gamma_k(\alpha)} \int_{D_2} g'(\|y\|)[g(\|y\|) - g(\|x\|)]^{(\alpha/k)-1} f(y)dy. \tag{45}$$

Thus,

$$T_{31}(f,x) = T_{32}(f,x) + T_{33}(f,x). \tag{46}$$

The fractional area balance operator with respect to the function g on E_n is defined by

$$T_{34}(f,x) = T_{33}(f,x) - T_{32}(f,x). \tag{47}$$

In particular, if $n = \beta = 1$ in Definition 13, then Definition 13 reduces to Definitions 7 and 10.

3. Some Lemmas

We require the following lemmas to prove our main results.

Lemma 14 ([15]). If $a_k, b_k, p_k > 0, 1 \le k \le n$, and f is a measurable function on $(0,\infty)$, then

$$\int_{B(r_1,r_2)} f\left(\sum_{k=1}^{n}\left(\frac{x_k}{a_k}\right)^{b_k}\right) x_1^{p_1-1}\cdots x_n^{p_n-1}dx_1\cdots dx_n$$

$$= \frac{\prod_{k=1}^{n} a_k^{p_k}}{\prod_{k=1}^{n} b_k} \cdot \frac{\prod_{k=1}^{n}\Gamma\left(\frac{p_k}{b_k}\right)}{\Gamma\left(\sum_{k=1}^{n}\frac{p_k}{b_k}\right)} \int_{r_1}^{r_2} f(t)t^{\left(\sum_{k=1}^{n}\frac{p_k}{b_k}-1\right)}dt,$$

where $B(r_1,r_2) = \{x \in E_n : 0 \le r_1 < \|x\| < r_2\}$.

We get Lemma 2 by taking $a_k = 1, b_k = \beta > 0, p_k = 1,\ 1 \le k \le n, r_1 = 0, r_2 = \infty$ in Lemma 1.

Lemma 15. *Let f be a measurable function on $(0,\infty)$, then*

$$\int_{E_n} f(\|x\|)dx = \frac{(\Gamma(1/\beta))^n}{\beta^n\Gamma(n/\beta)} \int_0^\infty f(t^{1/\beta})t^{(n/\beta)-1}dt. \tag{48}$$

4. Some Norm Inequalities for Operator T_{27} and T_{31}

Theorem 16. *Let* $1 < p, q < \infty$, $\frac{1}{p} + \frac{1}{q} \leq 1$, $0 < \frac{\lambda}{n} = 2 - \frac{1}{p} - \frac{1}{q}$, $\omega(x) = \|x\|^{c_0}$, *where*

$$c_0 = p\left(1 - n - \frac{\alpha(r+s)}{k}\right), x \in E_n. \tag{49}$$

If

$$c_1 = \frac{(\Gamma(1/\beta))^n}{\beta^{n-1}\Gamma(n/\beta)} \int_0^\infty |1 - u^{r+s}|^{\frac{n}{\lambda}(\frac{\alpha}{k}-1)p(1-\frac{1}{q})} \times u^{\frac{\alpha(r+s)}{k}[\frac{n}{\lambda}(p-1)-p]-\frac{n(p-1)}{\lambda}[1+n(1-\frac{1}{q})]+p-1} du < \infty \tag{50}$$

and

$$c_2 = \frac{(\Gamma(1/\beta))^n}{\beta^{n-1}\Gamma(n/\beta)} \times \int_0^\infty |1 - u^{r+s}|^{\frac{n}{\lambda}(\frac{\alpha}{k}-1)} u^{\frac{n}{\lambda}\left(1+n\left(1-\frac{1}{q}\right)-\frac{\alpha(r+s)}{k}\right)-n-1} du < \infty, \tag{51}$$

then the generalized conformable fractional integral operator T_{27} *is defined by* (38): $T_{27} : L^p(E_n) \to L^p_\omega(E_n)$ *exists as a bounded operator and*

$$\|T_{27}(f)\|_{p,\omega} \leq c_1^{1/p} c_2^{1-(1/p)} \frac{(r+s)^{1-(\alpha/k)}}{k\Gamma_k(\alpha)} \|f\|_p. \tag{52}$$

This implies that

$$\|T_{27}\| = \sup_{f\neq 0} \frac{\|T_{27}(f)\|_{p,\omega}}{\|f\|_p} \leq c_1^{1/p} c_2^{1-(1/p)} \frac{(r+s)^{1-(\alpha/k)}}{k\Gamma_k(\alpha)}. \tag{53}$$

Proof. Let $p_1 = \frac{p}{p-1}, q_1 = \frac{q}{1-q}$; it follows that

$$\frac{1}{p_1} + \frac{1}{q_1} + \left(1 - \frac{\lambda}{n}\right) = 1, \frac{p}{q_1} + p\left(1 - \frac{\lambda}{n}\right) = 1.$$

By Hölder's inequality, we get

$$\begin{aligned}
&T_{27}(f,x)\\
&\quad=\frac{(r+s)^{1-(\alpha/k)}}{k\Gamma_k(\alpha)}\int_{E_n}|\|x\|^{r+s}-\|y\|^{r+s}|^{(\alpha/k)-1}\|y\|^{r+s-1}f(y)dy\\
&\quad=\frac{(r+s)^{1-(\alpha/k)}}{k\Gamma_k(\alpha)}\int_{E_n}\left\{\|y\|^{(\frac{n^2}{\lambda p_1})}\right.\\
&\qquad\left.\times[|\|x\|^{r+s}-\|y\|^{r+s}|^{(\alpha/k)-1}\|y\|^{r+s-1}]^{n/\lambda}f^p(y)\right\}^{1/q_1}\\
&\qquad\times\left\{\|y\|^{-(\frac{n^2}{\lambda q_1})}[|\|x\|^{r+s}-\|y\|^{r+s}|^{(\alpha/k)-1}\|y\|^{r+s-1}]^{n/\lambda}\right\}^{1/p_1}\\
&\qquad\times\{f(y)\}^{p(1-\frac{\lambda}{n})}dy\le\frac{(r+s)^{1-(\alpha/k)}}{k\Gamma_k(\alpha)}\left\{\int_{E_n}\|y\|^{(\frac{n^2}{\lambda p_1})}\right.\\
&\qquad\left.\times[|\|x\|^{r+s}-\|y\|^{r+s}|^{(\alpha/k)-1}\|y\|^{r+s-1}]^{n/\lambda}|f(y)|^pdy\right\}^{1/q_1}\\
&\qquad\times\left\{\int_{E_n}\|y\|^{-(\frac{n^2}{\lambda q_1})}[|\|x\|^{r+s}-\|y\|^{r+s}|^{(\alpha/k)-1}\|y\|^{r+s-1}]^{n/\lambda}dy\right\}^{1/p_1}\\
&\qquad\times\|f\|_p^{p(1-\frac{\lambda}{n})}\\
&\quad=\frac{(r+s)^{1-(\alpha/k)}}{k\Gamma_k(\alpha)}I_1^{1/q_1}\times I_2^{1/p_1}\times\|f\|_p^{p(1-\frac{\lambda}{n})}.
\end{aligned}\tag{54}$$

□

In I_2, by using Lemma 2 and letting $u=\|x\|t^{-(1/\beta)}$, and using (51), we get

$$\begin{aligned}
I_2&=\int_{E_n}\|y\|^{-(\frac{n^2}{\lambda q_1})}[|\|x\|^{r+s}-\|y\|^{r+s}|^{(\alpha/k)-1}\|y\|^{r+s-1}]^{\frac{n}{\lambda}}dy\\
&=\int_{E_n}|1-(\|x\|\cdot\|y\|^{-1})^{r+s}|^{\frac{n}{\lambda}(\frac{\alpha}{k}-1)}\|y\|^{\frac{n}{\lambda}(\frac{\alpha(r+s)}{k}-\frac{n}{q_1}-1)}dy\\
&=\frac{(\Gamma(1/\beta))^n}{\beta^n\Gamma(n/\beta)}\int_0^\infty|1-(\|x\|t^{-(1/\beta)})^{r+s}|^{\frac{n}{\lambda}(\frac{\alpha}{k}-1)}\\
&\quad\times t^{\frac{n}{\beta\lambda}(\frac{\alpha(r+s)}{k}-\frac{n}{q_1}-1)+\frac{n}{\beta}-1}dt
\end{aligned}$$

$$= \frac{(\Gamma(1/\beta))^n}{\beta^{n-1}\Gamma(n/\beta)}\|x\|^{\frac{n}{\lambda}(\frac{\alpha(r+s)}{k}-\frac{n}{q_1}-1)+n}$$
$$\times \int_0^\infty |1-u^{r+s}|^{\frac{n}{\lambda}(\frac{\alpha}{k}-1)}u^{\frac{n}{\lambda}[1+\frac{n}{q_1}-\frac{\alpha(r+s)}{k}]-n-1}du$$
$$= c_2\|x\|^{\frac{n}{\lambda}(\frac{\alpha(r+s)}{k}-\frac{n}{q_1}-1)+n}. \tag{55}$$

Note that $\frac{1}{p}+\frac{1}{q}\leq 1$ implies $\frac{p}{q_1}\geq 1$, thus, by (54), (55), and Minkowski's inequality for integrals,

$$\left\{\int_X\left(\int_Y |f(x,y)|dy\right)^p\omega(x)dx\right\}^{1/p}$$
$$\leq \int_Y\left\{\int_X |f(x,y)|^p\omega(x)dx\right\}^{1/p}dy, 1\leq p<\infty, \tag{56}$$

and letting $u=\|y\|^{-1}t^{1/\beta}$, we conclude that

$$\|T_{27}(f)\|_{p,\omega}$$
$$= \left(\int_{E_n}|T_{27}(f,x)|^p\omega(x)dx\right)^{1/p}$$
$$\leq \frac{(r+s)^{1-(\alpha/k)}}{k\Gamma_k(\alpha)}\left(\int_{E_n} I_1^{p/q_1}I_2^{p/p_1}\|f\|_p^{p^2(1-\frac{\lambda}{n})}\omega(x)dx\right)^{1/p}$$
$$= \frac{(r+s)^{1-(\alpha/k)}}{k\Gamma_k(\alpha)}c_2^{1/p_1}\|f\|_p^{p(1-\frac{\lambda}{n})}\left\{\int_{E_n}\|x\|^{[\frac{n}{\lambda}(\frac{\alpha(r+s)}{k}-\frac{n}{q_1}-1)+n]\frac{p}{p_1}}\right.$$
$$\times\left[\int_{E_n}\|y\|^{\frac{n^2}{p_1\lambda}}(|\|x\|^{r+s}-\|y\|^{r+s}|^{(\alpha/k)-1}\|y\|^{r+s-1})^{\frac{n}{\lambda}}\right.$$
$$\left.\left.\times\ |f(y)|^p dy\right]^{\frac{p}{q_1}}\omega(x)dx\right\}^{1/p}$$
$$\leq \frac{(r+s)^{1-(\alpha/k)}}{k\Gamma_k(\alpha)}c_2^{1/p_1}\|f\|_p^{p(1-\frac{\lambda}{n})}\left\{\int_{E_n}\|y\|^{\frac{n}{\lambda}(\frac{n}{p_1}+r+s-1)}|f(y)|^p\right.$$
$$\times\left[\int_{E_n}\|x\|^{[\frac{n}{\lambda}(\frac{\alpha(r+s)}{k}-\frac{n}{q_1}-1)+n]\frac{p}{p_1}}|\|x\|^{r+s}\right.$$

$$- \|y\|^{r+s}|^{\frac{np}{\lambda q_1}(\frac{\alpha}{k}-1)}\omega(x)dx\Big]^{\frac{q_1}{p}} dy\Big\}^{1/q_1}$$

$$\leq \frac{(r+s)^{1-(\alpha/k)}}{k\Gamma_k(\alpha)} c_2^{1/p_1} \|f\|_p^{p(1-\frac{\lambda}{n})} \Big\{\int_{E_n} \|y\|^{\frac{n}{\lambda}(\frac{n}{p_1}+\frac{\alpha(r+s)}{k}-1)}|f(y)|^p$$

$$\times \Big[\int_{E_n} \|x\|^{[\frac{n}{\lambda}(\frac{\alpha(r+s)}{k}-\frac{n}{q_1}-1)+n]\frac{p}{p_1}}|(\|x\|\cdot\|y\|^{-1})^{r+s}$$

$$- 1|^{\frac{np}{\lambda q_1}(\frac{\alpha}{k}-1)}\omega(x)dx\Big]^{\frac{q_1}{p}} dy\Big\}^{1/q_1}$$

$$\leq \frac{(r+s)^{1-(\alpha/k)}}{k\Gamma_k(\alpha)} c_2^{1/p_1} \|f\|_p^{p(1-\frac{\lambda}{n})} \left(\frac{(\Gamma(1/\beta))^n}{\beta^n\Gamma(n/\beta)}\right)^{1/p}$$

$$\times \Big\{\int_{E_n} \|y\|^{\frac{n}{\lambda}(\frac{n}{p_1}+\frac{\alpha(r+s)}{k}-1)}|f(y)|^p \Big[\int_0^\infty |1-(t^{1/\beta}\|y\|^{-1})^{r+s}|^{\frac{np}{\lambda q_1}(\frac{\alpha}{k}-1)}$$

$$\times t^{\frac{pn}{\beta p_1\lambda}(\frac{\alpha(r+s)}{k}-\frac{n}{q_1}-1)+\frac{pn}{\beta p_1}+\frac{c_0+n}{\beta}-1} dt\Big]^{q_1/p} dy\Big\}^{1/q_1}$$

$$= \frac{(r+s)^{1-(\alpha/k)}}{k\Gamma_k(\alpha)} c_2^{1/p_1} \|f\|_p^{p(1-\frac{\lambda}{n})} \left(\frac{(\Gamma(1/\beta))^n}{\beta^n\Gamma(n/\beta)}\right)^{1/p}$$

$$\times \Big\{\int_{E_n} |f(y)|^p dy\Big\}^{1/q_1} \Big\{\int_0^\infty |1-u^{r+s}|^{\frac{np}{\lambda}(1-\frac{1}{q})(\frac{\alpha}{k}-1)}$$

$$\times u^{\frac{\alpha(r+s)}{k}[\frac{n(p-1)}{\lambda}-p]-\frac{n(p-1)}{\lambda}[1+n(1-\frac{1}{q})]+p-1} du\Big\}^{1/p}$$

$$= c_1^{1/p} c_2^{1-(1/p)} \frac{(r+s)^{1-(\alpha/k)}}{k\Gamma_k(\alpha)} \|f\|_p.$$

Thus,

$$\|T_{27}(f)\|_{p,\omega} \leq c_1^{1/p} c_2^{1-(1/p)} \frac{(r+s)^{1-(\alpha/k)}}{k\Gamma_k(\alpha)} \|f\|_p.$$

This implies that

$$\|T_{27}\| = \sup_{f\neq 0} \frac{\|T_{27}(f)\|_{p,\omega}}{\|f\|_p} \leq c_1^{1/p} c_2^{1-(1/p)} \frac{(r+s)^{1-(\alpha/k)}}{k\Gamma_k(\alpha)}.$$

The proof is completed. In the conjugate case: $\lambda = n, \frac{1}{p} + \frac{1}{q} = 1$, then by (50) and (51), we have

$$c_1 = c_2 = c = \frac{(\Gamma(1/\beta))^n}{\beta^{n-1}\Gamma(n/\beta)} \int_0^\infty |1 - u^{r+s}|^{\frac{\alpha}{k}-1} u^{-\frac{\alpha(r+s)}{k} - \frac{n}{q}} du,$$

thus, by Theorem 1, we get the following:

Corollary 17. *Let* $1 < p < \infty, \frac{1}{p} + \frac{1}{q} = 1, \omega(x) = \|x\|^{c_0}, x \in E_n$, *where* c_0 *is defined by* (49). *If*

$$c = \frac{(\Gamma(1/\beta))^n}{\beta^{n-1}\Gamma(n/\beta)} \int_0^\infty |1 - u^{r+s}|^{(\frac{\alpha}{k}-1)} u^{-\frac{\alpha(r+s)}{k} - \frac{n}{q}} du < \infty,$$

then the generalized conformable fractional integral operator T_{27} *is defined by* (38): $T_{27} : L^p(E_n) \to L^p_\omega(E_n)$ *exists as a bounded operator and*

$$\|T_{27}(f)\|_{p,\omega} \leq c\frac{(r+s)^{1-(\alpha/k)}}{k\Gamma_k(\alpha)}\|f\|_p.$$

This implies that

$$\|T_{27}\| = \sup_{f\neq 0} \frac{\|T_{27}(f)\|_{p,\omega}}{\|f\|_p} \leq c\frac{(r+s)^{1-(\alpha/k)}}{k\Gamma_k(\alpha)}.$$

Theorem 18. *Let* $1 < p, q < \infty, \frac{1}{p}+\frac{1}{q} \leq 1, 0 < \frac{\lambda}{n} = 2-\frac{1}{p}-\frac{1}{q}, \omega(x) = \|x\|^{c_0}$, *where*

$$c_0 = n\left[\frac{n}{\lambda} + \frac{p}{\lambda}\left(1 - n - \frac{\alpha}{k}\right) - 1\right], x \in E_n. \tag{57}$$

Let $g : (0,\infty) \to (0,\infty)$ *be an increasing function, and* $|g'(x)| \leq M$, *a.e.* $x \in E_n$. *If*

$$c_1 = \frac{(\Gamma(1/\beta))^n}{\beta^{n-1}\Gamma(n/\beta)} \int_0^\infty |1-u|^{\frac{np}{\lambda}(1-\frac{1}{q})(\frac{\alpha}{k}-1)} u^{\frac{n}{\lambda}[1-\frac{\alpha}{k}+n(1-\frac{1}{q})]-n-1} du < \infty \tag{58}$$

and

$$c_2 = \frac{(\Gamma(1/\beta))^n}{\beta^{n-1}\Gamma(n/\beta)} \int_0^\infty |1-u|^{\frac{n}{\lambda}(\frac{\alpha}{k}-1)} u^{\frac{n}{\lambda}[1+n(1-\frac{1}{q})-\frac{\alpha}{k}]-n-1} du < \infty, \tag{59}$$

then the generalized fractional integral operator T_{31} is defined by (43): $T_{31}: L^p(E_n) \to L^p_\omega(E_n)$ exists as a bounded operator and

$$\|T_{31}(f)\|_{p,\omega} \le c_1^{1/p} c_2^{1-(1/p)} M^{\frac{n\alpha}{k\lambda}} \frac{c}{k\Gamma_k(\alpha)} \|f\|_p.$$

This implies that

$$\|T_{31}\| = \sup_{f\neq 0} \frac{\|T_{31}(f)\|_{p,\omega}}{\|f\|_p} \le c_1^{1/p} c_2^{1-(1/p)} M^{\frac{n\alpha}{k\lambda}} \frac{c}{k\Gamma_k(\alpha)}.$$

Proof. Let $p_1 = \frac{p}{p-1}, q_1 = \frac{q}{q-1}$; it follows that

$$\frac{1}{p_1} + \frac{1}{q_1} + \left(1 - \frac{\lambda}{n}\right) = 1, \frac{p}{q_1} + p\left(1 - \frac{\lambda}{n}\right) = 1.$$

Note that $|g'(\|y\|)| \le M, a.e. y \in E_n$; it follows that

$$|g(\|x\|) - g(\|y\|)| \le M|\|x\| - \|y\||. \tag{60}$$

By Hölder's inequality, we get

$$\begin{aligned}
&T_{31}(f,x)\\
&\quad = \frac{c}{k\Gamma_k(\alpha)} \int_{E_n} g'(\|y\|)|g(\|\|x\|) - g(\|y\|)|^{(\alpha/k)-1} f(y)dy\\
&\quad = \frac{c}{k\Gamma_k(\alpha)} \int_{E_n} \left\{\|y\|^{(\frac{n^2}{\lambda p_1})}[g'(\|y\|)|g(\|x\|)\right.\\
&\qquad \left. - g(\|y\|)|^{(\alpha/k)-1}]^{\frac{n}{\lambda}} f^p(y)\right\}^{1/q_1} \left\{\|y\|^{-(\frac{n^2}{q_1\lambda})}[g'(\|y\|)|g(\|x\|)\right.\\
&\qquad \left. - g(\|y\|)|^{(\alpha/k)-1}]^{n/\lambda}\right\}^{1/p_1} (f(y))^{p(1-\frac{n}{\lambda})} dy\\
&\quad \le \frac{c}{k\Gamma_k(\alpha)} \left\{\int_{E_n} \|y\|^{\frac{n^2}{p_1\lambda}} [g'(\|y\|)|g(\|x\|)\right.\\
&\qquad \left. - g(\|y\|)|^{(\alpha/k)-1}]^{n/\lambda} |f(y)|^p dy\right\}^{1/q_1} \left\{\int_{E_n} \|y\|^{-(\frac{n^2}{q_1\lambda})} [g'(\|y\|)|g(\|x\|)\right.\\
&\qquad \left. - g(\|y\|)|^{(\alpha/k)-1}]^{n/\lambda} dy\right\}^{1/p_1} \|f\|_p^{p(1-\frac{\lambda}{n})}\\
&\quad = \frac{c}{k\Gamma_k(\alpha)} I_1^{1/q_1} \times I_2^{1/p_1} \times \|f\|_p^{p(1-\frac{\lambda}{n})}.
\end{aligned} \tag{61}$$

In I_2, by using Lemma 15 and letting $u = \|x\|t^{-(1/\beta)}$, and using (59) and (60), we get

$$\begin{aligned}
I_2 &= \int_{E_n} \|y\|^{-(\frac{n^2}{\lambda q_1})}[g'(\|y\|)|g(\|x\|) - g(\|y\|)|^{(\alpha/k)-1}]^{\frac{n}{\lambda}}dy \\
&\leq M^{\frac{n\alpha}{k\lambda}} \int_{E_n} \|y\|^{-(\frac{n^2}{q_1\lambda})}|\|x\| - \|y\||^{\frac{n}{\lambda}(\frac{\alpha}{k}-1)}dy \\
&= M^{\frac{n\alpha}{k\lambda}} \int_{E_n} |\|x\| \cdot \|y\|^{-1} - 1|^{\frac{n}{\lambda}(\frac{\alpha}{k}-1)}\|y\|^{\frac{n}{\lambda}(\frac{\alpha}{k}-1)-\frac{n^2}{q_1\lambda}}dy \\
&= M^{\frac{n\alpha}{k\lambda}} \times \frac{(\Gamma(1/\beta))^n}{\beta^n\Gamma(n/\beta)} \int_0^\infty |1 - (\|x\|t^{-(1/\beta)})|^{\frac{n}{\lambda}(\frac{\alpha}{k}-1)} \\
&\quad \times t^{\frac{n}{\beta\lambda}(\frac{\alpha}{k}-\frac{n}{q_1}-1)+\frac{n}{\beta}-1}dt \\
&= M^{\frac{n\alpha}{k\lambda}} \times \frac{(\Gamma(1/\beta))^n}{\beta^{n-1}\Gamma(n/\beta)}\|x\|^{\frac{n}{\lambda}(\frac{\alpha}{k}-\frac{n}{q_1}-1)+n} \\
&\quad \times \int_0^\infty |1 - u|^{\frac{n}{\lambda}(\frac{\alpha}{k}-1)}u^{\frac{n}{\lambda}(1+\frac{n}{q_1}-\frac{\alpha}{k})-n-1}du \\
&= c_2M^{\frac{n\alpha}{k\lambda}}\|x\|^{\frac{n}{\lambda}(\frac{\alpha}{k}-\frac{n}{q_1}-1)+n}.
\end{aligned} \tag{62}$$

Note that $\frac{1}{p} + \frac{1}{q} \leq 1$ implies that $\frac{p}{q_1} \geq 1$, thus, by (61), (62), and Minkowski's inequality for integrals (56), and letting $u = \|y\|^{-1}t^{1/\beta}$, we conclude that

$$\begin{aligned}
&\|T_{31}(f)\|_{p,\omega} \\
&\quad = \left(\int_{E_n} |T_{31}(f,x)|^p\omega(x)dx\right)^{1/p} \\
&\quad \leq \frac{c}{k\Gamma_k(\alpha)}\left(\int_{E_n} I_1^{p/q_1}I_2^{p/p_1}\|f\|_p^{p^2(1-(\lambda/n))}\omega(x)dx\right)^{1/p} \\
&\quad = \frac{c}{k\Gamma_k(\alpha)}c_2^{1/p_1}M^{\frac{n\alpha}{k\lambda p_1}}\|f\|_p^{p(1-\frac{\lambda}{n})}\left\{\int_{E_n} \|x\|^{[\frac{n}{\lambda}(\frac{\alpha}{k}-\frac{n}{q_1}-1)+n]\frac{p}{p_1}}\right. \\
&\qquad \times \left[\int_{E_n} \|y\|^{\frac{n^2}{p_1\lambda}}(g'(\|y\|)|g(\|x\|)\right.
\end{aligned}$$

$$
\begin{aligned}
&\quad - g(\|y\|)|^{(\alpha/k)-1)\frac{n}{\lambda}}|f(y)|^p dy\Big]^{\frac{p}{q_1}}\omega(x)dx\Bigg\}^{1/p} \\
&\le \frac{c}{k\Gamma_k(\alpha)} M^{\frac{n\alpha}{k\lambda}} c_2^{1/p_1}\|f\|_p^{p(1-\frac{\lambda}{n})}\Bigg\{\int_{E_n}\|y\|^{\frac{n^2}{\lambda p_1}}|f(y)|^p \\
&\quad\times\Bigg[\int_{E_n}\|x\|^{(\frac{n}{\lambda}(\frac{\alpha}{k}-\frac{n}{q_1}-1)+n)\frac{p}{p_1}}|\|x\|-\|y\||^{\frac{np}{\lambda q_1}(\frac{\alpha}{k}-1)}\omega(x)dx\Bigg]^{\frac{q_1}{p}}dy\Bigg\}^{1/q_1} \\
&\le \frac{c}{k\Gamma_k(\alpha)} M^{\frac{n\alpha}{k\lambda}} c_2^{1/p_1}\|f\|_p^{p(1-\frac{n}{\lambda})}\Bigg\{\int_{E_n}\|y\|^{\frac{n}{\lambda}(\frac{n}{p_1}+\frac{\alpha}{k}-1)}|f(y)|^p \\
&\quad\times\Bigg[\int_{E_n}\|x\|^{(\frac{n}{\lambda}(\frac{\alpha}{k}-\frac{n}{q_1}-1)+n)\frac{p}{p_1}}|(\|x\|\cdot\|y\|^{-1}) \\
&\quad - 1|^{\frac{np}{\lambda q_1}(\frac{\alpha}{k}-1)}\omega(x)dx\Bigg]^{\frac{q_1}{p}}dy\Bigg\}^{1/q_1} \\
&\le \frac{c}{k\Gamma_k(\alpha)} M^{\frac{n\alpha}{k\lambda}} c_2^{1/p_1}\|f\|_p^{p(1-\frac{\lambda}{n})}\left(\frac{(\Gamma(1/\beta))^n}{\beta^n\Gamma(n/\beta)}\right)^{1/p} \\
&\quad\times\Bigg\{\int_{E_n}\|y\|^{\frac{n}{\lambda}\frac{n}{p_1}+\frac{\alpha}{k}-1)}|f(y)|^p\Bigg[\int_0^\infty|1-(t^{1/\beta}\|y\|^{-1})|^{\frac{np}{\lambda q_1}(\frac{\alpha}{k}-1)} \\
&\quad\times t^{\frac{pn}{\beta p_1\lambda}(\frac{\alpha}{k}-\frac{n}{q_1}-1)+\frac{np}{\beta p_1}+\frac{c_0+n}{\beta}-1}dt\Bigg]^{\frac{q_1}{p}}dy\Bigg\}^{1/q_1} \\
&= \frac{c}{k\Gamma_k(\alpha)} M^{\frac{n\alpha}{k\lambda}} c_2^{1/p_1}\|f\|_p^{p(1-\frac{\lambda}{n})}\left(\frac{(\Gamma(1/\beta))^n}{\beta^n\Gamma(n/\beta)}\right)^{1/p} \\
&\quad\times\left\{\int_{E_n}|f(y)|^p dy\right\}^{1/q_1}\int_0^\infty|1-u|^{\frac{np}{\lambda}(1-\frac{1}{q})(\frac{\alpha}{k}-1)} \\
&\quad\times u^{\frac{n}{\lambda}[1-\frac{\alpha}{k}+n(1-\frac{1}{q})]-n-1}du \\
&= \frac{c}{k\Gamma_k(\alpha)} c_1^{1/p} c_2^{1-(1/p)} M^{\frac{n\alpha}{k\lambda}}\|f\|_p.
\end{aligned}
$$

Thus,

$$\|T_{31}(f)\|_{p,\omega} \leq \frac{c}{k\Gamma_k(\alpha)} c_1^{1/p} c_2^{1-(1/p)} M^{\frac{n\alpha}{k\lambda}} \|f\|_p.$$

This implies that

$$\|T_{31}\| = \sup_{f\neq 0} \frac{\|T_{31}(f)\|_{p,\omega}}{\|f\|_p} \leq \frac{c}{k\Gamma_k(\alpha)} c_1^{1/p} c_2^{1-(1/p)} M^{\frac{n\alpha}{k\lambda}}.$$

The proof is completed. □

In the conjugate case, $\lambda = n, \frac{1}{p} + \frac{1}{q} = 1$, then by (58) and (59), we get

$$c_1 = c_2 = \frac{(\Gamma(1/\beta))^n}{\beta^{n-1}\Gamma(n/\beta)} \int_0^\infty |1-u|^{(\frac{\alpha}{k}-1)} u^{-\frac{\alpha}{k}-\frac{n}{q}} du.$$

Thus, by Theorem 18, we get the following.

Corollary 19. *Let* $1 < p < \infty, \frac{1}{p} + \frac{1}{q} = 1, \omega(x) = \|x\|^{c_0}$, *where* c_0 *is defined by* (57). *If*

$$c_1 = \frac{(\Gamma(1/\beta))^n}{\beta^{n-1}\Gamma(n/\beta)} \int_0^\infty |1-u|^{(\frac{\alpha}{k}-1)} u^{-(\frac{\alpha}{k}+\frac{n}{q})} du < \infty,$$

then the generalized fractional integral operator T_{31} *is defined by* (43): $T_{31} : L^p(E_n) \to L^p_\omega(E_n)$ *exists as a bounded operator and*

$$\|T_{31}(f)\|_{p,\omega} \leq \frac{c}{k\Gamma_k(\alpha)} c_1 M^{\frac{\alpha}{k}} \|f\|_p.$$

This implies that

$$\|T_{31}\| = \sup_{f\neq 0} \frac{\|T_{31}(f)\|_{p,\omega}}{\|f\|_p} \leq \frac{c}{k\Gamma_k(\alpha)} c_1 M^{\frac{\alpha}{k}}.$$

References

[1] Kuang J.C. *Applied Inequalities*, 5th.edu. Shangdong Science and Technology Press, Jinan (2021) (in Chinese).

[2] Kilbas, A.A., Srivgstava, H.M. and Trujillo, J.J., Theory and Applications of Fractional Differential Equations, *North-Holland Mathematics Studies*, Elsevier, New York, 204 (2006).

[3] Mubeen, S. and Habibullah, G.M. k-fractional integrals and applications, *Int. J. Contemp. Math. Sci.*, 7, 89–94 (2012).

[4] Mubeen, S. and Iqbal, S. Grüss type integral inequalities for generalized Riemann-Liouville $k-$ fractional integrals, *J. Inequal. Appl.*, 109 (2016).

[5] Sarikaya, M.Z., Dahmani, Z., Kiris, M.E. and Ahmad, F. (k, s)-Riemann-Liouville fractional integral and applications, *Hacet. J. Math. Stat.*, 45(1), 77–89 (2016).

[6] Set.E., Tomar, M. and Sarkaya, M.Z. On generalized Grüss type inequalities for $k-$ fractional integrals, *Appl. Math. Comp.*, 8(269), 29–34 (2015).

[7] Abbas, G., Khan K.A., Farid, G. and Rehman, A.U. Generalizations of some fractional integral inequalities via generalized Mittag-Leffer function, *J. Inequal. Appl.*, 121 (2017).

[8] da Sousa, J.V. and de Oliveira, E.C., On the $\psi-$ Hilfer fractional derivative, *Commun. Nonlinear Sci. Numer. Simul.*, 60, 72–91 (2018).

[9] Zhao Yong, Sang Haiwei, Xiong weicheng and Cui Zhongwei, Hermite-Hadamard type inequalities involving $\psi-$ Riemann-Liouville fractional integrals via s-convex functions, *J. Inequal. Appl.*, 128 (2020).

[10] Dragomir, S.S., Inequalities for the area balance of absolutely continuous functions, *Stud. Univ. Babes-Bolyai Math.,* 63(1), 37–57 (2018).

[11] Khan, J.U. and Khan, M.A. Generalized conformable fractional integral operators, *J. Comput. Appl. Math.*, 346, 378–389 (2019).

[12] Rashid, S., Akdemir, A.O., Nisar, K.S., Abdeljawad T. and Rahman G. New generalized reverse Minkowski and related integral inequalities involving generalized fractional conformable integrals, *J. Inequal. Appl.*, 177 (2020).

[13] Çelik, B., Gürbüz, M.Ç., Özdemir M.E. and Set E. On integral inequalities related to the weighted and the extended Chebyshev functionals involving different fractional operators, *J. Inequal. Appl.*, 246 (2020).

[14] Iscan I., Jensen-Mercer inequality for $GA-$ convex functions and some related inequalities, *J. Inequal. Appl.*, 212 (2020).

[15] Kuang, J.C., Norm inequalities for generalized Laplace transforms. In: Raigorodskii, A. and Rassias, M.Th. (eds.), *Trigonometric Sums and Their Applications*, Springer, (2020).

[16] Kuang, J.C., Some new inequalities for fractional integral operators, In: Rassias, Th.M. (ed.), *Approximation and Computation in Science and Engineering*, Springer (2021).

[17] Sarikaya, M,Z. On the Hermite-Hadarmard — Type inequalities for co-ordinated convex function via fractional integrals, *Integral Transforms Spec. Funct.*, 25(2), 134–147 (2014).

[18] Erden, S., Budak, H., Sarikaya, M.Z., Iftikhar, S. and Kumam, P. Fractional Ostrowski type inequalities for bounded functions, *J. Inequal. Appl.*, 123 (2020).

[19] Erden, S., Budak, H. and Sarikaya, M.Z. Fractional Ostrowski type inequalities for functions of bounded variation with two variables, *Miskolc Math. Notes,* 21(1), 171–188 (2020).

https://doi.org/10.1142/9789811293009_0022

Chapter 22

Generalized Fractional Hilbert-type Integral Inequalities

Jichang Kuang*,‡ and Michael Th. Rassias†,§,¶

*Department of Mathematics, Hunan Normal University, Changsha, Hunan, P.R. China

†Department of Mathematics and Engineering Sciences, Hellenic Military Academy, Athens, Greece

‡jckuang@163.com

§mrassias@sse.gr

¶mthrassias@yahoo.com

In this chapter, we introduce some new generalized fractional Hilbert-type integral operators. We also establish the norm inequalities for these operators. The corresponding fractional Hilbert-type integral inequalities with the best possible constant factor are also provided. The results proved within this chapter constitute a significant improvement and generalization of many known and new classes of fractional integral operators.

1. Introduction

It is well known that fractional integral operators are an essential category of operators in harmonic analysis associated also with partial differential equations. In the recent past, different versions of fractional integral operators have been developed which are useful in the study of various classes of differential and integral equations. Researchers have identified that various fractional integral operators with different singular or non-singular kernels need to be studied in greater depth due to their real-world applications.

Many mathematicians have established a variety of inequalities for those fractional integral and derivative operators, some of which have turned out to be useful in analyzing solutions of certain fractional integral and differential equations. Hence, fractional integral inequalities are very important in the theory and applications of differential equations. Such inequalities are also of great importance in the mathematical modeling of fractional boundary value problems (cf. [1–15] and the references given therein).

We initially recall the following definitions and some related results.

Definition 1 (cf. [1, 2, 6]). Let $f \in L[a,b]$. Then the Riemann–Liouville fractional integrals of $f$of order $\alpha > 0$ with $a \geq 0$ are defined by

$$T_1(f,x) = \frac{1}{\Gamma(\alpha)} \int_a^x (x-t)^{\alpha-1} f(t)dt, \quad x > a, \tag{1}$$

and

$$T_2(f,x) = \frac{1}{\Gamma(\alpha)} \int_x^b (t-x)^{\alpha-1} f(t)dt, \quad x < b, \tag{2}$$

respectively, where

$$\Gamma(\alpha) = \int_0^\infty t^{\alpha-1} e^{-t} dt \tag{3}$$

stands for the gamma function and when $\alpha = 0$, $T_1(f,x) = T_2(f,x) = f(x)$.

Definition 2 (cf. [3]). Let $f \in L[a,b]$. Then the Riemann–Liouville k-fractional integrals of f of order $\alpha > 0$ with $a \geq 0$ are defined by

$$T_3(f,x) = \frac{1}{k\Gamma_k(\alpha)} \int_a^x (x-t)^{(\alpha/k)-1} f(t)dt, \quad x > a, \tag{4}$$

and

$$T_4(f,x) = \frac{1}{k\Gamma_k(\alpha)} \int_x^b (t-x)^{(\alpha/k)-1} f(t)dt, \quad x < b, \tag{5}$$

respectively, where

$$\Gamma_k(\alpha) = \int_0^\infty t^{\alpha-1} e^{-\left(\frac{t^k}{k}\right)} dt \quad (\alpha > 0) \tag{6}$$

stands for the k-gamma function. Also,

$$\Gamma(x) = \lim_{k\to 1} \Gamma_k(x), \quad \Gamma_k(\alpha) = k^{(\alpha/k)-1}\Gamma\left(\frac{\alpha}{k}\right),$$

and

$$\Gamma_k(\alpha + k) = \alpha\Gamma_k(\alpha).$$

It is well known that the Mellin transform of f is defined by

$$M(f,\alpha) = \int_0^\infty t^{\alpha-1} f(t)dt, \quad (\alpha > 0).$$

Therefore, the Mellin transform of the exponential function $\exp\{-\frac{t^k}{k}\}$ is the k-gamma function.

Definition 3 (cf. [4, 5]). Let $t^r f(t) \in L^1[a,b]$, $a \geq 0$. Then the generalized Riemann–Liouville fractional integrals of f of order $\alpha \geq 0$ and $r \geq 0$ are defined by

$$T_5(f,x) = \frac{(r+1)^{1-\alpha}}{\Gamma(\alpha)} \int_a^x (x^{r+1} - t^{r+1})^{\alpha-1} t^r f(t)dt, \quad x > a, \tag{7}$$

$$T_6(f,x) = \frac{(r+1)^{1-\alpha}}{\Gamma(\alpha)} \int_x^b (t^{r+1} - x^{r+1})^{\alpha-1} t^r f(t)dt, \quad x < b, \tag{8}$$

and there are also (k,r) fractional operators of Riemann type:

$$T_7(f,x) = \frac{(r+1)^{1-(\alpha/k)}}{k\Gamma_k(\alpha)} \int_a^x (x^{r+1} - t^{r+1})^{(\alpha/k)-1} t^r f(t)dt, \quad x > a, \tag{9}$$

$$T_8(f,x) = \frac{(r+1)^{1-(\alpha/k)}}{k\Gamma_k(\alpha)} \int_x^b (t^{r+1} - x^{r+1})^{(\alpha/k)-1} t^r f(t)dt, \quad x < b, \tag{10}$$

respectively, where $\alpha, k > 0, r \geq 0, x \in [a,b]$.

In particular, if $r = 0$, then definition 3 reduces to definitions 1 and 2.

Definition 4 (cf. [11, 12]). Let f be a conformable integrable function on $[a,b] \subset [0,\infty)$. The left-sided and right-sided generalized conformable fractional integrals T_9 and T_{10} of f of order $\alpha > 0$ are defined by

$$T_9(f,x) = \frac{1}{\Gamma(\alpha)} \int_a^x \left(\frac{x^{r+s} - t^{r+s}}{r+s} \right)^{\alpha-1} t^{r+s-1} f(t)dt, \quad x > a, \quad (11)$$

and

$$T_{10}(f,x) = \frac{1}{\Gamma(\alpha)} \int_x^b \left(\frac{t^{r+s} - x^{r+s}}{r+s} \right)^{\alpha-1} t^{r+s-1} f(t)dt, \quad x < b, \quad (12)$$

respectively, where $r \in R^1, s \in (0,1], r+s \neq 0$.

In particular, if $s = 1$, then T_9, T_{10} reduce to T_5, T_6, respectively.

Definition 5 (cf. [8, 9]). Let $f \in L[a,b]$, $g : [a,b] \to (0,\infty)$ be an increasing function, and $g' \in C[a,b], \alpha > 0$. Then the g-Riemann–Liouville fractional integrals of f with respect to the function g on $[a,b]$ are defined by

$$T_{11}(f,x) = \frac{1}{\Gamma(\alpha)} \int_a^x g'(t)[g(x) - g(t)]^{\alpha-1} f(t)dt, \quad x > a, \quad (13)$$

and

$$T_{12}(f,x) = \frac{1}{\Gamma(\alpha)} \int_x^b g'(t)[g(t) - g(x)]^{\alpha-1} f(t)dt, \quad x < b, \quad (14)$$

respectively.

In 2020, Kuang [15] introduced a new notion of the generalized fractional integral operators and fractional area balance operators:

Definition 6 ([15]). Let $f \in L[a,b]$, $g : [a,b] \to (0,\infty)$ be an increasing function, and $g \in AC[a,b], k, c, \alpha > 0$, $a \geq 0$. Then the generalized fractional integrals operator T_{13} with respect to the function g on $[a,b]$ is defined by

$$T_{13}(f,x) = \frac{c}{k\Gamma_k(\alpha)} \int_a^b g'(t)|g(x) - g(t)|^{(\alpha/k)-1} f(t)dt, \quad (15)$$

where $\Gamma_k(\alpha)$ is defined by (6).

Let

$$T_{14}(f,x) = \frac{c}{k\Gamma_k(\alpha)} \int_a^x g'(t)[g(x) - g(t)]^{(\alpha/k)-1} f(t)dt, \quad x > a, \tag{16}$$

and

$$T_{15}(f,x) = \frac{c}{k\Gamma_k(\alpha)} \int_x^b g'(t)[g(t) - g(x)]^{(\alpha/k)-1} f(t)dt, \quad x < b, \tag{17}$$

then

$$T_{13}(f,x) = T_{14}(f,x) + T_{15}(f,x). \tag{18}$$

The fractional area balance operator T_a with respect to the function g on $[a,b]$ is defined by

$$T_a(f,x) = T_{15}(f,x) - T_{14}(f,x).$$

Note that for a suitable and appropriate choice of parameters and functions, one can obtain various new and classical results. Hence, definition 6 unified and generalized many known and new classes of fractional integral operators.

Fractional integral operators were utilized extensively to study classical inequalities. The generalization of classical inequalities by means of fractional integral operators is considered as an interesting subject area. For instance, one of the best-known inequalities, the Hermite–Hadamard inequality, that is,

$$f\left(\frac{a+b}{2}\right) \leq \frac{1}{b-a}\int_a^b f(x)dx \leq \frac{f(a)+f(b)}{2}, \tag{19}$$

is generalized by means of several fractional integral operators (cf. [16–25, 32] and the references given therein).

Another famous inequality is the Hilbert inequality:

$$\int_0^\infty \int_0^\infty \frac{f(x)g(y)}{x+y} dxdy \leq \frac{\pi}{\sin(\pi/p)} \|f\|_p \|g\|_q, \tag{20}$$

where

$$\|f\|_p = \left(\int_0^\infty |f(x)|^p dx\right)^{1/p}, \quad 1 < p < \infty, \quad \frac{1}{p} + \frac{1}{q} = 1,$$

and the constant $\frac{\pi}{\sin(\pi/p)}$ is the best possible. The corresponding Hilbert operator

$$T_0(f,x) = \int_0^\infty \frac{f(y)}{x+y}dy \tag{21}$$

satisfies

$$\|T_0 f\|_p \le \|T_0\| \cdot \|f\|_p, \tag{22}$$

where

$$\|T_0\| = \frac{\pi}{\sin(\pi/p)}. \tag{23}$$

The Hilbert inequality is one of the most interesting inequalities in analysis and its applications. Hilbert-type inequalities have been generalized by many researchers. In 2008 and 2012, the author [26, 27] obtained some new inequalities related to the generalized Hilbert operator

$$T_{16}(f,x) = \int_0^\infty K(x,y)f(y)dy \tag{24}$$

with the general kernel $\mathrm{K}(x,y)$ on the Herz spaces and the weighted Morrey–Herz space, respectively (see [26–31] and the references given therein).

The aim of this chapter is to introduce some new generalized fractional Hilbert-type integral operators and establish some new fractional Hilbert-type integral inequalities. In Section 2, we define generalized fractional Hilbert-type integral operators. In Section 3, some new fractional Hilbert-type integral inequalities with the best possible constant factor are derived. The corresponding norm inequalities are established in Section 4. These constitute a significant improvement and generalization of many known and new classes of fractional integral operators.

2. Generalized Fractional Hilbert–Type Integral Operators

Definition 7. Let $f \in L(0,\infty)$, $\mathrm{K}(x,y)$ be a non-negative measurable function on $(0,\infty) \times (0,\infty)$, $c > 0$, then the generalized fractional Hilbert-type integral operator T_{17} is defined by

$$T_{17}(f,x) = \frac{c}{k\Gamma_k(\alpha)} \int_0^\infty (K(x,y))^{(\alpha/k)-1} f(y)dy, \tag{25}$$

where $\Gamma_k(\alpha)$ is defined by (6).

If K satisfies the following conditions:

$$K(tx,y) = t^{-\lambda_1} K\left(x, t^{-\left(\frac{\lambda_1}{\lambda_2}\right)} y\right); \quad K(x,ty) = t^{-\lambda_2} K\left(x t^{-\left(\frac{\lambda_2}{\lambda_1}\right)}, y\right),$$

where $t, \lambda_1, \lambda_2 > 0$, $\lambda_1 \neq \lambda_2$, then we call K a non-homogeneous kernel.

In particular, if $K(x,y) = \frac{1}{x^{\lambda_1}+y^{\lambda_2}}$, then

$$K(x,y) = x^{-\lambda_1} K\left(1, x^{-\left(\frac{\lambda_1}{\lambda_2}\right)} y\right) = y^{-\lambda_2} K\left(x y^{-\left(\frac{\lambda_2}{\lambda_1}\right)}, 1\right);$$

if $\lambda_1 = \lambda_2 = \lambda$, then

$$K(tx,ty) = t^{-\lambda} K(x,y),$$

that is, K is a homogeneous kernel of degree $-\lambda$.

If $K(x,y) = \frac{1}{(x^{\lambda_1}+y^{\lambda_2})^\lambda}$, where $\lambda, \lambda_1, \lambda_2 > 0$, then

$$T_{18}(f,x) = \frac{c}{k\Gamma_k(\alpha)} \int_0^\infty \left\{ \frac{1}{(x^{\lambda_1}+y^{\lambda_2})^\lambda} \right\}^{(\alpha/k)-1} f(y)dy \tag{26}$$

is said to be the fractional Hilbert-type integral operator with the non-singular kernel and

$$T(f,x) = \frac{c}{k\Gamma_k(\alpha)} \int_0^\infty \left\{ \frac{1}{|x^{\lambda_1}-y^{\lambda_2}|^\lambda} \right\}^{(\alpha/k)-1} f(y)dy \tag{27}$$

is said to be the fractional Hilbert-type integral operator with the singular kernel.

If $k = 1, \alpha = 2$, then (25) reduces to (24). Hence, the generalized Hilbert-type integral operators are extensions of the generalized Hilbert-type operators.

3. Main Results

Theorem 8. *Let* $1 < p < \infty, \frac{1}{p} + \frac{1}{q} = 1, \lambda, \alpha, \lambda_k > 0,\ k = 1, 2, 3, 4,$ *and*

$$\omega_1(x) = x^{\frac{\lambda_1}{\lambda_2}(\lambda_3+1)-\lambda_1\lambda(\frac{\alpha}{k}-1)-\lambda_4(p-1)}, \tag{28}$$

$$\omega_2(x) = x^{\frac{\lambda_2}{\lambda_1}(\lambda_4+1)-\lambda_2\lambda(\frac{\alpha}{k}-1)-\lambda_3(q-1)}. \tag{29}$$

Let f *and* g *be non-negative measurable functions on* $(0, \infty)$, *and* $f \in L^p_{\omega_1}(0, \infty)$, $g \in L^q_{\omega_2}(0, \infty)$. *If*

$$c_1 = \frac{1}{\lambda_2}\int_0^\infty \left(\frac{1}{|1-u|^\lambda}\right)^{\frac{\alpha}{k}-1} u^{\frac{\lambda_3+1}{\lambda_2}-1} du < \infty \tag{30}$$

and

$$c_2 = \frac{1}{\lambda_1}\int_0^\infty \left(\frac{1}{|1-u|^\lambda}\right)^{\frac{\alpha}{k}-1} u^{\frac{\lambda_4+1}{\lambda_1}-1} du < \infty, \tag{31}$$

then

$$\int_0^\infty\int_0^\infty \left(\frac{1}{|x^{\lambda_1}-y^{\lambda_2}|^\lambda}\right)^{\frac{\alpha}{k}-1} f(x)g(y)dxdy \le c_1^{1/p} c_2^{1/q} \|f\|_{p,\omega_1}\|g\|_{q,\omega_2}. \tag{32}$$

Proof. By Hölder's inequality, we obtain

$$\begin{aligned}
&\int_0^\infty\int_0^\infty \left(\frac{1}{|x^{\lambda_1}-y^{\lambda_2}|^\lambda}\right)^{\frac{\alpha}{k}-1} f(x)g(y)dxdy \\
&= \int_0^\infty\int_0^\infty \left(\frac{1}{|x^{\lambda_1}-y^{\lambda_2}|^\lambda}\right)^{\frac{\alpha}{k}-1} \left(\frac{y^{\lambda_3/p}}{x^{\lambda_4/q}} f(x)\right)\left(\frac{x^{\lambda_4/q}}{y^{\lambda_3/p}} g(y)\right) dxdy \\
&\le \left\{\int_0^\infty\int_0^\infty \left(\frac{1}{|x^{\lambda_1}-y^{\lambda_2}|^\lambda}\right)^{\frac{\alpha}{k}-1} \left(\frac{y^{\lambda_3}}{x^{\lambda_4(p-1)}} f^p(x)\right) dxdy\right\}^{1/p} \\
&\quad \times \left\{\int_0^\infty\int_0^\infty \left(\frac{1}{|x^{\lambda_1}-y^{\lambda_2}|^\lambda}\right)^{\frac{\alpha}{k}-1} \left(\frac{x^{\lambda_4}}{y^{\lambda_3(q-1)}} g^q(y)\right) dxdy\right\}^{1/q}
\end{aligned}$$

$$
= \left\{ \int_0^\infty \left[\int_0^\infty \left(\frac{1}{|x^{\lambda_1} - y^{\lambda_2}|^\lambda} \right)^{\frac{\alpha}{k}-1} \left(\frac{y^{\lambda_3}}{x^{\lambda_4(p-1)}} \right) dy \right] f^p(x) dx \right\}^{1/p}
$$

$$
\times \left\{ \int_0^\infty \left[\int_0^\infty \left(\frac{1}{|x^{\lambda_1} - y^{\lambda_2}|^\lambda} \right)^{\frac{\alpha}{k}-1} \left(\frac{x^{\lambda_4}}{y^{\lambda_3(q-1)}} \right) dx \right] g^q(y) dy \right\}^{1/q}
$$

$$
= \left\{ \int_0^\infty I_1 f^p(x) dx \right\}^{1/p} \times \left\{ \int_0^\infty I_2 g^q(y) dy \right\}^{1/q}, \tag{33}
$$

where

$$
I_1 = \int_0^\infty \left(\frac{1}{|x^{\lambda_1} - y^{\lambda_2}|^\lambda} \right)^{\frac{\alpha}{k}-1} \frac{y^{\lambda_3}}{x^{\lambda_4(p-1)}} dy \tag{34}
$$

and

$$
I_2 = \int_0^\infty \left(\frac{1}{|x^{\lambda_1} - y^{\lambda_2}|^\lambda} \right)^{\frac{\alpha}{k}-1} \frac{x^{\lambda_4}}{y^{\lambda_3(q-1)}} dx. \tag{35}
$$

With the change of variable $u = y^{\lambda_2} x^{-\lambda_1}$ and using (28) and (30), we derive that

$$
\begin{aligned}
I_1 &= \left\{ \int_0^\infty \frac{1}{|1 - y^{\lambda_2} x^{-\lambda_1}|^{\lambda((\alpha/k)-1)}} y^{\lambda_3} dy \right\} x^{-\lambda_1 \lambda((\alpha/k)-1) - \lambda_4(p-1)} \\
&= \left\{ \frac{1}{\lambda_2} \int_0^\infty \frac{1}{|1 - u|^{\lambda((\alpha/k)-1)}} u^{\frac{\lambda_3+1}{\lambda_2}-1} du \right\} \\
&\quad \times x^{\frac{\lambda_1}{\lambda_2}(\lambda_3+1) - \lambda_1 \lambda(\frac{\alpha}{k}-1) - \lambda_4(p-1)} \\
&= c_1 \omega_1(x).
\end{aligned} \tag{36}
$$

Similarly, by the change of variable $u = y^{-\lambda_2} x^{\lambda_1}$ and using (29) and (31), we obtain that

$$
I_2 = c_2 \omega_2(y). \tag{37}
$$

Combining (33), (36), and (37), we get the desired result. The proof is thus completed. □

Remark 9. Inequality (32) is a very general basic result because it contains six parameters $\lambda, \alpha, \lambda_k > 0, k = 1, 2, 3, 4$, that are all independent of each other. Note that for suitable and appropriate choice of these parameters, one can obtain various new and classical results.

Theorem 10. *Under the assumptions of Theorem 8, if*

$$\max\left\{\frac{\lambda_3+1}{\lambda_2}, \frac{\lambda_4+1}{\lambda_1}\right\} < \lambda\left(\frac{\alpha}{k}-1\right) < 1, \tag{38}$$

then

$$\begin{aligned}\int_0^\infty\int_0^\infty & \left(\frac{1}{|x^{\lambda_1}-y^{\lambda_2}|^{\lambda}}\right)^{\frac{\alpha}{k}-1} f(x)g(y)dxdy \\ & \le \frac{1}{\lambda_1^{1/q}\lambda_2^{1/p}}\left\{B\left(\frac{\lambda_3+1}{\lambda_2}, 1-\lambda\left(\frac{\alpha}{k}-1\right)\right)\right. \\ & \left. + B\left(\lambda\left(\frac{\alpha}{k}-1\right)-\frac{\lambda_3+1}{\lambda_2}, 1-\lambda\left(\frac{\alpha}{k}-1\right)\right)\right\}^{1/p} \\ & \times\left\{B\left(\frac{\lambda_4+1}{\lambda_1}, 1-\lambda\left(\frac{\alpha}{k}-1\right)\right) + B\left(\lambda\left(\frac{\alpha}{k}-1\right)\right.\right. \\ & \left.\left. -\frac{\lambda_4+1}{\lambda_1}, 1-\lambda\left(\frac{\alpha}{k}-1\right)\right)\right\}^{1/q} \|f\|_{p,\omega_1}\|g\|_{q,\omega_2}. \end{aligned} \tag{39}$$

Proof. Using Theorem 8 and the integral formula:

$$\int_0^\infty\left(\frac{1}{|1-u|^{\beta}}\right)^{\alpha} u^{\gamma-1}du = B(\gamma, 1-\alpha\beta) + B(\alpha\beta-\gamma, 1-\alpha\beta),$$

where $\gamma < \alpha\beta < 1$, the desired result is obtained. This completes the proof. □

Considering $\lambda_1 = \lambda_2, \lambda_3 = \frac{\lambda\lambda_1}{p}\left(\frac{\alpha}{k}-1\right)-1, \lambda_4 = \frac{\lambda\lambda_1}{q}\left(\frac{\alpha}{k}-1\right)-1$ in Theorem 10, we derive Theorem 11:

Theorem 11. *Let* $1 < p < \infty, \frac{1}{p}+\frac{1}{q} = 1, \lambda, \alpha, \lambda_1 > 0,$ *and*

$$\omega_1(x) = x^{\left(\lambda\lambda_1\left(1-\frac{\alpha}{k}\right)+1\right)(p-1)} \tag{40}$$

and

$$\omega_2(x) = x^{\left(\lambda\lambda_1\left(1-\frac{\alpha}{k}\right)+1\right)(q-1)}. \tag{41}$$

Let f and g be non-negative measurable functions on $(0,\infty)$, and $f \in L^p_{\omega_1}(0,\infty)$, $g \in L^q_{\omega_2}(0,\infty)$. Then,

$$\int_0^\infty \int_0^\infty \left(\frac{1}{|x^{\lambda_1} - y^{\lambda_1}|^\lambda}\right)^{\frac{\alpha}{k}-1} f(x)g(y)dxdy \leq c(p)\|f\|_{p,\omega_1}\|g\|_{q,\omega_2}, \tag{42}$$

where the constant factor

$$c(p) = \frac{1}{\lambda_1}\left\{B\left(\frac{\lambda}{p}\left(\frac{\alpha}{k}-1\right), 1-\lambda\left(\frac{\alpha}{k}-1\right)\right)\right.$$
$$\left. + B\left(\frac{\lambda}{q}\left(\frac{\alpha}{k}-1\right), 1-\lambda\left(\frac{\alpha}{k}-1\right)\right)\right\} \tag{43}$$

is the best possible.

Proof. Considering $\lambda_1 = \lambda_2, \lambda_3 = \frac{\lambda\lambda_1}{p}\left(\frac{\alpha}{k}-1\right) - 1, \lambda_4 = \frac{\lambda\lambda_1}{q}\left(\frac{\alpha}{k}-1\right) - 1$ in Theorem 10, we have

$$\omega_1(x) = x^{\frac{\lambda\lambda_1}{p}\left(\frac{\alpha}{k}-1\right) - \lambda\lambda_1\left(\frac{\alpha}{k}-1\right) - \left[\frac{\lambda\lambda_1}{q}\left(\frac{\alpha}{k}-1\right)-1\right](p-1)} = x^{\left(\lambda\lambda_1\left(1-\frac{\alpha}{k}\right)+1\right)(p-1)}$$

and

$$\omega_2(x) = x^{\frac{\lambda\lambda_1}{q}\left(\frac{\alpha}{k}-1\right) - \lambda\lambda_1\left(\frac{\alpha}{k}-1\right) - \left[\frac{\lambda\lambda_1}{p}\left(\frac{\alpha}{k}-1\right)-1\right](q-1)}$$
$$= x^{\left(\lambda\lambda_1\left(1-\frac{\alpha}{k}\right)+1\right)(q-1)},$$

$$c_1 = \frac{1}{\lambda_2}\left\{B\left(\frac{\lambda_3+1}{\lambda_2}, 1-\lambda\left(\frac{\alpha}{k}-1\right)\right) + B\left(\lambda\left(\frac{\alpha}{k}-1\right)\right.\right.$$
$$\left.\left. - \frac{\lambda_3+1}{\lambda_2}, 1-\lambda\left(\frac{\alpha}{k}-1\right)\right)\right\}$$
$$= \frac{1}{\lambda_1}\left\{B\left(\frac{\lambda}{p}\left(\frac{\alpha}{k}-1\right), 1-\lambda\left(\frac{\alpha}{k}-1\right)\right)\right.$$
$$\left. + B\left(\frac{\lambda}{q}\left(\frac{\alpha}{k}-1\right), 1-\lambda\left(\frac{\alpha}{k}-1\right)\right)\right\}.$$

Similarly, we have

$$c_2 = \frac{1}{\lambda_1}\left\{B\left(\frac{\lambda}{p}\left(\frac{\alpha}{k}-1\right), 1-\lambda\left(\frac{\alpha}{k}-1\right)\right)\right.$$
$$\left. + B\left(\frac{\lambda}{q}\left(\frac{\alpha}{k}-1\right), 1-\lambda\left(\frac{\alpha}{k}-1\right)\right)\right\}.$$

By Theorem 10, we obtain the desired result (42).

To prove that the constant factor

$$c(p) = \frac{1}{\lambda_1}\left\{B\left(\frac{\lambda}{p}\left(\frac{\alpha}{k}-1\right), 1-\lambda\left(\frac{\alpha}{k}-1\right)\right)\right.$$
$$\left. + B\left(\frac{\lambda}{q}\left(\frac{\alpha}{k}-1\right), 1-\lambda\left(\frac{\alpha}{k}-1\right)\right)\right\}$$

is the best possible, we set f_ε and g_ε as follows:

$$f_\varepsilon(t) = \begin{cases} 0, & 0 < t \le 1, \\ t^{\frac{\lambda\lambda_1}{q}\left(\frac{\alpha}{k}-1\right)-\frac{\varepsilon}{p}-1}, & 1 < t < \infty, \end{cases}$$

$$g_\varepsilon(t) = \begin{cases} 0, & 0 < t \le 1, \\ t^{\frac{\lambda\lambda_1}{p}\left(\frac{\alpha}{k}-1\right)-\frac{\varepsilon}{q}-1}, & 1 < t < \infty. \end{cases}$$

We get

$$\|f_\varepsilon\|_{p,\omega_1}^p = \int_1^\infty t^{\frac{p\lambda\lambda_1}{q}\left(\frac{\alpha}{k}-1\right)-\varepsilon-p} t^{(\lambda\lambda_1(1-\frac{\alpha}{k})+1)(p-1)} dt$$
$$= \int_1^\infty t^{-\varepsilon-1} dt = \frac{1}{\varepsilon}.$$

Similarly, $\|g_\varepsilon\|_{q,\omega_2}^q = \frac{1}{\varepsilon}$. If there exists a positive constant M, $M \le c(p)$, such that (42) holds true, it follows that

$$I = \int_0^\infty \int_0^\infty \left\{\frac{1}{|x^{\lambda_1} - y^{\lambda_1}|^\lambda}\right\}^{\frac{\alpha}{k}-1} f_\varepsilon(x) g_\varepsilon(y) dx dy$$
$$\le M\|f_\varepsilon\|_{p,\omega_1}\|g_\varepsilon\|_{q,\omega_2} = M\left(\frac{1}{\varepsilon}\right)^{1/p}\left(\frac{1}{\varepsilon}\right)^{1/q} = M\frac{1}{\varepsilon}. \tag{44}$$

With the change of variable $u = y^{\lambda_1}x^{-\lambda_1}$, we derive that

$$I = \int_0^\infty \int_0^\infty \left\{\frac{1}{|x^{\lambda_1} - y^{\lambda_1}|^\lambda}\right\}^{\frac{\alpha}{k}-1} f_\varepsilon(x) g_\varepsilon(y) dx dy$$
$$= \int_1^\infty \left\{\int_1^\infty \frac{1}{|x^{\lambda_1} - y^{\lambda_1}|^{\lambda((\alpha/k)-1)}} y^{\frac{\lambda\lambda_1}{p}\left(\frac{\alpha}{k}-1\right)-\frac{\varepsilon}{q}-1} dy\right\}$$
$$\times x^{\frac{\lambda\lambda_1}{q}\left(\frac{\alpha}{k}-1\right)-\frac{\varepsilon}{p}-1} dx$$
$$= \frac{1}{\lambda_1}\int_1^\infty x^{-\varepsilon-1}\left\{\int_{x^{-\lambda_1}}^\infty \frac{1}{|1-u|^{\lambda((\alpha/k)-1)}} u^{\frac{\lambda}{p}\left(\frac{\alpha}{k}-1\right)-\frac{\varepsilon}{q\lambda_1}-1} du\right\} dx$$

$$
\begin{aligned}
&= \frac{1}{\lambda_1}\int_1^{\infty} x^{-\varepsilon-1}\left\{\int_{x^{-\lambda_1}}^{1} \frac{1}{|1-u|^{\lambda((\alpha/k)-1)}} u^{\frac{\lambda}{p}\left(\frac{\alpha}{k}-1\right)-\frac{\varepsilon}{q\lambda_1}-1}du\right\}dx \\
&\quad + \frac{1}{\lambda_1}\int_1^{\infty} x^{-\varepsilon-1}\left\{\int_{1}^{\infty} \frac{1}{|1-u|^{\lambda((\alpha/k)-1)}} u^{\frac{\lambda}{p}\left(\frac{\alpha}{k}-1\right)-\frac{\varepsilon}{q\lambda_1}-1}du\right\}dx \\
&= I_1 + I_2,
\end{aligned} \tag{45}
$$

where

$$
\begin{aligned}
I_2 &= \frac{1}{\lambda_1}\int_1^{\infty} x^{-\varepsilon-1}\left\{\int_{1}^{\infty} \frac{1}{|1-u|^{\lambda((\alpha/k)-1)}} u^{\frac{\lambda}{p}\left(\frac{\alpha}{k}-1\right)-\frac{\varepsilon}{q\lambda_1}-1}du\right\}dx \\
&= \frac{1}{\lambda_1\varepsilon}\left\{\int_{1}^{\infty} \frac{1}{|1-u|^{\lambda((\alpha/k)-1)}} u^{\frac{\lambda}{p}\left(\frac{\alpha}{k}-1\right)-\frac{\varepsilon}{q\lambda_1}-1}du\right\}.
\end{aligned} \tag{46}
$$

Using Fubini's theorem, we get

$$
\begin{aligned}
I_1 &= \frac{1}{\lambda_1}\int_1^{\infty} x^{-\varepsilon-1}\left\{\int_{x^{-\lambda_1}}^{1} \frac{1}{|1-u|^{\lambda((\alpha/k)-1)}} u^{\frac{\lambda}{p}\left(\frac{\alpha}{k}-1\right)-\frac{\varepsilon}{q\lambda_1}-1}du\right\}dx \\
&= \frac{1}{\lambda_1}\int_0^{1} \frac{1}{|1-u|^{\lambda((\alpha/k)-1)}} u^{\frac{\lambda}{p}\left(\frac{\alpha}{k}-1\right)-\frac{\varepsilon}{q\lambda_1}-1}\left\{\int_{u^{-(1/\lambda_1)}}^{\infty} x^{-\varepsilon-1}dx\right\}du \\
&= \frac{1}{\lambda_1\varepsilon}\int_0^{1} \frac{1}{|1-u|^{\lambda((\alpha/k)-1)}} u^{\frac{\lambda}{p}\left(\frac{\alpha}{k}-1\right)+\frac{\varepsilon}{p\lambda_1}-1}du.
\end{aligned} \tag{47}
$$

Thus, by (44), (45), (46), and (47), we conclude that

$$
\begin{aligned}
&\frac{1}{\lambda_1}\int_0^{1} \frac{1}{|1-u|^{\lambda((\alpha/k)-1)}} u^{\frac{\lambda}{p}\left(\frac{\alpha}{k}-1\right)+\frac{\varepsilon}{p\lambda_1}-1}du \\
&\quad + \frac{1}{\lambda_1}\left\{\int_{1}^{\infty} \frac{1}{|1-u|^{\lambda((\alpha/k)-1)}} u^{\frac{\lambda}{p}\left(\frac{\alpha}{k}-1\right)-\frac{\varepsilon}{q\lambda_1}-1}du\right\} \le M.
\end{aligned} \tag{48}
$$

Letting $\varepsilon \to 0^+$ in (48), and using Fatou's lemma, we get

$$
\frac{1}{\lambda_1}\left\{\int_{0}^{\infty} \frac{1}{|1-u|^{\lambda((\alpha/k)-1)}} u^{\frac{\lambda}{p}\left(\frac{\alpha}{k}-1\right)-1}du\right\}
$$

$$= \frac{1}{\lambda_1}\left\{B\left(\frac{\lambda}{p}\left(\frac{\alpha}{k}-1\right), 1-\lambda\left(\frac{\alpha}{k}-1\right)\right)\right.$$
$$\left. + B\left(\frac{\lambda}{q}\left(\frac{\alpha}{k}-1\right), 1-\lambda\left(\frac{\alpha}{k}-1\right)\right)\right\} \le M.$$

Hence,

$$M = \frac{1}{\lambda_1}\left\{B\left(\frac{\lambda}{p}\left(\frac{\alpha}{k}-1\right), 1-\lambda\left(\frac{\alpha}{k}-1\right)\right)\right.$$
$$\left. + B\left(\frac{\lambda}{q}\left(\frac{\alpha}{k}-1\right), 1-\lambda\left(\frac{\alpha}{k}-1\right)\right)\right\}$$

is the best possible constant factor of (42). This completes the proof of Theorem 11. □

4. Norm Inequalities for Fractional Hilbert-Type Operators

Theorem 12. *Under the assumptions of Theorem 8, let*

$$\omega(y) = \omega_2(y)^{1/(1-q)}. \tag{49}$$

The integral operator T defined in (27), $T : L^p_{\omega_1}(0,\infty) \to L^p_{\omega}(0,\infty)$, exists as a bounded operator and

$$\|Tf\|_{p,\omega} \le \frac{c}{k\Gamma_k(\alpha)} c_1^{1/p} c_2^{1/q} \|f\|_{p,\omega_1}. \tag{50}$$

This implies that

$$\|T\| = \sup_{f\neq 0} \frac{\|Tf\|_{p,\omega}}{\|f\|_{p,\omega_1}} \le \frac{c}{k\Gamma_k(\alpha)} c_1^{1/p} c_2^{1/q}, \tag{51}$$

where c_1, c_2, ω_1, and ω_2 are defined in (30), (31), (28), and (29), respectively.

Proof. Let

$$g(y) = \operatorname{sgn} T(f,y)|T(f,y)|^{p-1}\omega(y).$$

By applying Theorem 8, we have

$$\begin{aligned}\|Tf\|_{p,\omega}^{p} &= \int_0^\infty T(f,y)\operatorname{sgn} T(f,y)|T(f,y)|^{p-1}\omega(y)dy \\ &= \int_0^\infty T(f,y)g(y)dy \\ &= \frac{c}{k\Gamma_k(\alpha)}\int_0^\infty\int_0^\infty \left(\frac{1}{|x^{\lambda_1}-y^{\lambda_2}|^{\lambda}}\right)^{(\alpha/k)-1} f(x)g(y)dxdy \\ &\le \frac{c}{k\Gamma_k(\alpha)} c_1^{1/p} c_2^{1/q}\|f\|_{p,\omega_1}\|g\|_{q,\omega_2}. \end{aligned} \tag{52}$$

Note that

$$\begin{aligned}\|g\|_{q,\omega_2} &= \left(\int_0^\infty |g(y)|^q\omega_2(y)dy\right)^{1/q} \\ &= \left\{\int_0^\infty (|T(f,y)|^{p-1}\omega(y))^q\omega_2(y)dy\right\}^{1/q} \\ &= \left\{\left(\int_0^\infty |T(f,y)|^p\omega(y)dy\right)^{1/p}\right\}^{p/q} = \|Tf\|_{p,\omega}^{p/q}. \end{aligned} \tag{53}$$

Hence, by (52) and (53), we get

$$\|Tf\|_{p,\omega}^{p} \le \frac{c}{k\Gamma_k(\alpha)} c_1^{1/p} c_2^{1/q}\|f\|_{p,\omega_1}\|Tf\|_{p,\omega}^{p/q}.$$

It follows that

$$\|Tf\|_{p,\omega} \le \frac{c}{k\Gamma_k(\alpha)} c_1^{1/p} c_2^{1/q}\|f\|_{p,\omega_1}. \tag{54}$$

This completes the proof. □

Theorem 13. *Under the assumptions of Theorem* 11*, the integral operator* T *defined in* (27)*,* $T : L^p_{\omega_1}(0,\infty) \to L^p_{\omega}(0,\infty)$*, exists as a bounded operator and*

$$\|Tf\|_{p,\omega} \le \frac{c}{k\Gamma_k(\alpha)} c(p)\|f\|_{p,\omega_1}, \tag{55}$$

where $c(p)$ is defined in (43). Thus,

$$\|T\| = \frac{c}{k\Gamma_k(\alpha)}c(p) = \frac{c}{\lambda_1 k\Gamma_k(\alpha)}\left\{B\left(\frac{\lambda}{p}\left(\frac{\alpha}{k}-1\right), 1-\lambda\left(\frac{\alpha}{k}-1\right)\right)\right.$$
$$\left. + B\left(\frac{\lambda}{q}\left(\frac{\alpha}{k}-1\right), 1-\lambda\left(\frac{\alpha}{k}-1\right)\right)\right\} \tag{56}$$

is the best constant factor and ω_1 is defined in (40) and

$$\omega(y) = y^{\lambda\lambda_1(\frac{\alpha}{k}-1)-1}.$$

Remark 14. If we set $\alpha = 2, k = c = 1$, $\lambda_1 = \lambda_2$, then we reduce to the classical case, that is,

(I) Inequality (42) reduces to

$$\int_0^\infty \int_0^\infty \left\{\frac{1}{|x^{\lambda_1} - y^{\lambda_1}|^\lambda}\right\} f(x)g(y)dxdy \le c(p)\|f\|_{p,\omega_1}\|g\|_{q,\omega_2}, \tag{57}$$

where $\omega_1(x) = x^{(1-\lambda\lambda_1)(p-1)}$, $\omega_2(x) = x^{(1-\lambda\lambda_1)(q-1)}$, and the constant factor

$$c(p) = \frac{1}{\lambda_1}\left\{B\left(\frac{\lambda}{p}, 1-\lambda\right) + B\left(\frac{\lambda}{q}, 1-\lambda\right)\right\} \tag{58}$$

is the best possible.

(II) Relation (27) reduces to

$$T(f, x) = \int_0^\infty \left\{\frac{1}{|x^{\lambda_1} - y^{\lambda_1}|^\lambda}\right\} f(y)dy, \tag{59}$$

and (55) reduces to

$$\|Tf\|_{p,\omega} \le c(p)\|f\|_{p,\omega_1}, \tag{60}$$

where $\omega(x) = x^{\lambda\lambda_1 - 1}$ and

$$c(p) = \|T\| = \frac{1}{\lambda_1}\left\{B\left(\frac{\lambda}{p}, 1-\lambda\right) + B\left(\frac{\lambda}{q}, 1-\lambda\right)\right\} \tag{61}$$

is the best constant factor.

References

[1] Kuang, J.C., *Applied Inequalities*. 5th edu. Shangdong Science and Technology Press, Jinan (2021) (in Chinese).

[2] Kilbas, A.A., Srivastava, H.M. and Trujillo, J.J. *Theory and Applications of Fractional Differential Equations*, North-Holland Mathematics Studies, Elsevier, New York, 204 (2006).

[3] Mubeen, S., and Habibullah, G.M, k-fractional integrals and applications, *Int. J. Contemp. Math. Sci.*, 7, 89–94 (2012).

[4] Mubeen, S., and Iqbal, S., Grüss type integral inequalities for generalized Riemann–Liouville k-fractional integrals, *J. Inequal. Appl.*, 109 (2016).

[5] Sarikaya, M.Z., Dahmani, Z., Kiris, M.E., and Ahmad, F., (k, s)-Riemann–Liouville fractional integral and applications, *Hacet. J. Math. Stat.*, 45(1), 77–89 (2016).

[6] Set. E., Tomar, M., and Sarkaya, M.Z., On generalized Grüss type inequalities for k-fractional integrals, *Appl. Math. Comp.*, 8(269), 29–34 (2015).

[7] Abbas, G., Khan, K.A., Farid, G. and Rehman, A.U. Generalizations of some fractional integral inequalities via generalized Mittag-Leffler function, *J. Inequal. Appl.*, 121 (2017).

[8] da Sousa, J.V. and de Oliveira, E.C., On the ψ-Hilfer fractional derivative, *Commun. Nonlinear Sci. Numer. Simul.*, 60, 72–91 (2018).

[9] Khan, J.U. and Khan, M.A., Generalized conformable fractional integral operators, *J. Comput. Appl. Math*, 346, 378–389 (2019).

[10] Rashid, S., Akdemir, A.O., Nisar, K.S. Abdeljawad Thabet and Rahman Gauhar, New generalized reverse Minkowski and related integral inequalities involving generalized fractional conformable integrals, *J. Inequal. Appl.*, 177 (2020).

[11] Çelik B., Gürbüz, M.Ç., Özdemir M.E., and Set E., On integral inequalities related to the weighted and the extended Chebyshev functionals involving different fractional operators, *J. Inequal. Appl.*, 246 (2020).

[12] Kashuri, A. and Rassias, Th.M. Fractional trapezium Type inequalities for strongly exponentially generalized preinvex functions with applications, *Appl. Anal. discrete Math.*, 14(3), 560–57 (2020).

[13] Awan, M.U., Noor, M.A., Talib, S., Noor, K.I., and Rassias, Th. M., New k-conformable fractional integral inequalities, In: Rassias, Th. M. (eds.), Approximation Theory and Analytic Inequalities, Springer (2021).

[14] Slimane, I., Damani, Z., ian, S. and Agarwal, P. Further results on continuous random variables via fractional integrals, In: Rassias, Th.M. (ed.), *Approximation Theory and Analytic Inequalities*, Springer (2021).

[15] Kuang, J.C., Some new inequalities for fractional integral. In: Rassias, Th. M. (ed.), *Approximation and Computation in Science and Engineering*, Springer (2021).

[16] Sarikaya, M.Z., On the Hermite-Hadamard-type inequalities for co-ordinated convex function via fractional integrals, *Integral Transforms Spec. Funct.*, 25(2), 134–147 (2014).

[17] Noor, M.A. and Rassias, Th.M. Some quantum Hermite-Hadamard-type inequalities for general convex functions. In: Pardalos, P.M. and Rassias, Th. M. (eds.), Contributions in Mathematics and Engineering in Honor of Constantin Carathéodory, Springer (2016).

[18] Zhao Yong, Sang Haiwei, Xiong weicheng and Cui Zhongwei, Hermite–Hadamard type inequalities involving ψ — Riemann-Liouville k-fractional integrals via s-convex functions, *J. Inequal. Appl.*, 128 (2020).

[19] Awan, M.U., Mihai, M.V., and Noor, K.I., Harmonic Hermite-Hadamard inequalities involving Mittag-Leffler function. In: Rassias, Th. M. (Editor), *Approximation Theory and Analytic Inequalities*, Springer (2021).

[20] Khan, M.B., Mohammed, P. O., Noor, M.A. and Hamed, Y.S. New Hermite-Hadamard inequalities in fuzzy-interval fractional calculus and related inequalities, *Symmetry*, 13(4), 673–684 (2021).

[21] Farid, G., Rehman, A.U., Bibi, S. and Chu, Y.M. Refinements of two fractional versions of Hadamard inequalities for Caputo fractional derivatives and related results, *Open J. Math. Sci*, 5(1), 1–10 (2021).

[22] Mohammed, P.O., Abdeljawad, T., Baleanu, D., Kashuri, A., Hamasalh, F. and Agarwal, P. New fractional inequalities of Hermite-Hadamard type involving the incomplete gamma functions, *J. Inequal. Appl.*, 263 (2020).

[23] Gürbüz, M., Akdemir, A.O., Rashid, S. and Set, E. Hermite-Hadamard inequality for fractional integrals of Caputo-Fabrizio type and related inequalities, *J. Inequal. Appl.*, 172 (2020).

[24] Kwun, Y.C., Saleem, M.S., Ghafoor, M., Nazeer, W. and Kang, S.M. Hermite-Hadarmard-type inequalities for functions whose derivatives are η-convex via fractional integrals, *J. Inequal. Appl.*, 44 (2019).

[25] Wu, S., Iqbal, S., Aamir, M., Samraiz, M. and Younus, A. On some Hermite-Hadamard inequalities involving k-fractional operators, *J. Inequal. Appl.*, 32 (2021).

[26] Kuang J.C., Generalized Hilbert integral operators on the Herz spaces, *Tamkang J. Math.*, 40(2), 193–200 (2009).

[27] Kuang J.C., Generalized Hilbert operators on weighted Morrey-Herz spaces, *J. Math. Inequal.*, 6(1), 69–77 (2012).

[28] Adiyasuren, V., Batbold, T. and Azar, L.E. A new discrete Hilbert-type inequality involving partial sums, *J. Inequal. Appl.*, 127 (2019).
[29] Krnić, M., Pečarić, J., Perić, I. and Vuković, P. Recent Advances in Hilbert-type inequality involving, Element, Zagreb (2012).
[30] Batbold, Ts., Krnić, M., Pečarić, J., Perić, I. and Vuković, P. Further Develoment of Hilbert-type inequalities, Element, Zagreb (2017).
[31] Yang, B., Liao, J. and Agarwal, R.P. *Hilbert-Type Inequalities: Operators, Compositions and Extensions*, Scientific Research Publishing, Inc., USA (2020).
[32] Kara, H., Budak, H., Alp, N., Kalsoom, H. and Sarikaya, M.Z. On new generalized quantum integrals and related Hermite-Hadamard inequalities, *J. Inequal. Appl.*, 180 (2021).

https://doi.org/10.1142/9789811293009_0023

Chapter 23

Hyers–Ulam–Rassias Stability of Additive (ρ_1, ρ_2)-Functional Inequalities and hom-Derivations in Banach Algebras

Jung Rye Lee[*,§], Choonkil Park[†,¶] and Themistocles M. Rassias[‡,||]

[*]*Department of Data Science, Daejin University, Kyunggi 11159, Korea*

[†]*Department of Mathematics, Hanyang University, Seoul 04763, Korea*

[‡]*Department of Mathematics, National Technical University of Athens, Zografou Campus, 15780 Athens, Greece*

[§]*jrlee@daejin.ac.kr*

[¶]*baak@hanyang.ac.kr*

[||]*trassias@math.ntua.gr*

In this chapter, we introduce and solve the following additive (ρ_1, ρ_2)-functional inequalities:

$$\begin{aligned} &\|f(x+y+z) - f(x) - f(y) - f(z)\| \\ &\quad \le \|\rho_1(f(x+y-z) - f(x) - f(y) + f(z))\| \\ &\qquad + \|\rho_2(f(x-y-z) - f(x) + f(y) + f(z))\|, \end{aligned} \tag{1}$$

$$\begin{aligned} &\|f(x+y-z)-f(x)-f(y)+f(z)\| \\ &\quad \leq \|\rho_1(f(x+y+z)-f(x)-f(y)-f(z))\| \\ &\qquad + \|\rho_2(f(x-y-z)-f(x)+f(y)+f(z))\|, \end{aligned} \tag{2}$$

where ρ_1 and ρ_2 are fixed nonzero complex numbers with $|\rho_1|+|\rho_2|<1$. Using the direct method, we prove the Hyers–Ulam stability of the additive (ρ_1, ρ_2)-functional inequalities (1) and (2) in complex Banach spaces.

1. Introduction and Preliminaries

Let A be a complex Banach algebra and $h : A \to A$ be a $\mathbb{C}$-linear mapping. Mirzavaziri and Moslehian [17, 18] introduced the concept of h-derivation as follows:

$$f(xy) = f(x)h(y) + h(x)f(y)$$

for all $x, y \in A$.

Let A be a complex Banach algebra, $h : A \to A$ be a homomorphism, and $\delta : A \to A$ be a derivation. Then,

$$\begin{aligned} h \circ \delta(xyz) &= h \circ \delta(x)h(y)h(z) + h(x)h \circ \delta(y)h(z) \\ &\quad + h(x)h(y)h \circ \delta(z) \\ &= h \circ \delta(x)h(yz) + h(x)h \circ \delta(y)h(z) + h(xy)h \circ \delta(z), \\ \delta \circ h(xyz) &= \delta \circ h(x)h(y)h(z) + h(x)\delta \circ h(y)h(z) \\ &\quad + h(x)h(y)\delta \circ h(z) \\ &= \delta \circ h(x)h(yz) + h(x)\delta \circ h(y)h(z) + h(xy)\delta \circ h(z) \end{aligned}$$

for all $x, y, z \in A$.

In this chapter, we introduce and investigate hom-derivations in a Banach algebra.

Definition 1. Let A be a complex Banach algebra and $h : A \to A$ be a homomorphism. A $\mathbb{C}$-linear mapping $D : A \to A$ is called a *hom-derivation* on A if D satisfies

$$D(xyz) = D(x)h(yz) + h(x)D(y)h(z) + h(xy)D(z)$$

for all $x, y, z \in A$.

The stability problem of functional equations originated from a question of Ulam [25] concerning the stability of group homomorphisms.

The functional equation $f(x + y) = f(x) + f(y)$ is called the *Cauchy equation*. In particular, every solution of the Cauchy equation is said to be an *additive mapping*. Hyers [12] gave a first affirmative partial answer to the question of Ulam for Banach spaces. Hyers' theorem was generalized by Aoki [1] for additive mappings and by Rassias [23] for linear mappings by considering an unbounded Cauchy difference. A generalization of the Rassias theorem was obtained by Găvruta [11] by replacing the unbounded Cauchy difference by a general control function in the spirit of Rassias' approach. The stability of quadratic functional equation was proved by Skof [24] for mappings $f : E_1 \to E_2$, where E_1 is a normed space and E_2 is a Banach space. Cholewa [7] noted that the theorem of Skof is still true if the relevant domain E_1 is replaced by an Abelian group.

Park [20, 21] defined additive ρ-functional inequalities and proved the Hyers–Ulam stability of the additive ρ-functional inequalities in Banach spaces and non-Archimedean Banach spaces. The stability problems of various functional equations have been extensively investigated by a number of authors [2, 6, 9, 10, 15, 26].

This chapterr is organized as follows: In Sections 2 and 3, we solve the additive (ρ_1, ρ_2)-functional inequalities (1) and (2) and prove the Hyers–Ulam stability of the additive (ρ_1, ρ_2)-functional inequalities (1) and (2) in Banach spaces by using the direct method. In Sections 4 and 5, using the direct method, we prove the Hyers–Ulam stability of hom-derivations on complex Banach algebras associated with the additive (ρ_1, ρ_2)-functional inequalities (1) and (2).

Throughout this chapter, let X be a real or complex normed space and Y a complex Banach space. Assume that A is a complex Banach algebra and that ρ_1 and ρ_2 are fixed nonzero complex numbers with $|\rho_1| + |\rho_2| < 1$.

2. Additive (ρ_1, ρ_2)-Functional Inequality (1): A Direct Method

In this section, we solve and investigate the additive (ρ_1, ρ_2)-functional inequality (1) in complex Banach spaces.

Lemma 2. *If a mapping $f : X \to Y$ satisfies*

$$\begin{aligned}\|f(x+y+z) - f(x) - f(y) - f(z)\| \\ \leq \|\rho_1(f(x+y-z) - f(x) - f(y) + f(z))\| \\ + \|\rho_2(f(x-y-z) - f(x) + f(y) + f(z))\|\end{aligned} \tag{3}$$

for all $x, y, z \in X$, then $f : X \to Y$ is additive.

Proof. Assume that $f : X \to Y$ satisfies (3).

Letting $x = y = z = 0$ in (3), we get $\|f(0)\| \leq \|\rho_2 f(0)\|$ and so $f(0) = 0$ since $|\rho_2| < 1$.

Letting $z = 0$ and $y = x$ in (3), we get

$$\|f(2x) - 2f(x)\| \leq \|\rho_1(f(2x) - 2f(x))\|$$

for all $x \in X$. Thus, $f(2x) = 2f(x)$ for all $x \in X$ since $|\rho_1| < 1$.

Letting $z = 0$ and $y = -x$ in (3), we get

$$\begin{aligned}\|f(x) + f(-x)\| &\leq |\rho_1|\|f(x) + f(-x)\| + |\rho_2|\|f(2x) - f(x) + f(-x)\| \\ &\leq (|\rho_1| + |\rho_2|)\|f(x) + f(-x)\|\end{aligned}$$

for all $x \in X$. So $f(-x) = -f(x)$ for all $x \in X$ since $|\rho_1| + |\rho_2| < 1$.

Letting $y = 0$ in (3), we get

$$\begin{aligned}\|f(x+z) - f(x) - f(z)\| \\ \leq (|\rho_1| + |\rho_2|)\|f(x-z) - f(x) + f(z)\|\end{aligned} \tag{4}$$

for all $x, z \in X$. Replacing z by $-z$ in (4), we get

$$\begin{aligned}\|f(x-z) - f(x) + f(z)\| \\ \leq (|\rho_1| + |\rho_2|)\|f(x+z) - f(x) - f(z)\|\end{aligned} \tag{5}$$

for all $x, z \in X$. It follows from (4) and (5) that

$$\|f(x+z) - f(x) - f(z)\| \leq (|\rho_1| + |\rho_2|)^2\|f(x+z) - f(x) - f(z)\|$$

for all $x, z \in X$. So $f(x+z) = f(x) + f(z)$ for all $x, z \in X$ since $|\rho_1| + |\rho_2| < 1$. Thus, f is additive. □

Now, we prove the Hyers–Ulam stability of the additive (ρ_1, ρ_2)-functional inequality (3) in complex Banach spaces by using the direct method.

Theorem 3. *Let $\varphi : X^3 \to [0, \infty)$ be a function such that*

$$\Psi(x, y, z) := \sum_{j=1}^{\infty} 2^j \varphi\left(\frac{x}{2^j}, \frac{y}{2^j}, \frac{z}{2^j}\right) < \infty \tag{6}$$

for all $x, y, z \in X$. Let $f : X \to Y$ be a mapping satisfying $f(0) = 0$ and

$$\begin{aligned} &\|f(x+y+z) - f(x) - f(y) - f(z)\| \\ &\quad \le \|\rho_1(f(x+y-z) - f(x) - f(y) + f(z))\| \\ &\qquad + \|\rho_2(f(x-y-z) - f(x) + f(y) + f(z))\| + \varphi(x, y, z) \end{aligned} \tag{7}$$

for all $x, y, z \in X$. Then there exists a unique additive mapping $A : X \to Y$ such that

$$\|f(x) - A(x)\| \le \frac{1}{2}\Psi(x, 0, x) \tag{8}$$

for all $x \in X$.

Proof. Letting $z = x$ and $y = 0$ in (7), we get

$$\|f(2x) - 2f(x)\| \le \varphi(x, 0, x) \tag{9}$$

and so

$$\left\|f(x) - 2f\left(\frac{x}{2}\right)\right\| \le \varphi\left(\frac{x}{2}, 0, \frac{x}{2}\right)$$

for all $x \in X$. Thus,

$$\begin{aligned} \left\|2^l f\left(\frac{x}{2^l}\right) - 2^m f\left(\frac{x}{2^m}\right)\right\| &\le \sum_{j=l}^{m-1} \left\|2^j f\left(\frac{x}{2^j}\right) - 2^{j+1} f\left(\frac{x}{2^{j+1}}\right)\right\| \\ &\le \sum_{j=l}^{m-1} 2^j \varphi\left(\frac{x}{2^{j+1}}, 0, \frac{x}{2^{j+1}}\right) \end{aligned} \tag{10}$$

for all non-negative integers m and l with $m > l$ and all $x \in X$. It follows from (10) that the sequence $\{2^k f(\frac{x}{2^k})\}$ is Cauchy for all

$x \in X$. Since Y is a Banach space, the sequence $\{2^k f(\frac{x}{2^k})\}$ converges. So, one can define the mapping $A : X \to Y$ by

$$A(x) := \lim_{k\to\infty} 2^k f\left(\frac{x}{2^k}\right)$$

for all $x \in X$. Moreover, letting $l = 0$ and passing the limit $m \to \infty$ in (10), we get (8).

It follows from (7) and (6) that

$$\begin{aligned}
&\|A(x+y+z) - A(x) - A(y) - A(z)\| \\
&= \lim_{n\to\infty} 2^n \left\| f\left(\frac{x+y+z}{2^n}\right) - f\left(\frac{x}{2^n}\right) - f\left(\frac{y}{2^n}\right) - f\left(\frac{z}{2^n}\right) \right\| \\
&\le \lim_{n\to\infty} 2^n |\rho_1| \left\| f\left(\frac{x+y-z}{2^n}\right) - f\left(\frac{x}{2^n}\right) - f\left(\frac{y}{2^n}\right) + f\left(\frac{z}{2^n}\right) \right\| \\
&\quad + \lim_{n\to\infty} 2^n |\rho_2| \left\| f\left(\frac{x-y-z}{2^n}\right) - f\left(\frac{x}{2^n}\right) + f\left(\frac{y}{2^n}\right) + f\left(\frac{z}{2^n}\right) \right\| \\
&\quad + \lim_{n\to\infty} 2^n \varphi\left(\frac{x}{2^n}, \frac{y}{2^n}, \frac{z}{2^n}\right) \\
&= \|\rho_1(A(x+y-z) - A(x) - A(y) + A(z))\| \\
&\quad + \|\rho_2(A(x-y-z) - A(x) + A(y) + A(z))\|
\end{aligned}$$

for all $x, y, z \in X$. So,

$$\begin{aligned}
&\|A(x+y+z) - A(x) - A(y) - A(z)\| \\
&\quad \le \|\rho_1(A(x+y-z) - A(x) - A(y) + A(z))\| \\
&\qquad + \|\rho_2(A(x-y-z) - A(x) + A(y) + A(z))\|
\end{aligned}$$

for all $x, y, z \in X$. By Lemma 2, the mapping $A : X \to Y$ is additive.

Now, let $T : X \to Y$ be another additive mapping satisfying (8). Then we have

$$\begin{aligned}
\|A(x) - T(x)\| &= \left\| 2^q A\left(\frac{x}{2^q}\right) - 2^q T\left(\frac{x}{2^q}\right) \right\| \\
&\le \left\| 2^q A\left(\frac{x}{2^q}\right) - 2^q f\left(\frac{x}{2^q}\right) \right\| + \left\| 2^q T\left(\frac{x}{2^q}\right) - 2^q f\left(\frac{x}{2^q}\right) \right\| \\
&\le 2^q \Psi\left(\frac{x}{2^q}, 0, \frac{x}{2^q}\right),
\end{aligned}$$

which tends to zero as $q \to \infty$ for all $x \in X$. So, we can conclude that $A(x) = T(x)$ for all $x \in X$. This proves the uniqueness of A. $\square$

Corollary 4. *Let $r > 1$ and θ be non-negative real numbers, and let $f : X \to Y$ be a mapping satisfying $f(0) = 0$ and*

$$\begin{aligned} &\|f\,(x+y+z) - f(x) - f(y) - f(z)\| \\ &\quad \le \|\rho_1(f(x+y-z) - f(x) - f(y) + f(z))\| \\ &\qquad + \|\rho_2\,(f(x-y-z) - f(x) + f(y) + f(z))\| \\ &\qquad + \theta(\|x\|^r + \|y\|^r + \|z\|^r) \end{aligned} \tag{11}$$

for all $x, y, z \in X$. Then there exists a unique additive mapping $A : X \to Y$ such that

$$\|f(x) - A(x)\| \le \frac{2\theta}{2^r - 2}\|x\|^r$$

for all $x \in X$.

Theorem 5. *Let $\varphi : X^3 \to [0, \infty)$ be a function and let $f : X \to Y$ be a mapping satisfying $f(0) = 0$, (7) and*

$$\Psi(x, y, z) := \sum_{j=0}^{\infty} \frac{1}{2^j}\varphi(2^j x, 2^j y, 2^j z) < \infty \tag{12}$$

for all $x, y, z \in X$. Then there exists a unique additive mapping $A : X \to Y$ such that

$$\|f(x) - A(x)\| \le \frac{1}{2}\Psi(x, 0, x)$$

for all $x \in X$.

Proof. It follows from (9) that

$$\left\|f(x) - \frac{1}{2}f(2x)\right\| \le \frac{1}{2}\varphi(x, 0, x)$$

for all $x \in X$. Hence,

$$\begin{aligned} \left\|\frac{1}{2^l}f(2^l x) - \frac{1}{2^m}f(2^m x)\right\| &\le \sum_{j=l}^{m-1} \left\|\frac{1}{2^j}f\,(2^j x) - \frac{1}{2^{j+1}}f\,(2^{j+1}x)\right\| \\ &\le \sum_{j=l}^{m-1} \frac{1}{2^{j+1}}\varphi(2^j x, 0, 2^j x) \end{aligned} \tag{13}$$

for all non-negative integers m and l with $m > l$ and all $x \in X$. It follows from (13) that the sequence $\{\frac{1}{2^n} f(2^n x)\}$ is a Cauchy sequence for all $x \in X$. Since Y is complete, the sequence $\{\frac{1}{2^n} f(2^n x)\}$ converges. So, one can define the mapping $A : X \to Y$ by

$$A(x) := \lim_{n\to\infty} \frac{1}{2^n} f(2^n x)$$

for all $x \in X$. Moreover, letting $l = 0$ and passing the limit $m \to \infty$ in (13), we get (13).

The rest of the proof is similar to the proof of Theorem 3. □

Corollary 6. *Let $r < 1$ and θ be positive real numbers, and let $f : X \to Y$ be a mapping satisfying $f(0) = 0$ and (11). Then there exists a unique additive mapping $A : X \to Y$ such that*

$$\|f(x) - A(x)\| \le \frac{2\theta}{2 - 2^r} \|x\|^r$$

for all $x \in X$.

3. Additive (ρ_1, ρ_2)-Functional Inequality (2): A Direct Method

In this section, we solve and investigate the additive (ρ_1, ρ_2)-functional inequality (2) in complex Banach spaces.

Lemma 7. *If a mapping $f : X \to Y$ satisfies $f(0) = 0$ and*

$$\begin{aligned} &\|f(x + y - z) - f(x) - f(y) + f(z)\| \\ &\quad \le \|\rho_1(f(x + y + z) - f(x) - f(y) - f(z))\| \\ &\qquad + \|\rho_2(f(x - y - z) - f(x) + f(y) + f(z))\| \end{aligned} \tag{14}$$

for all $x, y, z \in X$, then $f : X \to Y$ is additive.

Proof. Assume that $f : X \to Y$ satisfies (14).

Letting $z = 0$ and $y = x$ in (14), we get

$$\|f(2x) - 2f(x)\| \le |\rho_1| \|f(2x) - 2f(x)\|$$

for all $x \in X$. Thus, $f(2x) = 2f(x)$ for all $x \in X$, since $|\rho_1| < 1$.

Letting $z = 0$ and $y = -x$ in (14), we get

$$\begin{aligned}&\|f(x) + f(-x)\|\\&\quad \leq |\rho_1|\|f(x) + f(-x)\| + |\rho_2|\|f(2x) - f(x) + f(-x)\|\\&\quad \leq (|\rho_1| + |\rho_2|)\|f(x) + f(-x)\|\end{aligned}$$

for all $x \in X$. So, $f(-x) = -f(x)$ for all $x \in X$ since $|\rho_1| + |\rho_2| < 1$.

Letting $z = 0$ in (14), we get

$$\begin{aligned}\|f(x+y) - f(x) - f(y)\| \leq |\rho_1|\|f(x+y) - f(x) - f(y)\|\\ + |\rho_2|\|f(x-y) - f(x) + f(y)\|\end{aligned}$$

and so

$$\begin{aligned}&(1 - |\rho_1|)\|f(x+y) - f(x) - f(y)\|\\&\quad \leq |\rho_2|\|f(x-y) - f(x) + f(y)\| \end{aligned}\tag{15}$$

for all $x, y \in X$. Replacing y by $-y$ in (15), we get

$$\begin{aligned}&(1 - |\rho_1|)\|f(x-y) - f(x) + f(y)\|\|\|\\&\quad \leq |\rho_2|\|f(x+y) - f(x) - f(y)\| \end{aligned}\tag{16}$$

for all $x, y \in X$. It follows from (15) and (16) that

$$\|f(x+y) - f(x) - f(y)\| \leq \frac{|\rho_2|^2}{(1 - |\rho_1|)^2}\|f(x+y) - f(x) - f(y)\|$$

for all $x, y \in X$. So $f(x+y) = f(x) + f(y)$ for all $x, y \in X$ since $|\rho_1| + |\rho_2| < 1$. Thus, f is additive. □

Now, we prove the Hyers–Ulam stability of the additive (ρ_1, ρ_2)-functional inequality (14) in complex Banach spaces by using the direct method.

Theorem 8. *Let $\varphi : X^3 \to [0, \infty)$ be a function such that*

$$\Psi(x, y, z) := \sum_{j=1}^{\infty} 2^j \varphi\left(\frac{x}{2^j}, \frac{y}{2^j}, \frac{z}{2^j}\right) < \infty \tag{17}$$

for all $x, y, z \in X$. Let $f : X \to Y$ be a mapping satisfying $f(0) = 0$ and

$$\begin{aligned} &\|f(x+y-z) - f(x) - f(y) + f(z)\| \\ &\quad \le \|\rho_1(f(x+y+z) - f(x) - f(y) - f(z))\| \\ &\qquad + \|\rho_2(f(x-y-z) - f(x) + f(y) + f(z))\| + \varphi(x,y,z) \end{aligned} \tag{18}$$

for all $x, y, z \in X$. Then there exists a unique additive mapping $A : X \to Y$ such that

$$\|f(x) - A(x)\| \le \frac{1}{2(1-|\rho_1|)}\Psi(x,x,0) \tag{19}$$

for all $x \in X$.

Proof. Letting $y = x$ and $z = 0$ in (18), we get

$$\|f(2x) - 2f(x)\| \le \frac{1}{1-|\rho_1|}\varphi(x,x,0) \tag{20}$$

and so

$$\left\|f(x) - 2f\left(\frac{x}{2}\right)\right\| \le \frac{1}{1-|\rho_1|}\varphi\left(\frac{x}{2},\frac{x}{2},0\right)$$

for all $x \in X$. Thus,

$$\begin{aligned} \left\|2^l f\left(\frac{x}{2^l}\right) - 2^m f\left(\frac{x}{2^m}\right)\right\| &\le \sum_{j=l}^{m-1} \left\|2^j f\left(\frac{x}{2^j}\right) - 2^{j+1} f\left(\frac{x}{2^{j+1}}\right)\right\| \\ &\le \sum_{j=l}^{m-1} \frac{2^j}{1-|\rho_1|}\varphi\left(\frac{x}{2^{j+1}},\frac{x}{2^{j+1}},0\right) \end{aligned} \tag{21}$$

for all non-negative integers m and l with $m > l$ and all $x \in X$. It follows from (28) that the sequence $\{2^k f(\frac{x}{2^k})\}$ is Cauchy for all $x \in X$. Since Y is a Banach space, the sequence $\{2^k f(\frac{x}{2^k})\}$ converges. So, one can define the mapping $A : X \to Y$ by

$$A(x) := \lim_{k\to\infty} 2^k f\left(\frac{x}{2^k}\right)$$

for all $x \in X$. Moreover, letting $l = 0$ and passing the limit $m \to \infty$ in (28), we get (26).

It follows from (28) and (25) that

$$\begin{aligned}
&\|A(x+y-z) - A(x) - A(y) + A(z)\| \\
&\quad = \lim_{n\to\infty} 2^n \left\| f\left(\frac{x+y-z}{2^n}\right) - f\left(\frac{x}{2^n}\right) - f\left(\frac{y}{2^n}\right) + f\left(\frac{z}{2^n}\right) \right\| \\
&\quad \le \lim_{n\to\infty} 2^n |\rho_1| \left\| f\left(\frac{x+y+z}{2^n}\right) - f\left(\frac{x}{2^n}\right) - f\left(\frac{y}{2^n}\right) - f\left(\frac{z}{2^n}\right) \right\| \\
&\qquad + \lim_{n\to\infty} 2^n |\rho_2| \left\| f\left(\frac{x-y-z}{2^n}\right) - f\left(\frac{x}{2^n}\right) + f\left(\frac{y}{2^n}\right) + f\left(\frac{z}{2^n}\right) \right\| \\
&\qquad + \lim_{n\to\infty} 2^n \varphi\left(\frac{x}{2^n}, \frac{y}{2^n}, \frac{z}{2^n}\right) \\
&\quad = \|\rho_1(A(x+y+z) - A(x) - A(y) - A(z))\| \\
&\qquad + \|\rho_2(A(x-y-z) - A(x) + A(y) + A(z))\|
\end{aligned}$$

for all $x, y, z \in X$. So,

$$\begin{aligned}
&\|A(x+y-z) - A(x) - A(y) + A(z)\| \\
&\quad \le \|\rho_1(A(x+y+z) - A(x) - A(y) - A(z))\| \\
&\qquad + \|\rho_2(A(x-y-z) - A(x) + A(y) + A(z))\|
\end{aligned}$$

for all $x, y, z \in X$. By Lemma 7, the mapping $A : X \to Y$ is additive.

Now, let $T : X \to Y$ be another additive mapping satisfying (26). Then, we have

$$\begin{aligned}
\|A(x) - T(x)\| &= \left\| 2^q A\left(\frac{x}{2^q}\right) - 2^q T\left(\frac{x}{2^q}\right) \right\| \\
&\le \left\| 2^q A\left(\frac{x}{2^q}\right) - 2^q f\left(\frac{x}{2^q}\right) \right\| + \left\| 2^q T\left(\frac{x}{2^q}\right) - 2^q f\left(\frac{x}{2^q}\right) \right\| \\
&\le \frac{2^q}{1 - |\rho_1|} \Psi\left(\frac{x}{2^q}, \frac{x}{2^q}, 0\right),
\end{aligned}$$

which tends to zero as $q \to \infty$ for all $x \in X$. So, we can conclude that $A(x) = T(x)$ for all $x \in X$. This proves the uniqueness of A.

□

Corollary 9. *Let $r > 1$ and θ be non-negative real numbers, and let $f : X \to Y$ be a mapping satisfying $f(0) = 0$ and*

$$\begin{aligned}\|f(x+y-z) - f(x) - f(y) + f(z)\| &\le \|\rho_1(f(x+y+z) - f(x) - f(y) - f(z))\| \\ &\quad + \|\rho_2(f(x-y-z) - f(x) + f(y) + f(z))\| \\ &\quad + \theta(\|x\|^r + \|y\|^r + \|z\|^r) \end{aligned} \tag{22}$$

for all $x, y, z \in X$. Then there exists a unique additive mapping $A : X \to Y$ such that

$$\|f(x) - A(x)\| \le \frac{2\theta}{(2^r - 2)(1 - |\rho_1|)}\|x\|^r$$

for all $x \in X$.

Theorem 10. *Let $\varphi : X^3 \to [0, \infty)$ be a function and let $f : X \to Y$ be a mapping satisfying $f(0) = 0$, (18) and*

$$\Psi(x, y, z) := \sum_{j=0}^{\infty} \frac{1}{2^j}\varphi(2^j x, 2^j y, 2^j z) < \infty \tag{23}$$

for all $x, y, z \in X$. Then there exists a unique additive mapping $A : X \to Y$ such that

$$\|f(x) - A(x)\| \le \frac{1}{2(1 - |\rho_1|)}\Psi(x, x, 0)$$

for all $x \in X$.

Proof. It follows from (27) that

$$\left\|f(x) - \frac{1}{2}f(2x)\right\| \le \frac{1}{2(1 - |\rho_1|)}\varphi(x, x, 0)$$

for all $x \in X$. Hence,

$$\begin{aligned}\left\|\frac{1}{2^l}f(2^l x) - \frac{1}{2^m}f(2^m x)\right\| &\le \sum_{j=l}^{m-1}\left\|\frac{1}{2^j}f(2^j x) - \frac{1}{2^{j+1}}f(2^{j+1}x)\right\| \\ &\le \sum_{j=l}^{m-1}\frac{1}{2^{j+1}(1 - |\rho_1|)}\varphi(2^j x, 2^j x, 0) \end{aligned} \tag{24}$$

for all non-negative integers m and l with $m > l$ and all $x \in X$. It follows from (30) that the sequence $\{\frac{1}{2^n}f(2^n x)\}$ is a Cauchy sequence for all $x \in X$. Since Y is complete, the sequence $\{\frac{1}{2^n} f(2^n x)\}$ converges. So, one can define the mapping $A : X \to Y$ by

$$A(x) := \lim_{n\to\infty} \frac{1}{2^n} f(2^n x)$$

for all $x \in X$. Moreover, letting $l = 0$ and passing the limit $m \to \infty$ in (30), we get (29).

The rest of the proof is similar to the proof of Theorem 8. □

Corollary 11. *Let $r < 1$ and θ be positive real numbers, and let $f : X \to Y$ be a mapping satisfying $f(0) = 0$ and (22). Then there exists a unique additive mapping $A : X \to Y$ such that*

$$\|f(x) - A(x)\| \le \frac{2\theta}{(2-2^r)(1-|\rho_1|)}\|x\|^r$$

for all $x \in X$.

4. Stability of hom-Derivations in Banach Algebras Associated with (1)

Using the direct method, we prove the Hyers–Ulam stability of hom-derivations in Banach algebras, associated with the additive (ρ_1, ρ_2)-functional inequality (1).

Theorem 12. *Let $\varphi : A^3 \to [0, \infty)$ be a function such that*

$$\sum_{j=1}^{\infty} 8^j \varphi\left(\frac{x}{2^j}, \frac{y}{2^j}, \frac{z}{2^j}\right) < \infty \tag{25}$$

for all $x, y, z \in A$. Let $f, g : A \to A$ be mappings satisfying $f(0) = g(0) = 0$ and

$$\begin{aligned}
&\|f(\lambda(x+y+z)) - \lambda(f(x) + f(y) + f(z))\| \\
&\quad \le \|\rho_1(f(x+y-z) - f(x) - f(y) + f(z))\| \\
&\qquad + \|\rho_2(f(x-y-z) - f(x) + f(y) + f(z))\| + \varphi(x,y,z), \quad (26)\\
&\|g(\lambda(x+y+z)) - \lambda(g(x) + g(y) + g(z))\| \\
&\quad \le \|\rho_1(g(x+y-z) - g(x) - g(y) + g(z))\| \\
&\qquad + \|\rho_2(g(x-y-z) - g(x) + g(y) + g(z))\| + \varphi(x,y,z), \quad (27)
\end{aligned}$$

$$\|g(xy) - g(x)g(y)\| \le \varphi(x, y, 0), \tag{28}$$

$$\|f(xyz) - f(x)g(yz) - g(x)f(y)g(z) - g(xy)f(z)\| \le \varphi(x, y, z) \tag{29}$$

for all $\lambda \in \mathbb{T}^1 := \{\mu \in \mathbb{C} : |\mu| = 1\}$ *and all* $x, y, z \in A$. *Then there exist a unique homomorphism* $h : A \to A$ *and a unique hom-derivation* $D : A \to A$ *such that*

$$\|f(x) - D(x)\| \le \frac{1}{2}\Psi(x, 0, x), \tag{30}$$

$$\|g(x) - h(x)\| \le \frac{1}{2}\Psi(x, 0, x), \tag{31}$$

$$D(xyz) = D(x)h(yz) + h(x)D(y)h(z) + h(xy)D(z) \tag{32}$$

for all $x, y, z \in A$, *where*

$$\Psi(x, y, z) := \sum_{j=1}^{\infty} 2^j \varphi\left(\frac{x}{2^j}, \frac{y}{2^j}, \frac{z}{2^j}\right) < \infty$$

for all $x, y, z \in A$.

Proof. Let $\lambda = 1$ in (26) and (27). By the same reasoning as in the proof of Theorem 3, there are unique additive mappings $h, D : A \to A$ satisfying (30) and (31), respectively, which are given by

$$h(x) = \lim_{n\to\infty} 2^n g\left(\frac{x}{2^n}\right),$$

$$D(x) = \lim_{n\to\infty} 2^n f\left(\frac{x}{2^n}\right)$$

for all $x \in A$.

Letting $y = z = 0$ in (26), we get

$$\|f(\lambda x) - \lambda f(x)\| \le \varphi(x, 0, 0)$$

for all $\lambda \in \mathbb{T}^1$ and all $x \in A$. So,

$$\begin{aligned}\|D(\lambda x) - \lambda D(x)\| &\le 2^n \left\|f\left(\lambda\frac{x}{2^n}\right) - \lambda f\left(\frac{x}{2^n}\right)\right\| \\ &\le 2^n \varphi\left(\frac{x}{2^n}, 0, 0\right) \le \frac{2^n L^n}{8^n}\varphi(x, 0, 0)\end{aligned}$$

and so $D(\lambda x) = \lambda D(x)$ for all $\lambda \in \mathbb{T}^1$ and all $x \in A$. By the same reasoning as in the proof of [19, Theorem 2.1], the mapping $D : A \to A$ is $\mathbb{C}$-linear.

Similarly, one can show that the additive mapping $h : A \to A$ is $\mathbb{C}$-linear.

It follows from (28) that

$$\begin{aligned}
&\|h(xy) - h(x)h(y)\| \\
&= \lim_{n\to\infty} 4^n \left\| g\left(\frac{xy}{4^n}\right) - g\left(\frac{x}{2^n}\right) g\left(\frac{y}{2^n}\right)\right\| \\
&\leq \lim_{n\to\infty} 4^n \varphi\left(\frac{x}{2^n}, \frac{y}{2^n}, 0\right) \leq \lim_{n\to\infty} \frac{4^n L^n}{8^n} \varphi(x, y, 0) = 0
\end{aligned}$$

for all $x, y \in A$. So, $h(xy) = h(x)h(y)$ for all $x, y \in A$. Thus, the $\mathbb{C}$-linear mapping $h : A \to A$ is a homomorphism satisfying (31).

It follows from (29) that

$$\begin{aligned}
&\|D(xyz) - D(x)h(yz) - h(x)D(y)h(z) - h(xy)D(z)\| \\
&= \lim_{n\to\infty} 8^n \left\| f\left(\frac{xyz}{8^n}\right) - f\left(\frac{x}{2^n}\right) g\left(\frac{yz}{4^n}\right) - g\left(\frac{x}{2^n}\right) f\left(\frac{y}{2^n}\right) g\left(\frac{z}{2^n}\right)\right. \\
&\quad \left. - g\left(\frac{xy}{4^n}\right) f\left(\frac{z}{2^n}\right)\right\| \\
&\leq \lim_{n\to\infty} 8^n \varphi\left(\frac{x}{2^n}, \frac{y}{2^n}, \frac{z}{2^n}\right) \leq \lim_{n\to\infty} \frac{8^n L^n}{8^n} \varphi(x, y, z) = 0
\end{aligned}$$

for all $x, y, z \in A$. So $D(xyz) = D(x)h(yz) + h(x)D(y)h(z) + h(xy)D(z)$ for all $x, y, z \in A$. Thus, the $\mathbb{C}$-linear mapping $D : A \to A$ is a hom-derivation satisfying (30) and (32). □

Corollary 13. *Let $r > 3$ and θ be non-negative real numbers, and let $f, g : A \to A$ be mappings satisfying $f(0) = g(0) = 0$ and*

$$\begin{aligned}
&\|f(\lambda(x + y + z)) - \lambda(f(x) + f(y) + f(z))\| \\
&\quad \leq \|\rho_1(f(x + y - z) - f(x) - f(y) + f(z))\| \\
&\qquad + \|\rho_2(f(x - y - z) - f(x) + f(y) + f(z))\| \\
&\qquad + \theta(\|x\|^r + \|y\|^r + \|z\|^r), \qquad (33) \\
&\|g(\lambda(x + y + z)) - \lambda(g(x) + g(y) + g(z))\| \\
&\quad \leq \|\rho_1(g(x + y - z) - g(x) - g(y) + g(z))\|
\end{aligned}$$

$$+\|\rho_2(g(x-y-z)-g(x)+g(y)+g(z))\|$$
$$+\theta(\|x\|^r+\|y\|^r+\|z\|^r), \tag{34}$$
$$\|g(xy)-g(x)g(y)\| \le \theta(\|x\|^r+\|y\|^r), \tag{35}$$
$$\|f(xyz)-f(x)g(yz)-g(x)f(y)g(z)-g(xy)f(z)\|$$
$$\le \theta(\|x\|^r+\|y\|^r+\|z\|^r) \tag{36}$$

for all $\lambda \in \mathbb{T}^1$ *and all* $x, y, z \in A$. *Then there exist a unique homomorphism* $h : A \to A$ *and a unique hom-derivation* $D : A \to A$ *satisfying* (32) *and*

$$\|f(x)-D(x)\| \le \frac{2\theta}{2^r-2}\|x\|^r,$$
$$\|g(x)-h(x)\| \le \frac{2\theta}{2^r-2}\|x\|^r$$

for all $x \in A$.

Proof. The proof follows from Theorem 12 by taking $\varphi(x,y,z) = \theta(\|x\|^r+\|y\|^r+\|z\|^r)$ for all $x, y, z \in A$. Choosing $L = 2^{1-r}$, we obtain the desired result. □

Theorem 14. *Let* $\varphi : A^3 \to [0,\infty)$ *be a function such that*

$$\Psi(x,y,z) := \sum_{j=0}^{\infty} \frac{1}{2^j}\varphi(2^j x, 2^j y, 2^j z) < \infty \tag{37}$$

for all $x, y, z \in X$. *Let* $f, g : A \to A$ *be mappings satisfying* $f(0) = g(0) = 0$, (26), (27), (28), *and* (29). *Then there exist a unique homomorphism* $h : A \to A$ *and a unique hom-derivation* $D : A \to A$ *satisfying* (32) *and*

$$\|f(x)-D(x)\| \le \frac{1}{2}\Psi(x,0,x),$$
$$\|g(x)-h(x)\| \le \frac{1}{2}\Psi(x,0,x)$$

for all $x \in A$.

Proof. The proof is similar to the proof of Theorem 12. □

Corollary 15. *Let $r < 1$ and θ be positive real numbers, and let $f, g : A \to A$ be mappings satisfying $f(0) = g(0) = 0$, (33), (34), (35), and (36). Then there exist a unique homomorphism $h : A \to A$ and a unique hom-derivation $D : A \to A$ satisfying* (32) *and*

$$\|f(x) - D(x)\| \leq \frac{2\theta}{2 - 2^r}\|x\|^r,$$

$$\|g(x) - h(x)\| \leq \frac{2\theta}{2 - 2^r}\|x\|^r$$

for all $x \in A$.

Proof. The proof follows from Theorem 14 by taking $\varphi(x, y, z) = \theta(\|x\|^r + \|y\|^r + \|z\|^r)$ for all $x, y, z \in X$. Choosing $L = 2^{r-1}$, we obtain the desired result. □

5. Stability of hom-Derivations in Banach Algebras Associated with (2)

Using the fixed point method, we prove the Hyers–Ulam stability of hom-derivations in Banach algebras, associated with the additive (ρ_1, ρ_2)-functional inequality (1).

Theorem 16. *Let $\varphi : A^3 \to [0, \infty)$ be a function such that*

$$\sum_{j=1}^{\infty} 8^j \varphi\left(\frac{x}{2^j}, \frac{y}{2^j}, \frac{z}{2^j}\right) < \infty$$

for all $x, y, z \in A$. Let $f, g : A \to A$ be mappings satisfying $f(0) = g(0) = 0$, (28), (29), and

$$\begin{aligned}
&\|f(\lambda(x + y - z)) - \lambda(f(x) + f(y) - f(z))\| \\
&\quad \leq \|\rho_1(f(x + y + z) - f(x) - f(y) - f(z))\| \\
&\qquad + \|\rho_2(f(x - y - z) - f(x) + f(y) + f(z))\| + \varphi(x, y, z), \quad (38)\\
&\|g(\lambda(x + y - z)) - \lambda(g(x) + g(y) - g(z))\| \\
&\quad \leq \|\rho_1(g(x + y + z) - g(x) - g(y) - g(z))\| \\
&\qquad + \|\rho_2(g(x - y - z) - g(x) + g(y) + g(z))\| + \varphi(x, y, z) \quad (39)
\end{aligned}$$

for all $\lambda \in \mathbb{T}^1$ and all $x, y, z \in A$. Then there exist a unique homomorphism $h : A \to A$ and a unique hom-derivation $D : A \to A$ satisfying (32) and

$$\|f(x) - D(x)\| \leq \frac{1}{2(1 - |\rho_1|)}\Psi(x, x, 0), \tag{40}$$

$$\|g(x) - h(x)\| \leq \frac{1}{2(1 - |\rho_1|)}\Psi(x, x, 0) \tag{41}$$

for all $x \in A$, where

$$\Psi(x, y, z) := \sum_{j=1}^{\infty} 2^j \varphi\left(\frac{x}{2^j}, \frac{y}{2^j}, \frac{z}{2^j}\right)$$

for all $x, y, z \in A$.

Proof. Let $\lambda = 1$ in (38) and (39). By the same reasoning as in the proof of Theorem 12, there are unique additive mappings $h, D : A \to A$ satisfying (40) and (41), respectively, which are given by

$$h(x) = \lim_{n\to\infty} 2^n g\left(\frac{x}{2^n}\right),$$

$$D(x) = \lim_{n\to\infty} 2^n f\left(\frac{x}{2^n}\right)$$

for all $x \in A$.

Letting $y = z = 0$ in (38), we get

$$\|f(\lambda x) - \lambda f(x)\| \leq \varphi(x, 0, 0)$$

for all $\lambda \in \mathbb{T}^1$ and all $x \in A$. So,

$$\begin{aligned}\|D(\lambda x) - \lambda D(x)\| &\leq 2^n \left\| f\left(\lambda \frac{x}{2^n}\right) - \lambda f\left(\frac{x}{2^n}\right)\right\| \\ &\leq 2^n \varphi\left(\frac{x}{2^n}, 0, 0\right) \leq \frac{2^n L^n}{8^n}\varphi(x, 0, 0)\end{aligned}$$

and so $D(\lambda x) = \lambda D(x)$ for all $\lambda \in \mathbb{T}^1$ and all $x \in A$. By the same reasoning as in the proof of [19, Theorem 2.1], the mapping $D : A \to A$ is $\mathbb{C}$-linear.

Similarly, one can show that the additive mapping $h : A \to A$ is $\mathbb{C}$-linear.

The rest of the proof is similar to the proof of Theorem 12. □

Corollary 17. *Let $r > 3$ and θ be non-negative real numbers, and let $f, g : A \to A$ be mappings satisfying $f(0) = g(0) = 0$, (35), (36), and*

$$\begin{aligned}
&\|f(\lambda(x+y-z)) - \lambda(f(x)+f(y)-f(z))\| \\
&\quad \le \|\rho_1(f(x+y+z) - f(x) - f(y) - f(z))\| \\
&\qquad + \|\rho_2(f(x-y-z) - f(x) + f(y) + f(z))\| \\
&\qquad + \theta(\|x\|^r + \|y\|^r + \|z\|^r), \qquad (42) \\
&\|g(\lambda(x+y-z)) - \lambda(g(x)+g(y)-g(z))\| \\
&\quad \le \|\rho_1(g(x+y+z) - g(x) - g(y) - g(z))\| \\
&\qquad + \|\rho_2(g(x-y-z) - g(x) + g(y) + g(z))\| \\
&\qquad + \theta(\|x\|^r + \|y\|^r + \|z\|^r) \qquad (43)
\end{aligned}$$

for all $\lambda \in \mathbb{T}^1$ and all $x, y, z \in A$. Then there exist a unique homomorphism $h : A \to A$ and a unique hom-derivation $D : A \to A$ satisfying (32) *and*

$$\|f(x) - D(x)\| \le \frac{2\theta}{(2^r - 2)(1 - |\rho_1|)}\|x\|^r,$$

$$\|g(x) - h(x)\| \le \frac{2\theta}{(2^r - 2)(1 - |\rho_1|)}\|x\|^r$$

for all $x \in A$.

Proof. The proof follows from Theorem 16 by taking $\varphi(x, y, z) = \theta(\|x\|^r + \|y\|^r + \|z\|^r)$ for all $x, y, z \in A$. Choosing $L = 2^{1-r}$, we obtain the desired result. □

Theorem 18. *Let $\varphi : A^3 \to [0, \infty)$ be a function such that there exists an $L < 1$ with*

$$\Psi(x, y, z) := \sum_{j=0}^{\infty} \frac{1}{2^j}\varphi(2^j x, 2^j y, 2^j z) < \infty$$

for all $x, y, z \in X$. Let $f, g : A \to A$ be mappings satisfying $f(0) = g(0) = 0$, (28), (29), (38), *and* (39). *Then there exist a unique* *homomorphism $h : A \to A$ and a unique hom-derivation $D : A \to A$ satisfying* (32) *and*

$$\|f(x) - D(x)\| \leq \frac{1}{2(1 - |\rho_1|)} \Psi(x, x, 0),$$

$$\|g(x) - h(x)\| \leq \frac{1}{2(1 - |\rho_1|)} \Psi(x, x, 0)$$

for all $x \in A$.

Proof. The proof is similar to the proof of Theorem 16. □

Corollary 19. *Let $r < 1$ and θ be positive real numbers, and let $f, g : A \to A$ be mappings satisfying $f(0) = g(0) = 0$,* (35), (36), (42), *and* (43). *Then there exist a unique homomorphism $h : A \to A$ and a unique hom-derivation $D : A \to A$ satisfying* (32) *and*

$$\|f(x) - D(x)\| \leq \frac{2\theta}{(2 - 2^r)(1 - |\rho_1|)} \|x\|^r,$$

$$\|g(x) - h(x)\| \leq \frac{2\theta}{(2 - 2^r)(1 - |\rho_1|)} \|x\|^r$$

for all $x \in A$.

Proof. The proof follows from Theorem 18 by taking $\varphi(x, y, z) = \theta(\|x\|^r + \|y\|^r + \|z\|^r)$ for all $x, y, z \in X$. Choosing $L = 2^{r-1}$, we obtain the desired result. □

References

[1] Aoki, T. On the stability of the linear transformation in Banach spaces, *J. Math. Soc. Japan* 2, 64–66 (1950).

[2] Cădariu, L., Găvruta, L., and Găvruta, P. On the stability of an affine functional equation, *J. Nonlinear Sci. Appl.* 6, 60–67 (2013).

[3] Cădariu, L. and Radu, V. Fixed points and the stability of Jensen's functional equation, *J. Inequal. Pure Appl. Math.* 4(1), 1–7 (2003).

[4] Cădariu, L. and Radu, V. On the stability of the Cauchy functional equation: a fixed point approach, *Grazer Math. Ber.* 346, 43–52 (2004).

[5] Cădariu, L., and Radu, V. Fixed point methods for the generalized stability of functional equations in a single variable, *Fixed Point Theory Appl.* 2008, 749392, 1–15 (2008).

[6] Chahbi, A. and Bounader, N. On the generalized stability of d'Alembert functional equation, *J. Nonlinear Sci. Appl.* 6, 198–204 (2013).

[7] Cholewa, P.W. Remarks on the stability of functional equations, *Aequationes Math.* 27, 76–86 (1984).

[8] Diaz, J. and Margolis, B. A fixed point theorem of the alternative for contractions on a generalized complete metric space, *Bull. Am. Math. Soc.* 74, 305–309 (1968).

[9] Eghbali, N., Rassias, J.M. and Taheri, M. On the stability of a k-cubic functional equation in intuitionistic fuzzy n-normed spaces, *Results Math.* 70, 233–248 (2016).

[10] Eskandani, G.Z. and Găvruta, P. Hyers-Ulam-Rassias stability of pexiderized Cauchy functional equation in 2-Banach spaces, *J. Nonlinear Sci. Appl.* 5, 459–465 (2012).

[11] Găvruta, P. A generalization of the Hyers-Ulam-Rassias stability of approximately additive mappings, *J. Math. Anal. Appl.* 184, 431–436 (1994).

[12] Hyers, D.H. On the stability of the linear functional equation, *Proc. Nat. Acad. Sci. U.S.A.* 27, 222–224 (1941).

[13] Isac, G. and Rassias, Th. M. Stability of ψ-additive mappings: Applications to nonlinear analysis, *Int. J. Math. Math. Sci.* 19, 219–228 (1996).

[14] Kadison, R.V. and Ringrose, J.R. *Fundamentals of the Theory of Operator Algebras: Elementary Theory*. Academic Press, New York (1983).

[15] Khodaei, H. On the stability of additive, quadratic, cubic and quartic set-valued functional equations, *Results Math.* 68, 1–10 (2015).

[16] Miheţ, D. and Radu, V. On the stability of the additive Cauchy functional equation in random normed spaces, *J. Math. Anal. Appl.* 343, 567–572 (2008).

[17] Mirzavaziri, M. and Moslehian, M.S. Automatic continuity of σ-derivations on C^*-algebras, *Proc. Am. Math. Soc.* 134, 3319–3327 (2006).

[18] Mirzavaziri, M. and Moslehian, M.S. σ-Derivations in Banach algebras, *Bull. Iran. Math. Soc.* 32, 65–78 (2007).

[19] Park, C. Homomorphisms between Poisson JC^*-algebras, *Bull. Braz. Math. Soc.* 36, 79–97 (2005).

[20] Park, C. Additive ρ-functional inequalities and equations, *J. Math. Inequal.* 9, 17–26 (2015).

[21] Park, C. Additive ρ-functional inequalities in non-Archimedean normed spaces, *J. Math. Inequal.* 9, 397–407 (2015).

[22] Radu, V. The fixed point alternative and the stability of functional equations, *Fixed Point Theory,* 4, 91–96 (2003).

[23] Rassias, Th. M. On the stability of the linear mapping in Banach spaces, *Proc. Am. Math. Soc.* 72, 297–300 (1978).
[24] Skof, F. Propriet locali e approssimazione di operatori, *Rend. Sem. Mat. Fis. Milano* 53, 113–129 (1983).
[25] Ulam, S.M. *A Collection of the Mathematical Problems.* Interscience Publ., New York (1960).
[26] Wang, Z. Stability of two types of cubic fuzzy set-valued functional equations, *Results Math.* 70, 1–14 (2016).

https://doi.org/10.1142/9789811293009_0024

Chapter 24

An Additive (ρ_1, ρ_2)-Functional Inequality in Complex Banach Spaces

Jung Rye Lee[*,§], Choonkil Park[†,¶], and Michael Th. Rassias[‡,||]

[*] *Department of Data Science, Daejin University, Kyunggi 11159, Korea*

[†] *Department of Mathematics, Hanyang University, Seoul 04763, Korea*

[‡] *Department of Mathematics and Engineering Sciences, Hellenic Military Academy, 16673 Vari Attikis, Greece*

[§] *jrlee@daejin.ac.kr*

[¶] *baak@hanyang.ac.kr*

[||] *mthrassias@yahoo.com*

In this chapter, we introduce and solve the following additive (ρ_1, ρ_2)-functional inequality

$$\|f(x+y) - f(x) - f(y)\| \leq \|\rho_1(f(x+z) - f(x) - f(z))\| + \|\rho_2(f(y+z) - f(y) - f(z))\|, \qquad (1)$$

where ρ_1 and ρ_2 are fixed nonzero complex numbers with $|\rho_1| + |\rho_2| < 1$. Using the fixed point method and the direct method, we prove the Hyers–Ulam stability of the additive (ρ_1, ρ_2)-functional inequality (1) in complex Banach spaces.

1. Introduction and Preliminaries

The stability problem of functional equations originated from a question of Ulam [39] concerning the stability of group homomorphisms. The functional equation $f(x+y) = f(x) + f(y)$ is called the *Cauchy equation*. In particular, every solution of the Cauchy equation is said to be an *additive mapping*. Hyers [17] gave the first affirmative partial answer to the question of Ulam for Banach spaces. Hyers' theorem was generalized by Aoki [4] for additive mappings and by Rassias [36] for linear mappings by considering an unbounded Cauchy difference. A generalization of the Rassias theorem was obtained by Găvruta [16] by replacing the unbounded Cauchy difference by a general control function in the spirit of Rassias' approach. The stability of quadratic functional equation was proved by Skof [37] for mappings $f : E_1 \to E_2$, where E_1 is a normed space and E_2 is a Banach space. Cholewa [9] noted that the theorem of Skof is still true if the relevant domain E_1 is replaced by an Abelian group.

Park [32,33] defined additive ρ-functional inequalities and proved the Hyers–Ulam stability of the additive ρ-functional inequalities in Banach spaces and non-Archimedean Banach spaces. The stability problems of various functional equations have been extensively investigated by a number of authors [1–3, 5, 10, 11, 13–15, 18, 19, 21–29, 31, 34, 38].

We recall a fundamental result in fixed point theory.

Theorem 1 ([6, 12]). *Let (X, d) be a complete generalized metric space and let $J : X \to X$ be a strictly contractive mapping with Lipschitz constant $\alpha < 1$. Then for each given element $x \in X$, either*

$$d(J^n x, J^{n+1} x) = \infty$$

for all non-negative integers n or there exists a positive integer n_0 such that

(1) $d(J^n x, J^{n+1} x) < \infty, \qquad \forall n \geq n_0;$
(2) *the sequence $\{J^n x\}$ converges to a fixed point y^* of J;*
(3) *y^* is the unique fixed point of J in the set $Y = \{y \in X \mid d(J^{n_0} x, y) < \infty\}$;*
(4) $d(y, y^*) \leq \frac{1}{1-\alpha} d(y, Jy)$ *for all $y \in Y$.*

In 1996, Isac and Rassias [20] were the first to provide applications of stability theory of functional equations for the proof of new fixed point theorems with applications. By using fixed point methods, the stability problems of several functional equations have been extensively investigated by a number of authors [7, 8, 35].

In Section 2, we solve the additive (ρ_1, ρ_2)-functional inequality (1) and prove the Hyers–Ulam stability of the additive (ρ_1, ρ_2)-functional inequality (1) in Banach spaces by using the fixed point method. In Section 3, we prove the Hyers–Ulam stability of the additive (ρ_1, ρ_2)-functional inequality (1) in Banach spaces by using the direct method.

Throughout this chapter, let X be a real or complex normed space with norm $\|\cdot\|$ and Y a complex Banach space with norm $\|\cdot\|$. Assume that ρ_1 and ρ_2 are fixed nonzero complex numbers with $|\rho_1|+|\rho_2| < 1$.

2. Additive (ρ_1, ρ_2)-Functional Inequality (1): A Fixed Point Method

In this section, we solve and investigate the additive (ρ_1, ρ_2)-functional inequality (1) in complex Banach spaces.

Lemma 2. *If a mapping $f : X \to Y$ satisfies*

$$\begin{aligned} \|f(x+y) - f(x) - f(y)\| \le \|\rho_1(f(x+z) - f(x) - f(z))\| \\ + \|\rho_2(f(y+z) - f(y) - f(z))\| \end{aligned} \tag{2}$$

for all $x, y, z \in X$, then $f : X \to Y$ is additive.

Proof. Assume that $f : X \to Y$ satisfies (2).

Letting $x = y = z = 0$ in (2), we get $\|f(0)\| \le (|\rho_1| + |\rho_2|)\|f(0)\|$ and so $f(0) = 0$, since $|\rho_1| + |\rho_2| < 1$.

Letting $z = 0$ in (2), we get

$$\|f(x+y) - f(x) - f(y)\| \le 0$$

for all $x, y \in X$. So, $f(x+y) = f(x) + f(y)$ for all $x, y \in X$. Thus, f is additive. □

Using the fixed point method, we prove the Hyers–Ulam stability of the additive (ρ_1, ρ_2)-functional inequality (2) in complex Banach spaces.

Theorem 3. *Let $\varphi : X^3 \to [0, \infty)$ be a function such that there exists an $L < 1$ with*

$$\varphi\left(\frac{x}{2}, \frac{y}{2}, \frac{z}{2}\right) \leq \frac{L}{2}\varphi(x, y, z) \tag{3}$$

for all $x, y, z \in X$. Let $f : X \to Y$ be a mapping satisfying $f(0) = 0$ and

$$\begin{aligned}\|f(x+y) - f(x) - f(y)\| \leq{} & \|\rho_1(f(x+z) - f(x) - f(z))\| \\ & + \|\rho_2(f(y+z) - f(y) - f(z))\| \\ & + \varphi(x, y, z) \end{aligned} \tag{4}$$

for all $x, y, z \in X$. Then there exists a unique additive mapping $A : X \to Y$ such that

$$\|f(x) - A(x)\| \leq \frac{L}{2(1-L)}\varphi(x, x, 0)$$

for all $x \in X$.

Proof. Letting $y = x$ and $z = 0$ in (4), we get

$$\|f(2x) - 2f(x)\| \leq \varphi(x, x, 0) \tag{5}$$

for all $x \in X$.

Consider the set

$$S := \{h : X \to Y, \quad h(0) = 0\}$$

and introduce the generalized metric on S:

$$d(g, h) = \inf\{\mu \in \mathbb{R}_+ : \|g(x) - h(x)\| \leq \mu\varphi(x, x, 0), \quad \forall x \in X\},$$

where, as usual, $\inf \phi = +\infty$. It is easy to show that (S, d) is complete (see [30]).

Now, we consider the linear mapping $J : S \to S$ such that

$$Jg(x) := 2g\left(\frac{x}{2}\right)$$

for all $x \in X$.

Let $g, h \in S$ be given such that $d(g, h) = \varepsilon$. Then,

$$\|g(x) - h(x)\| \le \varepsilon\varphi(x, x, 0)$$

for all $x \in X$. Hence,

$$\begin{aligned}\|Jg(x) - Jh(x)\| &= \left\|2g\left(\frac{x}{2}\right) - 2h\left(\frac{x}{2}\right)\right\| \le 2\varepsilon\varphi\left(\frac{x}{2}, \frac{x}{2}, 0\right) \\ &\le 2\varepsilon\frac{L}{2}\varphi(x, x, 0) = L\varepsilon\varphi(x, x, 0)\end{aligned}$$

for all $x \in X$. So, $d(g, h) = \varepsilon$ implies that $d(Jg, Jh) \le L\varepsilon$. This means that

$$d(Jg, Jh) \le Ld(g, h)$$

for all $g, h \in S$.

It follows from (5) that

$$\left\|f(x) - 2f\left(\frac{x}{2}\right)\right\| \le \varphi\left(\frac{x}{2}, \frac{x}{2}, 0\right) \le \frac{L}{2}\varphi(x, x, 0)$$

for all $x \in X$ So $d(f, Jf) \le \frac{L}{2}$.

By Theorem 1, there exists a mapping $A : X \to Y$ satisfying the following:

(1) A is a fixed point of J, i.e.,

$$A(x) = 2A\left(\frac{x}{2}\right) \tag{6}$$

for all $x \in X$. The mapping A is a unique fixed point of J in the set

$$M = \{g \in S : d(f, g) < \infty\}.$$

This implies that A is a unique mapping satisfying (6) such that there exists a $\mu \in (0, \infty)$ satisfying

$$\|f(x) - A(x)\| \le \mu\varphi(x, x, 0)$$

for all $x \in X$.

(2) $d(J^l f, A) \to 0$ as $l \to \infty$. This implies the equality

$$\lim_{l\to\infty} 2^n f\left(\frac{x}{2^n}\right) = A(x)$$

for all $x \in X$.

(3) $d(f, A) \le \frac{1}{1-L} d(f, Jf)$, which implies

$$\|f(x) - A(x)\| \le \frac{L}{2(1-L)}\varphi(x, x, 0)$$

for all $x \in X$.

It follows from (3) and (4) that

$$\begin{aligned}
&\|A(x+y) - A(x) - A(y)\| \\
&= \lim_{n\to\infty} 2^n \left\| f\left(\frac{x+y}{2^n}\right) - f\left(\frac{x}{2^n}\right) - f\left(\frac{y}{2^n}\right)\right\| \\
&\le \lim_{n\to\infty} 2^n |\rho_1| \left\| f\left(\frac{x+z}{2^n}\right) - f\left(\frac{x}{2^n}\right) - f\left(\frac{z}{2^n}\right)\right\| \\
&\quad + \lim_{n\to\infty} 2^n |\rho_2| \left\| f\left(\frac{y+z}{2^n}\right) - f\left(\frac{y}{2^n}\right) - f\left(\frac{z}{2^n}\right)\right\| \\
&\quad + \lim_{n\to\infty} 2^n \varphi\left(\frac{x}{2^n}, \frac{y}{2^n}, \frac{z}{2^n}\right) \\
&= \|\rho_1(A(x+z) - A(x) - A(z))\| + \|\rho_2(A(y+z) - A(y) - A(z))\|
\end{aligned}$$

for all $x, y, z \in X$. So,

$$\begin{aligned}
\|A(x+y) - A(x) - A(y)\| \le \|\rho_1(A(x+z) - A(x) - A(z))\| \\
+ \|\rho_2(A(y+z) - A(y) - A(z))\|
\end{aligned}$$

for all $x, y, z \in X$. By Lemma 2, the mapping $A : X \to Y$ is additive. □

Corollary 4. *Let $r > 1$ and θ be non-negative real numbers, and let $f : X \to Y$ be a mapping satisfying $f(0) = 0$ and*

$$\begin{aligned}\|f(x+y) - f(x) - f(y)\| \leq \|\rho_1(f(x+z) - f(x) - f(z))\| \\ + \|\rho_2(f(y+z) - f(y) - f(z))\| \\ + \theta(\|x\|^r + \|y\|^r + \|z\|^r) \qquad (7)\end{aligned}$$

for all $x, y, z \in X$. Then there exists a unique additive mapping $A : X \to Y$ such that

$$\|f(x) - A(x)\| \leq \frac{2\theta}{2^r - 2}\|x\|^r$$

for all $x \in X$.

Proof. The proof follows from Theorem 3 by taking $L = 2^{1-r}$ and $\varphi(x, y, z) = \theta(\|x\|^r + \|y\|^r + \|z\|^r)$ for all $x, y, z \in X$. □

Theorem 5. *Let $\varphi : X^3 \to [0, \infty)$ be a function such that there exists an $L < 1$ with*

$$\varphi(x, y, z) \leq 2L\varphi\left(\frac{x}{2}, \frac{y}{2}, \frac{z}{2}\right) \qquad (8)$$

for all $x, y, z \in X$. Let $f : X \to Y$ be a mapping satisfying $f(0) = 0$ and (4). Then there exists a unique additive mapping $A : X \to Y$ such that

$$\|f(x) - A(x)\| \leq \frac{1}{2(1-L)}\varphi(x, x, 0)$$

for all $x \in X$.

Proof. Let (S, d) be the generalized metric space defined in the proof of Theorem 3.

Now, we consider the linear mapping $J : S \to S$ such that

$$Jg(x) := \frac{1}{2}g(2x)$$

for all $x \in X$.

It follows from (5) that

$$\left\|f(x) - \frac{1}{2}f(2x)\right\| \leq \frac{1}{2}\varphi(x, x, 0)$$

for all $x \in X$.

The rest of the proof is similar to the proof of Theorem 3. □

Corollary 6. *Let $r < 1$ and θ be positive real numbers, and let $f : X \to Y$ be a mapping satisfying $f(0) = 0$ and (7). Then there exists a unique additive mapping $A : X \to Y$ such that*

$$\|f(x) - A(x)\| \leq \frac{2\theta}{2 - 2^r}\|x\|^r$$

for all $x \in X$.

Proof. The proof follows from Theorem 5 by taking $L = 2^{r-1}$ and $\varphi(x, y, z) = \theta(\|x\|^r + \|y\|^r + \|z\|^r)$ for all $x, y, z \in X$. □

Theorem 7. *Let $\varphi : X^3 \to [0, \infty)$ be a function such that there exists an $L < 1$ satisfying (3). Let $f : X \to Y$ be a mapping satisfying $f(0) = 0$ and (4). Then there exists a unique additive mapping $A : X \to Y$ such that*

$$\|f(x) - A(x)\| \leq \frac{L}{2(1 - L)(1 - |\rho_1| - |\rho_2|)}\varphi(x, x, x)$$

for all $x \in X$.

Proof. Letting $y = z = x$ in (4), we get

$$\|f(2x) - 2f(x)\| \leq (|\rho_1| + |\rho_2|)\|f(2x) - 2f(x)\| + \varphi(x, x, x)$$

and so

$$\|f(2x) - 2f(x)\| \leq \frac{1}{1 - |\rho_1| - |\rho_2|}\varphi(x, x, x) \tag{9}$$

for all $x \in X$.

Consider the set

$$S := \{h : X \to Y, \quad h(0) = 0\}$$

and introduce the generalized metric on S:

$$d(g, h) = \inf\{\mu \in \mathbb{R}_+ : \|g(x) - h(x)\| \leq \mu\varphi(x, x, x), \quad \forall x \in X\},$$

where, as usual, $\inf \phi = +\infty$. It is easy to show that (S, d) is complete (see [30]).

Now, we consider the linear mapping $J : S \to S$ such that

$$Jg(x) := 2g\left(\frac{x}{2}\right)$$

for all $x \in X$.

Let $g, h \in S$ be given such that $d(g,h) = \varepsilon$. Then,

$$\|g(x) - h(x)\| \leq \varepsilon\varphi(x,x,x)$$

for all $x \in X$. Hence,

$$\begin{aligned}\|Jg(x) - Jh(x)\| &= \left\|2g\left(\frac{x}{2}\right) - 2h\left(\frac{x}{2}\right)\right\| \leq 2\varepsilon\varphi\left(\frac{x}{2}, \frac{x}{2}, \frac{x}{2}\right) \\ &\leq 2\varepsilon\frac{L}{2}\varphi(x,x,x) = L\varepsilon\varphi(x,x,x)\end{aligned}$$

for all $x \in X$. So, $d(g,h) = \varepsilon$ implies that $d(Jg, Jh) \leq L\varepsilon$. This means that

$$d(Jg, Jh) \leq Ld(g,h)$$

for all $g, h \in S$.

It follows from (9) that

$$\begin{aligned}\left\|f(x) - 2f\left(\frac{x}{2}\right)\right\| &\leq \frac{1}{1 - |\rho_1| - |\rho_2|}\varphi\left(\frac{x}{2}, \frac{x}{2}, \frac{x}{2}\right) \\ &\leq \frac{L}{2(1 - |\rho_1| - |\rho_2|)}\varphi(x,x,x)\end{aligned}$$

for all $x \in X$. So, $d(f, Jf) \leq \frac{L}{2(1-|\rho_1|-|\rho_2|)}$.

By Theorem 1, there exists a mapping $A : X \to Y$ satisfying the following:

(1) A is a fixed point of J, i.e.,

$$A(x) = 2A\left(\frac{x}{2}\right) \tag{10}$$

for all $x \in X$. The mapping A is a unique fixed point of J in the set

$$M = \{g \in S : d(f,g) < \infty\}.$$

This implies that A is a unique mapping satisfying (10) such that there exists a $\mu \in (0, \infty)$ satisfying

$$\|f(x) - A(x)\| \leq \mu\varphi(x,x,x)$$

for all $x \in X$.

(2) $d(J^l f, A) \to 0$ as $l \to \infty$. This implies the equality

$$\lim_{l\to\infty} 2^n f\left(\frac{x}{2^n}\right) = A(x)$$

for all $x \in X$.

(3) $d(f, A) \leq \frac{1}{1-L} d(f, Jf)$, which implies

$$\|f(x) - A(x)\| \leq \frac{L}{2(1-L)(1-|\rho_1|-|\rho_2|)} \varphi(x, x, x)$$

for all $x \in X$.

It follows from (3) and (4) that

$$\begin{aligned}
&\|A(x+y) - A(x) - A(y)\| \\
&= \lim_{n\to\infty} 2^n \left\| f\left(\frac{x+y}{2^n}\right) - f\left(\frac{x}{2^n}\right) - f\left(\frac{y}{2^n}\right) \right\| \\
&\leq \lim_{n\to\infty} 2^n |\rho_1| \left\| f\left(\frac{x+z}{2^n}\right) - f\left(\frac{x}{2^n}\right) - f\left(\frac{z}{2^n}\right) \right\| \\
&+ \lim_{n\to\infty} 2^n |\rho_2| \left\| f\left(\frac{y+z}{2^n}\right) - f\left(\frac{y}{2^n}\right) - f\left(\frac{z}{2^n}\right) \right\| \\
&+ \lim_{n\to\infty} 2^n \varphi\left(\frac{x}{2^n}, \frac{y}{2^n}, \frac{z}{2^n}\right) \\
&= \|\rho_1(A(x+z) - A(x) - A(z))\| + \|\rho_2(A(y+z) - A(y) - A(z))\|
\end{aligned}$$

for all $x, y, z \in X$. So,

$$\begin{aligned}
\|A(x+y) - A(x) - A(y)\| \leq{}& \|\rho_1(A(x+z) - A(x) - A(z))\| \\
&+ \|\rho_2(A(y+z) - A(y) - A(z))\|
\end{aligned}$$

for all $x, y, z \in X$. By Lemma 2, the mapping $A : X \to Y$ is additive. □

Corollary 8. *Let $r > 1$ and θ be non-negative real numbers, and let $f : X \to Y$ be a mapping satisfying $f(0) = 0$ and (7). Then there exists a unique additive mapping $A : X \to Y$ such that*

$$\|f(x) - A(x)\| \leq \frac{3\theta}{(2^r - 2)(1-|\rho_1|-|\rho_2|)} \|x\|^r$$

for all $x \in X$.

Proof. The proof follows from Theorem 7 by taking $L = 2^{1-r}$ and $\varphi(x, y, z) = \theta(\|x\|^r + \|y\|^r + \|z\|^r)$ for all $x, y, z \in X$. □

Theorem 9. *Let $\varphi : X^3 \to [0, \infty)$ be a function such that there exists an $L < 1$ satisfying (8). Let $f : X \to Y$ be a mapping satisfying $f(0) = 0$ and (4). Then there exists a unique additive mapping $A : X \to Y$ such that*

$$\|f(x) - A(x)\| \leq \frac{1}{2(1-L)(1-|\rho_1|-|\rho_2|)}\varphi(x, x, x)$$

for all $x \in X$.

Proof. Let (S, d) be the generalized metric space defined in the proof of Theorem 7.

Now we consider the linear mapping $J : S \to S$ such that

$$Jg(x) := \frac{1}{2}g(2x)$$

for all $x \in X$.

It follows from (9) that

$$\left\| f(x) - \frac{1}{2}f(2x) \right\| \leq \frac{1}{2(1-|\rho_1|-|\rho_2|)}\varphi(x, x, x)$$

for all $x \in X$.

The rest of the proof is similar to the proof of Theorem 7. □

Corollary 10. *Let $r < 1$ and θ be positive real numbers, and let $f : X \to Y$ be a mapping satisfying $f(0) = 0$ and (7). Then there exists a unique additive mapping $A : X \to Y$ such that*

$$\|f(x) - A(x)\| \leq \frac{3\theta}{(2-2^r)(1-|\rho_1|-|\rho_2|)}\|x\|^r$$

for all $x \in X$.

Proof. The proof follows from Theorem 9 by taking $L = 2^{r-1}$ and $\varphi(x, y, z) = \theta(\|x\|^r + \|y\|^r + \|z\|^r)$ for all $x, y, z \in X$. □

Remark 11. If ρ is a real number such that $|\rho_1| + |\rho_2| < 1$ and Y is a real Banach space, then all the assertions in this section remain valid.

3. Additive (ρ_1, ρ_2)-Functional Inequality (1): A Direct Method

In this section, we prove the Hyers–Ulam stability of the additive (ρ_1, ρ_2)-functional inequality (2) in complex Banach spaces by using the direct method.

Theorem 12. *Let $\varphi : X^3 \to [0, \infty)$ be a function such that*

$$\Psi(x, y, z) := \sum_{j=1}^{\infty} 2^j \varphi\left(\frac{x}{2^j}, \frac{y}{2^j}, \frac{z}{2^j}\right) < \infty \tag{11}$$

for all $x, y, z \in X$. Let $f : X \to Y$ be a mapping satisfying $f(0) = 0$ and (4). Then there exists a unique additive mapping $A : X \to Y$ such that

$$\|f(x) - A(x)\| \leq \frac{1}{2}\Psi(x, x, 0) \tag{12}$$

for all $x \in X$.

Proof. Letting $y = x$ and $z = 0$ in (4), we get

$$\|f(2x) - 2f(x)\| \leq \varphi(x, x, 0) \tag{13}$$

and so

$$\left\|f(x) - 2f\left(\frac{x}{2}\right)\right\| \leq \varphi\left(\frac{x}{2}, \frac{x}{2}, 0\right)$$

for all $x \in X$. Thus,

$$\begin{aligned}\left\|2^l f\left(\frac{x}{2^l}\right) - 2^m f\left(\frac{x}{2^m}\right)\right\| &\leq \sum_{j=l}^{m-1} \left\|2^j f\left(\frac{x}{2^j}\right) - 2^{j+1} f\left(\frac{x}{2^{j+1}}\right)\right\| \\ &\leq \sum_{j=l}^{m-1} 2^j \varphi\left(\frac{x}{2^{j+1}}, \frac{x}{2^{j+1}}, 0\right)\end{aligned} \tag{14}$$

for all non-negative integers m and l with $m > l$ and all $x \in X$. It follows from (14) that the sequence $\{2^k f(\frac{x}{2^k})\}$ is Cauchy for all

$x \in X$. Since Y is a Banach space, the sequence $\{2^k f(\frac{x}{2^k})\}$ converges. So, one can define the mapping $A : X \to Y$ by

$$A(x) := \lim_{k\to\infty} 2^k f\left(\frac{x}{2^k}\right)$$

for all $x \in X$. Moreover, letting $l = 0$ and passing the limit $m \to \infty$ in (14), we get (12).

It follows from (4) and (11) that

$$\begin{aligned}
&\|A(x+y) - A(x) - A(y)\| \\
&\quad = \lim_{n\to\infty} 2^n \left\| f\left(\frac{x+y}{2^n}\right) - f\left(\frac{x}{2^n}\right) - f\left(\frac{y}{2^n}\right)\right\| \\
&\quad \le \lim_{n\to\infty} 2^n |\rho_1| \left\| f\left(\frac{x+z}{2^n}\right) - f\left(\frac{x}{2^n}\right) - f\left(\frac{z}{2^n}\right)\right\| \\
&\quad + \lim_{n\to\infty} 2^n |\rho_2| \left\| f\left(\frac{y+z}{2^n}\right) - f\left(\frac{y}{2^n}\right) - f\left(\frac{z}{2^n}\right)\right\| \\
&\quad + \lim_{n\to\infty} 2^n \varphi\left(\frac{x}{2^n}, \frac{y}{2^n}, \frac{z}{2^n}\right) \\
&\quad = \|\rho_1(A(x+z) - A(x) - A(z))\| + \|\rho_2(A(y+z) - A(y) - A(z))\|
\end{aligned}$$

for all $x, y, z \in X$. So,

$$\begin{aligned}
\|A(x+y) - A(x) - A(y)\| \le \|\rho_1(A(x+z) - A(x) - A(z))\| \\
+ \|\rho_2(A(y+z) - A(y) - A(z))\|
\end{aligned}$$

for all $x, y, z \in X$. By Lemma 2, the mapping $A : X \to Y$ is additive.

Now, let $T : X \to Y$ be another additive mapping satisfying (12). Then we have

$$\begin{aligned}
\|A(x) - T(x)\| &= \left\|2^q A\left(\frac{x}{2^q}\right) - 2^q T\left(\frac{x}{2^q}\right)\right\| \\
&\le \left\|2^q A\left(\frac{x}{2^q}\right) - 2^q f\left(\frac{x}{2^q}\right)\right\| + \left\|2^q T\left(\frac{x}{2^q}\right) - 2^q f\left(\frac{x}{2^q}\right)\right\| \\
&\le 2^q \Psi\left(\frac{x}{2^q}, \frac{x}{2^q}, 0\right),
\end{aligned}$$

which tends to zero as $q \to \infty$ for all $x \in X$. So, we can conclude that $A(x) = T(x)$ for all $x \in X$. This proves the uniqueness of A. □

Corollary 13. *Let $r > 1$ and θ be non-negative real numbers, and let $f : X \to Y$ be a mapping satisfying $f(0) = 0$ and (7). Then there exists a unique additive mapping $A : X \to Y$ such that*

$$\|f(x) - A(x)\| \leq \frac{2\theta}{2^r - 2}\|x\|^r$$

for all $x \in X$.

Theorem 14. *Let $\varphi : X^3 \to [0, \infty)$ be a function and let $f : X \to Y$ be a mapping satisfying $f(0) = 0$, (4) and*

$$\Psi(x, y, z) := \sum_{j=0}^{\infty} \frac{1}{2^j}\varphi(2^j x, 2^j y, 2^j z) < \infty \tag{15}$$

for all $x, y, z \in X$. Then there exists a unique additive mapping $A : X \to Y$ such that

$$\|f(x) - A(x)\| \leq \frac{1}{2}\Psi(x, x, 0)$$

for all $x \in X$.

Proof. It follows from (13) that

$$\left\|f(x) - \frac{1}{2}f(2x)\right\| \leq \frac{1}{2}\varphi(x, x, 0)$$

for all $x \in X$. Hence,

$$\begin{aligned}\left\|\frac{1}{2^l}f(2^l x) - \frac{1}{2^m}f(2^m x)\right\| &\leq \sum_{j=l}^{m-1}\left\|\frac{1}{2^j}f\left(2^j x\right) - \frac{1}{2^{j+1}}f\left(2^{j+1}x\right)\right\| \\ &\leq \sum_{j=l}^{m-1}\frac{1}{2^{j+1}}\varphi(2^j x, 2^j x, 0)\end{aligned} \tag{16}$$

for all non-negative integers m and l with $m > l$ and all $x \in X$. It follows from (16) that the sequence $\{\frac{1}{2^n}f(2^n x)\}$ is a Cauchy sequence for all $x \in X$. Since Y is complete, the sequence $\{\frac{1}{2^n}f(2^n x)\}$ converges. So, one can define the mapping $A : X \to Y$ by

$$A(x) := \lim_{n\to\infty}\frac{1}{2^n}f(2^n x)$$

for all $x \in X$. Moreover, letting $l = 0$ and passing the limit $m \to \infty$ in (16), we get (16).

The rest of the proof is similar to the proof of Theorem 12. □

Corollary 15. *Let $r < 1$ and θ be positive real numbers, and let $f : X \to Y$ be a mapping satisfying $f(0) = 0$ and (7). Then there exists a unique additive mapping $A : X \to Y$ such that*

$$\|f(x) - A(x)\| \leq \frac{2\theta}{2 - 2^r}\|x\|^r$$

for all $x \in X$.

Theorem 16. *Let $\varphi : X^3 \to [0, \infty)$ be a function satisfying (11). Let $f : X \to Y$ be a mapping satisfying $f(0) = 0$ and (4). Then there exists a unique additive mapping $A : X \to Y$ such that*

$$\|f(x) - A(x)\| \leq \frac{1}{2(1 - |\rho_1| - |\rho_2|)}\Psi(x, x, x) \tag{17}$$

for all $x \in X$.

Proof. Letting $y = z = x$ in (4), we get

$$\|f(2x) - 2f(x)\| \leq (|\rho_1| + |\rho_2|)\|f(2x) - 2f(x)\| + \varphi(x, x, x)$$

and so

$$\|f(2x) - 2f(x)\| \leq \frac{1}{1 - |\rho_1| - |\rho_2|}\varphi(x, x, x) \tag{18}$$

for all $x \in X$. Thus,

$$\begin{aligned}\left\|2^l f\left(\frac{x}{2^l}\right) - 2^m f\left(\frac{x}{2^m}\right)\right\| &\leq \sum_{j=l}^{m-1}\left\|2^j f\left(\frac{x}{2^j}\right) - 2^{j+1} f\left(\frac{x}{2^{j+1}}\right)\right\| \\ &\leq \sum_{j=l}^{m-1}\frac{2^j}{1 - |\rho_1| - |\rho_2|}\varphi\left(\frac{x}{2^{j+1}}, \frac{x}{2^{j+1}}, \frac{x}{2^{j+1}}\right)\end{aligned} \tag{19}$$

for all non negative integers m and l with $m > l$ and all $x \in X$. It follows from (19) that the sequence $\{2^k f(\frac{x}{2^k})\}$ is Cauchy for all $x \in X$. Since Y is a Banach space, the sequence $\{2^k f(\frac{x}{2^k})\}$ converges. So, one can define the mapping $A : X \to Y$ by

$$A(x) := \lim_{k\to\infty} 2^k f\left(\frac{x}{2^k}\right)$$

for all $x \in X$. Moreover, letting $l = 0$ and passing the limit $m \to \infty$ in (19), we get (17).

It follows from (4) and (11) that

$$\begin{aligned}
&\|A(x+y) - A(x) - A(y)\| \\
&\quad = \lim_{n\to\infty} 2^n \left\| f\left(\frac{x+y}{2^n}\right) - f\left(\frac{x}{2^n}\right) - f\left(\frac{y}{2^n}\right) \right\| \\
&\quad \le \lim_{n\to\infty} 2^n |\rho_1| \left\| f\left(\frac{x+z}{2^n}\right) - f\left(\frac{x}{2^n}\right) - f\left(\frac{z}{2^n}\right) \right\| \\
&\qquad + \lim_{n\to\infty} 2^n |\rho_2| \left\| f\left(\frac{y+z}{2^n}\right) - f\left(\frac{y}{2^n}\right) - f\left(\frac{z}{2^n}\right) \right\| \\
&\qquad + \lim_{n\to\infty} 2^n \varphi\left(\frac{x}{2^n}, \frac{y}{2^n}, \frac{z}{2^n}\right) \\
&\quad = \|\rho_1(A(x+z) - A(x) - A(z))\| + \|\rho_2(A(y+z) - A(y) - A(z))\|
\end{aligned}$$

for all $x, y, z \in X$. So,

$$\begin{aligned}
\|A(x+y) - A(x) - A(y)\| \le{}& \|\rho_1(A(x+z) - A(x) - A(z))\| \\
&+ \|\rho_2(A(y+z) - A(y) - A(z))\|
\end{aligned}$$

for all $x, y, z \in X$. By Lemma 2, the mapping $A : X \to Y$ is additive.

Now, let $T : X \to Y$ be another additive mapping satisfying (17). Then we have

$$\begin{aligned}
\|A(x) - T(x)\| &= \left\| 2^q A\left(\frac{x}{2^q}\right) - 2^q T\left(\frac{x}{2^q}\right) \right\| \\
&\le \left\| 2^q A\left(\frac{x}{2^q}\right) - 2^q f\left(\frac{x}{2^q}\right) \right\| + \left\| 2^q T\left(\frac{x}{2^q}\right) - 2^q f\left(\frac{x}{2^q}\right) \right\| \\
&\le \frac{2^q}{1 - |\rho_1| - |\rho_2|} \Psi\left(\frac{x}{2^q}, \frac{x}{2^q}, \frac{x}{2^q}\right),
\end{aligned}$$

which tends to zero as $q \to \infty$ for all $x \in X$. So, we can conclude that $A(x) = T(x)$ for all $x \in X$. This proves the uniqueness of A. □

Corollary 17. *Let $r > 1$ and θ be non-negative real numbers, and let $f : X \to Y$ be a mapping satisfying $f(0) = 0$ and (7). Then there exists a unique additive mapping $A : X \to Y$ such that*

$$\|f(x) - A(x)\| \le \frac{3\theta}{(2^r - 2)(1 - |\rho_1| - |\rho_2|)} \|x\|^r$$

for all $x \in X$.

Theorem 18. *Let $\varphi : X^3 \to [0, \infty)$ be a function and let $f : X \to Y$ be a mapping satisfying $f(0) = 0$, (4) and (15). Then there exists a unique additive mapping $A : X \to Y$ such that*

$$\|f(x) - A(x)\| \leq \frac{1}{2(1 - |\rho_1| - |\rho_2|)} \Psi(x, x, x) \tag{20}$$

for all $x \in X$.

Proof. It follows from (18) that

$$\left\| f(x) - \frac{1}{2} f(2x) \right\| \leq \frac{1}{2(1 - |\rho_1| - |\rho_2|)} \varphi(x, x, x)$$

for all $x \in X$. Hence,

$$\begin{aligned} \left\| \frac{1}{2^l} f(2^l x) - \frac{1}{2^m} f(2^m x) \right\| &\leq \sum_{j=l}^{m-1} \left\| \frac{1}{2^j} f\left(2^j x\right) - \frac{1}{2^{j+1}} f\left(2^{j+1} x\right) \right\| \\ &\leq \sum_{j=l}^{m-1} \frac{1}{2^{j+1}(1 - |\rho_1| - |\rho_2|)} \varphi(2^j x, 2^j x, 2^j x) \end{aligned} \tag{21}$$

for all non-negative integers m and l with $m > l$ and all $x \in X$. It follows from (21) that the sequence $\{\frac{1}{2^n} f(2^n x)\}$ is a Cauchy sequence for all $x \in X$. Since Y is complete, the sequence $\{\frac{1}{2^n} f(2^n x)\}$ converges. So, one can define the mapping $A : X \to Y$ by

$$A(x) := \lim_{n \to \infty} \frac{1}{2^n} f(2^n x)$$

for all $x \in X$. Moreover, letting $l = 0$ and passing the limit $m \to \infty$ in (21), we get (20).

The rest of the proof is similar to the proof of Theorem 16. □

Corollary 19. *Let $r < 1$ and θ be positive real numbers, and let $f : X \to Y$ be a mapping satisfying $f(0) = 0$ and (7). Then there exists a unique additive mapping $A : X \to Y$ such that*

$$\|f(x) - A(x)\| \leq \frac{3\theta}{(2 - 2^r)(1 - |\rho_1| - |\rho_2|)} \|x\|^r$$

for all $x \in X$.

References

[1] Abdollahpour, M.R., Aghayaria, R. and Rassias, M.Th. Hyers-Ulam stability of associated Laguerre differential equations in a subclass of analytic functions, *J. Math. Anal. Appl.* 437, 605–612 (2016).

[2] Abdollahpour, M.R. and Rassias, M.Th. Hyers-Ulam stability of hypergeometric differential equations, *Aequationes Math.* 93, 691–698 (2019).

[3] Aczel, J. and Dhombres, J. *Functional Equations in Several Variables.* Cambridge University Press, Cambridge (1989).

[4] Aoki, T. On the stability of the linear transformation in Banach spaces, *J. Math. Soc. Japan* 2, 64–66 (1950).

[5] Bourgin, D.G. Classes of transformations and bordering transformations, *Bull. Am. Math. Soc.* 57, 223–237 (1951).

[6] Cădariu, L. and Radu, V. Fixed points and the stability of Jensen's functional equation, *J. Inequal. Pure Appl. Math.* 4(1), 1–7 (2003).

[7] Cădariu, L. and Radu, V. On the stability of the Cauchy functional equation: A fixed point approach, *Grazer Math. Ber.* 346, 43–52 (2004).

[8] Cădariu, L. and Radu, V. Fixed point methods for the generalized stability of functional equations in a single variable, *Fixed Point Theory Appl.* 2008, 749392, 1–15 (2008).

[9] Cholewa, P.W. Remarks on the stability of functional equations, *Aequationes Math.* 27, 76–86 (1984).

[10] Czerwik, S. On the stability of the quadratic mapping in normed spaces, *Abh. Math. Sem. Univ. Hamburg,* 62, 59–64 (1992).

[11] S. Czerwik, *Functional Equations and Inequalities in Several Variables.* World Scientific, Singapore (2002).

[12] Diaz, J. and Margolis, B. A fixed point theorem of the alternative for contractions on a generalized complete metric space, *Bull. Am. Math. Soc.* 74, 305–309 (1968).

[13] Eghbali, N., and Rassias, J.M., Taheri, M. On the stability of a k-cubic functional equation in intuitionistic fuzzy n-normed spaces, *Results Math.* 70, 233–248 (2016).

[14] Eskandani, G.Z., and Găvruta, P. Hyers-Ulam-Rassias stability of pexiderized Cauchy functional equation in 2-Banach spaces, *J. Nonlinear Sci. Appl.* 5, 459–465 (2012).

[15] Gajda, Z. On stability of additive mappings, *Int. J. Math. Math. Sci.* 14, 431–434 (1991).

[16] Găvruta, P. A generalization of the Hyers-Ulam-Rassias stability of approximately additive mappings, *J. Math. Anal. Appl.* 184, 431–436 (1994).

[17] Hyers, D.H. On the stability of the linear functional equation, *Proc. Nat. Acad. Sci. U.S.A.* 27, 222–224 (1941).

[18] Hyers, D.H., Isac, G. and Rassias, Th.M. *Stability of Functional Equations in Several Variables*. Birkhäuser, New York (1998).

[19] Hyers, D.H. Rassias, Th.M. Approximate homomorphisms, *Aequationes Math.* 44, 125–153 (1992).

[20] Isac, G. and Rassias, Th. M. Stability of ψ-additive mappings: Applications to nonlinear analysis, *Int. J. Math. Math. Sci.* 19, 219–228 (1996).

[21] Jung, S. Hyers-Ulam-Rassias, *Stability of Functional Equations in Nonlinear Analysis*. Springer, New York (2011).

[22] Jung, S., Mortici, C. and Rassias, M.Th. On a functional equation of trigonometric type, *Appl. Math. Comput.* 252, 294–303 (2015).

[23] Jung, S., Popa, D. and Rassias, M. Th. On the stability of the linear functional equation in a single variable on complete metric groups, *J. Global Optim.* 59, 165–171 (2014).

[24] Jung, S. and Rassias, M.Th. A linear functional equation of third order associated to the Fibonacci numbers, *Abstr. Appl. Anal.* 2014, 137468, 1–7 (2014).

[25] Kannappan, Pl. *Functional Equations and Inequalities with Applications*. Springer, New York (2009).

[26] Khodaei, H. On the stability of additive, quadratic, cubic and quartic set-valued functional equations, *Results Math.* 68, 1–10 (2015).

[27] Lee, Y., Jung, S. and Rassias, M. Th. On an n-dimensional mixed type additive and quadratic functional equation, *Appl. Math. Comput.* 228, 13–16 (2014).

[28] Lee, Y., Jung, S. and Rassias, M.Th. Uniqueness theorems on functional inequalities concerning cubic-quadratic-additive equation, *J. Math. Inequal.* 12, 43–61 (2018).

[29] Lee, Y. and Kim, G. Generalized Hyers-Ulam stability of the Pexider functional equation, *Math.* 7, 280 (2019).

[30] Miheţ, D. and Radu, V. On the stability of the additive Cauchy functional equation in random normed spaces, *J. Math. Anal. Appl.* 343, 567–572 (2008).

[31] Mortici, C., Jung, S. and Rassias, M. Th. On the stability of a functional equation associated with the Fibonacci numbers, *Abstr. Appl. Anal.* 2014, 546046, 1–6 (2014).

[32] Park, C. Additive ρ-functional inequalities and equations, *J. Math. Inequal.* 9, 17–26 (2015).

[33] Park, C. Additive ρ-functional inequalities in non-Archimedean normed spaces, *J. Math. Inequal.* 9, 397–407 (2015).

[34] Park, C. and Rassias, M.Th. Additive functional equations and partial multipliers in C^*-algebras, *Rev. R. Acad. Cien. Exactas Fís. Nat. Ser. A Mat. RACSAM* 113, 2261–2275.

[35] Radu, V. The fixed point alternative and the stability of functional equations, *Fixed Point Theory* 4, 91–96 (2003).

[36] Rassias, Th. M. On the stability of the linear mapping in Banach spaces, *Proc. Am. Math. Soc.* 72, 297–300 (1978).

[37] Skof, F. Propriet locali e approssimazione di operatori, *Rend. Sem. Mat. Fis. Milano* 53, 113–129 (1983).

[38] Székelyhidi, L. Superstability of functional equations related to spherical functions, *Open Math.* 15, 427–432 (2017).

[39] Ulam, S.M. *A Collection of the Mathematical Problems.* Interscience Publ., New York (1960).

https://doi.org/10.1142/9789811293009_0025

Chapter 25

The Effect of Two Exterior Masspoints on Bounds for Orthogonal Polynomials

D. S. Lubinsky

School of Mathematics, Georgia Institute of Technology, Atlanta, GA 30332-0160, USA

lubinsky@math.gatech.edu

Let ν be a symmetric positive measure supported on $[-1,1]$, with infinitely many points in its support. Let $\{p_n(\nu,x)\}_{n\geq 0}$ be its sequence of orthonormal polynomials. Suppose we add masspoints at r,s with $r\geq 1, |s|\geq 1$, and $r\neq s$, giving a new measure $\mu=\nu+M\delta_r+N\delta_s$. How much larger can $|p_n(\mu,0)|$ be than $|p_n(\nu,0)|$? We study this question for symmetric measures ν and obtain bounds that are uniform in n, M, N, r, s. In an earlier paper, we studied the case where $r=1, s=-1$, and investigated asymptotics.

Research supported by NSF grant DMS1800251

1. Results

Let μ be a finite positive Borel measure on the real line with infinitely many points in its support, and all finite moments

$$\int t^j d\mu(t), \, j = 0, 1, 2, \ldots.$$

Then we may define orthonormal polynomials

$$p_n(\mu, x) = \gamma_n(\mu) x^n + \ldots, \gamma_n(\mu) > 0,$$

$n = 0, 1, 2, \ldots$, satisfying the orthonormality conditions

$$\int p_n(\mu, x) p_m(\mu, x) d\mu(x) = \delta_{mn}.$$

The nth reproducing kernel for μ is

$$\begin{aligned} K_n(\mu, x, t) &= \sum_{j=0}^{n-1} p_j(\mu, x) p_j(\mu, t) \\ &= \frac{\gamma_{n-1}}{\gamma_n}(\mu) \frac{p_n(\mu, x) p_{n-1}(\mu, t) - p_{n-1}(\mu, x) p_{n-1}(\mu, t)}{x - t}. \end{aligned}$$

The three-term recurrence relation has the form

$$(x - b_n(\mu)) p_n(\mu, x) = a_{n+1}(\mu) p_{n+1}(\mu, x) + a_n(\mu) p_{n-1}(\mu, x),$$

where

$$a_n(\mu) = \frac{\gamma_{n-1}}{\gamma_n}(\mu).$$

Note that when μ is symmetric about 0, then $b_n(\mu) = 0$.

A central problem in the theory of orthonormal polynomials is to establish bounds on $p_n(\mu, x)$. Indeed, one of the most celebrated questions in this subject is the Steklov conjecture, posed in 1920 and solved by E. A. Rakhmanov in 1979 [15,16]. Steklov conjectured that if μ has support $[-1, 1]$ and is absolutely continuous there, while μ' is bounded below, then $\{p_n(\mu, x)\}_{n \geq 0}$ is uniformly bounded in $[-r, r]$, $0 < r < 1$, that is,

$$\sup_{n \geq 0} \sup_{x \in [-r, r]} |p_n(\mu, x)| < \infty.$$

Rakhmanov showed that Steklov's conjecture is false, and in fact, given $\varepsilon > 0$, one can find a measure satisfying its hypotheses such that for infinitely many n,

$$|p_n(\mu, 0)| \geq C n^{1/2-\varepsilon}.$$

There have since been a host of developments, due to among others, Ambroladze, Aptekarev, Denissov, Rush, and Tulyakov. In particular, Aptekarev, Denisov, and Tulyakov [3] showed that $|p_n(\mu, 0)|$ may grow faster than $n^{1/2}/\varepsilon_n$, where $\{\varepsilon_n\}$ is any sequence with limit 0, and this is the best one can aim for. Ambroladze showed [1] that even if μ' is continuous and satisfies some weak smoothness condition, Steklov's conjecture may still be false. See [7] for further results and references.

What about positive results? Most uniform bounds on orthonormal polynomials are consequences of much deeper asymptotics for orthonormal polynomials. Perhaps, the most general result for measures supported on $[-1, 1]$ is that of Badkov [5]. He proved that if μ satisfies Szegő's condition

$$\int_{-1}^{1} \frac{\log \mu'(x)}{\sqrt{1-x^2}} dx > -\infty$$

and μ' satisfies a local Dini condition, that is, in some interval $[a, b] \subset (-1, 1)$, the modulus of continuity

$$\omega(\delta) = \sup\left\{\left|\mu'(s) - \mu'(t)\right| : s, t \in [a, b] \text{ and } |s - t| \leq \delta\right\}, \delta > 0,$$

satisfies

$$\int_{0}^{1} \frac{\omega(t)}{t} dt < \infty,$$

then the corresponding orthonormal polynomials are uniformly bounded in compact subsets of (a, b) and admit local Szegő asymptotics. For non-Szegő weights, Rakhmanov [17] showed that orthogonal polynomials for Dini smooth weights on the unit circle admit uniform bounds. For exponential weights, supported on finite or infinite intervals, Eli Levin and the author established [13] global bounds, using extensions of a technique that goes back to Stan Bonan, a student of Paul Nevai. In many of these results, the bounds are then superseded by asymptotics.

In a recent paper, we investigated how adding masspoints at ± 1 can increase the size of the orthonormal polynomial at the origin. Consider a fixed positive measure ν supported on $[-1,1]$ with infinitely many points in its support, and that is symmetric about 0, so that $\nu([-b,-a]) = \nu([a,b])$ for all $[a,b] \subset [-1,1]$. Let $\mathcal{M}(\nu)$ denote the class of all measures

$$\mu = \nu + M\delta_1 + N\delta_{-1},$$

where $M, N \geq 0$. Also, let

$$U_n = K_n(\nu, 1, 1) - K_n(\nu, -1, 1),$$
$$V_n = K_n(\nu, 1, 1) + K_n(\nu, -1, 1).$$

We proved the following:

Theorem A. *Let ν be a positive measure with support in $[-1,1]$ and with infinitely many points in its support. Assume also that ν is symmetric so that $\nu([-b,-a]) = \nu([a,b])$ for all subintervals $[a,b]$ of $[-1,1]$. Let $n \geq 2$ be even. Then*

$$\sup_{\mu \in \mathcal{M}(\nu)} \left(\frac{p_n(\mu, 0)}{p_n(\nu, 0)} \right)^2 = \max \left\{ 1, \frac{U_n^2}{V_n V_{n+1}} \right\}.$$

We deduced the following:

Corollary B. *Assume in addition to the hypotheses of Theorem* 1.1 *that ν lies in the Nevai class so that the recurrence coefficients satisfy*

$$\lim_{n\to\infty} a_n(\nu) = \frac{1}{2}; \lim_{n\to\infty} b_n(\nu) = 0. \tag{1}$$

Then,

$$\lim_{n\to\infty} \left(\sup_{\mu \in \mathcal{M}(\nu)} \left(\frac{p_n(\mu, 0)}{p_n(\nu, 0)} \right)^2 \right) = 1.$$

Further, more detailed results were established in [14], with precise asymptotics for ultraspherical weights.

We noted there, as we do here, that there is an extensive literature on the effect of adding masspoints to orthogonal polynomials. In particular, various types of differential and other equations have been established [6, 11, 12], especially when the underlying measure comes from a Jacobi weight. Moreover, various asymptotic relations and estimates have been established in [2,4,8,10]. However, the focus is different from that in this chapter.

Here we consider what happens when we add masspoints at two arbitrary points exterior to $(-1,1)$. We prove the following:

Theorem 1. *Let ν be a positive measure with support in $[-1,1]$ and with infinitely many points in its support. Assume also that ν is symmetric so that $\nu([-b,-a]) = \nu([a,b])$ for all subintervals $[a,b]$ of $[-1,1]$. Let $n \geq 2$ be even. Let $M, N \geq 0$, $r \geq 1, s \leq -1$, and*

$$\mu = \nu + M\delta_r + N\delta_s.$$

Then,

$$\left|\frac{p_n(\mu,0)}{p_n(\nu,0)}\right| \leq 1 + \sqrt{20}\frac{\gamma_{n-1}}{\gamma_n}(\nu) \Big/ \frac{\gamma_{n-2}}{\gamma_{n-1}}(\nu).$$

For the case $r, s \geq 1$, we prove the following:

Theorem 2. *Assume the hypotheses of Theorem 1, except that we let $r, s \geq 1$. Then,*

$$\left|\frac{p_n(\mu,0)}{p_n(\nu,0)}\right| \leq 1 + \sqrt{10}\frac{\gamma_{n-1}}{\gamma_n}(\nu) / \frac{\gamma_{n-2}}{\gamma_{n-1}}(\nu).$$

Corollary 3. *Let ν be a positive measure with support in $[-1,1]$ and with infinitely many points in its support. Assume also that ν is symmetric so that $\nu([-b,-a]) = \nu([a,b])$ for all subintervals $[a,b]$ of $[-1,1]$. Assume also that either that (I) ν lies in the Nevai class so that the recurrence coefficients satisfy* (1), *or (II) ν is absolutely continuous in some subinterval J of $(-1,1)$ and $\nu' > 0$ a.e. there and is bounded above in J. Then there exists $C > 0$ depending only on ν such that for even $n \geq 2, M, N \geq 0$, and $r \geq 1$, $|s| \geq 1$, and μ as above,*

$$\left|\frac{p_n(\mu,0)}{p_n(\nu,0)}\right| \leq C.$$

Thus, we have an explicit bound that does not depend on the size of the masspoints or their location. We establish a basic identity in Section 2 and prove the theorems and corollary in Section 3.

In the sequel, $C, C_1, C_2, \ldots$ denote constants independent of n, x, t. The same symbol does not necessarily denote the same constant in different occurrences.

2. The Basic Identity

Throughout this section, we let $M, N > 0$ and $r \geq 1$, $|s| \geq 1, s \neq r$, and

$$\mu = \nu + M\delta_r + N\delta_s. \tag{2}$$

We assume that ν is a symmetric measure supported on $[-1, 1]$. We fix even $n \geq 2$. We let

$$X = \begin{bmatrix} p_n(\nu, s) \\ p_n(\nu, r) \end{bmatrix}; Y = \frac{\gamma_{n-1}}{\gamma_n}(\nu) \begin{bmatrix} \frac{p_{n-1}(\nu,s)}{s} \\ \frac{p_{n-1}(\nu,r)}{r} \end{bmatrix}, \tag{3}$$

$$\mathbb{K} = \begin{bmatrix} K_n(\nu, s, s) & K_n(\nu, r, s) \\ K_n(\nu, r, s) & K_n(\nu, r, r) \end{bmatrix}, \tag{4}$$

$$D = \begin{bmatrix} N & 0 \\ 0 & M \end{bmatrix}. \tag{5}$$

We prove the following identity. For the case $r = 1, s = -1$, it appeared in a slightly different form in [14]. Various identities of this type have been established for adding the effect of masspoints and most likely it is not new (although we could not find it in the literature).

Theorem 4.

$$\left(\frac{p_n(\mu, 0)}{p_n(\nu, 0)} \right)^2 = \frac{\left\{ 1 + X^T \left(D^{-1} + \mathbb{K} \right)^{-1} Y \right\}^2}{\left\{ 1 + X^T \left(D^{-1} + \mathbb{K} \right)^{-1} X \right\}}. \tag{6}$$

We begin the proof with the following.

Lemma 5.

(a) *Let*

$$\pi_{n-1}(y) = p_n(\mu, y) - \frac{\gamma_n(\mu)}{\gamma_n(\nu)} p_n(\nu, y). \tag{7}$$

Then,

$$p_n(\mu, y) = \frac{\gamma_n(\mu)}{\gamma_n(\nu)} \left\{ p_n(\nu, y) + \begin{bmatrix} -NK_n(\nu, y, s) \\ -MK_n(\nu, y, r) \end{bmatrix}^T (I + \mathbb{K}D)^{-1} X \right\}. \tag{8}$$

(b)

$$\left(\frac{\gamma_n(\mu)}{\gamma_n(\nu)} \right)^2 \{1 + X^T (D^{-1} + \mathbb{K})^{-1} X\} = 1. \tag{9}$$

Proof.

(a) Using the reproducing kernel property and orthogonality, we see that

$$\begin{aligned} \pi_{n-1}(y) &= \int_{-1}^{1} K_n(\nu, y, t)\, \pi_{n-1}(t)\, d\nu(t) \\ &= \int_{-1}^{1} K_n(\nu, y, t)\, p_n(\mu, t)\, d\nu(t) \\ &= -MK_n(\nu, y, r)\, p_n(\mu, r) - NK_n(\nu, y, s)\, p_n(\mu, s). \end{aligned} \tag{10}$$

Taking $y = s$ and $y = r$ gives

$$\begin{aligned} p_n(\mu, s) - \frac{\gamma_n(\mu)}{\gamma_n(\nu)} p_n(\nu, s) &= -MK_n(\nu, s, r)\, p_n(\mu, r) \\ &\quad - NK_n(\nu, s, s)\, p_n(\mu, s), \\ p_n(\mu, r) - \frac{\gamma_n(\mu)}{\gamma_n(\nu)} p_n(\nu, r) &= -MK_n(\nu, r, r)\, p_n(\mu, r) \\ &\quad - NK_n(\nu, r, s)\, p_n(\mu, s), \end{aligned}$$

and gathering the terms involving $p_n(\mu, r)$, $p_n(\mu, s)$ gives the matrix equation

$$\begin{bmatrix} 1 + NK_n(\nu, s, s) & MK_n(\nu, s, r) \\ NK_n(\nu, s, r) & 1 + MK_n(\nu, r, r) \end{bmatrix} \begin{bmatrix} p_n(\mu, s) \\ p_n(\mu, r) \end{bmatrix} = \frac{\gamma_n(\mu)}{\gamma_n(\nu)} \begin{bmatrix} p_n(\nu, s) \\ p_n(\nu, r) \end{bmatrix} \tag{11}$$

or, equivalently,

$$(I+\mathbb{K}D)\begin{bmatrix} p_n(\mu,s) \\ p_n(\mu,r) \end{bmatrix} = \frac{\gamma_n(\mu)}{\gamma_n(\nu)}X.$$

The matrix $I+\mathbb{K}D$ is non-singular, as can be seen from calculating its determinant:

$$\begin{aligned}\det(I+\mathbb{K}D) &= 1+NK_n(\nu,s,s)+MK_n(\nu,r,r)\\ &\quad +MN\left(K_n(\nu,r,r)\,K_n(\nu,s,s)-K_n(\nu,r,s)^2\right).\end{aligned}$$

The coefficients of M, N, MN are all non-negative. Solving the matrix equation (11) gives

$$\begin{bmatrix} p_n(\mu,s) \\ p_n(\mu,r) \end{bmatrix} = \frac{\gamma_n(\mu)}{\gamma_n(\nu)}(I+\mathbb{K}D)^{-1}X. \tag{12}$$

From (10) and this last identity,

$$\begin{aligned} & p_n(\mu,y)-\frac{\gamma_n(\mu)}{\gamma_n(\nu)}p_n(\nu,y)\\ &= \begin{bmatrix} -NK_n(\nu,y,s) \\ -MK_n(\nu,y,r) \end{bmatrix}^T \begin{bmatrix} p_n(\mu,s) \\ p_n(\mu,r) \end{bmatrix}\\ &= \frac{\gamma_n(\mu)}{\gamma_n(\nu)}\begin{bmatrix} -NK_n(\nu,y,s) \\ -MK_n(\nu,y,r) \end{bmatrix}^T (I+\mathbb{K}D)^{-1}X. \end{aligned}$$

Then, (8) follows.

(b) We obtain equations for $\frac{\gamma_n(\mu)}{\gamma_n(\nu)}$ in two ways: from the definition (7) of π_{n-1} and orthogonality,

$$\begin{aligned} &\int_{-1}^{1}\pi_{n-1}^2(y)\,d\nu(y)\\ &= \int_{-1}^{1}p_n^2(\mu,y)^2\,d\nu(y)-2\left(\frac{\gamma_n(\mu)}{\gamma_n(\nu)}\right)^2+\left(\frac{\gamma_n(\mu)}{\gamma_n(\nu)}\right)^2\\ &= 1-Mp_n(\mu,r)^2-Np_n(\mu,s)^2-\left(\frac{\gamma_n(\mu)}{\gamma_n(\nu)}\right)^2. \end{aligned}$$

Also, from (10),

$$\begin{aligned}
&\int_{-1}^{1} \pi_{n-1}^2(y)\, d\nu(y) \\
&= \int_{-1}^{1} \left(-NK_n(\nu, y, s)\, p_n(\mu, s) - MK_n(\nu, y, r)\, p_n(\mu, r)\right)^2 d\nu(y) \\
&= N^2 p_n^2(\mu, s)\, K_n(\nu, s, s) + M^2 p_n^2(\mu, r)\, K_n(\nu, r, r) \\
&\quad + 2MN p_n(\mu, s)\, p_n(\mu, r)\, K_n(\nu, r, s).
\end{aligned}$$

Then using the last two equations and solving for $1 - \left(\frac{\gamma_n(\mu)}{\gamma_n(\nu)}\right)^2$,

$$\begin{aligned}
&1 - \left(\frac{\gamma_n(\mu)}{\gamma_n(\nu)}\right)^2 \\
&= p_n^2(\mu, s)\left\{N + N^2 K_n(\nu, s, s)\right\} + p_n^2(\mu, r) \\
&\quad \times \left\{M + M^2 K_n(\nu, r, r)\right\} + 2MN p_n(\mu, s)\, p_n(\mu, r)\, K_n(\nu, r, s) \\
&= \begin{bmatrix} p_n(\mu, s) \\ p_n(\mu, r) \end{bmatrix}^T \begin{bmatrix} N & 0 \\ 0 & M \end{bmatrix} \\
&\quad \times \begin{bmatrix} 1 + NK_n(\nu, s, s) & MK_n(\nu, s, r) \\ NK_n(\nu, s, r) & 1 + MK_n(\nu, r, r) \end{bmatrix} \begin{bmatrix} p_n(\mu, s) \\ p_n(\mu, r) \end{bmatrix} \\
&= \begin{bmatrix} p_n(\mu, s) \\ p_n(\mu, r) \end{bmatrix}^T D(I + \mathbb{K}D) \begin{bmatrix} p_n(\mu, s) \\ p_n(\mu, r) \end{bmatrix}.
\end{aligned}$$

Using (12) gives

$$1 - \left(\frac{\gamma_n(\mu)}{\gamma_n(\nu)}\right)^2 = \left(\frac{\gamma_n(\mu)}{\gamma_n(\nu)}\right)^2 X^T \left((I + \mathbb{K}D)^{-1}\right)^T DX.$$

Here

$$\begin{aligned}
&\left((I + \mathbb{K}D)^{-1}\right)^T D \\
&= \left(D\,(I + \mathbb{K}D)^{-1}\right)^T \\
&= \left(\left((I + \mathbb{K}D)\, D^{-1}\right)^{-1}\right)^T \\
&= (D^{-1} + \mathbb{K})^{-1},
\end{aligned}$$

as both $D^{-1}, \mathbb{K}$ are symmetric. So,

$$1 = \left(\frac{\gamma_n(\mu)}{\gamma_n(\nu)}\right)^2 \{1 + X^T(D^{-1} + \mathbb{K})^{-1}X\}.$$

□

Proof of Theorem 4. Setting $y = 0$, and using $p_{n-1}(\nu, 0) = 0$, the Christoffel–Darboux formula gives

$$K_n(\nu, 0, y) = -\frac{\gamma_{n-1}(\nu)}{\gamma_n(\nu)} p_n(\nu, 0) \frac{p_{n-1}(\nu, y)}{y}.$$

Then setting $y = 0$ in (8),

$$\begin{aligned} p_n(\mu, 0) &= \frac{\gamma_n(\mu)}{\gamma_n(\nu)} \left\{ p_n(\nu, 0) + \begin{bmatrix} -NK_n(\nu, 0, s) \\ -MK_n(\nu, 0, r) \end{bmatrix}^T (I + \mathbb{K}D)^{-1}X \right\} \\ &= \frac{\gamma_n(\mu)}{\gamma_n(\nu)} p_n(\nu, 0) \\ &\quad \times \left\{ 1 + \frac{\gamma_{n-1}(\nu)}{\gamma_n(\nu)} \begin{bmatrix} N\frac{p_{n-1}(\nu,s)}{s} \\ M\frac{p_{n-1}(\nu,r)}{r} \end{bmatrix}^T (I + \mathbb{K}D)^{-1}X \right\} \\ &= \frac{\gamma_n(\mu)}{\gamma_n(\nu)} p_n(\nu, 0) \{1 + Y^T D(I + \mathbb{K}D)^{-1}X\}. \end{aligned}$$

As above,

$$\begin{aligned} &D(I + \mathbb{K}D)^{-1} \\ &\quad = \left((I + \mathbb{K}D)D^{-1}\right)^{-1} \\ &\quad = \left(D^{-1} + \mathbb{K}\right)^{-1}, \end{aligned}$$

so from (9),

$$\begin{aligned} &\left(\frac{p_n(\mu, 0)}{p_n(\nu, 0)}\right)^2 \{1 + X^T(D^{-1} + \mathbb{K})^{-1}X\} \\ &= \left\{1 + Y\left(D^{-1} + \mathbb{K}\right)^{-1} X\right\}^2. \end{aligned}$$

□

3. Proof of Theorems 1 and 2 and Corollary 3

We begin with a lemma that applies to both the situations of Theorems 1 and 2. Throughout we assume the hypotheses of Theorem 1 on ν and μ, though initially we assume that $M, N > 0$. We also assume notations (3)–(5).

Lemma 6. *For all* $Z = \begin{bmatrix} z_1 \\ z_2 \end{bmatrix} \in \mathbb{R}^2$, *we have*

$$Z^T \left(D^{-1} + \mathbb{K}\right)^{-1} Z \le Z^T \mathbb{K}^{-1} Z.$$

Proof. Now,

$$D^{-1} = \begin{bmatrix} N^{-1} & 0 \\ 0 & M^{-1} \end{bmatrix},$$

so

$$\begin{aligned}
&\left(D^{-1} + \mathbb{K}\right)^{-1} - \mathbb{K}^{-1} \\
&= \frac{1}{\det\left(D^{-1} + \mathbb{K}\right)} \begin{bmatrix} M^{-1} + K_n(\nu, r, r) & -K_n(\nu, r, s) \\ -K_n(\nu, r, s) & N^{-1} + K_n(\nu, s, s) \end{bmatrix} \\
&\quad - \frac{1}{\det \mathbb{K}} \begin{bmatrix} K_n(\nu, r, r) & -K_n(\nu, r, s) \\ -K_n(\nu, r, s) & K_n(\nu, s, s) \end{bmatrix} \\
&= \frac{1}{\det\left(D^{-1} + \mathbb{K}\right)} \begin{bmatrix} M^{-1} & 0 \\ 0 & N^{-1} \end{bmatrix} \\
&\quad + \left(\frac{1}{\det\left(D^{-1} + \mathbb{K}\right)} - \frac{1}{\det \mathbb{K}}\right) \begin{bmatrix} K_n(\nu, r, r) & -K_n(\nu, r, s) \\ -K_n(\nu, r, s) & K_n(\nu, s, s) \end{bmatrix}.
\end{aligned}$$

Then,

$$\begin{aligned}
&\left\{Z^T \left(\left(D^{-1} + \mathbb{K}\right)^{-1} - \mathbb{K}^{-1}\right) Z\right\} \det\left(D^{-1} + \mathbb{K}\right) \det(\mathbb{K}) \\
&= z_1^2 M^{-1} \det(\mathbb{K}) + z_2^2 N^{-1} \det(\mathbb{K}) + \left(\det K - \det\left(D^{-1} + K\right)\right) \\
&\quad \times \left\{z_1^2 K_n(\nu, r, r) + z_2^2 K_n(\nu, s, s) - 2 z_1 z_2 K_n(\nu, r, s)\right\}. \qquad (13)
\end{aligned}$$

Next, a straightforward calculation shows that

$$\begin{aligned}
&\det\left(D^{-1} + \mathbb{K}\right) \\
&= N^{-1} M^{-1} + M^{-1} K_n(\nu, s, s) + N^{-1} K_n(\nu, r, r) + \det(\mathbb{K}).
\end{aligned}$$

Then substituting in (13),

$$
\begin{aligned}
&\{Z^T((D^{-1}+\mathbb{K})^{-1}-\mathbb{K}^{-1})Z\}\det(D^{-1}+\mathbb{K})\det(\mathbb{K})\\
&\quad = z_1^2 M^{-1}\det(\mathbb{K}) + z_2^2 N^{-1}\det(\mathbb{K})\\
&\qquad - \left\{N^{-1}M^{-1} + M^{-1}K_n(\nu,s,s) + N^{-1}K_n(\nu,r,r)\right\}\\
&\qquad \times \left\{z_1^2 K_n(\nu,r,r) + z_2^2 K_n(\nu,s,s) - 2z_1z_2K_n(\nu,r,s)\right\}\\
&\quad = z_1^2\Big\{M^{-1}\det(\mathbb{K}) - N^{-1}M^{-1}K_n(\nu,r,r)\\
&\qquad - M^{-1}K_n(\nu,s,s)K_n(\nu,r,r) - N^{-1}K_n(\nu,r,r)^2\Big\}\\
&\qquad + z_2^2\Big\{N^{-1}\det(\mathbb{K}) - N^{-1}M^{-1}K_n(\nu,s,s)\\
&\qquad - M^{-1}K_n(\nu,s,s)^2 - N^{-1}K_n(\nu,s,s)K_n(\nu,r,r)\Big\}\\
&\qquad + 2z_1z_2K_n(\nu,r,s)\left\{N^{-1}M^{-1} + M^{-1}K_n(\nu,s,s)\right.\\
&\qquad \left. + N^{-1}K_n(\nu,r,r)\right\}\\
&\quad = z_1^2\Big\{-M^{-1}K_n(\nu,r,s)^2 - N^{-1}M^{-1}K_n(\nu,r,r)\\
&\qquad - N^{-1}K_n(\nu,r,r)^2\Big\} + z_2^2\Big\{-N^{-1}K_n(\nu,r,s)^2\\
&\qquad - N^{-1}M^{-1}K_n(\nu,s,s) - M^{-1}K_n(\nu,s,s)^2\Big\} + 2z_1z_2K_n(\nu,r,s)\\
&\qquad \times \left\{N^{-1}M^{-1} + M^{-1}K_n(\nu,s,s) + N^{-1}K_n(\nu,r,r)\right\}\\
&\quad = N^{-1}M^{-1}\left\{-z_1^2K_n(\nu,r,r) - z_2^2K_n(\nu,s,s) + 2z_1z_2K_n(\nu,r,s)\right\}\\
&\qquad + N^{-1}\Big\{-z_1^2K_n(\nu,r,r)^2 - z_2^2K_n(\nu,r,s)^2\\
&\qquad + 2z_1z_2K_n(\nu,r,r)K_n(\nu,r,s)\Big\}\\
&\qquad + M^{-1}\Big\{-z_1^2K_n(\nu,r,s)^2 - z_2^2K_n(\nu,s,s)^2\\
&\qquad + 2z_1z_2K_n(\nu,s,s)K_n(\nu,r,s)\Big\}\\
&\quad = N^{-1}M^{-1}\left\{\begin{array}{l} -\left(|z_1|\sqrt{K_n(\nu,r,r)} - |z_2|\sqrt{K_n(\nu,s,s)}\right)^2\\ +2\Big(z_1z_2K_n(\nu,r,s)\\ -|z_1||z_2|\sqrt{K_n(\nu,s,s)}\sqrt{K_n(\nu,r,r)}\Big)\end{array}\right\}
\end{aligned}
$$

$$- N^{-1} \left(z_1 K_n (\nu, r, r) - z_2 K_n (\nu, r, s) \right)^2$$
$$- M^{-1} \left(z_1 K_n (\nu, r, s) - z_2 K_n (\nu, s, s) \right)^2 \leq 0$$

since $K_n (\nu, r, s) \leq \sqrt{K_n (\nu, s, s)} \sqrt{K_n (\nu, r, r)}$. □

In the sequel, we let

$$C = \left(D^{-1} + \mathbb{K} \right)^{-1}.$$

Lemma 7. *Assume that* $M, N > 0$. *For* $r \geq 1, |s| \geq 1$,

$$\left(\frac{p_n (\mu, 0)}{p_n (\nu, 0)} \right)^2 \leq \left(1 + \left(Y^T \mathbb{K}^{-1} Y \right)^{1/2} \right)^2. \tag{14}$$

Proof. Note that $C = \left(D^{-1} + \mathbb{K} \right)^{-1}$ is positive definite, as both D and $\mathbb{K}$ are, so that

$$\left| X^T C Y \right| \leq \left(\left(X^T C X \right) \left(Y^T C Y \right) \right)^{1/2}.$$

Indeed, this follows from the Cauchy–Schwarz inequality as $< X, Y >= X^T C Y$ defines an inner product. Then, from (6),

$$\begin{aligned}
\left(\frac{p_n (\mu, 0)}{p_n (\nu, 0)} \right)^2 &= \frac{\left\{ 1 + X^T C Y \right\}^2}{\left\{ 1 + X^T C X \right\}} \\
&= \frac{1 + 2 X^T C Y + \left(X^T C Y \right)^2}{1 + X^T C X} \\
&\leq 1 + \frac{2 \left(\left(X^T C X \right) \left(Y^T C Y \right) \right)^{1/2} + \left(X^T C X \right) \left(Y^T C Y \right)}{1 + X^T C X} \\
&\leq 1 + 2 \left(\frac{Y^T C Y}{1 + X^T C X} \right)^{1/2} + Y^T C Y \\
&\leq 1 + 2 \left(Y^T C Y \right)^{1/2} + Y^T C Y = \left(1 + \left(Y^T C Y \right)^{1/2} \right)^2.
\end{aligned}$$

Finally, by Lemma 6,

$$Y^T C Y \leq Y^T \mathbb{K}^{-1} Y.$$

□

Lemma 8. *Let* $r \geq 1$ *and* $s \leq -1$. *Then,*

$$W = \det (\mathbb{K}) \geq \frac{1}{5} \max \{ K_n (\nu, r, r) K_{n-1} (\nu, s, s),$$
$$K_n (\nu, s, s) K_{n-1} (\nu, r, r) \}.$$

Proof. We use the Lagrange "error" formula for the Cauchy–Schwarz inequality:

$$\begin{aligned} W = \det(\mathbb{K}) &= K_n(\nu,r,r)\,K_n(\nu,s,s) - K_n(\nu,r,s)^2 \\ &= \frac{1}{2}\sum_{j,k=0}^{n-1}\left(p_j(\nu,r)\,p_k(\nu,s) - p_j(\nu,s)\,p_k(\nu,r)\right)^2. \end{aligned} \tag{15}$$

Using that p_j is odd or even according as j is and that $s<0$,

$$\begin{aligned} W &= K_n(\nu,r,r)\,K_n(\nu,s,s) - K_n(\nu,r,s)^2 \\ &\geq \frac{1}{2}\sum_{j\text{ odd},\,k\text{ even}}\left(p_j(\nu,r)\,p_k(\nu,|s|) + p_j(\nu,|s|)\,p_k(\nu,r)\right)^2 \\ &\quad + \frac{1}{2}\sum_{j\text{ even},\,k\text{ odd}}\left(p_j(\nu,r)\,p_k(\nu,|s|) + p_j(\nu,|s|)\,p_k(\nu,r)\right)^2 \\ &\geq \frac{1}{2}\left\{\begin{array}{l} \sum_{j\text{ odd}} p_j^2(\nu,r)\sum_{k\text{ even}} p_k^2(\nu,|s|) \\ \quad + \sum_{k\text{ even}} p_k^2(\nu,r)\sum_{j\text{ odd}} p_j^2(\nu,|s|) \\ \quad + \sum_{j\text{ even}} p_j^2(\nu,r)\sum_{k\text{ odd}} p_k^2(\nu,|s|) \\ \quad + \sum_{j\text{ even}} p_j^2(\nu,|s|)\sum_{k\text{ odd}} p_k^2(\nu,r) \end{array}\right\} \\ &\geq K_n^o(r)\,K_n^e(|s|) + K_n^e(r)\,K_n^o(|s|), \end{aligned} \tag{16}$$

where

$$K_n^e(x) = \sum_{0\leq j\leq n-1,\, j\text{ even}} p_j^2(x),$$

$$K_n^o(x) = \sum_{0\leq j\leq n-1,\, j\text{ odd}} p_j^2(x).$$

Next, from the recurrence relation,

$$xp_j(\nu,x) = \frac{\gamma_j}{\gamma_{j+1}}(\nu)\,p_{j+1}(\nu,x) + \frac{\gamma_{j-1}}{\gamma_j}(\nu)\,p_{j-1}(\nu,x)$$

so as $\frac{\gamma_j}{\gamma_{j+1}}(\nu)\leq 1$ (since $\operatorname{supp}[\nu]\subseteq[-1,1]$, [9]), we have

$$p_j(\nu,x)^2 \leq 2\left\{\left(\frac{p_{j+1}(\nu,x)}{x}\right)^2 + \left(\frac{p_{j-1}(\nu,x)}{x}\right)^2\right\}.$$

Then (recall n is even), for $|x| \geq 1$,

$$\begin{aligned} K_n(\nu, x, x) &= K_n^o(x) + K_n^e(x) \\ &\leq K_n^o(x) + \frac{2}{x^2}\{K_n^o(x) + K_n^o(x)\} \leq 5K_n^o(x). \end{aligned}$$

Similarly,

$$K_{n-1}(\nu, x, x) \leq 5K_n^e(x).$$

Hence, from (16),

$$W \geq \frac{1}{5}K_{n-1}(\nu, r, r)\left(K_n^o(|s|) + K_n^e(s)\right) = \frac{1}{5}K_{n-1}(\nu, r, r)\, K_n(\nu, s, s).$$

A similar estimate holds when we swap r and s. □

Proof of Theorem 1 Recalling (4) and the formula for the inverse of a 2×2 matrix,

$$\begin{aligned} & Y^T \mathbb{K}^{-1} Y \\ &= \frac{1}{W}\left(\frac{\gamma_{n-1}}{\gamma_n}(\nu)\right)^2 \left\{ \begin{array}{c} K_n(\nu, r, r)\left(\frac{p_{n-1}(\nu,s)}{s}\right)^2 \\ +K_n(\nu, s, s)\left(\frac{p_{n-1}(\nu,r)}{r}\right)^2 \\ -2K_n(\nu, r, s)\frac{p_{n-1}(\nu,s)}{s}\frac{p_{n-1}(\nu,r)}{r} \end{array} \right\} \\ &\leq \frac{1}{W}\left(\frac{\gamma_{n-1}}{\gamma_n}(\nu)\right)^2 \left\{ \sqrt{K_n(\nu, r, r)}\left|\frac{p_{n-1}(\nu, s)}{s}\right| \right. \\ &\quad \left. + \sqrt{K_n(\nu, s, s)}\left|\frac{p_{n-1}(\nu, r)}{r}\right| \right\}^2, \end{aligned}$$

by Cauchy–Schwarz so that from Lemma 8,

$$\begin{aligned} & Y^T \mathbb{K}^{-1} Y \\ &\leq 5\left(\frac{\gamma_{n-1}}{\gamma_n}(\nu)\right)^2 \\ &\quad \times \left[\frac{\left\{ \sqrt{K_n(\nu, r, r)}\left|\frac{p_{n-1}(\nu,s)}{s}\right| + \sqrt{K_n(\nu, s, s)}\left|\frac{p_{n-1}(\nu,r)}{r}\right| \right\}^2}{\max\{K_n(\nu, r, r)\, K_{n-1}(\nu, s, s), K_n(\nu, s, s)\, K_{n-1}(\nu, r, r)\}} \right] \\ &\leq 5\left(\frac{\gamma_{n-1}}{\gamma_n}(\nu)\right)^2 \left[\left\{ \left| \frac{p_{n-1}(\nu, s)}{s\sqrt{K_{n-1}(\nu, s, s)}} \right| + \left| \frac{p_{n-1}(\nu, r)}{r\sqrt{K_{n-1}(\nu, r, r)}} \right| \right\}^2 \right]. \end{aligned}$$

Here the recurrence relation gives

$$\begin{aligned}\left|sp_{n-2}(\nu,s)\right| &= \left|\frac{\gamma_{n-2}}{\gamma_{n-1}}(\nu)\,p_{n-1}(\nu,s)+\frac{\gamma_{n-3}}{\gamma_{n-2}}(\nu)\,p_{n-3}(\nu,s)\right| \\ &\geq \frac{\gamma_{n-2}}{\gamma_{n-1}}(\nu)\left|p_{n-1}(\nu,s)\right| \end{aligned} \tag{17}$$

as p_{n-1}, p_{n-3} have the same sign at s, so

$$\left|\frac{p_{n-1}(\nu,s)}{s\sqrt{K_{n-1}(\nu,s,s)}}\right| \leq \left(\frac{\gamma_{n-2}}{\gamma_{n-1}}(\nu)\right)^{-1}.$$

A similar inequality holds when we replace s by r, so

$$Y^T\mathbb{K}^{-1}Y \leq 20\left\{\frac{\gamma_{n-1}}{\gamma_n}(\nu)\Big/\frac{\gamma_{n-2}}{\gamma_{n-1}}(\nu)\right\}^2.$$

From Lemma 7,

$$\left(\frac{p_n(\mu,0)}{p_n(\nu,0)}\right)^2 \leq \left(1+\sqrt{20}\left\{\frac{\gamma_{n-1}}{\gamma_n}(\nu)\Big/\frac{\gamma_{n-2}}{\gamma_{n-1}}(\nu)\right\}\right)^2,$$

at least if $M, N > 0$. The case where $M = 0$ or $N = 0$ can be handled by letting M or $N \to 0$ and using continuity of the orthogonal polynomial in the measure, as follows from the classical determinantal representation for orthogonal polynomials. □

Proof of Theorem 2. We again need to estimate $Y^T\mathbb{K}^{-1}Y$. Now,

$$\begin{aligned} &WY^T\mathbb{K}^{-1}Y\Big/\left(\frac{\gamma_{n-1}}{\gamma_n}(\nu)\right)^2 \\ &= K_n(\nu,r,r)\left(\frac{p_{n-1}(\nu,s)}{s}\right)^2 + K_n(\nu,s,s)\left(\frac{p_{n-1}(\nu,r)}{r}\right)^2 \\ &\quad - 2K_n(\nu,r,s)\,\frac{p_{n-1}(\nu,s)}{s}\,\frac{p_{n-1}(\nu,r)}{r} \\ &= \sum_{j=0}^{n-1}\left\{p_j(\nu,r)\,\frac{p_{n-1}(\nu,s)}{s} - p_j(\nu,s)\,\frac{p_{n-1}(\nu,r)}{r}\right\}^2 \end{aligned}$$

$$= \frac{1}{(rs)^2} \sum_{j=0}^{n-2} \{rp_j(\nu,r)\,p_{n-1}(\nu,s) - sp_j(\nu,s)\,p_{n-1}(\nu,r)\}^2$$

$$+ \frac{(r-s)^2}{(rs)^2} (p_{n-1}(r)\,p_{n-1}(\nu,s))^2 .$$

Using the recurrence relation and $\frac{\gamma_j}{\gamma_{j+1}}(\nu) \le 1$, we continue this as

$$= \frac{1}{(rs)^2} \sum_{j=0}^{n-2} \left\{ \begin{array}{l} \left[\frac{\gamma_j}{\gamma_{j+1}}(\nu)\,p_{j+1}(\nu,r) + \frac{\gamma_{j-1}}{\gamma_j}(\nu)\,p_{j-1}(\nu,r)\right] p_{n-1}(\nu,s) \\ -\left[\frac{\gamma_j}{\gamma_{j+1}}(\nu)\,p_{j+1}(\nu,s) + \frac{\gamma_{j-1}}{\gamma_j}(\nu)\,p_{j-1}(\nu,s)\right] p_{n-1}(\nu,r) \end{array} \right\}^2$$

$$+ \frac{(r-s)^2}{(rs)^2} (p_{n-1}(r)\,p_{n-1}(\nu,s))^2$$

$$\le \frac{2}{(rs)^2} \sum_{j=0}^{n-2} \left\{ \begin{array}{l} \{p_{j+1}(\nu,r)\,p_{n-1}(\nu,s) - p_{j+1}(\nu,s)\,p_{n-1}(\nu,r)\}^2 \\ + \{p_{j-1}(\nu,r)\,p_{n-1}(\nu,s) - p_{j-1}(\nu,s)\,p_{n-1}(\nu,r)\}^2 \end{array} \right\}$$

$$+ \frac{(r-s)^2}{(rs)^2} (p_{n-1}(\nu,r)\,p_{n-1}(\nu,s))^2$$

$$\le \frac{4}{(rs)^2} \sum_{j,k=0}^{n-1} (p_j(r)\,p_k(s) - p_j(s)\,p_k(r))^2$$

$$+ \frac{(r-s)^2}{(rs)^2} (p_{n-1}(\nu,r)\,p_{n-1}(\nu,s))^2$$

$$= \frac{8W}{(rs)^2} + \frac{(r-s)^2}{(rs)^2} (p_{n-1}(\nu,r)\,p_{n-1}(\nu,s))^2, \tag{18}$$

recall (15). Next, from (15) again, and the Christoffel–Darboux formula,

$$W \ge \frac{1}{2} (p_{n-2}(\nu,r)\,p_{n-1}(\nu,s) - p_{n-2}(\nu,s)\,p_{n-1}(\nu,r))^2$$

$$= \frac{1}{2} \left(\frac{\gamma_{n-2}}{\gamma_{n-1}}(\nu)\right)^{-2} (K_{n-1}(\nu,r,s)\,(r-s))^2$$

$$\ge \frac{1}{2} \left(\frac{\gamma_{n-2}}{\gamma_{n-1}}(\nu)\right)^{-2} (p_{n-2}(\nu,r)\,p_{n-2}(\nu,s)\,(r-s))^2.$$

This last step uses that $r, s \geq 1$ so that all $p_j(\nu, r), p_j(\nu, s) \geq 0$. Then,

$$\begin{aligned}
&\frac{(r-s)^2}{(rs)^2}\left(p_{n-1}(\nu, r)\, p_{n-1}(\nu, s)\right)^2 \\
&= \left(p_{n-2}(\nu, r)\, p_{n-2}(\nu, s)(r-s)\right)^2 \frac{\left(p_{n-1}(\nu, r)\, p_{n-1}(\nu, s)\right)^2}{\left(r p_{n-2}(\nu, r)\, s p_{n-2}(\nu, s)\right)^2} \\
&\leq 2W\left(\frac{\gamma_{n-2}}{\gamma_{n-1}}(\nu)\right)^2 \frac{\left(p_{n-1}(\nu, r)\, p_{n-1}(\nu, s)\right)^2}{\left(r p_{n-2}(\nu, r)\, s p_{n-2}(\nu, s)\right)^2}.
\end{aligned}$$

As at (17), we continue this as

$$\frac{(r-s)^2}{(rs)^2}\left(p_{n-1}(\nu, r)\, p_{n-1}(\nu, s)\right)^2 \leq 2W\left(\frac{\gamma_{n-2}}{\gamma_{n-1}}(\nu)\right)^{-2}.$$

Substituting in (18) gives

$$\begin{aligned}
Y^T \mathbb{K}^{-1} Y \left(\frac{\gamma_{n-1}}{\gamma_n}(\nu)\right)^{-2} &\leq \frac{8}{(rs)^2} + 2\left(\frac{\gamma_{n-2}}{\gamma_{n-1}}(\nu)\right)^{-2} \\
&\leq 10\left(\frac{\gamma_{n-2}}{\gamma_{n-1}}(\nu)\right)^{-2},
\end{aligned}$$

so from Lemma 7, if $M, N > 0$,

$$\left(\frac{p_n(\mu, 0)}{p_n(\nu, 0)}\right)^2 \leq \left(1 + \sqrt{10}\left(\frac{\gamma_{n-1}}{\gamma_n}(\nu) \Big/ \frac{\gamma_{n-2}}{\gamma_{n-1}}(\nu)\right)\right)^2.$$

As above, the case where $M = 0$ or $N = 0$ can be handled by continuity. □

Proof of Corollary 3 If ν lies in the Nevai class, we have (1) so that for some $C > 0$ and all $n \geq 1$,

$$\frac{\gamma_{n-1}}{\gamma_n}(\nu) \Big/ \frac{\gamma_{n-2}}{\gamma_{n-1}}(\nu) \leq C. \tag{19}$$

Then Theorems 1 and 2 give the result. Next, suppose that in an interval J, $\nu' > 0$ a.e. while $\nu' \leq C$ for some constant C. Let I be a proper subinterval of J of positive length such that the distance

d from I to $(-1,1)\backslash J$ is positive. We use the reproducing property and Christoffel–Darboux formula: for $x \in J$,

$$\begin{aligned}1 &= \int_{-1}^{1} K_n(\nu, x, t)\, d\nu(t) \\ &= \frac{\gamma_{n-1}}{\gamma_n}(\nu)\left(p_n(\nu, x)\int_{-1}^{1}\frac{p_{n-1}(\nu, t)}{x-t}d\nu(t) - p_{n-1}(\nu, x)\right. \\ &\quad \left.\times \int_{-1}^{1}\frac{p_n(\nu, t)}{x-t}d\nu(t)\right),\end{aligned}$$

where the integrals are Cauchy principal value integrals. Then,

$$\begin{aligned}&\int_I \sqrt{\nu'(x)}dx \\ &\quad \leq \frac{\gamma_{n-1}}{\gamma_n}(\nu)\begin{pmatrix} \int_I |p_n(\nu, x)|\sqrt{\nu'(x)}\left|\int_{-1}^{1}\frac{p_{n-1}(\nu,t)}{x-t}d\nu(t)\right|dx \\ +\int_I |p_{n-1}(\nu, x)|\sqrt{\nu'(x)}\left|\int_{-1}^{1}\frac{p_n(\nu,t)}{x-t}d\nu(t)\right|dx\end{pmatrix}.\end{aligned} \tag{20}$$

Here using boundedness of the Hilbert transform in L_2 and Cauchy–Schwarz,

$$\begin{aligned}&\int_I |p_n(\nu, x)|\sqrt{\nu'(x)}\left|\int_{-1}^{1}\frac{p_{n-1}(\nu, t)}{x-t}d\nu(t)\right|dx \\ &\quad \leq \int_I |p_n(\nu, x)|\sqrt{\nu'(x)} \\ &\qquad \times\left\{\left|\int_J \frac{p_{n-1}(\nu, t)}{x-t}\nu'(t)\,dt\right| + \frac{1}{d}\int_{(-1,1)\backslash J}|p_{n-1}(\nu, t)|\,d\nu(t)\right\}dx \\ &\quad \leq \left(\int_J p_n^2(\nu, x)\,\nu'(x)\,dx\right)^{1/2}\left(\int_J\left|\int_J\frac{p_{n-1}(\nu, t)}{x-t}\nu'(t)\,dt\right|^2 dt\right)^{1/2} \\ &\qquad +\frac{1}{d}\left(\int_{-1}^{1}p_n^2(\nu, x)\,d\nu(x)\right)^{1/2}\left(\int_{-1}^{1}p_{n-1}^2(\nu, t)\,d\nu(t)\right)^{1/2} \\ &\qquad \times\left(\int_{-1}^{1}d\nu(t)\right) \\ &\quad \leq C\int_J p_{n-1}^2(\nu, t)\,\nu'(t)^2\,dt + C \leq C,\end{aligned}$$

where we are also using that ν' is bounded above in J. The other term in (20) may be similarly estimated. So, we obtain

$$\frac{\gamma_{n-1}}{\gamma_n}(\nu) \geq C > 0$$

and again (19) holds. □

References

[1] Ambroladze, M.U. On the possible rate of growth of polynomials that are orthogonal with a continuous positive weight. *Math. USSR-Sb.* 72, 311–331 (1992).

[2] Alvarez-Nodarse, R., Marcellán, F. and Petronilho, J. WKB approximation and Krall-type orthogonal polynomials, *Acta Applicandae Mathematicae*, 54(1), 27–58 (1998).

[3] Aptekarev, A., Denisov, S. and Tulyakov, D. On a problem by Steklov, *J. Amer. Math. Soc.* 29, 1117–1165 (2016).

[4] Arvesu, J., Marcellan, F. and Alvarez-Nodarse, R. On a Modification of the Jacobi Linear Functional: Asymptotic Properties and Zeros of the Corresponding Orthogonal Polynomials, *Acta Applicandae Mathematicae*, 71, 127–158 (2002).

[5] Badkov, V.M. The asymptotic behavior of orthogonal polynomials. (Russian) *Mat. Sb.* (N.S.) 109(151), 1, 46–59 (1979).

[6] Chihara, T.S. Orthogonal Polynomials and Measures with End Point Masses, *Rocky Mountain Journal of Math.*, 15, 705–712 (1985).

[7] Denisov, S. On the growth of polynomials orthogonal on the unit circle with a weight w that satisfies $w, w^{-1} \in L_\infty(T)$, *Math. Sbornik*, 209, 71–105 (2018).

[8] Fejzullahu, B. Xh. A Cohen type inequality for polynomial expansions associated with the measure $(1-x)^\alpha (1+x)^\beta dx + M\delta_{-1} + N\delta_1$, *J. Math. Sci. Univ. Tokyo* 15(2), 243–255 (2008).

[9] Freud, G. *Orthogonal Polynomials*, Pergamon Press, Akademiai Kiado, Budapest (1971).

[10] Guadalupe, J.J., Perez, M., Ruiz, F.J. and Varona, J.L. Asymptotic Behavior of Orthogonal Polynomials Relative to Measures with Mass Points, *Mathematika*, 40, 331–344 (1993).

[11] Kiesel H. and Wimp, J. A note on Koornwinder's polynomials with weight function $(1-x)^\alpha (1+x)^\beta + M\delta(x+1) + N\delta(x-1)$, *Numerical Algorithms*, 11, 229–241 (1996).

[12] Koornwinder, T.H. Orthogonal polynomials with weight function $(1-x)^{\alpha}(1+x)^{\beta}+M\delta(x+1)+N\delta(x-1)$, *Canad. Math. Bull.*, 27, 205–214 (1984).

[13] Eli Levin, and Lubinsky, *D.S. Orthogonal Polynomials for Exponential Weights*, Springer, New York (2001).

[14] Lubinsky, D.S. The effect of adding endpoint masspoints on bounds for orthogonal polynomials, *Dolomites Research Notes on Approximation*, 14, 42–55 (2021).

[15] Rahmanov, E.A. Steklov's conjecture in the theory of orthogonal polynomials, (Russian) *Mat. Sb.* (N.S.), 108(150), 581–608 (1979).

[16] Rakhmanov, E.A. Estimates of the growth of orthogonal polynomials whose weight is bounded away from zero, (Russian) *Mat. Sb.* (N.S.) 114(156), 269–298 (1981).

[17] Rakhmanov, E.A. On asymptotic properties of polynomials orthogonal on the circle with weights not satisfying Szegő's condition, *Math. USSR. Sbornik*, 58, 149–167 (1987).

https://doi.org/10.1142/9789811293009_0026

Chapter 26

The Stability of a Functional Equation in Two Variables

Laura Manolescu[*,§] and Michael Th. Rassias[†,‡,¶,||]

[*] *Department of Mathematics,*
Politehnica University of Timişoara,
Timişoara, Romania

[†] *Department of Mathematics and Engineering Sciences,*
Hellenic Military Academy, 16673 Vari Attikis, Greece

[‡] *Institute for Advanced Study,*
Program in Interdisciplinary Studies,
1 Einstein Dr, Princeton, NJ 08540, USA

[§] *laura.manolescu@upt.ro*

[¶] *mthrassias@yahoo.com*

[||] *mrassias@sse.gr*

In this chapter, we prove the generalized Hyers–Ulam stability of the following functional equation:

$$F(x+x', y+y') + F(x-x', y-y') - 2F(x,y) - 2F(x,y') = 0,$$

without any supplementary conditions.

1. Introduction

The starting point for the study of stability problems for functional equations is considered to be a question concerning the stability of group homomorphisms raised in 1940 by Ulam [36] while delivering a talk at the University of Wisconsin: Under what conditions does there exist a homomorphism near an approximate homomorphism? More generally, when a solution of an equation differing slightly from a given one must be somehow near to the solution of the given equation?

The first answer to Ulam's question, concerning the additive Cauchy equation, was given by Hyers [20] in 1941. The method used by Hyers to produce the additive mapping is called the direct method.

Subsequently, starting with the papers of Aoki [1], who gave a generalized solution to Ulam's problem for additive mappings, and Rassias [32], who generalized the theorem of Hyers for approximately linear mappings, a new direction of research was opened in stability theory of functional equations (cf. [10–12, 25, 28–31].

A further generalization was obtained by Găvruţa [17] in 1994, who provided a generalization of Rassias' theorem for the unbounded Cauchy difference and introduced the concept of generalized Hyers–Ulam–Rassias stability. See also [18] for more generalizations. For other results and generalizations, the reader is referred to [5, 16, 21, 24], and the references therein.

An interesting application of the stability in the sense of Hyers–Ulam was pointed out by Hyers, Isac, and Rassias in the study of complementarity problems (see [21]).

In 1991, Baker [2] used the Banach fixed point theorem to establish Hyers–Ulam stability results for a nonlinear functional equation. See also [8] and [9].

In [22], Hwang and Park presented some results concerning the generalized Hyers–Ulam–Rassias stability of the following functional equation:

$$F(x + x', y + y') + F(x - x', y - y') - 2F(x, y) - 2F(x, y') = 0. \tag{1}$$

They proved that if $F : X \times Y \to Z$ and $F(0, y) = F(x, 0) = 0$, $(\forall)$ $x \in X, y \in Y$, then F is additive in the first argument and quadratic in the second argument. Additionally, they showed that if $f : X \times Y \to Z$ is a mapping satisfying approximately (1) and

$$f(x, 0) = f(0, y) = 0, \quad (\forall)\ x \in X \text{ and } y \in Y, \tag{2}$$

then there exists a solution of equation (1) near f, which is unique under certain conditions.

We prove that if $F : X \times Y \to Z$ is the solution of equation (1), the condition $F(x, 0) = 0$, $(\forall)$ $x \in X$ always holds, and if $F(0, y) = 0$, then $F \equiv 0$.

In [14], Ciepliński presented a generalization of Hyers–Ulam stability for a general functional equation in four variables, also under condition (2). This equation contains equation (1) as a particular case and also the functional bilinear equation. Furthermore, K. Ciepliński [14] posed the following open problem: *Is there a way to prove the stability of this equation without the supplementary condition* (2)?

Bahyrycz and Sikorska [3] proved the Hyers–Ulam stability for a functional equation that generalizes the bilinear equation, without the supplementary condition (2).

In this chapter, we establish Hyers–Ulam–Rassias stability results for equation (1), without the supplementary hypothesis (2).

2. Results

Initially, let X, Y, Z be three linear spaces and $F : X \times Y \to Z$ be a mapping.

Theorem 1.

(a) *If F is a solution of the functional equation* (1), *then*

$$F(x, 0) = 0,\ (\forall)\ x \in X \text{ and if } F(0, y) = 0,\ (\forall)\ y \in Y \Longrightarrow F \equiv 0.$$

(b) *F is a solution of equation* (1) *if and only if there exists a quadratic function $G : Y \to Z$ such that*

$$F(x, y) = G(y), \quad (\forall)\ x \in X,\ y \in Y.$$

Proof. We set $x' = -x$ and $y' = y$ in (1) to obtain

$$F(0, 2y) + F(2x, 0) - 4F(x, y) = 0.$$

Hence,

$$F(x, y) = \frac{1}{4}F(2x, 0) + \frac{1}{4}F(0, 2y). \tag{3}$$

If in (1) we consider $x' = y = y' = 0$, we get

$$2F(x, 0) - 4F(x, 0) = 0 \Longrightarrow F(x, 0) = 0, \quad (\forall)x \in X.$$

If in (1) we set $x = x' = 0$, we get

$$F(0, y + y') + F(0, y - y') = 2F(0, y) + 2F(0, y'),$$

hence, $y \mapsto F(0, y)$ is quadratic.

If F constitutes a solution of equation (1), it follows from (3) that

$$F(x, y) = \frac{1}{4}F(0, 2y), \quad \text{i.e., } F(x, y) = G(y),$$

where G is quadratic mapping.

Conversely, if $G : Y \to Z$ is a quadratic mapping, then the two-variable function

$$F : X \times Y \to Z, \ F(x, y) = G(y), \quad (\forall)x \in X, \ y \in Y$$

is clearly a solution of equation (1). □

In the following, we denote by X and Y two linear spaces and by $(Z, \|\cdot\|)$ a Banach space. We also consider the following two functions:

$$f : X \times Y \to Z \quad \text{and} \quad \varphi : X \times X \times Y \times Y \to [0, \infty),$$

as well as the condition

$$\begin{aligned} &\|f(x + x', y + y') + f(x - x', y - y') - 2f(x, y) - 2f(x, y')\| \\ &\leq \varphi(x, x', y, y'), \end{aligned} \tag{4}$$

for all $x, x' \in X$ and $y, y' \in Y$.

We use the following lemma:

Lemma 2. *If f satisfies (4), then the function*

$$g(x,y) := f(x,y) - \frac{f(0,0)}{3}$$

satisfies

$$\|g(2x,2y) - 4g(x,y)\| \leq \varphi(x,x,y,y), \quad \text{for all } x \in X \text{ and } y \in Y. \tag{5}$$

Proof. We consider $x' = x$ and $y' = y$ in (4) and obtain

$$\|f(2x,2y) + f(0,0) - 4f(x,y)\| \leq \varphi(x,x,y,y)$$

or

$$\left\| f(2x,2y) - \frac{f(0,0)}{3} - 4\left(f(x,y) - \frac{f(0,0)}{3} \right) \right\| \leq \varphi(x,x,y,y),$$

hence, we deduce (5). □

Theorem 3. *Let us suppose that f satisfies (4) and the function φ satisfies the following conditions:*

$$\Phi_1(x,y) := \sum_{k=1}^{\infty} 4^{k-1} \varphi\left(\frac{x}{2^k}, \frac{x}{2^k}, \frac{y}{2^k}, \frac{y}{2^k} \right) < \infty \tag{6}$$

and

$$\lim_{k\to\infty} 4^k \varphi\left(\frac{x}{2^k}, \frac{x'}{2^k}, \frac{y}{2^k}, \frac{y'}{2^k} \right) = 0, \tag{7}$$

for any $x, x' \in X$ and $y, y' \subset Y$. Then the function

$$F_1(x,y) := \lim_{n\to\infty} 4^n f\left(\frac{x}{2^n}, \frac{y}{2^n} \right)$$

is well defined and is the unique function which satisfies (1) and

$$\left\| f(x,y) - F_1(x,y) \right\| \leq \Phi_1(x,y), \quad \text{for any } x \in X \text{ and } y \in Y. \tag{8}$$

Proof. From Lemma 2, it follows that the function

$$g(x,y) := f(x,y) - \frac{f(0,0)}{3}$$

satisfies (5). In (5), we set $\frac{x}{2}$ instead of x, and $\frac{y}{2}$ instead of y, to obtain

$$\left\| g(x,y) - 4g\left(\frac{x}{2}, \frac{y}{2}\right) \right\| \leq \varphi\left(\frac{x}{2}, \frac{x}{2}, \frac{y}{2}, \frac{y}{2}\right). \tag{9}$$

From the above inequality, we derive that

$$\left\| g\left(\frac{x}{2}, \frac{y}{2}\right) - 4g\left(\frac{x}{2^2}, \frac{y}{2^2}\right) \right\| \leq \varphi\left(\frac{x}{2^2}, \frac{x}{2^2}, \frac{y}{2^2}, \frac{y}{2^2}\right). \tag{10}$$

From (9) and (10), we get

$$\left\| g(x,y) - 4^2 g\left(\frac{x}{2^2}, \frac{y}{2^2}\right) \right\| \leq 4\varphi\left(\frac{x}{2^2}, \frac{x}{2^2}, \frac{y}{2^2}, \frac{y}{2^2}\right) + \varphi\left(\frac{x}{2}, \frac{x}{2}, \frac{y}{2}, \frac{y}{2}\right). \tag{11}$$

By induction, we deduce that

$$\left\| g(x,y) - 4^n g\left(\frac{x}{2^n}, \frac{y}{2^n}\right) \right\| \leq \sum_{k=1}^{n} 4^{k-1} \varphi\left(\frac{x}{2^k}, \frac{x}{2^k}, \frac{y}{2^k}, \frac{y}{2^k}\right). \tag{12}$$

Furthermore, we claim that the sequence $\left\{4^n g\left(\frac{x}{2^n}, \frac{y}{2^n}\right)\right\}$ is a Cauchy sequence in Z. Indeed, for $n > m$, we have

$$\begin{aligned}
&\left\| 4^m g\left(\frac{x}{2^m}, \frac{y}{2^m}\right) - 4^n g\left(\frac{x}{2^n}, \frac{y}{2^n}\right) \right\| \\
&\quad = 4^m \left\| g\left(\frac{x}{2^m}, \frac{y}{2^m}\right) - 4^{n-m} g\left(\frac{x}{2^n}, \frac{y}{2^n}\right) \right\| \\
&\quad = 4^m \left\| g\left(\frac{x}{2^m}, \frac{y}{2^m}\right) - 4^{n-m} g\left(\frac{x}{2^m} \cdot \frac{1}{2^{n-m}}, \frac{y}{2^m} \cdot \frac{1}{2^{n-m}}\right) \right\|.
\end{aligned}$$

In (12), we set $\frac{x}{2^m}$ instead of x, $\frac{y}{2^m}$ instead of y, and $n-m$ instead of m. We obtain

$$\begin{aligned}
&\left\| g\left(\frac{x}{2^m}, \frac{y}{2^m}\right) - 4^{n-m} g\left(\frac{x}{2^m} \cdot \frac{1}{2^{n-m}}, \frac{y}{2^m} \cdot \frac{1}{2^{n-m}}\right) \right\| \\
&\quad \leq \sum_{k=1}^{n-m} 4^{k-1} \varphi\left(\frac{x}{2^{k+m}}, \frac{x}{2^{k+m}}, \frac{y}{2^{k+m}}, \frac{y}{2^{k+m}}\right).
\end{aligned}$$

Hence,

$$\left\|4^m g\left(\frac{x}{2^m},\frac{y}{2^m}\right) - 4^n g\left(\frac{x}{2^n},\frac{y}{2^n}\right)\right\|$$

$$\leq \sum_{k=1}^{n-m} 4^{k+m-1}\varphi\left(\frac{x}{2^{k+m}},\frac{x}{2^{k+m}},\frac{y}{2^{k+m}},\frac{y}{2^{k+m}}\right)$$

$$= \sum_{p=m+1}^{n} 4^{p-1}\varphi\left(\frac{x}{2^p},\frac{x}{2^p},\frac{y}{2^p},\frac{y}{2^p}\right).$$

Using relation (6), it follows that

$$\lim_{m,n\to\infty}\left\|4^m g\left(\frac{x}{2^m},\frac{y}{2^m}\right) - 4^n g\left(\frac{x}{2^n},\frac{y}{2^n}\right)\right\| = 0.$$

Since Z is a Banach space, it follows that the sequence $\{4^n g(\frac{x}{2^n},\frac{y}{2^n})\}$ converges.

In order to prove (8), we consider in (12) the limit as $n\to\infty$ and derive that

$$\|g(x,y) - F_1(x,y)\| \leq \Phi_1(x,y), \quad \text{for any } x\in X \text{ and } y\in Y. \tag{13}$$

Recall that

$$F_1(x,y) = \lim_{n\to\infty} 4^n g\left(\frac{x}{2^n},\frac{y}{2^n}\right) = \lim_{n\to\infty} 4^n\left[f\left(\frac{x}{2^n},\frac{y}{2^n}\right) - \frac{f(0,0)}{3}\right].$$

From (6), it follows that $\varphi(0,0,0,0) = 0$. Then, from (4), it follows that $f(0,0) = 0$.

Hence,

$$F_1(x,y) = \lim_{n\to\infty} 4^n f\left(\frac{x}{2^n},\frac{y}{2^n}\right).$$

We now prove that the function F_1 constitutes a solution of (1). In (4), we consider

$$x\to\frac{x}{2^n},\quad x'\to\frac{x'}{2^n},\quad y\to\frac{y}{2^n},\quad y'\to\frac{y'}{2^n}$$

and derive that

$$\begin{aligned}&\left\| f\left(\frac{x}{2^n}+\frac{x'}{2^n},\frac{y}{2^n}+\frac{y'}{2^n}\right)+f\left(\frac{x}{2^n}-\frac{x'}{2^n},\frac{y}{2^n}-\frac{y'}{2^n}\right)\right.\\&\qquad\left. -2f\left(\frac{x}{2^n},\frac{y}{2^n}\right)-2f\left(\frac{x}{2},\frac{y'}{2^n}\right)\right\|\\&\quad\leq\varphi\left(\frac{x}{2^n},\frac{x'}{2^n},\frac{y}{2^n},\frac{y'}{2^n}\right).\end{aligned}$$

We multiply the above inequality by 4^n, then we let $n\to\infty$, and using (7), we get

$$\|F_1(x+x',y+y')+F_1(x-x',y-y')-2F_1(x,y)-2F_1(x,y')\|=0.$$

It remains to show that there exists a unique solution of equation (1) that satisfies (8). Hence, we suppose that there exits another solution F_1' of equation (1) such that

$$\|f(x,y)-F_1'(x,y)\|\leq\Phi_1(x,y).$$

From the above inequality and (8), it follows that

$$\|F_1(x,y)-F_1'(x,y)\|\leq 2\Phi_1(x,y).$$

We replace in the above inequality x by $\frac{x}{2^n}$ and y by $\frac{y}{2^n}$ and obtain that

$$\left\| F_1\left(\frac{x}{2^n},\frac{y}{2^n}\right)-F_1'\left(\frac{x}{2^n},\frac{y}{2^n}\right)_1\right\|\leq 2\Phi_1\left(\frac{x}{2^n},\frac{y}{2^n}\right).$$

We multiply the above inequality by 4^n and let $n\to\infty$ to obtain

$$\|F_1(x,y)-F_1'(x,y)\|=0.$$

Therefore, we proved that F_1 is uniquely defined. □

Corollary 4. *We suppose that f satisfies* (4) *and* $f(0,y)=0$, $(\forall)\ y\in Y$. *Then it follows that*

$$\|f(x,y)\|\leq\Phi_1(x,y)\quad(\forall)x\in X,\ y\in Y.$$

Proof. The desired result follows from Theorems 1 and 3 since the solution F_1 satisfies $f(0,y)=0,\ (\forall)y\in Y$. □

We consider now the normed spaces $(X, \|\cdot\|_1)$ and $(Y, \|\cdot\|_2)$. We apply Theorem 3 in the particular case when

$$\varphi(x, x', y, y') = \theta(\|x\|_1^r + \|x'\|_1^r)(\|y\|_2^r + \|y'\|_2^r).$$

We have

$$\begin{aligned} 4^k \varphi\left(\frac{x}{2^k}, \frac{x'}{2^k}, \frac{y}{2^k}, \frac{y'}{2^k}\right) &= 4^k \theta\left(\frac{\|x\|_1^r}{2^{kr}} + \frac{\|x'\|_1^r}{2^{kr}}\right)\left(\frac{\|y\|_1^r}{2^{kr}} + \frac{\|y'\|_1^r}{2^{kr}}\right) \\ &= 4^k \theta(\|x\|_1^r + \|x'\|_1^r)(\|y\|_2^r + \|y'\|_2^r)\frac{1}{4^{kr}} \\ &= \theta(\|x\|_1^r + \|x'\|_1^r)(\|y\|_2^r + \|y'\|_2^r)\left(\frac{4}{4^r}\right)^k \xrightarrow[k\to\infty]{} 0, \quad \text{if } r > 1. \end{aligned}$$

In this case,

$$\begin{aligned} \Phi_1(x, y) &= \sum_{k=1}^{\infty} 4^{k-1}\theta \cdot 4 \cdot \|x\|_1^r \cdot \|y\|_2^r \frac{1}{(4^r)^k} = \theta \cdot \|x\|_1^r \|y\|_2^r \sum_{k=1}^{\infty} (4^{1-r})^k \\ &= \theta \cdot \|x\|_1^r \|y\|_2^r \frac{4^{1-r}}{1 - 4^{1-r}} = \theta \|x\|_1^r \|y\|_2^r \frac{1}{4^{r-1} - 1}. \end{aligned}$$

We obtain the following corollary:

Corollary 5. *If $r > 1$ and*

$$\begin{aligned} &\|f(x + x', y + y') + f(x - x', y - y') - 2f(x, y) - 2f(x, y')\| \\ &\quad \le \theta(\|x\|_1^r + \|x'\|_1^r)(\|y\|_2^r + \|y'\|_2^r), \end{aligned}$$

then the function

$$F_1(x, y) = \lim_{n\to\infty} 4^n f\left(\frac{x}{2^n}, \frac{y}{2^n}\right)$$

is the unique solution of equation (1) which satisfies

$$\|f(x, y) - F_1(x, y)\| \le \frac{\theta}{4^{r-1} - 1}\|x\|_1^r \|y\|_2^r, \quad (\forall)\ x \in X,\ y \in Y.$$

We observe, by comparing our constant $\frac{\theta}{4^{r-1}-1}$ with the constant $\frac{2\theta}{2^r-2}$ obtained in [22], that our constant is smaller. Additionally, our results hold true for $r > 1$, whereas in [22], the results are proved for the case when $r > 2$. Furthermore, we apply Theorem 3 in the particular case when

$$\varphi(x, x', y, y') = \theta\|x\|_1^{\frac{r}{2}} \cdot \|x'\|_1^{\frac{r}{2}} \cdot \|y\|_2^{\frac{r}{2}} \cdot \|y'\|_2^{\frac{r}{2}}, \quad \text{where } r > 1.$$

We have

$$\begin{aligned}
&4^k\varphi\left(\frac{x}{2^k}, \frac{x'}{2^k}, \frac{y}{2^k}, \frac{y'}{2^k}\right)\\
&\quad = 4^k\theta\frac{\|x\|_1^{\frac{r}{2}}}{2^{\frac{kr}{2}}} \cdot \frac{\|x'\|_1^{\frac{r}{2}}}{2^{\frac{kr}{2}}} \cdot \frac{\|y\|_1^{\frac{r}{2}}}{2^{\frac{kr}{2}}} \cdot \frac{\|y'\|_1^{\frac{r}{2}}}{2^{\frac{kr}{2}}}\\
&\quad = 4^k\theta\|x\|_1^{\frac{r}{2}} \cdot \|x'\|_1^{\frac{r}{2}} \cdot \|y\|_2^{\frac{r}{2}} \cdot \|y'\|_2^{\frac{r}{2}}\frac{1}{4^{kr}}\\
&\quad = \theta\|x\|_1^{\frac{r}{2}} \cdot \|x'\|_1^{\frac{r}{2}} \cdot \|y\|_2^{\frac{r}{2}} \cdot \|y'\|_2^{\frac{r}{2}})\left(\frac{4}{4^r}\right)^k \xrightarrow[k\to\infty]{} 0\,, \quad \text{if } r > 1.
\end{aligned}$$

On the other hand,

$$\begin{aligned}
\Phi_1(x, y) &= \sum_{k=1}^{\infty} 4^{k-1}\varphi\left(\frac{x}{2^k}, \frac{x'}{2^k}, \frac{y}{2^k}, \frac{y'}{2^k}\right) = \frac{1}{4}\theta\sum_{k=1}^{\infty}(4^{1-r})^k\|x\|_1^r \cdot \|y\|_2^r\\
&= \theta \cdot \|x\|_1^r\|y\|_2^r\frac{4^{1-r}}{4^{1-r}-1} \cdot \frac{1}{4} = \theta\|x\|_1^r\|y\|_2^r\frac{1}{4^r-4}.
\end{aligned}$$

We obtain the following Corollary:

Corollary 6. *If* $r > 1$ *and*

$$\begin{aligned}
&\|f(x + x', y + y') + f(x - x', y - y') - 2f(x, y) - 2f(x, y')\|\\
&\quad \le \theta\|x\|_1^{\frac{r}{2}} \cdot \|x'\|_1^{\frac{r}{2}} \cdot \|y\|_2^{\frac{r}{2}} \cdot \|y'\|_2^{\frac{r}{2}},
\end{aligned}$$

then the function

$$F_1(x, y) = \lim_{n\to\infty} 4^n f\left(\frac{x}{2^n}, \frac{y}{2^n}\right)$$

is the unique solution of equation (1) *which satisfies*

$$\|f(x, y) - F_1(x, y)\| \le \frac{\theta}{4^r - 4}\|x\|_1^r\|y\|_2^r, \quad (\forall)\ x \in X,\ y \in Y.$$

We can also prove the following result:

Theorem 7. *Let us suppose that*

$$\lim_{n\to\infty}\frac{\varphi(2^n x, 2^n x', 2^n y, 2^n y')}{4^n} = 0 \quad \textit{and}$$

$$\Phi_2(x,y) := \sum_{k=0}^{\infty}\frac{\varphi(2^k x, 2^k x, 2^k y, 2^k y)}{4^{k+1}} < \infty.$$

If the function f satisfies (4), then the mapping

$$F_2(x,y) := \lim_{n\to\infty}\frac{f(2^n x, 2^n y)}{4^n}$$

is the unique solution of equation (1) with the property

$$\left\| f(x,y) - \frac{f(0,0)}{3} - F_2(x,y)\right\| \le \Phi_2(x,y), \quad \textit{for any } x\in X \textit{ and } y\in Y. \tag{14}$$

Proof. Recall that if

$$g(x,y) := f(x,y) - \frac{f(0,0)}{3},$$

then from Lemma 2, we have

$$\left\| g(x,y) - \frac{g(2x,2y)}{4}\right\| \le \frac{\varphi(x,x,y,y)}{4}. \tag{15}$$

In (15), we set $2x$ instead of x, and $2y$ instead of y, to obtain

$$\left\| \frac{g(2x,2y)}{4} - \frac{g(2^2x,2^2y)}{4^2}\right\| \le \frac{\varphi(2x,2x,2y,2y)}{4^2}. \tag{16}$$

From relations (15) and (16), we have

$$\left\| g(x,y) - \frac{g(2^2x,2^2y)}{4^2}\right\| \le \frac{\varphi(x,x,y,y)}{4} + \frac{\varphi(2x,2x,2y,2y)}{4^2}. \tag{17}$$

By induction, we obtain

$$\left\| g(x,y) - \frac{g(2^n x,2^n y)}{4^n}\right\| \le \sum_{k=0}^{n-1}\frac{\varphi(2^k x, 2^k x, 2^k y, 2^k y)}{4^{k+1}}. \tag{18}$$

Using relation (18) and similarly the method followed in Theorem 7, we can prove that $\{\frac{g(2^n x, 2^n y)}{4^n}\}$ is a Cauchy sequence. We set

$$F_2(x, y) = \lim_{n\to\infty} \frac{g(2^n x, 2^n y)}{4^n}.$$

We observe that

$$F_2(x, y) = \lim_{n\to\infty} \frac{f(2^n x, 2^n y)}{4^n}.$$

We prove that F_2 is a solution of equation (1). In (4), we set $2^n x$ instead of x, $2^n x'$ instead of x', $2^n y$ instead of y, and $2^n y'$ instead of y' and obtain that

$$\begin{aligned} &\left\| \frac{f(2^n x + 2^n x', 2^n y + 2^n y')}{4^n} + \frac{f(2^n x - 2^n x', 2^n y - 2^n y')}{4^n} \right. \\ &\qquad \left. - 2\frac{f(2^n x, 2^n y)}{4^n} - 2\frac{f(2^n x', 2^n y')}{4^n} \right\| \\ &\quad \le \frac{\varphi(2^n x, 2^n x', 2^n y, 2^n y')}{4^n}. \end{aligned}$$

Letting n go to infinity in the above inequality and from the hypothesis, it follows that

$$\|F_2(x + x', y + y') + F_2(x - x', y - y') - 2F_2(x, y) - 2F_2(x, y')\| = 0.$$

In order to prove (14), we consider the limit as $n \to \infty$ in (18) and we thus get that

$$\|g(x, y) - F_2(x, y)\| \le \Phi_2(x, y),$$

for any $x \in X$ and $y \in Y$.

It remains to show that there exists a unique solution of equation (1) that satisfies (14). We thus suppose that there exists another solution F_2' of (1) such that

$$\|f(x, y) - \frac{f(0, 0)}{3} - F_2'(x, y)\| \le \Phi_2(x, y).$$

From the above inequality and (14), it follows that

$$\|F_2(x,y) - F_2'(x,y)\| \leq 2\Phi_2(x,y).$$

In the above inequality, we replace x by $2^n x$ and y by $2^n y$, and we obtain

$$\|F_2(2^n x, 2^n y) - F_2'(2^n x, 2^n y)\| \leq 2\Phi_2(2^n x, 2^n y).$$

We divide the above inequality by 4^n and let $n \to \infty$ to obtain

$$\|F_2(x,y) - F_2'(x,y)\| = 0.$$

Hence, we have established that F_1 is uniquely defined. □

In Theorem 7, we consider the following particular cases:

- $(X, \|\cdot\|_1)$ and $(Y, \|\cdot\|_2)$ are two normed spaces;
- $\varphi(x,x',y,y') = \theta(\|x\|_1^r + \|x'\|_1^r)(\|y\|_2^r + \|y'\|_2^r)$, where $0 \leq r < 1$.

We have

$$\varphi(2^k x, 2^k x', 2^k y, 2^k y') = \theta 4^{kr}(\|x\|_1^r + \|x'\|_1^r)(\|y\|_2^r + \|y'\|_2^r).$$

Hence,

$$\frac{\varphi(2^k x, 2^k x', 2^k y, 2^k y')}{4^k} = \theta(4^{r-1})^k(\|x\|_1^r + \|x'\|_1^r)(\|y\|_2^r + \|y'\|_2^r).$$

Since $0 < 4^{r-1} < 1$, it follows that

$$\lim_{n\to k} \frac{\varphi(2^k x, 2^k x', 2^k y, 2^k y')}{4^k} = 0.$$

On the other hand,

$$\begin{aligned}\Phi_2(x,y) &= \sum_{k=0}^{\infty} \frac{\theta 4^{kr}(\|x\|_1^r + \|x'\|_1^r)(\|y\|_2^r + \|y'\|_2^r)}{4^{k+1}}\\ &= \theta\|x\|_1^r\|y\|_2^r \sum_{k=0}^{\infty}(4^{r-1})^k\\ &= \frac{\theta}{1-4^{r-1}}\|x\|_1^r\|y\|_2^r.\end{aligned}$$

Hence, we obtain the following result:

Corollary 8. *Let $0 \le r < 1$ and a mapping $f : X \times Y \to Z$ satisfying the relation*

$$\begin{aligned}&\|f(x+x', y+y') + f(x-x', y-y') - 2f(x,y) - 2f(x,y')\| \\ &\quad \le \theta(\|x\|_1^r + \|x'\|_1^r)(\|y\|_2^r + \|y'\|_2^r).\end{aligned}$$

Then, the function

$$F_2(x,y) := \lim_{n\to\infty} \frac{1}{4^n} f(2^n x, 2^n y)$$

is well defined and is the unique solution of equation (1) with the property

$$\|f(x,y) - F_2(x,y)\| \le \frac{\theta}{1-4^{r-1}} \|x\|_1 \|y\|_2, \quad \textit{for all } x \in X \textit{ and } y \in Y.$$

In [14], K. Ciepliński presented the following corollary.

Corollary 9 ([14], [15]). *Suppose that Z is a Banach space and $\varphi : X^4 \to [0,\infty)$ is a mapping such that*

$$\lim_{k\to\infty} 4^{-k} \varphi(2^k x, 2^k x', 2^k y, 2^k y') = 0, \quad (\forall) x, x', y, y' \in X$$

and

$$\varphi(2x, 2x, 2y, 2y) \le 4L\varphi(x,x,y,y)$$

for some $L \in (0,1)$. If $f : X^2 \to X$ is a function such that condition (2) is satisfied and

$$\begin{aligned}&\|f(x+x', y+y') + f(x-x', y-y') - 2f(x,y) - 2f(x,y')\| \\ &\quad \le \varphi(x, x', y, y'),\end{aligned}$$

$(\forall)$ $x, x', y, y' \in X$, then there exists a unique solution $F : X^2 \to Z$ of equation (1) satisfying (2) and

$$\|f(x,y) - F(x,y)\| \le \frac{1}{4(1-L)} \varphi(x,x,y,y), \quad x, y \in X.$$

The mapping F *is given by*

$$F(x, y) = \lim_{k\to\infty} 4^{-k} f(2^k x, 2^k y), \quad x, y \in X.$$

The following corollary follows immediately from Theorem 7:

Corollary 10. *We suppose that*

$$\lim_{n\to\infty} \frac{\varphi(2^n x, 2^n x', 2^n y, 2^n y')}{4^n} = 0$$

and

$$\varphi(2x, 2x, 2y, 2y) \leq 4L\varphi(x, x, y, y), \quad x, y \in X \tag{19}$$

for some $L \in (0, 1)$. *If the function* f *satisfies* (4), *then the mapping*

$$F(x, y) = \lim_{n\to\infty} \frac{f(2^n x, 2^n y)}{4^n}$$

is the unique solution of equation (1) *with the property*

$$\|f(x, y) - F(x, y)\| \leq \frac{1}{4(1 - L)}\varphi(x, x, y, y).$$

From Corollary 10, *we derive the following result, which improves the results from Corollary* 9:

Corollary 11. *We suppose that*

$$\lim_{n\to\infty} \frac{\varphi(2^n x, 2^n x', 2^n y, 2^n y')}{4^n} = 0$$

and

$$\varphi(2x, 2x, 2y, 2y) \leq 4L\varphi(x, x, y, y), \quad x, y \in X$$

for some $L \in (0, 1)$. *If the function* f *satisfies* (4) *and* $f(0, y) = 0$, $(\forall) y \in Y$, *then*

$$\|f(x, y)\| \leq \frac{1}{4(1 - L)}\varphi(x, x, y, y).$$

Proof. From Corollary 10, we have that F is the unique solution of equation (1), $F(0, y) = 0$. But, from Theorem 1, we have $F \equiv 0$. $\square$

References

[1] Aoki, T. On the stability of the linear transformation in Banach spaces. *J. M. Soc. Japan*, 2, 64–66 (1950).

[2] Baker, J.A. The stability of certain functional equations. *Proc. AMS*, 112(3), 729–732 (1991).

[3] Bahyrycz, A. and Sikorska, J. On stability of a general bilinear functional equation. *Results Math.*, 76, 143 (2021), https://doi.org/10.1007/s00025-021-01447-w.

[4] Brzdęk, J., Chudziak, J. and Páles Z. A fixed point approach to stability of functional equations. *Nonlinear Analysis — TMA*, 74, 6728–6732 (2011).

[5] Brzdęk, J., Popa, D., Raşa, I. and Xu, B. *Ulam Stability of Operators*, Academic Press, Cambridge, MA, USA (2018).

[6] Brzdęk, J., Popa, D. and Rassias Th.M. *Ulam Type Stability*, Springer (2019).

[7] Brzdęk, J., Cieplinski, K. and Rassias, Th.M. Developments in Functional Equations and Related Topics, Springer (2017).

[8] Cădariu, L., Găvruţa, L. and Găvruţa, P. Fixed points and generalized Hyers-Ulam stability. *Abstr. Appl. Anal.*, 2012, 712743, 10 (2012).

[9] Cădariu, L., Găvruţa, L. and Găvruţa, P. Weighted space method for the stability of some nonlinear equations. *Appl. Anal. Discrete Math.*, 6, 126–139 (2012).

[10] Cho, Y.J., Rassias, Th.M. and Saadati, R. *Stability of Functional Equations in Random Normed Spaces*, Springer (2013).

[11] Cho, Y.J., Park, C., Rassias, Th.M. and Saadati, R. *Stability of Functional Equations in Banach Algebras*, Springer (2015).

[12] Cho, Y.J., Rassias, Th.M. and Saadati, R. *Fuzzy Operator Theory in Mathematical Analysis*, Springer, New York (2018).

[13] Ciepliński, K. Applications of fixed point theorems to the Hyers-Ulam stability of functional equations — A survey. *Ann. Funct. Anal.*, 3(1), 151–164 (2012).

[14] Ciepliński, K. On the generalized Hyers-Ulam stability of a functional equation and Its consequences. *Results Math.*, 76, 60 (2021). https://doi.org/10.1007/s00025-021-01371-z.

[15] Ciepliński, K. On perturbations of two general equations in several variables. *Math. Ann.* (2022). https://doi.org/10.1007/s00208-022-02359-y.

[16] Czerwik, S. *Functional Equations and Inequalities in Several Variables*, World Scientific, New Jersey (2002).

[17] Găvruţa, P. A generalization of the Hyers-Ulam-Rassias stability of approximately additive mappings. *J. Math. Anal. Appl.*, 184, 431–436 (1994).

[18] Găvruţa, P. On a problem of G. Isac and Th. M. Rassias concerning the stability of mappings. *J. Math. Anal. Appl.*, 261, 543–553 (2001).

[19] Găvruţa, P. and Găvruţa, L. A new method for the generalized Hyers-Ulam-Rassias stability. *Int. J. Nonlinear Anal. Appl.*, 1(2), 11–18 (2010).

[20] Hyers, D.H. On the stability of the linear functional equation. *Proc. Natl. Acad. Sci.*, USA, 27, 222–224 (1941).

[21] Hyers, D.H., Isac, G. and Rassias, Th.M. *Stability of Functional Equations in Several Variables*, Birkhäuser, Boston (1998).

[22] Hwang, I. and Park, C. Ulam stability of additive-quadratic functional equation in Banach spaces. *J. Math. Ineq.*, 14(2), 421–436 (2020).

[23] Isac, G. Nonlinear analysis and complementarity theory. *J. Glob. Optim.*, 40, 129–146 (2008).

[24] Jung, S.-M. *Hyers-Ulam-Rassias Stability of Functional Equations in Nonlinear Analysis*, Series Springer Optimization and Its Applications 48, Springer (2011)

[25] Jung, S.-M., Popa, D. and Rassias, M.Th. On the stability of the linear functional equation in a single variable on complete metric groups. *Journal of Global Optimization*, 59 (2014), 165–171.

[26] Kannappan, Pl. *Functional Equations and Inequalities with Applications*, Springer (2009).

[27] Kuczma, M. *An Introduction to the Theory of Functional Equations and Inequalities: Cauchy's Equation and Jensen's Inequality*, Uniwersytet Slaski (1985).

[28] Lee, Y.-H., Jung, S.-M., and Rassias, M.Th. Uniqueness theorems on functional inequalities concerning cubic-quadratic-additive equation. *Journal of Mathematical Inequalities*, 12(1), 43–61 (2018).

[29] Lee, Y.-H., Jung, S.-M. and Rassias, M.Th. On an n-dimensional mixed type additive and quadratic functional equation. *Applied Mathematics and Computation*, 228, 13–16 (2014).

[30] Park, C. and Rassias, M.Th. Additive functional equations and partial multipliers in C*-algebras, Revista de la Real Academia de Ciencias Exactas. *Serie A. Matemáticas*, 113(3), 2261–2275 (2019).

[31] Rassias, J.M. Solution of a problem of Ulam. *J. Approximation Theory*, 57(3), 268–273 (1989).

[32] Rassias, Th.M. On the stability of the linear mapping in Banach spaces. *Proc. Amer. Math. Soc.*, 72, 297–300 (1978).

[33] Rassias, Th.M. On the stability of functional equations and a problem of Ulam. *Acta Appl. Math.*, 62, 23–130 (2000).
[34] Rassias, Th.M. and Brzdęk, J. *Functional Equations in Mathematical Analysis*, Springer (2012).
[35] Szekelyhidi, L. *Convolution Type Functional Equations on Topological Abelian Groups*, World Scientific Publishing Co. (1991).
[36] Ulam, S.M. *Problems in Modern Mathematics, Chapter VI*, Science Editors, Wiley, New York (1960).

https://doi.org/10.1142/9789811293009_0027

Chapter 27

Static Nuel Games with Terminal Payoff

Symeon Mastrakoulis* and Athanasios Kehagias†

Faculty of Engineering, Aristotle University, Thessaloniki, Greece

**symeonmastrakoulis@gmail.com*

†kehagiat@ece.auth.gr

In this chapter, we study a variant of the *Nuel* game (a generalization of the *duel*), which is played in turns by N players. In each turn, a single player *must* fire at one of the other players and has a certain probability of hitting and killing his target. The players shoot in a fixed sequence, and when a player is eliminated, the "move" passes to the next surviving player. The winner is the last surviving player. We prove that, for every $N \geq 2$, the Nuel has a stationary Nash equilibrium, and we provide algorithms for its computation.

1. Introduction

In this chapter, we study a variant of the *nuel* game (a generalization of the *duel*), which is played in turns by N players. In each turn, a single player *must* fire at one of the other players (in other words, *abstention* is not allowed) and has a certain probability of hitting and killing his target. The players shoot in a fixed sequence, and when a player is eliminated, the "move" passes to the next surviving player. The winner is the last surviving player.

In what follows, we use the term "N-uel" to describe the N-player game. Hence, the 2-uel, or *duel*, involves two players; the 3-uel, or *truel*, involves three players; etc.

Early works on the static 3-uel are [11, 16, 18–20], in which the postulated game rules guarantee the existence of *exactly one* survivor ("winner"). A more general analysis appears in [17], which considers the possibility of "cooperation" between the players. This idea is further studied in [13–15, 25]. Recent papers on the truel include [1–3, 5–9, 22–24]. Only a few of these papers [1–3, 25] make short remarks on the general N-uel (i.e., for $N \geq 3$), and of these, only [25] uses game theoretic analysis, but the matter is not pursued in depth.[1]

By *solving* the N-uel, we mean establishing the existence of one or more *Nash equilibria* and computing them. As will be explained in a later section, solving the N-uel involves solving a system of nonlinear equations, or, equivalently, minimizing the total *equilibrium error.* Hence, the problem is essentially one of global optimization.

The chapter is organized as follows. In Section 2, we present the rules of the game. In Section 3, we set up the N-uel *system of payoff equations* and prove that, for every N and under appropriate conditions, this system always has a unique solution, i.e., every N-uel has uniquely defined payoffs for all players. In Section 4, we prove that for every N, under appropriate conditions, the N-uel has a *stationary equilibrium strategy profile*, and we provide algorithms to compute this profile (equivalently, to solve the payoff equations system). In Section 5, we present illustrative experiments. In Section 6, we conclude and present some future research directions.

2. The N-uel Game

We now define the N-uel game rigorously for every $N \in \{2, 3, \ldots\}$. The game involves N players, who will be denoted by $P_1, \ldots, P_N$ or by $1, \ldots, N$, and evolves in discrete times (turns) $t \in \{0, 1, 2, \ldots\}$. For each $n \in \{1, \ldots, N\}$, P_n has a *marksmanship* p_n, which is the probability that he hits (and kills) the opponent whom he shoots.

[1] Let us also note the existence of an extensive literature on quite different types of duel games, which are essentially *games of timing* [4, 10, 12]. However, this literature is not relevant to the game studied in this chapter.

In what follows, we assume that, for all $n \in \{1, \ldots, N\}$, p_n lies *strictly* between 0 and 1. In the following sections, we define the components of the game.

2.1. *States and actions*

Each *game state* has the form $\mathbf{s} = s_0 s_1 \cdots s_N$, where $s_0 \in \{0, 1, \ldots, N\}$ and, for $n \in \{1, \ldots, N\}$, $s_n = 0$ (resp. 1) means that P_n is dead (resp. alive). Now, suppose that at time $t \in \{0, 1, 2, \ldots\}$, the game state is $\mathbf{s}(t) = s_0(t) \ldots s_N(t)$. We have the following possibilities:

(1) If $s_0(t) = n \in \{1, \ldots, N\}$, then the game is in progress, and P_n "has the move", i.e., P_n is the single player who will shoot in the current turn. In this case, we also have $s_1(t) \cdots s_N(t) \in \{0, 1\}^N$; $s_i(t) = 1$ means that P_i is alive, while $s_i(t) = 0$ means that he is dead.
(2) If $s_0(t) = 0$, then the game has terminated (i.e., there is a single alive player), and we must actually have $s_1(t) \cdots s_N(t) = 0 \cdots 010 \cdots 0$; i.e., there exists a single n such that $s_n(t) = 1$ (P_n is the single alive player) and, for $m \neq n$, $s_m(t) = 0$ (P_m is dead).

We exclude from consideration *inadmissible* states, i.e., states which will never be visited during a play of the game.[2] It will be useful to define the following sets of *admissible* states:

$$\forall k \in \{2, \ldots, N\}:$$

$$S_k = \left\{ \mathbf{s} = s_0 s_1 \cdots s_N : s_0 \neq 0 \text{ and } s_{s_0} = 1 \text{ and } \sum_{n=1}^{N} s_n = k \right\},$$

$$\forall k \in \{1, \ldots, N\}:$$

$$\widetilde{S}_k = \{\mathbf{s} = 0 s_1 \cdots s_N : \text{and } s_k = 1 \text{ and } s_n = 0 \text{ for } n \neq k\}.$$

S_k is the set of states in which k players are alive and P_{s_0} is the (alive) player who has the move; $\widetilde{S}_k$ is the set of states in which

[2]For example, with $N = 3$ players, the state $s_0 s_1 s_2 s_3 = 1011$ will never occur because it corresponds to the dead player P_1 having a move.

the sole surviving player is P_k and no player has the move. Letting $S_1 = \bigcup_{k=1}^{N} \widetilde{S}_k$, the set of all admissible states is $S = \bigcup_{k=1}^{N} S_k$.

Game actions have the form $a = m \in \{\lambda, 1, \ldots, N\}$. Suppose that at time $t \in \{1, 2, \ldots\}$, the game action is $a(t) = m$; if $m > 0$, then the player who has the move (as determined by the corresponding game state $\mathbf{s}$) will fire at P_m. Depending on the current state, an admissible action must satisfy the following: a player cannot fire either at himself or at an already dead player. We write $a = \lambda$ to denote the "*null action*", i.e., no shooting takes place.

We assume that the game has *perfect information*, i.e., at every t, all players know all previous states and actions.

2.2. *State transitions*

The game evolution is described in terms of state transitions:

(1) The game starts at some initial state $\mathbf{s}(0) = s_0(0)s_1(0)$ $s_2(0) \cdots s_N(0)$.

(2) Assume at $t \in \{1, 2, \ldots\}$, we have $\mathbf{s}(t-1) = s_0(t-1)s_1$ $(t-1) \cdots s_N(t-1)$ with $n = s_0(t-1) \neq 0$ and P_n performs action $a(t) = m \in \{1, \ldots, N\}$ where $s_m(t-1) = 1$. Then the next state is $\mathbf{s}(t) = s_0(t)s_1(t) \cdots s_N(t)$ and is obtained by the following rules:

   (a) For the $s_1(t) \cdots s_N(t)$ part of the state,we have the following:

   $$\Pr(s_m(t) = 0) = p_n \text{ (i.e., } P_n \text{ hit and killed } P_m\text{)},$$

   $$\Pr(s_m(t) = 1) = 1 - p_n \text{ (i.e., } P_n \text{ missed } P_m\text{)},$$

   $$\forall i \in \{1, \ldots, m-1, m+1, \ldots, N\} : \Pr(s_i(t) = s_i(t-1)) = 1,$$

   where we assume that the marksmanships $p_n \in (0, 1)$ *for all* $n \in \{1, \ldots, N\}$.

   (b) The $s_0(t)$ part of the state specifies the next player who has the move. This will be the "next alive player" and is best understood by some examples for the three-player case (analogously holds for $N > 3$ players):

      i. Suppose $\mathbf{s}(t-1) = 1111$, $a(t) = 2$ and $s_1(t)s_2(t)$ $s_3(t) = 111$. This means that P_1 had the move, fired at P_2, and missed him; hence, the next player to have the move is P_2, i.e., $s_0(t) = 2$.

ii. Suppose $\mathbf{s}(t-1) = 1111$, $a(t) = 2$ and $s_1(t)s_2(t)s_3(t) = 101$. This means that P_1 had the move, fired at P_2, and killed him; hence, the next player to have the move is P_3, i.e., $s_0(t) = 3$.
iii. Suppose $\mathbf{s}(t-1) = 3111$, $a(t) = 1$ and $s_1(t)s_2(t)s_3(t) = 111$. This means that P_3 had the move, fired at P_1, and missed him; hence, the next player to have the move is P_1, i.e., $s_0(t) = 1$.
iv. Suppose $\mathbf{s}(t-1) = 1101$, $a(t) = 3$ and $s_1(t)s_2(t)s_3(t) = 100$. This means that P_1 had the move, fired at P_3, and killed him; hence, there is no next player to have the move, i.e., the game has terminated and $s_0(t) = 0$.

(3) Finally, if $\mathbf{s}(t-1) = 0\cdots010\cdots0$ then the game has terminated, and there is neither a next action $a(t)$ nor a next state $\mathbf{s}(t)$.

Using the above rules, for every $N \in \{2, 3, \ldots\}$, we can construct a *controlled Markov chain* with state space S and transition probability matrix $\Pi(a)$, with

$$\forall \mathbf{s}' \in S \backslash S_1 : \Pi_{\mathbf{s}',\mathbf{s}''}(a) = \Pr(\mathbf{s}(t) = \mathbf{s}'' | \mathbf{s}(t-1) = \mathbf{s}', a(t) = a),$$
$$\forall \mathbf{s}' \in S_1 : \Pi_{\mathbf{s}',\mathbf{s}'}(\lambda) = 1.$$

The state sequence obtained from the controlled Markov chain above corresponds exactly to the state sequence of an N-uel, except for the fact that, in the Markov chain, every terminal state $\mathbf{s}' \in S_1$ loops back to itself with probability one, producing an infinite state sequence, while in the N-uel, as soon as a terminal state is entered, the game is over, resulting in a finite state sequence. Note that when $\mathbf{s}' \in S_1$, the only admissible action is $a = \lambda$ (the no-shoot action).

Because in each turn a player must shoot (unless he is the sole survivor) and $p_n > 0$ for all $n \in \{1, \ldots, N\}$, it is easy to prove the following.

Proposition 1. *For all N, the probability that the N-uel terminates in finite time equals one.*

The above completes the description of the N-uel, which we also denote by $\Gamma_N(\mathbf{p})$, where $\mathbf{p} = (p_1, \ldots, p_N)$.

2.3. *Histories, payoffs, and strategies*

We now proceed to define and discuss *histories*, *payoffs*, and *strategies* used in $\Gamma_N(\mathbf{p})$. In view of Proposition 1, we only need to consider finite-length histories. The game history at time t is $\mathbf{h}(t) = \mathbf{s}(0)a(1)\mathbf{s}(1)\cdots\mathbf{s}(t)$ and its *length* is t. The set of all *admissible terminal histories* is

$$H_0 = \{\mathbf{h}(t) = \mathbf{s}(0)a(1)\mathbf{s}(1)\cdots\mathbf{s}(t) : \\ \mathbf{h}(t) \text{ can occur in the game and } \mathbf{s}(t) \in S_1\}.$$

The set of all *admissible nonterminal histories* is

$$H_1 = \{\mathbf{h}(t) = \mathbf{s}(0)a(1)\mathbf{s}(1)\cdots\mathbf{s}(t) : \\ \mathbf{h}(t) \text{ can occur in the game } \mathbf{s}(t) \notin S_1\}.$$

The set of all *admissible histories* is $H = H_0 \cup H_1$.

For $n \in \{1,\ldots,N\}$, P_n's *payoff* function is $\mathsf{Q}_n : H \to \mathbb{R}$ and is defined as follows:

$$\forall n \in \{1,\ldots,N\}, \forall \mathbf{h} \in H :$$
$$\mathsf{Q}_n(\mathbf{h}) = \begin{cases} 1 & \text{iff } \mathbf{h} = \mathbf{s}(0)a(1)\cdots\mathbf{s}(t) \in H_0 \text{ and } s_n(t) = 1, \\ 0 & \text{otherwise.} \end{cases}$$

That is, upon game termination, the single surviving player receives one payoff unit.

Let Δ_N denote the set of N-long probability vectors $\mathbf{x} = (x_1,\ldots,x_N)$. A *strategy* is a function $\sigma : H_1 \to \Delta_N \cup \{(0,\ldots,0)\}$. Suppose $\mathbf{h} = \mathbf{s}(0)a(1)\cdots a(t-1)ns1\cdots sN$ and P_n uses the strategy $\mathbf{x}_n = (x_{n1},\ldots,x_{nN}) = \sigma_n(\mathbf{h})$, then

$$\forall m \in \{1,\ldots,N\} : x_{nm} = \Pr(\text{“}P_n \text{ shoots } P_m\text{”}).$$

Note that a strategy needs to be defined only for nonterminal histories since, at the end of a terminal history, no shooting takes place. We only consider *admissible strategies*, i.e., those which assign positive probability only to admissible actions.[3] In addition, when P_n

[3]For example, a strategy which assigns positive probability to shooting a dead player is inadmissible.

uses an admissible σ_n and $m \neq n$, we must have

$$\forall \mathbf{h} = \mathbf{s}(0)a(1)\cdots a(t-1)ms_1\cdots s_N : \sigma_n(\mathbf{h}) = (0,\ldots,0)$$

(since P_n does not have the move where he cannot fire at anybody).

The probabilities $\mathbf{x} = (\mathbf{x}_1,\ldots,\mathbf{x}_N) = \sigma_n(\mathbf{h})$ depend, in general, on the entire game history (up to the current time). A *stationary strategy* is one that only depends on the current state, and then we write $\sigma_n(\mathbf{s}) = \mathbf{x}_n = (x_{n1},\ldots,x_{nN})$. A *pure strategy* is a function that maps histories to probability vectors which concentrate all probabilities to a single component, equal to one; with a slight abuse of notation, we then also write $\sigma_n(\mathbf{h}) = m$, meaning that P_n shoots P_m with probability one ($\sigma_n(\mathbf{h}) = (0,\ldots,0)$ can be abbreviated as $\sigma_n(\mathbf{h}) = \lambda$, meaning that P_n does not shoot). A *pure stationary strategy* is a pure strategy which only depends on the current state.

A *strategy profile* is a vector $\sigma = (\sigma_1,\ldots,\sigma_N)$, where σ_n is the strategy used by P_n (for $n \in \{1,\ldots,N\}$). An initial state $\mathbf{s}(0)$ and a strategy profile σ define a probability measure on all histories, and hence the *expected payoff* to P_n is well defined by

$$Q_n(\mathbf{s}(0),\sigma) = \mathbb{E}_{\mathbf{s}(0),\sigma}(\mathsf{Q}_n(\mathbf{h})).$$

Note that $Q_n(\mathbf{s}(0),\sigma)$ equals the probability that P_n is the sole remaining survivor, or, equivalently, the winner of $\Gamma_N(\mathbf{p})$.

3. N-uel Payoff System

For each $N \in \{2,3,\ldots\}$ and each fixed strategy profile σ, the payoffs $(Q_i(\mathbf{s},\sigma))_{i\in\{1,\ldots,N\},\mathbf{s}\in S}$ satisfy a system of *payoff equations*. In this section, we formulate the *payoff system*, study its form and properties, and introduce appropriate algorithms for its solution. For clarity of presentation, we first study the 2-uel and 3-uel as special cases and then deal with the general case of the N-uel.

3.1. *The 2-uel*

With $N = 2$ players, we have

$$S_1 = \{010, 001\} \text{ and } S_2 = \{111, 211\},$$

$$S = S_1 \cup S_2 = \{010, 001, 111, 211\}.$$

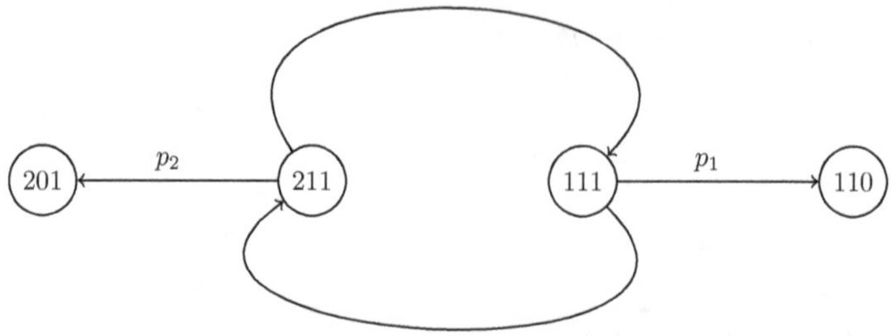

Fig. 1. The 2-uel state transition graph.

For each player, the only admissible strategy is to keep shooting at his opponent until one player is eliminated. The only nonzero state transition probabilities are

$$\Pr(\mathbf{s}(t) = 010 | \mathbf{s}(t-1) = 111, a(t) = 2) = p_1,$$
$$\Pr(\mathbf{s}(t) = 211 | \mathbf{s}(t-1) = 111, a(t) = 2) = 1 - p_1,$$
$$\Pr(\mathbf{s}(t) = 001 | \mathbf{s}(t-1) = 211, a(t) = 1) = p_2,$$
$$\Pr(\mathbf{s}(t) = 111 | \mathbf{s}(t-1) = 211, a(t) = 1) = 1 - p_2.$$

We can represent the above information compactly by a state transition graph, as shown in Figure 1, where actions and corresponding transition probabilities are written next to the respective edges.

The partition of states into the sets S_1 and S_2 is reflected in the structure of the state transition graph: each $\mathbf{s} \in S_2$ has two successors: a state $\mathbf{s}' \in S_2$ and a state $\mathbf{s}'' \in S_1$; each $\mathbf{s} \in S_1$ has no outgoing transitions.

Since the players have no choice of strategies, we do not have a strategic game here, but a *game of chance*. In fact, the 2-uel is essentially a Markov chain.[4] For every initial state, a player's payoff equals his winning probability, which we now compute. For notational brevity, we set

$$\forall i, n \in \{1, 2\}, \text{ for all admissible } ns_1, s_2 :$$
$$V_{i,ns_1s_2} = Q_i(ns_1s_2, (\sigma_1', \sigma_2')),$$

[4]Except for the fact that the terminal states have no associated state transitions.

i.e., the payoff to P_i when the game starts in state $\mathbf{s}(0) = ns_1s_2$, and both players use the same strategy $\sigma_1' = \sigma_2'$ of always shooting at the opponent. It is easily seen that the V_{1,ns_1s_2}'s satisfy the following system of equations:

$$\begin{aligned} V_{1,010} &= 1, \\ V_{1,001} &= 0, \\ V_{1,111} &= (1-p_1)V_{1,211} + p_1 V_{1,010}, \\ V_{1,211} &= (1-p_2)V_{1,111} + p_2 V_{1,200}. \end{aligned} \tag{1}$$

This system has the unique solution

$$\begin{aligned} &V_{1,010} = 1, \quad V_{1,001} = 0, \quad V_{1,111} = \frac{p_1}{p_1+p_2-p_1p_2}, \\ &V_{1,211} = \frac{p_1(1-p_2)}{p_1+p_2-p_1p_2}. \end{aligned} \tag{2}$$

A similar system can be set up for the variables V_{2,ns_1s_2} and has the unique solution

$$\begin{aligned} &V_{2,010} = 0, \quad V_{2,001} = 1, \quad V_{2,111} = \frac{p_2(1-p_1)}{p_1+p_2-p_1p_2}, \\ &V_{2,211} = \frac{p_2}{p_1+p_2-p_1p_2}. \end{aligned} \tag{3}$$

The formulas (2)–(3) provide the solution to the $\Gamma_2(\mathbf{p})$, i.e., the winning probability for each player and for each starting state.

3.2. *The 3-uel*

We limit our analysis to the case where all players use stationary strategies. Suppose $\sigma_n(\mathbf{s})$ is a stationary strategy used by P_n. This can be characterized as follows:

(1) For all $\mathbf{s} \in S_1$ (one player alive), there is no need to define $\sigma_n(\mathbf{s})$.
(2) For all $\mathbf{s} \in S_2$ (two players alive), the only admissible strategy $\sigma_n(\mathbf{s})$ is to shoot at the sole alive opponent.
(3) Finally, consider $\sigma_n(\mathbf{s}) = \mathbf{x}_n = (x_{n1}, x_{n2}, x_{n3})$ when $\mathbf{s} = s_0s_1 s_2s_3 \in S_3$:

(a) when $s_0 \neq n$, we will necessarily have $\mathbf{x}_n = (0, 0, 0)$;
(b) when $s_0 = n$, we will necessarily have $x_{nn} = 0$.

Hence, for every $n \in \{1, 2, 3\}$, an admissible stationary strategy σ_n for P_n is fully determined by the two positive numbers

$$\forall m \neq n : x_{nm} = \Pr(\text{“}P_n \text{ fires at } P_m\text{”} \,|\, \text{“the game state is } n111\text{”}),$$

which must satisfy $\sum_{m \neq n} x_{nm} = 1$. Using the x_{mn}'s, we can draw the state transition graph shown in Figure 2.

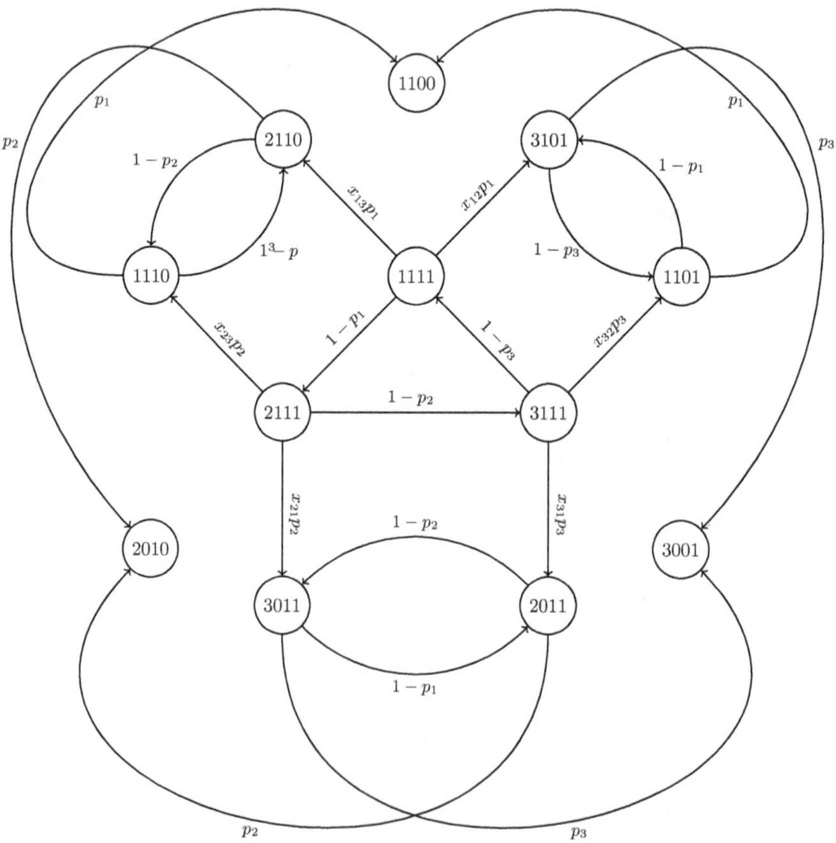

Fig. 2. The 3-uel state transition graph.

We now show the following.

Proposition 2. *For any p_1, p_2, p_3 such that $m \neq n \Rightarrow p_m \neq p_n$ and every admissible strategy profile σ, the 3-uel payoff system has a unique solution.*

Proof. Our goal is to compute each player's expected payoff (equivalently, his winning probability). To this end, similarly to the two-player game, we define

$$\forall i, n \in \{1, 2, 3\}, \text{for all admissible } ns_1s_2s_3 :$$
$$V_{i,ns_1s_2s_3} = Q_i(ns_1s_2s_3, \sigma),$$

i.e., the payoff to P_i when the game starts at state $ns_1s_2s_3$ and the strategy profile $\sigma = (\sigma_1, \sigma_2, \sigma_3)$ is used.[5]

Let us momentarily focus on P_1's payoffs. It is easily seen that the variables $(V_{1,ns_1s_2s_3})_{ns_1s_2s_3 \in S}$ must satisfy the following equations:

$$\begin{aligned}
V_{1,0100} &= 1,\\
V_{1,0010} &= 0,\\
V_{1,0001} &= 0,\\
V_{1,1101} &= (1-p_1)V_{1,3101} + p_1V_{1,0100},\\
V_{1,3101} &= (1-p_3)V_{1,1101} + p_3V_{1,0001},\\
V_{1,1110} &= (1-p_1)V_{1,2110} + p_1V_{1,0100},\\
V_{1,2110} &= (1-p_2)V_{1,1110} + p_2V_{1,0010},\\
V_{1,2011} &= (1-p_2)V_{1,3011} + p_2V_{1,0010},\\
V_{1,3011} &= (1-p_3)V_{1,2011} + p_3V_{1,0001},\\
V_{1,1111} &= (1-p_1)V_{1,2111} + x_{12}p_1V_{1,3101} + x_{13}p_1V_{1,2110},\\
V_{1,2111} &= (1-p_2)V_{1,3111} + x_{23}p_2V_{1,1110} + x_{21}p_2V_{1,3011},\\
V_{1,3111} &= (1-p_3)V_{1,1111} + x_{31}p_3V_{1,2011} + x_{32}p_3V_{1,1101}.
\end{aligned} \tag{4}$$

The above system can be solved in a stepwise fashion. The first three equations immediately yield the values of $V_{1,0100}$, $V_{1,0100}$, and $V_{1,0100}$.

[5]The dependence on σ is omitted from the notation for the sake of brevity.

The fourth and fifth equations can be solved to obtain the values of $V_{1,1101}$ and $V_{1,3101}$:

$$V_{1,1101} = \frac{p_1}{p_1 + p_3 - p_1 p_3}, \quad V_{1,3101} = \frac{p_1(1-p_3)}{p_1 + p_3 - p_1 p_3}; \tag{5}$$

naturally, these are exactly the payoffs for a duel between P_1 and P_3. Similarly, the sixth and seventh equations yield $V_{1,1110}$ and $V_{1,2110}$ (expressions for $V_{1,s_0 s_1 s_2 s_3}$ similar to those of (5)), and the eighth and ninth equations yield $V_{1,2011} = V_{1,3011} = 0$.

The final three equations involve the unknowns $V_{1,1111}$, $V_{1,2111}$, and $V_{1,3111}$ and the previously computed $V_{1,n s_1 s_2 s_3}$'s. The system can be rewritten as

$$\begin{aligned} V_{1,1111} - (1-p_1)V_{1,2111} &= x_{12} p_1 V_{1,3101} + x_{13} p_1 V_{1,2110}, \\ V_{1,2111} - (1-p_2)V_{1,3111} &= x_{23} p_2 V_{1,1110} + x_{21} p_2 V_{1,3011}, \\ V_{1,3111} - (1-p_3)V_{1,1111} &= x_{31} p_3 V_{1,2011} + x_{32} p_3 V_{1,1101}. \end{aligned}$$

Letting

$$\begin{aligned} A_1 &= x_{12} p_1 V_{1,3101} + x_{13} p_1 V_{1,2110}, \\ A_2 &= x_{23} p_2 V_{1,1110} + x_{21} p_2 V_{1,3011}, \\ A_3 &= x_{31} p_3 V_{1,2011} + x_{32} p_3 V_{1,1101}, \end{aligned} \tag{6}$$

the system becomes

$$\begin{bmatrix} 1 & -(1-p_1) & 0 \\ 0 & 1 & -(1-p_2) \\ -(1-p_3) & 0 & 1 \end{bmatrix} \begin{bmatrix} V_{1,1111} \\ V_{2,1111} \\ V_{3,1111} \end{bmatrix} = \begin{bmatrix} A_1 \\ A_2 \\ A_3 \end{bmatrix}.$$

The determinant of the system is $D = 1-(1-p_1)(1-p_2)(1-p_3) > 0$. Hence, the system has a unique solution, which is

$$V_{1,1111} = \frac{A_1 + A_2 + A_3 - p_1 A_2 - p_2 A_3 - p_1 A_3 + p_1 p_2 A_3}{1-(1-p_1)(1-p_2)(1-p_3)}, \tag{7}$$

$$V_{1,2111} = \frac{A_2 + A_3 + A_1 - p_2 A_3 - p_3 A_1 - p_2 A_1 + p_2 p_3 A_1}{1-(1-p_1)(1-p_2)(1-p_3)}, \tag{8}$$

$$V_{1,3111} = \frac{A_3 + A_1 + A_2 - p_3 A_1 - p_1 A_2 - p_3 A_2 + p_3 p_1 A_2}{1-(1-p_1)(1-p_2)(1-p_3)}. \tag{9}$$

In a similar manner, we can prove that the payoff systems for P_2 and P_3 have unique solutions, and this completes the proof. Note that $(V_{i,n111})_{i,n\in\{1,2,3\}}$ are actually functions of $\mathbf{x}_1$, $\mathbf{x}_2$, $\mathbf{x}_3$; for brevity, the dependence has been suppressed from the notation. □

The structure and stepwise solution of the payoff equations correspond to the structure of the state transition diagram. Namely, the vertices of the state transition graph are the game states, and the possible transitions are as follows: S_1 states are terminal, each S_2 state can transit either to a single other S_2 state or to a single S_1 state, and each S_3 state can transit either to a single other S_3 state or one of two S_2 states. Furthermore, the S_3 states form a cycle, i.e.,

$$\cdots \to (1111) \to (2111) \to (3111) \to (1111) \to \cdots .$$

These facts clearly correspond to the stepwise procedure for solving the payoff system. For each $n \in \{1, 2, 3\}$ separately, we first obtain the payoffs $(V_{n,\mathbf{s}})_{\mathbf{s}\in S_1}$, then $(V_{n,\mathbf{s}})_{\mathbf{s}\in S_2}$, and finally $(V_{n,\mathbf{s}})_{\mathbf{s}\in S_3}$.

3.3. *The N-uel*

We now write and solve the payoff system which, analogously to (1) and (4), governs the N-uel expected total payoffs. We use, for all $n \in \{1, \ldots, N\}$ and all $ms_1 \cdots s_N \in S$, the notation

$$V_{n,ms_1\cdots s_N} = Q_n(ms_1 \cdots s_N, \sigma).$$

In addition, we introduce the following notations. For all $\mathbf{s} = ms_1 \cdots s_N \in S$, we define

$$\text{the set of alive players} : L(\mathbf{s}) = \{n : s_n = 1\},$$

$$\text{the set of alive players other than } m : L_m(\mathbf{s}) = \{n : s_n = 1 \text{ and } n \neq m\}.$$

For all $\mathbf{s} = ms_1 \cdots s_N \in S$, we use $\mathbf{N}(\mathbf{s})$ to denote the state following $\mathbf{s}$ when no player is killed. For example, when $N = 4$, we have

$$\mathbf{N}(11111) = 21111, \quad \mathbf{N}(11011) = 31011, \quad \mathbf{N}(41111) = 11111, \ldots .$$

Furthermore, for all $\mathbf{s} = ms_1 \cdots s_N \in S$, we use $\mathbf{N}_i(\mathbf{s})$ to denote the the state following $\mathbf{s}$ when P_i is killed. For example, when $N = 4$, we have

$$\mathbf{N}_2(11111) = 31011, \quad \mathbf{N}_4(31111) = 11110, \ldots.$$

Finally, for all $n \in \{1, \ldots, N\}$ and all $\mathbf{s} = ms_1 \cdots s_N \in S$, the probability that P_m shoots P_n, when the state is $\mathbf{s}$ and P_m uses σ_m, is

$$x_{ms_1\cdots s_N,n} = \Pr(a = n | P_m \text{ uses strategy } \sigma_m, \text{ current state is } \mathbf{s} = ms_1 \cdots s_N).$$

Using the above notations and assuming a given strategy profile σ (which determines all the shooting probabilities $x_{ms_1\cdots s_N,n}$), the expected total payoffs for P_1 satisfy the following equations:

(1) At terminal states, we have

$$V_{1,0s_1\cdots s_N} = 1 \text{ when } s_1 = 1 \text{ and } V_{1,0s_1\cdots s_N} = 0 \text{ when } s_1 = 0. \tag{10}$$

(2) In all admissible states with two alive players P_1 and P_m, we have

$$V_{1,1s_1\cdots s_N} = \frac{p_1}{p_1 + p_m - p_1 p_m} \text{ and } V_{1,ms_1\cdots s_N} = \frac{p_1(1-p_m)}{p_1 + p_m - p_1 p_m}, \tag{11}$$

which are P_1's winning probabilities in a duel against P_m.

(3) At all admissible states $is_1 \cdots s_N$ with two alive players, both different from P_1, we have

$$V_{1,is_1\cdots s_N} = 0. \tag{12}$$

(4) At all admissible states with more than two alive players, for all $k \in \{3, \ldots, N\}$, $\mathbf{s} = ms_1 \cdots s_N \in S_k$, we have

$$\text{when } 1 \in L(\mathbf{s}) : V_{1,ms_1\cdots s_N} = (1 - p_m) V_{1,\mathbf{N}(\mathbf{s})} + \sum_{n \in L_m(\mathbf{s})} x_{ms_1\cdots s_N,n} p_m V_{1,\mathbf{N}_n(\mathbf{s})}, \tag{13}$$

$$\text{when } 1 \notin L(\mathbf{s}) : V_{1,ms_1\cdots s_N} = 0. \tag{14}$$

The payoff system, which must be satisfied by the $V_{1,ms_1\cdots s_N}$'s, consists of equations (10)–(14). Similar systems are satisfied by the $V_{n,ms_1\cdots s_N}$'s, for $n \in \{2,\ldots,N\}$.

Proposition 3. *For every $N \in \{2,3,\ldots\}$, for any $p_1,\ldots,p_N$ such that $m \neq n \Rightarrow p_m \neq p_n$, and every admissible strategy profile σ, the N-uel payoff system has a unique solution.*

Proof. We only consider the payoff systems regarding $(V_{1,\mathbf{s}})_{s\in S}$ (the cases $(V_{n,\mathbf{s}})_{s\in S}$ with $n \geq 2$ are treated similarly). The proof is by induction. Clearly, for $N = 2$, the payoff system has a unique solution, given by (2). Now, suppose the $(N-1)$-uel payoff system has a unique solution, and consider the N-uel payoff system (10)–(14).

Take any $K \in \{2,\ldots,N-1\}$; for any state $\mathbf{s} = s_0s_1\cdots s_N \in S_K$, we want to determine (for all $n \in \{1,\ldots,N\}$) the corresponding $V_{1,s_0s_1\cdots s_N} = Q_1(s_0s_1\cdots s_N,\sigma)$. This is P_1's payoff in an N-uel involving himself, $K-1$ other alive players, and $N-K$ dead players, which is the same as P_1's payoff in a K-uel involving himself and $K-1$ other alive players. Hence, $V_{n,s_0s_1\cdots s_N}$ can be computed by solving the respective K-uel with K alive players and relabeling the K-uel players and their payoffs so as to correspond with the K alive players of the N-uel. By the inductive assumption, the K-uel has a unique solution, and hence the value $V_{1,s_0s_1\cdots s_N}$ is also uniquely determined.

Consequently, the $V_{1,s_0s_1\cdots s_N}$'s are uniquely determined for all $\mathbf{s} = s_0s_1\cdots s_N \in \cup_{K=2}^{N-1}S_K$. It remains to show that the $V_{1,s_0s_1\cdots s_N}$'s with $\mathbf{s} = s_0s_1\cdots s_N \in S_N$ are also uniquely determined. Now, $\mathbf{s} \in S_N$ means there exist N alive players; hence, $s_1 = \cdots = s_N = 1$, and there exist exactly N such states:

$$S_N = \{11\cdots 1,\ldots,N1\cdots 1\}.$$

Also, since in such states all players are alive, we have

$$21\cdots 1 = \mathbf{N}(11\cdots 1),\quad 31\cdots 1 = \mathbf{N}(21\cdots 1),\ldots,$$
$$11\cdots 1\mathbf{N}(N1\cdots 1).$$

Each of the above states appears once on the left-hand side of an equation (13) and once on the right-hand side of another equation (13). Let us rename the corresponding variables $V_{1,s_0s_1\cdots s_N}$ as follows:

$$\forall m \in \{1,\ldots,N\} : Z_m = V_{1,ms_1\cdots s_N}.$$

Let us also define

$$\forall m \in \{1,\ldots,N\} : A_m = \sum_{i\in L_m(\mathbf{s})} x_{ms_1\cdots s_N,i} p_m V_{1,\mathbf{N}_i(ms_1\cdots s_N)}.$$

It follows that, for all $\mathbf{s} \in S_N$, equations (13) can be rewritten in the form

$$\begin{bmatrix} 1 & -(1-p_1) & 0 & \cdots & 0 \\ 0 & 1 & -(1-p_2) & \cdots & 0 \\ 0 & 0 & 1 & \cdots & 0 \\ \vdots & \vdots & \vdots & \ddots & \vdots \\ -(1-p_N) & 0 & 0 & \cdots & 1 \end{bmatrix} \begin{bmatrix} Z_1 \\ Z_2 \\ Z_3 \\ \vdots \\ Z_N \end{bmatrix} = \begin{bmatrix} A_1 \\ A_2 \\ A_3 \\ \vdots \\ A_N \end{bmatrix}. \tag{15}$$

Furthermore, since the states $\mathbf{N}_i(ms_1\cdots s_N) \in S_{N-1}$, the $V_{1,\mathbf{N}_i(ms_1\cdots s_N)}$'s are uniquely determined as solutions of an $(N-1)$-uel.

A necessary and sufficient condition for the system (15) to have a unique solution is that the determinant

$$D_N = \begin{vmatrix} 1 & -(1-p_1) & 0 & \cdots & 0 \\ 0 & 1 & -(1-p_2) & \cdots & 0 \\ 0 & 0 & 1 & \cdots & 0 \\ \vdots & \vdots & \vdots & \ddots & \vdots \\ -(1-p_N) & 0 & 0 & \cdots & 1 \end{vmatrix}$$

is different from zero. It can be easily proved that

$$D_N = 1 - (1-p_1)(1-p_2)\cdots(1-p_N).$$

Since, by assumption, for all n, we have $p_n \in (0,1)$, it follows that $D_N > 0$, and the inductive step is completed. □

Now, our problem is to solve the system of payoff equations, and the above proof suggests a solution method. In what follows, we define, for every $i \in \{1, \ldots, N\}$ and every $S' \subseteq S$, the *payoffs vector* $\mathbf{V}_{i,S'} = (V_{i,\mathbf{s}})_{\mathbf{s} \in S'}$, i.e., the vector of all payoffs indexed by players and states. For example, let $N = 3$; then, the sets of all states with one and two surviving players are, respectively,

$$S_1 = \{0100, 0010, 0001\}, \quad S_2 = \{1110, 2110, 1101, 3101, 2011, 3011\},$$

and the corresponding payoff vectors are

$$\begin{aligned}\mathbf{V}_{1,S_1} &= (V_{1,0100}, V_{2,0100}, V_{3,0100}, V_{1,0010}, V_{2,0010}, V_{3,0010}, V_{1,0001}, \\ &\qquad V_{2,0001}, V_{3,0001}), \\ \mathbf{V}_{1,S_2} &= (V_{1,1110}, V_{2,1110}, V_{3,1110}, V_{1,2110}, \ldots, V_{2,3011}, V_{3,3011}).\end{aligned}$$

Initialization of the $\mathbf{V}_{i,S_1}$'s, i.e., the set of payoffs (to all players) for states with a single surviving player, is immediate; for example, when $N = 3$, we have

$$\begin{aligned}V_{1,0100} &= 1, \quad V_{2,0100} = 0, \quad V_{3,0100} = 0, \\ V_{1,0010} &= 0, \quad V_{2,0010} = 1, \quad V_{3,0010} = 0, \\ V_{1,0001} &= 0, \quad V_{2,0001} = 0, \quad V_{3,0001} = 1,\end{aligned}$$

and similar values are obtained for $\mathbf{V}_{2,S_1}$ and $\mathbf{V}_{3,S_1}$.

The function SOLVENUEL, presented as follows in pseudocode, computes the $\mathbf{V}_{i,S}$ vectors (for $i \in \{1, \ldots, N\}$):

(1) The function inputs are:
 (a) the number of players N,
 (b) the player of interest i,
 (c) the vector of marksmanships $\mathbf{p} = (p_1, \ldots, p_N)$, and
 (d) the strategy profile σ.
(2) At initialization, the one-player payoff vector $\mathbf{V}_{i,S_1}$ is computed.
(3) In the outer loop of the function, K is the number of living players, from $K = 2$ to $K = N$; for each K, we create the set $\mathcal{C}$ of $\binom{N}{K}$ combinations of living players.
(4) In the inner loop, we solve a K-uel for each player set $C = \{n_1, n_2, \ldots, n_K\} \in \mathcal{C}$. This involves solving a system of the K unknown $\mathbf{V}_{S_C}$'s; when obtained, these are used to gradually populate the elements of the "full" $\mathbf{V}_{i,S}$ vector.

(5) On completion of both loops, all components of the $\mathbf{V}_{i,S}$ vector have been computed, and the function returns $\mathbf{V}_{i,S}$.

Algorithm 1. Function for recursive N-uel solution.

---

function SOLVENUEL($N, i, \mathbf{p}, \sigma$)
    Construct the state set S corresponding to the N-uel.
    Compute $\mathbf{V}_{i,S_1}$
    for $K = 2..N$ **do**
        Let $\mathcal{C}$ be the set of all combinations of K players out of $\{1, \ldots, N\}$
        for $C \in \mathcal{C}$ **do**
            Let S_C be the set of states corresponding to player subset C
            Compute $\mathbf{V}_{i,S_C}$ by solving a K-uel
        end for
    end for
    Return $\mathbf{V}_{i,S}$
end function

The $\mathbf{V}_{i,S_k}$ values, for $k \in \{3, \ldots, N\}$, are obtained by solving $\binom{N}{K}$ systems, each involving K unknowns. Exact values can be obtained using Cramer's rule or matrix inversion. However, we have found that implementation is simpler when the following iterative algorithm is used. We first present the algorithm and then prove its correctness:

(1) The function inputs are:
   (a) the number of alive players K,
   (b) the player of interest i,
   (c) the vector of marksmanships $\mathbf{p} = (p_1, \ldots, p_N)$,
   (d) the strategy profile σ,
   (e) the payoffs vector for $K - 1$ players $\mathbf{V}_{i,S_{K-1}}$, and
   (f) the termination parameter ε.
(2) We initialize, for all states $ms_1 \cdots s_N \in S_K$, at arbitrary values $V^{(0)}_{i,ms_1\cdots s_N}$.
(3) Then, we iterate, for $t \in \{0, 1, 2, \ldots\}$ and for each state $ms_1 \cdots s_N \in S_K$, to obtain new $V^{(t+1)}_{i,ms_1\cdots s_N}$ values using (16).
(4) If at some iteration t, we have $\max_{i\in\{1,\ldots,K\},\mathbf{s}\in S_K} \left| V^{(t+1)}_{i,\mathbf{s}} - V^{(t)}_{i,\mathbf{s}} \right| < \varepsilon$, the algorithm terminates and returns $\mathbf{V}_{i,S} = \mathbf{V}^{(t+1)}_{i,S}$.

Algorithm 2. Iterative solution of payoff system.

function ITERSOLVE($K, i, \mathbf{p}, \sigma, \mathbf{V}_{S_{K-1}}, \varepsilon$)
  for $\mathbf{s} = ms_1 \cdots s_N \in S_K$ **do**
    $V^{(0)}_{i,ms_1 \cdots s_N}$ arbitrary
  end for
  for $t \in \{0, 1, 2, \ldots\}$ **do**
    for $\mathbf{s} = ms_1 \cdots s_N \in S_K$ **do**

$$V^{(t+1)}_{i,\mathbf{s}} = (1 - p_m) V^{(t)}_{i,\mathbf{N}(\mathbf{s})} + \sum_{n \in L_m(\mathbf{s})} x_{\mathbf{s},n} p_m V_{i,\mathbf{N}_n(\mathbf{s})} \tag{16}$$

    end for
    if $\max_{\mathbf{s} \in S_K} \left| V^{(t+1)}_{i,\mathbf{s}} - V^{(t)}_{i,\mathbf{s}} \right| < \varepsilon$ **then**
      Break
    end if
  end for
  $\mathbf{V}_{i,S_K} = \mathbf{V}^{(t+1)}_{i,S_K}$
  Return $\mathbf{V}_{i,S_K}$
end function

The following proposition shows that, for a given strategy profile σ, the IPC algorithm yields in the limit the payoffs of the N-uel.

Proposition 4. *For every $K \in \{2, 3, \ldots\}$, $i \in \{1, 2, \ldots, K\}$, for any $p_1, \ldots, p_K$ such that $m \neq n \Rightarrow p_m \neq p_n$, and for every admissible strategy profile σ, the iterative solution of the payoff system always converges, and we have*

$$\forall \mathbf{s} \in \mathbf{S}_K : \lim_{t \to \infty} V^{(t+1)}_{i,\mathbf{s}} = V_{i,\mathbf{s}}.$$

Proof. Consider the same iteration starting from two different initial conditions: $V^{(0)}_{i,S_K}$ and $U^{(0)}_{i,S_K}$. Then, we have

$$\forall \mathbf{s} = ms_1 \cdots s_N \in S_K : V^{(t+1)}_{i,\mathbf{s}}$$
$$= (1 - p_m) V^{(t)}_{i,\mathbf{N}(\mathbf{s})} + \sum_{n \in L_m(\mathbf{s})} x_{\mathbf{s},n} p_m V_{i,\mathbf{N}_n(\mathbf{s})},$$

$$\forall \mathbf{s} = ms_1 \cdots s_N \in S_K : U_{i,\mathbf{s}}^{(t+1)}$$
$$= (1 - p_m) U_{i,\mathbf{N}(\mathbf{s})}^{(t)} + \sum_{n \in L_m(\mathbf{s})} x_{\mathbf{s},n} p_m V_{i,\mathbf{N}_n(\mathbf{s})}.$$

We then have

$$\forall \mathbf{s} = ms_1 \cdots s_N \in S_K : \left| V_{i,\mathbf{s}}^{(t+1)} - U_{i,\mathbf{s}}^{(t+1)} \right|$$
$$= (1 - p_m) \left| V_{i,\mathbf{N}(\mathbf{s})}^{(t)} - U_{i,\mathbf{N}(\mathbf{s})}^{(t)} \right|$$
$$\Rightarrow \sum_{\mathbf{s} \in S_K} \left| V_{i,\mathbf{s}}^{(t+1)} - U_{i,\mathbf{s}}^{(t+1)} \right| \le \sum_{\mathbf{s} \in S_K} \left(1 - \min_m p_m\right) \left| V_{i,\mathbf{N}(\mathbf{s})}^{(t)} - U_{i,\mathbf{N}(\mathbf{s})}^{(t)} \right|$$
$$\Rightarrow \sum_{m \in L(\mathbf{s})} \left| V_{i,\mathbf{s}}^{(t+1)} - U_{i,\mathbf{s}}^{(t+1)} \right| \le \left(1 - \min_m p_m\right) \sum_{\mathbf{s} \in S_K} \left| V_{i,\mathbf{N}(\mathbf{s})}^{(t)} - U_{i,\mathbf{N}(\mathbf{s})}^{(t)} \right|$$
$$\Rightarrow \sum_{m \in L(\mathbf{s})} \left| V_{i,\mathbf{s}}^{(t+1)} - U_{i,\mathbf{s}}^{(t+1)} \right| \le \left(1 - \min_m p_m\right)^{t+1} \sum_{\mathbf{s} \in S_K} \left| V_{i,\mathbf{N}(\mathbf{s})}^{(0)} - U_{i,\mathbf{N}(\mathbf{s})}^{(0)} \right|.$$

Since $|1 - \min_m p_m| \in (0, 1)$, we have

$$\lim_{t \to \infty} \sum_{m \in L(\mathbf{s})} \left| V_{i,\mathbf{s}}^{(t+1)} - U_{i,\mathbf{s}}^{(t+1)} \right| = 0.$$

This means that, for every i and $\mathbf{s}$, the iteration tends to a unique limit

$$\forall i, \mathbf{s} : \overline{V}_{i,\mathbf{s}} = \lim_{t \to \infty} \sum_{m \in L(\mathbf{s})} V_{i,\mathbf{s}}^{(t)},$$

and we have

$$\forall i, \mathbf{s} : \overline{V}_{i,\mathbf{s}} = (1 - p_m) \overline{V}_{i,\mathbf{N}(\mathbf{s})} + \sum_{n \in L_m(\mathbf{s})} x_{\mathbf{s},n} p_m V_{i,\mathbf{N}_n(\mathbf{s})}.$$

In other words, $\overline{V}_{i,S_K}$ satisfies the payoff equations, which means that the iteration yields the unique solution of the payoff system. □

It should be pointed out that Algorithms 1 and 2 are *guaranteed* to work (i.e., solve the N-uel) when the p_n marksmanships belong to $(0, 1)$; however, the algorithms *may* also work even when some of the p_n's are equal to zero or one.

4. N-uel Stationary Equilibria

We are now ready to study the existence of N-uel equilibria. For every N, we have to solve a separate system of *nonlinear* equations. For clarity of presentation, we first deal with the 3-uel and then consider the general N-uel.

Proposition 5. *For any p_1, p_2, p_3 such that $m \neq n \Rightarrow p_m \neq p_n$, the 3-uel has a unique stationary deterministic Nash equilibrium $\widehat{\sigma}$, which can be described as follows:*

$$\widehat{\sigma}(1111) = (0, \widehat{x}_{12}, 1 - \widehat{x}_{12}), \quad \widehat{\sigma}(2111) = (1 - \widehat{x}_{23}, 0, \widehat{x}_{23}),$$
$$\widehat{\sigma}(3111) = (\widehat{x}_{31}, 1 - \widehat{x}_{31}, 0),$$

where

$$\widehat{x}_{12} = \begin{cases} 1 & \text{iff } p_2 > p_3 \\ 0 & \text{else} \end{cases}, \quad \widehat{x}_{23} = \begin{cases} 1 & \text{iff } p_3 > p_1 \\ 0 & \text{else} \end{cases},$$

$$\widehat{x}_{31} = \begin{cases} 1 & \text{iff } p_1 > p_2 \\ 0 & \text{else} \end{cases}.$$

In other words, when in equilibrium, at every turn, each player shoots at his "strongest" opponent with probability one.

Proof. It suffices to consider P_1's point of view. His only strategy choice is when the game is in state $\mathbf{s} = 1111$ (in all other states, there exists a unique admissible strategy), i.e., P_1 must choose between x_{12} and x_{13} (subject to $x_{12} \geq 0$, $x_{13} \geq 0$, and $x_{12} + x_{13} = 1$) so as to maximize $V_{1,1111}$, $V_{1,2111}$, and $V_{1,3111}$. Recall that $V_{1,1111}$ is given by (7):

$$V_{1,1111} = \frac{A_1 + A_2 + A_3 - p_1 A_2 - p_2 A_3 - p_1 A_3 + p_1 p_2 A_3}{1 - (1 - p_1)(1 - p_2)(1 - p_3)},$$

where A_1, A_2, A_3 are given by (6). Note that x_{12} and x_{13} only appear in A_1. Hence, P_1 wants to maximize

$$\begin{aligned} F_1(x_{12}, x_{13}) &= \frac{A_1}{1 - (1 - p_1)(1 - p_2)(1 - p_3)} \\ &= \frac{x_{12} p_1 V_{1,3101} + x_{13} p_1 V_{1,2110}}{1 - (1 - p_1)(1 - p_2)(1 - p_3)} \end{aligned}$$

$$= \frac{x_{12}p_1\frac{p_1(1-p_3)}{p_1+p_3-p_1p_3} + x_{13}p_1\frac{p_1(1-p_2)}{p_1+p_2-p_1p_2}}{1-(1-p_1)(1-p_2)(1-p_3)}$$

$$= p_1^2\frac{\begin{array}{c}x_{12}(1-p_3)(p_1+p_2-p_1p_2)\\+x_{13}(1-p_2)(p_1+p_3-p_1p_3)\end{array}}{\begin{array}{l}(1-(1-p_1)(1-p_2)(1-p_3))(p_1+p_2-p_1p_2)\\ \quad(p_1+p_3-p_1p_3)\end{array}}$$

subject to the constraints $x_{12} + x_{13} = 1, x_{12} \geq 0, x_{13} \geq 0$. Since the denominator is positive, it suffices to choose x_{12} (and consequently $x_{13} = 1 - x_{12}$) so as to maximize

$$x_{12}(1-p_3)(p_1+p_2-p_1p_2) + x_{13}(1-p_2)(p_1+p_3-p_1p_3).$$

Finally, since

$$(1-p_3)(p_1+p_2-p_1p_2) - (1-p_2)(p_1+p_3-p_1p_3) = p_2 - p_3,$$

we have

$$p_2 > p_3 \Leftrightarrow (1-p_3)(p_1+p_2-p_1p_2) > (1-p_2)(p_1+p_3-p_1p_3),$$
$$p_2 < p_3 \Leftrightarrow (1-p_3)(p_1+p_2-p_1p_2) < (1-p_2)(p_1+p_3-p_1p_3).$$

Hence, the optimization rule for $V_{1,1111}$ is simple:

$$\begin{aligned}&\text{if } p_2 > p_3, \text{ then } \widehat{x}_{12} = 1, \widehat{x}_{13} = 0;\\&\text{if } p_2 < p_3, \text{ then } \widehat{x}_{12} = 0, \widehat{x}_{13} = 1.\end{aligned} \tag{17}$$

This rule also maximizes

$$\begin{aligned}V_{1,2111} &= \frac{A_2 + A_3 + A_1 - p_2A_3 - p_3A_1 - p_2A_1 + p_2p_3A_1}{1-(1-p_1)(1-p_2)(1-p_3)}\\&= \frac{A_1 - p_3A_1 - p_2A_1 + p_2p_3A_1 + A_2 + A_3 - p_2A_3}{1-(1-p_1)(1-p_2)(1-p_3)}\\&= \frac{(1-p_2)(1-p_3)A_1 + A_2 + A_3 - p_2A_3}{1-(1-p_1)(1-p_2)(1-p_3)}\end{aligned}$$

and

$$V_{1,3111} = \frac{A_3 + A_1 + A_2 - p_3A_1 - p_1A_2 - p_3A_2 + p_3p_1A_2}{1-(1-p_1)(1-p_2)(1-p_3)}$$
$$= \frac{A_1 - p_3A_1 + A_3 + A_2 - p_1A_2 - p_3A_2 + p_3p_1A_2}{1-(1-p_1)(1-p_2)(1-p_3)}$$
$$= \frac{(1-p_3)A_1 + A_3 + A_2 - p_1A_2 - p_3A_2 + p_3p_1A_2}{1-(1-p_1)(1-p_2)(1-p_3)}.$$

Hence, the rule (17) simultaneously maximizes $V_{1,1111}$, $V_{1,2111}$, and $V_{1,3111}$. This is the "strongest opponent" strategy and is P_1's best response to any of P_2's and P_3's strategies. Through a similar analysis, we can prove that the "strongest opponent" strategy is also P_2's and P_3's best response. This completes the proof. □

Proposition 6. *For every $N \in \{3, 4, \ldots, \}$, the N-uel has a stationary deterministic Nash equilibrium.*

Proof. It suffices to consider P_1's problem of determining his equilibrium strategy $\widehat{\sigma}_1$. The proof is inductive.

For all states $\mathbf{s} \in S_3$, P_1 can determine the value of his optimal strategy $\widehat{\sigma}_1(\mathbf{s})$ as follows. For every state in which he is not alive, he has no strategy choice. For every state in which he is alive, he must solve a 3-uel against the other two alive players; he can do this without considering the value of $\widehat{\sigma}_1(\mathbf{s})$ for states $\mathbf{s} \notin S_1 \cup S_2 \cup S_3$.

Suppose that P_1 has determined $\widehat{\sigma}_1(\mathbf{s})$ for all $\mathbf{s} \in \cup_{k=3}^{K-1} S_k$; now, he wants to determine $\widehat{\sigma}_1(\mathbf{s})$ for all states $\mathbf{s} \in S_K$. He has nothing to determine for states $\mathbf{s} \in S_K$ in which he is not alive. There exist $\binom{N-1}{K-1}$ sets of states where P_1 and $K-1$ other players are alive. Let S' be any such set, and let the alive players be $n_1, \ldots, n_K$; in particular, let $n_1 = 1$. With appropriate state reordering, we can write S' as

$$S' = \{\mathbf{s}_1, \ldots, \mathbf{s}_K\},$$

where P_{n_k} is the player having the move in $\mathbf{s}_k$; in particular, $\mathbf{s}_1$ is the state in which P_1 has the move. Now, letting

$$\forall k \in \{1, \ldots, K\} : \begin{cases} Z_k = V_{1,\mathbf{s}_k} \\ A_k = \sum_{i \in L_k(\mathbf{s}_k)} x_{\mathbf{s}_k,i} p_{n_k} V_{1,\mathbf{N}_i(\mathbf{s}_k)} \end{cases},$$

the following payoff system must be satisfied:

$$\begin{bmatrix} 1 & -(1-p_{n_1}) & \cdots & 0 \\ 0 & 1 & \cdots & 0 \\ \vdots & \vdots & \ddots & \vdots \\ -(1-p_{n_K}) & 0 & \cdots & 1 \end{bmatrix} \begin{bmatrix} Z_1 \\ Z_2 \\ \vdots \\ Z_K \end{bmatrix} = \begin{bmatrix} A_1 \\ A_2 \\ \vdots \\ A_K \end{bmatrix}. \tag{18}$$

The Z_k's are the unknowns, and since each state $\mathbf{N}_i(\mathbf{s}_k)$ belongs to S_{K-1}, the A_k's are known (but depending on the $x_{\mathbf{s}_k,i}$'s). P_1 wants to maximize $Z_k = V_{1,\mathbf{s}_k}$ (for all $k \in \{1,\ldots,K\}$). We can solve (18) using Cramer's rule. We have

$$\begin{vmatrix} 1 & 1-p_{n_1} & \cdots & 0 \\ 0 & 1 & \cdots & 0 \\ \vdots & \vdots & \ddots & \vdots \\ 1-p_{n_K} & 0 & \cdots & 1 \end{vmatrix} = 1-(1-p_{n_1})(1-p_{n_2})\cdots(1-p_{n_K}) > 0,$$

and expanding with respect to the first column, we have

$$V_{1,\mathbf{s}_1} = Z_1 = \frac{\begin{vmatrix} A_1 & 1-p_{n_1} & 0 & \cdots & 0 \\ A_2 & 1 & 1-p_{n_2} & \cdots & 0 \\ A_3 & 0 & 1 & \cdots & 0 \\ \vdots & \vdots & \vdots & \ddots & \vdots \\ A_K & 0 & 0 & \cdots & 1 \end{vmatrix}}{1-(1-p_{n_1})(1-p_{n_2})\cdots(1-p_{n_K})}, \tag{19}$$

or

$$\begin{aligned} V_{1,\mathbf{s}_1} = Z_1 &= \frac{A_1 \begin{vmatrix} 1 & 1-p_{n_2} & \cdots & 0 \\ 0 & 1 & \cdots & 0 \\ \vdots & \vdots & \ddots & \vdots \\ 0 & 0 & \cdots & 1 \end{vmatrix} - A_2 D_2 + A_3 D_3 - \cdots}{1-(1-p_{n_1})(1-p_{n_2})\cdots(1-p_{n_K})} \\ &= \frac{\sum_{i \in L_1(\mathbf{s}_1)} x_{\mathbf{s}_1,i} p_{n_1} V_{1,\mathbf{N}_i(\mathbf{s}_1)} + B}{1-(1-p_{n_1})(1-p_{n_2})\cdots(1-p_{n_K})}. \end{aligned}$$

In the above, D_k is the determinant of the submatrix obtained by removing the first column and the kth row ($k \in \{2, \ldots, K\}$) in (21) and $B = \sum_{k=2}^{K}(-1)^{k-1}A_k D_k$; neither the A_k's nor B contain P_1's shooting probabilities $x_{\mathbf{s}_1,i}$. Hence, P_1 only needs to maximize $\sum_{i \in L_1(\mathbf{s}_1)} x_{\mathbf{s}_1,i} p_{n_1} V_{1,\mathbf{N}_i(\mathbf{s}_1)}$. The rule to achieve this is simple:

$$\widehat{i} = \arg \max_{i \in L_1(\mathbf{s}_1)} V_{1,\mathbf{N}_i(\mathbf{s}_1)}, \tag{20}$$

$$\forall i \neq \widehat{i} : x_{\mathbf{s}_1,i} = 0 \text{ and } x_{\mathbf{s}_1,\widehat{i}} = 1. \tag{21}$$

In (20), $\arg\max_i$ is understood as the smallest i which achieves the maximum (there may exist more than one such, which may result in more than one Nash equilibria). After some additional algebra, it can be verified that this rule also maximizes $V_{1,\mathbf{s}_k}$ for all remaining $k \in \{2, \ldots, K\}$.

This completes the inductive proof for P_1's equilibrium strategy σ_1. The proof for all other players can be worked out in the same way. Let us define a *family* of rules $\mathbf{R}_k$ (for $K \in \{2, \ldots, N\}$):

> $\mathbf{R}_K$: When K players are alive, P_n shoots at some P_i, whose elimination results in a $(K-1)$-uel with the highest payoff to P_n.

What we have proved is that, for all $N \in \{2, 3, \ldots\}$ and all $n \in \{1, \ldots, N\}$, the family $(\mathbf{R}_K)_{K=1}^{N}$ yields a deterministic Nash equilibrium for the N-uel. □

The above proof also furnishes an algorithm, as follows, for computing the equilibrium $\widehat{\sigma} = (\widehat{\sigma}_1, \ldots, \widehat{\sigma}_N)$.

Algorithm 3. Computation of Nash equilibrium.

```
function NashCompute(N, p = (p_1, ..., p_N))
    Initialize σ̂_1(s), ..., σ̂_N(s) for all s ∈ S_1 ∪ S_2
    for K ∈ {3, ..., N} do
        Compute σ̂_1(s), ..., σ̂_N(s) for all s ∈ S_K
    end forK.
end function
```

If the original assumptions are violated, there is no guarantee that Algorithm 3 will yield the equilibrium of the N-uel. However, it is worth noting that if Algorithm 3 terminates, it will always yield an equilibrium, i.e., the algorithm *may work* for combinations of $p_1, \ldots, p_N$ values which violate some of our original assumptions (e.g., with some marksmanships being equal to zero or one).

As will be seen in the following section, the strongest opponent rule can result in rather interesting behaviors for certain combinations of $p_1, \ldots, p_N$.

5. Experiments

In this section, we use computer simulations to present some interesting cases of N-uels. In all the following tables, $\mathbf{s}_n$ denotes the state $(n, 1, 1, 1, \ldots, 1)$.

5.1. 3-*uels*

We begin with 3-uels in which, as mentioned, every player's optimal strategy is to shoot at his strongest opponent (and this holds for all states belonging to S_3).

5.1.1. *Strongest player has the highest/lowest winning probability*

In Table 1, we see a case where the strongest player $P_{\max} = P_1$, i.e., the one with the highest markmanship, has the greatest expected payoff regardless of who has the first move (note that, for each player, the optimal strategy is the same for every state: shoot at your strongest opponent). While this may seem natural, it is actually the exception and not the rule, as one might expect.

Table 1. Strongest player has the greatest expected payoff.

n	1	2	3
p_n	0.90	0.10	0.20
$\widehat{\sigma}_n(\mathbf{s}_n)$	3	1	1
$V_{n,1111}$	0.86	0.12	0.02
$V_{n,2111}$	0.62	0.18	0.20
$V_{n,3111}$	0.69	0.16	0.15

Table 2. The strongest player does not have the greatest expected payoff.

n	1	2	3
p_n	0.50	0.70	0.95
$\widehat{\sigma}_n(\mathbf{s}_n)$	3	3	2
$V_{n,1111}$	0.37	0.56	0.07
$V_{n,2111}$	0.56	0.30	0.14
$V_{n,3111}$	0.50	0.03	0.47

In Table 2, we see that the player $P_{\max} = P_3$ does *not* have the highest expected payoff. In fact, if he does not have the first move, he has the *lowest* expected payoff. This is because each player, $P_i \neq P_{\max}$, will shoot at $P_{\max}$ (by the optimal strategy of shooting at the strongest opponent), and he has a high probability of dying before he has a chance to shoot back. Hence, the "team" consisting of the two players with the lowest marksmanship has a better survival probability than $P_{\max}$ playing alone.

This is one of many cases in which the strongest player, P_3, has the lowest expected payoff when he does not have the first move. And even when he does have the first move, P_1 has the greatest expected payoff. On the other hand, P_1 has a higher expected payoff when P_2 has the move, and P_2 has a higher payoff when P_1 has the move.

5.1.2. *Zugzwang*

Consider the case of $p_1 = p_2 = p_3 = 1$, i.e., every player has perfect markmanship. Without loss of generality, we assume that $\hat{\sigma}_n = \mathbf{N}(n111)$, meaning that each P_n shoots at the next player (actually, it makes no difference to P_n which player he will shoot at). Let P_n be the player who has the first move. P_n will always lose, no matter what strategy he uses, since, after killing his first target, he will fight a 2-uel in which his opponent will have the first move *and* perfect markmanship and so will certainly kill P_n. These facts are illustrated in Table 3.

This resembles a *zugswang* position in chess, i.e., a position in which a player will necessarily lose if he moves *any* of his pieces,

Table 3. Player who has the first move loses.

n	1	2	3
p_n	1.00	1.00	1.00
$\widehat{\sigma}_n\,(n111)$	2	3	1
$V_{n,1111}$	0.00	0.00	1.00
$V_{n,2111}$	1.00	0.00	0.00
$V_{n,3111}$	0.00	1.00	0.00

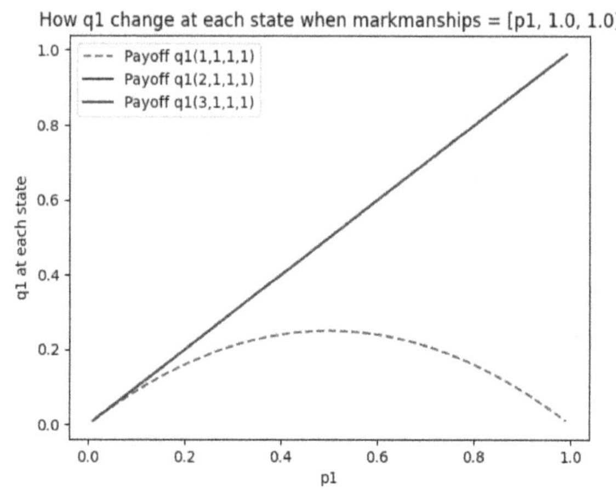

Fig. 3. P_1's payoff is a decreasing function of his marksmanship.

whereas he would not necessarily lose if he could pass (not move any piece).

5.1.3. *Being weaker may increase payoff*

This example is a continuation of the previous one. In Figure 3, we see that P_1's probability of winning is a *decreasing* function of marksmanship p_1 in the interval $[0.5, 1.0]$, when $p_2 = p_3 = 1$. In other words, having a lower marksmanship can increase one's probability of winning (and this is true regardless of which player has the first move).

5.2. *4-uels*

5.2.1. *Shooting the weakest opponent may yield the maximum payoff*

Adding one more player with $p_4 < p_1$ to the experiment in Section 5.1.3, we end up with a paradoxical 4-uel. Consider the example in Table 4.

P_1's optimal strategy at $\mathbf{s} = 11111$ is to shoot at the weakest player. Taking the possible 3-uels in which P_1 can end up if he shoots successfully, we have the following cases:

(1) With $\sigma_1(11111) = 2$, if P_1 kills P_2, he ends up in a truel, similar to that in Section 5.1.1, where P_3 with $p_3 = 1.00$ has the move and P_1 is the second-best player. Hence, the optimal strategy for P_3 is to shoot at P_1, and P_1 loses in the overall 4-uel, i.e., $V_{3,1011} = 0$.
(2) With $\sigma_1(11111) = 3$, P_1 ends up in a similar 3-uel where P_2 always shoots and kills P_1.
(3) With $\sigma_1(11111) = 4$, P_1 ends up in a truel similar to the one in Section 5.1.3, where he does not have the first move and achieves $V_{1,21110} = 0.66$. Hence, this strategy yields the best possible payoff to P_1.

Note that in this example, P_3's best strategy is also to shoot at the weakest player because the probability of P_1 shooting successfully at P_2 after that is high.

5.2.2. *Payoffs as functions of marksmanships p_1 and p_4*

In this example, we generalize the results of Section 5.2.1. In particular, we assume that $p_2 = p_3 = 1$, and we study the dependence of

Table 4. Optimal strategy is to shoot at the weakest player.

n	1	2	3	4
p_n	0.70	1.00	1.00	0.50
$\widehat{\sigma}_n(\mathbf{s}_n)$	4	3	4	2
$V_{n,11111}$	0.66	0.23	0.00	0.11

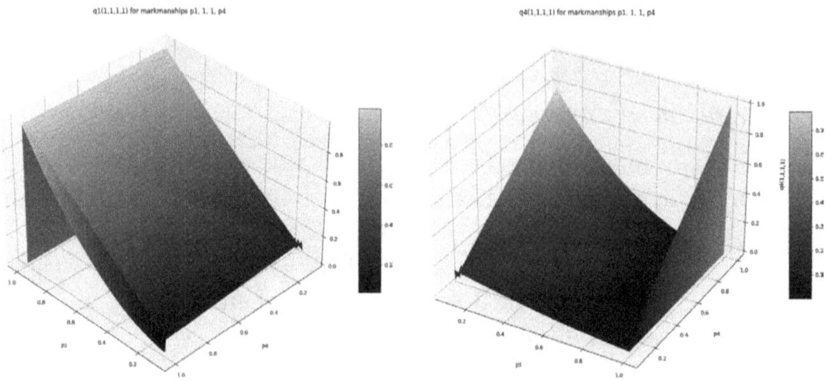

(a) (b)

Fig. 4. Payoffs (a) $V_{1,11111}$ and (b) $V_{4,11111}$ as functions of p_1 and p_4.

Table 5. An example of circular shooting and team formation.

n	1	2	3	4
p_n	0.80	0.40	0.85	0.50
$\widehat{\sigma}_n(\mathbf{s}_n)$	4	3	1	2

the payoffs for the "nonperfect" players P_1 and P_4 on their marksmanships p_1 and p_4.

The surface in Figure 4(a) (resp. Figure 4(b)) is $V_{1,11111}$ (resp. $V_{4,11111}$) as a function of p_1 and p_4, when $p_2 = p_3 = 1$. In both figures, we have a discontinuity at $p_1 = p_4$. This is due to the fact that players change strategies. Taking $p_4 > p_1$ in the 3-uel starting at, for example, $\mathbf{s} = 21101$, P_4 is P_2's new target as he is the next strongest player after P_2.

5.2.3. *Circular shooting sequences and formation of teams*

We now present two examples in which we focus on the players' strategies rather than their payoffs. Taking $\mathbf{p} = (0.80, 0.40, 0.85, 0.50)$, we get the optimal strategies shown in Table 5.

Taking $\mathbf{p} = (0.75, 0.25, 1.00, 0.50)$, we get the optimal strategies shown in Table 6.

Table 6. Another example of circular shooting and team formation.

n	1	2	3	4
p_n	0.75	0.25	1.00	0.50
$\widehat{\sigma}_n(\mathbf{s}_n)$	4	3	1	2

In both cases, the optimal strategy profile follows a *circular shooting order:*

$$P_1 \underset{\text{shoots at}}{\rightarrow} P_4 \underset{\text{shoots at}}{\rightarrow} P_2 \underset{\text{shoots at}}{\rightarrow} P_3 \underset{\text{shoots at}}{\rightarrow} P_1.$$

More generally, in all 4-uel experiments where we have a circular strategy, we can consider the players to be forming two teams. In this experiment, the teams are $\{P_1, P_2\}$ and $\{P_3, P_4\}$; the optimal strategy for each member of each team is to shoot at his teammate's shooter.

5.2.4. *Solidarity of the weakest*

In our final 4-uel example, taking $\mathbf{p} = (0.05, 0.10, 0.15, 0.70)$, we get a situation which resembles an economic model with three small businesses P_1, P_2, and P_3 and a large one P_4. The optimal strategy for the large business P_4 is to eliminate his strongest opponent. Hence, he shoots at P_3. The other three players are so weak that they do not want to lead the game to a truel similar to the experiment in Section 5.1.1, where the strongest player has the greatest payoff (for example, P_2 will not shoot P_1). Consequently, the three weak players (P_1, P_2, and P_3) cooperate to eliminate the strong player. The optimal strategies and respective payoffs are shown in Table 7. Note that P_4 has the greater expected payoff in each case, and in fact, his payoff is either very close or higher than 0.5.

In N-uels of this type, we see a *solidarity* effect between the weakest players. However, it must be noted that such N-uels are a small, non-representative subset of the possible cases, as we have seen from our previous examples.

Table 7. Solidarity of the weakest players.

n	1	2	3	4
p_n	0.05	0.10	0.15	0.70
$\widehat{\sigma}_n(\mathbf{s}_n)$	4	4	4	3
$V_{n,11111}$	0.18	0.20	0.13	0.49
$V_{n,21111}$	0.18	0.19	0.12	0.51
$V_{n,31111}$	0.16	0.18	0.09	0.57
$V_{n,41111}$	0.14	0.15	0.04	0.67

Table 8. An example of circular shooting.

n	1	2	3	4	5
p_n	0.25	0.20	0.10	0.06	0.04
$\widehat{\sigma}_n(\mathbf{s}_n)$	2	3	4	5	1

Table 9. An example of circular shooting.

n	1	2	3	4	5	6	7
p_n	0.40	0.26	0.18	0.12	0.08	0.06	0.04
$\widehat{\sigma}_n(\mathbf{s}_n)$	4	1	5	6	2	7	3

5.3. *N-uels*, $N \geq 5$

We conclude with two examples where, similar to the example in Section 5.2.3, we obtain circular shooting sequences. In the 5-uel with $\mathbf{p} = (0.25, 0.20, 0.10, 0.06, 0.04)$, computation shows that each player's optimal strategy is to shoot at the next player and the last to shoot at the first (Table 8).

Table 9 shows a 7-uel which has a similar effect.

While we have discovered many similar examples by brute-force computation, we have not been able to obtain a condition on the $\mathbf{p}$ values which guarantees the emergence of circular shooting orders. Also, it is not obvious what shooting order will emerge as soon as one player is eliminated; it is not usually the case that the resultant shooting order will again be circular. In the future, we intend to further study these questions.

6. Conclusion

We have studied an N-uel game (a generalization of the *duel*) in which finding a Nash equilibrium reduces to solving the system of nonlinear payoff equations. We have proved that this system has a solution (and hence the N-uel has a stationary Nash equilibrium), and we have provided algorithms for its computational solution. In the future, we want to study:

(1) the existence and computation of *nonstationary* Nash equilibria;
(2) the properties of the N-uel variant in which each player can *abstain*, i.e., not shoot at any of his opponents.

References

[1] Amengual, P. and Toral, R. Distribution of winners in truel games. *AIP Conference Proceedings*, vol. 1, 128–141 (2005).
[2] Amengual, P. and Toral, R. A Markov chain analysis of truels. *Proceedings of the 8th Granada Seminar on Computational Physics* (2005), vol. 1, pp. 7–11.
[3] Amengual, P. and Toral, R. Truels, or survival of the weakest. *Computing in Science & Engineering*, vol. 8, pp. 88–95 (2006).
[4] Barron, E. N. *Game Theory: An Introduction.* John Wiley & Sons, Hoboken, NJ (2013).
[5] Bossert, W., Brams S.J. and Kilgour, D.M. Cooperative vs non-cooperative truels: Little agreement, but does that matter? *Games and Economic Behavior*, 40, 185–202 (2002).
[6] Brams, S.J. and Kilgour, D.M. The truel. *Mathematics Magazine*, 70, 315–326 (1997).
[7] Brams, S.J. and Kilgour, D.M. Games that end in a bang or a whimper. Preprint, CV Starr Center for Applied Economics (2001).
[8] Brams, S.J., Kilgour, D.M., and Dawson, B. Truels and the future. *Math Horizons*, 10, 5–8 (2003).
[9] Dorraki, M., Allison, A., and Abbott, D. Truels and strategies for survival. *Scientific Reports*, 9, 1–7 (2019).
[10] Dresher, M. *Games of Strategy: Theory and Applications*, Prentice-Hall, Englewood Cliffs, NJ (1961).
[11] Gardner, M. *New Mathematical Puzzles and Diversions*, pp. 42–49, Simon and Schuster, New York (1966).
[12] Karlin, S. *Mathematical Methods and Theory in Games, Programming and Economics*, Vol. 2. Addison-Wesley, Reading, MA (1959).

[13] Kilgour, D.M. The simultaneous truel. *International Journal of Game Theory*, 1, 229–242 (1971).

[14] Kilgour, D.M. The sequential truel. *International Journal of Game Theory*, 4, 151–174 (1975).

[15] Kilgour, D.M. Equilibrium points of infinite sequential truels. *International Journal of Game Theory*, 6, 167–180 (1977).

[16] Kinnaird, C. *Encyclopedia of Puzzles and Pastimes*, Citadel Press, New York, NY, (1946).

[17] Knuth, D.E. The triel: A new solution. *Journal of Recreational Mathematics*, 6, 1–7 (1972).

[18] Larsen, H. D., and Moser L. *A Dart Game. American Mathematical Monthly*, vol. 55, pp. 640–641 (1948).

[19] Mosteller, F. *Fifty Challenging Problems in Probability with Solutions*, Addison-Wesley, Reading, MA (1987).

[20] Shubik, M. Does the fittest necessarily survive? *In Readings in Game Theory and Political Behavior*. Doubleday, New York, NY, (1954).

[21] Sobel, M.J. Noncooperative stochastic games. *The Annals of Mathematical Statistics*, 42, 1930–1935 (1971).

[22] Toral, R. and Amengual, P. Distribution of winners in truel games. *AIP Conference Proceedings (American Institute of Physics)*, Vol. 1, pp. 779 (2005).

[23] Wegener, M. and Mutlu, E. The good, the bad, the well-connected. *International Journal of Game Theory*, 50, 759–771 (2021).

[24] Xu, X. Game of the truel. *Synthese*, 185, 19–25 (2012).

[25] Zeephongsekul, P. Nash equilibrium points of stochastic N-uels. In *Recent Developments in Mathematical Programming*, pp. 425–452. CRC Press, Boca Raton, FL (1991).

https://doi.org/10.1142/9789811293009_0028

Chapter 28

Studying the Performance of LexOpt Algorithm for Unconstrained Optimization

Christina D. Nikolakakou[*,§], Theodoula N. Grapsa[†,¶] and George S. Androulakis[‡,||]

[*] *Department of Digital Systems, University of the Peloponnese, GR 23100 Sparta, Greece*

[†] *Department of Mathematics, University of Patras, GR 26504 Rio, Greece*

[‡] *Department of Business Administration, University of Patras, GR 265.04 Rio, Greece*

[§] *gnikolakak@uop.gr*

[¶] *grapsa@upatras.gr*

[||] *gandroul@upatras.gr*

An algorithm for unconstrained optimization problem, named lexicographic optimization algorithm, has been proposed. In this algorithm, under a sequential approach, approximations of the initial problem are constructed such that a new initial point is produced for the given problem. The objective functions of the produced problems are obtained from parts of the given objective function via a proper partitioning technique. In this chapter, first, a modified algorithm of LexOpt algorithm, is proposed. Additionally, two new algorithms based on using relaxation parameters on parts on the utilized objective functions in LexOpt and modified LexOpt algorithm correspondingly are developed. A study on the performance of all algorithms for a set of well-known test problems is presented.

1. Introduction

The general unconstrained optimization problem is considered as

$$\min_{x \in \Omega \subseteq \mathbb{R}^n} f(x), \tag{1}$$

where the function $f : \Omega \subseteq \mathbb{R}^n \to \mathbb{R}$ is called the objective function and $x = [x_1, x_2, \ldots, x_n] \in \Omega$ is the vector of unknown variables ([1, 2]).

A lot of iterative methods have been proposed for the above unconstrained optimization problem. The selection of initial point plays an important role in these class of methods [1–5].

An alternatively optimization strategy in order to optimize problem (1) has been proposed in [6–8] and its implementation has been done by the Lexicographic Optimization (LexOpt) algorithm. In this approach, a proper initial point for the optimization problem (1) via a preprocessing step is produced. The key idea is splitting the given objective function to elements, named cumulative elements (for more details, see definition 3.1 in [6]). Linear combinations of the produced elements are utilized for constructing problems to be approximations of the given one. Giving priorities on them through a sequential optimization process, a new initial point for the given problem is achieved.

In this chapter, a new way for constructing approximation problems is suggested and the Modified Lexicographic Optimization (MLexOpt) algorithm is proposed. In LexOpt algorithm, at least one common cumulative element is utilized for constructing the corresponding approximating objective functions, while in MLexOpt algorithm, no common cumulative elements are utilized for this reason.

Moreover, in this chapter, a study on how a relaxation on cumulative elements may effect the behavior of LexOpt and MLexOpt algorithms is done. Thus, each cumulative element is multiplied by a properly selected relaxation parameter (RelPar).

Finally, a study on the performance of all algorithms through a set of well-known test problems is presented. For this reason, ratios of the absolute value of largest to smallest eigenvalue of the given function, on the points produced from the utilized objective functions, are given. An improvement on the behavior of our algorithms seems, generally, to be succeeded by utilizing relaxation parameters.

The rest of this chapter is organized as follows: A brief description of LexOpt algorithm [6, 7] is introduced in Section 2. The new algorithms and the proofs for their convergence are given in Section 3. In Section 4, numerical results of the proposed algorithms on well-known test functions and a discussion on them are presented. Finally, conclusions and proposed directions for further research are summarized in Section 5.

2. LexOpt Algorithm: A Brief Description

The already proposed LexOpt algorithm [6–8] is based on producing functions, applying cumulative principle [6–8] on the problem given by relation (1).

According to the cumulative principle, the function $f : \Omega \subseteq \mathbb{R}^n \to \mathbb{R}$ is separated into some other functions $q_i(x)$, for $i = 1, \ldots, m$, such that its of them to be a part of the formula of the given objective function and such that their sum to be the given objective function.

In functions $q_i(x)$, named cumulative elements [6,7], priorities are given by the user. Then, problem (1) is transformed to the following one:

$$\min_{x\in\Omega\subseteq\mathbb{R}^n} H(x) = \left\{ \min_{x\in\Omega\subseteq\mathbb{R}^n} h_1(x), \min_{x\in\Omega\subseteq\mathbb{R}^n} h_2(x), \ldots, \min_{x\in\Omega\subseteq\mathbb{R}^n} h_m(x) \right\}, \tag{2}$$

where $h_i(x) : \Omega \subseteq \mathbb{R}^n \to \mathbb{R}$ and $i = 1, 2, \ldots, s-1$ with $s \in \mathbb{N} \setminus \{0, 1\}$ such that

$$\begin{aligned} h_1(x) &= q_1(x), \\ h_2(x) &= q_1(x) + q_2(x), \\ &\vdots \\ h_m(x) &= q_1(x) + q_2(x) + \cdots + q_m(x) = f(x). \end{aligned}$$

The function $h_1(x)$ has the lowest priority and the function $h_m(x)$, which is the objective function of the given problem, the highest one. Each function $h_i(x)$, for $i = 1, \ldots, m$, consists the approximation of the given objective one.

According to LexOpt algorithm, the problems given by relation (2) are optimized sequentially. Specifically, the point that each problem converges is utilized as the initial point for the next problem in the sequence. Hence, a kind of 'direct' walk is created and the last point of this walk is utilized as the initial point for the given problem, which is identical with the target problem given by relation (1). All the involved problems, except of the last one, are not optimized exhaustively [6, 7].

3. The New Proposed Algorithms

In this section, three new algorithms are presented. First, common parts and differences of these algorithms are as follows and then details for each of them are given.

For all new proposed algorithms, problem (1) is substituted by the problem given by relation (2), as in LexOpt algorithm. Following the cumulative principle, the vector

$$q(x) = \{q_1(x), q_2(x), \ldots, q_m(x)\}, m \in \mathbb{N}\backslash\{0, 1\} \tag{3}$$

is produced such that

$$\sum_{i=1}^{m} q_i(x) = f(x). \tag{4}$$

Giving priorities on the produced cumulative elements $q_i(x)$, $i = 1, \ldots, m$, they are ordered according to them. Thus, the function $q_1(x)$ has the lowest priority and the function $q_m(x)$ the highest one.

The key point for each proposed algorithm is the way the components $q_i(x)$, $i = 1, \ldots, m$ are utilized to pose the new problems $\min_{x \in \Omega \subseteq \mathbb{R}^n} h_i(x)$, $i = 1, \ldots, m$ for problem (2). Specifically, the difference between LexOpt and MLexOpt is the fact that in LexOpt at least one $q_i(x)$ is common in the constructed objective functions $h_i(x)$, $i = 1, \ldots, m$, while in MLexOpt algorithm, no common $q_i(x)$, $i = 1, \ldots, m$ is utilized.

3.1. *Modified Lexicographic Optimization (MLexOpt) Algorithm*

The problem given by relation (1) is proposed to be transformed as follows:

$$\min_{x\in\Omega\subseteq\mathbb{R}^n} H(x) = \left\{\min_{x\in\Omega\subseteq\mathbb{R}^n} h_1(x), \min_{x\in\Omega\subseteq\mathbb{R}^n} h_2(x), \ldots, \min_{x\in\Omega\subseteq\mathbb{R}^n} h_m(x)\right\}, \tag{5}$$

where $h_i(x) : \Omega \subseteq \mathbb{R}^n \to \mathbb{R}$ such that

$$\begin{aligned} h_1(x) &= q_1(x), \\ &\vdots \\ h_{m-1}(x) &= q_k(x) + \cdots + q_{(m)(m-k)}(x), \\ h_m(x) &= q_1(x) + \cdots + q_k(x) + \cdots + q_m(x) = f(x), \end{aligned}$$

where $k \in \mathbb{N} \setminus \{0,1\}$, such as $q_1(x) + \cdots + q_k(x) + \cdots + q_m(x) = f(x)$ and $q_i(x) : \Omega \subseteq \mathbb{R}^n \to \mathbb{R}$. Note that among all the constructed objective functions $h_i(x) : \Omega \subseteq \mathbb{R}^n \to \mathbb{R}$ and $i = 1, 2, \ldots, s-1$ with $s \in \mathbb{N}\setminus\{0,1\}$, except of the last one, there are no common cumulative elements.

The problem $\min_{x\in\Omega\subseteq\mathbb{R}^n} h_1(x)$ is set to have the lowest priority and the problem $\min_{x\in\Omega\subseteq\mathbb{R}^n} h_m(x)$ the highest one. The first $m-1$ problems are sequentially optimized, not exhaustively, utilizing a known optimization method. The point in which each problem converges is used as the initial one for the next problem in the sequence. Thus, a sequence of points is constructed and the last point (y^*) of that sequence is utilized as initial point for the last optimization problem, which is identical with the problem given by relation (1).

3.2. *Relaxation Lexicographic Optimization (RLexOpt) Algorithm*

The given problem, (1), is proposed to be transformed to the following one:

$$\min_{x\in\Omega\subseteq\mathbb{R}^n} H(x) = \left\{\min_{x\in\Omega\subseteq\mathbb{R}^n} h_1(x), \min_{x\in\Omega\subseteq\mathbb{R}^n} h_2(x), \ldots, \min_{x\in\Omega\subseteq\mathbb{R}^n} h_m(x)\right\}, \tag{6}$$

Algorithm 1. MLexOpt algorithm

Require: $\{f(x), x^0, \epsilon\}$ {*The initial objective function* $f(x)$, *the used optimization method(s) for each involved problem, the stopping criteria* ϵ, *for each one of the utilized optimization problem* (ϵ *may be a vector with as many components as the utilized optimization problems.*),}

STEP 1 Split the formula of the objective function $f(x)$ by Cumulative Principle ([6, 7]).

STEP 2 Give priorities on the produced Cumulative elements (Definition 3.1 in [6]).

STEP 3 Set m the quantity of Cumulative elements.

STEP 4 Define a Lexicographic Cumulative vector (Definition 3.4 in [6]) dimension m.

STEP 5 Define the utilized objective functions (See relation (5))).

STEP 6 $j \leftarrow 1$

STEP 7 $z^{(0)} \leftarrow x^0$

STEP 8 WHILE $(i < m)$

Use $z^{(i-1)}$ as initial point for the given optimization method

for $\min_{x \in \Omega \subseteq \mathbb{R}^n} h_i(x)$, $x \in \mathbb{R}^n$ problem in relation (5).

$z^{(i)} \leftarrow$ the point that termination criteria for $\min_{x \in \Omega \subseteq \mathbb{R}^n} h_i(x)$

problem is fulfilled.

$z^{(i-1)} \leftarrow z^{(i)}$

$i \leftarrow i + 1$

END WHILE

STEP 9 Use $z^{(m-1)}$ as initial point for the given optimization method for $\min_{x \in \Omega \subseteq \mathbb{R}^n} h_m(x)$, $x \in \mathbb{R}^n$ problem in relation (5) until the termination criteria to be fulfilled.

return x^*

where $h_i(x) : \Omega \subseteq \mathbb{R}^n \to \mathbb{R}$ such that

$$
\begin{aligned}
h_1(x) &= w_{11}q_1(x), \\
&\qquad 0 < w_{11} < 1, \\
h_2(x) &= w_{21}q_1(x) + w_{22}q_2(x), \\
&\qquad w_{21} + w_{22} = 1, \\
h_3(x) &= w_{31}q_1(x) + w_{32}q_2(x) + w_{33}q_3(x), \\
&\qquad w_{31} + w_{32} + w_{33} = 1, \\
&\ \vdots \\
h_{m-1}(x) &= w_{(m-1)1}q_1(x) + \cdots + w_{(m-1)(m-1)}q_{(m-1)(m-1)}(x), \\
&\qquad w_{(m-1)1} + \cdots + w_{(m-1)(m-1)} = 1, \\
h_m(x) &= q_1(x) + q_2(x) + \cdots + q_m(x) = f(x),
\end{aligned}
$$

where $k \in \mathbb{N} \setminus \{0, 1\}$, such as $q_1(x) + \cdots + q_k(x) + \cdots + q_m(x) = f(x)$ and $q_i(x) : \Omega \subseteq \mathbb{R}^n \to \mathbb{R}$. For each term w_{tj}, $t, j = 1, \ldots, (m-1)$, $m \in \mathbb{N} \setminus \{0, 1\}$, named relaxation parameter (RelPar), the relation $0 < w_{tj} < 1$ is holding. Moreover, at each used objective function $h_i(x) : \Omega \subseteq \mathbb{R}^n \to \mathbb{R}$ and $i = 1, 2, \ldots, s-1$ with $s \in \mathbb{N} \setminus \{0, 1\}$, the sum of the utilized relaxation parameters is equal to one unit.

As it has been already mentioned, the problem $\min_{x \in \Omega \subseteq \mathbb{R}^n} h_1(x)$ is set to have the lowest priority and the problem $\min_{x \in \Omega \subseteq \mathbb{R}^n} h_m(x)$ the highest one. The first $m-1$ problems are sequentially optimized, not exhaustively, utilizing a known optimization method. The point in which each problem converges is used as the initial one for the next problem in the sequence. Hence, a kind of a 'direct' walk is created and the last point of this walk is utilized as the initial point for the last optimization problem in the sequence, which is identical with the problem given by relation (1).

Algorithm 2. RLexOpt algorithm

Require: $\{f(x), x^0, \epsilon, w\}$ {*The initial objective function* $f(x)$, *the used optimization method(s) for each involved problem, the stopping criteria* ϵ, *for each one of the utilized optimization problem* (ϵ *may be a vector with as many components as the utilized optimization problems.*), *the relaxation parameters* w (w *may be a vector with as many components as the utilized relaxation parameters*) }

STEP 1 Split the formula of the given objective function $f(x)$ by Cumulative Principle ([6, 7]).

STEP 2 Give priorities on the produced Cumulative elements (Definition 3.1 in [6]).

STEP 3 Set m the quantity of Cumulative elements.

STEP 4 Define a Lexicographic Cumulative vector (Definition 3.4 in [6]) dimension m.

STEP 5 Define the utilized objective functions (See relation (6))).

STEP 6 $j \leftarrow 1$

STEP 7 $z^{(0)} \leftarrow x^0$

STEP 8 WHILE $(i < m)$

Use $z^{(i-1)}$ as initial point for the given optimization method
for $\min_{x\in\Omega\subseteq\mathbb{R}^n} h_i(x)$, $x \in \mathbb{R}^n$ problem in relation (6).
$z^{(i)} \leftarrow$ the point that termination criteria for $\min_{x\in\Omega\subseteq\mathbb{R}^n} h_i(x)$
problem is fulfilled.
$z^{(i-1)} \leftarrow z^{(i)}$
$i \leftarrow i + 1$
END WHILE

STEP 9 Use $z^{(m-1)}$ as initial point for the given optimization method for $\min_{x\in\Omega\subseteq\mathbb{R}^n} h_m(x)$, $x \in \mathbb{R}^n$ problem in relation (6) until the termination criteria to be fulfilled.

return x^*

3.3. *Modified Relaxation Lexicographic Optimization (MRLexOpt) Algorithm*

The problem given by relation (1) is proposed to be transformed as following:

$$\min_{x\in\Omega\subseteq\mathbb{R}^n} H(x) = \left\{\min_{x\in\Omega\subseteq\mathbb{R}^n} h_1(x), \min_{x\in\Omega\subseteq\mathbb{R}^n} h_2(x), \ldots, \min_{x\in\Omega\subseteq\mathbb{R}^n} h_m(x)\right\}, \tag{7}$$

where $h_i(x) : \Omega \subseteq \mathbb{R}^n \to \mathbb{R}$ such that

$$\begin{aligned} h_1(x) &= w_{11}q_1(x), \\ &\qquad 0 < w_{11} < 1, \\ &\vdots \\ h_{m-1}(x) &= w_{(m-1)k}q_k(x) + \cdots + w_{(m-1)(m)}q_{(m)(m-k)}(x), \\ &\qquad w_{(m-1)k} + \cdots + w_{(m-1)(m-k)} = 1, \\ h_m(x) &= q_1(x) + \cdots + q_k(x) + \cdots + q_m(x) = f(x), \end{aligned}$$

where $k \in \mathbb{N} \setminus \{0,1\}$, such as $q_1(x) + \cdots + q_k(x) + \cdots + q_m(x) = f(x)$ and $q_i(x) : \Omega \subseteq \mathbb{R}^n \to \mathbb{R}$. The relation $0 < w_{tj} < 1$ is holding, for each term $w_{tj}, t, j \in 1, \ldots, (m-1)$, $m \in \mathbb{N} \setminus \{0,1\}$ and at each used objective function $h_i(x) : \Omega \subseteq \mathbb{R}^n \to \mathbb{R}$ and $i = 1, 2, \ldots, s-1$ with $s \in \mathbb{N} \setminus \{0,1\}$, the sum of the utilized relaxation parameters is equal to one unit. Note that among all the utilized objective functions $h_i(x) : \Omega \subseteq \mathbb{R}^n \to \mathbb{R}$ and $i = 1, 2, \ldots, s-1$ with $s \in \mathbb{N} \setminus \{0,1\}$, except of the last one, no common cumulative elements exist.

As it has been already mentioned, the problem $\min_{x\in\Omega\subseteq\mathbb{R}^n} h_1(x)$ is set to have the lowest priority and the problem $\min_{x\in\Omega\subseteq\mathbb{R}^n} h_m(x)$ the highest one. Moreover, the first $m-1$ problems are sequentially optimized, not exhaustively, utilizing a known optimization method. The point in which each problem converges is used as the initial one for the next problem in the sequence. Thus, a sequence of points is constructed and the last point of that sequence is utilized as the initial point for the last optimization problem, which is identical with the problem given by relation (1).

Algorithm 3. MRLexOpt algorithm

Require: $\{f(x), x^0, \epsilon, w\}$ {*The initial objective function* $f(x)$, *the used optimization method(s) for each involved problem, the stopping criteria* ϵ, *for each one of the utilized optimization problem (*ϵ *may be a vector with as many components as the utilized optimization problems.), the relaxation parameters* w *(*w *may be a vector with as many components as the utilized relaxation parameters)*}

STEP 1 Split the formula of the objective function $f(x)$ by Cumulative Principle ([6, 7]).

STEP 2 Give priorities on the produced Cumulative elements (Definition 3.1 in [6]).

STEP 3 Set m the quantity of Cumulative elements.

STEP 4 Define a Lexicographic Cumulative vector (Definition 3.4 in [6]) dimension m.

STEP 5 Define the utilized objective functions (See relation (7))).

STEP 6 $j \leftarrow 1$

STEP 7 $z^{(0)} \leftarrow x^0$

STEP 8 WHILE $(i < m)$

Use $z^{(i-1)}$ as initial point for the given optimization method

for $\min_{x\in\Omega\subseteq\mathbb{R}^n} h_i(x)$, $x \in \mathbb{R}^n$ problem in relation (7).

$z^{(i)} \leftarrow$ the point that termination criteria for $\min_{x\in\Omega\subseteq\mathbb{R}^n} h_i(x)$

problem is fulfilled.

$z^{(i-1)} \leftarrow z^{(i)}$

$i \leftarrow i + 1$

END WHILE

STEP 9 Use $z^{(m-1)}$ as initial point for the given optimization method for $\min_{x\in\Omega\subseteq\mathbb{R}^n} h_m(x)$, $x \in \mathbb{R}^n$ problem in relation (7) until the termination criteria to be fulfilled.

return x^*

3.4. *The convergence theorem for MLexOpt, RLexOpt, and MRLexOpt Algorithm*

The following theorem refers to the convergence of the above three proposed algorithms.

Theorem 1. *The sequence $x^{(p)}$ generated by Algorithms* 1, 2, *and* 3 *converges to minimum value x^* of the optimization problem given by equation* (1).

Proof. The proof is similar to convergence theorem given in [6] by taking $h_i(x) : \Omega \subseteq \mathbb{R}^n \to \mathbb{R}$ for $i = 1, \ldots, m$ as given in Algorithms 1, 2, and 3 correspondingly. □

4. Numerical Results

In this section, numerical results of well-known test problems [9–12], for LexOpt (LX), MLexOpt (MLX), RLexOpt (RLX), and MRLexOpt (MRLX) algorithms, utilizing the well-known optimization method of Fletcher–Reeves (FR) [1,2] are presented. Implementation has been done using a new R program [13] and an IBM-compatible PC. Moreover, numerical results were analyzed with the statistical software R-program [13]. There are given numerical results in order to see the effectiveness of the proposed algorithms. Each of them was tested in a dense grid of initial points in areas of the domain of the given objective function.

The termination criteria for the given problem using FR method, which is identical with the last problem of LX, MLX, RLX, and MRLX algorithms, are as follows:

(1) $\left|f(x^{p+1}) - f(p^k)\right| \leq 10^{-8}$,
(2) $(\|\nabla f(x^{p+1})\|_2 \leq 10^{-4}$,
(3) $MFE = 1000$, where MFE is the maximum number of function evaluations.

The termination criteria for all problems in LX, MLX, RLX, and MRLX algorithms, except of the last one, are as follows:

(1) $\left|f(x^{p+1}) - f(x^p)\right| \leq retol_i$, where i denotes the corresponding utilized optimization problem, $i = 1, \ldots, m-1$;
(2) $\left\|\nabla f(x^{p+1})\right\|_2 \leq 10^{-4}$;
(3) $maxit_i$, the maximum number of iterations for the utilized FR method, for ith optimization problem, $i = 1, \ldots, m-1$;
(4) $MFE = 1000$, where MFE is the maximum number of function evaluations.

For the given numerical results, the tolerance is 0.00001 ($retol_i = 0.00001$) for each $\min_{x\in\Omega\subseteq\mathbb{R}^n} h_i(x)$, $i = 1, 2, 3$, given by relations (5), (6), or (7). Note that the numerical results are taken into account, all the RelPars combinations of $0 < w_{11}, w_{21}, w_{22} < 1$, with step 0.1.

In the captions of tables, there are given the following implementation details:

- The rectangle from which are extracted 25 initial points,
- the number of iterations ($maxit_i$) used by the FR method to optimize problem $\min_{x\in\Omega\subseteq\mathbb{R}^n} h_i(x)$, given by relations (5), (6), or (7), for $i = 1, \ldots, m-1$. The value of $maxit_i$ is user-defined,
- r_i: the ratio of the absolute value of the largest to the smallest eigenvalue of Hessian of the objective function at the point x_i. Thus, $r_i = \frac{|\lambda_{max,i}|}{|\lambda_{min,i}|}$, whereas $\lambda_{max,i}$ is the largest eigenvalue at the point x_i and $\lambda_{min,i}$ is the smallest eigenvalue at the point x_i,
- fs: the average function evaluations, required for all the corresponding utilized initial points,
- gs: the average gradient evaluations, required for all the corresponding utilized initial points,
- $NumMin$: the average number of minima that each method converges for the utilized initial points,
- p_i: the utilized permutation of cumulative elements, named partition.

For each test function, partitions, where the ratios $\frac{r_0}{r_1}$ and $\frac{r_0}{r_2}$ are close to or greater than 0.50, are given. This is meaning that more than 50% of the new initial points are creating better convergence preconditions from the given initial points.

Before each table, the objective functions of the utilized optimization problems and the utilized permutation of cumulative elements for each algorithm are given.

4.1. *Six hump camel back function*

This function [10, 11] has two global minima $x_1^* = (0.089842, -0.712656)$, $x_2^* = (-0.089842, 0.712656)$, four local minima, $x_3^* = (-1.704, 0.796)$, $x_4^* = (1.704, -0.796)$, $x_5^* = (-1.607, -0.569)$, $x_6^* = (1.607, 0.569)$, and is given by $f(x_1, x_2) = \left(4 - 2.1x_1^2 + \frac{x_1^4}{3}\right) x_1^2 + x_1x_2 + \left(-4 + 4x_2^2\right) x_2^2$.

The utilized cumulative elements for this function are obtained as

$$q_1(x_1, x_2) = \left(4 - 2.1x_1^2 + x_1^4/3\right) x_1^2,$$
$$q_2(x_1, x_2) = x_1 x_2,$$
$$q_3(x_1, x_2) = \left(-4 + 4x_2^2\right) x_2^2.$$

Numerical results are given in Tables 1 and 2.

The presented partitions for the LX algorithm are given by the following relations:

$$(p1) : h_1(x) = q_1(x), h_2(x) = q_1(x) + q_2(x),$$
$$h_3(x) = q_1(x) + q_2(x) + q_3(x),$$
$$(p2) : h_1(x) = q_1(x), h_2(x) = q_1(x) + q_3(x),$$
$$h_3(x) = q_1(x) + q_2(x) + q_3(x)$$

Table 1. Six hump camel back function, $x \in [-2.7, 2.7] \times [-1.8, 1.8]$, $maxit_1 = 5$, $maxit_2 = 10$.

RLX	eigenvalue ratio	
partition	r_0/r_1	r_0/r_2
p2	0.64	0.96
p4	0.48	0.56
p5	0.48	0.96
LX	eigenvalue ratio	
partition	r_0/r_1	r_0/r_2
p1	0.64	0.64
p2	0.64	0.64

MRLX	eigenvalue ratio	
permutation	r_0/r_1	r_0/r_2
p1	0.64	0.96
p2	0.48	0.88
MLX	eigenvalue ratio	
permutation	r_0/r_1	r_0/r_2
p3	0.48	0.96

Table 2. Six hump camel back function, $x \in [-2.7, 2.7] \times [-1.8, 1.8]$, $maxit_1 = 5$, $maxit_2 = 10$.

RLX			
partition	fs	gs	NumMin
p2	29.30	9.68	4
p4	43.24	13.96	6
p5	28.62	9.53	4
LX			
partition	fs	gs	NumMin
p1	29.72	10.44	4
p2	29.72	10.44	4

MRLX			
partition	fs	gs	NumMin
p1	35.73	12.47	4
p2	27.56	9.90	6
MLX			
partition	fs	gs	NumMin
p3	39.40	11.88	4

For the RLX algorithm, they are given by the following relations:

$$(p2): h_1(x) = w_{11}q_1(x), h_2(x) = w_{21}q_1(x) + w_{22}q_3(x),$$
$$h_3(x) = q_1(x) + q_2(x) + q_3(x),$$
$$(p4): h_1(x) = w_{11}q_2(x), h_2(x) = w_{21}q_2(x) + w_{22}q_3(x),$$
$$h_3(x) = q_1(x) + q_2(x) + q_3(x),$$
$$(p5): h_1(x) = w_{11}q_3(x), h_2(x) = w_{21}q_3(x) + w_{22}q_1(x),$$
$$h_3(x) = q_1(x) + q_2(x) + q_3(x).$$

For the MLX algorithm, they are given by the following relations:

$$(p3): h_1(x) = q_3(x), h_2(x) = q_1(x) + q_2(x),$$
$$h_3(x) = q_1(x) + q_2(x) + q_3(x).$$

For the MRLX algorithm, they are given by the following relations:

$$(p1): h_1(x) = w_{11}q_1(x), h_2(x) = w_{21}q_2(x) + w_{22}q_3(x),$$
$$h_3(x) = q_1(x) + q_2(x) + q_3(x),$$
$$(p2): h_1(x) = w_{11}q_2(x), h_2(x) = w_{21}q_1(x) + w_{22}q_3(x),$$
$$h_3(x) = q_1(x) + q_2(x) + q_3(x),$$

whereas $x = (x_1, x_2)$, $0 < w_{11}, w_{21}, w_{22} < 1$.

4.2. *Himmelblau function*

This function [9] has four global minima $x_1^* = (3, 2)$, $x_2^* = (-2.805118, 3.131312)$, $x_3^* = (3.584428, -1.848126)$, and $x_4^* = (-3.779310, -3.283186)$ with $f(x_i^*) = 0$, $i = 1, 2, 3, 4$ and is given by $f(x_1, x_2) = x_1^4 + 2x_1^2x_2 - 21x_1^2 - 13x_2^2 - 22x_2 + 170 + 2x_1x_2^2 - 14x_1 + x_2^4$.

The utilized cumulative elements for this function are obtained as

$$q_1(x_1, x_2) = -22x_2 + 170 - 14x_1,$$
$$q_2(x_1, x_2) = 2x_1^2x_2 - 21x_1^2 - 13x_2^2 + 2x_1x_2^2,$$
$$q_3(x_1, x_2) = x_1^4 + x_2^4.$$

Numerical results are given in Tables 3 and 4.

Table 3. Himmelblau function, $x \in [-5,5] \times [-5,5]$, $maxit_1 = 5$, $maxit_2 = 10$.

RLX	eigenvalue ratio	
partition	r_0/r_1	r_0/r_2
p2	0.52	0.60
p3	0.64	0.60
p4	0.64	0.56
LX	eigenvalue ratio	
partition	r_0/r_1	r_0/r_2
p3	0.64	0.64
p4	0.64	0.64

MRLX	eigenvalue ratio	
partition	r_0/r_1	r_0/r_2
p1	0.52	0.52
MLX	eigenvalue ratio	
partition	r_0/r_1	r_0/r_2
p1	0.48	0.44

Table 4. Himmelblau function, $x \in [-5,5] \times [-5,5]$, $maxit_1 = 5$, $maxit_2 = 10$.

RLX			
partition	fs	gs	NumMin
p2	54.14	13.28	4
p3	77.41	23.11	1
p4	60.47	18.21	4
LX			
partition	fs	gs	NumMin
p3	71.00	21.00	1
p4	56.96	17.08	4

MRLX			
partition	fs	gs	NumMin
p1	56.67	13.99	4
MLX			
partition	fs	gs	NumMin
p1	69.40	16.20	3

The presented partitions for the LX algorithm are given by the following relations:

$$(p3): h_1(x) = q_2(x), h_2(x) = q_2(x) + q_1(x),$$
$$h_3(x) = q_1(x) + q_2(x) + q_3(x),$$
$$(p4): h_1(x) = q_2(x), h_2(x) = q_2(x) + q_3(x),$$
$$h_3(x) = q_1(x) + q_2(x) + q_3(x).$$

For the RLX algorithm, they are given by the following relations:

$$(p2): h_1(x) = w_{11}q_1(x), h_2(x) = w_{21}q_1(x) + w_{22}q_3(x),$$
$$h_3(x) = q_1(x) + q_2(x) + q_3(x),$$
$$(p3): h_1(x) = w_{11}q_2(x), h_2(x) = w_{21}q_2(x) + w_{22}q_1(x),$$
$$h_3(x) = q_1(x) + q_2(x) + q_3(x),$$
$$(p4): h_1(x) = w_{11}q_2(x), h_2(x) = w_{21}q_2(x) + w_{22}q_3(x),$$
$$h_3(x) = q_1(x) + q_2(x) + q_3(x).$$

For the MRLX algorithm, they are given by the following relations:

$$(p1): h_1(x) = w_{11}q_1(x), h_2(x) = w_{21}q_2(x) + w_{22}q_3(x),$$
$$h_3(x) = q_1(x) + q_2(x) + q_3(x),$$

whereas $x = (x_1, x_2)$, $0 < w_{11}, w_{21}, w_{22} < 1$.

The presented partition for the MLX algorithm is the same as that for MRLX subtracting RelPars w_{ij}, for $i, j = 1, 2$.

4.3. *Trecanni function*

This function [12] has two global minima $f(x_1^*) = f(x_2^*) = 0$, at the points $x_1* = (0, 0)$ and $x_2* = (-2, 0)$, and is given by $f(x_1, x_2) = x_1^4 + 4x_1^3 + 4x_1^2 + x_2^2$.

The utilized cumulative elements for this function are obtained as

$$q_1(x) = x_1^4,$$

$$q_2(x) = 4x_1^3,$$

$$q_3(x) = 4x_1^2 + x_2^2.$$

Numerical results are given in Tables 5 and 6.

Table 5. Trecanni function, $x \in [-2, 2] \times [-2, 2]$, $maxit_1 = 2$, $maxit_2 = 3$.

RLX	eigenvalue ratio	
partition	r_0/r_1	r_0/r_2
p2	0.60	0.68
p3	0.40	0.80
p5	0.60	0.52
LX	eigenvalue ratio	
partition	r_0/r_1	r_0/r_2
p2	0.80	0.52
p5	0.68	0.60

MRLX	eigenvalue ratio	
partition	r_0/r_1	r_0/r_2
p1	0.50	0.58

MLX	eigenvalue ratio	
partition	r_0/r_1	r_0/r_2
p3	0.68	0.28

Table 6. Trecanni function, $x \in [-2,2] \times [-2,2]$, $maxit_1 = 2$, $maxit_2 = 3$.

RLX			
partition	fs	gs	NumMin
p2	15.84	7.01	1
p3	23.09	9.59	2
p5	15.54	6.77	1

LX			
partition	fs	gs	NumMin
p2	17.32	7.40	1
p5	14.52	6.60	1

MRLX			
partition	fs	gs	NumMin
p1*	29.00	10.31	2
MLX			
partition	fs	gs	NumMin
p3	23.20	9.72	2

Note: *MRLX diverges on 100 of initial points

The presented partitions for the LX algorithm are given by the following relations:

$$
\begin{aligned}
(p2): h_1(x) &= q_1(x), h_2(x) = q_1(x) + q_3(x),\\
h_3(x) &= q_1(x) + q_2(x) + q_3(x),\\
(p5): h_1(x) &= q_3(x), h_2(x) = q_3(x) + q_1(x),\\
h_3(x) &= q_1(x) + q_2(x) + q_3(x).
\end{aligned}
$$

For the RLX algorithm, they are given by the following relations:

$$
\begin{aligned}
(p2): h_1(x) &= w_{11}q_1(x), h_2(x) = w_{21}q_1(x) + w_{22}q_3(x),\\
h_3(x) &= q_1(x) + q_2(x) + q_3(x),\\
(p3): h_1(x) &= w_{11}q_2(x), h_2(x) = w_{21}q_2(x) + w_{22}q_1(x),\\
h_3(x) &= q_1(x) + q_2(x) + q_3(x),\\
(p5): h_1(x) &= w_{11}q_3(x), h_2(x) = w_{21}q_3(x) + w_{22}q_1(x),\\
h_3(x) &= q_1(x) + q_2(x) + q_3(x),
\end{aligned}
$$

whereas $x = (x_1, x_2)$.

For the MLX algorithm, they are given by the following relations:

$$
\begin{aligned}
(p3): h_1(x) &= q_3(x), h_2(x) = q_1(x) + q_2(x),\\
h_3(x) &= q_1(x) + q_2(x) + q_3(x).
\end{aligned}
$$

For the MRLX algorithm, they are given by the following relations:

$$(p1): h_1(x) = w_{11}q_1(x), h_2(x) = w_{21}q_2(x) + w_{22}q_3(x),$$
$$h_3(x) = q_1(x) + q_2(x) + q_3(x),$$

whereas $x = (x_1, x_2)$, $0 < w_{11}, w_{21}, w_{22} < 1$.

4.4. *Matyas function*

This function [12] has one global minimum $f(x^*) = 0$ at the point $x = (0,0)$ and is given by $f(x_1, x_2) = 0.26x_1^2 + 0.26x_2^2 - 0.48x_1x_2$.

The utilized cumulative elements for this function are obtained as

$$q_1(x) = 0.26x_1^2,$$
$$q_2(x) = 0.26x_2^2,$$
$$q_3(x) = -0.48x_1x_2.$$

Numerical results are given in Tables 7 and 8.

Table 7. Matyas function, $x \in [-1,2] \times [-1,2]$, $maxit_1 = 2$, $maxit_2 = 3$.

RLX	eigenvalue ratio	
partition	r_0/r_1	r_0/r_2
p1	0.36	0.52
p3	0.32	0.52
LX	**eigenvalue ratio**	
partition	r_0/r_1	r_0/r_2
p1	0.52	0.60
p2	0.52	0.60
p3	0.48	0.56
p4	0.48	0.56

MRLX	eigenvalue ratio	
partition	r_0/r_1	r_0/r_2
p3	0.32	0.48

MLX	eigenvalue ratio	
partition	r_0/r_1	r_0/r_2
p2	0.48	0.56

Table 8. Matyas function, $x \in [-1,2] \times [-1,2]$, $maxit_1 = 2$, $maxit_2 = 3$.

RLX			
partition	fs	gs	NumMin
p1	133.28	67.57	1
p3	133.28	67.57	1

LX			
partition	fs	gs	NumMin
p1	150.44	76.20	1
p2	150.44	76.20	1
p3	150.44	76.20	1
p4	150.44	76.20	1

MRLX			
partition	fs	gs	NumMin
p3	139.98	70.92	1

MLX			
partition	fs	gs	NumMin
p2	125.16	63.48	1

The presented partitions for the LX algorithm are given by the following relations:

$$
\begin{aligned}
(p1) : h_1(x) &= q_1(x), h_2(x) = q_1(x) + q_2(x),\\
h_3(x) &= q_1(x) + q_2(x) + q_3(x),\\
(p2) : h_1(x) &= q_1(x), h_2(x) = q_1(x) + q_3(x),\\
h_3(x) &= q_1(x) + q_2(x) + q_3(x),\\
(p3) : h_1(x) &= q_2(x), h_2(x) = q_2(x) + q_1(x),\\
h_3(x) &= q_1(x) + q_2(x) + q_3(x),\\
(p4) : h_1(x) &= q_2(x), h_2(x) = q_2(x) + q_3(x),\\
h_3(x) &= q_1(x) + q_2(x) + q_3(x).
\end{aligned}
$$

For the RLX algorithm, they are given by the following relations:

$$
\begin{aligned}
(p1) : h_1(x) &= w_{11}q_1(x), h_2(x) = w_{21}q_1(x) + w_{22}q_2(x),\\
h_3(x) &= q_1(x) + q_2(x) + q_3(x),\\
(p3) : h_1(x) &= w_{11}q_2(x), h_2(x) = w_{21}q_2(x) + w_{22}q_1(x),\\
h_3(x) &= q_1(x) + q_2(x) + q_3(x).
\end{aligned}
$$

For the MLX algorithm, they are given by the following relation:

$$(p2) : h_1(x) = q_2(x), h_2(x) = q_1(x) + q_3(x),$$
$$h_3(x) = q_1(x) + q_2(x) + q_3(x).$$

For the MRLX algorithm, they are given by the following relations:

$$(p3) : h_1(x) = w_{11}q_3(x), h_2(x) = w_{21}q_1(x) + w_{22}q_2(x),$$
$$h_3(x) = q_1(x) + q_2(x) + q_3(x),$$

whereas $x = (x_1, x_2)$, $0 < w_{11}, w_{21}, w_{22} < 1$.

Note that the numerical results, the number of minima in which a method converges, depend on the utilized partition. For example, in Table 2, the RLX algorithm leads to more number of minima by using partition $p4$ than $p2$ and $p5$, while the MRLX algorithm leads to more number of minima by using partition $p2$ than $p1$.

The same partition may lead to different number of minima, depending on the utilized algorithm. For example, based on Table 4, using partition $p1$ MRLX leads to more number of minima than the MLX algorithm, and, moreover, it is taking place with less computational cost.

Numerical results indicate that a proper selection of partition and algorithm may lead to more number of minima with fewer computational cost. It is shown in Table 2, by taking the algorithm MRLX and partition $p2$ instead of the MLX algorithm and partition $p3$.

There are partitions, whereas utilizing relaxation parameters in LX and MLX algorithms leads to the same number of minima with less computational cost than utilizing the corresponding algorithms without parameters. Thus, it is valid between RLX, LX and MLX, MRLX. It is shown in different test functions between LX and RLX algorithms. For example, in Table 2, utilizing partition $p2$, in Table 6, for partition $p2$, and, in Table 8, utilizing partitions $p1$ and $p3$. LX and RLX algorithms lead to the same number of minima, but by using the RLX algorithm, this happens with less computational cost.

Finally, noticing Tables 1, 3, and 5, more than 50% of initial points through anyone of the proposed algorithms lead to a new initial point for the given problem, where a reduction of the ratio of eigenvalues occurs.

5. Conclusions

In this chapter, three new algorithms are presented. They are based on the strategy proposed in [6, 7], to optimize a given problem by using other ones, which may be considered as approximations of it and thus a path of points to be build such that the last of them is to be used as the initial point for the given problem. The involved problems are constructed by utilizing objective functions which result by partitioning the objective function, giving priorities on the arisen parts and combining them.

Numerical results indicate that the FR optimization method may through the proposed algorithms reinforce its convergence features. Furthermore, utilizing relaxation parameters may lead to more minima and in some cases with less computational cost. It is worth pointing out that for each test function and for all the proposed algorithms, there is at least one partition, whereas more that 50% of initial points lead to a better initial point, as the ratio of the absolute value of the largest to the smallest eigenvalue of Hessian of the objective function at the given initial point and the new initial point for the given problem indicates.

In a future work, a study on how to take advantage of the ratio of the above eigenvalues such that an improvement of the proposed algorithms may be done. Also, the behavior of other optimization methods on the proposed algorithms may be studied. Finally, the significance of nonlinearity may be investigated under the view of defining rules for doing the partition of the objective function and for giving priorities whenever it is obligatory.

References

[1] Patrizi. G., Dennis, J.E. and Schnabel, R.B. *Numerical Methods for Unconstrained Optimization and Nonlinear Equations*, Prentice-Hall, Englewood Cliffs (1983), XII, Prentice-Hall Series in Computational Mathematics (1985).

[2] Nocedal, J. and Wright, S.J. *Numerical Optimization*, 2nd edition, Springer, New York (2006).

[3] Polak, E. *Computational Methods in Optimization: A Unified Approach.* Academic Press, 77 (1971).

[4] Horst, R. and Pardalos, P.M. *Handbook of Global Optimization.* Springer Science & Business Media, 2, (2013).

[5] Rao, S.S. *Optimization, Theory and Applications*, 2nd edition, 7th reprint, Wiley Eastern Limited, New Delhi (1992).

[6] Nikolakakou, C.D., Grapsa, T.N., Nikas, I. and Androulakis, G.S. A sequential approach for unconstrained optimization via a partitioning technique. *International Journal of Computer Mathematics*, 92(4), 722–741 (2015).

[7] Nikolakakou, C.D., Grapsa, T.N. and Androulakis, G.S. A lexicographic approach on unconstrainted optimization. In *23rd European Conference on Operational Research*, p. 90, Bonn (2009).

[8] Christina. N.G. *Techniques for Improving the Efficiency of Unconstrained Optimization Algorithms.* PhD thesis, University of Patras (2018).

[9] Himmelblau, D. M., *Applied Nonlinear Programming*, McGraw-Hill, New York (1972).

[10] Molga, M. and Smutnicki, C. Test functions for optimization needs. *Test Functions for Optimization Needs*, 101, 48 (2005).

[11] Pohlheim, H. Examples of objective functions. *Retrieved*, 4(10), 2012 (2007).

[12] Ernesto, P. and Diliman, U. *Mvf — Multivariate Test Functions Library in c for Unconstrained Global. Optimization*, University of the Philippines Diliman, Quezon City (2005).

[13] Team, R.D.C. A language and environment for statistical computing, *R Foundation for Statistical Computing.* 1, https://www.R–project.org, (2020). URL http://www.r-project.org.

https://doi.org/10.1142/9789811293009_0029

Chapter 29

Strongly Biconvex Functions and Bivariational Inequalities

Muhammad Aslam Noor*,‡, Khalida Inayat Noor*,§
and Michael Th. Rassias†,¶

**Department of Mathematics,*
COMSATS University Islamabad, Park Road,
Islamabad, Pakistan
†Institute of Mathematics, University of Zurich,
8057 Zurich, Switzerland
‡noormaslam@gmail.com
§khalidan@gmail.com
¶michail.rassias@math.uzh.ch

In this chapter, we introduce and study some new classes of biconvex functions involving an arbitrary bifunction, which are called strongly biconvex functions. Some new relationships among various concepts of strongly biconvex functions have been established. We have shown that the optimality conditions for the biconvex functions can be characterized by a class of bivariational inequalities. The auxiliary principle technique is used to propose proximal point methods for solving strongly bivariational inequalities. We also discuss the convergence criteria for the suggested methods under pseudomonotonicity. Several special cases are discussed as applications of our main concepts and results. Our method of proof is very simple compared to other methods.

1. Introduction

Convexity theory is a branch of mathematics with a wide range of applications in industry, physical, social, and engineering sciences. The general theory of the convexity started soon after the introduction of differential and integral calculus by Newton and Leibnitz, although some individual optimization problems had been investigated before that. Motivated by geometrical considerations, Euler deduced his first principle which is now referred to as Euler's differential equation for the determination of maximizing and minimizing arcs. By variational principles, we mean maximum and minimum problems arising in game theory, mechanics, geometrical optics, general relativity theory, economics, transportation, differential geometry, data analysis, machine learning, and related areas. It is worth mentioning that variational inequalities represent the optimality conditions for the differentiable convex functions on the convex sets in normed spaces. Variational inequality theory provides us with a simple, natural, unified, novel, and general framework to study an extensive range of unilateral, obstacle, free, moving, and equilibrium problems arising in fluid flow through porous media, elasticity, circuit analysis, transportation, oceanography, operations research, finance, economics, and optimization. Variational inequalities were introduced and studied in the early 1960s by Stampacchia [22]. Variational inequalities combine both theoretical and algorithmic advances with a plethora of applications. Analysis of these problems requires a blend of techniques from convex analysis, functional analysis, and numerical analysis.

Inspired by the research work conducted in this field, we introduce and study another class of non-convex functions with respect to an arbitrary bifunction. This class of non-convex functions is called the class of strongly biconvex functions. We have shown that the biconvex functions satisfy some nice properties. Several new concepts of monotonicity are introduced and are discussed. We establish the relationship between these classes and derive some new results under some mild conditions. It is shown that the optimality conditions of the differentiable biconvex functions can be characterized by a class of variational inequalities, called bivariational inequalities. Some iterative methods are suggested for solving bivariational inequalities using the auxiliary principle technique [4, 13, 18, 19, 24] involving Bregman distance functions. Convergence criteria are also discussed using the pseudomonotonicity which is a weaker condition than monotonicity. It is hoped that the ideas and techniques introduced in this chapter may stimulate further research in this field.

2. Preliminary Results

Let K be a non-empty closed set in a real Hilbert space H. We denote by $\langle \cdot, \cdot \rangle$ and $\| \cdot \|$ the inner product and norm, respectively. Let $F : K_\beta \to R$ be a continuous function and let $\beta(. - .) : K_\beta \times K_\beta \to R$ be an arbitrary continuous bifunction.

Definition 1. The set K_β in H is said to be a biconvex set with respect to an arbitrary bifunction $\beta(\cdot - \cdot)$ if

$$u + \lambda\beta(v - u) \in K_\eta, \quad \forall u, v \in K_\beta, \lambda \in [0, 1].$$

The biconvex set K_β is also called the β-connected set. Note that the biconvex set with $\beta(v, u) = v - u$ is a convex set K, but the converse is not true. For example, the set $K_\beta = R - (-\frac{1}{2}, \frac{1}{2})$ is a biconvex set with respect to β, where

$$\beta(v - u) = \begin{cases} v - u, & \text{for } v > 0, u > 0 \quad \text{or} \quad v < 0, u < 0 \\ u - v, & \text{for } v < 0, u > 0 \quad \text{or} \quad v < 0, u < 0. \end{cases}$$

It is clear that biconvex set K_β is not a convex set.

Remark 2. We would like to emphasize that, if $u + \lambda\beta(v - u) = v, \forall u, v \in K_\beta$, then $\beta((v - u) = v - u$. Consequently, the biconvex set K_β reduces to the convex set K. Thus, $K_\beta \subset K$. This implies that every convex set is a biconvex set and the converse is not true.

In the sequel, K_β is a non-empty closed biconvex set in H with respect to the bifunction $\beta(\cdot - \cdot)$, unless otherwise specified.

We now introduce some new concepts of biconvex functions and their variant forms, which is the main motivation of this chapter.

Definition 3. The function F on the biconvex set K_β is said to be a strongly biconvex with respect to the bifunction $\beta(\cdot - \cdot)$ if there exists a constant $\mu \geq 0$ such that

$$F(u + \lambda\beta(v - u)) \leq (1 - \lambda)F(u) + \lambda F(v) - \mu\lambda(1 - \lambda)\|\beta(v - u)\|^2, \forall u, v \in K_\beta, \lambda \in [0, 1]. \tag{1}$$

The function F is said to be strongly biconcave if and only if $-F$ is a strongly biconvex function. Consequently, we have a new concept.

Definition 4. A function F is said to be strongly affine biconvex involving an arbitrary bifunction $\beta(-)$ if there exists a constant $\mu \geq 0$ such that

$$F(u + \lambda\beta(v - u)) = (1 - \lambda)F(u) + \lambda F(v) - \mu\lambda(1 - \lambda)\|\beta(v - u)\|^2, \quad \forall u, v \in K_\beta, \lambda \in [0, 1]. \tag{2}$$

If $\beta(v - u) = v - u$, then the strongly biconvex function becomes a strongly convex function, that is,

$$F(u + \lambda(v - u)) \leq (1 - \lambda)F(u) + \lambda F(v) - \mu\lambda(1 - \lambda)\|\beta(v - u)\|^2, \quad \forall u, v \in K, \lambda \in [0, 1],$$

which were introduced by Polyak [21] and Karmardian [7] who used these strongly convex functions in the study of complementarity problems. For the properties of the convex functions in variational inequalities and equilibrium problems, cf. Noor [10, 11, 13–15] and Noor *et al.* [17–19]. Note that every strongly convex function is a strongly biconvex function, but the converse is not true.

Definition 5. The function F on the biconvex set K_β is said to be strongly quasi biconvex with respect to the bifunction $\beta(\cdot - \cdot)$ if there exists a constant $\mu \geq 0$ such that

$$F(u + \lambda\beta(v - u)) \leq \max\{F(u), F(v)\} - \mu\lambda(1 - \lambda)\|\beta(v - u)\|^2, \quad \forall u, v \in K_\beta, \lambda \in [0, 1].$$

Definition 6. The function F on the biconvex set K_β is said to be strongly log-biconvex with respect to the bifunction $\beta(\cdot - \cdot)$ if there exists a constant $\mu \geq 0$ such that

$$F(u + \lambda\beta(v-u)) \leq (F(u))^{1-\lambda}(F(v))^{\lambda} - \mu\lambda(1-\lambda)\|\beta(v-u)\|^2, \quad \forall u, v \in K_\beta, \lambda \in [0,1],$$

where $F(\cdot) > 0$.

We can rewrite Definition 6 in the following equivalent form.

Definition 7. The function F on the biconvex set K_β is said to be strongly log-biconvex with respect to the bifunction $\beta(\cdot - \cdot)$ if there exists a constant $\mu \geq 0$ such that

$$\begin{aligned}\log F(u + \lambda\beta(v-u)) &\leq (1-\lambda)\log F(u) + \lambda \log F(v) \\ &\quad - \mu\lambda(1-\lambda)\|\beta(v-u)\|^2, \quad \forall u, v \in K_\beta, \lambda \in [0,1],\end{aligned}$$

where $F(\cdot) > 0$.

This equivalent definition can be used to discus the properties of the differentiable strongly log-biconvex functions.

From the above definitions, we have

$$\begin{aligned}F(u + \lambda\beta(v-u)) &\leq (F(u))^{1-\lambda}(F(v))^{\lambda} - \mu\lambda(1-\lambda)\|\beta(v-u)\|^2 \\ &\leq (1-\lambda)F(u) + \lambda F(v) - \mu\lambda(1-\lambda)\|\beta(v-u)\|^2 \\ &\leq \max\{F(u), F(v)\} - \mu\lambda(1-\lambda)\|\beta(v-u)\|^2.\end{aligned}$$

This shows that every strongly log-biconvex function is a strongly biconvex function and every strongly biconvex function is a strongly quasi-biconvex function. However, the converse is not true.

Definition 8. The differentiable function F is said to be a strongly biconvex function if there exists a constant $\mu \geq 0$ such that

$$F(v) - F(u) \geq \langle F'(u), \beta(v-u)\rangle + \mu\|\beta(v-u)\|^2, \forall u, v \in K_\beta. \quad (3)$$

For $\lambda = 1$, Definitions 3 and 6 reduce to the following condition:

Condition A. The bifunction $\beta(,.,)$ satisfies the assumption

$$F(u + \beta(v-u)) \leq F(v), \quad \forall v \in K_\beta,$$

which is known as Condition A. This condition plays an important role in the derivation of the results.

We note that if $\mu = 0$, then Definition 3 reduces to the following:

Definition 9. The function F on the biconvex set K_β is said to be biconvex with respect to the bifunction $\beta(\cdot - \cdot)$ if

$$F(u + \lambda\beta(v - u)) \leq (1 - \lambda)F(u) + \lambda F(v), \quad \forall u, v \in K_\beta, \lambda \in [0, 1], \tag{4}$$

which were introduced and studied by Noor and Noor [16].

We now define the biconvex functions on the interval $K_\beta = I_\beta = [a, a + \beta(b - a)]$.

Definition 10. Let $I = [a, a + \beta(b - a)]$. Then F is a biconvex function if and only if

$$\begin{vmatrix} 1 & 1 & 1 \\ a & x & a + \beta(b - a) \\ F(a) & F(x) & F(b) \end{vmatrix} \geq 0; \quad a \leq x \leq a + \beta(b - a).$$

One can easily show that the following are equivalent:

(1) F is a biconvex function.

(2) $F(x) \leq F(a) + \frac{F(b)-F(a)}{\beta(b-a)}(x - a)$.

(3) $\frac{F(x)-F(a)}{x-a} \leq \frac{F(b)-F(a)}{\beta(b-a)}$.

(4) $\frac{F(a)}{(\beta(b-a))(a-x)} + \frac{F(x)}{(x-a-\beta(b-a))(a-x)} + \frac{F(b)}{\beta(b-a)(x-b)} \leq 0$,

where $x = a + \lambda\beta(b - a) \in [a, a + \beta(b - a)]$.

3. Properties of Biconvex Functions

In this section, we consider some basic properties of strongly biconvex functions and their variant forms. For this purpose, we need the following assumption regarding the bifunction $\beta(\cdot - \cdot)$.

Condition M. The bifunction $\beta(, -,)$ is said to be homogeneous if

$$\beta(\gamma(v - u)) = \gamma\beta(v - u), \quad \forall u, v \in K_\beta, \quad \gamma \in R^n.$$

Remark 11. Let $\beta(\cdot - \cdot) : K_\beta \times K_\beta \to H$ satisfy the assumption

$$\beta(v-u) = \beta(v-z) + \beta(z-u), \quad \forall u, v, z \in K_\beta. \tag{5}$$

One can easily show that $\beta(v-u) = 0, \quad \forall u, v \in K_\beta$. Consequently, $\beta(0) = 0$, for $v = u \in K_\beta$. Also, $\beta(v-u) + \beta(u-v) = 0$. This implies that the bifunction $\beta(. - .)$ is skew symmetric.

Theorem 12. *Let F be a differentiable function on the biconvex set K_β in H and let the condition M hold. Then the function F is a strongly biconvex function if and only if F is a strongly biconvex function, that is,*

$$F(v) - F(u) \geq \langle F'(u), \beta(v-u)) \rangle - \mu \|\beta(v-u)\|^2, \quad \forall u, v \in K_\beta. \tag{6}$$

Proof. Let F be a strongly biconvex function on the biconvex set K_β. Then,

$$F(u + \lambda\beta(v-u)) \leq (1-\lambda)F(u) + \lambda F(v) - \mu\lambda(1-\lambda)\|\beta(v-u)\|^2$$
$$\forall u, v \in K_\beta, \lambda \in [0,1],$$

which can be written as

$$F(v) - F(u) \geq \left\{ \frac{F(u + \lambda\beta(v-u)) - F(u)}{\lambda} \right\} - \mu(1-\lambda)\|\beta(v-u)\|^2.$$

Taking the limit in the above inequality as $\lambda \to 0$, we have

$$F(v) - F(u) \geq \langle F'(u), \beta(v-u)) \rangle - \mu\|\beta(v-u)\|^2.$$

This shows that F is a strongly biconvex function.

Conversely, let F be a strongly biconvex function on the biconvex set K_β. Then, $\forall u, v \in K_\beta, \lambda \in [0,1]$, $v_t = u + \lambda\beta(v-u) \in K_\beta$. Taking $v = v_t$ in (6) and using the condition M, we have

$$\begin{aligned} &F(v) - F(u + \lambda\beta(v-u)) \\ &\quad \geq \langle F'(u + \lambda\beta(v-u)), \beta(v - u + \lambda\beta(v-u))\rangle + \mu\|\beta(v_t - u)\|^2 \\ &\quad = (1-\lambda)F'(u + \lambda\beta(v-u)), \beta(v-u) + (-\lambda)^2\mu\|\beta(v_t - u)\|^2\rangle. \end{aligned} \tag{7}$$

In a similar way, we have

$$\begin{aligned} &F(u) - F(u + \lambda\beta(v-u)) \\ &\quad \geq \langle F'(u + \lambda\beta(v-u)), \beta(u, u + \lambda\beta(v-u))\rangle + \mu\|\beta(u - v_t) \\ &\quad = -\lambda F'(u + \lambda\beta(v-u)), \beta(v-u)\rangle + \mu\lambda^2\|\beta(v-u)\|^2. \end{aligned} \tag{8}$$

Multiplying (7) by λ and (8) by $(1-\lambda)$ and adding the resultant, we have

$$F(u + \lambda\beta(v-u)) \leq (1-\lambda)F(u) + \lambda F(v) - \mu\lambda(1-\lambda)\|\beta(v-u)\|^2,$$

showing that F is a strongly biconvex function. □

Theorem 13. *Let F be a differentiable strongly biconvex function on the biconvex set K_β. Then F is a strongly biconvex function if and only if*

$$\begin{aligned} &\langle F'(u), \beta(v-u))\rangle + \langle F'(v), \beta(u-v)\rangle \\ &\quad \leq -\mu\{\beta(v-u) + \beta(u-v)\}, \quad \forall u, v \in K_\beta. \end{aligned} \tag{9}$$

Proof. Let F be a biconvex function on the biconvex set K_β. Then,

$$F(v) - F(u) \geq \langle F'(u), \beta(v-u))\rangle + \mu\|\beta(v-u)\|^2, \quad \forall u, v \in K_\beta. \tag{10}$$

Changing the role of u and v in (10), we have

$$F(u) - F(v) \geq \langle F'(v), \beta(u-v)\rangle + \mu\|\beta(u-v)\|^2, \quad \forall u, v \in K_\beta. \tag{11}$$

Adding (10) and (11), we have

$$\begin{aligned} &\langle F'(u), \beta(v-u))\rangle + \langle F'(v), \beta(u-v)\rangle \\ &\quad \leq -\mu\{\|\beta(v-u)\|^2 + \|\beta(u-v)\|^2\}, \quad \forall u, v \in K_\beta, \end{aligned}$$

which is the required result (10).

Let $F'(.)$ satisfy inequality (10). Since K_β is an biconvex set, $\forall u, v \in K_\beta$, $\lambda \in [0,1]$ $v_t = u + \lambda\beta(v-u) \in K_\beta$. Taking $v = v_\lambda$ in (10) and using Condition M, we have

$$\begin{aligned}\langle F'(v_\lambda), \beta(-\lambda\beta(v-u))\rangle &\le \langle F'(u), \eta(\lambda\beta(v-u)))\rangle \\ &= -\lambda\langle F'(u), \beta(v-u)\rangle - 2\lambda^2\|\beta(v-u)\|^2,\end{aligned}$$

which implies that

$$\langle F'(v_\lambda), \beta(v-u)\rangle \ge \langle F'(u), \beta(v-u)\rangle + 2\lambda\|\beta(v-u)\|^2. \quad (12)$$

We now consider the auxiliary function

$$\xi(\lambda) = F(u + \lambda\beta(v-u)) = F(v_\lambda).$$

Then, from (12), we have

$$\begin{aligned}\xi'(\lambda) &= \langle F'(u+\lambda\beta(v-u)), \beta(v-u)\rangle \\ &\ge \langle F'(u), \beta(v-u)\rangle + 2\mu\|\beta(v-u)\|^2. \end{aligned} \quad (13)$$

Integrating (13) between 0 and 1, we have

$$\xi(1) - \xi(0) \ge \langle F'(u), \beta(v-u)\rangle + 2\mu \int_0^1 \lambda\|\beta(b-u)\|^2 d\lambda,$$

that is,

$$F(u + \beta(v-u)) - F(u) \ge \langle F'(u), \beta(v-u)\rangle + \mu\|\beta(v-u)\|^2.$$

By using Condition A, we have

$$F(v) - F(u) \ge \langle F'(u), \beta(v-u)\rangle + \mu\|\beta(v-u)\|^2,$$

which is the required result. □

4. Strongly Bivariational Inequalities

In this section, we consider strongly bivariational inequalities and suggest some iterative methods by using the auxiliary principle techniques involving the Bregman distance functions.

For the reader's convenience, we recall some basic properties of the Bregman [2] convex functions. For a strongly convex function F, we define the Bregman distance function as

$$B(v,u) = F(v) - F(u) - \langle F'(u), v-u\rangle \geq \alpha\|v-u\|^2, \quad \forall u,v \in K. \tag{14}$$

It is important to emphasize that various types of function F give different Bregman distances. We give the following important examples of some practical essential types of function F and their corresponding Bregman distances, cf. [6, 23].

Examples

(1) If $f(v) = \|v\|^2$, then $B(v,u) = \|v-u\|$, which is the squared Euclidean distance (SE).

(2) If $f(v) = \sum_{i=1}^{n} a_i \log(v_i)$, which is known as Shannon's entropy, then its corresponding Bregman distance is given by

$$B(v,u) = \sum^{n} i = 1 \left(v_i \log\left(\frac{v_i}{u+i}\right) + u_i - v_i \right).$$

This distance is called Kullback–Leibler distance (KL) and has become a very important tool in several areas of applied mathematics such as machine learning.

(3) If $f(v) = -\sum_{i=1}^{n} \log(v_i)$, which is called Burg entropy, then its corresponding Bregman distance is given by

$$B(v,u) = \sum_{i=1}^{n} \left(\log\left(\frac{v_i}{u_i}\right) + \frac{v_i}{u_i} - 1 \right).$$

This is called Itakura–Saito (IS) distance, which is very important in information theory, data analysis, and machine learning.

Remark 14. It is a challenging problem to explore the applications of Bregman distance for other types of non-convex functions, such as biconvex, k-convex functions, and harmonic functions.

We now discuss the optimality conditions for differentiable strongly biconvex functions.

Theorem 15. *Let F be a differentiable strongly biconvex function with modulus $\mu > 0$. If $u \in K_\beta$ is the minimum of the function F, then $u \in K_\beta$ satisfies the following:*

$$F(v) - F(u) \geq \mu\|\beta(v-u)\|^2, \quad \forall u, v \in K_\beta. \tag{15}$$

Proof. Let $u \in K_\beta$ be a minimum of the function F. Then,

$$F(u) \leq F(v), \quad \forall v \in K_\beta. \tag{16}$$

Since K_β is a biconvex set, we have, $\forall u, v \in K_\beta, \quad \lambda \in [0,1]$,

$$v_\lambda = u + \lambda\beta(v-u) \in K_\beta.$$

Taking $v = v_\lambda$ in (16), we have

$$0 \leq \lim_{\lambda \to 0}\left\{\frac{F(u+\lambda\beta((v-u)) - F(u)}{\lambda}\right\} = \langle F'(u), \beta(v-u)\rangle. \tag{17}$$

Since F is a differentiable biconvex function, we have

$$\begin{aligned} F(u+\lambda\beta(v-u)) \leq\ & F(u) + \lambda(F(v) - F(u)) \\ & - \mu\lambda(1-\lambda)\|\beta(v-u)\|^2, \quad \forall u, v \in K_\beta, \end{aligned}$$

from which, using (17), we derive that

$$\begin{aligned} F(v) - F(u) &\geq \lim_{\lambda \to 0}\left\{\frac{F(u+\lambda\beta(v-u)) - F(u)}{\lambda}\right\} \\ &\quad - \mu(1-\lambda)\|\beta(v-u)\|^2 \\ &= \langle F'(u), \beta(v,u)\rangle + \lambda\|\beta(v-u)\|^2 \\ &\geq \mu\|\beta(v-u)\|^2, \end{aligned}$$

which is the required result (15). □

Remark: We would like to mention that, if $u \in K_\beta$ satisfies the inequality

$$\langle F'(u), \beta(v,u)\rangle + \mu\|\beta(v-u)\|^2 \geq 0, \quad \forall u, v \in K_\beta, \tag{18}$$

then $u \in K_\beta$ is the minimum of the differentiable strongly biconvex function F.

The inequality of type (18) is called the bivariational inequality and appears to be new. It is worth mentioning that inequalities of type (18) may not arise as the minimization of the biconvex functions. This motivated us to consider a more general bivariational inequality of which (18) is a special case.

For a given operator T and a bifunction $\beta(. - .)$, consider the problem of finding $u \in K_\beta$ such that

$$\langle Tu, \beta(v-u)\rangle + \mu\|\beta(v-u)\|^2 \geq 0, \quad \forall v \in K_\beta, \tag{19}$$

which is called the strongly bivariational inequality.

It is worth mentioning that for a suitable and appropriate choice of the operators, biconvex sets, and spaces, one can obtain a wide class of variational inequalities and optimization problems. This shows that the strongly bivariational inequalities are quite flexible and unified.

We would like to mention that the projection method and its variant form cannot be used to suggest the iterative methods for solving these bivariational inequalities due to the nature of the problem. To overcome these drawback, one may use the auxiliary principle technique of Glowinski *et al.* [4] as developed by Noor [10,11,13–15] and Noor *et al.* [17–19] to suggest and analyze some iterative methods for solving the bivariational inequalities (19). This technique does not involve the concept of the projection, which is the main advantage of this technique. We again use the auxiliary principle technique coupled with Bergman functions. These applications are based on the type of convex function associated with the Bregman distance. We now suggest and analyze some iterative methods for bivariational inequalities (19) using the auxiliary principle technique coupled with Bregman functions as developed by Noor [14,15].

For given $u \in K_\beta$ satisfying the bivariational inequality (19), we consider the auxiliary problem of finding a $w \in K_\beta$ such that

$$\langle \rho Tw, \beta(v-w)\rangle + \langle E'(w) - E'(u), \beta(v-w)\rangle + \rho\mu\|\beta(v-w)\|^2 \geq 0, \quad \forall v \in K_\beta, \tag{20}$$

where $\rho > 0$ is a constant and $E'(u)$ is the differential of a strongly biconvex function $E(u)$ at $u \in K_\beta$. Since $E(u)$ is a strongly biconvex function, this implies that its differential E' is strongly β-monotone. Consequently, it follows that the problem (19) has an unique solution.

Remark 16. The function

$$\begin{aligned} B(w,u) &= E(w) - E(u) - \langle E'(u), \beta(w-u)\rangle \\ &\geq \mu\|\beta(w-u)\|^2, \forall u, w \in K_\beta, \end{aligned}$$

associated with the strongly biconvex function $E(u)$ is called the generalized Bregman distance function. By the strongly biconvexity of the function $E(u)$, the Bregman function $B(.,.)$ is non-negative and $B(w,u) = 0$ if and only if $u = w, \forall u, w \in K_\beta$. For the applications of the Bregman function in solving variational inequalities and complementarity problems, cf. [14, 15, 17–19, 24].

We note that, if $w = u$, then clearly w is solution of the strongly bivariational inequality (19). This observation enables us to suggest and analyze the following iterative method for solving (19).

Algorithm 1. For a given $u_0 \in H$, compute the approximate solution u_{n+1} by the iterative scheme

$$\begin{aligned} &\langle \rho T u_{n+1}, \beta(v-u_{n+1})\rangle + \langle E'(u_{n+1}) - E'(u_n), \beta(v-u_{n+1})\rangle \\ &\quad + \rho\mu\|\beta(v-u_{n+1})\|^2 \geq 0, \quad \forall v \in K_\beta, \end{aligned} \tag{21}$$

where $\rho > 0$ is a constant. Algorithm 1 is called the proximal method for solving bivariational inequalities (19). In passing, we remark that the proximal point method was suggested in the context of convex programming problems as a regularization technique.

If $\beta(v-u) = v-u$, then Algorithm 1 collapses to the following:

Algorithm 2. For a given $u_0 \in H$, compute the approximate solution u_{n+1} by the iterative scheme

$$\begin{aligned} &\langle \rho T(u_{n+1}), v-u_{n+1}\rangle + \langle E'(u_{n+1}) - E'(u), v-u_{n+1}\rangle \\ &\quad + \rho\mu\|v-u_{n+1}\|^2 \geq 0, \quad \forall v \in K, \end{aligned}$$

for solving the strongly variational inequality, which appears to be new.

For an appropriate choice of the operators and the spaces, one can obtain a number of known and new algorithms for solving variational inequalities and related problems.

For the convergence analysis of Algorithm 1, we need the following concept.

Definition 17. The operator T is said to be strongly pseudo β-monotone with respect to $\mu\|\beta(v-u)\|^2$ if

$$\langle Tu, \beta(v-u)\rangle + \mu\|\beta(v-u)\|^2 \geq 0$$

$$\Rightarrow$$

$$-\langle Tv, \beta(u-v)\rangle - \mu\|\beta(v-u)\|^2 \geq 0, \quad \forall v, u \in K_\beta.$$

Theorem 18. *Let the operator T be strongly β pseudomonotone with respect to $\mu\|beta(v-u)\|^2$. If E is a differentiable higher-order strongly biconvex function with module $\beta > 0$ and Condition M holds, then the approximate solution u_{n+1} obtained from Algorithm* 1 *converges to a solution $u \in K$ satisfying the strongly bivariational inequality* (19).

Proof. Let $u \in K_\beta$ be a solution of bivariational inequality (19). Then,

$$\langle Tu, \beta(v-u)\rangle + \mu\|\beta(v-u)\|^2 \geq 0, \quad \forall v \in K_\beta,$$

implies that

$$-\langle Tv, \beta(u-v))\rangle - \mu\|\beta(u-v)\|^2 \geq 0, \quad \forall v \in K_\beta, \tag{22}$$

since T is strongly β-pseudomonotone.

Taking $v = u$ in (21) and $v = u_{n+1}$ in (22), we have

$$\begin{aligned}&\langle \rho T(u_{n+1}), \beta(u, u-n+1)\rangle + \langle E'(u_{n+1}) - E'_k(u_n, \beta(u-u_{n+1}))\rangle \\ &\quad + \rho\mu\|\beta(u-u_{n+1}\|^2 \geq 0\end{aligned} \tag{23}$$

and

$$-\langle Tu_{n+1}, \beta(u-u_{n+1})\rangle - \mu\|\beta(u-un+1\|^2 \geq 0. \tag{24}$$

We now consider the Bregman function

$$\begin{aligned}B(u,w) &= E(u) - E(w) - \langle E'(w, \beta(u-w))\rangle \\ &\geq \mu\|\beta(u-w)\|^2,\end{aligned} \tag{25}$$

using the higher-order strongly biconvexity of E.

Now, combining (23), (24), and (25) and using the condition M, we have

$$\begin{aligned}
&B(u, u_n) - B(u, u_{n+1}) \\
&\quad = E(u_{n+1}) - E(u_n) - \langle E'(u_n), \beta(u - u_n)\rangle \\
&\qquad + \langle E'(u_{n+1}), \beta(u - u_{n+1})\rangle \\
&\quad = E(u_{n+1}) - E(u_n) - \langle E'(u_n) - E'(u_{n+1}, \beta(u - u_{n+1})\rangle \\
&\qquad - \langle E'(u_n, u_{n+1} - u_n\rangle \\
&\quad \geq \beta\|\beta(u_{n+1} - u_n)\|^2 + \langle E'(u_{n+1}) - E'(u_n), \beta(u - u_{n+1})\rangle \\
&\quad \geq \beta\|\beta(u_{n+1} - u_n)\|^2 - \rho\langle T(u_{n+1}), \beta(u - u_{n+1})\rangle \\
&\qquad - \rho\mu\|\beta(u - u_{n+1})\|^2 \\
&\quad \geq \beta\|\beta(u_{n+1} - u_n)\|^2.
\end{aligned}$$

If $u_{n+1} = u_n$, then clearly u_n is a solution of problem (19). Otherwise, it follows that $B(u, u_n) - B(u, u_{n+1})$ is non-negative and we must have

$$\lim_{n\to\infty} \|\beta(u_{n+1} - u_n)\| = 0,$$

from which, we have

$$\lim_{n\to\infty} \|u_{n+1} - u_n\| = 0.$$

It follows that the sequence $\{u_n\}$ is bounded. Let $\bar{u}$ be a cluster point of the subsequence $\{u_{n_i}\}$, and let $\{u_{n_i}\}$ be a subsequence converging to $\bar{u}$. Now using the technique of Zhu and Marcotte [24], it can be shown that the entire sequence $\{u_n\}$ converges to the cluster point $\bar{u}$ satisfying the bivariational inequality (19). □

It is well known that to implement the proximal point methods, one has to find the approximate solution implicitly, which is itself a difficult problem. To overcome this drawback, we now consider another method for solving the bivariational inequality (19) using the auxiliary principle technique.

For a given $u \in K_\beta$ satisfying bivariational inequality (19), find $w \in K_\beta$ such that

$$\langle \rho T(u, \beta(v-w)\rangle + \langle E'(w) - E', \beta(v-w)\rangle$$
$$+\mu\|\beta(v-w)\|^2 \geq 0, \quad \forall v \in K_\beta, \tag{26}$$

where $E'(u)$ is the differential of a biconvex function $E(u)$ at $u \in K_\beta$. Problem (19) has a unique solution since E is a strongly biconvex function. Note that problems (26) and (21) are quite different. It is clear that for $w = u$, w is a solution of (19). This fact allows us to suggest and analyze another iterative method for solving bivariational inequality (19).

Algorithm 3. For a given $u_0 \in H$, compute the approximate solution u_{n+1} by the iterative scheme

$$\langle \rho T u_n, \beta(v - u_{n+1})\rangle + \langle E'(u_{n+1}) - E'(u_n), \beta(v - u_{n+1})\rangle$$
$$+ \mu\|\beta(v - u_{n+1})\|^2 \geq 0, \quad \forall v \in K_\beta, \tag{27}$$

for solving the strongly bivariational inequality (19).

If $\beta(v, u)) = v - y$, Algorithm 3 collapses to the following:

Algorithm 4. For a given $u_0 \in H$, compute the approximate solution u_{n+1} by the iterative schemes

$$\rho\langle T u_n, v - u_{n+1}\rangle + \langle E'(u_{n+1}) - E'(u_n), v - u_{n+1}\rangle$$
$$+ \mu\|v - u_{n+1}\|^2 \geq 0, \quad \forall v \in K,$$

for solving the strongly variational inequalities, which appears to be new.

Remark 19. For suitable choice of the operators and the spaces, one can obtain various known and new algorithms for solving the strongly bivariational inequality (19) and related optimization problems. We have only presented a glimpse of the applications of the auxiliary principle techniques. It is an interesting problem from both analytical and numerical point of view.

5. Conclusion

In this chapter, we have introduced and studied some new classes of strongly biconvex functions. These concepts are more general and unifying. Several new properties of these strongly biconvex functions are discussed and their relations with previously known results are highlighted. It is shown that the optimality conditions of the differentiable strongly biconvex functions can be characterized by a class of bivariational inequalities. This result is used to introduce a more general class of higher-order strongly bivariational inequalities (19). Auxiliary principle techniques are used to suggest and analyze some iterative methods for solving strongly bivariational inequalities. Convergence analysis of the proposed methods is a condition using pseudomonotonicity, which is a weaker condition than monotonicity. It is itself an interesting problem to develop some efficient numerical methods for solving strongly bivariational inequalities along with applications in pure and applied sciences. Despite the current research in these fields, clearly a lot remains to be done. It is hoped that the ideas and techniques of this chapter may constitute a starting point for further future research.

Acknowledgments

We wish to express our deepest gratitude to our colleagues, students, collaborators, and friends, who have directly or indirectly contributed in the process of preparation of this chapter. The authors would like to thank the Rector, COMSATS University Islamabad, Islamabad, Pakistan, for providing excellent research and academic environments.

References

[1] Ben-Isreal, A. and Mond, B. What is invexity? *J. Austral. Math. Soc. Ser. B*, 28(1), 1–9 (1986).

[2] Bregman, L.M. The relaxation method for finding common points of convex sets and its application to the solution of problems in convex programming. *USSR Comput. Math. Math. Phys.*, 7, 200–217 (1967).

[3] Cristescu, G. and Lupsa, L. *Non-Connected Convexities and Applications.* Kluwer Academic Publisher, Dordrechet (2002).

[4] Glowinski, R., Lions, J.L. and Tremolieres, R. *Numerical Analysis of Variational Inequalities.* North-Holland, Amsterdam (1981).

[5] Hanson, M.A. On sufficiency of the Kuhn-Tucker conditions. *J. Math. Anal. Appl.*, 80, 545–550 (1981).

[6] Jolaoso, L.O., Aphane, M, and Khan, S.H., Two Bregman projection methods for solving variational inequality problems in Hilbert spaces with applications to signal processing. *Symmetry*, 12, 2007 (2020). doi:10.3390/sym12122007.

[7] Karamardian, S. The nonlinear complementarity problems with applications, Part 2. *J. Optim. Theory Appl.*, 4(3), 167–181 (1969).

[8] Lions, J.L. and Stampacchia, G. Variational inequalities. *Commun. Pure Apl. Math.*, 20, 493–519 (1967).

[9] Niculescu, C.P. and Persson, L.E. *Convex Functions and Their Applications.* Springer-Verlag, New York (2018).

[10] Noor, M.A. *On Variational Inequalities*, PhD Thesis, Brunel University, London, U.K. (1975).

[11] Noor, M.A. Variational like inequalities. *Optimization*, 30, 323–333 (1994)

[12] Noor, M.A. New approximation schemes for general variational inequalities. *J. Math. Anal. Appl.*, 151, 217–229 (2000).

[13] Noor, M.A. Some developments in general variational inequalities. *Appl. Math. Comput.*, 251, 199–277 (2004).

[14] Noor, M.A. Invex equilibrium problems. *J. Math. Anal. Appl.*, 302, 463–475 (2005).

[15] Noor, M.A. Fundamentals of equilibrium problems. *Math. Inequal. Appl.*, 9(3), 529–566 (2006).

[16] Noor, M.A. and Noor, K.I. *Prebinvex Functions and their Characterizations*, COMSATS University Islamabad, Islamabad, Pakistan, Preprint (2021).

[17] Noor, M.A., Noor, K.I. and Al-Said, E. Auxiliary principle technique for solving bifunction variational inequalities. *J. Optim. Theory Appl.*, 149, 441–445 (2011).

[18] Noor, M.A., Noor, K.I. and Rassias, M.Th. New trends in general variational inequalities. *Acta Appl. Mathematica*, 170(1), 981–1046 (2020).

[19] Noor, M.A., Noor, K.I. and Rassias, Th.M. Some aspects of variational inequalities. *J. Appl. Math. Comput.*, 47, 485–512 (1993).

[20] Pecaric, J., Proschan, F. and Tong, Y.L. *Convex Functions, Partial Orderings and Statistical Applications.* Academic Press, New York (1992).

[21] Polyak, B.T. Existence theorems and convergence of minimizing sequences in extremum problems with restrictions. *Soviet Math. Dokl.*, 7, 2–75 (1966).

[22] Stampacchia, G. Formes bilineaires coercivites sur les ensembles convexes. *C. R. Acad. Paris*, 258, 4413–4416 (1964).

[23] Sunthrayuth, P. and Cholamjiak, P. Modified extragradient method with Bregman distance for variational inequalities. *Appl. Anal.*, 2020, https://doi.org/10.1080/00036811.2020.1757078.

[24] Zhu, D.L. and Marcotte, P. Co-coercvity and its role in the convergence of iterative schemes for solving variational inequalities. *SIAM J. Optim.*, 6, 714–726 (1966).

https://doi.org/10.1142/9789811293009_0030

Chapter 30

System of General Variational Inclusions

Muhammad Aslam Noor*,‡, Khalida Inayat Noor*,§ and Themistocles M. Rassias†,¶

**Department of Mathematics, COMSATS University Islamabad, Park Road, Islamabad, Pakistan*

†Department of Mathematics, National Technical University of Athens, Zografou Campus, 157 80, Athens, Greece

‡noormaslam@gmail.com

§khalidan@gmail.com

¶trassias@math.ntua.gr

In this chapter, we introduce and consider a new system of variational inclusions involving three different operators, which is called the system of general variational inequalities. We prove that system of general variational inequalities is equivalent to the fixed point problem using the resolvent operator technique. This equivalent formulation is used to suggest and analyze some new explicit iterative methods for solving general variational inclusions. We also study the convergence analysis of the new iterative method under certain mild conditions. Since this new system includes the general variational inequalities involving the two operators, variational inequalities, and related optimization problems as special cases, results obtained in this chapter continue to hold for these problems. Our results can be viewed as a refinement and improvement of the previously known results for variational inclusions.

1. Introduction

Multi-valued variational inclusions, which were introduced and studied by Noor [17, 21], are being used as mathematical programming models to study a wide class of equilibrium problems arising in industry, finance, economics, transportation, optimization, operations research, and engineering sciences in a unified and general frame work. It has been shown that variational inclusions contain variational inequalities, which were introduced and studied by Stampacchia [43]. For the formulation, numerical methods, generalizations, dynamical systems, sensitivity analysis, and other aspects of variational inequalities and related optimization problems, see [1–46] and the references therein. Variational inclusions can provide new insight regarding problems being studied and can stimulate new and innovative ideas for problem solving. These activities have motivated to generalize and extend the variational inequalities and related optimization problems in several directions using new and novel techniques.

Inspired and motivated by research going on in this area, we introduce and consider a new system of general variational inclusions involving three different nonlinear operators. This class of system includes the system of general variational inequalities involving three operators and the classical variational inequalities as special cases. It is shown that odd order and non-symmetric boundary vale problems can be studied in the frame work of general variational inequality. Using the resolvent operator technique, we prove that the system of general variational inequalities is equivalent to the fixed point problem. This alternative equivalent formulation is used to suggest and analyze some iterative methods for solving the system of general variational inclusions. We also prove the convergence of the proposed iterative methods under weaker conditions. Since the new system of variational inclusions includes the system of variational inequalities and related optimization problems as special cases, results proved in this chapter continue to hold for these problems. Our result can be viewed as refinement and improvement of the previous results in this field. The comparison of these methods with other methods is a subject of future research.

2. Preliminaries

Let H be a real Hilbert space whose inner product and norm are denoted by $\langle \cdot, \cdot \rangle$ and $\|.\|$, respectively. Let K be a closed and convex set in H. Let $T_1, T_2, A, g : H \times H \to H$ be nonlinear different operators and let $\varphi : H \longrightarrow R \cup \{+\infty\}$ be a continuous function.

We now consider the problem of finding $x^*, y^* \in H$ such that

$$\left.\begin{array}{ll} 0 \in \rho T_1(y^*, x^*) + \rho A(g(x^*)) - g(y^*) + g(x^*), & \rho > 0 \\ 0 \in \eta T_2(x^*, y^*) + \eta A(g(y^*)) + g(y^*) - g(x^*), & \eta > 0 \end{array}\right\}, \tag{1}$$

which is called the system of general variational inclusions. We now discuss some special of problem (1), which appear to new ones.

Special cases

(I) If $T_1 = T_2 = T$, problem (1) is equivalent to finding $x^*, y^* \in H$ such that

$$\left.\begin{array}{ll} 0 \in \rho T(y^*, x^*) + \rho A(g(x^*)) - g(y^*) + g(x^*), & \rho > 0 \\ 0 \in \eta T(x^*, y^*) + \eta A(g(y^*)) + g(y^*) - g(x^*), & \eta > 0 \end{array}\right\}. \tag{2}$$

(II) If T_1, T_2 are univariate operators, then problem (1) is equivalent to finding $x^*, y^* \in H$ such that

$$\left.\begin{array}{ll} 0 \in \rho T_1(y^*) + \rho A(g(x^*)) - g(y^*) + g(x^*), & \rho > 0 \\ 0 \in \eta T_2(x^*) + \eta A(g(y^*)) + g(y^*) - g(x^*), & \eta > 0 \end{array}\right\}, \tag{3}$$

which appears to be a new one.

(III) If $T_1 = T_2 = T, \quad \rho = \eta, \quad x = x^* = y^*$, then problem (1) is equivalent to finding $x \in H$ such that

$$0 \in \rho T(x) + g(x) - g(x) + \rho A(g(x)), \tag{4}$$

which is known as the variational inclusion problem or finding the zero of the sum of two (more) monotone operators. It is well known that a wide class of linear and nonlinear problems can be studied via variational inclusion problems.

(IV) If $A(.) = \partial\varphi(.)$, the subdifferential of a proper, convex, and lower-semicontinuous function, then the system of variational inclusions (1) is equivalent to finding $x^*, y^* \in H$ such that

$$\left.\begin{array}{ll} 0 \in \rho T_1(y^*, x^*) + \rho\partial\varphi(g(x^*)) - g(y^*) + g(x^*), & \rho > 0 \\ 0 \in \eta T_2(x^*, y^*) + \eta\partial\varphi(g(y^*)) + g(y^*) - g(x^*), & \eta > 0 \end{array}\right\}, \tag{5}$$

or equivalently the problem of finding $x^*, y^* \in H$ such that

$$\left.\begin{array}{l} \langle \rho T_1(y^*, x^*) + g(x^*) - g(y^*), g(x) - g(x^*)\rangle \\ \quad \geq \rho\varphi(g(x^*)) - \rho\varphi(g(x)), \quad \forall x \in H, \quad \rho > 0 \\ \langle \eta T_2(x^*, y^*) + g(y^*) - g(x^*), g(x) - g(y^*)\rangle \\ \quad \geq \rho\varphi(g(y^*)) - \rho\varphi(g(x)), \quad \forall x \in H, \quad \eta > 0 \end{array}\right\}, \tag{6}$$

which is called the system of nonlinear mixed variational inequalities involving two different nonlinear operators, which has been introduced and considered by Noor [18, 28, 29].

(V) If $T_1 = T_2 = T$, then problem (6) reduces to the following system of mixed variational inequalities of finding $x^*, y^* \in H$ such that

$$\left.\begin{array}{l} \langle \rho T(y^*, x^*) + g(x^*) - g(y^*), g(x) - g(x^*)\rangle \\ \quad \geq \rho\varphi(g(x^*)) - \rho\varphi(g(x)), \quad \forall x \in H, \quad \rho > 0 \\ \langle \eta T(x^*, y^*) + g(y^*) - g(x^*), g(x) - g(y^*)\rangle \\ \quad \geq \rho\varphi(g(y^*)) - \rho\varphi(g(x)), \quad \forall x \in H, \quad \eta > 0 \end{array}\right\}. \tag{7}$$

(VI) If $T_1, T_2 : H \to H$ are univariate mappings, then problem (6) reduces to finding $x^*, y^* \in H$ such that

$$\left.\begin{array}{l} \langle \rho T_1(y^*) + g(x^*) - g(y^*), g(x) - g(x^*)\rangle \\ \quad \geq \rho\varphi(g(x^*)) - \rho\varphi(g(x)), \quad \forall x \in H, \quad \rho > 0 \\ \langle \eta T_2(x^*) + g(y^*) - g(x^*), g(x) - g(y^*)\rangle \\ \quad \geq \rho\varphi(g(y^*)) - \rho\varphi(g(x)), \quad \forall x \in H, \quad \eta > 0 \end{array}\right\}, \tag{8}$$

which appears to be a new one.

(VII) If $T_1 = T_1 = T$ is univariate operator and $\rho = \eta$, $x = x^* = y^*$, then problem (6) is equivalent to finding $x \in H$ such that

$$\langle Tx, g(y) - g(x)\rangle \geq \varphi(g(x)) - \varphi(g(y)), \quad \forall y \in H, \tag{9}$$

which is known as the mixed general variational inequality or general variational inequality of the second type. For the applications and numerical methods for solving the mixed general variational inequalities, see Noor [18, 27–29, 32] and Noor *et al.* [33, 35].

(VIII) If φ is an indicator function of a closed convex set K in H, then problem (6) is equivalent to finding $x^*, y^* \in K$ such that

$$\left.\begin{array}{ll}\langle \rho T_1(y^*, x^*) + g(x^*) - g(y^*), g(x) - g(x^*)\rangle \geq 0, \quad \forall x \in K, & \rho > 0 \\ \langle \eta T_2(x^*, y^*) + g(y^*) - g(x^*), g(x) - g(y^*)\rangle \geq 0, \quad \forall x \in K, & \eta > 0\end{array}\right\} \tag{10}$$

is called the system of variational inequalities and has been considered and studied by Huang and Noor [9].

(IX) If $T_1 = T_2 = T$, then problem (10) is equivalent to the following system of variational inequalities (SVI) of finding $x^*, y^* \in K$ such that

$$\left.\begin{array}{ll}\langle \rho T(y^*, x^*) + g(x^*) - g(y^*), g(x) - g(x^*)\rangle \geq 0, \quad \forall x \in K, & \rho > 0 \\ \langle \eta T(x^*, y^*) + g(y^*) - g(x^*), g(x) - g(y^*)\rangle \geq 0, \quad \forall x \in K, & \eta > 0\end{array}\right\}, \tag{11}$$

which has been considered and studied in [9, 33].

(X) If $T_1, T_2 :\longrightarrow H$ are univariate operators, then problem (10) reduces to finding $x^*, y^* \in K$ such that

$$\left.\begin{array}{ll}\langle \rho T_1(y^*) + g(x^*) - g(y^*), g(x) - g(x^*)\rangle \geq 0, \quad \forall x \in K, & \rho > 0 \\ \langle \eta T_2(x^*) + g(y^*) - g(x^*), g(x) - g(y^*)\rangle \geq 0, \quad \forall x \in K, & \eta > 0\end{array}\right\}, \tag{12}$$

which has been considered and studied in [9].

(XI) If $\varphi(.)$ is the indicator function of a closed convex set K, then problem (9) is equivalent to finding $x^* \in K$ such that

$$\langle Tx^*, g(x) - g(x^*)\rangle \geq 0, \quad \forall x \in K, \tag{13}$$

which is known as the general variational inequality introduced and studied by Noor [12–14,20,23]. For the formulation, numerical methods, error bounds, dynamical system, and other aspects of the general variational inequalities, see [12–16, 16–32] and the references therein.

It is known [6,44] that if the operator is not symmetric and non-positive, then it can be made symmetric and positive with respect to an arbitrary operator. For the sake of completeness and to convey the idea, we include all the details.

Definition 1 ([3,31]). An operator $T : H \longrightarrow H$ with respect to an arbitrary operator g is said to be

(a) *g-symmetric*, if and only if,

$$\langle Tu, g(v)\rangle = \langle g(u), Tv\rangle, \quad \forall u, v \in H,$$

(b) g-positive, if and only if,

$$\langle Tu, g(u)\rangle \geq 0, \quad \forall u, v \in H,$$

(c) *g-coercive (g-elliptic)*, if there exists a constant $\alpha > 0$ such that

$$\langle Tu, g(u)\rangle \geq \alpha\|g(u)\|^2, \quad \forall u, v \in H.$$

Note that g-coercivity implies g-positivity, but the converse is not true. It is also worth mentioning that there are operators which are not g-symmetric but g-positive. On the other hand, there are g-positive but not g-symmetric operators. Furthermore, it is well know [6, 44] that, for a linear operator T, there exists an inverse operator T^{-1} on $R(T)$ with $\overline{R(T)} = H$, then one can find an infinite set of auxiliary operators g such that the operator T is both g-symmetric and g-positive.

We now show that the third-order boundary value problems can be studied via problem (11).

Example 2. Consider the third-order absolute boundary value problem of finding u such that

$$\left.\begin{array}{ll} -u'''(x(\geq f(x) & \text{on } \Omega = [a,b] \\ u(x) \geq \phi(x) & \text{on } \Omega = [a,b] \\ [-u'''(x) - f(x)][u - \phi(x)] = 0 & \text{on } \Omega = [a,b] \\ u(a) = 0, \quad u'(a) = 0, \quad u'(b) = 0, & \end{array}\right\} \tag{14}$$

where $f(x)$ is a continuous function and $\phi(x)$ is the cost (obstacle) function. Problem (14) can be studied in the framework of the general variational inequality (11). To do so, let

$$H_0^2[a,b] = \{u \in H, u(a) = 0, \quad u'(a) = 0, \quad u'(b) = 0\}$$

be a Hilbert space, see [6]. One can easily show that the energy functional associated with (11) is

$$\begin{aligned} I[v] &= -\int_a^b \frac{d^3v}{dx^3} v dx - 2\int_a^b f\frac{dv}{dx} dx, \quad \forall \frac{dv}{dx} \in H_0^2[a,b] \\ &= \int_a^b \left(\frac{d^2v}{dx^2}\right)^2 \quad 2\int_a^b f\frac{dv}{dx} vdx \\ &= \langle Lv, g(v)\rangle - 2\langle f, g(v)\rangle, \end{aligned}$$

where

$$\langle Lu, g(v)\rangle = -\int_a^b \frac{d^3u}{dx^3}\frac{dv}{dx} dx = \int_a^b \left(\frac{d^2u}{dx^2}\right)\left(\frac{d^2v}{dx^2}\right) dx, \tag{15}$$

and

$$\langle f, g(v)\rangle = \int_a^b f\frac{dv}{dx} dx,$$

where $g = \frac{d}{dx}$ is linear operator. It is clear that the operator L defined by (15) is linear, g-symmetric, and g-positive and g is a linear operator.

Thus, the minimum of the functional $I[v]$ defined on the Hilbert space $H_0^1[a,b]$ can be characterized by inequality (11). This shows that the third-order absolute boundary value problems can be studied in the framework of the general variational inequality (11).

(XII) If $g = I$, the identity operator, then problem (13) is equivalent to finding $u \in K$ such that

$$\langle Tx^*, x - x^* \rangle \geq 0, \quad \forall x \in K, \tag{16}$$

is called the classical variational inequality, introduced and studied by Stampacchia [43] in potential theory. It has been shown that a wide class of unrelated problems, which arise in various branches of pure and applied sciences, can be studied in the unified and general framework of variational inequalities. Clearly, the system of general variational inclusions (1) is more general and includes several classes of variational inequalities and related optimization problems as special cases. For the recent applications, numerical methods, and formulations of variational inequalities and variational inclusions. For more details, see [1–46] and the references therein.

3. Iterative Methods

In this section, we suggest some explicit iterative algorithms for solving the system of general variational inclusion (1). First of all, we establish the equivalence between the system of variational inclusions and fixed point problems. For this purpose, we recall the following well-known result.

Definition 3. For any maximal operator T, the resolvent operator associated with T, for any $\rho > 0$, is defined as

$$J_T(u) = (I + \rho T)^{-1}(u), \quad \forall u \in H.$$

It is well known that an operator T is maximal monotone if and only if its resolvent operator J_T is defined everywhere. It is single-valued and non-expansive, that is,

$$\|J_A u - J_A v\| \leq \|u - v\|, \quad \forall u, v \in H.$$

If $\varphi(.)$ is a proper, convex, and lower-semicontinuous function, then its subdifferential $\partial\varphi(.)$ is a maximal monotone operator. In this case, we can define the resolvent operator

$$J_\varphi(u) = (I + \rho\partial\varphi)^{-1}(u), \quad \forall u \in H,$$

associated with the subdifferential $\partial\varphi(.)$. The resolvent operator J_φ has the following useful characterization.

Lemma 4 ([2]). *For a given $z \in H$, $u \in H$ satisfies the inequality*

$$\langle u - z, v - u\rangle + \rho\varphi(v) - \rho\varphi(u) \geq 0, \quad \forall v \in H,$$

if and only if

$$u = J_\varphi(z),$$

where $J_\varphi = (I + \rho\partial\varphi)^{-1}$ is the resolvent operator. It is well known the resolvent operator J_φ is non-expansive.

We now show that the system of general variational inclusions (1) is equivalent to the fixed-point problem and this is the motivation of our next result.

Lemma 5. *If the operator A is maximal monotone, then $(x^*, y^*) \in H$ is a solution of the system of general variational inclusions* (1) *if and only if $x^*, y^* \in H$ satisfies*

$$g(x^*) = J_A[g(y^*) - \rho T_1(y^*, x^*)], \tag{17}$$

$$g(y^*) = J_A[g(x^*) - \eta T_2(x^*, y^*)]. \tag{18}$$

Proof. Let $(x^*, y^*) \in H$ be a solution of (1). Then,

$$\left.\begin{array}{l} g(y^*) - \rho T_1(y^*, x^*) \in (I + \rho A)(g(x^*)) \\ g(x^*) - \eta T_2(x^*, y^*) \in (I + \eta A)(g(y^*)) \end{array}\right\}$$

which implies that

$$g(x^*) = J_A[g(y^*) - \rho T_1(y^*, x^*)],$$

$$g(y^*) = J_A[g(x^*) - \eta T_2(x^*, y^*)],$$

which is the required result. □

Lemma 5 shows that the system of general variational inequalities to the fixed problems. This alternative equivalence formulation enables us to suggest the following explicit iterative method for solving the system of general variational inclusions (1).

Algorithm 1. For initial points $x_0, y_0 \in H$, compute the sequence $\{x_n\}$ and $\{y_n\}$ by

$$g(x_{n+1}) = (1 - a_n)g(x_n) + a_n J_A[g(y_n) - \rho T_1(y_n, x_n)]$$

$$g(y_{n+1}) = J_A[g(x_{n+1}) - \eta T_2(x_{n+1}, y_n)],$$

where $a_n \in [0, 1]$ for all $n \geq 0$.

If $T_1, T_2 : H \to H$ are univariate mappings and $g = I$ is the identity operator, then Algorithm 1 reduces to the following iterative method for solving the system of variational inclusion.

Algorithm 2. For initial points $x_0, y_0 \in K$, compute the sequence $\{x_n\}$ and $\{y_n\}$ by

$$x_{n+1} = (1 - a_n)x_n + a_n J_A[y_n - \rho T_1(y_n)], \tag{19}$$

$$y_{n+1} = J_A[x_{n+1} - \eta T_2(x_{n+1})], \tag{20}$$

where $a_n \in [0, 1]$ for all $n \geq 0$ satisfies some suitable conditions.

If $T_1 = T_2 = T$, then Algorithm 1 reduces to the following method for solving (2).

Algorithm 3. For initial points $x_0, y_0 \in H$, compute the sequence $\{x_n\}$ and $\{y_n\}$ by

$$g(x_{n+1}) = (1 - a_n)g(x_n) + a_n J_A[g(y_n) - \rho T(y_n, x_n)],$$
$$g(y_{n+1}) = J_A[g(x_{n+1}) - \eta T(x_{n+1}, y_n)],$$

where $a_n \in [0, 1]$ for all $n \geq 0$.

If $A(.) = \partial\varphi(.)$, the subdifferential of a proper, convex, and lower semicontinuous function φ, then $J_A = J_\varphi$. In this case, Algorithms 1–3 collapse to the following iterative methods for solving system of mixed general variational inequalities (6)–(8):

Algorithm 4. For initial points $x_0, y_0 \in H$, compute the sequence $\{x_n\}$ and $\{y_n\}$ by

$$g(x_{n+1}) = (1 - a_n)g(x_n) + a_n J_\varphi[g(y_n) - \rho T_1(y_n, x_n)]$$
$$g(y_{n+1}) = J_\varphi[g(x_{n+1}) - \eta T_2(x_{n+1}, y_n)],$$

where $a_n \in [0, 1]$ for all $n \geq 0$.

If $T_1, T_2 : H \to H$ are univariate mappings, then Algorithm 3.4 reduces to the following iterative method for solving the system of mixed general variational inequalities (8).

Algorithm 5. For initial points $x_0, y_0 \in K$, compute the sequence $\{x_n\}$ and $\{y_n\}$ by

$$\begin{aligned} g(x_{n+1}) &= (1-a_n)g(x_n) + a_n J_{\varphi}[g(y_n) - \rho T_1(y_n)], \\ g(y_{n+1}) &= J_{\varphi}[g(x_{n+1}) - \eta T_2(x_{n+1})], \end{aligned}$$

where $a_n \in [0,1]$ for all $n \geq 0$ satisfies some suitable conditions.

If $T_1 = T_2 = T$, then Algorithm 4 reduces to the following method for solving the system of general variational inequalities (7).

Algorithm 6. For initial points $x_0, y_0 \in H$, compute the sequence $\{x_n\}$ and $\{y_n\}$ by

$$\begin{aligned} g(x_{n+1}) &= (1-a_n)g(x_n) + a_n J_{\varphi}[y_n - \rho T(y_n, x_n)], \\ g(y_{n+1}) &= J_{\varphi}[g(x_{n+1}) - \eta T(x_{n+1}, y_n)], \end{aligned}$$

where $a_n \in [0,1]$ for all $n \geq 0$.

If φ is the indicator function of a closed convex set K in H, then $J_{\varphi} = P_K$, the projection of H onto the closed convex set K. Then Algorithms 4–6 collapse to the following iterative projection method for solving systems of general variational inequalities (10)–(12).

Algorithm 7. For initial points $x_0, y_0 \in K$, compute the sequence $\{x_n\}$ and $\{y_n\}$ by

$$\begin{aligned} g(x_{n+1}) &= (1-a_n)g(x_n) + a_n P_K[g(y_n) - \rho T_1(y_n, x_n)] \\ g(y_{n+1}) &= P_K[g(x_{n+1}) - \eta T_2(x_{n+1}, y_n)], \end{aligned}$$

where $a_n \in [0,1]$ for all $n \geq 0$.

If $T_1, T_2 : K \to H$ are univariate mappings, then Algorithm 7 reduces to the following iterative method for solving the system of general variational inequalities (12).

Algorithm 8. For initial points $x_0, y_0 \in K$, compute the sequence $\{x_n\}$ and $\{y_n\}$ by

$$\begin{aligned} x_{n+1} &= (1-a_n)x_n + a_n P_K[y_n - \rho T_1(y_n)], \\ y_{n+1} &= P_K[x_{n+1} - \eta T_2(x_{n+1})]. \end{aligned}$$

If $T_1 = T_2 = T$, then Algorithm 7 is reduced to the following.

Algorithm 9. For initial points $x_0, y_0 \in K$, compute the sequence $\{x_n\}$ and $\{y_n\}$ by

$$g(x_{n+1}) = (1 - a_n)g(x_n) + a_n P_K[y_n - \rho T(y_n, x_n)],$$
$$g(y_{n+1}) = P_K[g(x_{n+1}) - \eta T(x_{n+1}, y_n)],$$

where $a_n \in [0, 1]$ for all $n \geq 0$.

We again use Lemma 4 to suggest another method for solving the system of general variational inclusions (1):

Algorithm 10. For initial points $x_0, y_0 \in H$, compute the sequence $\{x_n\}$ and $\{y_n\}$ by

$$g(x_{n+1}) = (1 - a_n)g(x_n) + a_n J_A[g(y_{n+1}) - \rho T_1(y_{n+1}, x_n)]$$
$$g(y_{n+1}) = J_A[g(x_n) - \eta T_2(x_n, y_n)],$$

where $a_n \in [0, 1]$ for all $n \geq 0$. Algorithm 10 is called the Gauss–Seidel-type algorithm for a class of solving system of general variational inequalities.

If $g = I$, then Algorithm 10 reduces to the following method for solving the system of variational inclusions.

Algorithm 11. For initial points $x_0, y_0 \in H$, compute the sequence $\{x_n\}$ and $\{y_n\}$ by

$$x_{n+1} = (1 - a_n)x_n + a_n J_A[y_{n+1} - \rho T_1(y_{n+1}, x_n)]$$
$$y_{n+1} = J_A[x_n - \eta T_2(x_n, y_n)],$$

where $a_n \in [0, 1]$ for all $n \geq 0$. Using the technique of Noor *et al.* [33], one can investigate the convergence analysis of Algorithm 11.

Remark 6. For appropriate and suitable choices of the operators T_1, T_2, g, and the spaces, one can suggest several new iterative methods for solving the system of general variational inclusions. This shows that Algorithms 1 and 10 are quite flexible and general and includes various known and new algorithms for solving general variational inequalities and related optimization problems as special cases.

Definition 7. A mapping $T : H \to H$ is called r-strongly monotone if there exists a constant $r > 0$ such that

$$\langle Tx - Ty, x - y\rangle \geq r||x - y||^2, \quad \forall x, y \in H.$$

Definition 8. A mapping $T : H \to H$ is called relaxed γ-cocoercive if there exists a constant $\gamma > 0$ such that

$$\langle Tx - Ty, x - y\rangle \geq -\gamma||Tx - Ty||^2, \quad \forall x, y \in H.$$

Definition 9. A mapping $T : H \to H$ is called relaxed (γ, r)-cocoercive if there exists constants $\gamma > 0, r > 0$ such that

$$\langle Tx - Ty, x - y\rangle \geq -\gamma||Tx - Ty||^2 + r||x - y||^2, \quad \forall x, y \in H.$$

The class of relaxed (γ, r)-cocoercive mappings is more general than the class of strongly monotone mappings.

Definition 10. A mapping $T : H \to H$ is called μ-Lipschitzian if there exists a constant $\mu > 0$ such that

$$||Tx - Ty|| \leq \mu||x - y||, \quad \forall x, y \in H.$$

Lemma 11. *Suppose* $\{\delta_n\}_{n=0}^{\infty}$ *is a non-negative sequence satisfying the following inequality:*

$$\delta_{n+1} \leq (1 - \lambda_n)\delta_n + \sigma_n, \quad \forall \quad n \geq 0,$$

with $\lambda_n \in [0, 1]$, $\sum_{n=0}^{\infty} \lambda_n = \infty$, *and* $\sigma_n = o(\lambda_n)$. *Then,* $\lim_{n\to\infty} \delta_n = 0$.

4. Convergence Analysis

In this section, we consider the convergence criteria of Algorithm 2 under some suitable mild conditions and this is the main motivation of this chapter. In a similar way, one can consider the convergence analysis of other algorithms.

Theorem 12. *Let* (x^*, y^*) *be the solution of* (1). *If* $T_1 : H \to H$ *is relaxed* (γ_1, r_1)*-cocoercive and* μ_1*-Lipschitzian, and* $T_2 : H \times H \to H$

is relaxed (γ_2, r_2)*-cocoercive and* μ_2*-Lipschitzian with conditions*

$$0 < \rho < \min\{2(r_1 - \gamma_1\mu_1^2)/\mu_1^2,\ 2(r_2 - \gamma_2\mu_2^2)/\mu_2^2\}, \quad r_1 > \gamma_1\mu_1^2, \tag{21}$$

$$0 < \eta < \min\{2(r_1 - \gamma_1\mu_1^2)/\mu_1^2,\ 2(r_2 - \gamma_2\mu_2^2)/\mu_2^2\}, \quad r_2 > \gamma_2\mu_2^2, \tag{22}$$

and $a_n \in [0,1]$, $\sum_{n=0}^{\infty} a_n = \infty$, *then for arbitrarily initial points* $x_0, y_0 \in H$, x_n *and* y_n *obtained from Algorithm* 2 *converge strongly to* x^* *and* y^*, *respectively.*

Proof. From (17), (19), and the non-expansive property of the resolvent operator J_A, we have

$$\begin{aligned}
&||x_{n+1} - x^*|| \\
&\quad = ||(1 - a_n)x_n + a_n J_A[y_n - \rho T_1(y_n)] \\
&\qquad - (1 - a_n)x^* - a_n J_A[y^* - \rho T_1(y^*)]|| \\
&\quad \le (1 - a_n)||x_n - x^*|| + a_n||J_A[y_n - \rho T_1(y_n)] - J_A[y^* - \rho T_1(y^*)]|| \\
&\quad \le (1 - a_n)||x_n - x^*|| + a_n||[y_n - \rho T_1(y_n)] - [y^* - \rho T_1(y^*)]|| \\
&\quad = (1 - a_n)||x_n - x^*|| + a_n||y_n - y^* - \rho[T_1(y_n) - T_1(y^*)]||.
\end{aligned} \tag{23}$$

From the relaxed (γ_1, r_1)-cocoercive and μ_1-Lipschitzian definition of T_1, we have

$$\begin{aligned}
&||y_n - y^* - \rho[T_1(y_n) - T_1(y^*)]||^2 \\
&\quad = ||y_n - y^*||^2 - 2\rho\langle T_1(y_n) - T_1(y^*), y_n - y^*\rangle \\
&\qquad + \rho^2||T_1(y_n) - T_1(y^*)||^2 \\
&\quad \le ||y_n - y^*||^2 - 2\rho[-\gamma_1||T_1(y_n) - T_1(y^*)||^2 + r_1||y_n - y^*||^2] \\
&\qquad + \rho^2||T_1(y_n) - T_1(y^*)||^2 \\
&\quad \le ||y_n - y^*||^2 + 2\rho\gamma_1\mu_1^2||y_n - y^*||^2 \\
&\qquad - 2\rho r_1||y_n - y^*||^2 + \rho^2\mu_1^2||y_n - y^*||^2 \\
&\quad = [1 + 2\rho\gamma_1\mu_1^2 - 2\rho r_1 + \rho^2\mu_1^2]||y_n - y^*||^2.
\end{aligned} \tag{24}$$

Set $\theta_1 = [1+2\rho\gamma_1\mu_1^2 - 2\rho r_1 + \rho^2\mu_1^2]^{1/2}$. It is clear from condition (21) that $0 \le \theta_1 < 1$. Hence, from (24), it follows that

$$||y_n - y^* - \rho[T_1(y_n, x_n) - T_1(y^*, x^*)]|| \le \theta_1||y_n - y^*||. \quad (25)$$

Similarly, from the relaxed (γ_2, r_2)-cocoercive and μ_2-Lipschitzian definition of T_2, we obtain

$$\begin{aligned}
&||x_{n+1} - x^* - \eta[T_2(x_{n+1}) - T_2(x^*)]||^2 \\
&\quad = ||x_{n+1} - x^*||^2 - 2\eta\langle T_2(x_{n+1}) - T_2(x^*), x_{n+1} - x^*\rangle \\
&\qquad + \eta^2||T_2(x_{n+1}) - T_2(x^*)||^2 \\
&\quad \le ||x_{n+1} - x^*||^2 - 2\eta[-\gamma_2||T_2(x_{n+1}) - T_2(x^*)||^2 + r_2||x_{n+1} - x^*||^2] \\
&\qquad + \eta^2||T_2(x_{n+1}) - T_2(x^*)||^2 \\
&\quad = ||x_{n+1} - x^*||^2 + 2\eta\gamma_2||T_2(x_{n+1}) - T_2(x^*)||^2 - 2\eta r_2||x_{n+1} - x^*||^2 \\
&\qquad + \eta^2||T_2(x_{n+1}) - T_2(x^*)||^2 \\
&\quad \le ||x_{n+1} - x^*||^2 + 2\eta\gamma_2\mu_2^2||x_{n+1} - x^*||^2 \\
&\qquad - 2\eta r_2||x_{n+1} - x^*||^2 + \eta^2\mu_2^2||x_{n+1} - x^*||^2 \\
&\quad = [1 + 2\eta\gamma_2\mu_2^2 - 2\eta r_2 + \eta^2\mu_2^2]||x_{n+1} - x^*||^2. \qquad (26)
\end{aligned}$$

Set $\theta_2 = [1 + 2\eta\gamma_2\mu_2^2 - 2\eta r_2 + \eta^2\mu_2^2]^{1/2}$. It is clear from the condition (22) that $0 \le \theta_2 < 1$. Then, from (26), it follows that

$$||x_{n+1} - x^* - \eta[T_2(x_{n+1}, y_n) - T_2(x^*, y^*)]|| \le \theta_2||x_{n+1} - x^*||. \quad (27)$$

Hence, from (23), (25), (27), and the non-expansive property of the resolvent operator J_A, we have

$$\begin{aligned}
&||y_{n+1} - y^*|| \\
&\quad = ||J_A[x_{n+1} - \eta T_2(x_{n+1}, y_n)] - J_A[x^* - \eta T_2(x^*, y^*)]|| \\
&\quad \le ||[x_{n+1} - \eta T_2(x_{n+1}, y_n)] - [x^* - \eta T_2(x^*, y^*)]|| \\
&\quad = ||x_{n+1} - x^* - \eta[T_2(x_{n+1}, y_n) - T_2(x^*, y^*)]|| \\
&\quad \le \theta_2||x_{n+1} - x^*||, \qquad (28)
\end{aligned}$$

which implies that for all $n \geq 1$,

$$||y_n - y^*|| \leq \theta_2 ||x_n - x^*||. \tag{29}$$

Then, from (23), (28), and (29), we obtain that

$$\begin{aligned}
&||x_{n+1} - x^*|| \\
&\quad \leq (1 - a_n)||x_n - x^*|| + a_n||y_n - y^* - \rho[T_1(y_n, x_n) - T_1(y^*, x^*)]|| \\
&\quad \leq (1 - a_n)||x_n - x^*|| + a_n\theta_1||y_n - y^*|| \\
&\quad \leq (1 - a_n)||x_n - x^*|| + a_n\theta_1 \cdot \theta_2||x_n - x^*|| \\
&\quad = [1 - a_n(1 - \theta_1\theta_2)]||x_n - x^*||.
\end{aligned}$$

Since the constant $(1 - \theta_1\theta_2) \in (0, 1]$, and $\sum_{n=0}^{\infty} a_n(1 - \theta_1\theta_2) = \infty$, from Lemma 2.3, we have $\lim_{n\to\infty} ||x_n - x^*|| = 0$. Hence, the result $\lim_{n\to\infty} ||y_n - y^*|| = 0$ is from (29). □

If $\varphi(.)$ is the indicator function of a closed convex set K in H, then $J_\varphi = P_K$, and $g = I$, the identity operator, then the projection of H onto K. Consequently, Theorem 4.1 reduces to the result for solving systems of variational inequalities (10)–(12), which is mainly due to Huang and Noor [9].

5. Conclusion

In this chapter, we have introduced a new system of variational inclusions involving three different operators. Using the resolvent operator technique, we have established the equivalence between the system of variational inclusions and the fixed point problems. We have used this alternative equivalent formulation to suggest and analyze a number of iterative resolvent methods for solving this new system. Convergence analysis is also considered. Several special cases of our results are also discussed. It is interested to compare the idea and technique of this chapter with other techniques. These research areas do not appeared to have developed to an extent that these provide a complete framework for studying these problems. These problems can be very useful in practice. It is worth mentioning that the concept of fuzzy mappings can be extended for a system of general variational inclusions (1) by using the technique of Noor [33]. Much more work is needed in all these areas to develop a sound basis for applications in engineering and physical sciences. The interested reader is advised to explore this field further and discover new and interesting applications of the system of variational inclusions and related optimization problems.

Acknowledgments

The authors wish to express their deepest gratitude to their teachers, students, colleagues, collaborators, and friends, who have directly or indirectly contributed in the preparation of this chapter.

References

[1] Bnouhachem, A., Noor, M.A. and Khalfaoui, M. Modified descent-projection method for solving variational inequalities. *Appl. Math. Computation*, 190, 1691–1700 (2007).
[2] Brezis, H. *Operateurs Maximaux Monotone et Semigroupes de Contractions dans les Espace d'Hilbert*. North-Holland, Amsterdam, Holland (1973).
[3] Chang, S.S., Lee, H.W.J. and Chan, C.K. Generalized system for relaxed cocoercive variational inequalities in Hilbert spaces. *Appl. Math. Lett.*, 20, 329–334 (2006).
[4] Ding, X.P. Perturbed proximal point algorithms for generalized quasi variational inclusions. *J. Math. Anal. Appl.*, 210, 88–101 (1997).
[5] Giannessi, F. and Maugeri, A. *Variational Inequalities and Network Equilibrium Problems*. Plenum Press, New York (1995).
[6] Filippov, V.M. *Variational Principles for Nonpotential Operators*. Amer. Math. soc. Providence, Rode Island, USA (1989).
[7] Giannessi, F., Maugeri, A. and Pardalos, P.M. *Equilibrium Problems, Nonsmooth Optimization and Variational Inequalities Problems*. Kluwer Academic Publishers, Dordrecht, Holland (2001).
[8] Glowinski, R., Lions, J.L. and Tremolieres, R. *Numerical Analysis of Variational Inequalities*. North-Holland, Amsterdam, Holland (1981).
[9] Huang, Z. and Noor, M.A. An explicit projection method for a system of nonlinear variational inequalities with different (γ, r)-cocoercive mappings. *Appl. Math. Comput.*, 190, 356–361 (2007).
[10] Lions, J.L. and Stampacchia, G. Variational inequalities, *Commun. Pure Appl. Math.*, 20, 493–512 (1967).

[11] Noor, M.A. On Variational Inequalities, PhD Thesis, Brunel University, London, UK (1975).

[12] Noor, M.A. General variational inequalities. *Appl. Math. Letters*, 1, 119–121 (1988).

[13] Noor, M.A. Wiener-Hopf equations and variational inequalities. *J. Optim. Theory Appl.*, 79, 197–206 (1993).

[14] Noor, M.A. Variational inequalities in physical oceanography, in: M. Rahman (ed.), *Ocean Wave Engineering*, Computational Mechanics Publications, Southampton, UK, 201–226 (1994).

[15] Noor, M.A. Some recent advances in variational inequalities, Part I, basic concepts, New Zealand. *J. Math.*, 26, 53–80 (1997).

[16] Noor, M.A. Some recent advances in variational inequalities, Part II, other concepts, New Zealand. *J. Math.*, 26, 229–255 (1997).

[17] Noor, M.A. Generalized set-valued variational inclusions and resolvent equations. *J. Math. Anal. Appl.*, 228, 206–220 (1998).

[18] Noor, M.A. Some Algorithms for general monotone mixed variational inequalities. *Math. Comput. Modell.*, 29, 1–9 (1999).

[19] Noor, M.A. Variational inequalities for fuzzy mappings, III. *Fuzzy Sets and Systems*, 110, 101–108 (2000).

[20] Noor, M.A. New approximation schemes for general variational inequalities. *J. Math. Anal. Appl.*, 251, 217–229 (2000).

[21] Noor, M.A. Three-step iterative algorithms for multivalued quasi variational inclusions. *J. Math. Anal. Appl.*, 255, 589–604 (2001).

[22] Noor, M.A. New extragradient-type methods for general variational inequalities. *J. Math. Anal. Appl.*, 277, 379–395 (2003).

[23] Noor, M.A. Some developments in general variational inequalities. *Appl. Math. Computation*, 152, 199–277 (2004).

[24] Noor, M.A. Mixed quasi variational inequalities. *Appl. Math. Computation*, 146, 553–578 (2003).

[25] Noor, M.A. Fundamentals of mixed quasi variational inequalities. *Inter. J. Pure Appl. Math.*, 15, 137–258 (2004).

[26] Noor, M.A. Fundamentals of equilibrium problems. *Math. Inequal. Appl.*, 9, 529–566 (2006).

[27] Noor, M.A. Merit functions for general variational inequalities. *J. Math. Anal. Appl.*, 316, 736–752 (2006).

[28] Noor, M.A. On iterative methods for solving a system of mixed variational inequalities. *Appl. Anal.*, 87, 99–108 (2008).

[29] Noor, M.A. Differentiable nonconvex functions and general variational inequalities. *Appl. Math. Computation*, 199, 623–630 (2008).

[30] Noor, M.A. Extended general variational inequalities. *Appl. Math. Letters*, 22, 182–186 (2009).

[31] Noor, M.A. On a class of general variational inequalities. *J. Adv. Math. Stud.*, 1, 75–86 (2008).

[32] Noor, M.A. On a system of general mixed variational inequalities. *Optim. Letters*, 3, 437–457 (2009).

[33] Noor, M.A,, Khan, A.G., Noor, K.I. and Pervez, A. Gauss-Seidel type algorithms for a class of variational inequalities. *Filomat*, 32(2), 395–407 (2018).

[34] Noor, M.A. and Noor, K.I. Sensitivity analysis of quasi variational inclusions. *J. Math. Anal. Appl.*, 236, 290–299 (1999).

[35] Noor, M.A. and Noor, K.I. Projection algorithms for solving system of general variational inequalities. *Nonl. Anal.*, 70, 2700–2706 (2009).

[36] Noor, M.A., Noor, K.I. and Yaqoob, H. On general mixed variational inequalities. *Acta Appl. Math.*, 110, 227–246 (2010). https://doi.org/10.1007/s10440-008-9402-4.

[37] Noor, M.A. and Noor, K.I. Self-adaptive projection algorithms for general variational inequalities. *Appl. Math. Computation*, 151, 659–670 (2004).

[38] Noor, M.A. Noor, K.I. and Rassias, Th.M. Some aspects of variational inequalities. *J. Comput. Appl. Math.*, 47, 285–312 (1993).

[39] Noor, M.A., Noor, K.I. and Rassias, M.Th. New trends in general variational inequalities. *Acta Math. Applicandae*, 170(1), 981–1046 (2020).

[40] Noor, M.A., Noor, K.I. and Rassias, T.M. Set-valued resolvent equations and mixed variational inequalities. *J. Math. Anal. Appl.*, 220(1998), 741–759.

[41] Patriksson, M. *Nonlinear Programming and Variational Inequalities: A Unified Approach*. Kluwer Academic Publishers, Dordrecht (1998).

[42] Robinson, S.M. Generalized equations and their solutions. *Math. Program. Stud.*, 10, 128–141 (1979).

[43] Stampacchia, G. Formes bilineaires coercivities sur les ensembles convexes. *C.R. Acad. Sci. Paris*, 258, 4413–4416 (1964).

[44] Tonti, E. Variational formulation for every nonlinear problem. *Intern. J. Eng. Sci.*, 22(11–12), 1343–1371 (1984).

[45] Uko, L.U. Strongly nonlinear generalized equations. *J. Math. Anal. Appl.*, 220, 65–76 (1998)

[46] Weng, X.L. Fixed point iteration for local strictly pseudocontractive mappings. *Proc. Amer. Math. Soc.*, 113, 727–731 (1991).

https://doi.org/10.1142/9789811293009_0031

Chapter 31

Higher-Order Bihemivariational Inequalities

Muhammad Aslam Noor[*,§], Khalida Inayat Noor[*,¶], Themistocles M. Rassias[†,||] and Muhammad Uzair Awan[‡,**]

[*] *Department of Mathematics, COMSATS University Islamabad, Park Road, Islamabad, Pakistan*

[†] *Department of Mathematics, National Technical University of Athens, Zografou Campus, 157 80, Athens, Greece*

[‡] *Department of Mathematics, Government College University, Faisalabad, Pakistan*

[§] *noormaslam@gmail.com*

[¶] *khalidan@gmail.com*

[||] *trassias@math.ntua.gr*

[**] *awan.uzair@gmail.com*

In this chapter, we introduce and study a new class of variational inequalities, known as higher-order bihemivariational inequality. It is shown that the optimality conditions of the sum of differentiable strongly biconvex functions and directionally differentiability can be characterized by the higher-order bihemivariational inequalities. The auxiliary principle technique is applied to suggest and analyze some iterative schemes for solving higher-order bihemivariational inequalities. We prove the convergence of these iterative methods under some weak conditions. Our method of proof of the convergence criteria is simple as compared with other techniques. Several important special cases are discussed as applications. Results obtained in this chapter continue to hold for new and known classes of variational inequalities and related optimization problems.

1. Introduction

Variational inequality theory, which was introduced by Stampachia [42] in potential theory, provides us with a simple, general, and unified framework to study a wide class of problems arising in pure and applied sciences. Variational inequalities have been extended and generalized in several directions using novel and innovative techniques. Hemivariational inequalities [39, 40] involving the nonlinear Lipschitz continuous functions can be viewed as novel and important generalization of variational inequalities and variational principles and were introduced and studied by Panagiotopoulos [39, 40] in the structural analysis. It have been shown [39,40] that, if a non-smooth and non-convex superpotential of a structure is quasi-differentiable, then these problems can be studied via the hemivariational inequalities. In passing, we remark that, if the nonlinear Lipschitz continuous function is a differentiable convex function, then hemivariational inequalities coincide with the mildly nonlinear variational inequalities introduced and studied by Noor [11] in 1975.

Stampaccchia [42] proved that the minimum of the differentiable convex function can be characterized by the variational inequalities. To be more precise, let K be a convex set in the Hilbert space H. Then the minimum $u \in K$ of the differentiable convex function F is equivalent to finding $u \in K$ such that

$$\langle F'(u), v - u \rangle \geq 0, \quad \forall v \in K, \tag{1}$$

which is called the variational inequality (1). Here $F'(u)$ is the Frechet derivative of the convex function F at $u \in K$ in the direction $v - u$. Noor and Noor [12] introduced the convex of biconvex set and biconvex functions involving the biconfunction. The biconvex function can be viewed as an important generalization of the convex functions. Noor and Noor [21–23, 35] proved that the optimality conditions of the differentiable biconvex function is equivalent to finding $u \in K_\beta$, where K_β is a biconvex set, such that

$$\langle F'(u), \beta(v-u)\rangle \geq 0, \quad \forall v \in K_\beta, \quad p \geq 1, \tag{2}$$

which is called the bivariational inequality. For the properties, numerical methods, and applications of the bivariational inequalities, see [21–23] and the references therein.

We would like to emphasize that the hemivariational inequalities and bivariational inequalities are quite different generalizations of the variational inequalities and related optimizations problems. It is natural to study these different problems in a unified framework. This motivated us to introduce and consider some new classes of higher-order bihemivariational inequality. These higher-order bihemivariational inequalities are more general and include variational inequalities and related optimization problems as special case. Due to structure of these inequalities, it is not possible to extend the usual projection and resolvent techniques for solving bihemivariational inequalities. To overcome these difficulties, one can use the auxiliary principle technique, which is mainly due to Lions *et al.* [6] and Glowinski *et al.* [4]. Noor [10, 14–16] and Noor *et al.* [18–24, 27, 28, 31–33] have used this technique to develop some iterative schemes for solving various classes of variational inequalities and equilibrium problems. The main idea involving this technique is to first consider an suitable auxiliary problem and then to show that the solution of the auxiliary problem is the solution of the original problem by using the fixed-point approach. We point out that this technique does not involve any projection and the resolvent of the operator and is flexible. It is shown that the auxiliary principle technique can be used to suggest and analyze a class of new inertial iterative methods for solving bihemivariational inequalities. It is worth mentioning that the inertial type method was suggested by Polyak [42] to speed up the convergence of iterative methods. We also prove that the convergence of these new methods requires pseudo bimonotonicity, which is a weaker condition than bimonotonicity. As special cases, one obtains several known and new results for bihemivariational inequalities, variational inequalities, and related optimization problems. Results obtained in this chapter represent an improvement and refinement of the known results for variational inequalities.

2. Preliminaries and Formulation

Let H be a real Hilbert space whose inner product and norm are denoted by $\langle\cdot,\cdot\rangle$ and $\|.\|$, respectively. Let K be a non-empty closed convex set in H. Let $j : H \longrightarrow R$ be a locally Lipschitz continuous function and et $\beta(. - .) : H \times H \to R$ be an arbitrary bifunction.

First of all, we recall the following concepts and results from non-smooth analysis, see [1].

Definition 1. [1] Let j be locally Lipschitz continuous at a given point $x \in H$ and v be any other vector in H. Clarke's generalized directional derivative of j at x in the direction v, denoted by $j^0(x;v)$, is defined as

$$f^0(x;v) = \lim_{t\to 0^+} \sup_{h\to 0} \frac{f(x+h+tv) - f(x+h)}{t}.$$

The generalized gradient of j at x, denoted $\partial j(x)$, is defined to be subdifferential of the function $j^0(x;v)$ at 0. That is,

$$\partial j(x) = \{w \in H : \langle w, v\rangle \le j^0(x;v), \quad \forall v \in H\}.$$

Lemma 2 ([1]). *Let j be locally Lipschitz continuous at a given point $x \in H$ with a constant L. Then,*

(i) $\partial j(x)$ is a non-empty compact subset of H and $\|\xi\| \le L$ for each $\xi \in \partial j(x)$,
(ii) for every $v \in H$, $\quad j^0(x;v) = max\{\langle\xi, c\rangle : \xi \in \partial j(x)\}$,
(iii) the function $v \longrightarrow j^0(x;v)$ is finite, positively homogeneous, subadditive, convex, and continuous,
(iv) $j^0(x;-v) = (-j)^0(x;v)$,
(v) $j^0(x;v)$ is upper semicontinuous as a function of $(x;v)$,
(vi) $\forall x \in H$, there exists a constant $\alpha > 0$ such that

$$|j^0(x;v)| \le \alpha\|v\|, \quad \forall v \in H.$$

If j is convex on K and locally Lipschitz continuous at $x \in K$, then $\partial j(x)$ coincides with the subdifferential $j'(x)$ of j at x in the sense of convex analysis, and $j^0(x;v)$ coincides with the directional derivative $j'(x;v)$ for each $v \in H$, that is, $j^0(x;v) = \langle j'(x), v\rangle$.

Also, we recall the following concepts and basic results of biconvex set and biconvex functions, which are mainly due to Noor *et al.* [21–23, 35].

Definition 3 ([21–23, 35]). The set K_β is said to be a biconvex set with respect to arbitrary bifunction $\beta(.-.)$ if there exists a bifunction $\beta(.-.)\colon H \times H \to R$ such that

$$u + t\beta(v-u) \in K_\beta, \quad \forall u, v \in K_\beta, \quad t \in [0,1].$$

For $\beta(v-u) = v-u$, the biconvex set $K_\beta = K$ becomes the convex set.

Definition 4. A function F on the convex set K_β is said to be higher-order strongly biconvex with respect to bifunction if there exists a constant $\eta > 0$ such that

$$\begin{aligned} F(u+t\beta(v-u)) \le (1-t)F(u) + tF(v) \\ -\eta\{t^p(1-t) + t(1-t)^p\}\|\beta(v-u)\|^p, \\ \forall u, v \in K_\beta, t \in [0,1]. \end{aligned}$$

A function F is said to higher-order biconcave function if and only if $-F$ is a higher-order biconvex function.

If $t = \frac{1}{2}$, then

$$\begin{aligned} F\left(\frac{2u+\beta(v-u)}{2}\right) &\le \frac{F(u)+F(v)}{2} \\ &\quad - \eta\frac{1}{2^p}\|\beta(v-u)\|^p, \quad \forall u, v \in K_\beta, t \in [0,1]. \end{aligned}$$

The function F is said to be a higher-order J-biconvex function. If $p = 2$, then we have the following:

Definition 5. A function F on the convex set K_β is said to be strongly biconvex with respect to bifunction $\beta(.-.) : H \times H \to R$ if there exists a constant $\eta > 0$ such that

$$\begin{aligned} F(u+t\beta(v-u)) &\le (1-t)F(u) + tF(v) \\ &\quad -\eta t(1-t)\|\beta(v-u)\|^2, \quad \forall u, v \in K_\beta, t \in [0,1]. \end{aligned}$$

We also need the following condition.

Assumption 6. The bifunction $\beta(,-,)$ is required to satisfy

$$\beta(u, u + t\beta(v-u)) = -t\beta(v-u), \quad \forall u, v \in K_\beta,$$
$$\beta(v, u + t\beta(v-u) = (1-t)\beta(v-u), \quad \forall u, v \in K_\beta.$$

Note that $\beta(v-u) = 0$ if and only if, $v = u \in K_\beta$. Also

$$\beta(v-u) + \beta(u-v) = 0, \quad \forall u, v \in K_\beta.$$

This implies that the bifunction $\beta(,-,)$ is skew symmetric.

Definition 7. An operator T with respect to the bifunction $\beta(,-,)$ said to

(i) strongly bimonotone if there exits a constant $\alpha > 0$ such that

$$\langle Tu, \beta(v-u)\rangle + \langle Tv, \beta(u-v)\rangle$$
$$\leq -\alpha\{\|\beta(v-u)\|^2 + \|\beta(u-v)\|^2, \quad \forall u, v \in K_\beta,$$

(ii) bimonotone if

$$\langle Tu, \beta(v-u)\rangle + \langle Tv, \beta(u-v)\rangle \leq 0, \quad \forall u, v \in K_\beta,$$

(iii) pseudo bimonotone if

$$\langle Tu, \beta(v-u)\rangle \geq 0 \quad \Rightarrow \quad -\langle Tv, \beta(u-v)\rangle \geq 0, \quad \forall u, v \in K_\beta.$$

Using the technique of Noor *et al.* [21–23], one can easily prove the following results:

Theorem 8. *Let F be a differentiable strongly biconvex function. If Assumption 6 holds, then*

$$\langle F'(u), \beta(v-u)\rangle + \langle F'(v), \beta(u-v)\rangle$$
$$\leq -\alpha\{\|\beta(v-u)\|^2 + \|\beta(u-v)\|^2, \quad \forall u, v \in K_\beta.$$

The converse is also true.

We now consider the problem of the sum of two biconvex functions and discuss its optimality conditions under some suitable conditions:

$$I[v] = F(v) + \phi(v), \tag{3}$$

where $F(v)$ and $\phi(v)$ are biconvex functions. The functional $I[v]$ is usually called the energy functional.

We prove that the optimality criteria of the energy function $I[v]$ define by (3) can be characterized by a class of variational inequalities.

Theorem 9. *Let F be a differentiable higher-order biconvex function with modulus $\eta > 0$ and let $\phi(v)$ be a directionally differentiable biconvex function on the convex set $K\beta$. If $u \in K_\beta$ is the minimum of the functional $I[v]$ defined by (3), then*

$$\begin{aligned} &F(v) + \phi(v) - (F(u) + \phi(u)) \\ &\quad \geq \langle F'(u), \beta(v-u)\rangle + \phi'(u; v-u) + \eta\|\beta(v-u)\|^p, \quad \forall v, u \in K_\beta. \end{aligned} \tag{4}$$

Proof. Let $u \in K_\beta$ be a minimum of the functional $I[v]$. Then,

$$I[u] \leq I[v], \quad \forall v \in K\beta,$$

which implies that

$$F(u) + \phi(u) \leq F(v) + \phi(v), \quad \forall v \in K_\beta.$$

Since K_β is a biconvex set, $\forall u, v \in K_\beta, \lambda \in [0,1]$, $v_t = (u + \lambda\beta(v-u) \in K_\beta$.

Taking $v = v_t$ in (5), we have

$$F(u) + \phi(u) \leq F(v_t) + \phi(v_t), \quad \forall v \in K_\beta. \tag{5}$$

Since F is differentiable strongly biconvex function,

$$\begin{aligned} F(u + \lambda\beta(v-u)) \leq F(u) &+ \lambda(F(v) - F(u)) \\ &- \eta\{\lambda^p(1-\lambda) + \lambda(1-\lambda)^p\}\|\beta(v-u)\|^p, \\ &\forall u, v \in K_\beta, \quad p \geq 1, \end{aligned}$$

from which, using (5), we have

$$\begin{aligned} F(v) - F(u) &\geq \lim_{\lambda \to 0}\left\{\frac{F(u + \lambda\beta(v-u)) - F(u)}{\lambda}\right\} + \eta\|\beta(v-u)\|^p \\ &= \langle F'(u), \beta(v-u)\rangle + \eta\|\beta(v-u)\|^p, \quad p \geq 1. \end{aligned} \tag{6}$$

In a similar way,

$$\phi(v)-\phi(u) \geq \lim_{\lambda\to 0}\left\{\frac{\phi(u+\lambda\beta(v-u))-\phi(u)}{\lambda}\right\}$$
$$= \phi'(u;\beta(v-u)). \tag{7}$$

From (7) and (6), we have

$$F(v)+\phi(v)-(F(u)+\phi(u))$$
$$\geq \langle F'(u),\beta(v-u)\rangle + \phi'(u;\beta(v-u)) + \eta\|\beta(v-u)\|^p,$$

which is the required result (4). □

Remark. We would like to point out that if $u \in K_\beta$ satisfies the inequality

$$\langle F'(u),\beta(v-u)\rangle$$
$$+\phi'(u;\beta(v-u)) + \eta\|\beta(v-u)\|^p \geq 0, \quad \forall u,v \in K_\beta, \tag{8}$$

then $u \in K_\beta$ is the minimum of the functional $I[v] = F(v)+\phi(v)$. The inequality of type (8) is called the higher-order bihemivariational inequality.

In many applications, the inequality of type (8) may not arise as the minimum of the sum of the two differentiable biconvex functions. These facts motivated us to consider more general bihemivariational inequality, which contains the inequality (8) as a special case.

For given nonlinear continuous operators $T, A : H \longrightarrow H$, we consider the problem of finding $u \in K$ such that

$$\langle Tu,\beta(v-u)\rangle$$
$$+A(u;\beta(v-u)) + \zeta\|\beta(v-u)\|^p \geq 0, \quad \geq 1, \quad \forall v \in K_\beta, \tag{9}$$

which is called the *higher-order bihemivariational inequality.*

Special Cases

We now discuss some new and known classes of variational inequalities and related optimization problems:

(i) If $A(u;\beta(v-u)) = \phi'(u;\beta(v-u))$ denotes directional derivative of the biconvex function ϕ. u in the direction $v-u$, then problem (9) reduces to finding $u \in K_\beta$ such that

$$\langle Tu, \beta(v-u)\rangle + \phi'(u; \beta(v-u)) \\ + \zeta\|\beta(v-u)\|^p \geq 0, \quad \geq 1 \quad \forall v \in K_\beta, \qquad (10)$$

which is also called the higher-order bihemivariational inequality.

(ii) For $\beta(v-u) = v-u$, $A(u; v-u) = J^0(u; v-u)$, and $\zeta = 0$, problem (9) reduces to finding $u \in K$ such that

$$\langle Tu, v-u\rangle + J^0(u; v-u) \geq 0, \quad \forall v \in K, \qquad (11)$$

which is known as hemivariational inequality introduced and studied by Panagiotopoulos [39, 40]. Hemivariational inequalities have important applications in superpotential analysis of elasticity and structural analysis.

(iii) If the operator T is linear, symmetric, and positive and $j(.)$ is locally Lipschitz continuous function, then one can easily show that problem (11) is equivalent to finding the minimum of the functional on the convex set K, where

$$I[v] = \langle Tv, v\rangle + 2j(v), \qquad (12)$$

is known as the nonlinear energy functional associated with hemivariational inequalities.

(iv) If $\phi(.)$ is a smooth and convex function and $\beta(v-u) = v-u$, then $\phi'(u; v-u) = \langle \phi'(u), v-u\rangle$, and consequently problem (10) is equivalent to finding $u \in K$ such that

$$\langle Tu, v-u\rangle + \langle \phi'(u), v-u\rangle \\ + \zeta\|v-u\|^p, \quad p \geq 1, \quad \forall v \in K, \quad \geq 0, \qquad (13)$$

which is called the higher-order mildly nonlinear variational inequality.

(v) If $\zeta = 0$, $\beta(v-u) = v-u$ and $A(u; v-u) = \langle Au, v-u\rangle$, then problem (9) reduces to finding $u \in K$ such that

$$\langle Tu, v-u\rangle + \langle A(u), v-u\rangle \geq 0, \quad \forall v \in K, \qquad (14)$$

which is called the mildly nonlinear variational inequality, introduced and studied by Noor [10] in 1975.

(vi) If $K^\star = \{u \in H : \langle u, v\rangle \geq 0, \quad \forall v \in K\}$ is a polar convex of the convex cone K, then problem (14) is equivalent to fining $u \in H$ such that

$$u \in K, \quad Tu + A(u) \in K^\star, \quad \langle Tu + A(u), u\rangle = 0, \tag{15}$$

is called the generalized complementarity problem, see Noor [11, 12]. For the applications, numerical methods, and other aspects of complementarity problems, see [2, 5, 9, 12, 14, 31, 32, 38] and the references therein.

(vii) If $K = H$, then problem (14) is equivalent to fining $u \in H$ such that

$$\langle Tu + A(u), v\rangle = 0, \quad \forall v \in H, \tag{16}$$

which is called the weak formulation of the mildly nonlinear boundary value problem, see [10].

From (16), we have

$$Tu + A(u) = b, \quad \forall v \in H, \tag{17}$$

where b is the given data and is called the system of equations, see [30, 36].

If $A(u) = A|u|$, then problem (18) is called the system of absolute value equations. That is,

$$Tu + A|u| = b, \quad \forall v \in H. \tag{18}$$

It is well known [7,8,17] that the system of absolute value equations is equivalent to the complementarity problems, which are equivalent to variational inequalities.

(viii) If $A(.;.) = 0$, then problem (9) reduces to finding $u \in K_\beta$ such that

$$\langle Tu, \beta(v-u)\rangle + \zeta\|\beta(v-u)\|^p \geq 0, \quad \forall v \in K_\beta, \quad p \geq 1, \tag{19}$$

is called the higher-order bivariational inequality introduced and studied in [21–23].

(ix) If $\zeta = 0, \quad \beta(v-u) = v - u$, then problem (19) is exactly the variational inequality problem of finding $u \in K$ such that

$$\langle Tu, v - u\rangle \geq 0, \quad \forall v \in K, \tag{20}$$

which is known as the classical variational inequality, introduced and studied by Stampacchia [43] in potential theory.

> For the recently applications, numerical methods, sensitivity analysis, dynamical systems, and local uniqueness of solutions of variational inequalities, see [2–44] and the references therein.

This shows that problem (9) is a quite and unified one. Due to the structure and nonlinearity involved, one has to consider its own. It is an open problem to develop unified numerical implementation numerical methods for solving the higher-order bihemivariational inequalities.

3. Main Results

The projection and resolvent methods cannot be applied to suggest the iterative methods for solving the hemivariational inequalities. To overcome this drawback, we use the auxiliary principle technique. This technique is mainly due to Lions *et al.* [6] and Glowinski *et al.* [4] as developed in [14–16, 27, 28, 31–34, 36, 43, 44] to suggest and analyze some inertial iterative methods for solving higher-order hemivariational inequalities (9).

For a given $u \in K_\beta$ satisfying (9), consider the problem of finding $w \in K_\beta$ such that

$$\begin{aligned}&\langle \rho T(w+\eta(w-u)), \beta(v-w)\rangle + \langle w-u, v-w\rangle \\ &\quad + A(w;\beta(v-w)) + \zeta\|\beta(v-w)\|^p \geq 0, \quad \forall v \in K_\beta, \ \ p \geq 1, \quad (21)\end{aligned}$$

where $\rho > 0, \eta, \zeta \in [0,1]$ are constants.

Inequality of type (21) is called the auxiliary higher-order hemivariational inequality.

If $w = u$, then w is a solution of (9). This simple observation enables us to suggest the following iterative method for solving (9).

Algorithm 10. For a given $u_0 \in K_\beta$, compute the approximate solution u_{n+1} by the iterative scheme

$$\begin{aligned}&\langle \rho T(u_{n+1}+\eta(u_n-u_{n+1})), \beta(v-u_{n+1}\rangle + \langle u_{n+1}-u_n, v-u_{n+1}\rangle \\ &\quad \geq -A(u_{n+1};\beta(v-u_{n+1})) - \zeta\|\beta(u_{n+1}-u_n)\|^p, \\ &\quad \forall v \in K_\beta, \ \ p \geq 1.\end{aligned}$$

Algorithm 10 is called the hybrid proximal point algorithm for solving higher-order bihemivariational inequalities (9).

Special Cases

We now consider some cases of Algorithm 10.

(I) For $\eta = 0$, Algorithm 10 reduces to the following:

Algorithm 11. For a given $u_0 \in K_\beta$, compute the approximate solution u_{n+1} by the iterative scheme

$$\begin{aligned}&\langle \rho T u_{n+1}, \beta(v - u_{n+1})\rangle + \langle u_n - u_{n+1}, v - u_{n+1}\rangle\\&\geq -A(u_{n+1}; \beta(v - u_{n+1})) - \zeta\|\beta(v - u_{n+1})\|^p,\\&\forall v \in K_\beta,\ \ p \geq 1.\end{aligned} \tag{22}$$

(II) If $\eta = 1$, then Algorithm 10 reduces to the following:

Algorithm 12. For a given $u_0 \in K_\beta$, compute the approximate solution u_{n+1} by the iterative scheme

$$\begin{aligned}&\langle \rho T u_n, \beta(v - u_{n+1})\rangle + \langle u_n - u_{n+1}, v - u_{n+1}\rangle\\&\geq -A(u_{n+1}; \beta(v - u_{n+1})) - \zeta\|\beta(u_n - u_{n+1})\|^p,\\&\forall v \in K_\beta,\ \ p \geq 1.\end{aligned}$$

(III) If $\eta = \frac{1}{2}$, then Algorithm 10 collapses to the following:

Algorithm 13. For a given $u_0 \in K_\beta$, compute the approximate solution u_{n+1} by the iterative scheme

$$\begin{aligned}&\left\langle \rho T\left(\frac{u_{n+1} + u_n}{2}\right), \beta(v - u_{n+1})\right\rangle + \langle u_n - u_{n+1}, v - u_{n+1}\rangle\\&\geq -A(u_{n+1}; \beta(v - u_{n+1})) - \zeta\|\beta(u_n - u_{n+1})\|^p,\\&\forall v \in K_\beta,\ \ p \geq 1,\end{aligned}$$

which is called the midpoint proximal method for solving problem (9).

If $A(.;.) = 0$, then Algorithm 10 reduces to the following:

Algorithm 14. For a given $u_0 \in H$, compute the approximate solution u_{n+1} by the iterative scheme

$$\begin{aligned}&\langle \rho T(u_{n+1} + \eta(u_n - u_{n+1})), \beta(v - u_{n+1}\rangle + \langle u_{n+1} - u_n, v - u_{n+1}\rangle\\&\geq -\zeta\|\beta(u_{n+1} - u_n)\|^p, \quad \forall v \in K_\beta,\ \ p \geq 1,\end{aligned}$$

for solving higher-order bivariational inequality.

For the convergence analysis of Algorithm 11, we recall the following concepts and results.

Definition 15. An operator $T : H \to H$ is said to be pseudo bimonotone with respect to $A(.;\beta(.-.))+\eta\|\beta(.-.)\|^p,\}$ if and only if

$$\langle Tu,\beta(v-u)\rangle + A(u;\beta(v-u)) + \zeta\|\beta(v-u)\|^p \geq 0, \quad \forall u,v,z \in H$$

$$\Longrightarrow$$

$$\langle Tv,\beta(v-u)\rangle - A(v;\beta(v-u)) - \zeta\|\beta(v-u)\|^p \geq 0,$$
$$\forall u,v,z \in H.$$

Definition 16. The operator $A(.;.)$ is called bimonotone iff

$$A(u;\beta(v-u)) + A(v;\beta(u-v)) \leq 0, \quad \forall u,v \in H.$$

Lemma 17. $\forall u,v \in H$,

$$2\langle u,v\rangle = \|u+v\|^2 - \|u\|^2 - \|v\|^2. \tag{23}$$

We now consider the convergence criteria of Algorithm 11. The analysis is in the spirit of Noor [14]. We include the proof for the sake of completeness and to convey an idea of the technique involved.

Theorem 18. *Let $u \in K_\beta$ be a solution of* (9) *and let u_{n+1} be the approximate solution obtained from Algorithm* 11. *If the operator T is pseudo bimonotone with respect to $A(.;.)+\zeta\|\beta(v-u)\|^p$, then*

$$\|u_{n+1}-u\|^2 \leq \|u_n-u\|^2 - \|u_{n+1}-u_n\|^2. \tag{24}$$

Proof. Let $u \in K_\beta$ be a solution of (2.1). Then,

$$\langle Tv,\beta(v-u)\rangle - A(v;\beta(v-u)) - \zeta\|\beta(v-u)\|^p \geq 0 \quad \forall v \in K_\beta, \quad p \geq 1, \tag{25}$$

since T is a pseudomonotone operator with respect to $A(.;.)+\eta\|v-u\|^p$.

Now taking $v = u_{n+1}$ in (25), we have

$$\langle Tu_{n+1},\beta(u_{n+1}-u)\rangle - A(u_{n+1};\beta(u_{n+1}-u)) - \zeta\|\beta(u_{n+1}-u)\|^p \geq 0. \tag{26}$$

Taking $v = u$ in (22), we get

$$\begin{aligned}&\langle \rho T u_{n+1}, \beta(u - u_{n+1})\rangle + \langle u_n - u_{n+1}, u - u_{n+1}\rangle\\ &\quad \geq -A(u_{n+1}; \beta(u_{n+1}) - u) - \zeta\|\beta(u_{n+1} - u)\|^p, \quad \forall v \in K_\beta, p \geq 1,\end{aligned}$$

which can be written as

$$\begin{aligned}&\langle u_{n+1} - u_n, u - u_{n+1}\rangle\\ &\quad \geq \langle \rho T u_{n+1}, \beta(u_{n+1} - u)\rangle\\ &\qquad - \rho A(u_{n+1}; \beta(u_{n+1} - u)) - \rho\zeta\|\beta(u_{n+1} - u)\|^p\\ &\quad \geq 0, \end{aligned} \tag{27}$$

where we have used (26).

Setting $u = u - u_{n+1}$ and $v = u_{n+1} - u_n$ in (23), we obtain

$$\begin{aligned}&2\langle u_{n+1} - u_n, u - u_{n+1}\rangle\\ &\quad = \|u - u_n\|^2 - \|u - u_{n+1}\|^2 - \|u_{n+1} - u_n\|^2.\end{aligned} \tag{28}$$

Combining (27) and (28), we have

$$\|u_{n+1} - u\|^2 \leq \|u_n - u\|^2 - \|u_{n+1} - u_n\|^p,$$

which is the required result (24). □

Theorem 19. *Let H be a finite dimensional space and all the assumptions of Theorem* 18 *hold. Then the sequence $\{u_n\}_1^\infty$ given by Algorithm* 11 *converges to a solution u of* (9).

Proof. Let $u \in K_\beta$ be a solution of (9). From (24), it follows that the sequence $\{||u - u_n||\}$ is non-increasing and consequently $\{u_n\}$ is bounded. Furthermore, we have

$$\sum_{n=0}^{\infty} \|u_{n+1} - u_n\|^2 \leq \|u_0 - u\|^2,$$

which implies that

$$\lim_{n\to\infty} \|u_{n+1} - u_n\| = 0. \tag{29}$$

Let $\hat{u}$ be the limit point of $\{u_n\}_1^\infty$; a subsequence $\{u_{n_j}\}_1^\infty$ of $\{u_n\}_1^\infty$ converges to $\hat{u} \in H$. Replacing w_n by u_{n_j} in (3.2), taking the limit $n_j \longrightarrow \infty$, and using (29), we have

$$\langle T\hat{u}, \beta(v - \hat{u})\rangle + A(\hat{u}; \beta(v - \hat{u})) + \zeta\|\beta(v - \hat{u})\|^p \geq 0, \quad \forall v \in K_\beta,$$

which implies that $\hat{u}$ solves the higher-order bihemivariational inequality (9) and

$$\|u_{n+1} - u\|^2 \leq \|u_n - u\|^2.$$

Thus, it follows from the above inequality that $\{u_n\}_1^\infty$ has exactly one limit point $\hat{u}$ and

$$\lim_{n\to\infty}(u_n) = \hat{u},$$

which is the required result. □

We again consider the auxiliary principle technique to suggest some hybrid inertial proximal point methods for solving problem (9).

For a given $u \in K$ satisfying (9), consider the problem of finding $w \in K$ such that

$$\begin{aligned}&\langle \rho T(w + \eta(w - u)), \beta(v - w)\rangle + w - u + \alpha(u - u), v - w\rangle\\&\quad + A((w + \xi(w - u)); \beta(v - w)) + \zeta\|\beta(v - w)\|^p \geq 0,\\&\quad \forall v \in K_\beta, p \geq 1,\end{aligned} \tag{30}$$

where $\rho > 0, \alpha, \xi, \eta, \zeta \in [0, 1]$ are constants.

Clearly, for $w = u, w$ is a solution of (9). This fact motivated us to suggest the following iterative method for solving (9).

Algorithm 20. For given $u_0, u_1 \in K_\beta$, compute the approximate solution u_{n+1} by the iterative scheme

$$\begin{aligned}&\langle \rho T(u_{n+1} + \eta(u_n - u_{n+1})), \beta(v - u_{n+1})\rangle\\&\quad + u_{n+1} - u_n + \alpha(u_n - u_{n-1}, v - u_{n+1}\rangle\\&\geq -A((u_{n+1} + \xi(u_n - u_{n+1})); \beta(v - u_{n+1}))\\&\quad - \zeta\|\beta(v - u_{n+1})\|^p, \quad \forall v \in K_\beta, p \geq 1,\end{aligned}$$

which is known as the hybrid inertial iterative method.

Note that for $\alpha = 0$, $\xi = 0$, Algorithm 20 is exactly Algorithm 10. Using essentially the technique of Theorem 18 and Noor [14], one can study the convergence analysis of Algorithm 20.

If $\eta = \frac{1}{2}$ and $\xi = \frac{1}{2}$, then Algorithm 20 reduces to the following.

Algorithm 21. For given $u_0, u_1 \in K_\beta$, compute the approximate solution u_{n+1} by the iterative scheme

$$\begin{aligned}&\left\langle \rho T\left(\frac{u_{n+1}+u_n}{2}\right), \beta(v-u_{n+1})\right\rangle\\ &\qquad + u_{n+1} - u_n + \alpha(u_n - u_{n-1}, v - u_{n+1})\\ &\qquad \geq -A\left(\frac{u_{n+1}+u_n}{2}\right);\ \beta(v-u_{n+1})\\ &\qquad -\zeta\|\beta(v-u_{n+1})\|^p, \quad \forall v \in K_\beta, p \geq 1,\end{aligned}$$

which is known as the hybrid inertial iterative method.

We now suggest some iterative methods by using the auxiliary principle techniques involving the Bregman distance functions.

For the readers' convenience, we recall some basic properties of the Bregman convex functions. For strongly biconvex function F, we define the Bregman distance function as

$$\begin{aligned}B(v,u) &= F(v) - F(u) - \langle F'(u), \beta(v-u)\rangle\\ &\geq \alpha\|\beta(v-u)\|^2, \quad \forall u, v \in K_\beta. \qquad (31)\end{aligned}$$

It is important to emphasize that various types of function F give different Bregman distances. We give the following important examples of some practical important types of convex function F and their corresponding Bregman distances:

(1) If $f(v) = \|v\|^2$, then $B(v,u) = \|v-u\|$, which is the squared Euclidean distance.

(2) If $f(v) = \sum_{i=1}^n a_i \log(v_i)$, which is known as Shannon entropy, then its corresponding Bregman distance function is given as

$$B(v,u) = \sum^n i = 1\left(v_i \log\left(\frac{v_i}{u_i}\right) + u_i - v_i\right).$$

This Bregman distance function is called Kullback–Leibler distance (KL) and has become a very important tool in several areas of applied mathematics, such as machine learning.

(3) If $f(v) = -\sum_{i=1}^{n} \log(v_i)$, which is called Burg entropy, then its corresponding Bregman distance function is given as

$$B(v,u) = \sum_{i=1}^{n} \left(\log\left(\frac{v_i}{u_i}\right) + \frac{v_i}{u_i} - 1 \right).$$

This is called Itakura–Saito (IS) distance function, which is very important in the information theory, data analysis, and machine learning.

Remark 22. It is a challenging problem to explore the applications of Bregman distance function for other types of non-convex functions as biconvex, k-convex functions and harmonic functions.

For a given $u \in K_\beta$, consider the auxiliary problem of finding $w \in K_\beta$ such that

$$\begin{aligned} &\rho\langle Tw, \beta(v-w)\rangle + \langle E'(w) - E'(u), \beta(v-w)\rangle \\ &\quad + \rho A(w; \beta(v-w)) + \zeta\|\beta(v-w)\|^p \geq 0, \quad \forall v \in K_\beta, \end{aligned} \tag{32}$$

where $\rho > 0$ is a constant and $E'(u)$ is the differential of a strongly biconvex function $E(u)$ at $u \in K_\beta$.

We note that if $w = u$, then clearly w is solution of problem (9). This observation enables us to suggest and analyze the following iterative method for solving (9).

Algorithm 23. For a given $u_0 \in H$, compute the approximate solution u_{n+1} by the iterative scheme

$$\begin{aligned} &\rho\langle Tu_{n+1}, \beta(v-u_{n+1})\rangle + \langle E'(u_{n+1}) - E'(u), \beta(v-u_{n+1})\rangle \\ &\quad + \rho A(u_{n+1}; \beta(v-u_{n+1})) + \zeta\|\beta(v-u_{n+1})\|^p \geq 0, \quad \forall v \in K_\beta. \end{aligned} \tag{33}$$

Algorithm 23 is called the proximal method for solving problem (9).

It is well known that to implement the proximal point methods, one has to find the approximate solution implicitly, which is itself a difficult problem. To overcome this drawback, we now consider another method for solving (9) using the auxiliary principle technique.

For a given $u \in K_\beta$, consider the auxiliary problem of finding $w \in K_\beta$ such that

$$\rho\langle Tu, \beta(v-w)\rangle + \langle E'(w) - E'(u), \beta(v-w)\rangle \\ + \rho A(u; \beta(v-w)) + \zeta\|\beta(v-w)\|^p \geq 0, \quad \forall v \in K_\beta, \quad (34)$$

where $\rho > 0$ is a constant and $E'(u)$ is the differential of a strongly biconvex function $E(u)$ at $u \in K_\beta$. Since $E(u)$ is a strongly convex function, problem (3.1) has a unique solution. We note that if $w = u$, then clearly w is the solution of problem (9). This observation enables us to suggest and analyze the following iterative method for solving (9).

Algorithm 24. For a given $u_0 \in H$, compute the approximate solution u_{n+1} by the iterative scheme

$$\rho\langle Tu_n, \beta(v-u_{n+1})\rangle + \langle E'(u_{n+1}) - E'(u), \beta(v-u_{n+1})\rangle \\ + \rho A(u_n; \beta(v-u_n)) + \zeta\|\beta(v-u_{n+1})\|^p \geq 0, \quad \forall v \in K_\beta, \quad (35)$$

where $\rho > 0$ is a constant. Algorithm 24 is called the proximal method for solving problem (9). For different and appropriate values of the parameters, ξ, η, ζ, α, the operators T, A and spaces, one can obtain a wide class of inertial-type iterative methods for solving the higher-order bihemivariational inequalities and related optimization problems.

4. Conclusion

In this chapter, we have shown that the auxiliary principle technique can be applied for considering some inertial-type iterative methods for solving higher-order bihemivariational inequalities with suitable modifications. We note that this technique is independent of the projection and the resolvent of the operator. Moreover, we have studied the convergence analysis of these new methods under weaker conditions. We have only considered the theoretical aspects of the hybrid inertial iterative methods. It is an interesting open problem to compare the idea and technique of this paper with other techniques. These research areas do not appeared to have developed to an extent that these provide a complete framework for studying these problems.

These problems can be very useful in practice. It is worth mentioning that the concept of fuzzy mappings and quantum calculus can be extended for higher-order bihemivariational inequalities, exponentially equilibrium, and related optimization problems. Much more work is needed in all these areas to develop a sound basis for applications in engineering and physical sciences. The interested reader is advised to explore this field further and discover new and interesting applications of the system of variational inclusions and related optimization problems. It is an interesting problem to implement these methods numerically and compare with other iterative schemes.

Acknowledgments

The authors wish to express their deepest gratitude to their respected teachers, students, colleagues, collaborators, referees, and friends, who have directly or indirectly contributed in the preparation of this chapter.

References

[1] Clarke, F.H. *Optimization and Nonsmooth Analysis*, John Wiley and Sons, New York (1983).

[2] Cottle, R.W., Pang, Jong-S. and Stone, R.E. The Linear Complementarity Problem, SIAM Publication (1992).

[3] Giannessi, F. Maugeri, A. and Pardalos, M.S. Equilibrium Problems: Nonsmooth Optimization and Variational Inequality Models, Kluwer Academic, Dordrecht, Holland (2001).

[4] Glowinski, R., Lions, J.-L., and Tremolieres, R. *Numerical Analysis of Variational Inequalities*, North-Holland, Amsterdam, Holland (1981).

[5] Karamardian, S. Generalized complementarity problems, *J. Opt. Theory Appl.* 8, 161–168 (1971).

[6] Lions, J.L. and Stampacchia, G. Variational inequalities, Commun. *Pure Appl. Math.*, 20, 493–519 (1967).

[7] Mangasarian, O.L. Absolute value equation solution via dual complementarity, *Optim. Lett.* 7, 625–630 (2013).

[8] Mangasarian, O.L. and Meyer, R.R. Absolute value equations, *Lin. Alg. Appl.* 419(2), 359–367 (2006).

[9] Mohsen, B.B., Noor, M.A., Noor, K.I. and Postolcahe, M. Strongly convex functions of higher order involving bifunction, *Mathematics*, 7, 1028 (2019). https://doi.org/10.3390/math7111028.

[10] Noor, M.A. *On Variational Inequalities*, PhD Thesis, Brunel University, London, UK (1975).

[11] Noor, M.A. General variational inequalities, *Appl. Math. Letters*, 1, 119–121 (1988).

[12] Noor, M.A. Fixed point approach for complementaritty problems, *J. Math. Anal. Appl.* 133, 437–448 (1988).

[13] Noor, M.A. New approximation schemes for general variational inequalities, *J. Math. Anal. Appl.* 251, 217–229 (2000).

[14] Noor, M.A. Some developments in general variational inequalities, *Appl. Math. Comput.* 152, 199–277 (2004).

[15] Noor, M.A. Hemiequilibrium problems, *J. Appl. Math. Stoschast. Anal.* 2004(3), 235–244 (2004).

[16] Noor, M.A. Hemivariational inequalities, *J. Appl. Math. Computing*, 17, 59–72 (2005).

[17] Noor, M.A., Iqbal, J., Noor K.I. and Al-Said, E. On an iterative method for solving absolute value equations, *Optim. Lett.* 6, 1027–1033 (2012).

[18] Noor, M.A. and Noor, K.I. Higher order strongly general convex functions and variational inequalities, *AIMS Math.*, 5(4), 3646–3663 (2020).

[19] Noor, M.A. and Noor, K.I. Some classes of higher order general convex functions and variational inequalities, *Montes Taurus J. Pure Appl. Math.* 55(3), 1–15 (2023).

[20] Noor, M.A., Noor, K.I. and Al-Bayatti, H.M.Y. Higher order variational inequalities, *Inform. Sci. Lett.*, 11(1), 1–5 (2022).

[21] Noor, M.A. and Noor, K.I. Hihger order strongly exponentially biconvex functions and bivariational inequalities, *J. Math. Anal.*, 12(2), 23–43 (2021).

[22] Noor, M.A. and Noor, K.I. Higher order strongly biconvex functions and biequilibrium problems, *Advanc. Lin. Algebr. Matrix Theory*, 11(2), 31–53 (2021). doi:10.4236/alamt.2021.112004.

[23] Noor, M.A. and Noor, K.I. General biconvex functions and bivariational inequalities, *Numerical Algeb. Control Optim.*, 13(1), 11–27 (2023). doi:10.3934/naco.2021041.

[24] Noor, M.A. and Noor, K.I. Some classes of higher order general convex functions and variational inequalities, *Montes Taurus J. Pure Appl. Math.* 5(3), 1–15 (2023).

[25] Noor, M.A. and Noor, K.I. Dynamical system technique for solving quasi variational inequalities, *U.P.B. Sci. Bull.*, Series A, 84(4), 55–66 (2022).

[26] Noor, M.A. and Noor, K.I. New inertial approximation schemes for general quasi variational inclusions, *Filomat*, 36(18), 6071–6084 (2022).

[27] Noor, M.A. and Noor, K.I. Higher order generalized variational inequalites and noncovex optimization, *U.P.B. Sci. Bull.*, Series A, 83(2), 77–88 (2023).

[28] Noor, M.A. and Noor, K.I. Some new trends in mixed variational inequalities, *J. Advanc. Math. Studies*, 15(2), 105–140 (2022).

[29] Noor, M.A. and Noor, K.I. Iterative methods and sensitivity analysis for exponential general variational inclusions, *Earthhline J. Math. Sci.* 12(1), 53–107 (2023).

[30] Noor, M.A. and Noor, K.I. Some new systems of exponentially general equations, *Advanc. Linear Algeb. Matrix Theory*, 12(3), 67–86 (2022).

[31] Noor, M.A., Noor, K.I. and Rassias, M.T. New trends in general variational inequalities, *Acta. Appl. Mathematicae*, 170(1), 981-1-64 (2020).

[32] Noor, M.A., Noor, K.I. and Rassias, M.TH. Characterizations of higher order strongly generalized convex functions, in: Themistocles M. Rassias (ed.), *Nonlinear Analysis, Differential Equations, and Applications*, Springer Optimization and Its Applications, Springer, pp. 341–364 (2021).

[33] Noor, M.A., Noor, K.I., Mohesen, B.B., Rassias, M.Th. and Raigorodskii, A. General preinvex functions and variational-like inequalities, in: Nicholas J. Daras and Themistocles M. Rassias (eds.), *Approximation and Computation in Science and Engineering*, Springer Optimization and Its Applications, Springer, pp. 643–666 (2022).

[34] Noor, M.A., Noor, K.I. and Rassias, M.Th. Some aspects of variational inequalities, *J. Comput. Appl. Math.* 47, 285–312 (1993).

[35] Noor, M.A. and Noor, K.I. Approximate exponentially bivariational inequalities, *J. Advanc. Math. Studies*, 16(3), 243–252 (2022).

[36] Noor, M.A. and Noor, K.I. Some new systems of exponentially general equations, *Advanc. Linear Algeb. Matrix Theory*, 12(3), 67–86 (2022).

[37] Noor, M.A., Noor, K.I., Hamdi, A. and El-Shemas, E.H. On difference of two monotone operators, *Optim. Letters*, 3, 329–335 (2009).

[38] Noor, M.A. and Oettli, W. On general nonlinear complementarity problems and quasi equilibria, *Le Math.* (Catania), 49, 313–331 (1994).

[39] Panagiotopoulos, P.D. Nonconvex energy functions, Hemivariational inequalities and substationary principles, *Acta Mechanica*, 42, 160–183 (1983).

[40] Panagiotopolous, P.D. Hemivariational inequalities: Applications to Mechanics and Engineering, Springer Verlag, Berlin, Germany (1995).

[41] Patriksson, M. Nonlinear Programming and Variational Inequalities: A Unified Approach, Kluwer Academic Publishers, Dordrecht, Holland (1999).

[42] Polyak, B.T. Some methods of speeding up the convergence of iterative methods, *USSR Comput. Math. Math. Phys.* 4, 1–17 (1964).

[43] Stampacchia, G. Formes bilineaires coercitives sur les ensembles convexes, *Comptes Rendus de l'Academie des Sciences*, Paris, 258, 4413–4416 (1964).

[44] Zu, D.L. and Marcotte, P. Co-coercivity and its role in the convergence of iterative schemes for solving variational inequalities, *SIAM J. Optim.*, 6(3), 714–726 (1996).

https://doi.org/10.1142/9789811293009_0032

Chapter 32

On the Ranking of Players in Network Games with Local Average

Mauro Passacantando*,‡ and Fabio Raciti†,§,¶

*Department of Business and Law,
University of Milano-Bicocca,
Via Bicocca degli Arcimboldi 8

†Department of Mathematics and Computer Science,
University of Catania, Viale
Doria 6, 95125 Catania, Italy

‡mauro.passacantando@unimib.it

§fabio.raciti@unict.it

We consider a network Nash equilibrium problem where players are represented as nodes of the graph of social relationships and each player's action is influenced by the actions of her social contacts. Within this model, we investigate the problem of ranking the importance of different players. To this end, we first consider some classic topological measures, commonly used to assess the importance of nodes of a graph, and combine them using the entropy weighting method to obtain a single topological measure. We then consider a different measure of importance, based on the social welfare function, computed at the Nash equilibrium, for the original network, as well as for all the networks obtained from

¶Corresponding author.

the initial one by removing a single player. The two measures are then illustrated and compared with the help of small size test problems.

1. Introduction

Games played on networks are Nash equilibrium problems where players are modeled as nodes of a graph, and the social or economic relationship between any two players is represented by a (possibly direct) link connecting them. A main feature of this topic is that the graph-theoretical properties are deeply intermingled with the game-theoretical concepts. Indeed, in the seminal paper [3], the unique Nash equilibrium of a network game was expressed by using the power expansion of the adjacency matrix of the graph following the method in [6]. In the last fifteen years, many scholars have devoted a large number of papers to theoretically extend the original model in [3] and provided a wealth of social and economic applications, see, for instance, [2, 4, 12, 13]. Most authors confined themselves to treat the case of quadratic utility functions and interior solutions in order to obtain closed-form expressions that could be analyzed in detail from the social and economic point of view. Only recently, the powerful theory of variational inequalities (VIs) has been used to analyze and solve network games, although the VI approach to Nash equilibrium was developed many decades ago in [11]. In this respect, we mention the interesting paper [20], where the authors investigate some models of network games and provide a detailed analysis of monotonicity properties and stability of solutions, and the papers [21–23] where the authors use the VI approach to extend some results previously obtained in the case of interior solution.

It is important to point out that, in order to carry out a detailed investigation of how the structure of the graph of relationships influences the Nash equilibrium, it is useful to consider game classes where all players are influenced by their neighbors in the same direction, giving rise to the classes of games with strategic complements and games with strategic substitutes, which are defined precisely in the following. Specific forms of the utility functions are also very appropriate to model conformity effects, where a player tends to imitate the actions of her peers, represented as her graph neighbors. This kind of conformity effect has been analyzed in a general economic model in [9] and framed in the network game theory in [17].

In this chapter, we apply the variational inequality approach to the game described in [17] and provide a ranking of players according to their contribution to the overall network utility, which is computed using the social welfare function at the Nash equilibrium. Since players are represented as nodes in the graph of relationships, they can also be ranked according to some classic topological measures of node importance, such as *degree, betweenness*, and *closeness*. We combine these three topological measures by using the entropy method of multi-criterion analysis and compare the single topological measure thus obtained with the social welfare based measure, with the help of two small scale examples.

This chapter is structured as follows. Section 2 covers some basic concepts on network games and variational inequalities. In Section 3, we first present the model with the conformity (or local average) term and then investigate the monotonicity properties and the social welfare function in two subsections. In Section 4, we describe the four ranking criteria for players and, in Section 5, we outline the entropy method used to weight the three topological criteria. Section 6 is then devoted to illustrate the abovementioned concepts with the help of two small size test problems. In the last concluding section, we summarize our results and mention possible developments.

2. Basics on Network Games and Variational Inequalities

In network games, players are represented by the nodes (or vertices) of a graph (V, E), where $V = \{v_1, \ldots, v_n\}$ is the sets of nodes and E is the set of directed edges (also called arcs or links), formed by ordered pairs of nodes (v_i, v_j). There are neither multiple arcs connecting the same pair of nodes nor loops in our model. If (v, w) and (w, v) are the same, the graph is said to be undirected. Two nodes v_i and v_j are said to be adjacent if they are connected by the edge (v_i, v_j). The information about the adjacency of nodes can be stored in the adjacency matrix G whose elements g_{ij} are equal to 1 if (v_i, v_j) is an edge, 0 otherwise. G is thus a zero-diagonal matrix. Given a node v, the nodes connected to v with an edge are called the *neighbors* of v and are grouped in the set $N_v(G)$. The number of elements of $N_v(G)$ is the *degree* $d_v(G)$ of v. A *walk* in the graph is a finite sequence of the form $v_{i_0}, e_{j_1}, v_{i_1}, e_{j_2}, \ldots, e_{j_k}, v_{j_k}$, which consists of alternating nodes and edges of the graph such that $v_{i_{t-1}}$ and v_{i_t} are end nodes of edge e_{j_t}. The *length* of a walk is the number of its edges. Let us remark that it is allowed to visit a node or go through an edge more than once. The indirect connections between any two nodes in the graph are described by means of the powers of the adjacency matrix G. Indeed, it can be proved that the element $g_{ij}^{[k]}$ of G^k gives the number of walks of length k between nodes v_i and v_j.

In the sequel, the set of players is denoted by $\{1, 2, \ldots, n\}$ instead of $\{v_1, v_2, \ldots, v_n\}$. We denote with $A_i \subset \mathbb{R}$ the action space of player i, while $A = A_1 \times \cdots \times A_n$ is called the space of action profiles. For each $a = (a_1, \ldots, a_n)$, $a_{-i} = (a_1, \ldots, a_{i-1}, a_{i+1}, \ldots, a_n)$ and the notation $a = (a_i, a_{-i})$ is used when we want to distinguish the action of player i from the action of all the other players. Each player i is endowed with a payoff function $u_i : A \to \mathbb{R}$ that she wishes to maximize. The notation $u_i(a, G)$ is often utilized when one wants to emphasize that the utility of player i also depends on the actions taken by her neighbors in the graph.

The solution concept that we consider here is the Nash equilibrium of the game, that is, we seek an element $a^* \in A$ such that for each $i \in \{1, \ldots, n\}$:

$$u_i(a_i^*, a_{-i}^*) \geq u_i(a_i, a_{-i}^*), \qquad \forall\, a_i \in A_i. \tag{1}$$

Depending on how variations of the actions of player's i neighbors affect her marginal utility, the game can be classified as follows.

Definition 1. The network game has the property of strategic complements if

$$\frac{\partial^2 u_i}{\partial a_j \partial a_i}(a) > 0, \quad \forall\, (i,j) : \; g_{ij} = 1, \; \forall\, a \in A.$$

Definition 2. The network game has the property of strategic substitutes if

$$\frac{\partial^2 u_i}{\partial a_j \partial a_i}(a) < 0, \quad \forall \, (i, j) : \, g_{ij} = 1, \, \forall \, a \in A.$$

For the subsequent development, it is important to recall (see, e.g., [19]) that if the u_i are continuously differentiable functions on A and $u_i(\cdot, a_{-i})$ are concave, the Nash equilibrium problem is equivalent to the variational inequality $VI(F, A)$: find $a^* \in A$ such that

$$[F(a^*)]^\top (a - a^*) \geq 0, \quad \forall \, a \in A, \tag{2}$$

where

$$[F(a)]^\top := -\left(\frac{\partial u_1}{\partial a_1}(a), \ldots, \frac{\partial u_n}{\partial a_n}(a)\right) \tag{3}$$

is also called the pseudogradient of the game. For an account of variational inequalities, the interested reader can refer to [10, 16]. We recall here some useful monotonicity properties.

Definition 3. A map $T : \mathbb{R}^n \to \mathbb{R}^n$ is said to be monotone on A iff

$$[T(x) - T(y)]^\top (x - y) \geq 0, \quad \forall \, x, y \in A.$$

If the equality holds only when $x = y$, then T is said to be strictly monotone.

T is said to be β-strongly monotone on A iff there exists $\beta > 0$ such that

$$[T(x) - T(y)]^\top (x - y) \geq \beta \|x - y\|^2, \quad \forall \, x, y \in A.$$

For linear operators on $\mathbb{R}^n$, the two concepts of strict and strong monotonicity coincide and are equivalent to the positive definiteness of the corresponding matrix. Conditions that ensure the unique solvability of a variational inequality problem are given by the following theorem (see, e.g., [10, 15]).

Theorem 4. *If $K \subset \mathbb{R}^n$ is a compact convex set and $T : \mathbb{R}^n \to \mathbb{R}^n$ is continuous on K, then the variational inequality problem $VI(F, K)$ admits at least one solution. In the case K is unbounded, the existence of a solution may be established under the following coercivity condition:*

$$\lim_{\|x\| \to +\infty} \frac{[T(x) - T(x_0)]^\top (x - x_0)}{\|x - x_0\|} = +\infty,$$

for $x \in K$ and some $x_0 \in K$.

Furthermore, if T is strictly monotone on K, the solution is unique.

3. The Network Game with Local Average

We consider a social network with n players with symmetric relationships, whose (symmetric) adjacency matrix G has elements $g_{ij} \in \{0, 1\}$. Let $d_i(G)$ represent the degree of player i, which we assume greater than zero for every i, i.e., there are no disconnected players in the network. We then introduce a second matrix $\hat{G}$ with elements:

$$\hat{g}_{ij} = \frac{g_{ij}}{d_i(G)}.$$

We then have that $0 \leq \hat{g}_{ij} \leq 1$, and $\sum_{j=1}^{n} \hat{g}_{ij} = 1$ for any $i = 1, \ldots, n$. Let us note that, in general, $\hat{G}$ is not a symmetric matrix even if G is so. For each player i, we define the average action of her neighbors as

$$\bar{a}_i(G) = \frac{1}{d_i(G)} \sum_{j=1}^{n} g_{ij} a_j = \sum_{j=1}^{n} \hat{g}_{ij} a_j.$$

Each player is endowed with a utility function given by

$$u_i(a, G) = \alpha_i a_i - \frac{1}{2} a_i^2 - \frac{\delta}{2} \left(a_i - \frac{1}{d_i(G)} \sum_{j=1}^{n} g_{ij} a_j \right)^2, \qquad (4)$$

where $\delta > 0$, and $\alpha_i > 0$ for any $i = 1, \ldots, n$. We also assume that $a_i \in [0, L_i]$, for any $i = 1, \ldots, n$. The first term in expression (4) represents the gain of player i when her action is a_i and the second term is the corresponding intrinsic cost. The third term models a conformity effect and, since $\delta > 0$, penalizes the utility of player i when her action differs from the average action of her neighbors. The pseudogradient's components are

$$F_i(a) = -\frac{\partial u_i}{\partial a_i} = -\alpha_i + (1+\delta)a_i - \delta \sum_{j=1}^{n} \hat{g}_{ij} a_j, \quad \forall\, i = 1, \ldots, n,$$

which, in compact form, can be written as

$$F(a) = (1+\delta)\left[I - \frac{\delta}{1+\delta}\hat{G}\right] a - \alpha, \tag{5}$$

where $\alpha = (\alpha_1, \ldots, \alpha_n)$.

In the case of interior solution, the first-order conditions applied to the utility functions (4) are (see [17])

$$a_i = \frac{\alpha_i}{1+\delta} + \frac{\delta}{1+\delta}\sum_{j=1}^{n} \hat{g}_{ij} a_j, \quad \forall\, i = 1, \ldots, n.$$

The system in compact form is written as

$$\left[I - \frac{\delta}{1+\delta}\hat{G}\right] a = \frac{1}{1+\delta}\alpha,$$

which can be uniquely solved if the matrix $[I - \frac{\delta}{1+\delta}\hat{G}]$ is non-singular. In the general case, we have to solve the variational inequality $VI(F, A)$.

3.1. *Monotonicity properties*

The map F is strictly monotone iff the matrix $I - \frac{\delta}{1+\delta}\hat{G}$ is positive definite. We then consider the symmetric part of $I - \frac{\delta}{1+\delta}\hat{G}$, that is, $I - \frac{\delta}{1+\delta}(\frac{\hat{G}+\hat{G}^\top}{2})$, and check conditions for its smallest eigenvalue to be positive. Since the eigenvalues of $I - \frac{\delta}{1+\delta}(\frac{\hat{G}+\hat{G}^\top}{2})$ can be written as $1 - \frac{\delta}{1+\delta}\lambda_i(\frac{\hat{G}+\hat{G}^\top}{2})$, where $\lambda_i(\frac{\hat{G}+\hat{G}^\top}{2})$ denotes the ith eigenvalue of

$\frac{\hat{G}+\hat{G}^\top}{2}$, the condition $\min_i\{1 - \frac{\delta}{1+\delta}\lambda_i(\frac{\hat{G}+\hat{G}^\top}{2})\} > 0$ can be written in an equivalent manner as

$$\frac{\delta}{1+\delta}\lambda_{max}\left(\frac{\hat{G}+\hat{G}^\top}{2}\right) < 1. \tag{6}$$

Let us recall that for a symmetric matrix, whose entries are all non-negative real numbers, the Perron–Frobenius theorem entails that its greater eigenvalue coincides with the spectral radius. We then get that

$$\frac{\delta}{1+\delta}\lambda_{max}\left(\frac{\hat{G}+\hat{G}^\top}{2}\right) < 1 \Longleftrightarrow \frac{\delta}{1+\delta}\rho\left(\frac{\hat{G}+\hat{G}^\top}{2}\right) < 1. \tag{7}$$

Now, let us note that the condition which ensures the invertibility of $I - \frac{\delta}{1+\delta}\hat{G}$ is $\frac{\delta}{1+\delta}\rho(\hat{G}) < 1$. We then have to compare $\rho(\hat{G})$ with $\rho(\frac{\hat{G}+\hat{G}^\top}{2})$.

Lemma 5. *Let M be a square matrix with non-negative entries. We then have*

$$\rho(M) \leq \rho\left(\frac{M+M^\top}{2}\right),$$

where the equality holds if and only if M and $M^\top$ share an eigenvalue corresponding to $\rho(M)$.

Proof. See, e.g., [5] ex. 6.5, page 53, or [25]. □

3.2. *Social Welfare*

As usual, in game theory, the social welfare W is defined as the sum of the utilities of all players:

$$\begin{aligned} W(a,\hat{G}) &:= \sum_{i=1}^{n} u_i(a,\hat{G}) \\ &= \sum_{i=1}^{n} \alpha_i a_i - \frac{1}{2}\sum_{i=1}^{n} a_i^2 - \frac{\delta}{2}\sum_{i=1}^{n}\left(a_i - \sum_{j=1}^{n}\hat{g}_{ij}a_j\right)^2. \end{aligned} \tag{8}$$

To obtain an expression of W which can be easily analyzed, we observe that

$$\sum_{i=1}^{n}\left(\sum_{j=1}^{n}\hat{g}_{ij}a_j\right)^2 = \left(\hat{G}a\right)^\top \hat{G}a = a^\top \hat{G}^\top \hat{G}a$$

and

$$\delta\sum_{i=1}^{n}a_i\sum_{j=1}^{n}\hat{g}_{ij}a_j = \delta a^\top \hat{G}a = \delta a^\top\left(\frac{\hat{G}+\hat{G}^\top}{2}\right)a,$$

whence

$$W(a,\hat{G}) = -\frac{1}{2}a^\top[(1+\delta)I - \delta(\hat{G}+\hat{G}^\top - \hat{G}^\top \hat{G})]a + \alpha^\top a. \tag{9}$$

It is interesting to provide conditions which ensure the strict concavity of W. For this, let us note that the quadratic form in W is negative definite iff the matrix

$$(1+\delta)I - \delta(\hat{G}+\hat{G}^\top - \hat{G}^\top \hat{G})$$

is positive definite and, because $\hat{G}^\top \hat{G}$ is positive definite, it is enough that $(1+\delta)I - \delta(\hat{G}+\hat{G}^\top)$ is positive definite which amounts to requiring that

$$\rho(\hat{G}+\hat{G}^\top) < 1 + \frac{1}{\delta}.$$

Given that the uniqueness of the Nash equilibrium holds if $\rho(\hat{G}+\hat{G}^\top) < 2(1+\frac{1}{\delta})$, we get that if $\rho(\hat{G}+\hat{G}^\top) < 1+\frac{1}{\delta}$, there exist a unique Nash equilibrium and a unique social optimum.

For the reader's convenience, we report here the expressions of the first and second derivatives of the social welfare, namely,

$$\frac{\partial W}{\partial a_l} = -a_l + \alpha_l - \delta\sum_{i=1}^{n}\left(a_i - \sum_{j=1}^{n}\hat{g}_{ij}a_j\right)\left(\delta_{il} - \sum_{j=1}^{n}\hat{g}_{ij}\delta_{lj}\right)$$

$$= \alpha_l - (1+\delta)a_l + \delta\sum_{i=1}^{n}(\hat{g}_{il}+\hat{g}_{li})a_i - \delta\sum_{i=1}^{n}(\hat{g}_{il})^2 a_l$$

$$- \delta\sum_{i=1}^{n}\sum_{j\neq i}\hat{g}_{ij}\hat{g}_{il}a_j,$$

$$\frac{\partial^2 W}{\partial a_l^2} = -\left[1+\delta+\delta\sum_{i=1}^{n}(\hat{g}_{il})^2\right],$$

$$\frac{\partial^2 W}{\partial a_k \partial a_l} = \delta\left(\hat{g}_{kl}+\hat{g}_{lk}-\sum_{i=1}^{n}\hat{g}_{ik}\hat{g}_{il}\right), \quad l \neq k.$$

4. Ranking Criteria

We now define some common topological measures used to assess the importance of a node i in a network:

- *degree centrality*: $DC_i = d_i(G)$;
- *closeness centrality*: $CC_i = \frac{1}{\sum_{j\neq i} d_{ij}}$, where d_{ij} is the shortest path length between i and j;
- *betweenness centrality*: $BC_i = \sum_{s\neq i}\sum_{t\neq i}\frac{n_{st}(i)}{n_{st}}$, where n_{st} is the number of shortest paths between s and t and $n_{st}(i)$ is the number of such paths that pass through node i.

Together with the topological measure of importance of a node, we also consider a different measure based on the Nash equilibrium of the game.

Specifically, for any node $i = 1, \ldots, n$, we measure the importance of i as the relative variation of the social welfare computed at the Nash equilibrium after i is removed from the network, that is,

$$\textit{social welfare centrality}:\ SWC_i = \frac{W(NE(G)) - W(NE(G\setminus\{i\}))}{W(NE(G))}, \tag{10}$$

where $NE(G)$ is the Nash equilibrium in the network G, $NE(G\setminus\{i\})$ is the Nash equilibrium in the network G, where node i has been removed, and W is the social welfare function (9). Note that $SWC(i)$ can be negative if the total social welfare of the network increases after removing the node i (see Example 7). This situation can be compared to the well-known Braess paradox [7], where the efficiency of a network improves due to the removal of a link. The use of a mix of topological indices to measure the importance of a nodes in real transport network has been recently applied to the case of the railway network of Shanghai in [26], using the entropy method described in the following.

5. The Entropy Method in Multi-Criterion Analysis

In multi-criterion analysis, a decision-maker has to make a choice among different alternatives, taking into account multiple, possibly conflicting, and objectives. For instance, in a road network, there are multiple paths which connect an origin and a destination, and a driver usually considers travel time and travel cost as the two main criteria which influence her decision. How to compare the importance of travel time and travel cost can depend on personal preferences as well as on objective facts. In our model, we are given a social network, and we wish to assess the importance of its nodes. Since there can be different criteria to measure the importance of a node, an aggregate measure would be useful.

A large number of problems from economics or other fields of applied sciences present the feature of the above examples, of making a choice considering multiple objectives. In those cases, a simple aggregation method consists of assigning weights to each criterion and then computing the weighted sum. The process of assigning weights is often subjective and can lead to unsatisfactory results. In his influential work, M. Zeleny [27] proposed a method to assess the weights in an objective manner, still allowing for the possible intervention of a decision-maker. To this end, he first specified the concept of weight as follows:

"A weight, assigned to the i-th attribute as a measure of its relative importance for a given decision problem, is directly related to the average intrinsic information generated by a given set of alternatives through the i-th attribute, as well as to its subjective assessment."

Thus, a procedure to generate weights requires an operational definition of what the *average intrinsic information* is. To this end, Zeleny made use of the concept of Shannon entropy. We now recall the basic version of the entropy method (see also [24]).

Assume that we are given m alternatives, $a_1, a_2, \ldots, a_m$, and n criteria, $C_1, C_2, \ldots, C_n$. The utility of applying criterion C_j to the alternative a_i is denoted with a_{ij}. In our model, the alternatives are the nodes of the network, and the criteria are the three topological measures of their importance. The matrix whose entries are the elements a_{ij} is called the decision matrix. The coefficients a_{ij} must be normalized so as to take on values in the interval $[0, 1]$. Once the weights w_j are computed as explained in the following, for each alternative a_i, we evaluate the number

$$R(a_i) = \sum_{j=1}^{n} w_j \frac{a_{ij}}{\sum_{i=1}^{m} a_{ij}},$$

which gives the importance of a_i. The steps to compute weights are as follows:

(1) For each criterion C_j, compute the associated Shannon entropy:

$$E_j = -k \sum_{i=1}^{m} \frac{a_{ij}}{\sum_{i=1}^{m} a_{ij}} \log \left(\frac{a_{ij}}{\sum_{i=1}^{m} a_{ij}} \right),$$

where $k = 1/\log m$, to have $E_j \in [0, 1]$.

(2) It is well known that, for each j, the entropy is maximized when the values $\frac{a_{ij}}{\sum_{i=1}^{m} a_{ij}}$, $i = 1, \ldots, m$, are all equal, which is the case where no information is provided about the ranking of the alternatives. To rank the alternatives, we thus compute the following:

$$D_j = 1 - E_j, \quad \forall\, j = 1, \ldots, n.$$

(3) Since we want that weights to take on values in the interval $[0, 1]$, we further normalize:

$$w_j = \frac{D_j}{\sum_{i=1}^{n} D_i}, \quad \forall\, j = 1, \ldots, n.$$

6. Numerical Examples

In this section, we show two numerical examples, in which the social welfare centrality measure is compared with three topological centrality measures (degree, closeness, and betweenness) and their weighted sum obtained by exploiting the entropy weighting method. To find Nash equilibria, we reformulated the affine variational inequality (2) as an equivalent convex quadratic optimization problem (see [1]) and solved it with the function `quadprog` from the MATLAB optimization toolbox.

Example 6. We consider a network game on a graph with 10 nodes, where the adjacency matrix G has been randomly generated (see Fig. 1), and the game parameters are $\delta = 0.9/[\rho(\hat{G}+\hat{G}^\top)-1]$, $\alpha_i = 5\,i$ and $L_i = 30$ for any $i = 1,\ldots,10$. Table 1 shows the normalized values of the degree, closeness, and betweenness centrality (col. 2–4), their weighted sum with entropy weighting method (col. 5), and the social welfare centrality (col. 6). Table 2 shows the ranking of nodes according to the analyzed centrality measures. It is interesting noting that the ranking defined by the new measure is quite different from that provided by the other measures. In particular, node 1 is the most important according to topological measures, while it is the least important according to the social welfare measure. Moreover, Fig. 1 gives a graphical representation of the normalized values associated with the network nodes for each considered centrality measure.

Example 7. We consider a network game on a graph with 15 nodes, where the adjacency matrix G has been randomly generated (see Fig. 2), and the game parameters are $\delta = 0.9/[\rho(\hat{G} + \hat{G}^\top) - 1]$, $\alpha_i = 5\,i$ and $L_i = 40$ for any $i = 1,\ldots,15$. Tables 3 and 4 show the normalized values of the five centrality measures and the corresponding ranking of nodes. Note that nodes 1 and 2 have negative values according to the social welfare centrality measure. Moreover, node 1 is the most important according to degree and closeness centrality measures, while it is the least important according to the social welfare measure. Figure 2 gives a graphical representation of the normalized values associated with the network nodes for each considered centrality measure.

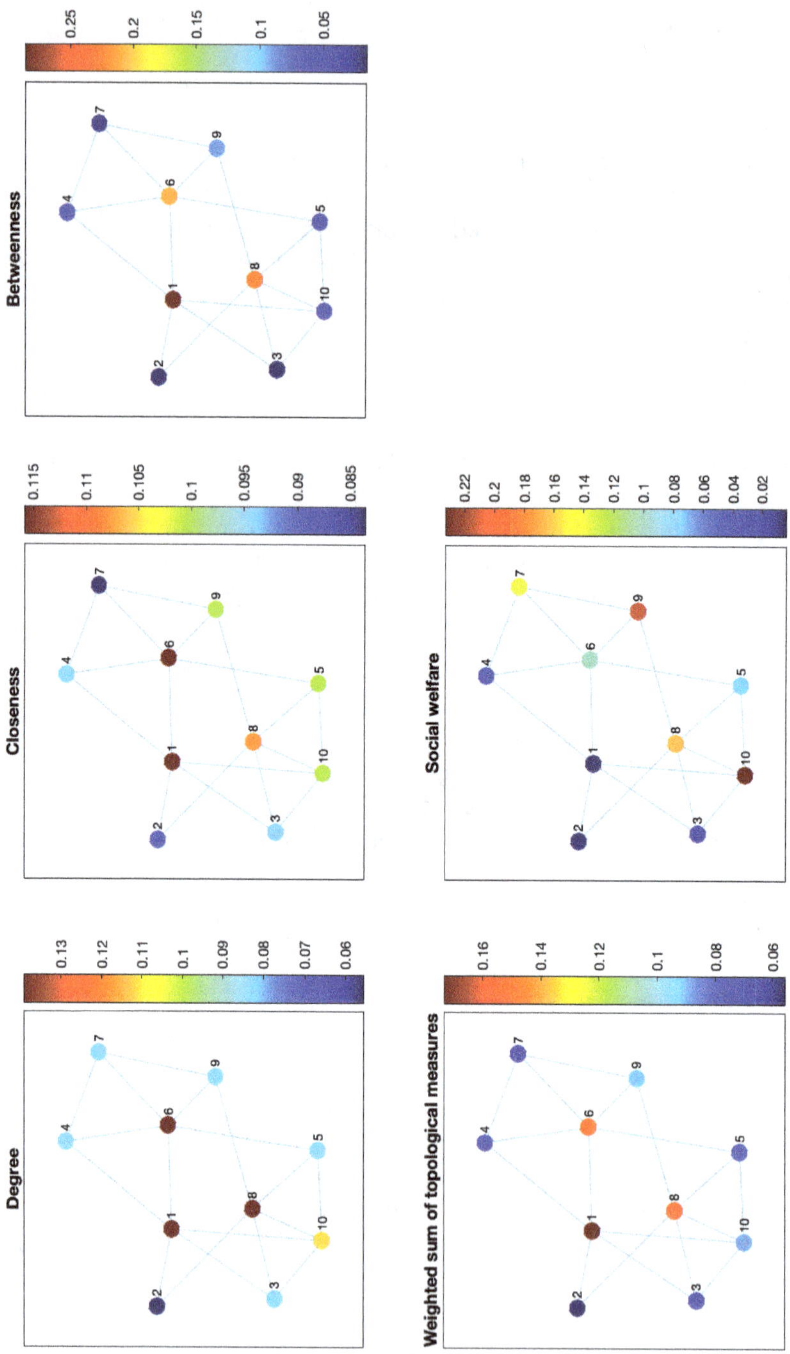

Fig. 1. Example 6: Comparison between degree, closeness, and betweenness centrality, their weighted sum, and social welfare centrality.

Table 1. Example 6: Normalized values of degree, closeness, and betweenness centrality, their weighted sum, and social welfare centrality.

	Topological measures				
Nodes	Degree	Closeness	Betweenness	Weighted sum of topological measures	Social welfare
1	0.1389	0.1157	0.2849	0.1733	0.0039
2	0.0556	0.0885	0.0161	0.0558	0.0062
3	0.0833	0.0940	0.0161	0.0675	0.0205
4	0.0833	0.0940	0.0457	0.0762	0.0449
5	0.0833	0.1003	0.0457	0.0784	0.0752
6	0.1389	0.1157	0.2097	0.1513	0.1001
7	0.0833	0.0836	0.0215	0.0654	0.1377
8	0.1389	0.1075	0.2231	0.1523	0.1654
9	0.0833	0.1003	0.0887	0.0910	0.2132
10	0.1111	0.1003	0.0484	0.0890	0.2328

Table 2. Example 6: Ranking of nodes according to degree, closeness, and betweenness centrality, their weighted sum, and social welfare centrality.

	Topological measures				
Rank	Degree	Closeness	Betweenness	Weighted sum of topological measures	Social welfare
1	1	1	1	1	10
2	6	6	8	8	9
3	8	8	6	6	8
4	10	5	9	9	7
5	3	9	10	10	6
6	4	10	4	5	5
7	5	3	5	4	4
8	7	4	7	3	3
9	9	2	2	7	2
10	2	7	3	2	1

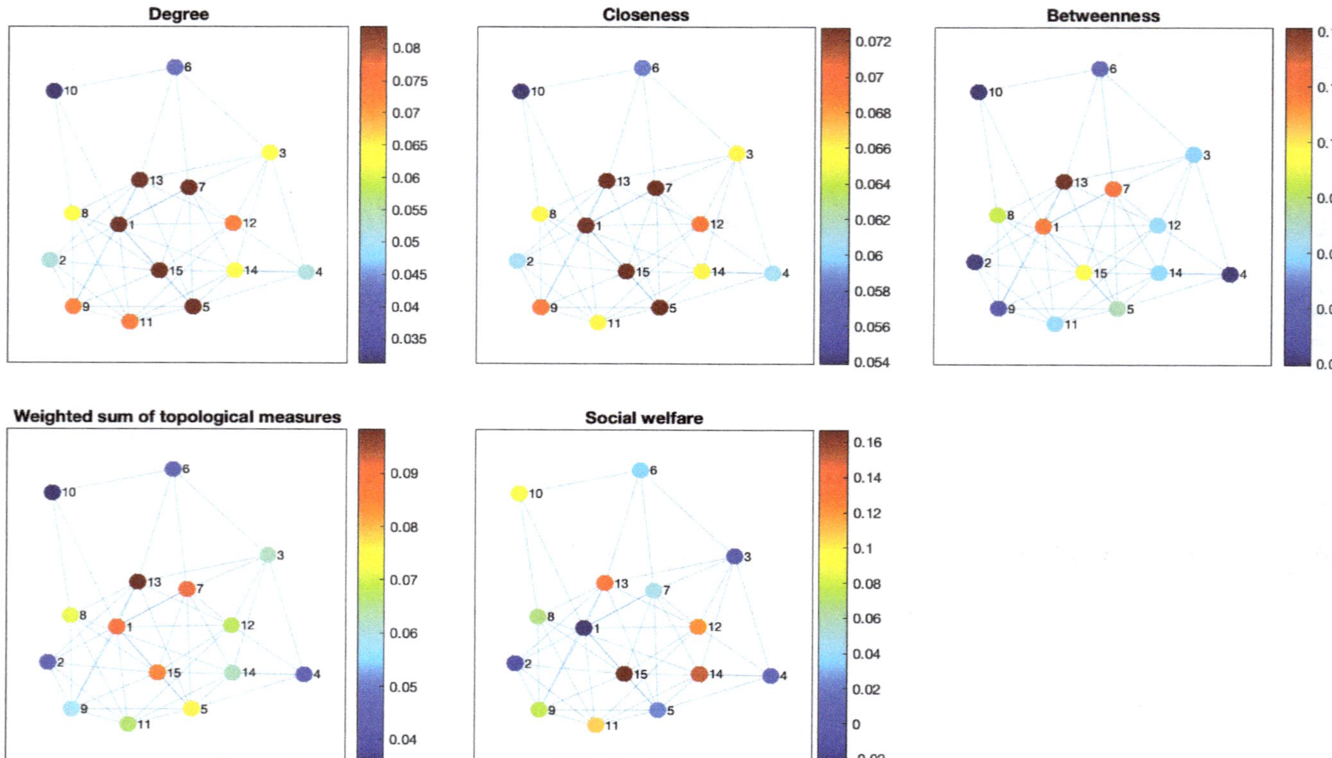

Fig. 2. Example 7: Comparison between degree, closeness, and betweenness centrality, their weighted sum, and social welfare centrality.

Table 3. Example 7: Normalized values of degree, closeness, and betweenness centrality, their weighted sum, and social welfare centrality.

	Topological measures			Weighted sum of	Social
Nodes	Degree	Closeness	Betweenness	topological measures	welfare
1	0.0833	0.0727	0.1170	0.0905	−0.0244
2	0.0521	0.0606	0.0208	0.0450	−0.0135
3	0.0625	0.0661	0.0569	0.0620	0.0009
4	0.0521	0.0606	0.0208	0.0450	0.0116
5	0.0833	0.0727	0.0728	0.0763	0.0230
6	0.0417	0.0582	0.0391	0.0465	0.0353
7	0.0833	0.0727	0.1208	0.0918	0.0496
8	0.0625	0.0661	0.0847	0.0709	0.0685
9	0.0729	0.0693	0.0300	0.0579	0.0769
10	0.0312	0.0539	0.0194	0.0352	0.0891
11	0.0729	0.0661	0.0608	0.0667	0.1076
12	0.0729	0.0693	0.0600	0.0675	0.1218
13	0.0833	0.0727	0.1412	0.0983	0.1356
14	0.0625	0.0661	0.0581	0.0623	0.1513
15	0.0833	0.0727	0.0975	0.0843	0.1669

Table 4. Example 7: Ranking of nodes according to degree, closeness, and betweenness centrality, their weighted sum, and social welfare centrality.

	Topological measures			Weighted sum of	Social
Rank	Degree	Closeness	Betweenness	topological measures	welfare
1	1	1	13	13	15
2	5	5	7	7	14
3	7	7	1	1	13
4	13	13	15	15	12
5	15	15	8	5	11
6	9	9	5	8	10
7	11	12	11	12	9
8	12	3	12	11	8
9	3	8	14	14	7
10	8	11	3	3	6
11	14	14	6	9	5
12	2	2	9	6	4
13	4	4	2	2	3
14	6	6	4	4	2
15	10	10	10	10	1

7. Conclusions

In this chapter, we considered the problem of ranking the importance of players in a network game with conformity effect, where each player tends to use a strategy close to the average strategy of her peers. Our ranking was based on the use of the social welfare function computed at the Nash equilibrium for the original network and for all the networks resulting from the removal of a single player. This ranking was compared with the one obtained combining three standard topological measures of node importance in a graph with the entropy weighting method. In future work, we aim to apply this model to a class of environmental games described in [14]. Another interesting research direction is the introduction of additional shared constraints, which would yield to a generalized Nash equilibrium problem (see, e.g., [18] for some recent results on this topic).

Acknowledgments

The authors are members of the Gruppo Nazionale per l'Analisi Matematica, la Probabilità e le loro Applicazioni (GNAMPA — National Group for Mathematical Analysis, Probability and their Applications) of the Instituto Nazionale di Alta Matematica (INdAM — National Institute of Higher Mathematics). This research was partially supported by the research project "Programma ricerca di ateneo UNICT 2020-22 linea 2-OMNIA" of the University of Catania. This support is gratefully acknowledged.

References

[1] Aghassi M., Bertsimas D. and Perakis G. Solving asymmetric variational inequalities via convex optimization. *Oper. Res. Lett.* 34, 481–490 (2006).

[2] Belhaj M., Bramoulleé Y. and Deroian F. Network games under strategic complementarities. *Games Econon. Behav.* 88, 310–319 (2014).

[3] Ballester C., Calvo-Armengol A. and Zenou Y. Who's who in networks. Wanted: The key player. *Econometrica* 74, 1403–1417 (2006).

[4] Ballester C., Calvo-Armengol A. and Zenou Y. Delinquent networks. *J. Eur. Econ. Assoc.* 8, 34–61 (2010).

[5] Berman A. and Plemmons R.J. Nonnegative Matrices in Mathematical Sciences, SIAM, Society for Industrial and Applied Mathematics, Philadelphia (1994).

[6] Bonacich P. Power and centrality: A family of measures. *Am. J. Sociol.* 92, 1170–1182 (1987).

[7] Braess D. Über ein Paradoxon aus der Verkehrsplanung. Unternehmenforschung 12, 258–268 (1968).

[8] Calvo-Armengol A., Patacchini E. and Zenou Y. Peer effects and social networks in education. *Rev. Econ. Stud.* 76, 1239–1267 (2009).

[9] Clark A.E. and Oswald A. Comparison-concave utility and following behaviour in social and economic settings. *J. Publ. Econ.* 70, 133–175 (1998).

[10] Facchinei F. and Pang J.-S. *Finite-Dimensional Variational Inequalities and Complementarity Problems.* Springer, Berlin (2003).

[11] Gabay D. and Moulin H. On the uniqueness and stability of Nash equilibria in noncooperative games. In: Bensoussan A., Kleindorfer P. and Tapiero C.S. (eds.), *Applied Stochastic Control in Econometrics and Management Science.* North-Holland, Amsterdam, 271–294 (1980).

[12] Helsley R.W. and Zenou Y. Social networks and interactions in cities, *J. Economic Theory* 150, 426–466 (2014).

[13] Jackson M.O. and Zenou Y. Games on Networks, in: *Handbook of Game Theory with Economic Applications*, Elsevier, 95–163 (2015).

[14] Jadamba B. and Raciti F. On the modelling of some environmental games with uncertain data. *J. Optim. Theory Appl.* 167, 959–968 (2015).

[15] Kinderleher D. and Stampacchia G. *An Introduction to Variational Inequalities and their Applications*, Academic Press (1980).

[16] Konnov I. Equilibrium Models and Variational Inequalities, Elsevier (2007).

[17] Liu X., Patacchini E. and Zenou Y. Endogenous peer effects: Local aggregate or local average? *Journal of Economic Behavior & Organization*, 103, 39–59 (2014).

[18] Mastroeni G., Pappalardo M. and Raciti F. Generalized Nash equilibrium problems and variational inequalities in Lebesgue spaces. *Minimax Theory Appl.* 5, 47–64 (2020).

[19] Nagurney A. *Network Economics. A Variational Inequality Approach*, Springer (1999).

[20] Parise F. and Ozdaglar A. A variational inequality framework for network games: Existence, uniqueness, convergence and sensitivity analysis. *Games Econ. Behav.* 114, 47–82 (2019).

[21] Passacantando M. and Raciti F., A note on generalized Nash games played on networks, in: M.Th. Rassias (ed.), *Nonlinear Analysis,*

Differential Equations, and Applications, Springer Optimization and Its Applications, Springer, 173, 365–380 (2021)

[22] Passacantando M. and Raciti F. A Variational Inequality Approach to a Class of Network Games with Local Complementarities and Global Congestion, Amorosi, Dell'Olmo, Lari (eds.), in: *Optimization in Artificial Intelligence and Data Sciences*, AIRO Springer Series, 1–11 (2022).

[23] Passacantando M. and Raciti F. A finite convergence algorithm for solving linear-quadratic network games with strategic complements and bounded strategies, Optimization Methods and Software, published online: 19 May 2023, https://doi.org/10.1080/10556788.2023.2205644.

[24] Pomerol J.-C. and Barba-Romero S. *Multicriterion Decision in Management*, Kluwer Academic Publishers, (2000).

[25] Schwenk A.J. Tight bounds on the spectral radius of asymmetric nonnegative matrices, *Linear Algebra and its applications,* 75, 257–265 (1986).

[26] Shi J., Wen S., Zhao X. and Wu G. Sustainable development of urban rail transit networks: A vulnerability perspective, *Sustainability,* 11, 1335 (2019).

[27] Zeleny M. Linear multiobjective programming, *Lecture Notes in Economics and Mathematical Systems*, Springer (1974).

https://doi.org/10.1142/9789811293009_0033

Chapter 33

A Two-Stage Game Theoretical Model of Social Interactions and Location Choice in City Areas

Mauro Passacantando*,¶, Fabio Raciti†,‖
and Michael Th. Rassias‡,§,**,††

*Department of Business and Law, University of Milano-Bicocca, Via Bicocca degli Arcimboldi 8, 20126 Milan, Italy
†Department of Mathematics and Computer Science, University of Catania, Viale Doria 6, 95125 Catania, Italy
‡Department of Mathematics and Engineering Sciences Hellenic Military Academy 16673 Vari Attikis, Greece
§Institute for Advanced Study, Program in Interdisciplinary Studies, 1 Einstein Dr, Princeton, NJ 08540, USA
¶mauro.passacantando@unipi.it
‖fabio.raciti@unict.it
**mrassias@sse.gr
††mthrassias@yahoo.com

In this chapter, we further investigate a previous game theoretical model of social and economic interactions where players are considered according to their position in a social network and also to their geographic location. Our model allows for intermediate geographic areas, between the city center and the periphery. Moreover, an analysis of some simple but paradigmatic networks shows that the game whose solution yields the players' location can have multiple equilibria.

1. Introduction

In this chapter, we extend the game theoretical model introduced in [8] to describe the economic interactions among various agents (players) situated in two different geographic locations: the city center and the periphery. Agents are also considered as members of a social network, hence the utility function of each of them depends on two different groups of variables: the geographic variables and the social variables. Specifically, in [8], it is assumed that all interactions occur in the city center and the authors first assume that each player has a prescribed geographic position and there is a cost of commuting for those who reside in the periphery. In this setting, the variable of each player measures how often he/she decides to visit the city center. Each visit implies economic interactions so that the aggregate number of visits has the meaning of overall social or economic activity. The modeling and analysis of this game is performed according to the general framework described in the influential paper [2] (see also [4]) where players are considered as nodes of a social network and the utility function of each player depends on his/her action as well as on the actions of his/her neighbor nodes. The interesting feature of this model is that it shows how, at the Nash equilibrium, the action of a given player can be influenced by the actions of players who do not correspond to any of her adjacent nodes in the network structure. This effect is described by means of a series expansion of the equilibrium solutions which uses the powers of the network adjacency matrix, along the same lines as in [5]. Other interesting applications of this methodology can be found, for instance, in [3] and [6].

In this chapter, we contribute to extend and clarify the two-stage model of [8] where the geographic locations of players are not assigned, but must be chosen, given that living in the city center has an intrinsic cost (not to be confused with the transport cost). Our contribution is twofold: (i) we generalize the model by considering a partition of the geographic space in a finite number of areas (not necessarily two) and (ii) we show that even in the case of two geographic areas, the game can have multiple equilibria, thus contrasting with the results in [8].

This chapter is organized as follows. In Section 2, we describe the model and provide a numerical example showing non-uniqueness of the Nash equilibrium. We then study, in Section 3, two simple classic social networks (complete and star-shaped networks) in the case of two geographic locations and derive exact results, which prove that also in the case of two locations one can have multiple equilibria. A short concluding section describes future research directions, and an appendix summarizes the basic concepts and notations of network games used throughout this chapter.

2. Two-Stage Game Theoretical Model

We consider n economic agents in a city, which is divided into $k+1$ areas: a center (where all interactions among agents occur) and k peripheral areas. Each agent can choose the geographic location in which to live: $x_i \in \{0, \ldots, k\}$ represents the location of agent i and $x_i = 0$ means that i chooses to locate at the city center. The social connections among agents are represented by an undirected graph (V, E), where $V = \{1, \ldots, n\}$ is the set of agents and each arc $(i, j) \in E$ means that agents i and j are directly connected. In addition to the geographical location, each agent has to decide on the number of visits to the center: $v_i \geq 0$ represents the number of visits that agent i makes to the center. Denote the vectors $x = (x_1, \ldots, x_n)$ and $v = (v_1, \ldots, v_n)$.

According to [8], we assume that the utility function of player i is defined as follows:

$$U_i(x, v) = y_i - t_i(x_i)v_i - c_i(x_i) + \alpha_i v_i - \frac{1}{2}v_i^2 + \theta \sum_{j=1}^{n} g_{ij} v_i v_j,$$

where y_i represents the income due to the interaction with other agents at the center, $t_i(x_i)$ is the marginal transport cost between the location chosen by agent i and the city center, $c_i(x_i)$ is the cost associated with the location chosen by player i, while the last three terms represent the utility function of interactions, with $\alpha_i > 0$ and $\theta > 0$.

In order to model the interactions among agents, we consider a two-stage game theoretical model: in the first stage, the agents (players) decide where to locate (x), while in the second stage, they decide their optimal effort in the network (v). The aim of each agent is to maximize its own utility function, which depends on the number of visits v (of all the agents) that, in turn, depend on the location choices x (of all the agents). We remark that once the agents' locations x have been chosen, the Nash equilibrium in visit choices is related to the Katz–Bonacich centrality measure (see Appendix A). The following example shows that multiple Nash equilibria (in locations) can occur.

Example 1. Consider a game with four agents connected as in Figure 1. The adjacency matrix of the network is

$$G = \begin{pmatrix} 0 & 1 & 1 & 1 \\ 1 & 0 & 1 & 0 \\ 1 & 1 & 0 & 0 \\ 1 & 0 & 0 & 0 \end{pmatrix}.$$

We assume that agents have three possible locations in the city (i.e., $k = 2$) and the other parameters are defined as follows: $\theta = 0.3$, $\alpha_i = 1$ for any $i = 1, \ldots, 4$, $y_i = 5$ for any $i = 1, \ldots, 4$, the marginal transport cost is

$$t_i(x_i) = \begin{cases} 0 & \text{if } x_i = 0, \\ t/3 & \text{if } x_i = 1, \\ t & \text{if } x_i = 2, \end{cases}$$

for any $i = 1, \ldots, 4$, with $t \in (0, 1)$, and the location cost is

$$c_i(x_i) = \begin{cases} c & \text{if } x_i = 0, \\ 2c/3 & \text{if } x_i = 1, \\ c/2 & \text{if } x_i = 2, \end{cases}$$

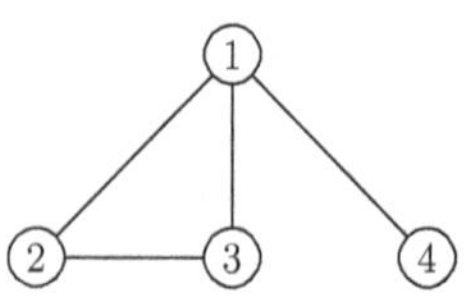

Fig. 1. Social network of the game in Example 1.

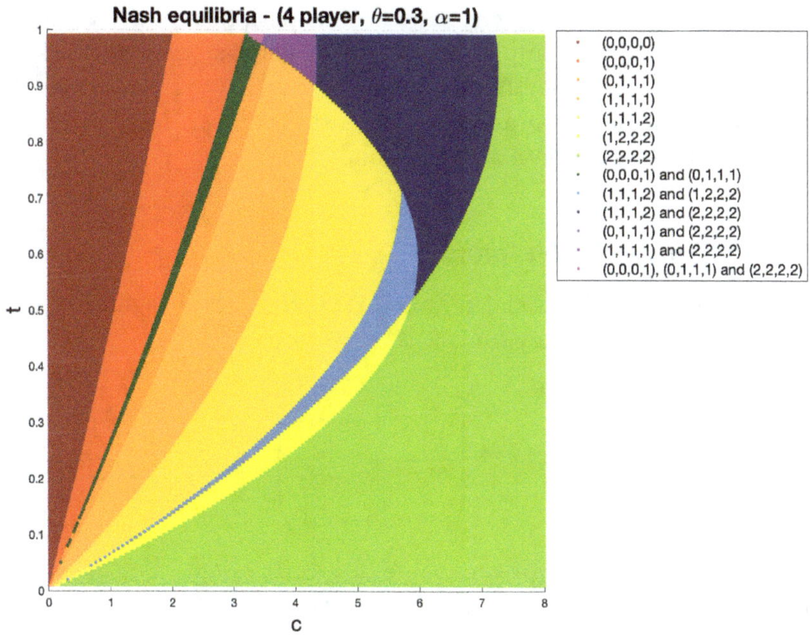

Fig. 2. Nash equilibria (in locations) of the game described in Example 1.

for any $i = 1, \ldots, 4$, with $c > 0$. Figure 2 shows the equilibrium locations according to the values of the parameters c and t. The Nash equilibria (in locations) were computed by means of a comprehensive analysis of the payoff matrices of the players, where the (unique) Nash equilibrium in visit choices was computed according to formula (A.3). Specifically, there are seven possible Nash equilibria (in locations): $x = (0,0,0,0)$, $x = (0,0,0,1)$, $x = (0,1,1,1)$, $x = (1,1,1,1)$, $x = (1,1,1,2)$, $x = (1,2,2,2)$, $x = (2,2,2,2)$. Note that there exist values of c and t such that the Nash equilibrium (in locations) is not unique. For instance, if $c = 3.4$ and $t = 0.99$, then there are three Nash equilibria (in locations): $x = (0,0,0,1)$, $x = (0,1,1,1)$, and $x = (2,2,2,2)$.

3. Equilibria for Two Special Social Networks

In this section, we provide an analytical study of the Nash equilibria (in locations) in the case $k = 1$ (i.e., the agents have only two possible locations in the city: center and periphery) for two special network structures: complete networks and star-shaped networks. In particular, we assume that $\alpha_i = \alpha$ for any $i = 1, \ldots, n$ and the marginal transport cost and location cost are linear: $t_i(x_i) = tx_i$ and $c_i(x_i) = c(1 - x_i)$, where $t \in (0, \alpha)$ and $c > 0$.

3.1. *Complete Networks*

We assume that the social network related to the n agents is complete, thus its adjacency matrix is

$$G = \begin{pmatrix} 0 & 1 & \dots & 1 \\ 1 & \ddots & \ddots & \vdots \\ \vdots & \ddots & \ddots & 1 \\ 1 & \dots & 1 & 0 \end{pmatrix},$$

where $\rho(G) = n - 1$. For any value of $\theta \in (0, 1/(n-1))$, the matrix $I - \theta G$ is invertible and its inverse is given by

$$M = (I - \theta G)^{-1} = \frac{1}{1 - (n-2)\theta - (n-1)\theta^2} \times \begin{pmatrix} 1-(n-2)\theta & \theta & \cdots & & \theta \\ \theta & & \ddots & \ddots & \vdots \\ \vdots & & \ddots & \ddots & \theta \\ \theta & & \cdots & \theta & 1-(n-2)\theta \end{pmatrix}.$$

Proposition 2. *In the case of a complete social network, only two Nash equilibria (in locations) exist:* $x = (0, \ldots, 0)$ *and* $x = (1, \ldots, 1)$. *In particular, if* $2c < s_1$, *where*

$$s_1 := \frac{t[1 - (n-2)\theta]}{[1 - (n-2)\theta - (n-1)\theta^2]^2} \times [(2\alpha - t)[1 - (n-2)\theta] + 2(\alpha - t)\theta(n-1)],$$

then $x = (0, \ldots, 0)$ *is the unique Nash equilibrium; if* $2c > s_2$, *where*

$$s_2 := \frac{t[1 - (n-2)\theta]}{[1 - (n-2)\theta - (n-1)\theta^2]^2} \times [(2\alpha - t)[1 - (n-2)\theta] + 2\alpha\theta(n-1)],$$

then $x = (1, \ldots, 1)$ *is the unique Nash equilibrium; if* $2c \in [s_1, s_2]$, *then* $x = (0, \ldots, 0)$ *and* $x = (1, \ldots, 1)$ *are both Nash equilibria.*

Proof. It follows from [8, Proof of Proposition 4] that $x = (0, \ldots, 0)$ is a Nash equilibrium if and only if

$$2c \leq tm_{ii} \left[(2\alpha - t)m_{ii} + 2\alpha \sum_{\substack{1 \leq j \leq n \\ j \neq i}} m_{ij} \right], \quad \forall\, i = 1, \ldots, n, \tag{1}$$

where the definition and interpretation of m_{ij} are given in (A.4) and (A.5). From now on, let us denote $A = 1 - (n-2)\theta$ and $B = 1 - (n-2)\theta - (n-1)\theta^2$. Then, conditions (1) are equivalent to the following inequality:

$$2c \leq \frac{tA}{B} \left[(2\alpha - t)\frac{A}{B} + 2\alpha(n-1)\frac{\theta}{B} \right] = s_2.$$

On the other hand, $x = (1, \ldots, 1)$ is a Nash equilibrium if and only if

$$2c \geq tm_{ii} \left[(2\alpha - t)m_{ii} + 2(\alpha - t) \sum_{\substack{1 \leq j \leq n \\ j \neq i}} m_{ij} \right], \quad \forall\, i = 1, \ldots, n, \tag{2}$$

that is, equivalent to

$$2c \geq \frac{tA}{B} \left[(2\alpha - t)\frac{A}{B} + 2(\alpha - t)(n-1)\frac{\theta}{B} \right] = s_1.$$

The last step is to prove that Nash equilibria different from $x = (0, \ldots, 0)$ and $x = (1, \ldots, 1)$ cannot exist. Assume, by contradiction, that there exists an integer k, with $1 \leq k \leq n-1$, such that $x = (\underbrace{0, \ldots, 0}_{k}, 1, \ldots, 1)$ is a Nash equilibrium. Then, the following conditions must hold:

$$\begin{cases} 2c \le t(2\alpha - t)m_{ii}^2 + 2t\alpha m_{ii} \sum\limits_{\substack{1\le j\le k \\ j\ne i}} m_{ij} + 2t(\alpha - t)m_{ii} \sum\limits_{j=k+1}^{n} m_{ij} \\ \quad \forall\, i = 1, \dots, k, \\ 2c \ge t(2\alpha - t)m_{ii}^2 + 2t\alpha m_{ii} \sum\limits_{j=1}^{k} m_{ij} + 2t(\alpha - t)m_{ii} \sum\limits_{\substack{k+1\le j\le n \\ j\ne i}} m_{ij} \\ \quad \forall\, i = k+1, \dots, n. \end{cases}$$

The above conditions are equivalent to the following system of inequalities:

$$\begin{cases} 2c \le t(2\alpha - t)\dfrac{A^2}{B^2} + 2t\alpha\dfrac{A}{B}(k-1)\dfrac{\theta}{B} + 2t(\alpha - t)\dfrac{A}{B}(n-k)\dfrac{\theta}{B} \\ 2c \ge t(2\alpha - t)\dfrac{A^2}{B^2} + 2t\alpha\dfrac{A}{B}k\dfrac{\theta}{B} + 2t(\alpha - t)\dfrac{A}{B}(n-k-1)\dfrac{\theta}{B} \end{cases}$$

that can be equivalently rewritten as follows:

$$\begin{cases} 2c \le t(2\alpha - t)\dfrac{A^2}{B^2} + 2t\theta\dfrac{A}{B^2}\left[\alpha(n-1) - t(n-k)\right] \\ 2c \ge t(2\alpha - t)\dfrac{A^2}{B^2} + 2t\theta\dfrac{A}{B^2}\left[\alpha(n-1) - t(n-k-1)\right] \end{cases}$$

that is impossible. Therefore, $x = (\underbrace{0, \dots, 0}_{k}, 1, \dots, 1)$ cannot be a Nash equilibrium and the proof is complete. □

Remark 3. Proposition 2 generalizes [8, Proposition 8] to the case of a complete network with n agents and provides correct sets of parameters for the corresponding Nash equilibria.

Example 4. Consider a game with $n = 3$ players connected by a complete social network (see Figure 3). According to Proposition 2, three possible cases can occur:

(a) if $2c < \frac{t(1-\theta)}{(1-\theta-2\theta^2)^2}\left[(2\alpha - t)(1-\theta) + 4(\alpha - t)\theta\right]$, then $x = (0, 0, 0)$ is the unique Nash equilibrium;

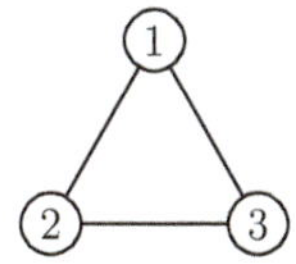

Fig. 3. Social network of the game in Example 4.

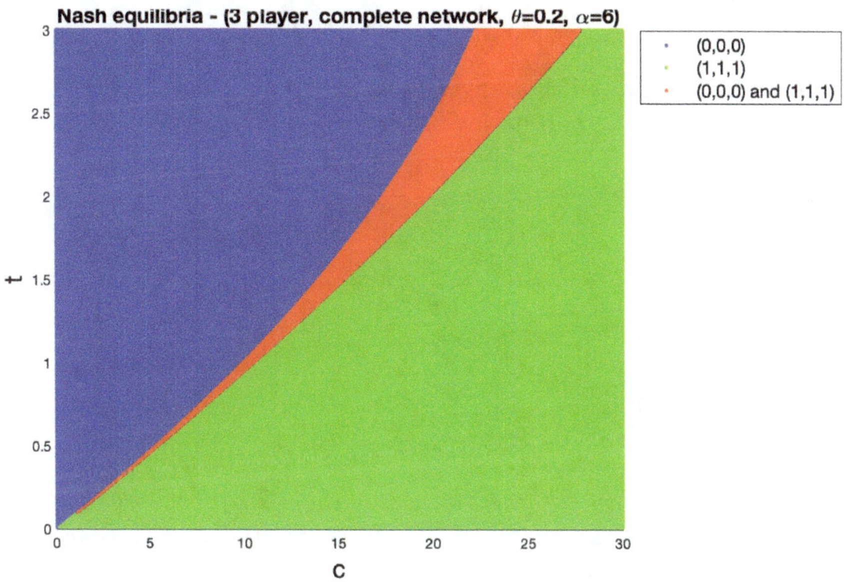

Fig. 4. Nash equilibria (in locations) of the game described in Example 4.

(b) if $2c > \frac{t(1-\theta)}{(1-\theta-2\theta^2)^2}\left[(2\alpha - t)(1-\theta) + 4\alpha\theta\right]$, then $x = (1,1,1)$ is the unique Nash equilibrium;

(c) if $\frac{t[1-\theta]}{[1-\theta-2\theta^2]^2}\left[(2\alpha - t)[1-\theta] + 4(\alpha - t)\theta\right] \leq 2c$ and $2c \leq \frac{t[1-\theta]}{[1-\theta-2\theta^2]^2}$ $\left[(2\alpha - t)[1-\theta] + 4\alpha\theta\right]$, then $x = (0,0,0)$ and $x = (1,1,1)$ are both Nash equilibria.

Figure 4 shows the Nash equilibria of the game according to the values of the parameters c and t in the case $\theta = 0.2$, $\alpha = 6$, and $y = 5$.

3.2. *Star-shaped networks*

We now consider a star-shaped social network with n agents, where node 1 is the center of the star. The adjacency matrix is

$$G = \left(\begin{array}{c|ccc} 0 & 1 & \cdots & 1 \\ \hline 1 & 0 & \cdots & 0 \\ \vdots & \vdots & \ddots & \vdots \\ 1 & 0 & \cdots & 0 \end{array}\right),$$

where $\rho(G) = \sqrt{n-1}$. For any value of $\theta \in (0, 1/\sqrt{n-1})$, the matrix $I - \theta G$ is invertible and its inverse is given by

$$M = (I - \theta G)^{-1} = \frac{1}{1-(n-1)\theta^2}$$

$$\times \left(\begin{array}{c|ccccc} 1 & \theta & & \cdots & \cdots & \theta \\ \hline \theta & 1-(n-2)\theta^2 & \theta^2 & \cdots & & \theta^2 \\ \vdots & \theta^2 & & \ddots & \ddots & \vdots \\ \vdots & \vdots & & \ddots & & \theta^2 \\ \theta & \theta^2 & & \cdots & \theta^2 & 1-(n-2)\theta^2 \end{array}\right).$$

Proposition 5. *In the case of a star-shaped social network, three Nash equilibria exist:* $x = (0, \ldots, 0)$, $x = (0, 1, \ldots, 1)$, *and* $x = (1, \ldots, 1)$. *In particular, if we denote*

$$s_1 := \frac{t[1-(n-2)\theta^2]}{[1-(n-1)\theta^2]^2}\left[(2\alpha - t)[1-(n-2)\theta^2] + 2\alpha\theta + 2\alpha\theta^2(n-2)\right],$$

$$s_2 := \frac{t}{[1-(n-1)\theta^2]^2}\,[2\alpha - t + 2(\alpha - t)\theta(n-1)],$$

$$s_3 := \frac{t[1-(n-2)\theta^2]}{[1-(n-1)\theta^2]^2}\Big[(2\alpha - t)[1-(n-2)\theta^2]$$
$$+ 2\alpha\theta + 2(\alpha - t)\theta^2(n-2)\Big],$$

then seven scenarios can occur:

(a) *If* $2c < \min\{s_2, s_3\}$, *then* $x = (0, \ldots, 0)$ *is the unique Nash equilibrium.*

(b) *If* $2c > \max\{s_1, s_2\}$, *then* $x = (1, \ldots, 1)$ *is the unique Nash equilibrium.*
(c) *If* $2c \in (s_1, s_2)$, *then* $x = (0, 1, \ldots, 1)$ *is the unique Nash equilibrium.*
(d) *If* $2c \in (s_2, s_1]$, *then there are two Nash equilibria:* $x = (0, \ldots, 0)$ *and* $x = (1, \ldots, 1)$.
(e) *If* $2c \in [s_3, \min\{s_1, s_2\}]$, *then there are two Nash equilibria:* $x = (0, \ldots, 0)$ *and* $x = (0, 1, \ldots, 1)$.
(f) *If* $2c = s_2 > s_1$, *then there are two Nash equilibria:* $x = (0, 1, \ldots, 1)$ *and* $x = (1, \ldots, 1)$.
(g) *If* $2c = s_2 \leq s_1$, *then there are three Nash equilibria:* $x = (0, \ldots, 0)$, $x = (1, \ldots, 1)$, *and* $x = (0, 1, \ldots, 1)$.

Proof. We know that $x = (0, \ldots, 0)$ is a Nash equilibrium if and only if conditions (1) hold. From now on, let us denote $A = 1 - (n - 2)\theta^2$ and $B = 1 - (n - 1)\theta^2$. Then, conditions (1) are equivalent to the following system of inequalities:

$$\begin{cases} 2c \leq \dfrac{t}{B^2}\left[2\alpha - t + 2\alpha\theta(n-1)\right] & \text{(for } i = 1\text{)}, \\[2ex] 2c \leq \dfrac{tA}{B^2}\left[(2\alpha - t)A + 2\alpha\theta + 2\alpha\theta^2(n-2)\right] & \text{(for } i > 1\text{)}. \end{cases}$$

Since $\theta^2 < \theta$ and $A < 1$, we get

$$\frac{tA}{B^2}\left[(2\alpha - t)A + 2\alpha\theta + 2\alpha\theta^2(n-2)\right] < \frac{t}{B^2}\left[2\alpha - t + 2\alpha\theta(n-1)\right],$$

hence conditions (1) are equivalent to

$$2c \leq \frac{tA}{B^2}\left[(2\alpha - t)A + 2\alpha\theta + 2\alpha\theta^2(n-2)\right] = s_1.$$

The vector $x = (1, \ldots, 1)$ is a Nash equilibrium if and only if conditions (2) hold that are equivalent to the following system of inequalities:

$$\begin{cases} 2c \geq \dfrac{t}{B^2}\left[2\alpha - t + 2(\alpha - t)\theta(n-1)\right] & \text{(for } i = 1\text{)}, \\[2ex] 2c \geq \dfrac{tA}{B^2}\left[(2\alpha - t)A + 2(\alpha - t)[\theta + \theta^2(n-2)]\right] & \text{(for } i > 1\text{)}. \end{cases}$$

Since $\theta^2 < \theta$ and $A < 1$, we get

$$\begin{aligned}&\frac{t}{B^2}\left[2\alpha - t + 2(\alpha - t)\theta(n-1)\right]\\&\quad > \frac{tA}{B^2}\left[(2\alpha - t)A + 2(\alpha - t)[\theta + \theta^2(n-2)]\right],\end{aligned}$$

hence conditions (2) are equivalent to

$$2c \geq \frac{t}{B^2}\left[2\alpha - t + 2(\alpha - t)\theta(n-1)\right] = s_2.$$

The vector $x = (0, 1, \ldots, 1)$ is a Nash equilibrium if and only if the following conditions hold:

$$\begin{cases} 2c \leq \dfrac{t}{B^2}\left[2\alpha - t + 2(\alpha - t)\theta(n-1)\right] & (\text{for } i = 1), \\ 2c \geq \dfrac{tA}{B^2}\left[(2\alpha - t)A + 2\alpha\theta + 2(\alpha - t)\theta^2(n-2)\right] & (\text{for } i > 1), \end{cases}$$

that is, $2c \in [s_3, s_2]$.

We now prove that Nash equilibria different from $x = (0, \ldots, 0)$, $x = (1, \ldots, 1)$, and $(0, 1, \ldots, 1)$ cannot exist. Let us consider three possible cases:

(a) Assume, by contradiction, that there exists an integer k, with $2 \leq k \leq n-1$, such that $x = (\underbrace{0, \ldots, 0}_{k}, 1, \ldots, 1)$ is a Nash equilibrium.

Then, the following conditions must hold:

$$\begin{cases} 2c \leq \dfrac{t}{B^2}\left[2\alpha - t + 2\alpha\theta(k-1) + 2(\alpha - t)\theta(n-k)\right], \\ 2c \leq \dfrac{tA}{B^2}\left[(2\alpha - t)A + 2\alpha\theta + 2\theta^2[\alpha(k-2) + (\alpha - t)(n-k)]\right], \\ 2c \geq \dfrac{tA}{B^2}\left[(2\alpha - t)A + 2\alpha\theta + 2\theta^2[\alpha(k-1) + (\alpha - t)(n-k-1)]\right]. \end{cases} \tag{3}$$

Since the following chain of equalities and inequalities holds:

$$\begin{aligned} \alpha(k-1) + (\alpha - t)(n-k-1) &= \alpha(n-2) - t(n-k-1) && (4)\\ &> \alpha(n-2) - t(n-k) && (5)\\ &= \alpha(k-2) + (\alpha - t)(n-k), && (6) \end{aligned}$$

we get from (3) that

$$\begin{aligned}2c &\geq \frac{tA}{B^2}\left[(2\alpha - t)A + 2\alpha\theta + 2\theta^2[\alpha(k-1) + (\alpha - t)(n-k-1)]\right]\\ &> \frac{tA}{B^2}\left[(2\alpha - t)A + 2\alpha\theta + 2\theta^2[\alpha(k-2) + (\alpha - t)(n-k)]\right]\\ &\geq 2c\end{aligned}$$

that is impossible. Therefore, $x = (\underbrace{0, \ldots, 0}_{k}, 1, \ldots, 1)$ cannot be a Nash equilibrium.

(b) Assume, by contradiction, that there exists an integer k, with $2 \leq k \leq n-1$, such that $x = (1, \underbrace{0, \ldots, 0}_{k-1}, 1, \ldots, 1)$ is a Nash equilibrium. Then, the following conditions must hold:

$$\begin{cases} 2c \geq \dfrac{t}{B^2}\left[2\alpha - t + 2\alpha\theta(k-1) + 2(\alpha - t)\theta(n-k)\right], \\ 2c \leq \dfrac{tA}{B^2}\left[(2\alpha - t)A + 2(\alpha - t)\theta \right. \\ \qquad \left. + 2\theta^2[\alpha(k-2) + (\alpha - t)(n-k)]\right], \\ 2c \geq \dfrac{tA}{B^2}\left[(2\alpha - t)A + 2(\alpha - t)\theta \right. \\ \qquad \left. + 2\theta^2[\alpha(k-1) + (\alpha - t)(n-k-1)]\right]. \end{cases} \tag{7}$$

It follows from (4) and (7) that

$$\begin{aligned}2c &\geq \frac{tA}{B^2}\left[(2\alpha - t)A + 2(\alpha - t)\theta \right.\\ &\qquad \left. + 2\theta^2[\alpha(k-1) + (\alpha - t)(n-k-1)]\right]\\ &> \frac{tA}{B^2}\left[(2\alpha - t)A + 2(\alpha - t)\theta + 2\theta^2[\alpha(k-2) + (\alpha - t)(n-k)]\right]\\ &\geq 2c\end{aligned}$$

that is impossible. Therefore, $x = (1, \underbrace{0, \ldots, 0}_{k-1}, 1, \ldots, 1)$ cannot be a Nash equilibrium.

(c) Assume, by contradiction, that $x = (1, 0, \ldots, 0)$ is a Nash equilibrium. Then, the following conditions must hold:

$$\begin{cases} 2c \geq \dfrac{t}{B^2}\left[2\alpha - t + 2\alpha\theta(n-1)\right], \\ 2c \leq \dfrac{tA}{B^2}\left[(2\alpha - t)A + 2(\alpha - t)\theta + 2\alpha\theta^2(n-2)\right]. \end{cases}$$

Thus, we have

$$\begin{aligned} 2c &\leq \frac{tA}{B^2}\left[(2\alpha - t)A + 2(\alpha - t)\theta + 2\alpha\theta^2(n-2)\right] \\ &< \frac{tA}{B^2}\left[(2\alpha - t)A + 2\alpha\theta + 2\alpha\theta^2(n-2)\right] \\ &< \frac{tA}{B^2}\left[(2\alpha - t)A + 2\alpha\theta + 2\alpha\theta(n-2)\right] \\ &< \frac{t}{B^2}\left[(2\alpha - t) + 2\alpha\theta + 2\alpha\theta(n-2)\right] \\ &\leq 2c \end{aligned}$$

that is impossible. Therefore, $x = (1, 0, \ldots, 0)$ cannot be a Nash equilibrium.

In summary, we proved that there exist three Nash equilibria and

$$x = (0, \ldots, 0) \text{ is a Nash equilibrium} \iff 2c \leq s_1, \tag{8}$$

$$x = (1, \ldots, 1) \text{ is a Nash equilibrium} \iff 2c \geq s_2, \tag{9}$$

$$x = (0, 1, \ldots, 1) \text{ is a Nash equilibrium} \iff 2c \in [s_3, s_2]. \tag{10}$$

Since $s_3 < s_1$ holds, the seven scenarios described in the statement can be easily derived. □

Remark 6. Proposition 5 generalizes [8, Proposition 7] to the case of a star-shaped network with n agents and provides correct sets of parameters for the corresponding Nash equilibria.

Example 7. Consider a game with $n = 3$ players connected by a star-shaped social network (see Figure 5). According to Proposition 5, seven possible scenarios can occur. Figure 6 shows the Nash equilibria of the game according to the values of parameters c and t in the case $\theta = 0.5$, $\alpha = 1$, and $y = 5$.

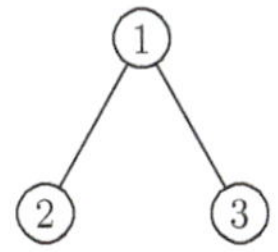

Fig. 5. Social network of the game in Example 7.

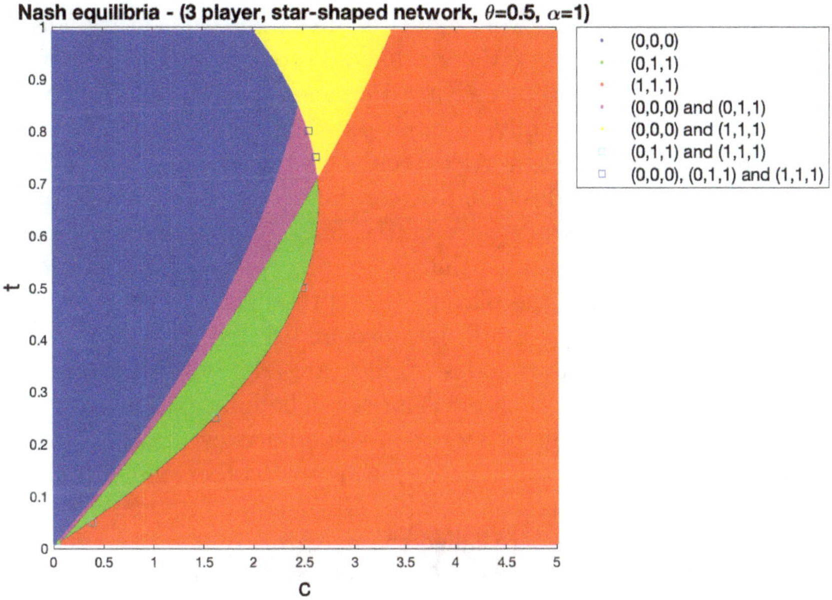

Fig. 6. Nash equilibria (in locations) of the game described in Example 7.

4. Conclusions

In this chapter, we contributed to improve the model of social interactions in city areas put forward in [8]. We generalized the model by considering any finite number of city areas instead of only two (i.e., the periphery and the city center). Moreover, we showed by means of simple test problems, for which we computed the solutions in closed form, that, even in the case of two geographic locations, multiple equilibrium locations can occur, thus contrasting with the results in [8]. Future research is needed to formulate a general theorem so as to generalize the analytic formulas of the examples and also clarify the content of Proposition 5 in [8]. Other possible extensions of this model are the introduction of uncertain data, along the same lines as in [10], the variational inequality approach (see, e.g., [13]) or the use of joint constraints (see, e.g., [7, 12]).

A.1. Basics about Network Games

The formulation of a network game requires some elementary notions of graph theory and game theory that we provide here. Moreover, we describe here the concept of Katz–Bonacich vector which is ubiquitous in this field. Formally, a graph g is a pair of sets (V, E), where V is the set of nodes (or vertices) and E is the set of arcs (or edges), formed by pairs of nodes (v, w). Arcs which have the same end nodes are called parallel, while arcs of the form (v, v) are called loops. We focus on *simple* graphs, that is, graphs with no parallel arcs or loops. In our setting, the players are represented by the n nodes in the graph. Moreover, a graph is called undirected if arcs (v, w) and (w, v) are the same. Two nodes v and w are adjacent if they are connected by an arc, i.e., if (v, w) is an arc. The information about the adjacency of nodes can be stored in the adjacency matrix G whose elements g_{ij} are equal to 1 if (v_i, v_j) is an arc, 0 otherwise. G is thus a symmetric and zero diagonal matrix. Given a node v, the nodes connected to v with an arc are called the *neighbors* of v and are grouped in the set $N_v(g)$. The number of elements of $N_v(g)$ is the *degree* of v. A *walk* in the graph g is a finite sequence of the form

$$v_{i_0}, e_{j_1}, v_{i_1}, e_{j_2}, \ldots, e_{j_k}, v_{j_k},$$

which consists of alternating nodes and arc of the graph such that $v_{i_{t-1}}$ and v_{i_t} are end nodes of e_{j_t}. The *length* of a walk is the number of its arcs. Let us remark that it is allowed to visit a node or go through an arc more than once. A *path* is a walk with all different nodes (except possibly the initial and terminal ones if the walk is closed). The indirect connections between any two nodes in the graph are described by means of the powers of the adjacency matrix G. Indeed, it can be proved that the element $g_{ij}^{[k]}$ of G^k gives the number of walks of length k between v_i and v_j.

We now provide the basic notions of game theory used throughout this chapter. For simplicity, the set of players will be denoted by $\{1, 2, \ldots, n\}$ instead of $\{v_1, v_2, \ldots, v_n\}$. We denote with $A_i \subset \mathbb{R}$

the action space of player i, while $A = A_1 \times \cdots \times A_n$ and the notation $a = (a_i, a_{-i})$ is used when we want to distinguish the action of player i from the action of all the other players. Each player i is endowed with a payoff function $u_i : A \to \mathbb{R}$ that he/she wishes to maximize. The notation $u_i(a, g)$ is often utilized when one wants to emphasize the influence of the graph structure (e.g., when studying perturbation with respect to the removal of an arc). The solution concept considered is the Nash equilibrium of the game, according to the following definition.

Definition 8. A Nash equilibrium is an element $a^* \in A$ such that for each $i \in \{1, \dots, n\}$,

$$u_i(a_i^*, a_{-i}^*) \geq u_i(a_i, a_{-i}^*), \quad \forall\, a_i \in A_i.$$

In our model, as in most network games, a_{-i} only consists of components a_j such that $j \in N_i(g)$, that is, j is a neighbor of i.

For any given player i, it is interesting to distinguish how variations of the actions of player's i neighbors affect his/her marginal utility. In the case where the utility functions are twice continuously differentiable, the following definitions clarify this point.

Definition 9. We say that the network game has the property of strategic substitutes if for each player i the following condition holds:

$$\frac{\partial^2 u_i(a_i, a_{-i})}{\partial a_j \partial a_i} < 0, \quad \forall (i, j) : g_{ij} = 1, \ \forall\, a \in A.$$

Definition 10. We say that the network game has the property of strategic complements if for each player i the following condition holds:

$$\frac{\partial^2 u_i(a_i, a_{-i})}{\partial a_j \partial a_i} > 0, \quad \forall (i, j) : g_{ij} = 1, \forall\, a \in A.$$

In most network games, an in-depth analytic study can be performed in the case of quadratic utility functions.

Let $A_i = \mathbb{R}_+$ for any $i \in \{1, \dots, n\}$, hence $A = \mathbb{R}^n_+$. The payoff of player i is given by

$$u_i(a, g) = -\frac{1}{2} a_i^2 + \delta_i a_i + \phi \sum_{j=1}^{n} g_{ij} a_i a_j, \quad \delta_i, \phi > 0. \tag{A.1}$$

The last term describes the interaction between neighbors, and since $\phi > 0$, this interaction falls in the class of strategic complements.

It is useful to introduce the so-called the pseudogradient of the game, according to the terminology introduced by Rosen:

$$[F(a)]^\top := -\left(\frac{\partial u_1}{\partial a_1}(a), \ldots, \frac{\partial u_n}{\partial a_n}(a)\right). \tag{A.2}$$

The pseudogradient's components of this game are easily computed as

$$F_i(a) = a_i - \delta_i - \phi \sum_{j=1}^{n} g_{ij} a_j, \quad i \in \{1, \ldots, n\},$$

which can be written in compact form as

$$F(a) = (I - \phi G)a - \delta.$$

Lemma 11 (see, e.g., [9]). *The matrix $I - \phi G$ is positive definite iff $\phi\rho(G) < 1$, where $\rho(G)$ is the spectral radius of G.*

Proof. The symmetric matrix $I - \phi G$ is positive definite if and only if $\lambda_{min}(I - \phi G) > 0$. On the other hand, $\lambda_{min}(I - \phi G) = 1 - \phi\lambda_{max}(G)$. Since G is a symmetric non-negative matrix, the Perron–Frobenius theorem guarantees that $\lambda_{max}(G) = \rho(G)$, hence $I - \phi G$ is positive definite if and only if $\phi\rho(G) < 1$. □

To be selfconsistent, in the following lemma, we recall the following result about series of matrices.

Lemma 12 (see, e.g., [1]). *Let T be a square matrix and consider the series*

$$I + T + T^2 + \cdots + T^k + \cdots.$$

The series converges provided that $\lim_k T^k = 0$, which is equivalent to $\rho(T) < 1$. In such a case, the matrix $I - T$ is non-singular and we have

$$(I - T)^{-1} = I + T + T^2 + \cdots + T^k + \cdots.$$

Theorem 13 (see, e.g., [9]). *If $\phi\rho(G) < 1$, then the unique Nash equilibrium is*

$$a^* = (I - \phi G)^{-1}\delta = \sum_{p=0}^{\infty} \phi^p G^p \delta. \tag{A.3}$$

Remark 14. The expansion in (A.3) suggests an interesting interpretation. Indeed, it can be shown that the (i, j) entry, $g_{ij}^{[p]}$, of the matrix G^p gives the number of walks of length p between nodes i and j. Based on this observation, a measure of centrality on the network was proposed by Katz and Bonacich (see, e.g., [5]). Specifically, for any weight $w \in \mathbb{R}_+^n$, the weighted vector of Katz–Bonacich is given by

$$b_w(G, \phi) = M(G, \phi) = (I - \phi G)^{-1} w = \sum_{p=0}^{\infty} \phi^p G^p w. \tag{A.4}$$

In the case where $w = \mathbf{1}$, the (non-weighted) centrality measure of Katz–Bonacich of node i is given by

$$b_{1,i}(G, \phi) = \sum_{j=1}^{n} m_{ij}(G, \phi) = m_{ii} + \sum_{j \neq i} m_{ij}(G, \phi) \tag{A.5}$$

and counts the total number of walks in the graph, which start at node i, exponentially damped by ϕ.

References

[1] Atkinson, K. and Han, W. *Theoretical Numerical Analysis*. Springer (2007).

[2] Ballester, C., Calvo-Armengol, A. and Zenou, Y. Who's who in networks. Wanted: The key player. *Econometrica*, 74, 1403–1417 (2006).

[3] Ballester, C., Calvo-Armengol, A. and Zenou, Y. Delinquent networks. *J. Eur. Econ. Assoc.*, 8, 34–61 (2010).

[4] Belhaj M., Bramoulleé Y. and Deroian F. Network games under strategic complementarities. *Games Econon. Behav.*, 88, 310–319 (2014).

[5] Bonacich, P. Power and centrality: A family of measures. *Am. J., Sociol.*, 92, 1170–1182 (1987).

[6] Calvo-Armengol, A., Patacchini, E. and Zenou, Y. Peer effects and social networks in education. *Rev. Econ. Stud.*, 76, 1239–1267 (2009).

[7] Faraci, F. and Raciti, F. On generalized Nash equilibrium problems in infinite dimension: The lagrange multipliers approach. *Optimization*, 64, 321–338 (2015).

[8] Helsley, R.W. and Zenou, Y. Social networks and interactions in cities. *J. Economic Theory*, 150, 426–466, (2014).

[9] Jackson, M.O. and Zenou, Y. Games on Networks, in: *Handbook of Game Theory with Economic Applications*, Elsevier, 95–163 (2015).

[10] Jadamba, B. and Raciti, F. On the modelling of some environmental games with uncertain data. *J. Optim. Theory Appl.*, 167, 959–968 (2015).

[11] Monderer, D. and Shapley, L.S. Potential games. *Games Econ. Behav.*, 14, 124–143 (1996).

[12] Passacantando M. and Raciti F. A note on generalized Nash games played on networks, in: Rassias T.M. (ed.), *Nonlinear Analysis, Differential Equations, and Applications.* Springer Optimization and Its Applications, Springer, Cham, 173 (2021).

[13] Parise, F. and Ozdaglar, A. A variational inequality framework for network games: Existence, uniqueness, convergence and sensitivity analysis. *Games Econ. Behav.*, 114, 47–82 (2019).

https://doi.org/10.1142/9789811293009_0034

Chapter 34

On a Theorem of Miron Nicolescu: Proofs, Generalizations, and Applications to Zeros Separation of Some Special Functions

Adrian Petruşel*,† and Ioan A. Rus*,‡

**Babeş-Bolyai University of Cluj-Napoca, Department of Mathematics, Kogălniceanu Street, No. 1, 400084 Cluj-Napoca, Romania*

†*adrian.petrusel@ubbcluj.ro*

‡*iarus@math.ubbcluj.ro*

In this chapter, we put first in the light a zeros separation theorem for a system of two linear and homogeneous differential equations, given by Miron Nicolescu (see M. Nicolesco, Sur le théorème de Sturm, *Mathematica*, 1, 111–114 (1929)). Several proofs of this result are considered. Then, some generalization of this result, as well as some applications to special function theory, is presented.

1. Introduction and Preliminaries

For the functions $p_i, q_i \in C[a,b]$ $(i \in \{1,2\})$, we consider the following system of differential equations:

$$\begin{cases} y' + p_1 y + q_1 z = 0 \\ z' + p_2 y + q_2 z = 0. \end{cases} \tag{1}$$

The following result is known as Nicolescu's theorem ([7, 8, 12, 14, 18, 21–28], ...):

Theorem 1. *If $q_1(x) \neq 0$, $p_2(x) \neq 0$, for all $x \in [a,b]$, then we have the following:*

(1) *If $(y_1, z_1), (y_2, z_2) \in C^1([a,b], \mathbb{R}^2)$ are two linearly independent solution of* (1), *then the zeros of y_1 and y_2, and of z_1 and z_2 separate each other.*
(2) *If $(y, z) \in C^1([a,b], \mathbb{R}^2)$ is a nontrivial solution of* (1), *then the zeros of y and z separate each other.*

Since in many papers on this subject the above result is not mentioned ([2, 15, 16, 20, 30, 31], ...), we visit first Nicolescu's theorem. After some proofs of it, we give some generalizations, as well as some applications to special function theory.

We first recall from [28] the following notations, notions, and results.

Let $a, b \in \mathbb{R}$, $a < b$ and $y, z \in C[a,b]$ be two linearly independent functions. We denote by $H(y,z)$ the linear hull of $\{y, z\}$ over $\mathbb{R}$.

For a function $y : [a,b] \to \mathbb{R}$, we denote by Z_y the zero (i.e., zero point) set of y.

Definition 2. By definition, the zeros of two functions $y, z \in C^1[a,b]$ separate each other iff

(1) $Z_y \cap Z_z = \emptyset$;
(2) all zeros of y and z are simple;
(3) if x_1 and x_2 are two consecutive zeros of y (from condition (2), the functions y and z have finite number of zeros!), then z has one and only one zero in $]x_1, x_2[$, and if x_1, x_2 are two consecutive zeros of z, then y has one and only one zero in $]x_1, x_2[$.

If $y, z \in C^1[a,b]$ are two given functions, then we denote by $W(y,z)$ the Wronskian associated with these functions, i.e.,

$$W(y,z)(x) := \begin{vmatrix} y & z \\ y' & z' \end{vmatrix}(x).$$

We have the following result (see Theorem 2.1 in [28]; see also [2] for the case y and $z \in C^2[a,b]$).

Theorem 3. (*Wronskian separation principle for scalar functions*). *Let $y, z \in C^1[a,b]$ be two functions. We suppose that $W(y,z)(x) \neq 0$, for all $x \in [a,b]$. Let $u, v \in H(y,z)$ be two linearly independent functions. Then the zeros of u and v separate each other.*

For $y, z \in C([a,b], \mathbb{R}^2)$, we denote by $H(y,z)$ the linear hull of $\{y,z\}$ over $\mathbb{R}$ and by

$$V(y,z) := \begin{vmatrix} y_1 & y_2 \\ z_1 & z_2 \end{vmatrix},$$

the Wronskian of $y = (y_1, y_2)$ and $z = (z_1, z_2)$.

In this case, we have (see Theorem 2.2 in [28]):

Theorem 4. (*Wronskian separation principle for vectorial functions*). *We suppose the following:*

(1) $V(y,z)(x) \neq 0$, *for all $x \in [a,b]$.*
(2) *The zeros of y_1 and y_2, and of z_1 and z_2 separate each other.*

Then the zeros of y_1 and z_1, and of y_2 and z_2 separate each other. Moreover, if the zeros of two linearly independent elements of $H(y_1,y_2)$ and of $H(z_1,z_2)$ separate each other, then if $u = (u_1,u_2) \in H(y,z)$ is a non-trivial element, then the zeros of u_1 and u_2 separate each other.

The structure of this chapter is as follows. Section 1 details the introduction and preliminaries. Section 2 provides some proofs of Miron Nicolescu theorem. In Section 3, generalizations in the nonlinear case are given. Section 4 provides the applications to some special functions. In Section 5, zeros separation results in the case of two second-order differential equations are given.

2. Some Proofs of Miron Nicolescu Theorem

There are various proofs of Nicolescu result (see [7, 8, 12, 14, 18, 22, 24, 26–28], ...). Here are some of them.

Nicolescu's proof. Nicolescu proves Theorem 1, when $p_i, q_i \in C^1[a,b]$, for each $i \in \{1,2\}$. Let $(y,z) \in C^1([a,b],\mathbb{R}^2)$ be a solution of (1). From the regularity of p_i, q_i, it follows that $(y,z) \in C^2([a,b],\mathbb{R}^2)$. From (1), we have that y is the solution of a second-order equation

$$y'' + \left(p_1 + q_2 - \frac{q_1'}{q_1}\right) y' + \left(p_1 q_2 - p_2 q_1 + \frac{p_1' q_1 - p_1 q_1'}{q_1}\right) y = 0 \quad (2)$$

and z is a solution of the equation

$$z'' + \left(p_1 + q_2 - \frac{p_2'}{p_2}\right) z' + \left(p_1 q_2 - p_2 q_1 + \frac{q_2' p_2 - p_2' q_2}{p_2}\right) z = 0. \quad (3)$$

If (y_1, z_1) and (y_2, z_2) are two linearly independent solution of (1), then y_1 and y_2 are linearly independent solutions of (2) and z_1 and z_2 are two linearly independent solutions of (3). So, conclusion (1) in Theorem 1 follows from zeros separation theorem of Sturm.

If (y,z) is a non-trivial solution of (1), then there exists a solution (Y,Z) of (1) such that

$$\begin{vmatrix} y & Y \\ z & Z \end{vmatrix}(x) \neq 0, \ \forall\, x \in [a,b].$$

It is clear that we can suppose that this determinant is positive. Let, for example, x_1 and x_2 be two consecutive zeros of y, then

$$z(x_1)Y(x_1) < 0 \text{ and } z(x_2)Y(x_2) < 0.$$

From conclusion (1), $Y(x_1)Y(x_2) < 0$. So, $z(x_1)z(x_2) < 0$. Conclusion (2) follows.

Butlewski's proof ([7]). System (1) is put in the following form:

$$\left(e^{\int_a^x p_1(s)ds} y(x)\right)' + q_1(x) e^{\int_a^x p_1(s)ds} z(x) = 0,$$

$$\left(e^{\int_a^x q_2(s)ds} z(x)\right)' + p_2(x) e^{\int_a^x q_2(s)ds} y(x) = 0. \quad (4)$$

For $c_1, c_2 \in \mathbb{R}$ and (y,z), a non-trivial solution of (4) (i.e., of (1)), we consider the function $\phi(c_1, c_2; y, z) \in C^1[a,b]$ defined by

$$\phi(c_1, c_2; y, z)(x) := c_1 e^{\int_a^x p_1(s)ds} y(x) + c_2 e^{\int_a^x q_2(s)ds} z(x).$$

For this function ϕ, we have the following:

Lemma 5. *If* c_1, c_2 *are such that*

$$D(c_1,c_2)(x) := c_1 q_1(x) e^{\int_a^x (p_1(s)-q_2(s))ds} - c_2 p_2(x) e^{-\int_a^x (p_1(s)-q_2(s))ds} \neq 0, \ \forall\, x \in [a,b],$$

then the zeros of $\phi(c_1,c_2;y,z)$ *are simple for each* (y,z) *non-trivial solution of* (1).

Indeed, if for $x_0 \in [a,b]$, $\phi(x_0) = \phi'(x_0) = 0$, we have that $y(x_0) = z(x_0) = 0$. But (y,z) is a non-trivial solution of (1).

Lemma 6. *If* $c_1, c_2, \widetilde{c}_1, \widetilde{c}_2 \in \mathbb{R}$ *are such that* $D(c_1,c_2)(x) \neq 0$, $D(\widetilde{c}_1, \widetilde{c}_2) \neq 0$, *for all* $x \in [a,b]$, *then we have the following:*

(a) *If* (y_1,z_1) *and* (y_2,z_2) *are linearly independent solution of* (1), *then the zeros of* $\phi(c_1,c_2;y_1,z_1)$ *and* $\phi(c_1,c_2;y_2,z_2)$ *separate each other.*

(b) *If* $c_1\widetilde{c}_2 - c_2\widetilde{c}_1 \neq 0$ *and* (y,z) *is a non-trivial solution of* (1), *then the zeros of* $\phi(c_1,c_2;y,z)$ *and* $\phi(\widetilde{c}_1,\widetilde{c}_2;y,z)$ *separate each other.*

Indeed, we remark that

$$W(\phi(c_1,c_2;y_1,z_2), \phi(c_1,c_2;y_2,z_2))(x) \neq 0$$

and

$$W(\phi(c_1,c_2;y,z), \phi(\widetilde{c}_1,\widetilde{c}_2;y,z))(x) \neq 0, \ \forall\, x \in [a,b].$$

Conclusions (a) and (b) follow from Theorem 3.

Now, if we take in $\phi(c_1,c_2;y_i,z_i)$, $i = 1,2$, $c_1 = 1$, $c_2 = 0$ and $c_1 = 0$, $c_2 = 1$, we have (1) in Theorem 1. If we take in $\phi(c_1,c_2;y,z)$ and $\phi(\widetilde{c}_1,\widetilde{c}_2;y,z)$, $c_1 = 1$, $c_2 = 0$, $\widetilde{c}_1 = 0$, $\widetilde{c}_2 = 1$, we have (2) in Theorem 1.

Let us present the proof given in [28]. First, if (y_1,z_1) and (y_2,z_2) are two linearly independent solutions of (1), then

$$W(y_1,y_2)(x) = -q_1(x) V((y_1,y_2),(z_1,z_2))(x) \neq 0, \ \forall\, x \in [a,b]$$

and

$$W(z_1,z_2)(x) = p_2(x) V((y_1,z_1),(y_2,z_2))(x) \neq 0, \ \forall\, x \in [a,b].$$

Now, the proof of conclusion (1) in Theorem 1 follows from Theorem 3 and of conclusion (2) follows from Theorem 4.

Remark 7. From the above considerations it is clear that we have that:

If (y_1, z_1) and (y_2, z_2) are two solutions of (1), with $q_1(x) \neq 0$, $p_2(x) \neq 0$, for all $x \in [a, b]$, then the following statements are equivalent:

(1) $V((y_1, z_1), (y_2, z_2))(x) \neq 0$, for all $x \in [a, b]$.
(1′) (y_1, z_1) and (y_2, z_2) are linearly independent.
(2) $W(y_1, y_2)(x) \neq 0$, for all $x \in [a, b]$.
(2′) y_1 and y_2 are linearly independent.
(3) $W(z_1, z_2)(x) \neq 0$, for all $x \in [a, b]$.
(3′) z_1 and z_2 are linearly independent.

Remark 8. For $P, Q \in C[a, b]$, we consider the differential equation

$$y'' + Py' + Qy = 0. \tag{5}$$

If we denote $y' = z$, then we have from (5) the following system (a particular case of (1)):

$$y' - z = 0,$$

$$z' + Qy + Pz = 0.$$

From Theorem 1, we have the following:

Theorem 9 ([18]). We suppose that $Q(x) \neq 0$, for all $x \in [a, b]$. If y is a non-trivial solution of *(5)*, then the zeros of y and y' separate each other.

3. Generalizations in the Nonlinear Case

There are some generalizations of Nicolescu's theorem for nonlinear systems ([5, 11, 12, 21, 23, 28], ...). For our considerations in this section, we need the following notion and result. There are various notions of zeros interlacing of two functions ([2, 9, 15–17, 19], ...).

In this chapter, we use the following one.

Definition 10. By definition, the zeros of two functions $y_1, y_2 \in C^1[a, b]$ interlace iff

(i) $Z_{y_i} \cap Z_{y_i'} = \emptyset$, for $i \in \{1, 2\}$;
(ii) if x_1 and x_2 are two consecutive zeros of y_1 (of y_2), then $Z_{y_2} \cap [x_1, x_2] \neq 0$ $(Z_{y_1} \cap [x_1, x_2] \neq \emptyset)$.

Tonelli Lemma. *Let* $y_1, y_2 \in C^1[a,b]$ *be such that*

(i) $y_1(a) = y_1(b) = 0$, $y_1(x) > 0$, *for all* $x \in]a,b[$;
(ii) $y_2(x) > 0$, *for all* $x \in [a,b]$.

Then there exists $x_0 \in]a,b[$ *and* $\lambda > 0$ *such that*

(1) $y_2(x_0) = \lambda y_1(x_0)$;
(2) $y_2'(x_0) = \lambda y_1'(x_0)$.

For $F, G : \mathbb{R}^3 \times [a,b] \to \mathbb{R}$, we consider the following system of differential equations:

$$\begin{aligned} F(y', y, z, x) &= 0, \\ G(z', y, z, x) &= 0. \end{aligned} \tag{6}$$

Our results for (6) are as the follows:

Theorem 11. *We suppose that*

(N_1) *the Cauchy problem for* (6) *has at most a solution;*
(N_2) $F(0,0,0,x) = 0$, $G(0,0,0,x) = 0$, *for all* $x \in [a,b]$;
(N_3) $x \in [a,b]$, $r \in \mathbb{R}$, $F(0,0,r,x) = 0 \Rightarrow r = 0$;
(N_4) $x \in [a,b]$, $r \in \mathbb{R}$, $G(0,r,0,x) = 0 \Rightarrow r = 0$.

If $(y,z) \in C^1([a,b], \mathbb{R}^2)$ *is a non-trivial solution of* (6), *then*

(1) $Z_y \cap Z_z = \emptyset$;
(2) $Z_y \cap Z_{y'} = \emptyset$;
(3) $Z_z \cap Z_{z'} = \emptyset$.

Proof. (1) From (N_2), the constant function, $x \mapsto (0,0)$, is a solution of (6). From (N_1), if there exist $x_0 \in Z_y \cap Z_z$, then $(y,z) = (0,0)$, but (y,z) is a non-trivial solution.

(2) + (3) Let, for example, $x_0 \in Z_y \cap Z_{y'}$. From (N_3), we have that $z(x_0) = 0$, and from (N_4), $y(x_0) = 0$. Conclusion (2) follow from (1). In a similar way, we have (3). □

Theorem 12. *In addition to the conditions in Theorem* 11, *we suppose that*

(N_5) *there exist* $k_i > 0$, $i = 1, 2$, *such that*

$$F(\lambda x_1, \lambda r_2, \lambda r_3, x) = \lambda^{k_1} F(r_1, r_2, r_3, x)$$

and

$$G(\lambda r_1, \lambda r_2, \lambda r_3, x) = \lambda^{k_2} G(r_1, r_2, r_3, x),$$

for all $\lambda > 0$, $r_1, r_2, r_3 \in \mathbb{R}$ *and* $x \in [a, b]$;

(N_6) *if* (y_1, z_1) *and* (y_2, z_2) *are two solution of* (6), *then:*

(a) *if for some* $x_0 \in [a, b]$, $y_1(x_0) = y_2(x_0)$ *and* $y_1'(x_0) = y_2'(x_0)$, *then* $(y_1, z_1) = (y_2, z_2)$;

(b) *if for some* $x_0 \in [a, b]$, $z_1(x_0) = z_2(x_0)$ *and* $z_1'(x_0) = z_2'(x_0)$, *then* $(y_1, z_1) = (y_2, z_2)$.

In the above conditions, if (y_1, z_1) *and* (y_2, z_2) *are two linearly independent solutions of* (6), *then the zeros of* y_1 *and* y_2, *and of* z_1 *and* z_2 *interlace each other.*

Proof. From Theorem 11, the zeros of y_i, z_i are simple. Let, for example, x_1, x_2 be two consecutive zeros of y_1. Let us suppose that $Z_{y_2} \cap [x_1, x_2] = \emptyset$. Since F, G are homogeneous, we can suppose that $y_1(x) > 0$, for all $x \in]x_1, x_2[$ and $y_2(x) > 0$, for all $x \in [x_1, x_2]$. From Tonelli lemma, there exist $x_0 \in]x_1, x_2[$ and $\lambda > 0$ such that $y_2(x_0) = \lambda y_1(x_0)$ and $y_2'(x_0) = \lambda y_1'(x_0)$. Since $(\lambda y_1, \lambda z_1)$ is a solution of (6), from (N_6), we have that $(y_2, z_2) = \lambda(y_1, z_1)$. This is a contradiction since (y_1, z_1), (y_2, z_2) are linearly independent. □

Theorem 13. *In addition to conditions in Theorem* 11, *we suppose that*

(N_7) $x \in [a, b]$, $r_1, r_2 \in \mathbb{R}$, $F(r_1, 0, r_2, x) = 0 \Rightarrow \text{sign} r_1 = \text{sign} r_2$;

(N_8) $x \in [a, b]$, $r_1, r_2 \in \mathbb{R}$, $G(r_1, r_2, 0, x) = 0 \Rightarrow \text{sign} r_1 = \text{sign} r_2$.

In these conditions, if $(y, z) \in C^1([a, b]), \mathbb{R}^2$ *is a non-trivial solution of* (6), *then the zeros of* y *and* z *separate each other.*

Proof. From Theorem 11, the zeros of y and of z are simple and $Z_y \cap Z_z = \emptyset$. Let, for example, x_1 and x_2 be two consecutive zeros of y. Then, $y'(x_1)y'(x_2) < 0$. From (N_7), since $F(y'(x_i), 0, z(x_i)) = 0$, $i = 1, 2$, it follows that $z(x_1)z(x_2) < 0$. So, $Z_z \cap]x_1, x_2[\neq \emptyset$. In a similar way, we prove the conclusion in the case of z. □

Remark 14. In Theorem 13, instead of conditions (N_7) and (N_8), we can put

(N_7') $x \in [a,b]$, $r_1, r_2 \in \mathbb{R}$, $F(r_1, 0, r_2, x) = 0 \Rightarrow \text{sign} r_1 = -\text{sign} r_2$;
(N_8') $x \in [a,b]$, $r_1, r_2 \in \mathbb{R}$, $G(r_1, r_2, 0, x) = 0 \Rightarrow \text{sign} r_1 = -\text{sign} r_2$.
Or (N_7') and (N_8), or (N_7) and (N_8').

Remark 15. For $F : \mathbb{R}^3 \times [a,b] \to \mathbb{R}$, we consider the second-order differential equation:

$$F(y'', y', y, x) = 0. \tag{7}$$

If we denote $y' := z$, then from (7), we have the following particular system of (6):

$$\begin{aligned} y' - z &= 0, \\ F(z', z, y, x) &= 0. \end{aligned} \tag{8}$$

From Theorem 11, 12, and 13, we have the following result for equation (7).

Theorem 16. *We suppose that*

$(\widetilde{N}_1)$ *the Cauchy problem for* (7) *has at most a solution;*
$(\widetilde{N}_2)$ $F(0, 0, x) = 0$, *for all* $x \in [a,b]$;
$(\widetilde{N}_3)$ $x \in [a,b]$, $r \in \mathbb{R}, F(0, 0, r, x) = 0 \Rightarrow r = 0$.

If $y \in C^2[a,b]$ *is a nontrivial solution of* (7), *then*

(1) $Z_y \cap Z_{y'} = \emptyset$, *i.e., the zeros of* y *are simple;*
(2) $Z_{y'} \cap Z_{y''} = \emptyset$, *i.e., the zeros of* y' *are simple.*

Theorem 17. *In addition to the conditions in Theorem* 16, *we suppose that*

$(\widetilde{N}_4)$ $F(\cdot, \cdot, \cdot, x) : \mathbb{R}^3 \to \mathbb{R}$ *is a homogeneous function, for all* $x \in [a,b]$;
$(\widetilde{N}_5)$ *if* y_1, y_2 *are two solutions of* (7), *then if for some* $x_0 \in [a,b]$, $y_1'(x_0) = y_2'(x_0)$ *and* $y_1''(x_0) = y_2''(x_0)$, *then* $y_1 = y_2$.

Then, if y_1, y_2 *are two linearly independent solutions of* (7), *then zeros of* y_1 *and* y_2, *and* y_1' *and* y_2' *interlace each other.*

Theorem 18. *In addition to the condition in Theorem* 16, *we suppose that*

$(\widetilde{N}_6)$ $x \in [a,b]$, $r_1, r_2 \in \mathbb{R}$, $F(r_1, 0, r_2, x) = 0 \Rightarrow \text{sign} r_1 = \text{sign} r_2$.

Then, if y is a non-trivial solution of (7), then the zeros of y and y' separate each other.

4. Applications to Some Classes of Special Functions

4.1. *Bessel functions*

Bessel functions, named after the German mathematician, physicist, astronomer, and geodesist Friedrich Bessel (1784–1846), are defined as solutions of the following Bessel differential equation:

$$x^2y''(x) + xy'(x) + (x^2 - \nu^2)y(x) = 0, \tag{9}$$

where ν is a given complex number. It is easy to see that $x = 0$ is a regular singular point.

The so-called Bessel functions of the first kind, denoted by $J_\nu(x)$, are solutions of Bessel's differential equation. The Bessel functions can be defined by the series

$$J_\nu(x) := \sum_{k=0}^{\infty} \frac{(-1)^k}{k!\Gamma(k+\nu+1)} \cdot \left(\frac{x}{2}\right)^{2k+\nu}, \tag{10}$$

where Γ denotes Euler's function.

For non-integer ν, the functions $J_\nu(x)$ and $J_{-\nu}(x)$ are linearly independent.

If ν is an integer (denoted by n), then the following relation holds:

$$J_{-n}(x) = (-1)^n J_n(x).$$

In this case, Bessel functions are also known as cylinder functions since they appear in the solution of Laplace's equation in cylindrical coordinates. If the argument x is real and the order ν is integer, the Bessel function is real, and its graph has the form of a damped vibration. Moreover, for this case (see [32]), considering the case $x > 0$, the following recurrence relations can be established:

$$\begin{cases} J'_{n-1}(x) - \frac{n-1}{x}J_{n-1}(x) + J_n(x) = 0 \\ J'_n(x) - J_{n-1}(x) + \frac{n+1}{x}J_n(x) = 0. \end{cases} \tag{11}$$

Hence, J_{n-1} and J_n are solutions of a system of differential equations of the following form:

$$\begin{cases} y'(x) + p_1(x)y(x) + q_1(x)z(x) = 0 \\ z'(x) + p_2(x)y(x) + q_2(x)z(x) = 0. \end{cases} \tag{12}$$

In this particular case, $q_1(x) = 1$ and $p_2(x) = -1$, while $p_1(x) = -\frac{n-1}{x}$ and $q_2(x) = \frac{n+1}{x}$ are C^1 on every interval $[a, b] \subset]0, +\infty[$. As a consequence, one can apply Miron Nicolescu's theorem and we have the following result.

Theorem 19. *In the above context, the zeros of J_{n-1} and J_n separate each other on any interval $[a, b] \subset]0, +\infty[$.*

4.2. *Legendre functions*

Legendre polynomials (functions) are solutions to a very important differential equation, the Legendre equation:

$$(1 - x^2)y''(x) - 2xy'(x) + n(n+1)y(x) = 0, \ x \in]-1, 1[, \tag{13}$$

where $n \in \mathbb{N}$.

The equation takes its name from the French mathematician Adrien Marie Legendre (1752–1833). Two well-known and important concepts in Mathematics, namely, Legendre polynomials and Legendre transformation, are named after him.

The Legendre functions are defined by Rodrigues formula

$$P_n(x) = \frac{1}{n!2^n} \frac{d^n}{dx^n}(x^2 - 1)^n, \ n \in \mathbb{N},$$

and have the representation

$$P_n(x) := \frac{1}{2^n} \sum_{k=0}^{[\frac{n}{2}]} \frac{(-1)^k (2n-2k)!}{k!(n-k)!(n-2k)!} \cdot x^{n-2k}.$$

The Legendre polynomial of order n satisfies the Legendre equation (13), which is related to the solution of Laplace's equation in spherical coordinates, when the method of separation of variables is used, see [3].

Moreover, the Legendre functions satisfy the following recurrence relations:

$$\begin{cases} (1-x^2)P'_{n-1}(x) - nxP_{n-1}(x) + n(x^2+x-1)P_n(x) = 0 \\ (1-x^2)P'_n(x) - nP_{n-1}(x) + nxP_n(x) = 0. \end{cases} \tag{14}$$

Hence, P_{n-1} and P_n are again solutions of a system of differential equations of the form (12). In this particular case, $q_1(x) = \frac{n(x^2+x-1)}{1-x^2}$ and $p_2(x) = -\frac{n}{1-x^2}$, while $p_1(x) = -\frac{n}{1-x^2}$ and $q_2(x) = \frac{nx}{1-x^2}$ are C^1 on every interval $[a,b] \subset]-1,1[$. As a consequence, one can apply Miron Nicolescu's theorem and we have the following result.

Theorem 20. *In the above context, the zeros of P_{n-1} and P_n separate each other on any interval $[a,b] \subset]-1,1[$.*

4.3. *Hermite functions*

Hermite polynomials (functions), named after the French mathematician Charles Hermite, satisfy the following differential equation of second order:

$$y''(x) - 2xy'(x) + 2ny(x) = 0, \ x \in \mathbb{R}, \tag{15}$$

where $n \in \mathbb{N}$.

Hermite polynomials were defined by the French mathematician Pierre-Simon Laplace in 1810 and studied in detail by the Russian mathematician Pafnuty Lvovich Chebyshev in 1859. Unfortunately, Chebyshev's work has been forgotten and they were named later after another French mathematician Charles Hermite, who wrote some works on the polynomials in 1864–1865.

The Hermite functions are defined by the Rodrigues formula

$$H_n(x) = (-1)^n e^{x^2} \frac{d^n}{dx^n} e^{-x^2}$$

and they can be also represented by

$$H_n(x) := \sum_{k=0}^{[\frac{n}{2}]} \frac{(-1)^k n!}{k!(n-2k)!} \cdot (2x)^{n-2k}.$$

Moreover, the Hermite functions satisfy the following recurrence relations:

$$\begin{cases} H'_{n-1}(x) - 2xH_{n-1}(x) + H_n(x) = 0 \\ H'_n(x) - 2nH_{n-1}(x) = 0. \end{cases} \tag{16}$$

Hence, H_{n-1} and H_n are solutions of a system of differential equations of the form (12). In this particular case, $q_1(x) = 1$ and $p_2(x) = -2n$, while $p_1(x) = -2x$ and $q_2(x) = 0$ are C^1 on every interval $[a,b] \subset \mathbb{R}$. As a consequence, one can apply Miron Nicolescu's theorem and we have the following result.

Theorem 21. *In the above context, the zeros of H_{n-1} and H_n separate each other on any interval $[a,b] \subset \mathbb{R}$.*

Remark 22. Another example that can be given in this context are the so-called parabolic cylinder functions, also named as Weber–Hermite functions, which are solutions of the differential equation

$$y''(x) + \left(n + \frac{1}{2} - \frac{x^2}{4}\right) y(x) = 0, \; x \in \mathbb{R},$$

where $n \in \mathbb{N}$.

For other details and considerations, see [1, 4, 6].

5. Zeros Separation Results in the Case of Two Second-Order Differential Equations

There are various comparison results in the theory of second-order differential equations ([5, 7, 20, 27, 29, 31, 33], ...). The results given in Section 4 give rise to the following problem (see also [28], Problem 6.3).

Problem 23. For $P_i, Q_i \in C[a,b] (i \in \{1,2\})$, we consider the following second-order differential equations:

$$y'' + P_1 y' + Q_1 y = 0, \tag{17}$$

$$z'' + P_2 z' + Q_2 z = 0. \tag{18}$$

Let $y \in C^2[a,b]$ be a non-trivial solution of (17) and $z \in C^2[a,b]$ be a non-trivial solution of (18). If $Z_y \cap Z_z = \emptyset$, in which conditions do the zeros of y and z separate each other?

From Theorem 4, we have the following result.

Theorem 24. *Let y_1 and y_2 be two linearly independent solutions of* (17) *and z_1 and z_2 be two linearly independent solutions of* (18). *If $V((y_1, z_1), (y_2, z_2))(x) \neq 0$, for every $x \in [a, b]$, then the zeros of $y := c_1 y_1 + c_2 y_2$ and $z := c_1 z_1 + c_2 z_2$ separate each other, for all $(c_1, c_2) \in \mathbb{R}^2 \backslash \{(0, 0)\}$.*

Proof. From Sturm's zero separation theorem, the zeros of y_1 and y_2, and z_1 and z_2 separate each other. Now, the conclusion follows by Theorem 4. □

Remark 25. The above theorem also follows from Miron Nicolescu's theorem, conclusion (2). Indeed, since $V((y_1, z_1), (y_2, z_2))(x) \neq 0$, for every $x \in [a, b]$, the pair (y, z) is a solution of the system

$$\begin{cases} y' \begin{vmatrix} y_1 & y_2 \\ z_1 & z_2 \end{vmatrix} + y \begin{vmatrix} z_1 & z_2 \\ y_1' & y_2' \end{vmatrix} + z \begin{vmatrix} y_1' & y_2' \\ y_1 & y_2 \end{vmatrix} = 0 \\ z' \begin{vmatrix} y_1 & y_2 \\ z_1 & z_2 \end{vmatrix} + y \begin{vmatrix} z_1 & z_2 \\ z_1' & z_2' \end{vmatrix} + z \begin{vmatrix} z_1' & z_2' \\ y_1 & y_2 \end{vmatrix} = 0. \end{cases} \tag{19}$$

Note that, in the condition of Theorem 24, this system verifies all the assumptions of Miron Nicolescu's theorem.

From Theorem 24, the following question naturally rises.

Problem 26. Let y_1 and y_2 be two linearly independent solutions of (17) and z_1 and z_2 be two linearly independent solutions of (18). In which conditions do we have that

$$V((y_1, z_1), (y_2, z_2))(x) \neq 0, \text{ for every } x \in [a, b]?$$

In this respect, the following example is illustrative. Consider the following equations:

$$y'' + y = 0, \tag{20}$$

$$z'' - 2z' + 2z = 0. \tag{21}$$

In this case, the solutions are

$$y(x) = c_1 \sin x + c_2 \cos x, x \in [a, b],$$

$$z(x) = c_1 e^x \sin\left(x + \frac{\pi}{4}\right) + c_2 e^x \cos\left(x + \frac{\pi}{4}\right), x \in [a, b].$$

Thus, the zeros of y and z separate each other. Indeed, we remark that

$$\begin{vmatrix} \sin x & \cos x \\ e^x \sin(x+\frac{\pi}{4}) & e^x \cos(x+\frac{\pi}{4}) \end{vmatrix} \neq 0. \text{ for any } x \in [a,b].$$

Moreover, this example suggests the following open problem.

Problem 27. Let $S_i \subset C^2[a,b]$ be the solution set of the equation $(5.i)$, for $i \in \{1,2\}$. The problem is to find a bijective linear operator $L : S_1 \to S_2$ such that the zeros of y and $L[y]$ separate each other, for all $y \in S_1$.

For example, if there exists a base $\{y_1, y_2\}$ of S_1 such that

$$V((y_1, L[y_1]), (y_2, L[y_2]))(x) \neq 0, \text{ for every } x \in [a,b],$$

then, for all $y \in S_1 \backslash \{0\}$, the zeros of y and $L[y]$ separate each other.

References

[1] Agarwal, R.P. and O'Regan, D. *Ordinary and Partial Differential Equations. with Special Functions, Fourier Series, and Boundary Value Problems*. Springer, New York, p. 410.

[2] Barrett, L.C. and Bendixen, G.E. Separation and interlacing theorems. *Quart. Appl. Math.* 23, 69–78 (1965).

[3] Beals, R. and Wong, R. *Special Functions*. Cambridge University Press, p. 456 (2010).

[4] Beals, R. and Wong, R. *Special functions and orthogonal polynomials*. Cambridge University Press, p. 474 (2016).

[5] Bihary, I. Extension of certain theorems of the Sturmian type to non-linear second order differential equations. *Math. Inst. Hung. Acad. Sci.* 3, 13–20 (1958).

[6] Brychkov, Y.A. *Handbook of Special Functions. Derivatives, Integrals, Series and Other Formulas*. Boca Raton, FL: Chapman & Hall CRC, p. 680 (2008).

[7] Butlewski, Z. Sur les zéros des intégrales réelles des équations différentielles linéaires. *Mathematica* 17, 85–110 (1941).

[8] Cuculescu, I. Zeros separation of a pair of solutions for a system of differential equations. *Gazeta Matematică şi Fizică (Bucureşti)* 5, 194–195 (1953).

[9] Dimitrov, D.K. A late report on interlacing of zeros of polynomials, in: Nikolov, G. and Uluchev, R. (eds.), *Constructive Theory of Functions.* Prof. Marin Drinov Academic Publishing House, 69–79 (2012).

[10] Foiaş, C., Gussi, G. and Poenaru, V. Sur le problème polylocal pour les équations différentielles linéaires du second ordre. (in Romanian) *Acad. R. P. Romîne. Bul. Şti. Secţ. Şti. Mat. Fiz.* 7, 699–721 (1955).

[11] Ilea, V., Otrocol, D. and Rus, I.A. Some properties of solutions of the homogeneous nonlinear second order differential equations. *Mathematica,* 57(80), (1–2), 38–43 (2015).

[12] Kamke, E. Bemerkungen zu einigen Trennungssätzen von M. Nicolesco und S. Takahashi für die Nullstellen der Lösungen von Differentialgleichungen. *Mathematica, Cluj* 15, 201–203 (1939).

[13] Konstantinesku, F. (Constantinescu, F.). A new proof of a theorem of V.A. Markov using a theorem of Sturm. *Uspehi Mat. Nauk (N.S.),* 12(6), (78), 147–148 (1957).

[14] Krasnoselskii, M.A., Perov, A.I., Povolotskiy, A.I. and Zabrejko, P.P. *Plane Vector Fields.* Academic Press New York and London, p. 42 (1966).

[15] Marx, I. On the structure of recurrence relations. *Michigan Math. J.* 2, 45–50 (1954).

[16] Marx, I. On the structure of recurrence relations. II. *Michigan Math. J.* 2, 99–103 (1954).

[17] Montel, P. Sur les fractions ratilonnelles à termes entrelacés. *Mathematica,* 5, 110–129 (1931).

[18] Nicolesco, M. Sur le théorème de Sturm. *Mathematica,* 1, 111–114 (1929).

[19] Pinkus, A. and Ziegler, Z. Interlacing properties of the zeros of the error functions in best L^p-approximations. *J. Approx. Theory,* 27(1), 1–18 (1979).

[20] Reid, W.T. *Sturmian Theory for Ordinary Differential Equations.* Springer-Verlag, New York-Heidelberg-Berlin, p. 559 (1980).

[21] Rus, I.A. Sturm type theorems. (in Romanian) *Studia Univ. Babeş-Bolyai Math.* 1, 131–136 (1961).

[22] Rus, I.A. On a Sturm type thorem. (in Romanian) *Studia Univ. Babeş-Bolyai Math.* 2, 33–36 (1962).

[23] Rus, I.A. On the roots of the component solutions for a system of two differential equations of first order. (in Romanian) *Studii şi Cercetări de Matematică (Cluj)* 14, 151–156 (1963).

[24] Rus, I.A. Families of functions with Sturm's property. (in Romanian) *Studia Univ. Babeş-Bolyai Math.* 6, 37–40 (1966).

[25] Rus, I.A. Some properties of the zeros of the component solutions for a system of two nonlinear differential equations of first order. (in Romanian) *Studii şi Cercetări Matematice (Bucureşti).* 18, 1549–1553 (1966).

[26] Rus, I.A. Separations theorems for the zeros of some real functions. *Mathematica,* 27, 43–46 (1985).

[27] Rus, I.A. *Differential Equations, Integral Equations and Dynamical Systems.* Transilvania Press Cluj-Napoca, p. 229 (1996).

[28] Rus, I.A. "Homogeneous" second order differential equation: zeros separation principles. (English) Zbl 1438.34122 *Stud. Univ. Babeş-Bolyai Math.* 63, 245–256 (2018).

[29] Sansone, G. *Equazioni differenziali nel campo reale. I, II.* (in Italian) Nicola Zanichelli Bologna, p. 375 & p. 400 (1940 & 1941).

[30] Segura, J. The zeros of special functions from a fixed point method. *SIAM J. Numer. Anal.* 40, 114–133 (2002).

[31] Swanson, C.A. *Comparison and Oscillation Theory of Linear Differential Equations.* New York - London, Academic Press, p. 227 (1968).

[32] Wang, Z.X. and Guo, D.R. *Special Functions.* World Scientific, Farrer Road, Singapore (1989).

[33] Willett, D. Classification of second order linear differential equations with respect to oscillation. *Advances in Mathematics* 3, 594–623 (1969).

https://doi.org/10.1142/9789811293009_0035

Chapter 35

Minimization of the Free Energy with an Expectation Constraint

Iosif Pinelis

Michigan Technological University,
Department of Mathematical Sciences,
Houghton, MI 49931 USA
ipinelis@mtu.edu

The Gibbs lower bound on the free energy is refined in the presence of an expectation constraint.

Let $(\Omega, \mathcal{F})$ be a measurable space. Let P and Q be probability measures defined on $\mathcal{F}$. The relative entropy of P with respect to Q is given by the formula

$$\mathcal{H}(P|Q) := \int_\Omega h\left(\frac{dP}{dQ}\right) dQ$$

if P is absolutely continuous with respect to Q, where $\frac{dP}{dQ}$ is the Radon–Nikodym derivative of P with respect to Q and

$$h(u) := u \ln u \text{ for real } u > 0, \text{ with } h(0) := 0.$$

If P is not absolutely continuous with respect to Q, then $\mathcal{H}(P|Q) := \infty$. It is well known that $\mathcal{H}(P|Q)$ may take only non-negative values. One may have $\mathcal{H}(P|Q) = \infty$ even when P is absolutely continuous with respect to Q. It is also well known that

$\mathcal{H}(P|Q)$ is a convex function of the pair (P, Q); see, e.g., [1, Proposition 15.5(d)]. Another name for the relative entropy is the Kullback–Leibler divergence/information [2].

Let now μ be any measure on $\mathcal{F}$ such that both P and Q are absolutely continuous with respect to μ. Such a measure μ always exists. For instance, one may take $\mu := P + Q$. Let

$$p := \frac{dP}{d\mu} \quad \text{and} \quad q := \frac{dQ}{d\mu}.$$

Then,

$$\mathcal{H}(P|Q) = \mathcal{H}(p||q) := \int_\Omega h\left(\frac{p}{q}\right) q\, d\mu \tag{1}$$

if $p = 0$ Q-almost everywhere on the set $\{\omega \in \Omega\colon q(\omega) = 0\}$, with $\mathcal{H}(p||q) := \infty$ otherwise. The expression for $\mathcal{H}(p||q)$ in (1) will hold in all cases assuming the following conventions:

$$\frac{p}{q} := \infty \text{ and } h\left(\frac{p}{q}\right) q := \infty \text{ on the set } \{\omega \in \Omega\colon q(\omega) = 0 < p(\omega)\}, \tag{2}$$

which will be indeed assumed henceforth. Also, on the set $\{\omega \in \Omega\colon q(\omega) = 0 = p(\omega)\}$, the values of $\frac{p}{q}$ and $h(\frac{p}{q})\, q$ will be chosen in whatever way we find suitable for us.

From now on, consider the measure μ and the density q fixed.

Let $H\colon \Omega \to \mathbb{R}$ be a P-integrable measurable function, which in this context is called a Hamiltonian. Then,

$$E(p) := \int_\Omega Hp\, d\mu = \int_\Omega H\, dP$$

is the energy and

$$V(p) := E(p) + \mathcal{H}(p||q) = \int_\Omega \left(Hp + h\left(\frac{p}{q}\right) q\right) d\mu \tag{3}$$

is the free energy corresponding to the (probability measure P with) density p, relative to density q.

Let $A\colon \Omega \to \mathbb{R}$ be another P-integrable measurable function, with

$$a_- := (Q)\operatorname*{essinf}_{\omega\in\Omega} A(\omega) \in [-\infty,\infty) \quad \text{and}$$

$$a_+ := (Q)\operatorname*{esssup}_{\omega\in\Omega} A(\omega) \in (-\infty,\infty],$$

where $(Q)\,\text{essinf}$ and $(Q)\,\text{esssup}$ denote, respectively, the essential infimum and the essential supremum with respect to the probability measure Q. Assume that

$$a_- < a_+ \tag{4}$$

and take any

$$a_* \in (a_-, a_+). \tag{5}$$

Consider the problem of minimization of the free energy $V(p)$ in p given the expectation constraint

$$\int_\Omega A\,p\,d\mu = a_*. \tag{6}$$

Note that, for any probability density p (with respect to μ), the value of $\int_\Omega A\,p\,d\mu$ must be in the interval $[a_-, a_+]$. So, the only feasible values of $\int_\Omega A\,p\,d\mu$ that condition (6) may exclude are the endpoints of the interval $[a_-, a_+]$, containing all feasible values of $\int_\Omega A\,p\,d\mu$. In particular, a_* must be finite. Note also that, in the extreme case when $\int_\Omega A\,p\,d\mu = a_+$ and $a_+ < \infty$, the function A must be constant P-almost everywhere. Similarly, when $\int_\Omega A\,p\,d\mu = a_-$ and $a_- > -\infty$, the function A must be constant P-almost everywhere.

We shall assume that for all real c

$$Z(c) := \int_\Omega e^{cA-H} q\,d\mu < \infty. \tag{7}$$

In the absence of restriction (6), $Z(0)$ is called the partition function; it is a function of the temperature, which is not explicitly present in our setting, due to a suitable choice of physically relevant units.

Remark 1. The absence of restriction (6) can be formally considered as the presence of (6) with $A = 0$ and $a_* = 0$. In this case, we have $Z(c) = Z(0)$ for all real c.

In general, one may refer to $Z(c)$ as the tilted (by A) partition function; cf., e.g., [3, §23.4].

Again for each real c, let

$$p_c := \frac{e^{cA-H}q}{Z(c)}, \tag{8}$$

which is clearly the density (with respect to μ) of a certain probability measure P_c.

Note that $Z(c) > 0$ for all real c and let

$$L(c) := \ln Z(c). \tag{9}$$

Lemma 2. *The function L is continuously differentiable, with*

$$L'(c) = \frac{\int_\Omega Ae^{cA-H}q\,d\mu}{\int_\Omega e^{cA-H}q\,d\mu} \tag{10}$$

for all real c. Moreover, the derivative L' of L is strictly and continuously increasing on $\mathbb{R}$ from a_- to a_+.

Lemma 2 is essentially well known in the large deviation theory; see, e.g., [4, Theorem 6.1]. For readers' convenience, the following is the proof:

Proof of Lemma 2. Using the dominated convergence theorem, one can differentiate the integral in (7) with respect to the parameter c to get

$$Z'(c) = \int_\Omega Ae^{cA-H}q\,d\mu.$$

So, (10) follows from (9) and (7).

Similarly,

$$\begin{aligned} L''(c) &= \frac{\int_\Omega A^2e^{cA-H}q\,d\mu}{\int_\Omega e^{cA-H}q\,d\mu} - \left(\frac{\int_\Omega Ae^{cA-H}q\,d\mu}{\int_\Omega e^{cA-H}q\,d\mu}\right)^2 \\ &= \mathsf{E}_c\,A^2 - (\mathsf{E}_c\,A)^2 = \mathsf{Var}_c\,A > 0, \end{aligned} \tag{11}$$

where E_c and Var_c are, respectively, the expectation and variance with respect to the probability measure P_c. The inequality in (11) is strict because of assumption (4). So, indeed L' is strictly and continuously increasing on $\mathbb{R}$.

To complete the proof of Lemma 2, it remains to show that $L'(\infty-) := \lim_{c\to\infty} L'(c) = a_+$ and $L'(-\infty+) := \lim_{c\to-\infty} L'(c) = a_-$. To do this, take any $a \in (a_-, a_+)$. Then, by (10), for real c,

$$(L'(c) - a)e^{-ca} \int_\Omega e^{cA-H} q\,d\mu$$
$$= \int_\Omega (A - a)e^{c(A-a)-H} q\,d\mu = I_-(c) + I_+(c),$$

where

$$I_-(c) := \int_\Omega (A - a)\,1(A < a)\,e^{c(A-a)-H} q\,d\mu \underset{c\to\infty}{\longrightarrow} 0$$

by dominated convergence (which follows by condition (7)) and

$$I_+(c) := \int_\Omega (A - a)\,1(A > a)\,e^{c(A-a)-H} q\,d\mu \underset{c\to\infty}{\longrightarrow} \infty$$

by (say) Fatou's lemma and condition $a < a_+$. So, for any $a \in (a_-, a_+)$ and all large enough real c, we will have $L'(c) - a > 0$. On the other hand, it is clear from (10) that $L'(c) \le a_+$ for all real c. So, $L'(\infty-) = a_+$. Similarly, $L'(-\infty+) = a_-$. This completes the proof of Lemma 2.

It follows from condition (5) and Lemma 2 that there exists a unique real c_* such that

$$L'(c_*) = a_*.$$

Let

$$p_* := p_{c_*},$$

with p_c defined by formula (8).

Note that p_* satisfies restriction (6); that is, (6) holds with p_* in place of p.

Note that the function h is convex and $h'(u) = 1 + \ln u$ for real $u > 0$. So, for any probability density p (with respect to μ),

$$\begin{aligned} h\left(\frac{p}{q}\right) q \ge \tilde{h}(p) :=& h\left(\frac{p_*}{q}\right) q + \left(1 + \ln\frac{p_*}{q}\right)(p - p_*) \\ =& h\left(\frac{p_*}{q}\right) q + (1 + c_*A - H - L(c_*))(p - p_*), \end{aligned} \tag{12}$$

by (8) and (9).

Note that the function h is bounded from below (by $1/e$).

Assume that the Hamiltonian function H is bounded from below as well.

Then, in view of (3), for any probability density p, the free energy $V(p)$ is well defined (even though may equal ∞). Moreover, in view of (3) and (8),

$$V(p_*) = c_* L'(c_*) - L(c_*) \in \mathbb{R}.$$

Therefore, and in view of (12), we may write

$$\begin{aligned} V(p) - V(p_*) &= \int_\Omega d\mu \left(H(p - p_*) + h\left(\frac{p}{q}\right) q - h\left(\frac{p_*}{q}\right) q \right) \\ &\geq \int_\Omega d\mu \left(H(p - p_*) + \tilde{h}(p) - h\left(\frac{p_*}{q}\right) q \right) \\ &= \int_\Omega d\mu \,(p - p_*)(1 + H + c_* A - H - L(c_*)) \\ &= \int_\Omega d\mu \,(p - p_*)(1 - L(c_*)) + c_* \int_\Omega d\mu \,(p - p_*) A \\ &= 0 + 0 = 0 \end{aligned} \tag{13}$$

if the probability density p satisfies condition (6).

Thus, we have proved the following theorem.

Theorem 3. *Suppose that the Hamiltonian function H is bounded from below and condition* (7) *holds for all real c. Let a_* be as in* (5). *Then for all probability densities p satisfying restriction* (6), *we have*

$$V(p) \geq V(p_*) = c_* L'(c_*) - L(c_*). \tag{14}$$

So, p_* minimizes the free energy $V(p)$ among all probability densities p satisfying restriction (6).

The corresponding inequality

$$V(p) \geq -L(0), \tag{15}$$

which holds for all probability densities p, without restriction (6), is the famous Gibbs inequality; see, e.g., [1, Chapter 15]. One may note that the Gibbs "unrestricted" lower bound $-L(0)$ on the free energy $V(p)$ in (15) equals the value of the "restricted" lower bound $c_* L'(c_*) - L(c_*)$ in (14) at $c_* = 0$. This observation is in agreement with Remark 1.

References

[1] Georgii, H.-O. *Gibbs measures and phase transitions. De Gruyter Studies in Mathematics*, Walter de Gruyter & Co., Berlin, 9, (2011), 2nd edition. https://doi-org.services.lib.mtu.edu/10.1515/9783110250329.

[2] Kullback, S. and Leibler, R.A. On information and sufficiency, *Ann. Math. Statistics.* 22, 79–86 (1951). https://doi-org.services.lib.mtu.edu/10.1214/aoms/1177729694.

[3] Klenke, A. *Probability theory—a comprehensive course.* Universitext, (Springer, Cham, [2020] ©2020). https://doi-org.services.lib.mtu.edu/10.1007/978-3-030-56402-5. Third edition [of 2372119].

[4] Daniels, H.E. Saddlepoint approximations in statistics, *Ann. Math. Statist.* 25, 631–650 (1954). https://doi-org.services.lib.mtu.edu/10.1214/aoms/1177728652.

https://doi.org/10.1142/9789811293009_0036

Chapter 36

An Extended Hardy–Hilbert's Integral Inequality with Internal Variables Involving Two Upper-Limit Functions

Michael Th. Rassias[*,†,‡,¶,||] and Bicheng Yang[§,**,††]

[*] *Institute of Mathematics, University of Zurich, CH-8057, Zurich, Switzerland*

[†] *Institute of Physics and Technology, 141700 Dolgoprudny, Institutskiy per, d. 9, Moscow, Russia*

[‡] *Institute for Advanced Study, Program in Interdisciplinary Studies, 1 Einstein Dr, Princeton, NJ 08540, USA*

[§] *School of Mathematics, Guangdong University of Education, Guangzhou, Guangdong 510303, P.R. China*

[¶] *mthrassias@yahoo.com*

[||] *michail.rassias@math.uzh.ch*

[**] *bcyang@gdei.edu.cn*

[††] *bcyang818@163.com*

By means of the weight functions, the idea of introduced parameters as well as internal variables and the technique of real analysis, a new extended Hardy–Hilbert's integral inequality with the homogeneous kernel as $\frac{1}{(U(x)+V(y))^{\lambda}}$ ($\lambda > 0$) involving two upper-limit functions is obtained. The equivalent statements of the best possible constant factor related to several parameters are considered. Some particular inequalities and the case of reverses are provided.

1. Introduction

Assuming that $p > 1, \frac{1}{p} + \frac{1}{q} = 1, a_m, b_n \geq 0, 0 < \sum_{m=1}^{\infty} a_m^p < \infty$, and $0 < \sum_{n=1}^{\infty} b_n^q < \infty$, the following Hardy–Hilbert's inequality with the best possible constant factor $\pi/\sin(\frac{\pi}{p})$ was provided (cf. [1], Theorem 315):

$$\sum_{n=1}^{\infty}\sum_{m=1}^{\infty} \frac{a_m b_n}{m+n} < \frac{\pi}{\sin(\pi/p)} \left(\sum_{m=1}^{\infty} a_m^p\right)^{\frac{1}{p}} \left(\sum_{n=1}^{\infty} b_n^q\right)^{\frac{1}{q}}. \tag{1}$$

If $f(x), g(y) \geq 0, 0 < \int_0^{\infty} f(x)dx < \infty$, and $0 < \int_0^{\infty} g(y)dy < \infty$, then we still have the integral analogue of (1) named in Hardy–Hilbert's integral inequality as follows (cf. [1], Theorem 316):

$$\int_0^{\infty}\int_0^{\infty} \frac{f(x)g(y)}{x+y} dxdy < \frac{\pi}{\sin(\pi/p)} \left(\int_0^{\infty} f^p(x)dx\right)^{\frac{1}{p}} \left(\int_0^{\infty} g^q(y)dy\right)^{\frac{1}{q}}, \tag{2}$$

where the same constant factor $\pi/\sin(\frac{\pi}{p})$ is still the best possible. Inequalities (1) and (2) played an important role in analysis and its applications (cf. [2–13]).

In 2006, by using Euler–Maclaurin summation formula, Krnic *et al.* [14] gave an extension of (1) with the kernel as $\frac{1}{(m+n)^{\lambda}}(0 < \lambda \leq 4)$. Applying the result of [14], in 2019, Adiyasuren *et al.* [15] considered an extension of (1) involving partial sums, and then in 2020, Mo *et al.* [16] gave an extension of (2) involving two upper-limit functions. In 2016–2017, Hong *et al.* [17, 18] provided several equivalent statements of the extensions of (1) and (2) related to multi-parameters and the best possible constant factors. Some further results are given in [19]–[41].

In this chapter, following the way of [15] and [17], by means of the weight functions, the idea of introduced parameters as well as internal variables, and the techniques of real analysis, a new extended Hardy–Hilbert-type integral inequality with the kernel as $\frac{1}{(V(x)+V(y))^{\lambda}}$ $(\lambda > 0)$ involving two upper-limit functions is provided, which is an extension of the result in [16]. The equivalent statements of the best possible constant factor related to several parameters are considered. A few particular inequalities and the case of reverses are obtained.

2. Some Lemmas

In what follows, we suppose that $p > 0$ $(p \neq 1)$, $\frac{1}{p}+\frac{1}{q} = 1, 0 < \lambda_i < \lambda$ $(i = 1, 2), \widehat{\lambda}_1 := \frac{\lambda-\lambda_2}{p} + \frac{\lambda_1}{q}$, $\widehat{\lambda}_2 := \frac{\lambda-\lambda_1}{q} + \frac{\lambda_2}{p}$, $\alpha, \beta \in \mathbf{R} = (-\infty, \infty)$, $U(x)$ $(x > \alpha)$ and $V(y)$ $(y > \beta)$ are positive functions with $U'(x) > 0, V'(y) > 0$, $U(\alpha^+) = V(\beta^+) = 0$, and $U(\infty) = V(\infty) = \infty$, $f(x)$ $(x > \alpha)$ and $g(y)$ $(y > \beta)$ are non-negative Lebesgue integrable functions, two upper-limit functions $F(x) := \int_{\alpha}^{x} f(t)dt$ $(x \in [\alpha, \infty))$ and $G(y) := \int_{\beta}^{y} g(t)dt$ $(y \in [\beta, \infty))$, such that

$$F(x) = o(e^{tU(x)}), \quad G(y) = o(e^{tV(y)}) \quad (t > 0; x, y \to \infty).$$

We also assume that

$$0 < \int_{\alpha}^{\infty} \frac{U'(x)}{(U(x))^{p\widehat{\lambda}_1+1}} F^p(x)dx < \infty \quad \text{and}$$

$$0 < \int_{\beta}^{\infty} \frac{V'(y)}{(V(y))^{q\widehat{\lambda}_2+1}} G^q(y)dy < \infty.$$

We indicate the gamma function as follows:

$$\Gamma(\gamma) := \int_{0}^{\infty} e^{-t}t^{\gamma-1}dt(\gamma > 0), \tag{3}$$

satisfying $\Gamma(\gamma + 1) = \gamma\Gamma(\gamma)$, and define the following beta function (cf. [42]):

$$B(u, v) := \int_{0}^{\infty} \frac{t^{u-1}dt}{(1 + t)^{u+v}} = \frac{\Gamma(u)\Gamma(v)}{\Gamma(u + v)}(u, v > 0). \tag{4}$$

By (3), for $\lambda > 0, x > \alpha, y > \beta$, we find the following expression:

$$\frac{1}{(U(x)+V(y))^{\lambda}} = \frac{1}{\Gamma(\lambda)} \int_0^{\infty} e^{-(U(x)+V(y))t} t^{\lambda-1} dt. \tag{5}$$

Lemma 1. *For $t > 0$, we have the following formulas:*

$$\int_{\alpha}^{\infty} e^{-tU(x)} f(x)dx = t \int_{\alpha}^{\infty} e^{-tU(x)} U'(x)F(x)dx, \tag{6}$$

$$\int_{\beta}^{\infty} e^{-tV(y)} g(y)dy = t \int_{\beta}^{\infty} e^{-tV(y)} V'(y)G(y)dy. \tag{7}$$

Proof. Since $F(\alpha) = 0$, $F(x) = o(e^{tU(x)})$ $(t > 0)$, integration by parts, we obtain

$$\int_{\alpha}^{\infty} e^{-tU(x)} f(x)dx = \int_{\alpha}^{\infty} e^{-tU(x)} dF(x) = e^{-tU(x)} F(x)|_{\alpha}^{\infty}$$

$$-\int_{\alpha}^{\infty} F(x)de^{-tU(x)} = t \int_{\alpha}^{\infty} e^{-tx} U'(x)F(x)dx.$$

Hence, we have (6). In the same way, we obtain (7). □

The lemma is proved.

Lemma 2. *For $0 < s_1 < s$, we set $k_s(s_1) := B(s_1, s - s_1)$. Assuming that $x \in (\alpha, \infty)$,$y \in (\beta, \infty)$, we define the following weight functions:*

$$\varpi(\lambda_2, x) := (U(x))^{\lambda+1-\lambda_2} \int_{\beta}^{\infty} \frac{(V(t))^{\lambda_2} V'(t)}{(U(x)+V(t))^{\lambda+2}} dt, \tag{8}$$

$$\omega(\lambda_1, y) := (V(y))^{\lambda+1-\lambda_1} \int_{\alpha}^{\infty} \frac{(U(t))^{\lambda_1} U'(t)}{(U(t)+V(y))^{\lambda+2}} dt. \tag{9}$$

We have the following expressions:

$$\varpi(\lambda_2, x) = k_{\lambda+2}(\lambda_2 + 1)(x \in (\alpha, \infty)), \tag{10}$$

$$\omega(\lambda_1, y) = k_{\lambda+2}(\lambda_1 + 1)(y \in (\beta, \infty)). \tag{11}$$

Proof. For fixed $x > \alpha$, setting $u = \frac{V(t)}{U(x)}$, we find $U(x)du = V'(t)dt$ and

$$\begin{aligned}\varpi(\lambda_2, x) &= (U(x))^{\lambda+1-\lambda_2} \int_0^\infty \frac{(uU(x))^{\lambda_2} U(x)}{(U(x) + uU(x))^{\lambda+2}} du \\ &= \int_0^\infty \frac{u^{\lambda_2}}{(1+u)^{\lambda+2}} du = k_{\lambda+2}(\lambda_2 + 1),\end{aligned}$$

namely, (10) follows. In the same way, we obtain (11). □

The lemma is proved.

Lemma 3. (i) *For $p > 1$ $(q > 1)$, we have the following extended Hardy–Hilbert's integral inequality:*

$$\begin{aligned}I_{\lambda+2} &:= \int_\beta^\infty \int_\alpha^\infty \frac{(U'(x)F(x))(V'(y)G(y))}{(U(x) + V(y))^{\lambda+2}} dxdy \\ &< k_{\lambda+2}^{\frac{1}{p}}(\lambda_2 + 1) B_{\lambda+2}^{\frac{1}{q}}(\lambda_1 + 1) \left[\int_\alpha^\infty \frac{U'(x)F^p(x)}{(U(x))^{p\widehat{\lambda}_1+1}} dx\right]^{\frac{1}{p}} \\ &\quad \times \left[\int_\beta^\infty \frac{V'(y)G^q(y)}{(V(y))^{q\widehat{\lambda}_2+1}} dy\right]^{\frac{1}{2}};\end{aligned} \tag{12}$$

(ii) *for $0 < p < 1$ $(q < 0)$, we have the reverse of* (12).

Proof. (i) By Hölder's inequality (cf. [43]), we obtain

$$\begin{aligned}I_{\lambda+2} &= \int_\beta^\infty \int_\alpha^\infty \frac{1}{(U(x) + V(y))^{\lambda+2}} \left[\frac{(V(y))^{\lambda_2/p}(V'(y))^{1/p}}{(U(x))^{\lambda_1/q}(U'(x))^{1/q}} U'(x)F(x)\right] \\ &\quad \times \left[\frac{(U(x))^{\lambda_1/q}(U'(x))^{1/q}}{(V(y))^{\lambda_2/p}(V'(y))^{1/p}} V'(y)G(y)\right] dxdy \\ &\le \left\{\int_\alpha^\infty \left[\int_\beta^\infty \frac{(U'(x))^{1-p}V'(y)}{(U(x) + V(y))^{\lambda+2}} \frac{(V(y))^{\lambda_2} dy}{(U(x))^{\lambda_1(p-1)}}\right] (U'(x)F(x))^p dx\right\}^{\frac{1}{p}} \\ &\quad \times \left\{\int_\beta^\infty \left[\int_\alpha^\infty \frac{(V'(y))^{1-q}U'(x)}{(U(x) + V(y))^{\lambda+2}} \frac{(U(x))^{\lambda_1} dx}{(V(y))^{\lambda_2(q-1)}}\right] (V'(y)G(y))^q dy\right\}^{\frac{1}{q}} \\ &= \left[\int_\alpha^\infty \varpi(\lambda_2, x) \frac{U'(x)F^p(x)}{(U(x))^{p\widehat{\lambda}_1+1}} dx\right]^{\frac{1}{p}} \left[\int_\beta^\infty \omega(\lambda_1, y) \frac{V'(y)G^q(y)}{(V(y))^{q\widehat{\lambda}_2+1}} dy\right]^{\frac{1}{q}}.\end{aligned} \tag{13}$$

□

If (13) protains the form of equality, then there exist constants A and B such that they are not both zero, satisfying

$$A\frac{(V(y))^{\lambda_2}V'(y)(U'(x)F(x))^p}{(U(x))^{\lambda_1(p-1)}(U'(x))^{p-1}} = B\frac{(U(x))^{\lambda_1}U'(x)(V'(y)G(y))^q}{(V(y))^{\lambda_2(q-1)}(V'(y))^{q-1}}$$

$$a.e. \text{ in } (\alpha,\infty)\times(\beta,\infty).$$

Assuming that $A \neq 0$, for fixed $a, e, y \in (\beta, \infty)$, we have

$$\frac{U'(x)F^p(x)}{(U(x))^{p\widehat{\lambda}_1+1}} = \frac{BG^q(y)U'(x)}{A(V(y))^{q\lambda_2}}(U(x))^{-1-b} \quad a.e. \text{ in } (\alpha,\infty),$$

where $b := \lambda - \lambda_1 - \lambda_2 \in \mathbf{R}$, and $\int_\alpha^\infty (U(x))^{-1-b}U'(x)dx = \infty$. Then the above expression contradicts the fact that

$$0 < \int_\alpha^\infty \frac{U'(x)}{(U(x))^{p\widehat{\lambda}_1+1}}F^p(x)dx < \infty.$$

Therefore, by (10), (11), and (13), we have (12).

(ii) By the reverse Hölder's inequality and in the same way, we obtain the reverse of (12).

The lemma is proved.

3. Main Results

Theorem 4. (i) *For $p > 1$, we have the following Hardy–Hilbert-type integral inequality involving two upper-limit functions:*

$$\begin{aligned} I &:= \int_\beta^\infty \int_\alpha^\infty \frac{f(x)g(y)}{(U(x)+V(y))^\lambda}dxdy \\ &< \frac{\Gamma(\lambda+2)}{\Gamma(\lambda)}k_{\lambda+2}^{\frac{1}{p}}(\lambda_2+1)B_{\lambda+2}^{\frac{1}{q}}(\lambda_1+1) \\ &\times \left[\int_\alpha^\infty \frac{U'(x)F^p(x)}{(U(x))^{p\widehat{\lambda}_1+1}}dx\right]^{\frac{1}{p}} \left[\int_\beta^\infty \frac{V'(y)G^q(y)}{(V(y))^{q\widehat{\lambda}_2+1}}dy\right]^{\frac{1}{q}}; \end{aligned} \tag{14}$$

(ii) *for $0 < p < 1$, we have the reverse of (14).*

In particular, for $p > 1$, $\lambda_1 + \lambda_2 = \lambda$, we reduce (14) to the following:

$$I = \int_\beta^\infty \int_\alpha^\infty \frac{f(x)g(y)dxdy}{(U(x)+V(y))^\lambda} < \lambda_1\lambda_2 B(\lambda_1,\lambda_2) \times \left[\int_\alpha^\infty \frac{U'(x)F^p(x)}{(U(x))^{p\lambda_1+1}}dx\right]^{\frac{1}{p}} \left[\int_\beta^\infty \frac{V'(y)G^q(y)}{(V(y))^{q\lambda_2+1}}dy\right]^{\frac{1}{q}}, \quad (15)$$

where the constant factor $\lambda_1\lambda_2 B(\lambda_1,\lambda_2)$ is the best possible.

Proof. (i) In view of (7), (8), and Fubini theorem (cf. [44]), we find

$$\begin{aligned}
I &= \frac{1}{\Gamma(\lambda)} \int_\beta^\infty \int_\alpha^\infty f(x)g(y)\left[\int_0^\infty t^{\lambda-1}e^{-(U(x)+V(y))t}dt\right]dxdy \\
&= \frac{1}{\Gamma(\lambda)} \int_0^\infty t^{\lambda-1}\left(\int_\alpha^\infty e^{-U(x)t}f(x)dx\right)\left(\int_\beta^\infty e^{-V(y)t}g(y)dy\right)dt \\
&= \frac{1}{\Gamma(\lambda)} \int_0^\infty t^{\lambda-1}\left(t\int_\alpha^\infty e^{-U(x)t}U'(x)F(x)dx\right) \\
&\quad \times \left(t\int_\beta^\infty e^{-V(y)t}V'(y)G(y)dy\right)dt \\
&= \frac{1}{\Gamma(\lambda)} \int_\beta^\infty \int_\alpha^\infty U'(x)F(x)V'(y)G(y) \\
&\quad \times \left[\int_0^\infty t^{\lambda+1}e^{-(U(x)+V(y))t}dt\right]dxdy \\
&= \frac{\Gamma(\lambda+2)}{\Gamma(\lambda)} \int_\beta^\infty \int_\alpha^\infty \frac{(U'(x)F(x))(V'(y)G(y))}{(U(x)+V(y))^{\lambda+2}}dxdy \\
&= \frac{\Gamma(\lambda+2)}{\Gamma(\lambda)} I_{\lambda+2}. \qquad (16)
\end{aligned}$$

Then by (12), we have (14).

(ii) In the same way, by the reverse of (12), we have the reverse of (14). □

For $p > 1, \lambda_1 + \lambda_2 = \lambda$ in (14), by simplification, we have (15).

For any $0 < \varepsilon < \min\{p\lambda_1, q\lambda_2\}$, we set the following functions:

$$\widetilde{f}(x) := \begin{cases} 0, \alpha < x < \alpha + 1, \\ (U(x))^{\lambda_1 - \frac{\varepsilon}{p} - 1} U'(x), x \geq \alpha + 1 \end{cases},$$

$$\widetilde{g}(y) := \begin{cases} 0, \beta < y \leq \beta + 1, \\ (V(y))^{\lambda_2 - \frac{\varepsilon}{q} - 1} V'(y), y > \beta + 1 \end{cases},$$

$$\widetilde{F}(x) = \int_{\alpha}^{x} \widetilde{f}(t)dt, \widetilde{G}(y) = \int_{\beta}^{y} \widetilde{g}(t)dt.$$

Then it follows that $\widetilde{F}(x) = 0$ $(\alpha \leq x < \alpha + 1)$, $\widetilde{G}(y) = 0$ $(\beta \leq y < \beta + 1)$,

$$\begin{aligned} \widetilde{F}(x) &= \int_{\alpha+1}^{x} (U(t))^{\lambda_1 - \frac{\varepsilon}{p} - 1} U'(t)dt \\ &= \frac{1}{\lambda_1 - \frac{\varepsilon}{p}} [(U(x))^{\lambda_1 - \frac{\varepsilon}{p}} - (U(\alpha+1))^{\lambda_1 - \frac{\varepsilon}{p}}] \ (x \geq \alpha + 1), \\ \widetilde{G}(y) &= \int_{\beta+1}^{y} (V(t))^{\lambda_2 - \frac{\varepsilon}{q} - 1} V'(t)dt \\ &= \frac{1}{\lambda_2 - \frac{\varepsilon}{q}} [(V(y))^{\lambda_{21} - \frac{\varepsilon}{q}} - (V(\beta+1))^{\lambda_2 - \frac{\varepsilon}{q}}] \ (y \geq \beta + 1). \end{aligned}$$

We observe that $\widetilde{F}(x) = o(e^{tU(x)})$, $\widetilde{G}(y) = o(e^{tV(y)})$ $(t > 0; x, y \to \infty)$.

If there exists a positive constant $M(\leq \lambda_1\lambda_2 B(\lambda_1, \lambda_2))$ such that (15) is valid when we replace $\lambda_1\lambda_2 B(\lambda_1, \lambda_2)$ by M, then in particular, since

$$\begin{aligned} \widetilde{J} &:= \left[\int_{\alpha}^{\infty} \frac{U'(x)\widetilde{F}^p(x)}{(U(x))^{p\lambda_1+1}} dx\right]^{\frac{1}{p}} \left[\int_{\beta}^{\infty} \frac{V'(y)\widetilde{G}^q(y)}{(V(y))^{q\lambda_2+1}} dy\right]^{\frac{1}{2}} \\ &< \frac{\left[\int_{\alpha+1}^{\infty} (U(x)^{-\varepsilon-1} U'(x)dx\right]^{\frac{1}{p}}}{(\lambda_1 - \frac{\varepsilon}{p})(\lambda_2 - \frac{\varepsilon}{q})} \left[\int_{\beta+1}^{\infty} (V(y))^{-\varepsilon-1} V'(y)dy\right]^{\frac{1}{q}} \\ &= \frac{1}{\varepsilon(\lambda_1 - \frac{\varepsilon}{p})(\lambda_2 - \frac{\varepsilon}{q})} (U(\alpha+1))^{-\frac{\varepsilon}{p}} (V(\beta+1))^{-\frac{\varepsilon}{q}}, \end{aligned}$$

we have

$$\widetilde{I} := \int_{\beta}^{\infty} \int_{\alpha}^{\infty} \frac{\widetilde{f}(x)\widetilde{g}(y)}{(U(x)+V(y))^{\lambda}} dxdy < M\widetilde{J}$$

$$< \frac{M}{\varepsilon(\lambda_1 - \frac{\varepsilon}{p})(\lambda_2 - \frac{\varepsilon}{q})} (U(\alpha+1))^{-\frac{\varepsilon}{p}} (V(\beta+1))^{-\frac{\varepsilon}{q}}. \quad (17)$$

In view of Fubini theorem (cf. [44]), setting $v = V(y)/U(x)$, it follows that

$$\widetilde{I} = \int_{\alpha+1}^{\infty} \left[\int_{\beta+1}^{\infty} \frac{(V(y))^{\lambda_2 - \frac{\varepsilon}{q} - 1} V'(y)}{(U(x)+V(y))^{\lambda}} dy \right] (U(x))^{\lambda_1 - \frac{\varepsilon}{p} - 1} U'(x) dx$$

$$= \int_{\alpha+1}^{\infty} (U(x))^{-\varepsilon-1} U'(x) \left[\int_{\frac{V(\beta+1)}{U(x)}}^{\infty} \frac{v^{\lambda_2 - \frac{\varepsilon}{q} - 1}}{(1+v)^{\lambda}} dv \right] dx$$

$$= \int_{\alpha+1}^{\infty} (U(x))^{-\varepsilon-1} U'(x) \left[\int_{\frac{V(\beta+1)}{U(x)}}^{\frac{V(\beta+1)}{U(\alpha+1)}} \frac{v^{\lambda_2 - \frac{\varepsilon}{q} - 1}}{(1+v)^{\lambda}} dv \right] dx$$

$$+ \int_{\alpha+1}^{\infty} (U(x))^{-\varepsilon-1} U'(x) \left[\int_{\frac{V(\beta+1)}{U(\alpha+1)}}^{\infty} \frac{v^{\lambda_2 - \frac{\varepsilon}{q} - 1}}{(1+v)^{\lambda}} dv \right] dx$$

$$= \int_{0}^{\frac{V(\beta+1)}{U(\alpha+1)}} \left[\int_{U^{-1}(\frac{V(\beta+1)}{v})}^{\infty} (U(x))^{-\varepsilon-1} U'(x) dx \right] \frac{v^{\lambda_2 - \frac{\varepsilon}{q} - 1}}{(1+v)^{\lambda}} dv$$

$$+ \frac{1}{\varepsilon} (U(\alpha+1))^{-\varepsilon} \int_{\frac{V(\beta+1)}{U(\alpha+1)}}^{\infty} \frac{v^{\lambda_2 - \frac{\varepsilon}{q} - 1}}{(1+v)^{\lambda}} dv$$

$$= \frac{1}{\varepsilon} (V(\beta+1))^{-\varepsilon} \int_{0}^{\frac{V(\beta+1)}{U(\alpha+1)}} \frac{v^{\lambda_2 + \frac{\varepsilon}{p} - 1}}{(1+v)^{\lambda}} dv$$

$$+ \frac{1}{\varepsilon} (U(\alpha+1))^{-\varepsilon} \int_{\frac{V(\beta+1)}{U(\alpha+1)}}^{\infty} \frac{v^{\lambda_2 - \frac{\varepsilon}{q} - 1}}{(1+v)^{\lambda}} dv.$$

Based on (17) and the above results, it follows that

$$(V(\beta+1))^{-\varepsilon}\int_0^{\frac{V(\beta+1)}{U(\alpha+1)}}\frac{v^{\lambda_2+\frac{\varepsilon}{p}-1}}{(1+v)^{\lambda}}dv$$

$$+(U(\alpha+1))^{-\varepsilon}\int_{\frac{V(\beta+1)}{U(\alpha+1)}}^{\infty}\frac{v^{\lambda_2-\frac{\varepsilon}{q}-1}}{(1+v)^{\lambda}}dv$$

$$<\varepsilon\widetilde{I}<\frac{M(U(\alpha+1))^{-\frac{\varepsilon}{p}}(V(\beta+1))^{-\frac{\varepsilon}{q}}}{(\lambda_1-\frac{\varepsilon}{p})(\lambda_2-\frac{\varepsilon}{q})}.$$

Putting $\varepsilon\to 0^+$ in the above inequality, in view of the continuity of the beta function, we find $B(\lambda_1,\lambda_2)\le\frac{M}{\lambda_1\lambda_2}$, namely, $\lambda_1\lambda_2 B(\lambda_1,\lambda_2)\le M$. Hence, $M=\lambda_1\lambda_2 B(\lambda_1,\lambda_2)$ is the best possible constant factor in (15).

The theorem is proved.

Remark 5. For $p>1$, $\widehat{\lambda}_1=\frac{\lambda-\lambda_2}{p}+\frac{\lambda_1}{q}$, $\widehat{\lambda}_2=\frac{\lambda-\lambda_1}{q}+\frac{\lambda_2}{p}$, it follows that $\widehat{\lambda}_1+\widehat{\lambda}_2=\lambda$. We find $0<\widehat{\lambda}_1<\frac{\lambda}{p}+\frac{\lambda}{q}=\lambda$ and then $0<\widehat{\lambda}_2=\lambda-\widehat{\lambda}_1<\lambda$. By Hölder's inequality (cf. [43]), we obtain

$$0<B(\widehat{\lambda}_1+1,\widehat{\lambda}_2+1)=\int_0^{\infty}\frac{u^{\widehat{\lambda}_1}}{(1+u)^{\lambda+2}}du$$

$$=\int_0^{\infty}\frac{u^{\frac{\lambda-\lambda_2}{p}+\frac{\lambda_1}{q}}}{(1+u)^{\lambda+2}}du=\int_0^{\infty}\frac{(u^{\frac{\lambda-\lambda_2}{p}})}{(1+u)^{\lambda+2}}\left(u^{\frac{\lambda_1}{q}}\right)du$$

$$\le\left[\int_0^{\infty}\frac{u^{(\lambda-\lambda_2+1)-1}}{(1+u)^{\lambda+2}}du\right]^{\frac{1}{p}}\left[\int_0^{\infty}\frac{u^{(\lambda_1+1)-1}}{(1+u)^{\lambda+2}}du\right]^{\frac{1}{q}}$$

$$\times\left[\int_0^{\infty}\frac{v^{(\lambda_2+1)-1}}{(1+v)^{\lambda+2}}dv\right]^{\frac{1}{p}}\left[\int_0^{\infty}\frac{u^{(\lambda_1+1)-1}}{(1+u)^{\lambda+2}}du\right]^{\frac{1}{q}}$$

$$=k_{\lambda+2}^{\frac{1}{p}}(\lambda_2+1)k_{\lambda+2}^{\frac{1}{q}}(\lambda_1+1). \tag{18}$$

Theorem 6. *For $p > 1$, if the constant factor*

$$\frac{\Gamma(\lambda+2)}{\Gamma(\lambda)} k_{\lambda+2}^{\frac{1}{p}}(\lambda_2+1) k_{\lambda+2}^{\frac{1}{q}}(\lambda_1+1)$$

in (14) *is the best possible, then we have $\lambda_1 + \lambda_2 = \lambda$.*

Proof. By (15) (for $\lambda_i = \widehat{\lambda}_i (i = 1, 2)$), we still have

$$I = \int_\beta^\infty \int_\alpha^\infty \frac{f(x)g(y)dxdy}{(U(x)+V(y))^\lambda} < \frac{\Gamma(\lambda+2)}{\Gamma(\lambda)} B(\widehat{\lambda}_1+1, \widehat{\lambda}_2+1)$$

$$\times \left[\int_\alpha^\infty \frac{U'(x)F^p(x)}{(U(x))^{p\widehat{\lambda}_1+1}} dx\right]^{\frac{1}{p}} \left[\int_\beta^\infty \frac{V'(y)G^q(y)}{(V(y))^{q\widehat{\lambda}_2+1}} dy\right]^{\frac{1}{q}}, \qquad (19)$$

Since

$$\frac{\Gamma(\lambda+2)}{\Gamma(\lambda)} k_{\lambda+2}^{\frac{1}{p}}(\lambda_2+1) k_{\lambda+2}^{\frac{1}{q}}(\lambda_1+1)$$

is the best possible constant factor in (14), by (19), we have

$$\frac{\Gamma(\lambda+2)}{\Gamma(\lambda)} k_{\lambda+2}^{\frac{1}{p}}(\lambda_2+1) k_{\lambda+2}^{\frac{1}{q}}(\lambda_1+1)$$
$$\leq \frac{\Gamma(\lambda+2)}{\Gamma(\lambda)} B(\widehat{\lambda}_1+1, \widehat{\lambda}_2+1) \ (\in \mathbf{R}_+),$$

namely,

$$B(\widehat{\lambda}_1+1, \widehat{\lambda}_2+1) \geq k_{\lambda+2}^{\frac{1}{p}}(\lambda_2+1) k_{\lambda+2}^{\frac{1}{q}}(\lambda_1+1).$$

It follows that (18) protains the form of equality.

We observe that (18) protains the form of equality if and only if there exist constants A and B such that they are not both zero and $Au^{\lambda-\lambda_2} = Bu^{\lambda_1}$ *a.e.* in $\mathbf{R}_+$ (cf. [43]). Assuming that $A \neq 0$, it follows that $u^{\lambda-\lambda_2-\lambda_1} = B/A$ *a.e.* in $\mathbf{R}_+$, namely, $\lambda - \lambda_2 - \lambda_1 = 0$, and then $\lambda_1 + \lambda_2 = \lambda$.

The theorem is proved. □

Remark 7. (i) For $U(x) = x^\gamma$ $(\gamma > 0; x > 0), V(y) = y^\rho$ $(\rho > 0; y > 0)$ in (15), we have the following extended Hardy–Hilbert's inequality involving two upper-limit functions:

$$\int_0^\infty \int_0^\infty \frac{f(x)g(y)}{(x^\gamma + y^\rho)^\lambda} dxdy < \gamma^{\frac{1}{p}} \rho^{\frac{1}{q}} \lambda_1 \lambda_2 B(\lambda_1, \lambda_2) \times \left(\int_0^\infty \frac{F^p(x)dx}{x^{p\gamma\lambda_1+1}} \right)^{\frac{1}{p}} \left(\int_0^\infty \frac{G^q(y)dy}{y^{q\rho\lambda_2+1}} \right)^{\frac{1}{q}}. \tag{20}$$

In particular, for $\gamma = \rho = 1$, we have the following Hardy–Hilbert's integral inequality which is provided by [16]:

$$\int_0^\infty \int_0^\infty \frac{f(x)g(y)}{(x+y)^\lambda} dxdy < \lambda_1 \lambda_2 B(\lambda_1, \lambda_2) \left(\int_0^\infty \frac{F^p(x)dx}{x^{p\lambda_1+1}} \right)^{\frac{1}{p}} \left(\int_0^\infty \frac{G^q(y)dy}{y^{q\lambda_2+1}} \right)^{\frac{1}{q}}. \tag{21}$$

(ii) For $U(x) = \ln x$ $(x > \alpha = 1), V(y) = \ln y$ $(y > \beta = 1)$ in (15), we have the following extended Mulholland's inequality involving two upper-limit functions:

$$\int_1^\infty \int_1^\infty \frac{f(x)g(y)}{\ln^\lambda xy} dxdy < \lambda_1 \lambda_2 B(\lambda_1, \lambda_2) \left(\int_1^\infty \frac{F^p(x)dx}{x \ln^{p\lambda_1+1} x} \right)^{\frac{1}{p}} \left(\int_1^\infty \frac{G^q(y)dy}{y \ln^{q\lambda_2+1} y} \right)^{\frac{1}{q}}. \tag{22}$$

The constant factors in the above inequalities are the best possible.

4. The Reverses

For $0 < p < 1 (q < 0)$, by Theorem 1, we have the following reverse of (14):

$$I := \int_\beta^\infty \int_\alpha^\infty \frac{f(x)g(y)}{(U(x)+V(y))^\lambda} dxdy > \frac{\Gamma(\lambda+2)}{\Gamma(\lambda)} k_{\lambda+2}^{\frac{1}{p}}(\lambda_2+1) B_{\lambda+2}^{\frac{1}{q}}(\lambda_1+1) \times \left[\int_\alpha^\infty \frac{U'(x)F^p(x)}{(U(x))^{p\widehat{\lambda}_1+1}} dx \right]^{\frac{1}{p}} \left[\int_\beta^\infty \frac{V'(y)G^q(y)}{(V(y))^{q\widehat{\lambda}_2+1}} dy \right]^{\frac{1}{q}}. \tag{23}$$

In particular, for $\lambda_1 + \lambda_2 = \lambda$, we reduce (24) to the following:

$$I = \int_\beta^\infty \int_\alpha^\infty \frac{f(x)g(y)dxdy}{(U(x)+V(y))^\lambda} > \lambda_1\lambda_2 B(\lambda_1,\lambda_2) \\ \times \left[\int_\alpha^\infty \frac{U'(x)F^p(x)}{(U(x))^{p\lambda_1+1}}dx\right]^{\frac{1}{p}} \left[\int_\beta^\infty \frac{V'(y)G^q(y)}{(V(y))^{q\lambda_2+1}}dy\right]^{\frac{1}{q}}. \quad (24)$$

Theorem 8. *For $0 < p < 1$, the constant factor $\lambda_1\lambda_2 B(\lambda_1,\lambda_2)$ in (24) is the best possible.*

Proof. For any $0 < \varepsilon < \lambda_1 \min\{p,|q|\}$, we set functions $\widetilde{f}(x), \widetilde{F}(x), \widetilde{g}(y)$, and $\widetilde{G}(y)$ as in Theorem 1. If there exists a positive constant $M \geq \lambda_1\lambda_2 B(\lambda_1,\lambda_2)$ such that (24) is valid when we replace $\lambda_1\lambda_2 B(\lambda_1,\lambda_2)$ by M, then, in particular, we have

$$\widetilde{I} = \int_\beta^\infty \int_\alpha^\infty \frac{\widetilde{f}(x)\widetilde{g}(y)}{(U(x)+V(y))^\lambda}dxdy > M\widetilde{J}, \quad (25)$$

where

$$\widetilde{J} := \left[\int_\alpha^\infty \frac{U'(x)\widetilde{F}^p(x)}{(U(x))^{p\lambda_1+1}}dx\right]^{\frac{1}{p}} \left[\int_\beta^\infty \frac{V'(y)\widetilde{G}^q(y)}{(V(y))^{q\lambda_2+1}}dy\right]^{\frac{1}{2}}.$$

Since

$$\lim_{x\to\infty} \frac{[1-U(\alpha+1)(U(x))^{-\lambda_1+\frac{\varepsilon}{p}}]^p - 1}{(U(x))^{-\lambda_1+\frac{\varepsilon}{p}}} \left(\frac{0}{0}\right) \\ = \lim_{x\to\infty} \frac{p[1-U(\alpha+1)(U(x))^{-\lambda_1+\frac{\varepsilon}{p}}]^{p-1}[-U(\alpha+1)(U(x))^{-\lambda_1+\frac{\varepsilon}{p}}]'}{[(U(x))^{-\lambda_1+\frac{\varepsilon}{p}}]'} \\ = -pU(\alpha+1),$$

we have

$$[1-U(\alpha+1)(U(x))^{-\lambda_1+\frac{\varepsilon}{p}}]^p = 1 + O_1((U(x))^{-\lambda_1+\frac{\varepsilon}{p}}).$$

In the same way, it follows that

$$[1-V(\beta+1)(V(y))^{-\lambda_2+\frac{\varepsilon}{q}}]^q = 1 + O_2((V(y))^{-\lambda_2+\frac{\varepsilon}{q}}) \ (y\to\infty).$$

We obtain

$$\widetilde{J}=\frac{\left\{\int_{\alpha+1}^{\infty}(U(x))^{-\varepsilon-1}U'(x)[1-U(\alpha+1)(U(x))^{-\lambda_1+\frac{\varepsilon}{p}}]^p dx\right\}^{\frac{1}{p}}}{(\lambda_1-\frac{\varepsilon}{p})(\lambda_2-\frac{\varepsilon}{q})}$$

$$\times\left\{\int_{\beta+1}^{\infty}(V(y))^{-\varepsilon-1}V'(y)[1-V(\beta+1)(V(y))^{-\lambda_2+\frac{\varepsilon}{q}}]^q dy\right\}^{\frac{1}{q}}$$

$$=\frac{\left[\int_{\alpha+1}^{\infty}(U(x))^{-\varepsilon-1}U'(x)dx+\int_{\alpha+1}^{\infty}O_1((U(x))^{-\lambda_1-\frac{\varepsilon}{q}-1})U'(x)dx\right]^{\frac{1}{p}}}{(\lambda_1-\frac{\varepsilon}{p})(\lambda_2-\frac{\varepsilon}{q})}$$

$$\times\left[\int_{\beta+1}^{\infty}(V(y))^{-\varepsilon-1}V'(y)dy+\int_{\beta+1}^{\infty}O_2((V(y))^{-\lambda_2-\frac{\varepsilon}{p}-1})V'(y)dy\right]^{\frac{1}{q}}$$

$$=\frac{[(U(\alpha+1))^{-\varepsilon}+\varepsilon O_3(1)]^{\frac{1}{p}}}{\varepsilon(\lambda_1-\frac{\varepsilon}{p})(\lambda_2-\frac{\varepsilon}{q})}[(V(\beta+1))^{-\varepsilon}+\varepsilon O_4(1)]^{\frac{1}{q}}.$$

In view of the proof in Theorem 1 and (25) with the above results, it follows that

$$(V(\beta+1))^{-\varepsilon}\int_0^{\frac{V(\beta+1)}{U(\alpha+1)}}\frac{v^{\lambda_2+\frac{\varepsilon}{p}-1}}{(1+v)^{\lambda}}dv$$

$$+(U(\alpha+1))^{-\varepsilon}\int_{\frac{V(\beta+1)}{U(\alpha+1)}}^{\infty}\frac{v^{\lambda_2-\frac{\varepsilon}{q}-1}}{(1+v)^{\lambda}}dv=\varepsilon\widetilde{I}$$

$$>\frac{M((U(\alpha+1))^{-\varepsilon}+\varepsilon O_3(1))^{\frac{1}{p}}(V(\beta+1))^{-\varepsilon}+\varepsilon O_4(1))^{\frac{1}{q}}}{(\lambda_1-\frac{\varepsilon}{p})(\lambda_2-\frac{\varepsilon}{q})}.$$

Putting $\varepsilon\to 0^+$ in the above inequality, in view of the continuity of the beta function, we find $B(\lambda_1,\lambda_2)\geq\frac{M}{\lambda_1\lambda_2}$, namely, $\lambda_1\lambda_2 B(\lambda_1,\lambda_2)\geq M$. Hence, $M=\lambda_1\lambda_2 B(\lambda_1,\lambda_2)$ is the best possible constant factor in (24).

The theorem is proved. □

Remark 9. For $0<p<1$, $\widehat{\lambda}_1=\frac{\lambda-\lambda_2}{p}+\frac{\lambda_1}{q}=\lambda_1+\frac{\lambda-\lambda_1-\lambda_2}{p}$, $\widehat{\lambda}_2=\frac{\lambda-\lambda_1}{q}+\frac{\lambda_2}{p}$, it follows that $\widehat{\lambda}_1+\widehat{\lambda}_2=\lambda$. For $\lambda-\lambda_1-\lambda_2\in(-p\lambda_1,$ $p(\lambda-\lambda_1))$, we find $0<\widehat{\lambda}_1<\lambda$ and then $0<\widehat{\lambda}_2=\lambda-\widehat{\lambda}_1<\lambda$.

By the reverse Hölder's inequality (cf. [43]), we obtain

$$\begin{aligned}\infty > B(\widehat{\lambda}_1 + 1, \widehat{\lambda}_2 + 1) &= \int_0^\infty \frac{u^{\frac{\lambda-\lambda_2}{p}+\frac{\lambda_1}{q}}}{(1+u)^{\lambda+2}} du \\ &= \int_0^\infty \frac{1}{(1+u)^{\lambda+2}} \left(u^{\frac{\lambda-\lambda_2}{p}}\right)\left(u^{\frac{\lambda_1}{q}}\right) du \\ &\geq \left[\int_0^\infty \frac{u^{\lambda-\lambda_2}}{(1+u)^{\lambda+2}} du\right]^{\frac{1}{p}} \left[\int_0^\infty \frac{u^{\lambda_1}}{(1+u)^{\lambda+2}} du\right]^{\frac{1}{q}} \\ &= k_{\lambda+2}^{\frac{1}{p}}(\lambda_2+1) B_{\lambda+2}^{\frac{1}{q}}(\lambda_1+1). \end{aligned} \tag{26}$$

Theorem 10. *For $0 < p < 1$, if the constant factor*

$$\frac{\Gamma(\lambda+2)}{\Gamma(\lambda)} k_{\lambda+2}^{\frac{1}{p}}(\lambda_2+1) B_{\lambda+2}^{\frac{1}{q}}(\lambda_1+1)$$

in (23) *is the best possible, then for $\lambda - \lambda_1 - \lambda_2 \in (-p\lambda_1, p(\lambda - \lambda_1))$, we have $\lambda_1 + \lambda_2 = \lambda$.*

Proof. By (24) (for $\lambda_i = \widetilde{\lambda}_i (i = 1, 2)$), since

$$\frac{\Gamma(\lambda+2)}{\Gamma(\lambda)} k_{\lambda+2}^{\frac{1}{p}}(\lambda_2+1) B_{\lambda+2}^{\frac{1}{q}}(\lambda_1+1)$$

is the best possible constant factor in (23), we have

$$\begin{aligned}&\frac{\Gamma(\lambda+2)}{\Gamma(\lambda)} k_{\lambda+2}^{\frac{1}{p}}(\lambda_2+1) B_{\lambda+2}^{\frac{1}{q}}(\lambda_1+1) \\ &\quad \geq \frac{\Gamma(\lambda+2)}{\Gamma(\lambda)} B(\widehat{\lambda}_1+1, \widehat{\lambda}_2+1) \ (\in \mathbf{R}_+),\end{aligned}$$

namely,

$$B(\widehat{\lambda}_1 + 1, \widehat{\lambda}_2 + 1) \leq k_{\lambda+2}^{\frac{1}{p}}(\lambda_2+1) B_{\lambda+2}^{\frac{1}{q}}(\lambda_1+1).$$

It follows that (26) protains the form of equality.

We observe that (26) protains the form of equality if and only if there exist constants A and B, such that they are not both zero and $Au^{\lambda-\lambda_2} = Bu^{\lambda_1}$ *a.e.* in R_+ (cf. [43]). Assuming that $A \neq 0$, it follows that $u^{\lambda-\lambda_2-\lambda_1} = B/A$ *a.e.* in R_+, namely, $\lambda - \lambda_2 - \lambda_1 = 0$, and then $\lambda_1 + \lambda_2 = \lambda$.

The theorem is proved. □

5. Conclusions

In this chapter, following the way of some published works, by means of the weight functions, the idea of introduced parameters as well as internal variables and the techniques of real analysis, a new extended Hardy–Hilbert's integral inequality with the kernel as $\frac{1}{(V(x)+V(y))^{\lambda}}$ $(\lambda > 0)$ involving two upper-limit functions is provided in Theorem 1. The equivalent statements of the best possible constant factor related to several parameters are considered in Theorems 2 and 3. A few particular inequalities are deduced in Remark 2, and the case of reverses is obtained in Theorems 4 and 5. The lemmas and theorems provide an extensive account of these types of inequalities.

Acknowledgments

The authors thank the referee for his useful propose to reform this chapter.

References

[1] Hardy, G.H., Littlewood, J.E. and Polya, G., *Inequalities.* Cambridge University Press, Cambridge (1934).
[2] Yang, B.C. *The Norm of Operator and Hilbert-Type Inequalities.* Science Press, Beijing, China (2009).
[3] Yang, B.C. *Hilbert-Type Integral Inequalities.* Bentham Science Publishers Ltd., The United Arab Emirates (2009).
[4] Yang, B.C. On the norm of an integral operator and applications. *J. Math. Anal. Appl.*, 321, 182–192 (2006).
[5] Xu, J.S. Hardy-Hilbert's inequalities with two parameters. *Advances in Mathematics*, 36(2), 63–76 (2007).
[6] Xie, Z.T., Zeng, Z. and Sun, Y.F. A new Hilbert-type inequality with the homogeneous kernel of degree −2. *Advances and Applications in Mathematical Sciences*, 12(7), 391–401 (2013).

[7] Zeng, Z., Raja, K., Gandhi, R. and Xie, Z.T. A new Hilbert-type inequality with the homogeneous kernel of degree −2 and with the integral. *Bulletin of Mathematical Sciences and Applications*, 3(1), 11–20 (2014).

[8] Xin, D.M. A Hilbert-type integral inequality with the homogeneous kernel of zero degree. *Mathematical Theory and Applications*, 30(2), 70–74 (2010).

[9] Azar, L.E. The connection between Hilbert and Hardy inequalities. *Journal of Inequalities and Applications*, 2013, 452 (2013).

[10] Batbold, T. and Sawano, Y. Sharp bounds for m-linear Hilbert-type operators on the weighted Morrey spaces. *Math. Inequal. Appl.*, 20, 263–283 (2017).

[11] Adiyasuren, V., Batbold, T. and Krnic, M. Multiple Hilbert-type inequalities involving some differential operators. *Banach Journal of Mathematical Analysis*, 10, 320–337 (2016).

[12] Adiyasuren, V., Batbold, T. and Krnic, M. Hilbert–type inequalities involving differential operators, the best constants and applications. *Math. Inequal. Appl.*, 18, 111–124 (2015).

[13] Batbold, T. and Azar, L.E. A new form of Hilbert integral inequality. *Math. Inequal. Appl.*, 12, 379–390 (2018).

[14] Krnic, M. and Pecaric, J. Extension of Hilbert's inequality. *J. Math. Anal. Appl.*, 324(1), 150–160 (2006).

[15] Adiyasuren, V., Batbold, T. and Azar, L.E. A new discrete Hilbert-type inequality involving partial sums. *Journal of Inequalities and Applications*, 2019, 127 (2019).

[16] Mo, H.M. and Yang, B.C. On a new Hilbert-type integral inequality involving the upper limit functions. *Journal of Inequalities and Applications*, 2020, 5 (2020).

[17] Hong, Y. and Wen, Y. A necessary and sufficient condition of that Hilbert type series inequality with homogeneous kernel has the best constant factor. *Annals Mathematica*, 37A(3), 329–336 (2016).

[18] Hong, Y. On the structure character of Hilbert's type integral inequality with homogeneous kernel and applications. *Journal of Jilin University* (Science Edition), 55(2), 189–194 (2017).

[19] Hong, Y., Huang, Q.L., Yang, B.C. and Liao J.L. The necessary and sufficient conditions for the existence of a kind of Hilbert-type multiple integral inequality with the non-homogeneous kernel and its applications. *Journal of Inequalities and Applications*, 2017, 316 (2017).

[20] Xin, D.M., Yang, B.C. and Wang, A.Z. Equivalent property of a Hilbert-type integral inequality related to the beta function in the whole plane. *Journal of Function Spaces*, 2018, 2691816, 8.

[21] Hong, Y., He, B. and Yang, B.C. Necessary and sufficient conditions for the validity of Hilbert type integral inequalities with a class of quasi-homogeneous kernels and its application in operator theory. *Journal of Mathematics Inequalities*, 12(3), 777–788 (2018).

[22] Liao, J.Q. Wu, S.H. and Yang, B.C. On a new half-discrete Hilbert-type inequality involving the variable upper limit integral and the partial sum. *Mathematics*, 8, 229 (2020), doi:10.3390/math8020229.

[23] Rassias, M.Th. and Yang, B.C. On an equivalent property of a reverse Hilbert-type integral inequality related to the extended Hurwitz-zeta function. *Journal of Mathematical Inequalities*, 13(2), 315–334 (2019).

[24] Rassias M.Th. and Yang, B.C. A reverse Mulholland-type inequality in the whole plane with multi-parameters. *Appl. Anal. Discrete Math.*, 13, 290–308 (2019).

[25] Rassias, M.Th., Yang, B.C. and Raigorodskii, A. On Hardy-type integral inequality in the whole plane related to the extended Hurwitz-zeta fanction. *Journal of Inequalities and Applications*, 2020, 94 (2020).

[26] Rassias, M.Th., Yang, B.C. and Raigorodskii, A. On the reverse Hardy-type integral inequalities in the whole plane with the extended Riemann-zeta function. *Journal of Mathematics Inequalities*, 14(2), 525–546 (2020).

[27] Rassias, M.Th., Yang, B.C. and Raigorodskii, A. On a more accurate reverse Hilbert-type inequality in the whole plane. *Journal of Mathematical Inequalities*, 14(4), 1359–1374 (2020).

[28] Rassias, M.Th., Yang, B.C. and Raigorodskii, A.A Hilbert-type integral inequality in the whole plane related to the Arc tangent function. *Symmetry*, 13, 351 (2021). https://doi.org/10.3390/sym13020351.

[29] Rassias, M.Th., Yang, B.C. and Raigorodskii, A. Equivalent properties of two kinds of Hardy-type integral inequalities. *Symmetry*, 13, 1006 (2021). https://doi.org/10.3390/sym13061006.

[30] Rassias, M.Th., Yang, B.C. and Meletiou, G.C. A more accurate half-discrete Hilbert-type inequality in the whole plane and the reverses. *Ann. Funct. Anal.*, 12(50) (2021).

[31] Yang, B.C., Rassias, M.Th. and Raigorodskii, A. On an extension of a Hardy-Hilbert-type inequality with multi-parameters. *Mathematics*, 9, 2432 (2021). https://doi.org/10.3390/math9192432.

[32] Rassias, M.Th., Yang, B.C. and Raigorodskii, A. A new Hardy–Mulholland-type inequality with a mixed kernel. *Advances in Operator Theory* (2021) 6:27. https://doi.org/10.1007/s43036-020-00123-0.

[33] Rassias, M.Th., Yang, B.C. and Raigorodskii, A. Equivalent conditions of a multiple Hilbert-type integral inequality with the non homogeneous kernel. *Rev. Real Acad. Cienc. Exact as Fis. Nat.Ser.A-Mat.*, 116, 107 (2022).

[34] Yang, B.C., Andrica, D., Bagdasar, O. and Rassias, M.Th. An equivalent property of a Hilbert-type integral inequality and its applications. *Appl. Anal. Discrete Math.*, 16, 548–563 (2022).

[35] Yang, B.C., Andrica, D., Bagdasar, O. and Rassias, M.Th. On a Hilbert-type integral inequality in the whole plane with the equivalent forms. *Rev. Real Acad. Cienc. Exactas Fis. Nat. Ser. A-Mat.*, 117, 57 (2023). https://doi.org/10.1007/s13398-023-01388-9.

[36] Yang, B.C and Rassias, M.Th. A new Hardy–Hilbert-type integral inequality involving one multiple upper limit function and one derivative function of higher order. *Axioms*, 12, 499 (2023).

[37] He, B., Hong, Y. and Li, Z. Conditions for the validity of a class of optimal Hilbert type multiple integral inequalities with non-homogeneous. *Journal of Inequalities and Applications*, 2021, 64 (2021).

[38] Chen, O. He, B. Hong, Y. and Li, Z. Equivalent parameter conditions for the validity of half-discrete Hilbert-type multiple integral inequality with generalized homogeneous kernel. *Journal of Function Spaces*, 2020, 7414861, 6.

[39] He, B., Hong, Y., Chen, Q., The equivalent parameter conditions for constructing multiple integral half-discrete Hilbert-type inequalities with a class of non-homogeneous kernels and their applications. *Open Mathematics*, 2021(19), 400–411 (2021).

[40] Hong, Y., Huang, Q.L. and Chen, Q. The parameter conditions for the existence of the Hilbert type multiple integral inequality and its best constant factor. *Annals of Functional Analysis*, https://doi.org/10.1007/s43034-020-00087-5.

[41] Hong, Y. and Chen, Q. Equivalent parameter conditions for the construction of Hilbert-type integral inequalities with a class of nonhomogeneous kernels. *Journal of South China Normal University* (Natural Science Edition). 52(5), 124–128 (2020).

[42] Wang, Z.X. and Guo, D.R. *Introduction to Special Functions.* Science Press, Beijing (1979).

[43] Kuang, J.C. *Applied Inequalities*. Shangdong Science and Technology Press, Jinan, China (2004).

[44] Kuang, J.C. *Real and Functional Analysis* (Continuation), Higher Education Press, Beijing, 2 (2015).

https://doi.org/10.1142/9789811293009_0037

Chapter 37

Implicit Sequential Contractions in Partially Ordered Metric Spaces

Mihai Turinici
A. Myller Mathematical Seminar, A. I. Cuza University, 700506 Iasi, Romania
mturi@uaic.ro

A fixed point result is established for a class of implicit sequential contractions on partially ordered metric spaces. The connections with some related statements in the area due to Jaggi [Indian J. Pure Appl. Math., 8 (1977), 223–230], Chandok *et al.* [J. Egyptian Math. Soc., 23 (2015), 95–101], and Harjani *et al.* [Abstr. Appl. Anal., Volume 2010, Article ID 190701] are also being established.

Subject Class [2010]. 47H17 (Primary), 54H25 (Secondary).

Keywords: Partially ordered metric space, fixed point, Picard operator, self-closeness, implicit sequential contraction, nright/npoint property, normal couple.

1. Introduction

Let X be a non-empty set. Call the subset Y of X, *almost-singleton* (in short, *asingleton*) provided $y_1, y_2 \in Y$ implies $y_1 = y_2$, and *singleton* if, in addition, Y is non-empty; note that in this case, $Y = \{y\}$, for some $y \in X$. Take a *metric* $d : X \times X \to R_+ := [0, \infty[$ over X, as well as a selfmap $T \in \mathcal{F}(X)$. [Here, for each couple A, B of non-empty sets, $\mathcal{F}(A, B)$ denotes the class of all functions from A to B; when $A = B$, we write $\mathcal{F}(A)$ in place of $\mathcal{F}(A, A)$]. Denote $\text{Fix}(T) = \{x \in X; x = Tx\}$; each point of this set is referred to as *fixed* under T. The determination of such points is carried out in the following context, comparable with the one in Rus [47, Ch 2, Sect 2.2]:

pic-(1) We say that T is a *Picard operator* (modulo d) if, for each $x \in X$, the iterative sequence $(T^n x; n \geq 0)$ is d-Cauchy.

pic-(2) We say that T is a *strong Picard operator* (modulo d) if, for each $x \in X$, $(T^n x; n \geq 0)$ is d-convergent with $\lim_n (T^n x) \in \text{Fix}(T)$.

pic-(3) We say that T is *fix-asingleton* if $\text{Fix}(T)$ is asingleton, and *fix-singleton*, provided $\text{Fix}(T)$ is singleton.

Concerning the existence and uniqueness of such points, a basic result (referred to as Banach fixed point theorem; in short, (B-fpt)) may be stated as follows. Call the selfmap T, *$(d; \alpha)$-contractive* (where $\alpha \geq 0$), if

$$\text{(con)} \quad d(Tx, Ty) \leq \alpha d(x, y), \text{ for all } x, y \in X.$$

Theorem 1. *Assume that T is Banach $(d; \alpha)$-contractive, for some $\alpha \in [0, 1[$. In addition, let X be d-complete. Then, T is a strong Picard operator and fix-asingleton (or, equivalently: fix-singleton); precisely,*

$$\text{(11-a)} \quad \text{Fix}(T) \textit{ is a singleton, } \{z\}$$

$$\text{(11-b)} \quad T^n x \xrightarrow{d} z \textit{ as } n \to \infty, \textit{ for each } x \in X.$$

This result, established in 1922 by Banach [5], found some important applications to the operator equations theory. Consequently, a multitude of extensions for it were proposed. From the perspective of this exposition, the set implicit ones are of interest. Denote, for $x, y \in X$,

$$Q_1(x, y) = d(x, Tx),\ Q_2(x, y) = d(x, y),\ \ Q_3(x, y) = d(x, Ty),$$
$$Q_4(x, y) = d(Tx, y),\ \ Q_5(x, y) = d(Tx, Ty),\ Q_6(x, y) = d(y, Ty),$$

$$\mathcal{Q}(x,y) = (Q_1(x,y), Q_2(x,y), Q_3(x,y), Q_4(x,y), Q_5(x,y), Q_6(x,y)).$$

Then, the underlying contractions may be written as

(i-s-con) $\mathcal{Q}(x,y) \in \Upsilon$, for all $x, y \in X$ with $x\mathcal{R}y$,

where $\Upsilon \subseteq R_+^6$ is a (non-empty) subset and $\mathcal{R}$ is a relation over X. In particular, when Υ is the zero section of a certain function $F : R_+^6 \to R$, i.e.,

$$\Upsilon = \{(t_1, \ldots, t_6) \in R_+^6; F(t_1, \ldots, t_6) \leq 0\},$$

the implicit contractive condition above has the functional form:

(i-f-con) $F(\mathcal{Q}(x,y)) \leq 0$, for all $x, y \in X$ with $x\mathcal{R}y$

(where $\mathcal{R}$ is the precise one). Four basic particular cases appear.

(I) Suppose that $\mathcal{R} = X \times X$ (the *trivial* relation over X); this is the natural case discussed in the immense majority of papers. Concerning the "genuine" implicit case, some recent contributions in the area may be found in Akkouchi [1], Berinde and Vetro [7], or Nashine *et al.* [37]. We stress that in almost all papers based on implicit techniques — including the ones we just quoted — it is claimed that the starting point in the area is represented by the 1999 contribution due to Popa [40]. Unfortunately, this claim is not true: fixed point results based on implicit techniques were obtained more than two decades ago in two papers by Turinici [53, 54]. But, we must note that some partial aspects of the set-implicit theory (as described above) have been discussed in the (classical by now) 1969 Meir–Keeler fixed point principle [32]. On the other hand, for the explicit case, we must quote the basic contributions due to Boyd and Wong [10], Leader [29], Reich [43], or Matkowski [31]; see also the survey paper by Rhoades [44].

(II) Suppose that $\mathcal{R}$ is reflexive, transitive, antisymmetric (hence, a *partial order* on X). A first couple of 1986 results was formulated — in the realm of Matkowski type contractions — by Turinici [55, 56]. Two decades later, this fixed point statement was re-discovered — over the Banach contractive setting — by Ran and Reurings [42]; see also Nieto and Rodriguez-Lopez [38], and since then, the number of papers devoted to the precise topic increased rapidly.

(III) Suppose that $\mathcal{R}$ is a relation over X (with certain specific properties). An early result of this type was obtained in Berzig [9]; see also Turinici [60]. Some up-to-date aspects concerning this direction may be found in Alam *et al.* [2].

(IV) Suppose that $\mathcal{R}$ is an *amorphous* relation over X (with no specific properties at all). The first result in the area was obtained in 2008 by Jachymski [23] over metric spaces endowed with a graph. Further, in 2012, Samet and Turinici [48] established a lot of Leader type results in the realm of relational metric spaces. In fact, as shown in Roldán and Shahzad [46], these approaches are, practically, identical.

Having these precise, it is our aim in the following to give — under the lines in Turinici [61, Paper-3-1] — a lot of sequential-type fixed point results — involving six-dimensional contractive conditions — for mappings acting upon partially ordered metric spaces. Note that, as a by-product of our developments, one gets a couple of rather distinct fixed point statements involving rational-type contractions on partially ordered metric spaces due to Harjani *et al.* [21] and Chandok *et al.* [14] We stress that the partial order setting of underlying results cannot be extended to the relational framework. Finally, by a little modification of these techniques, one may derive the Meir–Keeler contractive fixed point statements in Samet *et al.* [49]; see also Turinici [62]. Further aspects are delineated elsewhere.

2. Dependent Choice Principle

Throughout this exposition, the axiomatic system in use is Zermelo–Fraenkel's (abbreviated: (ZF)), as described by Cohen [17, Ch 2]. The notations and basic facts to be considered in this system are more or less standard. Some important ones are described as follows.

(A) Let X be a non-empty set. By a *relation* over X, we mean any (non-empty) part $\mathcal{R} \subseteq X \times X$; then, $(X, \mathcal{R})$ will be referred to as a *relational structure*. For simplicity, we sometimes write $(x, y) \in \mathcal{R}$ as $x\mathcal{R}y$. Note that $\mathcal{R}$ may be regarded as a mapping between X and $\exp[X]$ (=the class of all subsets in X). In fact, denote for $x \in X$

$X(x, \mathcal{R}) = \{y \in X; x\mathcal{R}y\}$ (the *section* of $\mathcal{R}$ through x);

then, the desired mapping representation is $[\mathcal{R}(x) = X(x, \mathcal{R}); x \in X]$. A basic example of such object is

$\mathcal{I} = \{(x, x); x \in X\}$ [the *identical relation* over X].

Given the relations $\mathcal{R}$, $\mathcal{S}$ over X, define their *product* $\mathcal{R} \circ \mathcal{S}$ as

$(x, z) \in \mathcal{R} \circ \mathcal{S}$, if there exists $y \in X$ with $(x, y) \in \mathcal{R}$, $(y, z) \in \mathcal{S}$.

Also, for each relation $\mathcal{R}$ in X, denote

$\mathcal{R}^{-1} = \{(x, y) \in X \times X; (y, x) \in \mathcal{R}\}$ (the *inverse* of $\mathcal{R}$).

Finally, given the relations $\mathcal{R}$ and $\mathcal{S}$ on X, let us say that $\mathcal{R}$ is *coarser* than $\mathcal{S}$ (or, equivalently, $\mathcal{S}$ is *finer* than $\mathcal{R}$), provided

$\mathcal{R} \subseteq \mathcal{S}$; i.e., $x\mathcal{R}y$ implies $x\mathcal{S}y$.

Given a relation $\mathcal{R}$ on X, the following properties are to be discussed here:

(P1) $\mathcal{R}$ is *reflexive*: $\mathcal{I} \subseteq \mathcal{R}$
(P2) $\mathcal{R}$ is *irreflexive*: $\mathcal{R} \cap \mathcal{I} = \emptyset$
(P3) $\mathcal{R}$ is *symmetric*: $\mathcal{R}^{-1} = \mathcal{R}$
(P4) $\mathcal{R}$ is *antisymmetric*: $\mathcal{R}^{-1} \cap \mathcal{R} \subseteq \mathcal{I}$
(P5) $\mathcal{R}$ is *transitive*: $\mathcal{R} \circ \mathcal{R} \subseteq \mathcal{R}$.

This yields the classes of relations to be used; the following ones are important for our developments:

(C0) $\mathcal{R}$ is *amorphous* (i.e.: it has no specific properties at all)
(C1) $\mathcal{R}$ is a *quasi-order* (reflexive and transitive)
(C2) $\mathcal{R}$ is a *strict order* (irreflexive and transitive)
(C3) $\mathcal{R}$ is an *equivalence* (reflexive, transitive, symmetric)
(C4) $\mathcal{R}$ is a *(partial) order* (reflexive, transitive, antisymmetric)
(C5) $\mathcal{R} = X \times X$ (the *trivial* relation over X).

(B) A basic example of relational structure is to be constructed as follows. Let $N = \{0, 1, 2, \ldots\}$ denote the set of *natural* numbers introduced as

$0 := \emptyset$, $1 := \{0\}$, $2 := \{0, 1\}$, and so on.

Technically speaking, the basic (algebraic and order) structures over N may be obtained by means of the *(immediate) successor* function suc : $N \to N$, and the following Peano properties (deductible in our system (ZF)):

(pea-1) $0 \in N$ and $0 \notin \text{suc}(N)$

(pea-2) suc(.) is injective ($\text{suc}(n) = \text{suc}(m)$ implies $n = m$)

(pea-3) if $M \subseteq N$ is such that $[0 \in M]$ and $[\text{suc}(M) \subseteq M]$, then $M = N$.

(Note that, in the absence of our axiomatic setting, these properties become the well-known Peano axioms, as described in Halmos [20, Ch 12].) In fact, starting from these properties, one may construct, in a recurrent way, an *addition* $(a, b) \mapsto a + b$ over N, according to

$$(\forall m \in N)\colon m + 0 = m;\ m + \text{suc}(n) = \text{suc}(m + n).$$

This, in turn, makes possible the introduction of a partial order $(\leq)$ and a strict order $(<)$ over N, as

(for each $m, n \in N$):

$m \leq n$ iff $m + p = n$, for some $p \in N$; $m < n$ iff $m \leq n$, and $m \neq n$.

Concerning the properties of this structure, the most important one writes

$(N, \leq)$ is well ordered:
any (non-empty) subset of N has a first element.

This tells us that *inductive* reasonings and constructions are allowed over this structure. Further, denote for $p, q \in N$, $p \leq q$,

$$N[p, q] = \{n \in N; p \leq n \leq q\},\ N]p, q[= \{n \in N; p < n < q\},$$
$$N[p, q[= \{n \in N; p \leq n < q\},\ N]p, q] = \{n \in N; p < n \leq q\},$$

as well as, for $r \in N$,

$$N[r, \infty[= \{n \in N; r \leq n\},\ N]r, \infty[= \{n \in N; r < n\}.$$

For each $r \geq 1$, $N[0, r[= N(r, >)$ is referred to as the *initial interval* (of N) induced by r. Any set P with $P \sim N$ (in the sense, there exists a bijection from P to N) is referred to as *effectively denumerable.*

In addition, given some natural number $n \geq 1$, any set Q with $Q \sim N(n, >)$ is said to be *n-finite*; when n is generic here, we say that Q is *finite*. Finally, the (non-empty) set Y is called (at most) *denumerable* iff it is either finite or effectively denumerable.

Let X be a non-empty set. By a *sequence* in X, we mean any mapping $x : N \to X$, where $N = \{0, 1, \ldots\}$ is the set of *natural* numbers. For simplicity reasons, it will be useful to denote it as $(x(n); n \geq 0)$, or $(x_n; n \geq 0)$; moreover, when no confusion can arise, we further simplify this notation as $(x(n))$ or (x_n), respectively. Also, any sequence $(y_n := x_{i(n)}; n \geq 0)$ with

$(i(n); n \geq 0)$ is strictly ascending (hence: $i(n) \to \infty$ as $n \to \infty$)

is referred to as a *subsequence* of $(x_n; n \geq 0)$. Note that, under such a convention, the relation "subsequence of" is transitive; i.e.,

(z_n) = subsequence of (y_n) and (y_n) = subsequence of (x_n)
imply (z_n) = subsequence of (x_n).

(C) Remember that, an outstanding part of (ZF) is the *Axiom of Choice* (abbreviated: (AC)), which, in a convenient manner, may be written as

(AC) For each couple (J, X) of nonempty sets and each function $F : J \to \exp(X)$, there exists a selective function $f : J \to X$ (in the sense: $f(\nu) \in F(\nu)$, for each $\nu \in J$).

(Here, $\exp(X)$ stands for the class of all non-empty elements in $\exp[X]$.) Sometimes, when the index set J is denumerable, the existence of such a selective function may be determined by using a weaker form of (AC), called *Dependent Choice* principle(in short, (DC)). Call the relation $\mathcal{R}$ over X *proper* when

$(X(x, \mathcal{R}) =)\mathcal{R}(x)$ is non-empty, for each $x \in X$.

Then, $\mathcal{R}$ is to be viewed as a mapping between X and $\exp(X)$, and the couple $(X, \mathcal{R})$ is referred to as a *proper relational structure*. Further, given $a \in X$, let us say that the sequence $(x_n; n \geq 0)$ in X is *$(a; \mathcal{R})$-iterative*, provided

$x_0 = a$, and $x_n \mathcal{R} x_{n+1}$ (i.e.: $x_{n+1} \in \mathcal{R}(x_n)$), for all n.

Proposition 2. *Let the relational structure $(X, \mathcal{R})$ be proper. Then, for each $a \in X$, there is at least an $(a, \mathcal{R})$-iterative sequence in X.*

This principle — proposed, independently, by Bernays [8] and Tarski [52] — is deductible from (AC) but not conversely; cf. Wolk [64]. Moreover, by the developments in Moskhovakis [34, Ch 8] and Schechter [50, Ch 6], the *reduced system* (ZF-AC+DC) is comprehensive enough so as to cover the "usual" mathematics; see also Moore [33, Appendix 2].

Let $(\mathcal{R}_n; n \geq 0)$ be a sequence of relations on X. Given $a \in X$, let us say that the sequence $(x_n; n \geq 0)$ in X is $(a; (\mathcal{R}_n; n \geq 0))$-*iterative* provided

$$x_0 = a, \text{ and } x_n \mathcal{R}_n x_{n+1} \text{ (i.e.: } x_{n+1} \in \mathcal{R}_n(x_n)), \text{ for all } n.$$

The following *Diagonal Dependent Choice* principle (in short, (DDC)) is available.

Proposition 3. *Let $(\mathcal{R}_n; n \geq 0)$ be a sequence of proper relations on X. Then, for each $a \in X$, there exists at least one $(a; (\mathcal{R}_n; n \geq 0))$-iterative sequence in X.*

Clearly, (DDC) includes (DC), to which it reduces when $(\mathcal{R}_n; n \geq 0)$ is constant. The reciprocal of this is also true. In fact, letting the premises of (DDC) hold, put $P = N \times X$, and let $\mathcal{S}$ be the relation over P introduced as

$$\mathcal{S}(i, x) = \{i + 1\} \times \mathcal{R}_i(x), \quad (i, x) \in P.$$

It will suffice applying (DC) to $(P, \mathcal{S})$ and $b := (0, a) \in P$ to get the conclusion in the statement; we do not give details.

A basic consequence of (DC) (or, equivalently: (DDC)) is the so-called *Denumerable Axiom of Choice* [in short, (AC(N))].

Proposition 4. *Let $F : N \to \exp(X)$ be a function. Then, for each $a \in F(0)$, there exists a function $f : N \to X$ with $f(0) = a$ and $f(n) \in F(n)$, $\forall n \geq 0$.*

Proof. Denote $Q = N \times X$, and let us introduce the (proper) relation $\mathcal{R}$ over it, according to

$$\mathcal{R}(n, x) = \{n + 1\} \times F(n + 1), \quad n \geq 0, x \in X.$$

By an application of (DC) to the proper relational structure $(Q, \mathcal{R})$, the conclusion follows; we do not give details. □

As a consequence of the above facts,

(DC) $\Longrightarrow$ (AC(N)) in the *strongly reduced system* (ZF-AC),

or, equivalently,

(AC(N)) is deductible in the reduced system (ZF-AC+DC).

The reciprocal of the written inclusion is not true; see Moskhovakis [34, Ch 8, Sect 8.25] for details.

3. Conv-Cauchy Structures

Let X be a non-empty set. Denote by $\mathcal{S}(X)$ the class of all sequences (x_n) in X. By a (sequential) *convergence structure* on X, we mean, as in Kasahara [25], any part $\mathcal{C}$ of $\mathcal{S}(X) \times X$ with the properties

(conv-1) $\mathcal{C}$ is hereditary:

$((x_n); x) \in \mathcal{C} \Longrightarrow ((y_n); x) \in \mathcal{C}$, for each subsequence (y_n) of (x_n)

(conv-2) $\mathcal{C}$ is *reflexive*: for each $a \in X$,

the constant sequence $(x_n = a; n \geq 0)$ fulfills $((x_n); a) \in \mathcal{C}$.

For simplicity, the relation $((x_n); x) \in \mathcal{C}$ is denoted as $x_n \xrightarrow{\mathcal{C}} x$ and reads

(x_n), $\mathcal{C}$-converges to x (also referred to as: x is the $\mathcal{C}$*-limit* of (x_n)).

The set of all such x is denoted as $\mathcal{C} - \lim_n(x_n)$; concerning this concept, there are two possible alternatives:

(div) (x_n) is $\mathcal{C}$*-divergent*, when $\mathcal{C} - \lim_n(x_n)$ is empty,

(conv) (x_n) is $\mathcal{C}$*-convergent*, when $\mathcal{C} - \lim_n(x_n)$ is non-empty.

The following condition about the introduced convergence structure $\mathcal{C}$ is to be optionally considered

(conv-3) $\mathcal{C}$ is *separated*:
$\mathcal{C} - \lim_n(x_n)$ is an asingleton, for each sequence $(x_n; n \geq 0)$ in X;

when it holds, $\mathcal{C} - \lim_n(x_n) = \{z\}$ is also written as $\mathcal{C} - \lim_n(x_n) = z$.

Likewise, by a (sequential) *Cauchy structure* on X, we mean, as in Turinici [57], any part $\mathcal{H}$ of $\mathcal{S}(X)$ with

(Cauchy-1) $\mathcal{H}$ is *hereditary*:
$(x_n) \in \mathcal{H} \Longrightarrow (y_n) \in \mathcal{H}$, for each subsequence (y_n) of (x_n)

(Cauchy-2) $\mathcal{H}$ is *reflexive*: for each $a \in X$,
the constant sequence $(x_n = a; n \geq 0)$ fulfills $(x_n) \in \mathcal{H}$.

Each element of $\mathcal{H}$ is referred to as a $\mathcal{H}$-*Cauchy* sequence in X.

Finally, the couple $(\mathcal{C}, \mathcal{H})$ will be referred to as a *conv-Cauchy structure* on X. The optional conditions about this object to be considered here are

(CC-1) $(\mathcal{C}, \mathcal{H})$ is *regular*: each $\mathcal{C}$-convergent sequence is $\mathcal{H}$-Cauchy,

(CC-2) $(\mathcal{C}, \mathcal{H})$ is *complete*: each $\mathcal{H}$-Cauchy sequence is $\mathcal{C}$-convergent.

(A) In the following, a basic example of conv-Cauchy structure is given.

Let X be some non-empty set. In what follows, by a *pseudometric* over X, we will mean any map $d : X \times X \to R_+$. Suppose that we fixed such an object, with, in addition,

(met-1) d is *triangular*: $d(x, z) \leq d(x, y) + d(y, z)$, for all $x, y, z \in X$

(met-2) d is *reflexive*: $d(x, x) = 0$, for each $x \in X$

(met-3) d is *sufficient*: $d(x, y) = 0$ implies $x = y$

(met-4) d is *symmetric*: $d(x, y) = d(y, x)$, for all $x, y \in X$.

Then, $d(.,.)$ is called a *metric* on X, and (X, d) is referred to as a *metric space*.

We introduce a d-convergence and a d-Cauchy structure on X as follows. Given the sequence (x_n) in X and the point $x \in X$, we say that (x_n), d-*converges* to x (written as $x_n \xrightarrow{d} x$) provided $d(x_n, x) \to 0$ as $n \to \infty$; i.e.,

$$\forall \varepsilon > 0,\ \exists i = i(\varepsilon):\ n \geq i \Longrightarrow d(x_n, x) < \varepsilon,$$

or, equivalently,
$\forall \varepsilon > 0, \exists i = i(\varepsilon)$: $n \geq i \Longrightarrow d(x_n, x) \leq \varepsilon$.

By this very definition, we have the following properties:

(d-conv-1) $(\xrightarrow{d})$ is *hereditary*:
$x_n \xrightarrow{d} x$ implies $y_n \xrightarrow{d} x$, for each subsequence (y_n) of (x_n)
(d-conv-2) $(\xrightarrow{d})$ is *reflexive*: for each $a \in X$,
the constant sequence $(x_n = a; n \geq 0)$ fulfills $x_n \xrightarrow{d} a$.

As a consequence, $(\xrightarrow{d})$ is a sequential convergence on X. The set of all such limit points of (x_n) is denoted as $d - \lim_n(x_n)$, or $\lim_n(x_n)$, when d is understood. Concerning this concept, there are two possible alternatives:

(d-div) (x_n) is *d-divergent*, when $\lim_n(x_n)$ is empty

(d-conv) (x_n) is *d-convergent*, when $\lim_n(x_n)$ is non-empty.

Remember that, by a previous definition,

(d-conv-3) $(\xrightarrow{d})$ is *separated* (referred to as: d is *separated*) iff $\lim_n(x_n)$ is an asingleton, for each sequence (x_n).

Note that this holds in our metrical setting, as it can be directly seen. Then, by the accepted notations, $\lim_n(x_n) = \{z\}$ is also written as $\lim_n(x_n) = z$.

The introduced concepts allow us giving a useful property. (Since the proof is immediate, we do not give the details.)

Proposition 5. *The mapping $(x, y) \mapsto d(x, y)$ is d-Lipschitz, in the sense*

(31-1) $|d(x, y) - d(u, v)| \leq d(x, u) + d(y, v)$, $\forall (x, y), (u, v) \in X \times X$.

As a consequence, this map is d-continuous; i.e.,

(31-2) $x_n \xrightarrow{d} x$, $y_n \xrightarrow{d} y$ *imply* $d(x_n, y_n) \to d(x, y)$.

Further, call the sequence (x_n), *d-Cauchy* when $d(x_n, x_m) \to 0$ as $n, m \to \infty$, $n < m$; i.e.,

$\forall \varepsilon > 0$, $\exists j = j(\varepsilon)$: $j \leq n < m \Longrightarrow d(x_n, x_m) < \varepsilon$,

or, equivalently,

$\forall \varepsilon > 0$, $\exists j = j(\varepsilon)$: $j \leq n < m \Longrightarrow d(x_n, x_m) \leq \varepsilon$;

the class of all such sequences is denoted as $Cauchy(d)$. Clearly,

(Cauchy-1) $Cauchy(d)$ is *hereditary*:

(x_n) is d-Cauchy implies (y_n) is d-Cauchy,

for each subsequence $(y_n; n \geq 0)$ of $(x_n; n \geq 0)$

(Cauchy-2) $Cauchy(d)$ is *reflexive*: for each $a \in X$,

the constant sequence $(x_n = a; n \geq 0)$ is d-Cauchy

so that $Cauchy(d)$ is a Cauchy structure on X.

Now, according to the general setting, call the couple $((\xrightarrow{d}), Cauchy(d))$ a *conv-Cauchy structure* induced by d. The following regularity conditions about this structure are to be considered:

(CC-1) d is *regular*: each d-convergent sequence is d-Cauchy

(CC-2) d is *complete*: each d-Cauchy sequence is d-convergent.

Note that the former of these holds in our setting; we do not give details.

(B) Let X be a non-empty set. Take a metric $d(.,.)$ on X, and let $(\leq)$ be a quasi-order on X; the triple $(X, d, \leq)$ is referred to as a *quasi-ordered metric space*. Let also $(<)$ stand for the attached relation

$x < y$ iff $x \leq y$ and $x \neq y$.

For the moment, $(<)$ is irreflexive; however, it is not in general transitive [hence, a *strict order*] on X unless $(\leq)$ is antisymmetric (i.e., a partial order on X).

Further, call the sequence $(x_n; n \geq 0)$, $(\leq)$-*ascending* provided [$i \leq j$ implies $x_i \leq x_j$]. In the following, some d-Cauchy criteria are stated for such objects.

Let us say that the sequence $(x_n; n \geq 0)$ is *d-asymptotic*, provided

$r_n := d(x_n, x_{n+1}) \to 0$ as $n \to \infty$.

Clearly, each d-Cauchy sequence is d-asymptotic too; the reciprocal of this is not in general true. This tells us that the d-Cauchy criteria we are looking for are to be sought in the class of $(\leq)$-ascending d-asymptotic sequences. To get concrete examples of such properties, we need some conventions and auxiliary facts.

Given the d-asymptotic sequence $(x_n; n \geq 0)$ and the number $\varepsilon > 0$, let us say that $i \in N$ is *ε-regular*, provided $[i \leq n$ implies $d(x_n, x_{n+1}) < \varepsilon]$. The class $\mathcal{Z}(\varepsilon)$ of all these ranks fulfills

$\mathcal{Z}(\varepsilon)$ is non-empty, hereditary ($i \in \mathcal{Z}(\varepsilon)$ and $i \leq j$ imply $j \in \mathcal{Z}(\varepsilon)$);

so that: $Z(\varepsilon) := \min \mathcal{Z}(\varepsilon)$ is well defined, as an element of N;

with, in addition: $d(x_n, x_{n+1}) < \varepsilon$, for all $n \geq Z(\varepsilon)$.

Define the subsets of $N \times N$

$(\leq; N) = \{(m,n) \in N \times N; m \leq n\}$, $(<; N) = \{(m,n) \in N \times N; m < n\}$;

these are just the graphs of the relations $(\leq)$ and $(<)$ over N, respectively.

Further, let us say that the subset Θ of $R^0_+ :=]0,\infty[$ is *$(>)$-cofinal* in R^0_+, when

for each $\varepsilon > 0$, there exists $\theta \in \Theta$ with $\varepsilon > \theta$;

note that, in this case, the following hereditary property holds

$\beta \in \Theta$ implies $\Theta_\beta := \Theta(\beta, >)$ is $(>)$-cofinal in R^0_+.

Given this subset, define (as before)

$(\geq; \Theta) = \{(\beta,\gamma) \in \Theta \times \Theta; \beta \geq \gamma\}$, $(>; \Theta) = \{(\beta,\gamma) \in \Theta \times \Theta; \beta > \gamma\}$;

these are just the graphs of the relations $(\geq)$ and $(>)$ over Θ, respectively. In addition, for each $(\beta,\gamma) \in (>; \Theta)$, denote

$B(\beta,\gamma) = \{n \in N(2,\leq); 3^{-n} < \min\{\beta - \gamma, \gamma\}$;

clearly, $B(\beta,\gamma)$ is non-empty and hereditary:

$i \in B(\beta,\gamma)$ and $i \leq j$ imply $j \in B(\beta,\gamma)$.

In this case, it makes sense denoting

$$b(\beta, \gamma) = \min B(\beta, \gamma),\ (\beta, \gamma) \in (>; \Theta).$$

If no confusion arises, we write $b(\beta, \gamma)$ as b, for simplicity; hence, by definition,

$$(b \geq 2 \text{ and}) \ k \in N(b, \leq) \text{ implies } 3^{-k} < \min\{\beta - \gamma, \gamma\}.$$

From the perspective of our initial problem, the points $\beta \in \Theta$ appearing here must be chosen according to some admissible properties. Some conventions are in order. Let (x_n) be a sequence in X. We say that $\beta \in \Theta$ is (x_n)-*admissible* when

(ad-be) $E(j; \beta) := \{(m, n) \in (<; N); j \leq m < n, d(x_m, x_n) > \beta\} \neq \emptyset,\ \forall j.$

The class of all these points is denoted as $\text{adm}(\Theta; (x_n))$.

Proposition 6. *Let (x_n) be a sequence in X that is not d-Cauchy and Θ be a $(>)$-cofinal part of R^0_+. Then, the following are valid:*

(32-1) $\text{adm}(\Theta; (x_n))$ *is non-empty*

(32-2) $\beta \in \text{adm}(\Theta; (x_n))$ *implies* $\Theta(\beta, >) \subseteq \Theta(\beta, \geq) \subseteq \text{adm}(\Theta; (x_n))$;

so that, $\text{adm}(\Theta; (x_n))$ *is* $(>)$-*cofinal in* R^0_+

(32-3) *for each* $\beta \in \text{adm}(\Theta; (x_n))$ *and each* $\gamma \in \Theta$, *we have* $(\beta, \gamma) \in (>; \Theta)$ *iff* $(\beta, \gamma) \in (>; \text{adm}(\Theta; (x_n)))$.

Proof. (i) By definition, the d-Cauchy property of our sequence writes

$$\forall \varepsilon \in R^0_+,\ \exists a \in N,\ \forall (m, n) \in (<; N){:}\ a \leq m < n \Longrightarrow d(x_m, x_n) \leq \varepsilon.$$

As Θ is a $(>)$-cofinal part in R^0_+, this property may be also written as

$$\forall \theta \in \Theta,\ \exists \alpha \in N,\ \forall (m, n) \in (<; N){:}\ \alpha \leq m < n \Longrightarrow d(x_m, x_n) \leq \theta.$$

The negation of this property means that there exists $\beta \in \Theta$ such that

$$E(j;\beta) := \{(m,n) \in (<;N); j \leq m < n, d(x_m, x_n) > \beta\} \neq \emptyset, \ \forall j,$$

which shows that $\beta \in \text{adm}(\Theta; (x_n))$ and proves our first conclusion. Moreover, as

$$(\forall j)\text{: } \beta_1 \geq \beta_2 \text{ implies } E(j;\beta_1) \subseteq E(j;\beta_2),$$

it is clear that the first half of our second conclusion holds; wherefrom (by a previous observation) the second part of underlying conclusion holds too.

(ii): Evident, by the preceding stage. □

Finally, the following conventions are needed. For each sequence of ranks $(\lambda(k); k \geq 0)$ in $N(1, \leq)$, define the property

$(\lambda(k); k \geq 0)$ is *strictly ascending*: $i < j$ implies $\lambda(i) < \lambda(j)$;

hence, $(\lambda(k))$ is divergent $(\lim_k \lambda(k) = \infty)$.

On the other hand, for each sequence (r_n) in R and each point $r \in R$, let us write

$r_n \to r+$ (or, $\lim_n(r_n) = r+$) if $r_n \to r$ and $(r_n > r, \ \forall n)$,

$r_n \to\to r+$ (or, $\lim_n(r_n) = r++$) if $r_n \to r$ and $(r_n > r, \ \forall\forall n)$.

Here, given a property $\pi(k)$ depending on $k \in N$, let us say that it holds for *nearly all* k [written: $(\pi(k), \ \forall\forall k)$], provided

there exists $c = c(\pi) \in N$ such that ($\pi(k)$ is true, for all $k \geq c$).

The following result, referred to as *Boyd–Wong non-Cauchy Criterion* (in short, (BW-n-CC)), is now available.

Theorem 7. *Let the sequence $(x_n; n \geq 0)$ in X be such that*

(31-i) *$(x_n; n \geq 0)$ is $(\leq)$-ascending ($i \leq j$ implies $x_i \leq x_j$)*

(31-ii) *$(x_n; n \geq 0)$ is d-asymptotic ($r_n := d(x_n, x_{n+1}) \to 0$ as $n \to \infty$)*

(31-iii) *$(x_n; n \geq 0)$ is not d-Cauchy.*

Further, let the subset Θ of R^0_+ be $(>)$-cofinal in R^0_+; hence (see above)

> $\mathrm{adm}(\Theta;(x_n))$ *(=the class of (x_n)-admissible points in Θ) is non-empty*

and take a strictly ascending rank sequence $(\lambda(k); k \geq 0)$ in $N(1,\leq)$. Then, for each (x_n)-admissible couple $(\beta,\gamma) \in (>; \mathrm{adm}(\Theta;(x_n)))$ (with the associated rank $b = b(\beta,\gamma)$), there exists a rank sequence $(J(k); k \geq 0)$ in $N(1,\leq)$ and a couple of rank sequences $(m(k); k \geq 0)$ and $(n(k); k \geq 0)$ in $N(1,\leq)$ so that

> (31-a) $k+1 \leq J(k) \leq m(k) < m(k)+3\lambda(k) < n(k)$,
>
> *and* $x_{m(k)} \leq x_{n(k)}$, $\forall k$
>
> (31-b) $J(k) \leq m(k) < m(k)+2\lambda(k) < n(k)-1 < n(k)$, *and*
>
> $d(x_{m(k)}, x_{n(k)}) > \gamma$, $d(x_{m(k)}, x_{n(k)-1}) \leq 3^{-b-k}+\gamma$, $\forall k$,
>
> (31-c) $U_k := d(x_{m(k)}, x_{n(k)}) \to \gamma+$ *as* $k \to \infty$
>
> (31-d) *for each couple of rank sequences $(\mu(k))$ and $(\nu(k))$ in N with $(\mu(k),\nu(k) \leq 3\lambda(k), \forall k)$, $V_k := d(x_{m(k)+\mu(k)}, x_{n(k)+\nu(k)}) \to \gamma+$ as $k \to \infty$*
>
> (31-e) *for each couple of ranks (i,j) in $N \times N$ with $(i,j \leq 3\lambda(0))$,*
>
> *we derive* $S_k := d(x_{m(k)+i}, x_{n(k)+j}) \to \gamma+$ *as* $k \to \infty$
>
> (31-f) *for each couple of ranks (i,j) in $N \times N$, one has*
>
> $T_k := d(x_{m(k)+i}, x_{n(k)+j}); k \geq 0) \to\to \gamma+$ *as* $k \to \infty$.

Proof. Let the (x_n)-admissible couple $(\beta,\gamma) \in (>; \mathrm{adm}(\Theta;(x_n)))$ be fixed in the sequel; note that, as a consequence of $\beta \in \mathrm{adm}(\Theta;(x_n))$, we have

> (rela-1) $E(j) := \{(m,n) \in (<; N); j \leq m < n, d(x_m, x_n) > \beta\} \neq \emptyset$, $\forall j$.

Further, take a strictly ascending rank sequence $(\lambda(k); k \geq 0)$ in $N(1,\leq)$. With the aid of these data, define the rank sequence $(J(k); k \geq 0)$ in $N(1,\leq)$ according to

$$(J(k) = 1 + k + Z(3^{-2b-k}/\lambda(k)); k \geq 0),$$

where $b = b(\beta, \gamma)$ (see above), and the mapping $\varepsilon \mapsto Z(\varepsilon)$ is introduced by the d-asymptotic property of (x_n). Then, denote

$(A(k) = E(J(k)); k \geq 0)$; hence, by definition,
$A(k) := \{(m, n) \in (<; N); J(k) \leq m < n, d(x_m, x_n) > \beta\}$, $k \geq 0$,
with, in addition, $A(k)$ is a non-empty relation over N, for each $k \geq 0$.

By the triangle inequality (and the choice of b)

$(\forall k)$: $(m, n) \in A(k)$ implies $(m, n) \in (<; N)$, $J(k) \leq m < n$, and
$d(x_{m+s}, x_{n+t}) \geq d(x_m, x_n) - d(x_m, x_{m+s}) - d(x_n, x_{n+t})$
$> \beta - 3^{-2b-k+1} - 3^{-2b-k+1} > \beta - 3^{-b-k} > \gamma, \forall s, t \in N[0, 3\lambda(k)]$,

which tells us that

$(\forall k)$: $B(k) := \{(m, n) \in (<; N); J(k) \leq m < n,$
$d(x_{m+s}, x_{n+t}) > \gamma, \forall s, t \in N[0, 3\lambda(k)]\}$ is a non-empty relation over N.

Having this precise, denote for each $k \geq 0$

$m(k) = \min \mathrm{Dom}(B(k))$, $n(k) = \min B(k)(m(k))$.

By this very definition, we get

(pro-1) $(\forall k)$: $k + 1 \leq J(k) \leq m(k) < n(k)$, $x_{m(k)} \leq x_{n(k)}$,
(pro-2) $(\forall k)$: $d(x_{m(k)+s}, x_{n(k)+t}) > \gamma$, $\forall s, t \in N[0, 3\lambda(k)]$.

We claim that the couple $(\beta, \gamma) \in (>; \mathrm{adm}(\Theta; (x_n)))$, the rank sequence $(J(k))$, and the couple of rank sequences $[(m(k)), (n(k))]$ fulfill all desired conclusions.

(i) By (pro-1), it is clear that the first, second, and third relation of (31-a) holds.

(ii) Suppose by contradiction that

$(m(k) <) n(k) \leq m(k) + 3\lambda(k)$, for some $k \geq 0$.

By $m(k) \geq J(k) \geq Z(3^{-2b-k}/\lambda(k))$, the triangle inequality, and the choice of b,

$d(x_{m(k)}, x_{n(k)}) \leq 3^{-2b-k+1} < 3^{-b-k} < \gamma$, in contradiction with (pro-2); whence, the fourth relation in (31-a) holds too.

(iii) The first and second parts of (31-b) are directly obtainable from the preceding stage and (pro-2), respectively. Concerning the third part of (31-b), let $k \geq 0$ be arbitrary fixed. By definition, $n(k)$ is the minimum of all ranks $p \in N$ with

$$(m(k), p) \in B(k), \text{ that is,}$$
$$J(k) \leq m(k) < p \text{ and } d(x_{m(k)+s}, x_{p+t}) > \gamma, \ \forall s, t \in N[0, 3\lambda(k)].$$

As $m(k) < m(k) + 2\lambda(k) < n(k) - 1$, we must have (by this minimal property)

(pro-3) $d(x_{m(k)+s}, x_{n(k)-1+t}) \leq \gamma$, for some $s, t \in N[0, 3\lambda(k)]$.

But, in view of (pro-2), once again,

(pro-4) $d(x_{m(k)+u}, x_{n(k)-1+v}) > \gamma$, $\forall u \in N[0, 3\lambda(k)]$, $\forall v \in N[1, 3\lambda(k)]$.

This, combined with (pro-3), tells us that, necessarily,

(pro-5) $d(x_{m(k)+s}, x_{n(k)-1}) \leq \gamma$, for some $s \in N[0, 3\lambda(k)]$.

By $m(k) \geq Z(3^{-2b-k}/\lambda(k))$, the triangle inequality, and $b \geq 2$, we then have

$$\begin{aligned} d(x_{m(k)}, x_{n(k)-1}) \leq\ & d(x_{m(k)}, x_{m(k)+s}) + d(x_{m(k)+s}, x_{n(k)-1}) \\ & \leq 3^{-2b-k+1} + \gamma \leq 3^{-b-k} + \gamma, \end{aligned}$$

and the last conclusion of (31-b) follows.

(iv) From these facts and triangular inequality,

$$\begin{aligned} \gamma < d(x_{m(k)}, x_{n(k)}) & \leq d(x_{m(k)}, x_{n(k)-1}) + r_{n(k)-1} \\ & \leq 3^{-b-k} + \gamma + r_{n(k)-1}, \ \forall k. \end{aligned}$$

Passing to limit as $k \to \infty$ in this double inequality gives (31-c).

(v) By the very definition of $[(\mu(k)), (\nu(k))]$, and (pro-2),

$V_k > \gamma$, for all $k \geq 0$, so, the first half of (31-d) holds.

Moreover, from a metrical property of d, the very definition of $(m(k); k \geq 0)$ and $(n(k)); k \geq 0)$, and the choice of b,

$$\begin{aligned}&|d(x_{m(k)}, x_{n(k)}) - d(x_{m(k)+\mu(k)}, x_{n(k)+\nu(k)})|\\ &\quad \leq d(x_{m(k)}, x_{m(k)+\mu(k)}) + d(x_{n(k)}, x_{n(k)+\nu(k)})\\ &\quad \leq 3^{-2b-k+1} + 3^{-2b-k+1} < 3^{-2b-k+2} \leq 3^{-b-k}, \quad \text{for all } k \geq 0.\end{aligned}$$

Passing to limit as $k \to \infty$ in the relation between the extremal members of this relation gives the second half of (31-d).

(vi) By the strict ascending property of $(\lambda(k); k \geq 0)$, the sequences $(\mu(k) = i; k \geq 0)$ and $(\nu(k) = j; k \geq 0)$ fulfill $(\mu(k), \nu(k) \leq 3\lambda(k), \forall k)$, and this, along with the preceding stage, yields the desired conclusion.

(vii) Let $(i, j) \in N \times N$ be given. By the strict ascending property of $(\lambda(k); k \geq 0)$, there exists an index $L = L(i, j) \in N$ with

$$i, j \leq 3\lambda(k), \text{ for all } k \geq L.$$

Then, define the couple of sequences $(\mu(k); k \geq 0)$ and $(\nu(k); k \geq 0)$ as

$$(\mu(k) = \nu(k) = 0;\ k \leq L);\ (\mu(k) = i,\ \nu(k) = j;\ k > L).$$

Clearly, $(\mu(k), \nu(k) \leq 3\lambda(k), \forall k)$, and this, along with the preceding stage, yields

$$T_k^* = d(x_{m(k)+\mu(k)}, x_{n(k)+\nu(k)}) \to \gamma+ \text{ as } n \to \infty.$$

It will now suffice observing that

$$T_k = T_k^*, \text{ for all } k > L \text{ (hence, for all } k \geq L+1)$$

to get the written conclusion. The proof is complete. □

Let us say that $[(\beta, \gamma); (J(k)); (m(k)); (n(k))]$ is a *Boyd–Wong* $[\Theta, (\lambda(k))]$*-system* attached to (x_n), provided

(BW-1) $k + 1 \leq J(k) \leq m(k) < m(k) + 3\lambda(k) < n(k), \forall k$

(BW-2) $\forall$ couple of rank sequences $(\mu(k))$ and $(\nu(k))$ in N with $(\mu(k), \nu(k) \leq 3\lambda(k), \forall k)$, $V_k := d(x_{m(k)+\mu(k)}, x_{n(k)+\nu(k)}) \to \gamma+$ as $k \to \infty$

(BW-3) for each couple of ranks (i, j) in $N \times N$ with $(i, j \leq 3\lambda(0))$, $S_k := d(x_{m(k)+i}, x_{n(k)+j}) \to \gamma+$ as $k \to \infty$

(BW-4) for each couple of ranks (i, j) in $N \times N$, one has
$T_k := d(x_{m(k)+i}, x_{n(k)+j}); k \geq 0) \to\to \gamma+$ as $k \to \infty$.

In this case, the result above yields the following characterization of the d-Cauchy property for $(\leq)$-ascending d-asymptotic sequences.

Theorem 8. *Let the sequence $(x_n; n \geq 0)$ in X be $(\leq)$-ascending, d-asymptotic. Then, the following conditions/properties are equivalent:*

(32-a) *$(x_n; n \geq 0)$ is not d-Cauchy;*

(32-b) *for each $(>)$-cofinal subset Θ of R^0_+ and each strictly ascending rank sequence $(\lambda(k); k \geq 0)$ in $N(1, \leq)$, there is at least one Boyd–Wong $[\Theta, (\lambda(k))]$-system attached to (x_n).*

Proof. By the preceding result, the first condition includes the second one. Conversely, if the second condition holds, then, under $\Theta = R^0_+$ and $(\lambda(k) = k + 1; k \geq 0)$, there must be a Boyd–Wong $[\Theta, (\lambda(k))]$-system $[(\beta, \gamma); (J(k)); (m(k)); (n(k))]$ attached to (x_n); hence, in particular,

$$d(x_{m(k)}, x_{n(k)}) \to \gamma+, \text{ as } k \to \infty.$$

This necessarily gives us that $(x_n; n \geq 0)$ is not d-Cauchy; otherwise, under a d-Cauchy condition upon our starting sequence, it follows that

$$d(x_{m(k)}, x_{n(k)}) \to 0, \text{ as } k \to \infty,$$

in contradiction with the limiting property above. □

In particular, when $\Theta = R^0_+$ and $(\lambda(k) = k + 1; k \geq 0)$, the obtained statement covers the 1969 one in Boyd and Wong [10], so, it is natural that this result be referred to in the proposed way. Further aspects may be found in Reich [43]; see also Khan *et al.* [26].

4. Statement of the Problem

Let $(X, d, \leq)$ be a partially ordered metric space. Given the relations $\mathcal{A}, \mathcal{B}$ on X, denote for $Z \in \exp(X)$

$$(\mathcal{A}; \mathcal{B}; Z) = \mathcal{A} \cap \mathcal{B} \cap (Z \times Z), \ (\mathcal{A}; Z) = (\mathcal{A}; \mathcal{A}; Z);$$

in this perspective, $(\mathcal{A};\mathcal{B})$ means $(\mathcal{A};\mathcal{B};X)$, and $\mathcal{A}$ is just $(\mathcal{A};X)$. Likewise, given the couple of sequences $(u_n; n \geq 0)$ and $(v_n; n \geq 0)$ in X, let us put

$$(u_n; n \geq 0)\mathcal{A}(v_n; n \geq 0) \text{ iff } u_i\mathcal{A}v_j, \text{ for all } i, j \in N.$$

Finally, call the subset Y of X, *$\mathcal{B}$-almost-singleton* (in short, *$\mathcal{B}$-asingleton*) provided $y_1, y_2 \in Y, y_1\mathcal{B}y_2 \Longrightarrow y_1 = y_2$, and *$\mathcal{B}$-singleton* when, in addition, Y is non-empty.

Further, let $T \in \mathcal{F}(X)$ be a selfmap of X. In the following, sufficient conditions are given for the existence and/or uniqueness of elements in $\mathrm{Fix}(T)$.

(4-I) Concerning the proposed question, we must say that the points of $\mathrm{Fix}(T)$ are to be reached via iterative processes that may act beyond the limit points of these. Some preliminaries are needed.

Let $\mathrm{aord}(X)$ stand for the class of all admissible (with respect to X) ordinals. Roughly speaking, this concept is to be introduced according to the well-known *Sierpiński inductive covering theorem* [51, Ch XVI, Sect 1]; this, in particular, means that one may arrange for $\mathrm{aord}(X)$ including ω (=the first transfinite ordinal) and Ω (=the first uncountable ordinal). Denote, for the arbitrary fixed $x \in X$,

$T^\omega(x) = \lim_n(T^n x)$, when this limit exists, and then

(for each $i \in N$): $T^{\omega+i}x = T^i z$, where $z := \lim_n(T^n x)$.

(4-I-1) The object of our developments is represented by the *iterative processes.*

Let x_0 be some point in X. By an *orbital* (or: *iterative*) sequence attached to x_0 (and T), we mean any sequence $X_0 := (x_n; n \geq 0)$ (or, simply, $X_0 = (x_n)$) defined as $(x_n = T^n x_0; n \geq 0)$. When x_0 is generic here, the resulting object is referred to as an *orbital* (or *iterative*) sequence on X. Then, we say that the orbital sequence $X_0 = (x_n)$ is *$(\leq)$-ascending* (in short, *ascending*), provided

(x_n) is $(\leq)$-ascending (in short, *ascending*): $x_n \leq x_m$, if $n \leq m$.

As before, when x_0 is generic here, we talk about *ascending orbital sequences* (in short, *(a-o)-sequences*) on X; the class of all these is denoted as $\mathrm{ao}(T)$.

Now, let us consider the condition

(a-reg) T is *ascending regular*: $\text{ao}(T)$ is non-empty.

A basic problem is that of obtaining ascending iterative sequences by starting from global conditions upon T. For example, this is available under

(s-pro) T is $(\leq)$-*semiprogressive* $[X(T, \leq) := \{x \in X; x \leq Tx\} \neq \emptyset]$,

(incr) T is $(\leq)$-*increasing* $[x \leq y \Longrightarrow Tx \leq Ty]$.

We stress that such requirements are just an illustration of the ascending regular property being attainable, so, these conditions will be non-effective in the sequel.

By the ascending regular condition, it results that ascending orbital sequences exist. Fix one of these, $X_0 = (x_n)$, and denote

$[X_0] = \{x_n; n \geq 0\}$ (the *trajectory* attached to $X_0 = (x_n)$),

$[[X_0]] := \text{cl}([X_0])$ (the *complete trajectory* attached to $X_0 = (x_n)$).

Proposition 9. *Under these conventions,*

$$[[X_0]] = [X_0] \cup \{z\}, \textit{ whenever } z := \lim_n(x_n) \textit{ exists.}$$

Proof. Denote for simplicity $U_0 = [X_0]$, $V_0 = [[X_0]]$; hence, $V_0 = \text{cl}(U_0)$. Clearly, $V_0 \supseteq U_0 \cup \{z\}$. Suppose that there exists $v \in V_0$ that is outside $U_0 \cup \{z\}$. By the limit definition, there exists $\sigma > 0$ such that

$$X(v, \sigma) := \{x \in X; d(v, x) < \sigma\} \text{ is disjoint from } U_0 \cup \{z\}.$$

In particular, this tells us that v cannot belong to $\text{cl}(U_0) = V_0$; contradiction. Consequently, $V_0 \subseteq U_0 \cup \{z\}$, and we are done. □

(4-I-2) Passing to the basic part of our setting, denote for $x, y \in X$,

$$Q_1(x, y) = d(x, Tx),\ Q_2(x, y) = d(x, y),\ Q_3(x, y) = d(x, Ty),$$
$$Q_4(x, y) = d(Tx, y),\ Q_5(x, y) = d(Tx, Ty),\ Q_6(x, y) = d(y, Ty),$$
$$\mathcal{Q}(x, y) = (Q_1(x, y), Q_2(x, y), Q_3(x, y), Q_4(x, y), Q_5(x, y), Q_6(x, y)).$$

Further, let us construct the family of functions [for $x, y \in X$]:

$$B_0(x,y) = \min\{Q_2, Q_5\}(x,y),$$
$$B_1(x,y) = \min\{Q_1, Q_2, Q_5, Q_6\}(x,y),$$
$$B_2(x,y) = \min\{Q_2, Q_3, Q_4, Q_5\}(x,y),$$
$$B_3(x,y) = \min\{Q_1, Q_2, Q_3, Q_4, Q_5, Q_6\}(x,y)$$

and introduce the relations (over X)

(rela-B) $(B_i > 0) = \{(x,y) \in X \times X; B_i(x,y) > 0\}$, $i \in \{0,1,2,3\}$.

Finally, let Υ be a non-empty subset of R_+^6, and $\mathcal{S} = \mathcal{S}(\Upsilon)$ stand for the relation

$\mathcal{S} = \mathcal{Q}^{-1}(\Upsilon)$; hence, $\mathcal{Q}(\mathcal{S}) \subseteq \Upsilon$.

Given the relation $\mathcal{R}$ over X, let us introduce the contractive property

(R-contr) T is $(\leq, \mathcal{R}; \Upsilon)$-*contractive*, provided $(\leq; \mathcal{R}) \subseteq \mathcal{S}$;
that is, $\mathcal{Q}(x,y) \in \Upsilon$, $\forall (x,y) \in (\leq; \mathcal{R})$.

As usual, the starting point of our metrical fixed point theory for these objects is the construction of a $(\leq)$-ascending orbital sequence $X_0 = (x_n)$ with

(R-asc) $X_0 = (x_n)$ is $\mathcal{R}$-ascending: $x_n \mathcal{R} x_{n+1}$, for all n.

The general aspects of this method will be discussed elsewhere. Here, we consider a particular case of it — that, in fact, includes a large number of such contractions — based on the restrictive condition

(B1-adm) $\mathcal{R}$ is B_1-*admissible*: $(B_1 > 0) \subseteq \mathcal{R}$.

This tells us that, in place of our general contraction, we study its restrictive part

(B1-contr) T is $(\leq, (B_1 > 0); \Upsilon)$-*contractive*, provided $(\leq; B_1 > 0) \subseteq \mathcal{S}$;
that is, $\mathcal{Q}(x,y) \in \Upsilon$, $\forall (x,y) \in (\leq; B_1 > 0)$.

Another condition to be used, in certain moments of our exposition, is

(B2-adm) $\mathcal{R}$ is B_2-admissible: $(B_2 > 0) \subseteq \mathcal{R}$,

with the associated (stronger) contractive condition

(B2-contr) T is $(\leq, (B_2 > 0); \Upsilon)$-*contractive*, provided $(\leq; B_2 > 0) \subseteq \mathcal{S}$;
that is, $\mathcal{Q}(x, y) \in \Upsilon$, $\forall (x, y) \in (\leq; B_2 > 0)$.

This, as we will see, has an impact upon the uniqueness property. Note that both admissible conditions above hold under

(B0-adm) $\mathcal{R}$ is B_0-admissible: $(B_0 > 0) \subseteq \mathcal{R}$,

and both associated contractive conditions are fulfilled under

T is $(\leq, (B_0 > 0); \Upsilon)$-*contractive*, provided $(\leq; B_0 > 0) \subseteq \mathcal{S}$;
that is, $\mathcal{Q}(x, y) \in \Upsilon$, $\forall (x, y) \in (\leq; B_0 > 0)$.

(4-II) Returning to the B_1-admissible setting, we stress that, in this case, our initial objective of constructing a $\mathcal{R}$-ascending orbital sequence is directly attainable, in terms of the the coarser relation $(B_1 > 0)$. This may be carried out as follows. Call the orbital sequence $X_0 = (x_n)$, $(B_1 > 0)$-*ascending* provided

(x_n) is $(B_1 > 0)$-ascending: $B_1(x_n, x_{n+1}) > 0$, for all n,
referred to as (x_n) is B-*ascending*.

We then say that the (a-o) sequence $X_0 = (x_n)$ has the *B-property* or, equivalently, that $X_0 = (x_n)$ is (Ba-o). The possibility of reaching such a property is a consequence of the following reasoning. Given the ascending orbital sequence $X_0 = (x_n)$, we have two alternatives.

(Alt-1) The ascending orbital sequence $X_0 = (x_n)$ is *telescopic*, in the sense

there exists $h \geq 0$, such that $d(x_h, x_{h+1}) = 0$; i.e., $x_h = x_{h+1}$.

By the iterative definition of our sequence, one derives

$$x_n = x_h, \text{ for all } n \geq h; \text{ whence, } x_n \xrightarrow{d} x_h \in \text{Fix}(T).$$

Consequently, this case is completely clarified from the fixed point perspective.

(Alt-2) The ascending orbital sequence $X_0 = (x_n)$ is *non-telescopic*, in the sense

$$d(x_n, x_{n+1}) > 0 \text{ (hence, } x_n < x_{n+1}\text{), for all } n.$$

Note that, in this case,

$$(B_1(x_n, x_{n+1}) > 0, \forall n)\text{; hence, } X_0 = (x_n) \text{ is (Ba-o).}$$

A useful characterization of this property may be given as follows. (Since the verification is immediate, we do not give details.)

Proposition 10. *For the (Ba-o) sequence $X_0 = (x_n)$, the following are equivalent:*

(42-1) $X_0 = (x_n)$ *is non-telescopic (see above)*

(42-2) $X_0 = (x_n)$ *is strictly ascending: $i < j$ implies $x_i < x_j$*

(42-3) $X_0 = (x_n)$ *is full: $i \neq j$ implies $x_i \neq x_j$.*

Summing up, the ascending orbital sequences to be used in the sequel are, in addition, full; hence, (Ba-o-f) (in short). In this case, the basic directions under which the investigations be conducted are described in the following list:

(opic-1) We say that the (Ba-o-f) sequence $X_0 = (x_n)$ is *semi-Picard* (modulo $(\leq, \mathcal{R}; T)$) when (x_n) is d-asymptotic.

(opic-2) We say that the (Ba-o-f) sequence $X_0 = (x_n)$ is *Picard* (modulo $(\leq, \mathcal{R}; T)$) when (x_n) is d-Cauchy.

(opic-3) We say that the (Ba-o-f) sequence $X_0 = (x_n)$ is *strongly Picard* (modulo $(\leq, \mathcal{R}; T)$) when $T^\omega x := \lim_n(T^n x)$ exists with $T^\omega x \in \text{Fix}(T)$.

(opic-4) We say that the (Ba-o-f) sequence $X_0 = (x_n)$ is *Bellman Picard* (modulo $(\leq, \mathcal{R}; T)$) when $T^\omega x := \lim_n(T^n x)$ exists, $T^\omega x \in \text{Fix}(T)$, with, in addition, $T^n x \leq T^\omega x$, $\forall n \in N$.

(opic-5) We say that the (Ba-o-f) sequence $X_0 = (x_n)$ is *almost Bellman Picard* (modulo $(\leq, \mathcal{R}; T)$) when $T^\omega x := \lim_n(T^n x)$ exists, $T^{\omega+i} x \in \text{Fix}(T)$ for some $i \in N$ and $T^\beta x \leq T^{\omega+i} x$, for all $\beta \leq \omega + i$.

(opic-6) We say that T is *fix-($\leq$)-asingleton*, provided Fix(T) is ($\leq$)-asingleton, and *fix-($\leq$)-singleton*, in case Fix(T) is ($\leq$)-singleton.

In particular, when

T is ($\leq$)-semiprogressive and ($\leq$)-increasing,

these conventions are comparable with the ones in Turinici [58]. Note that, as $(\leq) = X \times X$ is not possible (by the antisymmetric property), these developments are not comparable with the ones in Rus [47, Ch 2, Sect 2.2].

Sufficient conditions for such properties are obtainable under the following lines:

(reg-1) Let Y be a (nonempty) subset of X. Then, we say that

(com-X) Y is *(Ba-o-f,d)-complete in X*, provided: each d-Cauchy

(Ba-o-f)-sequence in Y, d-converges towards some point of X.

(com-it) Y is *(Ba-o,d)-complete (in itself)*, provided: each d-Cauchy

(Ba-o-f)-sequence in Y, d-converges towards some point of Y.

(reg-2) Let us say that T is *(Ba-o-f,d)-continuous* when, for any (Ba-o-f) sequence $Y_0 = (y_n; n \geq 0)$, one has $y_n \overset{d}{\longrightarrow} z$ implies $Ty_n \overset{d}{\longrightarrow} Tz$.

(reg-3) Call the ambient partial order ($\leq$),

(Ba-o-f,d)-selfclosed when, for any (Ba-o-f) sequence $Y_0 = (y_n; n \geq 0)$,

one has $y_n \overset{d}{\longrightarrow} z$ implies $(y_n; n \geq 0) \leq z$,

(Ba-o-f,d)-sup-selfclosed when, for any (Ba-o-f) sequence

$Y_0 = (y_n; n \geq 0)$, one has $y_n \overset{d}{\longrightarrow} z$ implies $\sup(y_n; n \geq 0) = z$.

Here, as usual,

(bd) $(y_n; n \geq 0) \leq z$ means $y_n \leq z$, for each n

(sup-bd) $\sup(y_n; n \geq 0) = z$ means: $(y_n; n \geq 0) \leq z$ and $((y_n; n \geq 0) \leq w$ implies $z \leq w)$.

The following auxiliary fact involving these notions is in effect for us.

Proposition 11. *Suppose that T is $(\leq)$-semiprogressive, $(\leq)$-increasing, X is (Ba-o-f,d)-complete, and $(\leq)$ is (Ba-o-f,d)-sup-selfclosed. Then, the (non-empty) subset $X(T, \leq)$ is (Ba-o-f, d)-complete in itself.*

Proof. Let $(z_n; n \geq 0)$ be a d-Cauchy (Ba-o-f)-sequence in $X(T, \leq)$. As X is (Ba-o-f,d)-complete, $z_n \xrightarrow{d} z$ as $n \to \infty$, for some $z \in X$, and this, along with $(\leq)$ being (Ba-o,d)-sup-selfclosed, gives $\sup(z_n; n \geq 0) = z$. Combining with the increasing property of T gives

$$(\forall n)\colon z_n \leq z \Longrightarrow z_n \leq Tz_n \leq Tz$$

so that by the definition of supremum, $z \leq Tz$; i.e., $z \in X(T, \leq)$. □

(4-III) In addition to this, let us attach to each (Ba-o-f) sequence $X_0 = (x_n)$ an orbital sequence $Z_0 := \mathcal{L}(X_0)$, referred to as the *limit orbital sequence* of $X_0 = (x_n)$. Two alternatives of this functional construction must be discussed.

(FC-1) Suppose that

$X_0 = (x_n)$ is *d-divergent*: $\lim_n(x_n)$ does not exist.

By definition, we put $\mathcal{L}(X_0) = X_0$ (the initial (Ba-o-f) sequence).

(FC-2) Suppose that

$X_0 = (x_n)$ is *d-convergent*: $z_0 := \lim_n(x_n)$ exists.

In this case, one may construct in a unique way, an orbital sequence $Z_0 = (z_n)$, as $(z_n = T^n z_0; n \geq 0)$. By definition, we then put $\mathcal{L}(X_0) = Z_0$.

Now, let us say that the (Ba-o-f) sequence $X_0 = (x_n)$ is *regular*, provided

the associated limit orbital sequence $Z_0 = \mathcal{L}(X_0)$ is ascending; hence, an (a-o) sequence.

Note that, when $X_0 = (x_n)$ is d-divergent, this happens. Concerning the d-convergent case, the following regularity criterion is useful in practice.

Proposition 12. *Suppose that T is $(\leq)$-increasing, $(\leq)$ is (Ba-o-f,d)-sup-selfclosed, and let the (Ba-o-f) sequence $X_0 = (x_n)$ be d-convergent. Then,*

(44-1) *$X_0 = (x_n)$ is regular: $Z_0 := \mathcal{L}(X_0)$ is an (a-o) sequence,*

(44-2) *in addition, $X_0 \leq Z_0$ ($x_i \leq z_j$, for $i, j \in N$).*

Proof. Suppose that $z_0 := \lim_n(x_n)$ exists. As $(\leq)$ is (Ba-o-f)-sup-selfclosed, we must have $\sup(x_n; n \geq 0) = z_0$. In this case,

$$(x_n \leq z_0 \text{ (whence, } x_n \leq Tx_n \leq Tz_0), \forall n) \text{ so that } z_0 \leq Tz_0,$$

by the very definition of supremum. Combining with T=increasing, one derives that the sequence $Z_0 := (z_n; n \geq 0)$, where $(z_n = T^n z_0; n \geq 0)$ is ascending orbital, and, in addition, $X_0 \leq Z_0$. The proof is complete. □

(4-IV) Having these precise, we now state the working conditions upon our data. So, let $X_0 = (x_n)$ be a (Ba-o-f) sequence in X, and $Z_0 := \mathcal{L}(X_0)$ be the limit orbital sequence attached to X_0. Then, denote

$$[X_0] = \{x_n; n \geq 0\}, [[X_0]] = \text{cl}[x_0]), [Z_0] = \{z_n; n \geq 0\}.$$

Finally, define the couple of real sequences

$$(r_n(X_0) = d(x_n, x_{n+1}); n \geq 0), (p_n(X_0) = d(x_n, x_{n+2}); n \geq 0);$$

clearly, $(r_n := r_n(X_0))$ and $(p_n := p_n(X_0))$ and sequences in R^0_+ and R_+, respectively, with in addition

$$(r_n, r_n, p_n, 0, r_{n+1}, r_{n+1}) = \mathcal{Q}(x_n, x_{n+1}) \in \mathcal{Q}(\leq; B_1 > 0; [X_0])$$
$$\text{and } |p_n - r_n| \leq r_{n+1}, \text{ for all } n;$$

these are referred to as the *telescopic sequences* attached to X_0.

(I) The first condition is of *asymptotic* type. Two variants of it are of interest.

(I-a) Call Υ, *asymptotic* on $[X_0]$ when

(asy) the telescopic sequences $(r_n := r_n(X_0))$ and $(p_n := p_n(X_0))$

attached to X_0 satisfy $r_n \to 0$; hence, $p_n \to 0$ as well.

(I-b) Call Υ, *descending asymptotic* on $[X_0]$ when

(desc-asy) the telescopic sequences $(r_n := r_n(X_0))$ and $(p_n := p_n(X_0))$ attached to X_0 satisfy (r_n)=strictly descending and $r_n \to 0$; hence,

via $(0 < r_n - r_{n+1} \leq p_n \leq r_n + r_{n+1}, \forall n)$, (p_n) is a sequence in R^0_+ with (in addition) $p_n \to 0$.

Clearly, the following inclusion holds:

Υ is descending asymptotic on $[X_0]$ implies Υ is asymptotic on $[X_0]$.

However, as we will see, the immense majority of contractive conditions is based on the descending asymptotic hypothesis and not on the asymptotic one.

(II) The second condition to be considered is related to *right* properties. Take some point $c = (c_1, \ldots, c_6)$ in R^6_+. We say that the sequence $(t^n := (t^n_1, \ldots, t^n_6); n \geq 0)$ in R^6_+ is *right* at c if

(r-c) $(t^n_i \to c_i+)$ as $n \to \infty$, for all $i \in \{1, \ldots, 6\}$.

Given $b > 0$, let us say that Υ is *nright* at b on $[X_0]$ if

(nright) for each sequence $(t^n; n \geq 0)$ in $\mathcal{Q}(\leq; B_3 > 0; [X_0]) \cap \Upsilon$, the right property at $(0, b, b, b, b, 0)$ is not true.

The class of all these $b > 0$ is denoted as $\text{nright}(\Upsilon; [X_0])$. In this case, define

(a-n-r) Υ is *almost nright* on $[X_0]$ if $\Theta := \text{nright}(\Upsilon; [X_0])$

is $(>)$-cofinal in R^0_+ (for each $\varepsilon \in R^0_+$ there exists $\theta \in \Theta$ with $\varepsilon > \theta$)

(n-r) Υ is *nright* on $[X_0]$, if $\Theta := \text{nright}(\Upsilon; [X_0])$ is identical with R^0_+.

(III) The third condition is a bilateral version of the preceding one. Take some point $c = (c_1, \ldots, c_6)$ in R^6_+. We say that the sequence $(t^n := (t^n_1, \ldots, t^n_6); n \geq 0)$ in R^6_+ is *bilateral* at c if

(bila-c) $t^n_i \to c_i$, as $n \to \infty$, $\forall i \in \{1, \ldots, 6\}$.

Given $b > 0$, let us say that Υ is *nbilateral* at b on $[X_0] \cup [Z_0]$ if

(nbila) for each sequence $(t^n; n \geq 0)$ in $\mathcal{Q}(\leq; B_3 > 0; [X_0] \cup [Z_0]) \cap \Upsilon$,

the bilateral property at $(0, b, b, b, b, 0)$ is not true.

The class of all these $b > 0$ is denoted as $\text{nbila}(\Upsilon; [X_0] \cup [Z_0])$. Define

(n-b) Υ is *nbilateral* on $[X_0] \cup [Z_0]$ if

$\Theta := \text{nbila}(\Upsilon; [X_0] \cup [Z_0])$ is identical with R^0_+.

(IV) The fourth condition involves *point* properties. Take some point $c = (c_1, \ldots, c_6)$ in R^6_+. We say that the sequence $(t^n := (t^n_1, \ldots, t^n_6); n \geq 0)$ in R^6_+ is *point* at c if

(p-c) $[t^n_i \to c_i$ as $n \to \infty$, $\forall i]$, and $[t^n_6 = c_6$, $\forall n]$.

Given $b > 0$, let us say that Υ is *npoint* at b on $[[X_0]]$ if

(npoint) for each sequence $(t^n; n \geq 0)$ in $\mathcal{Q}(\leq; B_3 > 0; [[X_0]]) \cap \Upsilon$,

the point property at $(0, 0, b, 0, b, b)$ is not true.

The class of all these $b > 0$ is denoted as $\text{npoint}(\Upsilon; [[X_0]])$. Then, define

(n-p) Υ is *npoint* on $[[X_0]]$, if $\Theta := \text{npoint}(\Upsilon; [[X_0]])$ is identical with R^0_+.

(V) The fifth condition to be posed is *normality*. Given $a > 0$, let us say that

(norm) Υ is *normal* at a, when $(0, a, a, a, a, 0) \in \Upsilon$,
(n-norm) Υ is *nnormal* at a, when $(0, a, a, a, a, 0) \notin \Upsilon$.

The class of all nnormal $a > 0$ is denoted as $\text{nnorm}(\Upsilon)$. Then, define

(n-n) Υ is *nnormal*, if $\Theta := \text{nnorm}(\Upsilon)$ is identical with R^0_+.

Some concrete examples of such conditions are given a bit further.

5. Main Result

Let $(X, d, \leq)$ be a partially ordered metric space. Further, let $T \in \mathcal{F}(X)$ be a selfmap of X, with

(a-reg) T is *ascending regular*:
there are ascending orbital sequences in X.

In this case, the family $\text{ao}(T)$ of all these objects is non-empty; let $X_0 = (x_n)$ be an element of it; remember that, without loss, one may assume that

$X_0 = (x_n)$ is *non-telescopic*: hence, a (Ba-o-f) sequence.

Then, let $Z_0 := \mathcal{L}(X_0)$ be the limit orbital sequence attached to X_0.

As precise, we are interested to determine sufficient conditions for the (uniqueness and) existence of elements in $\text{Fix}(T)$, via contractive-type requirements involving the iterative processes $[X_0 = (x_n), Z_0 = (z_n)]$ and their associated sets:

$$[X_0] = \{x_n; n \geq 0\},\ [[X_0]] = \text{cl}([X_0]),\ [Z_0] = \{z_n; n \geq 0\}.$$

The basic directions and regularity conditions relative to X_0 under which the problem of determining fixed points of T is to be solved were already listed, and the contractive-type framework (involving X_0 and Z_0) was settled.

We are now in position to state our main result in this exposition.

Theorem 13. *Let the subset $\Upsilon \in \exp(R^6_+)$ and the relation $\mathcal{R}$ be such that*

(51-i) *T is $(\leq, \mathcal{R}; \Upsilon)$-contractive, and $\mathcal{R}$ is B_1-admissible*

(51-ii) *Υ is asymptotic on $[Y_0]$, for each (Ba-o-f) sequence $Y_0 = (y_n)$*

(51-iii) *Υ is almost nright on $[Y_0]$, for each (Ba-o-f) sequence $Y_0 = (y_n)$.*

Further, assume that X is (Ba-o-f,d)-complete. Then (with $X_0 = (x_n)$ and $Z_0 = (z_n)$ taken as before)

(51-a) *$X_0 = (x_n)$ is semi-Picard (modulo $(\leq, \mathcal{R}; T)$)*

(51-b) *$X_0 = (x_n)$ is Picard (modulo $(\leq, \mathcal{R}; T)$)*

(51-c) *$X_0 = (x_n)$ is strongly Picard (modulo $(\leq, \mathcal{R}; T)$), whenever T is (Ba-o-f,d)-continuous*

(51-d) *$X_0 = (x_n)$ is Bellman Picard (modulo $(\leq, \mathcal{R}; T)$), whenever $(\leq)$ is (Ba-o-f,d)-selfclosed and Υ is npoint on $[[X_0]]$*

(51-e) *$X_0 = (x_n)$ is almost Bellman Picard (modulo $(\leq, \mathcal{R}; T)$), whenever $(\leq)$ is (Ba-o-f,d)-sup-selfclosed, $X_0 = (x_n)$ is regular, and Υ is nbilateral on $[X_0] \cup [Z_0]$*

(51-f) *T is fix-$(\leq)$-asingleton (hence, fix-$(\leq)$-singleton, under either of these extra requirements) when, in addition to the conditions above, ($\mathcal{R}$ is B_2-admissible and Υ is nnormal).*

Proof. Before developing the argument, we need some technical facts to be derived, for a later use. The following auxiliary statement collects all of these.

Proposition 14. *Let the conditions of the main result be accepted and take some (Ba-o-f) sequence $Y_0 = (y_n)$. Then,*

(51-1) *$Y_0 = (y_n)$ is d-asymptotic; hence, semi-Picard (modulo $(\leq, \mathcal{R}; T)$)*

(51-2) *$Y_0 = (y_n)$ is d-convergent: $z_0 := \lim_n(y_n)$ exists*

(51-3) *$Y_0 < z_0$ (that is: $y_n < z_0$, for all n),*
whenever $(\leq)$ is (Ba-o-f,d)-selfclosed.

Proof. There are several parts to be passed.

Part 1. Let $(r_n := r_n(Y_0))$ and $(p_n := p_n(Y_0))$ be the telescopic sequences attached to $Y_0 = (y_n)$:

$$(r_n(Y_0) = d(y_n, y_{n+1}); n \geq 0), \ (p_n(Y_0) = d(y_n, y_{n+2}); n \geq 0);$$

as precise, these enjoy the properties

$(r_n, r_n, p_n, 0, r_{n+1}, r_{n+1}) = \mathcal{Q}(y_n, y_{n+1}) \in \mathcal{Q}(\leq; B_1 > 0; [Y_0])$
and $|p_n - r_n| \leq r_{n+1}$, for all n.

By the asymptotic property of Υ at $[Y_0]$, one derives

(sd-asy) $r_n \to 0$ (hence, $p_n \to 0$), as $n \to \infty$; proving that
$Y_0 = (y_n)$ is d-asymptotic; hence, semi-Picard (modulo $(\leq, \mathcal{R}; T)$).

Part 2. As Υ is almost nright at $[Y_0]$, $\Theta := \text{nright}(\Upsilon; [Y_0])$ appears as $(>)$-cofinal in R^0_+. We show that, under these conditions, $Y_0 = (y_n)$ is d-Cauchy. Suppose not, and let $(\lambda(k))$ be a strictly ascending sequence in $N(1, \leq)$. By a previous auxiliary fact, there exists a Boyd–Wong $[\Theta, (\lambda(k))]$-system $[(\beta, \gamma); (J(k)); (m(k)); n(k))]$ attached to (x_n) (where $(\beta, \gamma) \in (>; \text{adm}(\Theta; (x_n)))$, $(J(k))$ is a rank sequence in $N(1, \leq)$, and $[(m(k)), (n(k))]$ is a couple of rank-sequences), with the properties

(BW-1) $k + 1 \leq J(k) \leq m(k) < m(k) + 3\lambda(k) < n(k)$, $\forall k$

(BW-2) $\forall$ couple of rank sequences $(\mu(k))$ and $(\nu(k))$ in N with $(\mu(k), \nu(k) \leq 3\lambda(k), \forall k)$, $V_k := d(y_{m(k)+\mu(k)}, y_{n(k)+\nu(k)}) \to \gamma+$ as $k \to \infty$

(BW-3) for each couple of ranks (i, j) in $N \times N$ with $(i, j \leq 3\lambda(0))$, $S_k(i, j) := d(y_{m(k)+i}, y_{n(k)+j}) \to \gamma+$ as $k \to \infty$

(BW-4) for each couple of ranks (i, j) in $N \times N$, one has $T_k(i, j) := d(y_{m(k)+i}, y_{n(k)+j}); k \geq 0) \to\to \gamma+$ as $k \to \infty$.

By the d-asymptotic property of the (Ba-o-f) sequence $Y_0 = (y_n)$,

$(t^k_1 := r_{m(k)}; k \geq 0)$ and $(t^k_6 := r_{n(k)}; k \geq 0)$

are sequences in R^0_+ with $t^k_1, t^k_6 \to 0+$ as $k \to \infty$.

Moreover, taking these properties into account yields

$(t^k_2 := S_k(0, 0); k \geq 0)$, $(t^k_3 := S_k(0, 1); k \geq 0)$, and

$(t^k_4 := S_k(1, 0); k \geq 0)$, $(t^k_5 := S_k(1, 1); k \geq 0)$

are sequences in R^0_+ with $t^k_2, t^k_3, t^k_4, t^k_5 \to \gamma+$ as $k \to \infty$;

hence, putting these together,

(t-right) the vectorial sequence $(t^k := (t^k_1, t^k_2, t^k_3, t^k_4, t^k_5, t^k_6); k \geq 0)$ is right at $(0, \gamma, \gamma, \gamma, \gamma, 0)$.

In addition, by the choice of our system and $Y_0 = (y_n)$ being a (Ba-o-f) sequence,

$(\forall k)$: $y_{m(k)} < y_{n(k)}$, and $B_3(y_{m(k)}, y_{n(k)})$

$$= \min\{r_{m(k)}, S_k(0, 0), S_k(0, 1), S_k(1, 0), S_k(1, 1), r_{n(k)}\} > 0;$$

whence, $t^k \in \mathcal{Q}(\leq; B_3 > 0; [X_0]) = \mathcal{Q}(\leq; B_3 > 0; [X_0]) \cap \Upsilon$, where the last equality is a consequence of contractive condition. This, via (t-right), contradicts the choice of γ as an element of $\Theta :=$ nright$(\Upsilon; [Y_0])$. Hence, our working assumption is not acceptable, and the assertion follows.

Part 3. From these developments, $Y_0 = (y_n)$ is (Ba-o-f) and d-Cauchy. As X is (Ba-o-f,d)-complete, we must have

Y_0 is d-convergent: $y_n \xrightarrow{d} z_0$ as $n \to \infty$, for some $z_0 \in X$.

This tells us that, whenever $(\leq)$ is (Ba-o-f,d)-selfclosed, we must have

$Y_0 \leq z_0$ (that is: $y_n \leq z_0$, $\forall n$), and this yields
(by the (Ba-o-f) property): $Y_0 < z_0$ (that is: $y_n < z_0$, for all n).

The proof is thereby complete. □

With these preliminaries, we may now pass to the basic steps in the proof.

Step 1. By the above auxiliary fact, applied to the (Ba-o-f) sequence $X_0 = (x_n)$, we get the conclusions

(X-con-1) $X_0 = (x_n)$ is semi-Picard (modulo $(\leq, \mathcal{R}; T)$)
(X-con-2) $X_0 = (x_n)$ is Picard (modulo $(\leq, \mathcal{R}; T)$)
(X-con-2) $X_0 = (x_n)$ is d-convergent: $z_0 := \lim_n(x_n)$ exists.

Three possible cases — treated in the following steps — are to be discussed.

Step 2. Suppose that T is (Ba-o-f,d)-continuous. Then, $u_n := Tx_n \xrightarrow{d} Tz_0$ as $n \to \infty$. On the other hand, $(u_n = x_{n+1}; n \geq 0)$ is a subsequence of $(x_n; n \geq 0)$; whence, $u_n \xrightarrow{d} z_0$ as $n \to \infty$. Combining with d=separated yields $z_0 = Tz_0$.

Step 3. Suppose that $(\leq)$ is (Ba-o-f,d)-selfclosed and Υ is npoint on $[[X_0]]$. From the convergence property above, one has

(eva-z) $X_0 \leq z_0$ (that is: $x_n \leq z_0$, $\forall n$), and this yields
(by the (Ba-o-f) property): $X_0 < z_0$ (that is: $x_n < z_0$, for all n).

From the same (Ba-o-f) property of $X_0 = (x_n)$,

$H_1 := \{n \in N; Tx_n = Tz_0\}$, $H_2 := \{n \in N; x_n = Tz_0\}$ are asingletons, and this tells us that

$(\exists i = i(z_0) \in N$ such that
$n \geq i$ implies $Tx_n \neq Tz_0$ (hence, $x_n \neq z_0$) and $x_n \neq Tz_0$.

Denote for simplicity $(u_n = x_{n+i}; n \geq 0)$; clearly, by the preceding relation,

$(\forall n)$: $Tu_n \neq Tz_0$ (hence, $u_n \neq z_0$) and $u_n \neq Tz_0$.

Again combining with the (Ba-o-f) property of $X_0 = (x_n)$, one derives

(posi-1) $(\forall n)$: $Q_1(u_n, z_0) = d(u_n, Tu_n) > 0$,
$Q_2(u_n, z_0) = d(u_n, z_0) > 0$, $Q_3(u_n, z_0) = d(u_n, Tz_0) > 0$,
$Q_4(u_n, z_0) = d(Tu_n, z_0) > 0$, $Q_5(u_n, z_0) = d(Tu_n, Tz_0) > 0$.

Suppose by contradiction that

(posi-2) $b := d(z_0, Tz_0) > 0$ [whence, $Q_6(u_n, z_0) = b > 0$, $\forall n$].

We show that this is not compatible with Υ being npoint at b.

From the preceding observations, we have

(t-inclu) $(\forall n)$: $u_n < z_0$, and
$B_3(u_n, z) = \min\{Q_1(u_n, z_0), \ldots, Q_6(u_n, z_0)\} > 0$; whence,
$t^n := \mathcal{Q}(u_n, z_0) \in \mathcal{Q}(\leq; B_3 > 0; [[X_0]]) = \mathcal{Q}(\leq; B_3 > 0;$
$[[X_0]]) \cap \Upsilon$,

where the last equality is a consequence of contractive condition. Let us evaluate the convergence properties of (t^n). From the preceding facts,

$(t_1^n := Q_1(u_n, z_0); n \geq 0)$, $(t_2^n := Q_2(u_n, z_0); n \geq 0)$,
$(t_3^n := Q_3(u_n, z_0); n \geq 0)$, $(t_4^n := Q_4(u_n, z_0); n \geq 0)$,
$(t_5^n := Q_5(u_n, z_0); n \geq 0)$ are sequences in R_+^0 with
$(t_1^n, t_2^n, t_3^n, t_4^n, t_5^n) \to (0, 0, b, 0, b)$ as $n \to \infty$.

At the same time,

$(t_6^n := Q_6(u_n, z_0) = b; n \geq 0)$ is a constant sequence, so, putting these together,

(t-point) $(t^n := (t_1^n, t_2^n, t_3^n, t_4^n, t_5^n, t_6^n); n \geq 0)$ is point at $(0, 0, b, 0, b, b)$.

This, along with (t-inclu), contradicts the relation $b \in \text{npoint}(\Upsilon; [[X_0]])$. Hence, the assumption $b > 0$ cannot be accepted, and then, $b = 0$; that is, $z_0 \in \text{Fix}(T)$.

Step 4. Suppose that $(\leq)$ is (Ba-o-f,d)-sup-selfclosed, $X_0 = (x_n)$ is regular, and Υ is nbilateral on $[X_0] \cup [Z_0]$, where $Z_0 := \mathcal{L}(X_0)$. In view of $x_n \xrightarrow{d} z_0$ as $n \to \infty$, we have that $Z_0 := \mathcal{L}(X_0)$ is represented as $Z_0 = (z_n)$, where $(z_n = T^n z_0; n \geq 0)$. Moreover, under the regularity property, we derive

$Z_0 = (z_n)$ is ascending (hence, an (a-o) sequence) and

$X_0 \leq Z_0$; that is: $x_i \leq z_j$, for all $i, j \in N$.

Denote for simplicity $(\rho_n := d(z_n, z_{n+1}); n \geq 0)$. Two possible alternatives come into discussion.

(Alt-1) Suppose that

$Z_0 = (z_n)$ is telescopic: $H := \{n \in N; \rho_n = 0\}$ is non-empty.

Putting $i := \min(H)$, one gets

$$z_i = T^{\omega+i} x_0 \in \text{Fix}(T) \text{ and } T^\beta x_0 \leq T^{\omega+i} z_0, \text{ for all } \beta \leq \omega + i,$$

proving that $X_0 = (x_n)$ is almost Picard (modulo $(d, \leq; T)$).

(Alt-2) Suppose that

$Z_0 = (z_n)$ is non-telescopic: $\rho_n > 0$, for each n.

Note that, as a direct consequence of this, one has (by the ascending property of our iterative sequence)

(Con-1) $(z_n < z_{n+1}, \forall n)$, $(B_1(z_n, z_{n+1}) = \min\{\rho_n, \rho_{n+1}\} > 0, \forall n)$:

hence, $Z_0 = (z_n)$ is a (Ba-o-f) sequence

(Con-2) $X_0 < Z_0$ (that is: $x_i < z_j$, for all $i, j \in N$).

We prove that this alternative yields a contradiction.

By the conclusion (Con-1), the auxiliary fact above applies to the (Ba-o-f) sequence $Z_0 = (z_n)$ and yields

(Z-con-1) $Z_0 = (z_n)$ is Picard (modulo $(\leq, \mathcal{R}; T)$)

(Z-con-2) $Z_0 = (z_n)$ is d-convergent: $w_0 := \lim_n(z_n)$ exists

(Z-con-3) $Z_0 < w_0$ (that is: $z_n < w_0$, $\forall n$); hence, $b := d(z_0, w_0) > 0$.

Further, by the properties of the couple (X_0, Z_0),

(Posi) $(\forall n)$: $Q_1(x_n, z_n) = d(x_n, x_{n+1}) > 0$, $Q_2(x_n, z_n) = d(x_n, z_n) > 0$,
$Q_3(x_n, z_n) = d(x_n, z_{n+1}) > 0$, $Q_4(x_n, z_n) = d(x_{n+1}, z_n) > 0$,
$Q_5(x_n, z_n) = d(x_{n+1}, z_{n+1}) > 0$, $Q_6(x_n, z_n) = d(z_n, z_{n+1}) > 0$.

As a consequence of these,

(t-inclus) $(\forall n)$: $x_n < z_n$ and
$B_3(x_n, z_n) = \min\{Q_1(x_n, z_n), \ldots, Q_6(x_n, z_n)\} > 0$; whence,
$t^n := Q(x_n, z_n) \in \mathcal{Q}(\leq; B_3 > 0; [X_0] \cup [Z_0]) = \mathcal{Q}(\leq; B_3 > 0; [X_0] \cup [Z_0]) \cap \Upsilon$,

where the last equality is a consequence of contractive condition. Let us evaluate the convergence properties of (t^n). Clearly,

$(t_1^n := Q_1(x_n, z_n); n \geq 0)$, $(t_6^n := Q_6(x_n, z_n); n \geq 0)$
are sequences in R_+^0 with $(t_1^n, t_6^n) \to (0, 0)$ as $n \to \infty$.

At the same time (taking a metric property of d into account),

$(t_2^n := Q_2(x_n, z_n); n \geq 0)$, $(t_3^n := Q_3(x_n, z_n); n \geq 0)$,
$(t_4^n := Q_4(x_n, z_n); n \geq 0)$, $(t_5^n := Q_5(x_n, z_n); n \geq 0)$
are sequences in R_+^0 with $(t_2^n, t_3^n, t_4^n, t_5^n) \to (b, b, b, b)$ as $n \to \infty$,

so, putting these together,

(t-bila) $(t^n := (t_1^n, t_2^n, t_3^n, t_4^n, t_5^n, t_6^n); n \geq 0)$ is bilateral at $(0, b, b, b, b, 0)$.

This, via (t-inclus), contradicts the relation $b \in \text{nbila}(\Upsilon; [X_0] \cup [Z_0])$. Hence, the initial working alternative upon our data is not acceptable here, as claimed.

Step 5. Take the points $z_1, z_2 \in \text{Fix}(T)$ according to

$$z_1 \leq z_2, \text{ and } z_1 \neq z_2; \text{ hence, } z_1 < z_2,\ a := d(z_1, z_2) > 0.$$

Then,

$$B_2(z_1, z_2) = d(z_1, z_2) = a > 0; \text{ hence, } (z_1, z_2) \in \mathcal{R}$$

if one takes the B_2-admissible property of $\mathcal{R}$ into account. This tells us that the contractive condition applies to (z_1, z_2) and gives

$$\mathcal{Q}(z_1, z_2) \in \Upsilon; \text{ wherefrom } (0, a, a, a, a, 0) \in \Upsilon,$$

a contradiction with respect to the nnormal property of Υ. Hence, our working condition is not accepted, and the assertion follows. □

Note that multi-valued extensions of this result are possible, under the lines in Nadler [36]. On the other hand, an extended setting of these developments can be worked out under the lines discussed in the 2001 PhD Thesis by Hitzler [22, Ch 1, Sect 1.2]; see, for instance, Pasicki [39] and Turinici [60]. Further aspects may be found in Beg and Butt [6].

6. Analytic Methods in KSRDC Theory

In the following, as a by-product of our main result, some analytic methods in the Rhoades contractions theory are being developed.

Let $(X, d, \leq)$ be a partially ordered metric space. Further, let $T \in \mathcal{F}(X)$ be a selfmap of X, with

(a-reg) T is *ascending regular*:

there are ascending orbital sequences in X.

In this case, the family $\text{ao}(T)$ of all these objects is non-empty. Let $X_0 = (x_n)$ be an element of it; remember that, without loss, one may assume that

$B_1(x_n, x_{n+1}) > 0$, $\forall n$; proving that $X_0 = (x_n)$ is a (Ba-o-f) sequence.

Then, let $Z_0 = \mathcal{L}(X_0)$ be the limit orbital sequence attached to X_0. As precise, we are interested to determine sufficient conditions for existence/uniqueness of elements in $\mathrm{Fix}(T)$, via contractive-type requirements involving the iterative processes $X_0 = (x_n)$, $Z_0 = (z_n)$, and the associated sets

$$[X_0] = \{x_n; n \geq 0\},\ [[X_0]] = \mathrm{cl}([X_0]),\ [Z_0] = \{z_n; n \geq 0\}.$$

The basic directions and regularity conditions relative to X_0 under which the problem of determining fixed points of T is to be solved were already listed; and the contractive-type framework (involving X_0 and Z_0) was settled. As a by-product of these, we established our main result in this exposition, Theorem 13. It is our aim in the following to show that, starting from this principle, it is possible to describe the basic lines for a kind of analytical approach in treating fixed points of perturbed contractions over metric spaces of the type introduced by Krasnoselskii and Stetsenko [27], re-discovered by Rhoades [45], refined by Dutta and Choudhury [18], and extended over the partial order setting by Harjani *et al.* [21].

(A) Roughly speaking, the approach to be considered requires an appropriate description of contractive conditions to be used. Denote, for $x, y \in X$,

$$\begin{aligned}
&Q_1(x,y) = d(x,Tx),\ Q_2(x,y) = d(x,y),\\
&Q_3(x,y) = d(x,Ty),\ Q_4(x,y) = d(Tx,y),\\
&Q_5(x,y) = d(Tx,Ty),\ Q_6(x,y) = d(y,Ty),\\
&\mathcal{Q}(x,y) = (Q_1(x,y), Q_2(x,y), Q_3(x,y), Q_4(x,y), Q_5(x,y),\\
&Q_6(x,y)).
\end{aligned}$$

Then, let us construct the family of functions [for $x, y \in X$]

$$\begin{aligned}
&B_0(x,y) = \min\{Q_2, Q_5\}(x,y),\\
&B_1(x,y) = \min\{Q_1, Q_2, Q_5, Q_6\}(x,y),\\
&B_2(x,y) = \min\{Q_2, Q_3, Q_4, Q_5\}(x,y),\\
&B_3(x,y) = \min\{Q_1, Q_2, Q_3, Q_4, Q_5, Q_6\}(x,y)
\end{aligned}$$

and put, for simplicity,

$$(B_i > 0) = \{(x,y) \in X \times X; B_i(x,y) > 0\},\ i \in \{0,1,2,3\}.$$

Let (ψ, φ, χ) be a triple of functions over $\mathcal{F}(R_+, R)$. Further, letting (u, v, w, z) be a quadruple of functions over $\mathcal{F}(R_+^6, R_+)$, define the subsets

$$\Upsilon_0 = \{t \in R_+^6; u(t), v(t), w(t) > 0\},$$
$$\Upsilon = \{t \in \Upsilon_0; \psi(u(t)) \le \psi(v(t)) - \varphi(w(t)) + \chi(z(t))\}.$$

Finally, define the couple of relations over X

$$\mathcal{R} = \mathcal{Q}^{-1}(\Upsilon_0),\ \mathcal{S} = \mathcal{Q}^{-1}(\Upsilon);$$

note that, by this convention,

$$(x, y) \in \mathcal{R} \text{ iff } \mathcal{Q}(x, y) \in \Upsilon_0;\ (x, y) \in \mathcal{S} \text{ iff } \mathcal{Q}(x, y) \in \Upsilon.$$

Under this setting, we may introduce two contractive conditions upon our data. The former of these is a functional contractive condition, written as

(fct-contr) T is $(\le, \mathcal{R}; \psi, \varphi, \chi; u, v, w, z)$-*contractive* if
$\psi(u \circ \mathcal{Q}(x, y)) \le \psi(v \circ \mathcal{Q}(x, y)) - \varphi(w \circ \mathcal{Q}(x, y)) + \chi(z \circ \mathcal{Q}(x, y))$,
when $(x, y) \in (\le; \mathcal{R})$.

The latter of these appears as a set contractive condition and writes

(set-contr) T is $(\le, \mathcal{R}; \Upsilon)$-contractive:
$\mathcal{Q}(x, y) \in \Upsilon$, if $(x, y) \in (\le; \mathcal{R})$.

The connection between these is described as follows.

Proposition 15. *Under the above developments, one has*

T is $(\le, \mathcal{R}; \psi, \varphi, \chi; u, v, w, z)$-contractive iff T is $(\le, \mathcal{R}; \Upsilon)$-contractive.

Proof. Two steps must be passed.

Step 1. Suppose that the functional contractive property (fct-contr) is holding, and let the couple $(x, y) \in (\le; \mathcal{R})$ be arbitrary fixed. For the moment,

$(x, y) \in (\le; \mathcal{R})$ implies $x \le y$ and $t := \mathcal{Q}(x, y) \in \Upsilon_0$
(i.e., $u(t) > 0$, $v(t) > 0$, $w(t) > 0$).

Moreover, by the very definition of function contractive property,

$$\psi(u(t)) \leq \psi(v(t)) - \psi(w(t)) + \chi(z(t)); \text{ that is: } t := \mathcal{Q}(x, y) \in \Upsilon.$$

This, by the arbitrariness of the couple $(x, y) \in (\leq; \mathcal{R})$, shows that the set contractive property holds, and the left to right inclusion follows.

Step 2. Suppose that the set contractive property (set-contr) is holding, and let the couple $(x, y) \in (\leq; \mathcal{R})$ be arbitrary fixed. For the moment,

$(x, y) \in \mathcal{R}$ implies $t := \mathcal{Q}(x, y) \in \Upsilon_0$ (i.e.: $u(t) > 0$, $v(t) > 0$, $w(t) > 0$).

Moreover, by the very definition of Υ,

$$\psi(u(t)) \leq \psi(v(t)) - \psi(w(t)) + \chi(z(t)); \text{ where } t := \mathcal{Q}(x, y).$$

This tells us that the couple (x, y) fulfills the functional contractive property and proves the right to left inclusion. □

In other words, the functional contractive condition (fct-contr) is equivalent with the set-contractive condition (set-contr), to which the methods of the main result are applicable. Clearly, an effective application of the main result to the set contractive problem (set-contr) amounts to verifying the asymptotic, right, bilateral, point, and normality properties upon Υ, in terms of its components: the relation $\mathcal{R}$, the triple (ψ, φ, χ), and the quadruple (u, v, w, z). Some preliminaries are needed.

(B) Define, for a sequence (r_n) in R,

$\liminf_n(r_n) = \sup_m A(m)$, where $(A(m) = \inf\{r_i; i \geq m\}$; $m \geq 0)$,

$\limsup_n(r_n) = \inf_m B(m)$, where $(B(m) = \sup\{r_j; j \geq m\}$; $m \geq 0)$.

Clearly, $(A(m); m \geq 0)$ is ascending in $R \cup \{-\infty\}$, and $(B(m); m \geq 0)$ is descending in $R \cup \{\infty\}$. In addition, $[A(m) \leq B(m), \forall m]$, wherefrom

$$A(p) \leq B(q), \text{ if } p \leq q; \text{ so that, } \liminf_n(r_n) \leq \limsup_n(r_n).$$

Clearly, for each sequence (r_n), we have

$$\liminf_n(-r_n) = -\limsup_n r_n, \ \limsup_n(-r_n) = -\liminf_n r_n,$$

so, in the following, we may concentrate on the superior limit operator. In this direction, the following additive property of underlying concept is to be noted.

Proposition 16. *For each couple of real sequences (r_n) and (s_n),*

$$\limsup_n(r_n + s_n) \leq \limsup_n r_n + \limsup_n s_n,$$

whenever the right member exists.

Proof. Denote

$$(U_m = \sup\{r_j + s_j; j \geq m\}; m \geq 0),$$
$$(V_m = \sup\{r_j; j \geq m\}; m \geq 0), \ (W_m = \sup\{s_j; j \geq m\}; m \geq 0).$$

Clearly, (U_m), (V_m), (W_m) are descending sequences in $R \cup \{\infty\}$ and

$$U_m \leq V_m + W_m, \text{ for all } m.$$

Passing to limit as $m \to \infty$, we are done. □

(C) Let $\psi \in \mathcal{F}(R_+, R)$ be given. For any $s \in R_+$, put

$\Lambda^+\psi(s) = \inf_{0<\varepsilon<s} \Psi^+(s+)(\varepsilon)$, where $\Psi^+(s+)(\varepsilon) = \sup \psi(]s, s+\varepsilon[)$
(*right superior limit* of ψ at s),
$\Lambda_+\psi(s) = \sup_{0<\varepsilon<s} \Psi_+(s+)(\varepsilon)$, where $\Psi_+(s+)(\varepsilon) = \inf \psi(]s, s+\varepsilon[)$
(*right inferior limit* of ψ at s).

Then, for each $s \in R^0_+$, denote

$\Lambda^\pm\psi(s) = \inf_{0<\varepsilon<s} \Psi^\pm(s\pm)(\varepsilon)$, where $\Psi^\pm(s\pm)(\varepsilon) = \sup \psi(]s-\varepsilon, s+\varepsilon[)$
(*bilateral superior limit* of ψ at s),
$\Lambda_\pm\psi(s) = \sup_{0<\varepsilon<s} \Psi_\pm(s\pm)(\varepsilon)$, where $\Psi_\pm(s\pm)(\varepsilon) = \inf \psi(]s-\varepsilon, s+\varepsilon[)$
(*bilateral inferior limit* of ψ at s).

The following auxiliary fact is useful.

Proposition 17. *Let $\psi \in \mathcal{F}(R_+, R)$ be given. Then, we have the following:*

(63-1) *For each $s \in R_+$, we have*

(63-1-a) $\limsup_n(\psi(t_n)) \leq \Lambda^+\psi(s)$,
for each sequence (t_n) in R_+^0 with $t_n \to s+$
(63-1-b) *there exists a sequence (r_n) in R_+^0 with*
$r_n \to s+$ *and* $\psi(r_n) \to \Lambda^+\psi(s)$

(63-2) *For each $s \in R_+^0$, we have*

(63-2-a) $\limsup_n(\psi(t_n)) \leq \Lambda^\pm\psi(s)$,
for each sequence (t_n) in R_+^0 with $t_n \to s$
(63-2-b) *there exists a sequence (r_n) in R_+^0 with*
$r_n \to s$ *and* $\psi(r_n) \to \Lambda^\pm\psi(s)$.

Proof. It suffices to establish the first half (concerning right limits); the argument of the second half (concerning bilateral limits) is very similar.

(i) Given $\varepsilon \in]0, s[$, there exists a rank $p(\varepsilon) \geq 0$ such that $s < t_n < s + \varepsilon$, for all $n \geq p(\varepsilon)$; hence,

$$\limsup_n(\varphi(t_n)) \leq \sup\{\varphi(t_n); n \geq p(\varepsilon)\} \leq \Psi^+(s+)(\varepsilon).$$

It suffices to take the infimum over $\varepsilon > 0$ in this relation to get the desired fact.

(ii) Denote, for simplicity,

$$\alpha = \Lambda^+\psi(s); \text{ hence, } \alpha = \inf_{0<\varepsilon<s} \Psi^+(s+)(\varepsilon).$$

The sequence $(\beta_n := \Psi^+(s+)(s2^{-n-1}); n \geq 0)$ is descending in $R \cup \{\infty\}$, with

$$(\beta_n \geq \alpha, \ \forall n) \text{ and } \inf_n \beta_n = \alpha; \text{ hence } \lim_n \beta_n = \alpha.$$

By these properties, there may be constructed a sequence $(\gamma_n; n \geq 0)$ in R, with

$$\gamma_n < \beta_n, \ \forall n; \ \lim_n \gamma_n = \lim_n \beta_n = \alpha.$$

For example, we may take for each $n \geq 0$,

$$\gamma_n = 3^n, \text{ if } \beta_n = \infty, \ \gamma_n = \beta_n - 3^{-n}, \text{ if } \beta_n < \infty.$$

Let $n \geq 0$ be arbitrary fixed. By the supremum definition, there exists $r_n \in]s, s + s2^{-n-1}[$ such that $\psi(r_n) > \gamma_n$; moreover (again by definition), $\varphi(r_n) \leq \beta_n$. The obtained sequence $(r_n; n \geq 0)$ fulfills $r_n \to s+$ and $\varphi(r_n) \to \alpha$. □

Having these precise, we introduce two basic concepts. Let $\psi \in \mathcal{F}(R_+, R)$ be a function, and $b, c \in R_+$ be a couple of points. Define the *right oscillation* of ψ at (b, c) and, respectively, *point oscillation* of ψ at (b, c) as

(osc-r) $\text{osc}(+)(\psi; b, c)$ = the supremum of all $\limsup_n[\psi(t_n) - \psi(s_n)]$ where (t_n), (s_n) are sequences in R^0_+ with $t_n \to c+$, $s_n \to b+$,

(osc-p) $\text{osc}(\psi; b, c)$ = the supremum of all $\limsup_n(\psi(t_n) - \psi(s_n))$ where (t_n), (s_n) are sequences in R^0_+ with $t_n \to c$, $s_n \to b$.

Here, for each sequence (r_n) in R^0_+ and each $u \in R_+$,

$r_n \to u$ is the bilateral convergence, if $u > 0$

$r_n \to u$ means the right convergence $r_n \to u+$, if $u = 0$.

A global characterization of these oscillations is provided in the following statement. Define, for each $a \geq 0$,

$$\Lambda^*\psi(a) = \Lambda^{\pm}\psi(a), \text{ if } a > 0, \ \Lambda^*\psi(a) = \Lambda^{+}\psi(a), \text{ if } a = 0,$$
$$\Lambda_*\psi(a) = \Lambda_{\pm}\psi(a), \text{ if } a > 0, \ \Lambda_*\psi(a) = \Lambda_{+}\psi(a), \text{ if } a = 0.$$

Proposition 18. *For each $\psi \in \mathcal{F}(R_+, R)$ and each $b, c \in R_+$, we have (when the right member exists)*

(64-1) $\text{osc}(+)(\psi; b, c) = \Lambda^+\psi(c) - \Lambda_+\psi(b)$,

(64-2) $\text{osc}(\psi; b, c) = \Lambda^*\psi(c) - \Lambda_*\psi(b)$.

Proof. As before, it suffices to verify the first part; the remaining one is proved in a similar way.

(i) Let (t_n), (s_n) be sequences in R with $t_n \to c+$, $s_n \to b+$. By the previous auxiliary fact,

$$\begin{aligned}&\limsup_n[\psi(t_n) - \psi(s_n)]\\ &\quad \leq \limsup_n \psi(t_n) - \liminf_n \psi(s_n) \leq \Lambda^+\psi(c) - \Lambda_+\psi(b),\end{aligned}$$

and this gives the left to right inequality.

(ii) By the same fact, there are two sequences (u_n), (v_n) in R, with

$v_n \to c+$, $\Lambda^+\psi(c) = \lim_n \psi(v_n)$, as well as
$u_n \to b+$, $\Lambda^+(-\psi(b) = \lim_n(-\psi(u_n))$; so, $-\Lambda_+\psi(b) = -\lim_n \psi(u_n)$.

Adding these relations yield

$$\Lambda^+\psi(c) - \Lambda_+\psi(b) = \lim_n[\psi(v_n) - \psi(u_n)] \leq \text{osc}(+)(\psi; b, c),$$

and, from this, all is clear. □

A simplified form of these representations is available under special regularity assumptions about ψ. Some preliminaries are needed.

For each function $\psi \in \mathcal{F}(R_+, R)$, define

ψ is *right regulated* at $a \geq 0$: $\psi(a+0) = \lim_{t\to a+} \psi(t)$ (the *right limit* of ψ at a) exists (in $R \cup \{-\infty, \infty\}$)

ψ is *left regulated* at $a > 0$: $\psi(a-0) = \lim_{t\to a-} \psi(t)$ (the *left limit* of ψ at a) exists (in $R \cup \{-\infty, \infty\}$)

ψ is *regulated* at $a > 0$: both $\psi(a+0)$ and $\psi(a-0)$ exist (in $R \cup \{-\infty, \infty\}$)

ψ is *continuous* at $a > 0$: ψ is regulated at a and $\psi(a+0) = \psi(a-0) = \psi(a)$.

The class of all such points is indicated, respectively, as

rreg(ψ) = the right regulated domain of ψ,

lreg(ψ) = the left regulated domain of ψ,

reg(ψ) = the regulated domain of ψ,

cont(ψ) = the continuous domain of ψ.

Proposition 19. *Let the function $\psi \in \mathcal{F}(R_+, R)$ be given. Then,*

(65-1) $\Lambda^+\psi(b) = \psi(b+0)$, *whenever ψ is right regulated at $b \geq 0$*

(65-2) $\Lambda^\pm\psi(c) = \max\{\psi(c+0), \psi(c), \psi(c-0)\}$, *whenever ψ is regulated at $c > 0$.*

Proof. (i) By a previous observation, there exists a sequence (s_n) in R_+^0 with $s_n \to b+$ and $\psi(s_n) \to \Lambda^+\psi(b)$. On the other hand, by the right-regulated property, $\psi(s_n) \to \psi(b+0)$. Combining these, we are done.

(ii) Let (t_n) be a sequence in R^0_+ with $t_n \to c$. From a previous observation,

$\limsup_n \psi(t_n) \leq \Lambda^{\pm}\psi(c)$,
which yields $\max\{\psi(c+0), \psi(c), \psi(c-0)\} \leq \Lambda^{\pm}\psi(c)$,

by choosing (t_n) according to the supplementary alternatives

(alter) $(t_n > c, \forall n)$, $(t_n = c, \forall n)$, $(t_n < c, \forall n)$.

Moreover, by another observation, there exists a sequence (r_n) in R^0_+ with $r_n \to c$ and $\psi(r_n) \to \Lambda^{\pm}\psi(c)$. This, combined with the written alternatives (possibly on subsequences of (r_n)) tells us that

$\Lambda^{\pm}\psi(c) \in \{\psi(c+0), \psi(c), \psi(c-0)\}$ so that
$\Lambda^{\pm}\psi(c) \leq \max\{\psi(c+0), \psi(c), \psi(c-0)\}$;

wherefrom, all is clear. □

Note that, as a consequence of these, some simplified forms of the attached oscillation operators are available; we do not give details.

(E) In the following, some technical facts are developed for a class of binary operations to be used further.

Let $(a,b) \mapsto a * b$ be some binary relation over R_+; we call it *semiregular*, provided the following properties are valid:

(sreg-1) $(a \leq a * b)$, $(c * c = c)$, $\forall a, b, c \in R_+$

(sreg-2) $(a,b) \mapsto a * b$ is continuous in its arguments.

Note that, by the former condition,

(sreg-3) $a, b > 0$ implies $a * b > 0$; so, $(*)$ is a binary operation on R^0_+.

Given such an object, call it

(reg) *regular*, when $(b < a * b$ implies $b < a)$
or, equivalently, $(a \leq b$ implies $a * b \leq b)$

(s-reg) *strongly regular*, when ($b \neq a * b$ implies $b < a$) or, equivalently, ($a \leq b$ implies $a * b = b$).

A basic example of such an operation is

$a \top b = \max\{a, b\} = a + \max\{0, b - a\}$, $a, b \in R_+$;
clearly, ($\top$) is strongly regular.

Some other examples are provided by the following statement. Let $I(.)$ stand for the unitary function of $\mathcal{F}(R_+)$ defined as ($I(t) = t$; $t \geq 0$). Given the increasing $f \in \mathcal{F}(R_+)$, define the binary operations

$a(f)b = a + \max\{0, f(b) - f(a)\}$, $a, b \in R_+$,
$a[f]b = a + |f(b) - f(a)|$, $a, b \in R_+$,

Clearly, f and $f - \alpha$ (where $0 \leq \alpha \leq f(0)$) generate the same binary operations, so, without loss, one may assume that $f(0) = 0$.

Proposition 20. *Let the function $f \in \mathcal{F}(R_+)$ be continuous increasing with $f(0) = 0$. Then,*

(66-1) *the binary operation (f) is regular when $I - f$ is increasing,*

(66-2) *the binary operation $[f]$ is regular whenever $I - f$ is increasing,*

(66-3) *the binary operation (f) is strongly regular iff $f = I$; moreover, in this case, ($a \top b = a(f)b$, $\forall a, b \in R_+$),*

(66-4) *the binary operation $[f]$ is strongly regular iff $f = I$; moreover, in this case, ($a \top b \leq a[f]b$, $\forall a, b \in R_+$).*

Proof. (i) Clearly, (f) is semiregular. Further, let $a, b \in R_+$ be such that

$a \leq b$; hence, $f(b) \geq f(a)$; wherefrom $a(f)b = a + f(b) - f(a)$.

The posed relation $a(f)b \leq b$ becomes

$a - f(a) \leq b - f(b)$; evident, via $I - f =$ increasing.

(ii) As before, $[f]$ is semiregular. Further, let $a, b \in R_+$ be such that

$$a \leq b; \text{ hence, } f(b) \geq f(a); \text{ wherefrom } a(f)b = a + f(b) - f(a).$$

The posed relation $a[f]b \leq b$ becomes

$$a - f(a) \leq b - f(b); \text{ evident, via } I - f = \text{increasing}.$$

(iii) Let $a, b \in R_+$ be such that

$$a \leq b; \text{ hence, } f(b) \geq f(a); \text{ wherefrom } a(f)b = a + f(b) - f(a).$$

The posed relation $a(f)b = b$ becomes

$$a - f(a) = b - f(b) \text{ (for all such } (a, b)).$$

But then (combining with $f(0) = 0$), $f = I$, and the assertion follows.

(iv) Let $a, b \in R_+$ be such that

$$a \leq b; \text{ hence, } f(b) \geq f(a); \text{ wherefrom } a[f]b = a + f(b) - f(a).$$

The posed relation $a[f]b = b$ becomes

$$a - f(a) = b - f(b) \text{ (for all such } (a, b)).$$

But then (combining with $f(0) = 0$), $f = I$, and the assertion follows. □

An open problems is that of these operations being the only ones endowed with (strongly) regular properties. We conjecture that the answer is negative.

(F) The following completion of these facts is useful. Let $L : R_+ \times R_+ \to R_+$ and $\lambda : R_+ \to R_+$ be a couple of maps. (Actually, $(t, s) \mapsto L(t, s)$ is a binary operation over R_+, but this is not important for us.) We say that (L, λ) is *right compatible*, provided

(r-comp) for each sequence (r_n) in R_+^0 and each $r \in R_+^0$ with $r_n \to r+$,

we have $L(r_n, r_{n+1}) \to \lambda(r)+$.

The following right compatibility criterion is useful in practice. Denote, for each couple $\alpha, \beta \in R$,

co(α, β) = the convex cover of $\{\alpha, \beta\}$,

that is, the interval $[\min\{\alpha, \beta\}, \max\{\alpha, \beta\}]$.

Proposition 21. *Let $L \in \mathcal{F}(R_+ \times R_+, R_+)$ and $\lambda \in \mathcal{F}(R_+)$ be such that*

(67-I) *L is λ-subordinated: $L(t,s) \in \mathrm{co}(\lambda(t), \lambda(s))$, for each $t, s \in R_+^0$*

(67-II) *$\lambda(.)$ is continuous strictly increasing on R_+^0.*

Then, necessarily, (L, λ) is right compatible.

Proof. Let the sequence (r_n) in R_+^0 and the point $r \in R_+^0$ be such that $r_n \to r+$. By the subordination-type condition,

$(\forall n)$: $L(r_n, r_{n+1}) \in \mathrm{co}(\lambda(r_n), \lambda(r_{n+1}))$, that is,
$\min\{\lambda(r_n), \lambda(r_{n+1})\} \le L(r_n, r_{n+1}) \le \max\{\lambda(r_n), \lambda(r_{n+1})\}$.

On the other hand, by the regularity conditions imposed upon λ,

$\lambda(r_n) \to \lambda(r)+$ as $n \to \infty$; whence,
$\min\{\lambda(r_n), \lambda(r_{n+1})\} \to r+$, $\max\{\lambda(r_n), \lambda(r_{n+1})\} \to r+$, as $n \to \infty$.

Putting these together yields the desired fact. □

In particular, with λ taken as before, the mappings introduced as

$L_1(t,s) = a\lambda(t) + b\lambda(s)$, $t, s \in R_+$, where $a, b \in R_+$, $a + b = 1$,
$L_2(t,s) = \max\{\lambda(t), \lambda(s)\}$, $L_3(t,s) = \min\{\lambda(t), \lambda(s)\}$, $t, s \in R_+$

are λ-subordinated; hence, (L_i, λ) = right compatible, $i \in \{1, 2, 3\}$. On the other hand, let us consider the mapping

$L(t,s) = \mu(t) + |\mu(s) - \mu(t)|$, $t, s \in R_+$,

where $\mu \in \mathcal{F}(R_+)$ is continuous strictly increasing on R_+^0. It is not hard to see that (L, μ) is right compatible. However, this is not a consequence of the preceding criterion, in the sense

L is not λ-subordinated, for each $\lambda \in \mathcal{F}(R_+)$ with
λ = continuous and strictly increasing on R_+^0.

In fact, suppose that this would be true: there exists $\lambda \in \mathcal{F}(R_+)$ with $\lambda =$ continuous and strictly increasing on R^0_+ such that

> (sub) L is λ-*subordinated*: $L(t,s) \in \text{co}(\lambda(t),\lambda(s))$, for each t, $s \in R^0_+$;
>
> hence, in particular, $\lambda(s) \leq 2\mu(t) - \mu(s) \leq \lambda(t)$, $\forall t,s \in R^0_+$, $t > s$.

Passing to limit as $t \to s+$ yields

$$(\forall r > 0)\colon \lambda(r) \leq \mu(r) \leq \lambda(r);\ \text{hence},\ \lambda(r) = \mu(r).$$

Taking the preceding relation into account gives

$$\mu(s) \leq 2\mu(t) - \mu(s) \leq \mu(t),\ \text{for all}\ t,s \in R^0_+\ \text{with}\ t > s.$$

This, along with the underlying inequality giving

> $\mu(s) \leq \mu(t) \leq \mu(s)$ (hence, $\mu(t) = \mu(s)$), for all $t,s \in R^0_+$ with $t > s$

yields a contradiction with the initial choice of μ; hence, the claim.

(G) Let (ψ,φ) be taken as before. Remember that, we defined the right and point oscillation of ψ at some couple (b,c) over R_+. In this case, for each triple ($a \in R_+$, $b,c \in R_+$), we introduce the related properties:

> (n-r-osc) (ψ,φ) is *non right oscillating at* (b,c,a): for each sequence (τ_n) in R^0_+ with $\tau_n \to a+$, we have $\limsup_n \varphi(\tau_n) > \text{osc}(+)(\psi;b,c)$.
>
> (n-p-osc) (ψ,φ) is *non point oscillating at* (b,c,a): for each sequence (τ_n) in R^0_+ with $\tau_n \to a$, we have $\limsup_n \varphi(\tau_n) > \text{osc}(\psi;b,c)$.

In particular, let (ξ,η,ζ) be a triple of functions in $\mathcal{F}(R_+)$; for example, one may take $\xi = I$ [the identity function $(I(t) = t; t \in R_+)$], as well as $\eta = 0$ and/or $\zeta = 0$ [the identically zero function $(0(t) = 0; t \in R_+)$].

Given $b > 0$, let us say that the system $(\psi,\varphi;\xi,\eta,\zeta)$ is *non right oscillating* at b, provided

> (ψ,φ) is *non right oscillating* at $(\xi(b),\eta(b),\zeta(b))$ (see above).

The set of all such $b > 0$ will be denoted as $\text{nrosc}(\psi, \varphi; \xi, \eta, \zeta)$. Then, we say that

(a-nrosc) $(\psi, \varphi; \xi, \eta, \zeta)$ is *almost non right oscillating* if
$\Theta := \text{nrosc}(\psi, \varphi; \xi, \eta, \zeta)$ is $(>)$-cofinal in R_+^0,
(nrosc) $(\psi, \varphi; \xi, \eta, \zeta)$ is *non right oscillating* if
$\Theta := \text{nrosc}(\psi, \varphi; \xi, \eta, \zeta)$ is identical with R_+^0.

Given $b > 0$, let us say that the system $(\psi, \varphi; \xi, \eta, \zeta)$ is *non point oscillating* at b, provided

(ψ, φ) is *non point oscillating* at $(\xi(b), \eta(b), \zeta(b))$ (see above).

If this holds for all $b > 0$, then $(\psi, \varphi; \xi, \eta, \zeta)$ is called *non point oscillating.*

Finally, let (u, v, w, z) be a quadruple of functions over $\mathcal{F}(R_+^6, R_+)$. We say that (u, v, w, z) is

(r-asy) *right asymptotic of type* $(\xi, \eta, \zeta, 0)$ at $[X_0]$, when:
$\forall$ sequence $(t^n; n \geq 0)$ in $\mathcal{Q}(\leq; B_3 > 0; [X_0])$, and $\forall b > 0$:
$(t^n; n \geq 0)$ has the right property at $(0, b, b, b, b, 0)$
implies $u(t^n) \to \xi(b)+$, $v(t^n) \to \eta(b)+$, $w(t^n) \to \zeta(b)+$, $z(t^n) \to 0$
(b-asy) *bilateral asymptotic of type* $(\xi, \eta, \zeta, 0)$ at $[X_0] \cup [Z_0]$, when:
$\forall$ sequence $(t^n; n \geq 0)$ in $\mathcal{Q}(\leq; B_3 > 0; [X_0] \cup [Z_0])$, and $\forall b > 0$:
$(t^n; n \geq 0)$ has the bilateral property at $(0, b, b, b, b, 0)$
implies $u(t^n) \to \xi(b)$, $v(t^n) \to \eta(b)$, $w(t^n) \to \zeta(b)$, $z(t^n) \to 0$
(p-asy) *point asymptotic of type* $(\xi, \eta, \zeta, 0)$ at $[[X_0]]$, when:
$\forall$ sequence $(t^n; n \geq 0)$ in $\mathcal{Q}(\leq; B_3 > 0; [[X_0]])$, and $\forall b > 0$:
$(t^n; n \geq 0)$ has the point property at $(0, 0, b, 0, b, b)$
implies $u(t^n) \to \xi(b)$, $v(t^n) \to \eta(b)$, $w(t^n) \to \zeta(b)$, $z(t^n) \to 0$.

The following statement is an essential step toward the precise objective. Letting (ψ, φ, χ) and (u, v, w, z) be taken as above, remember that we defined the couple of subsets in R^6_+ as well as the couple of relations over X

(subs) $\Upsilon_0 = \{t \in R^6_+; u(t), v(t), w(t) > 0\}$,
$\Upsilon = \{t \in \Upsilon_0; \psi(u(t)) \le \psi(v(t)) - \varphi(w(t)) + \chi(z(t))\}$;
(rela) $\mathcal{R} = \mathcal{Q}^{-1}(\Upsilon_0)$, $\mathcal{S} = \mathcal{Q}^{-1}(\Upsilon)$.

Theorem 22. *Let the selfmap T be taken so as*

(61-i) *T is $(\le, \mathcal{R}; \psi, \varphi, \chi; u, v, w, z)$-contractive;*

or, equivalently: $(\le; \mathcal{R}; \Upsilon)$-contractive,

where the relation $\mathcal{R}$ over X, the triple (ψ, φ, χ) of functions over $\mathcal{F}(R_+, R)$, and the quadruple (u, v, w, z) of functions over $\mathcal{F}(R^6_+, R_+)$ fulfill (for the moment)

(61-ii) *$\mathcal{R}$ is B_1-admissible, and φ is strictly positive ($\varphi(R^0_+) \subseteq R^0_+$)*

(61-iii) *χ is null right continuous at origin ($\chi(0{+}0) = 0 = \chi(0)$).*

Then,

(I) *The following are valid, at the level of each arbitrary fixed (Ba-o-f) sequence $Y_0 = (y_n)$*

(61-a) *Υ is descending asymptotic on $[Y_0]$, when, for the telescopic sequences $(r_n := d(y_n, y_{n+1}); n \ge 0)$ and $(p_n := d(y_n, y_{n+2}))$ attached to $Y_0 = (y_n)$ there exists a strongly regular binary operation $(*)$ and a right compatible pair (L, λ) (where $L \in \mathcal{F}(R_+ \times R_+, R_+)$, $\lambda \in \mathcal{F}(R_+)$), with*

(61-a-i) *(u, v, w, z) is strongly $(*, L, \lambda)$-iterative, in the sense:*

$u^n := u(A_n) = r_{n+1}$, $v^n := v(A_n) = r_n * r_{n+1}$,

$w^n := w(A_n) = L(r_n, r_{n+1})$, $z^n := z(A_n) = 0$, *for all n,*

where $(A_n := (r_n, r_n, p_n, 0, r_{n+1}, r_{n+1}); n \ge 0)$

(61-a-ii) *$(\psi, \varphi; I, I, \lambda)$ is non right oscillating*

(61-b) *Υ is descending asymptotic on $[Y_0]$, when ψ is increasing and, for the telescopic sequences $(r_n := d(y_n, y_{n+1}); n \ge 0)$ and $(p_n := d(y_n, y_{n+2}))$ attached to $Y_0 = (y_n)$ there exists a regular binary operation $(*)$ and a compatible pair (L, λ) (where $L \in \mathcal{F}(R_+ \times R_+, R_+)$, $\lambda \in \mathcal{F}(R_+)$), with*

(61-b-i) (u, v, w, z) *is weakly* $(*, L, \lambda)$*-iterative, in the sense:* $u^n := u(A_n) \geq r_{n+1}$, $v^n := v(A_n) \leq r_n * r_{n+1}$, $w^n := w(A_n) = L(r_n, r_{n+1})$, $z^n := z(A_n) = 0$, *for all* n, *where* $(A_n := (r_n, r_n, p_n, 0, r_{n+1}, r_{n+1}); n \geq 0)$

(61-b-ii) $(\psi, \varphi; I, I, \lambda)$ *is non right oscillating*

(61-c) Υ *is almost nright on* $[Y_0]$, *provided there exists* $\mu \in \mathcal{F}(R_+)$ *such that*

(61-c-i) (u, v, w, z) *is right asymptotic of type* $(I, I, \mu, 0)$ *at* $[Y_0]$

(61-c-ii) $(\psi, \varphi; I, I, \mu)$ *is almost non right oscillating*

(II) *The following are valid, at the level of our (Ba-o-f) sequence* $X_0 = (x_n)$ *and its limit sequence* $Z_0 = (z_n)$

(61-d) Υ *is nbilateral on* $[X_0] \cup [Z_0]$, *provided there exists a triple* (ξ, η, ζ) *in* $\mathcal{F}(R_+)$ *such that*

(61-d-i) (u, v, w, z) *is bilateral asymptotic of type* $(\xi, \eta, \zeta, 0)$ *at* $[X_0] \cup [Z_0]$

(61-d-ii) $(\psi, \varphi; \xi, \eta, \zeta)$ *is non point oscillating*

(61-e) Υ *is npoint on* $[[X_0]]$, *provided there exists a triple* (ξ, η, ζ) *in* $\mathcal{F}(R_+)$, *with*

(61-e-i) (u, v, w, z) *is point asymptotic of type* $(\xi, \eta, \zeta, 0)$ *at* $[[X_0]]$

(61-e-ii) $(\psi, \varphi; \xi, \eta, \zeta)$ *is non point oscillating*

(III) *Finally, the following is valid at the level of* Υ *above*

(61-f) Υ *is nnormal, provided*

(u, v, w, z) *is invariant, in the sense:* $u(0, a, a, a, a, 0) = a$, $v(0, a, a, a, a, 0) = a$, $w(0, a, a, a, a, 0) = a$, $z(0, a, a, a, a, 0) = 0$, $\forall a > 0$.

Proof. The argument consists in a number of parts.

Part 1. [Υ is descending asymptotic on $[Y_0]$ under the first lot of conditions].

Let the telescopic sequences $(r_n := d(y_n, y_{n+1}); n \geq 0)$ and $(p_n := d(y_n, y_{n+2}))$ attached to $Y_0 = (y_n)$ be given. As (u, v, w, z) is strongly $(*, L, \lambda)$-iterative,

$$(\forall n)\colon u^n := u(A_n) = r_{n+1},\ v^n := v(A_n) = r_n * r_{n+1},$$
$$w^n := w(A_n) = L(r_n, r_{n+1}),\ z^n := z(A_n) = 0.$$

By the very representation of Υ (and the choice of χ)

(iter-1) $(\forall n)\colon \psi(r_{n+1}) \leq \psi(r_n * r_{n+1}) - \varphi(w^n)$,
where $w^n = L(r_n, r_{n+1})$.

Combining with the strict positivity of φ, and $(*)$ being strongly regular,

$\psi(r_{n+1}) < \psi(r_n * r_{n+1})$ (so that, $r_{n+1} \neq r_n * r_{n+1}$), $\forall n$;
whence $(r_{n+1} < r_n,\ \forall n)$, that is, (r_n) is strictly descending,

which tells us that $r := \lim_n(r_n)$ exists in R_+. Suppose by contradiction that $r > 0$. As (r_n) is strictly descending,

$$r_n \to r+;\ \text{hence}\ r_n * r_{n+1} \to r+,\ w^n = L(r_n, r_{n+1}) \to \lambda(r)+$$

(cf. the choice of $(*)$, (L, λ), and (w^n)). By (iter-1), we get

(iter-2) $\varphi(w^n) \leq \psi(r_n * r_{n+1}) - \psi(r_{n+1})$, $\forall n$.

Passing to $\limsup$ as $n \to \infty$, one gets

$$\limsup_n \varphi(w^n) \leq \limsup_n [\psi(r_n * r_{n+1}) - \psi(r_{n+1})]$$
$$\leq \mathrm{osc}(+)(\psi; r, r),$$

in contradiction with $(\psi, \varphi; I, I, \lambda)$ being non right oscillating. Hence, $r = 0$, and the assertion follows.

Part 2. [Υ is descending asymptotic on $[Y_0]$ under the second lot of conditions].

Let the telescopic sequences $(r_n := d(y_n, y_{n+1}); n \geq 0)$ and $(p_n := d(y_n, y_{n+2}))$ attached to $Y_0 = (y_n)$ be given. As (u, v, w, z) is weakly $(*, L, \lambda)$-iterative,

$$(\forall n)\colon u^n := u(A_n) \geq r_{n+1},\ v^n := v(A_n) \leq r_n * r_{n+1},$$
$$w^n := w(A_n) = L(r_n, r_{n+1}),\ z^n := z(A_n) = 0.$$

By the very representation of Υ (and ψ=increasing)

$$\text{(iter-3)} \quad (\forall n)\colon \psi(r_{n+1}) \leq \psi(u^n) \leq \psi(v^n) - \varphi(w^n)$$
$$\leq \psi(r_n * r_{n+1}) - \varphi(w^n), \text{ where } w^n = L(r_n, r_{n+1}).$$

Combining with the strict positive condition upon φ and ψ = increasing, this yields (by the regularity of our binary operation)

$$\psi(r_{n+1}) < \psi(r_n * r_{n+1}) \text{ (so that, } r_{n+1} < r_n * r_{n+1}), \forall n;$$
$$\text{whence } (r_{n+1} < r_n, \forall n)\text{; that is: } (r_n) \text{ is strictly descending,}$$

which tells us that $r := \lim_n(r_n)$ exists in R_+. Suppose by contradiction that $r > 0$. By the strict descending property,

$$r_n \to r+; \text{ hence } r_n * r_{n+1} \to r+,\ w^n = L(r_n, r_{n+1}) \to \lambda(r)+$$

(cf. the choice of $(*)$, (L, λ) and (w^n)). By (iter-3), we get

$$\text{(iter-4)} \quad \varphi(w^n) \leq \psi(r_n * r_{n+1}) - \psi(r_{n+1}), \forall n.$$

Passing to $\limsup$ as $n \to \infty$, one gets

$$\limsup_n \varphi(w^n) \leq \limsup_n[\psi(r_n * r_{n+1}) - \psi(r_{n+1})]$$
$$\leq \operatorname{osc}(+)(\psi; r, r),$$

in contradiction with $(\psi, \varphi; I, I, \lambda)$ being non right oscillating. Hence, $r = 0$, and the assertion follows.

Part 3. [Υ is almost nright on $[Y_0]$, under the precise conditions].

Take some point $c = (c_1, \ldots, c_6)$ in R^6_+. We say that the sequence $(t^n := (t^n_1, \ldots, t^n_6); n \geq 0)$ in R^6_+ is *right* at c, if

$$\text{(right-c)} \quad t^n_i \to c_i+, \text{ as } n \to \infty, \forall i.$$

Given $b > 0$, let us say that Υ is *nright* at b on $[Y_0]$, if

for each sequence $(t^n; n \geq 0)$ in $\mathcal{Q}(\leq; B_3 > 0; [Y_0]) \cap \Upsilon$,
the right property at $(0, b, b, b, b, 0)$ is not true.

The class of all these $b > 0$ will be denoted as $\mathrm{nright}(\Upsilon; [Y_0])$. In this case, we say that Υ is

(a-n-r) *almost nright* on $[Y_0]$, if $\Theta := \mathrm{nright}(\Upsilon; [Y_0])$ is $(>)$-cofinal in R^0_+ (for each $\varepsilon \in R^0_+$ there exists $\theta \in \Theta$ with $\varepsilon > \theta$)
(n-r) *nright* on $[Y_0]$, if $\Theta := \mathrm{nright}(\Upsilon; [Y_0])$ is identical with R^0_+.

We have to establish that the former property is retainable for our data. To do this, we start by noting that

$(\psi, \varphi; I, I, \mu)$ = almost non right oscillating implies
$\Theta := \mathrm{nrosc}(\psi, \varphi; I, I, \mu)$ is $(>)$-cofinal in R^0_+.

We now claim that

$\Theta \subseteq \mathrm{nright}(\Upsilon; [Y_0])$; wherefrom, $\mathrm{nright}(\Upsilon; [Y_0])$ is $(>)$-cofinal in R^0_+.

This amounts to establish that

for each sequence $(t^n := (t^n_1, \ldots, t^n_6); n \geq 0)$ in $\mathcal{Q}(\leq; B_3 > 0; [Y_0]) \cap \Upsilon$,
and each $b \in \Theta$, the right property at $(0, b, b, b, b, 0)$ is not true.

Suppose by absurd that there exists a (vectorial) sequence $(t^n := (t^n_1, \ldots, t^n_6); n \geq 0)$ in R^6_+ and some $b \in \Theta$, with

(right-b-1) $(\forall n)$: $t^n := (t^n_1, \ldots, t^n_6) \in \mathcal{Q}(\leq; B_3 > 0; [Y_0]) \cap \Upsilon$;
whence, $\psi(u(t^n)) \leq \psi(v(t^n)) - \varphi(w(t^n)) + \chi(z(t^n))$
(right-b-2) $(t^n := (t^n_1, \ldots, t^n_6); n \geq 0)$ has the right property at $(0, b, b, b, b, 0)$; i.e., $(t^n_i \to 0+,\ \forall i \in \{1, 6\})$, and $(t^n_i \to b+,\ \forall i \in \{2, 3, 4, 5\})$.

By the former of these properties,

$(\forall n)$: $\varphi(w(t^n)) - \chi(z(t^n)) \leq \psi(v(t^n)) - \psi(u(t^n))$.

As (u, v, w, z) is right asymptotic of type $(I, I, \mu, 0)$ at $[Y_0]$,

$$u(t^n) \to b+,\ v(t^n) \to b+,\ w(t^n) \to \mu(b)+,\ z(t^n) \to 0.$$

Passing to lim sup as $n \to \infty$ in this relation gives

$$\begin{aligned}\limsup_n \varphi(w(t^n)) &\leq \limsup_n[\psi(v(t^n)) - \psi(u(t^n))]\\ &\leq \mathrm{osc}(+)(\psi, b, b);\end{aligned}$$

in contradiction with $(\psi, \varphi; I, I, \mu)$ = non right oscillating at b. Hence, our working assumption cannot be true; and the assertion follows.

Part 4. [Υ is nbilateral on $[X_0] \cup [Z_0]$ under the described assumptions].

Take some point $c = (c_1, \ldots, c_6)$ in R^6_+. We say that the sequence $(t^n := (t^n_1, \ldots, t^n_6); n \geq 0)$ in R^6_+ is *bilateral* at c if

(bila-c) $t^n_i \to c_i$ as $n \to \infty$, $\forall i$.

Given $b > 0$, let us say that Υ is *nbilateral* at b on $[X_0] \cup [Z_0]$ if

(npoint) for each sequence $(t^n; n \geq 0)$ in $\mathcal{Q}(\leq; B_3 > 0; [X_0] \cup [Z_0]) \cap \Upsilon$,
the bilateral property at $(0, b, b, b, b, 0)$ is not true.

The class of all these $b > 0$ will be denoted as nbila$(\Upsilon; [X_0] \cup [Z_0])$. In this case, we define

(n-bila) Υ is *nbilateral* on $[X_0] \cup [Z_0]$ if
$\Theta :=$ nbila$(\Upsilon; [X_0] \cup [Z_0])$ is identical with R^0_+.

We have to establish that this property holds; that is,

for each sequence $(t^n; n \geq 0)$ in $\mathcal{Q}(\leq; B_3 > 0; [X_0] \cup [Z_0]) \cap \Upsilon$, and each $b > 0$, the bilateral property at $(0, b, b, b, b, 0)$ is not true.

Assume by contradiction that this assertion is false: there exists a (vectorial) sequence $(t^n = (t^n_1, \ldots, t^n_6); n \geq 0)$ in R^6_+ and some $b > 0$, with

(bila-b-1) $(\forall n)$: $t^n = (t^n_1, \ldots, t^n_6) \in \mathcal{Q}(\leq; B_3 > 0; [X_0] \cup [Z_0]) \cap \Upsilon$;
whence, $\psi(u(t^n)) \leq \psi(v(t^n)) - \varphi(w(t^n)) + \chi(z(t^n))$

(bila-b-2) $(t^n = (t_1^n, \ldots, t_6^n); n \geq 0)$ is bilateral at $(0, b, b, b, b, 0)$; that is,

$(t_i^n \to 0, \forall i \in \{1, 6\}), (t_i^n \to b, \forall i \in \{2, 3, 4, 5\})$.

By the former of these properties,

$$(\forall n)\colon \varphi(w(t^n)) - \chi(z(t^n)) \leq \psi(v(t^n)) - \psi(u(t^n)).$$

As (u, v, w, z) is bilateral asymptotic of type $(\xi, \eta, \zeta, 0)$ at $[X_0] \cup [Z_0]$,

$$u(t^n) \to \xi(b),\ v(t^n) \to \eta(b),\ w(t^n) \to \zeta(b).\ z(t^n) \to 0.$$

Passing to lim sup as $n \to \infty$ in this relation gives

$$\begin{aligned}\limsup_n \varphi(w(t^n)) &\leq \limsup_n [\psi(v(t^n)) - \psi(u(t^n))] \\ &\leq \operatorname{osc}(\psi, \xi(b), \eta(b)),\end{aligned}$$

in contradiction with $(\psi, \varphi; \xi, \eta, \zeta)$ = non point oscillating at b. Hence, our working assumption cannot be true, and the assertion follows.

Part 5. [Υ is npoint on $[[X_0]]$ under the described assumptions].

Take some point $c = (c_1, \ldots, c_6)$ in R_+^6. We say that the sequence $(t^n := (t_1^n, \ldots, t_6^n); n \geq 0)$ in R_+^6 is *point* at c if

(point-c) $(t_i^n \to c_i, \forall i)$ and $[t_6^n = c_6, \forall n]$.

Given $b > 0$, let us say that Υ is *npoint* at b on $[[X_0]]$ if

(npoint) for each sequence $(t^n; n \geq 0)$ in $\mathcal{Q}(\leq; B_3 > 0; [[X_0]]) \cap \Upsilon$,
the point property at $(0, 0, b, 0, b, b)$ is not true.

The class of all these $b > 0$ will be denoted as $\text{npoint}(\Upsilon; [[X_0]])$. In this case, define

(n-point) Υ is *npoint* on $[[X_0]]$ if
$\Theta := \text{npoint}(\Upsilon; [[X_0]])$ is identical with R_+^0.

We have to establish that this property holds; that is,

for each sequence $(t^n; n \geq 0)$ in $\mathcal{Q}(\leq; B_3 > 0; [[X_0]]) \cap \Upsilon$, and each $b > 0$, the point property at $(0, 0, b, 0, b, b)$ is not true.

Assume by contradiction that this assertion is false: there exists a (vectorial) sequence $(t^n = (t_1^n, \ldots, t_6^n); n \geq 0)$ in R_+^6 and some $b > 0$, with

(point-b-1) $(\forall n)$: $t^n = (t_1^n, \ldots, t_6^n) \in \mathcal{Q}(\leq; B_3 > 0; [[X_0]]) \cap \Upsilon$;
whence, $\psi(u(t^n)) \leq \psi(v(t^n)) - \varphi(w(t^n)) + \chi(z(t^n))$
(point-b-2) $(t^n = (t_1^n, \ldots, t_6^n); n \geq 0)$ is point at $(0, 0, b, 0, b, b)$; that is,

$(t_i^n \to 0, \forall i \in \{1, 2, 4\})$, $(t_i^n \to b, \forall i \in \{3, 5, 6\})$, and $(t_6^n = b, \forall n)$.

By the former of these properties,

$$(\forall n)\colon \varphi(w(t^n)) - \chi(z(t^n)) \leq \psi(v(t^n)) - \psi(u(t^n)).$$

As (u, v, w, z) is point asymptotic of type $(\xi, \eta, \zeta, 0)$ at $[[X_0]]$,

$$u(t^n) \to \xi(b),\ v(t^n) \to \eta(b),\ w(t^n) \to \zeta(b),\ z(t^n) \to 0.$$

Passing to lim sup as $n \to \infty$ in this relation gives

$$\begin{aligned}\limsup_n \varphi(w(t^n)) &\leq \limsup_n [\psi(v(t^n)) - \psi(u(t^n))] \\ &\leq \operatorname{osc}(\psi, \xi(b), \eta(b)),\end{aligned}$$

in contradiction with $(\psi, \varphi; \xi, \eta, \zeta)$ = non point oscillating at b. Hence, our working assumption cannot be true, and the assertion follows.

Part 6. [Υ is normal under the posed condition].
Remember that, this property means

$(0, a, a, a, a, 0) \in \Upsilon$ is impossible, for each $a > 0$.

Suppose by contradiction that

there exists $a > 0$ with $A := (0, a, a, a, a, 0) \in \Upsilon$.

In view of (u, v, w, z) = invariant,

$$u(A) = a,\ v(A) = a,\ w(A) = a,\ z(A) = 0.$$

This first means

$A := (0, a, a, a, a, 0) \in \Upsilon_0$ (because $a > 0$).

Second, by the complete definition of Υ,

$$\psi(u(A)) \le \psi(v(A)) - \varphi(w(A)) + \chi(z(A)).$$

So, again combining with $(u, v, w, z) =$ invariant, we must have

$$\psi(a) \le \psi(a) - \varphi(a); \text{ that is, } \varphi(a) \le 0;$$

absurd, by $\varphi =$ strictly positive, and our claim follows. □

Now, by simply combining the obtained fact with Theorem 13, one gets the following couple of fixed point statements with a practical value. The former of these statements (referred to as Rhoades–Dutta–Choudhury principle for general functions; in short, (RDC-gen)) is based on a general framework upon ψ.

Theorem 23. *Let the selfmap T be taken so as*

(62-i) T is $(\le, \mathcal{R}; \psi, \varphi, \chi; u, v, w, z)$-contractive

or, equivalently: $(\le; \mathcal{R}; \Upsilon)$-contractive,

where the relation $\mathcal{R}$ over X, the triple (ψ, φ, χ) of functions over $\mathcal{F}(R_+, R)$, and the quadruple (u, v, w, z) of functions over $\mathcal{F}(R_+^6, R_+)$ fulfill (for the moment)

(62-ii) $\mathcal{R}$ is B_1-admissible, and φ is strictly positive ($\varphi(R_+^0) \subseteq R_+^0$)

(62-iii) χ is null right continuous at origin ($\chi(0+0) = 0 = \chi(0)$).

Further, suppose that at the level of each (Ba-o-f) sequence $Y_0 = (y_n)$, we have

(62-iv) *for the telescopic sequences $(r_n := d(y_n, y_{n+1}); n \ge 0)$ and $(p_n := d(y_n, y_{n+2}))$ attached to $Y_0 = (y_n)$ there exists a strongly regular binary operation $(*)$ and and a compatible pair (L, λ) (where $L \in \mathcal{F}(R_+ \times R_+, R_+)$, $\lambda \in \mathcal{F}(R_+)$), with the properties*

(62-iv-1) *(u, v, w, z) is strongly $(*, L, \lambda)$-iterative, in the sense:*

$u^n := u(A_n) = r_{n+1}$, $v^n := v(A_n) = r_n * r_{n+1}$,

$w^n := w(A_n) = L(r_n, r_{n+1})$, $z^n := z(A_n) = 0$, *for all n,*

where $(A_n := (r_n, r_n, p_n, 0, r_{n+1}, r_{n+1}); n \ge 0)$

(62-iv-2) *$(\psi, \varphi; I, I, \lambda)$ is non right oscillating*

(62-v) *there exists* $\mu \in \mathcal{F}(R_+)$ *such that*

(62-v-1) (u, v, w, z) *is right asymptotic of type* $(I, I, \mu, 0)$ *at* $[Y_0]$

(62-v-2) $(\psi, \varphi; I, I, \mu)$ *is almost non right oscillating.*

Finally, let X *be (Ba-o-f,d)-complete,* $X_0 = (x_n)$ *be a (Ba-o-f) sequence, and* $Z_0 = (z_n)$ *be its associated limit sequence. Then,*

(62-a) $X_0 = (x_n)$ *is semi-Picard (modulo* $(\leq, \mathcal{R}; T)$*)*

(62-b) $X_0 = (x_n)$ *is Picard (modulo* $(\leq, \mathcal{R}; T)$*)*

(62-c) $X_0 = (x_n)$ *is strongly Picard (modulo* $(\leq, \mathcal{R}; T)$*), whenever* T *is (Ba-o-f,d)-continuous*

(62-d) $X_0 = (x_n)$ *is Bellman Picard (modulo* $(\leq, \mathcal{R}; T)$*), provided (in addition to the above)* $(\leq)$ *is (Ba-o-f,d)-selfclosed and there exists a triple* (ξ, η, ζ) *in* $\mathcal{F}(R_+)$ *such that*

(62-d-1) (u, v, w, z) *is point asymptotic of type* $(\xi, \eta, \zeta, 0)$ *at* $[[X_0]]$

(62-d-2) $(\psi, \varphi; \xi, \eta, \zeta)$ *is non point oscillating*

(62-e) $X_0 = (x_n)$ *is almost Bellman Picard (modulo* $(\leq, \mathcal{R}; T)$*), provided (in addition to the above)* $(\leq)$ *is (Ba-o-f,d)-sup-selfclosed,* $X_0 = (x_n)$ *is regular, and there exists a triple* (ξ, η, ζ) *in* $\mathcal{F}(R_+)$ *such that*

(62-e-1) (u, v, w, z) *is bilateral asymptotic of type* $(\xi, \eta, \zeta, 0)$ *at* $[X_0] \cup [Z_0]$

(62-e-2) $(\psi, \varphi; \xi, \eta, \zeta)$ *is non point oscillating*

(62-f) T *is fix-*$(\leq)$*-asingleton (hence, fix-*$(\leq)$*-singleton, under any of these extra requirements) when, in addition to the conditions above,*

$\mathcal{R}$ *is* B_2*-admissible and* $(u, v, w.z)$ *is invariant: for each* $a > 0$,

$$u(A) = v(A) = w(A) = a, \; z(A) = 0, \; \text{where } A = (0, a, a, a, a, 0).$$

The latter of these statements (referred to as Rhoades–Dutta–Choudhury principle for increasing functions; in short, (RDC-incr)) is based on an increasing condition upon ψ.

Theorem 24. *Let the selfmap T be taken so as*

(63-i) *T is $(\leq, \mathcal{R}; \psi, \varphi, \chi; u, v, w, z)$-contractive or, equivalently, $(\leq; \mathcal{R}; \Upsilon)$-contractive,*

where the relation $\mathcal{R}$ over X, the triple (ψ, φ, χ) of functions over $\mathcal{F}(R_+, R)$, and the quadruple (u, v, w, z) of functions over $\mathcal{F}(R_+^6, R_+)$ fulfill (for the moment)

(63-ii) *$\mathcal{R}$ is B_1-admissible, ψ is increasing and φ is strictly positive*

(63-iii) *χ is null right continuous at origin ($\chi(0{+}0) = 0 = \chi(0)$).*

Further, suppose that at the level of each (Ba-o-f) sequence $Y_0 = (y_n)$, we have

(63-iv) *for the telescopic sequences $(r_n := d(y_n, y_{n+1}); n \geq 0)$ and $(p_n := d(y_n, y_{n+2}))$ attached to $Y_0 = (y_n)$ there exists a regular binary operation $(*)$ and a compatible pair (L, λ) (where $L \in \mathcal{F}(R_+ \times R_+, R_+)$, $\lambda \in \mathcal{F}(R_+)$), with the properties*

(63-iv-1) *(u, v, w, z) is weakly $(*, L, \lambda)$-iterative, in the sense:*
$u^n := u(A_n) \geq r_{n+1}$, $v^n := v(A_n) \leq r_n * r_{n+1}$,
$w^n := w(A_n) = L(r_n, r_{n+1})$, $z^n := z(A_n) = 0$, for all n,
where $(A_n := (r_n, r_n, p_n, 0, r_{n+1}, r_{n+1}); n \geq 0)$

(63-iv-2) *$(\psi, \varphi; I, I, \lambda)$ is non right oscillating*

(63-v) *there exists $\mu \in \mathcal{F}(R_+)$ such that:*

(63-v-1) *(u, v, w, z) is right asymptotic of type $(I, I, \mu, 0)$ at $[Y_0]$*

(63-v-2) *$(\psi, \varphi; I, I, \mu)$ is almost non right oscillating.*

Finally, let X be (Ba-o-f,d)-complete, $X_0 = (x_n)$ be a (Ba-o-f) sequence, and $Z_0 = (z_n)$ be its associated limit sequence. Then,

(63-a) *$X_0 = (x_n)$ is semi-Picard (modulo $(\leq, \mathcal{R}; T)$)*

(63-b) *$X_0 = (x_n)$ is Picard (modulo $(\leq, \mathcal{R}; T)$)*

(63-c) *$X_0 = (x_n)$ is strongly Picard (modulo $(\le, \mathcal{R}; T)$), whenever T is (Ba-o-f,d)-continuous*

(63-d) *$X_0 = (x_n)$ is Bellman Picard (modulo $(\le, \mathcal{R}; T)$), provided (in addition to the above) $(\le)$ is (Ba-o-f,d)-selfclosed and there exists a triple (ξ, η, ζ) in $\mathcal{F}(R_+)$ such that*

(63-d-1) *(u, v, w, z) is point asymptotic of type $(\xi, \eta, \zeta, 0)$ at $[[X_0]]$*
(63-d-2) *$(\psi, \varphi; \xi, \eta, \zeta)$ is non point oscillating*

(63-e) *$X_0 = (x_n)$ is almost Bellman Picard (modulo $(\le, \mathcal{R}; T)$), provided (in addition to the above) $(\le)$ is (Ba-o-f,d)-sup-selfclosed, $X_0 = (x_n)$ is regular, and there exists a triple (ξ, η, ζ) in $\mathcal{F}(R_+)$ such that*

(63-e-1) *(u, v, w, z) is bilateral asymptotic of type $(\xi, \eta, \zeta, 0)$ at $[X_0] \cup [Z_0]$*

(63-e-2) *$(\psi, \varphi; \xi, \eta, \zeta)$ is non point oscillating*

(63-f) *T is fix-$(\le)$-asingleton (hence, fix-$(\le)$-singleton, under any of these extra requirements) when, in addition to the conditions above,*

$\mathcal{R}$ is B_2-admissible and $(u, v, w.z)$ is invariant: for each $a > 0$,
$u(A) = v(A) = w(A) = a$, $z(A) = 0$, where $A = (0, a, a, a, a, 0)$.

It is worth noting that an extended setting of these developments is possible, under the lines described in the 2001 PhD Thesis by Hitzler [22, Ch 1, Sect 1.2]; see also Turinici [60]. Further aspects are discussed elsewhere.

7. Particular Aspects

Let $(X, d, \le)$ be a partially ordered metric space. Further, let $T \in \mathcal{F}(X)$ be a selfmap of X, with

(a-reg) T is *ascending regular*:
there are ascending orbital sequences in X.

In this case, the family ao(T) of all these objects is nonempty. Let $X_0 = (x_n)$ be an element of it; remember that, without loss, one may assume that

> $B_1(x_n, x_{n+1}) > 0$, $\forall n$, proving that $X_0 = (x_n)$ is a (Ba-o-f) sequence.

Let $Z_0 = \mathcal{L}(X_0)$ be the limit orbital sequence attached to X_0. As precise, we are interested to determine sufficient conditions for existence/uniqueness of elements in Fix(T), via contractive-type requirements involving the iterative processes $X_0 = (x_n)$, $Z_0 = (z_n)$, and the associated sets

$$[X_0] = \{x_n; n \geq 0\},\ [[X_0]] = \mathrm{cl}([X_0]),\ [Z_0] = \{z_n; n \geq 0\}.$$

The specific part of this construction was already presented. Let (ψ, φ, χ) be a triple of functions over $\mathcal{F}(R_+, R)$; and (u, v, w, z) be a quadruple of functions over $\mathcal{F}(R_+^6, R_+)$. Define the couple of subsets in R_+^6

$$\Upsilon_0 = \{t \in R_+^6; u(t), v(t), w(t) > 0\},$$
$$\Upsilon = \{t \in \Upsilon_0; \psi(u(t)) \leq \psi(v(t)) - \varphi(w(t)) + \chi(z(t))\},$$

as well as the couple of relations over X

$$\mathcal{R} = \mathcal{Q}^{-1}(\Upsilon_0),\ \mathcal{S} = \mathcal{Q}^{-1}(\Upsilon).$$

We may introduce two contractive conditions upon our data. The former of these is a functional contractive condition written as

> (fct-contr) T is $(\leq, \mathcal{R}; \psi, \varphi, \chi; u, v, w, z)$-*contractive*:
> $\psi(u \circ \mathcal{Q}(x, y)) \leq \psi(v \circ \mathcal{Q}(x, y)) - \varphi(w \circ \mathcal{Q}(x, y)) + \chi(z \circ \mathcal{Q}(x, y))$, when $(x, y) \in (\leq; \mathcal{R})$.

The latter of these appears as a set contractive condition and writes

> (set-contr) T is $(\leq, \mathcal{R}; \Upsilon)$-contractive: $\mathcal{Q}(x, y) \in \Upsilon$, if $(x, y) \in (\leq; \mathcal{R})$.

The connection between these is described as

> T is $(\leq, \mathcal{R}; \psi, \varphi, \chi; u, v, w, z)$-contractive iff T is $(\leq, \mathcal{R}; \Upsilon)$-contractive.

For this set contractive condition, sufficient conditions were given so that the main result be applicable; these, naturally, are ultimately expressed in terms of $\mathcal{R}$, (ψ, φ, χ) and (u, v, w, z). The effective application of the quoted statement involves the couple of fixed point results we just exposed. It is our aim in the following to show that, as a basic particular case of these, one may get a couple of fixed point statements involving rational type contractions on partially ordered metric spaces including the ones in Harjani *et al.* [21] and Chandok *et al.* [14]. Further aspects involving these developments are also discussed.

(A) For the moment, some direct constructions are proposed.

(Constr-0) Define the auxiliary system of functions $(e_0, e_1, e_2, e_3, e_4)$ over the class $\mathcal{F}(R^6_+, R_+)$, as, for each $t = (t_1, \ldots, t_6) \in R^6_+$,

$e_0(t) = t_2 + |t_1 - t_6|$, $e_1(t) = t_6(1 + t_1)/(1 + t_2)$,
$e_2(t) = t_4(1 + t_3)/(1 + t_2)$, $e_3(t) = \max\{t_1, t_6\}$, $e_4(t) = (1/2)(t_3 + t_4)$,
$e_5(t) = t_1 t_6/t_2$, if $t_2 \neq 0$, and $e_5(t) = 0$, if $t_2 = 0$.

(Constr-1) Further, define the triple (u_1, u_2, u_3) over the class $\mathcal{F}(R^6_+, R_+)$, as, for each $t = (t_1, \ldots, t_6) \in R^6_+$,

(u-def) $u_1(t) = \max\{t_5, e_1(t)\}$, $u_2(t) = \max\{t_5, t_6\}$, $u_3(t) = t_5$.

(Constr-2) Then, let us construct the system of functions $(v_1, \ldots, v_5)$ over the class $\mathcal{F}(R^6_+, R_+)$, as, for each $t = (t_1, \ldots, t_6) \in R^6_+$,

(v-def) $v_1(t) = \max\{t_2, e_1(t), e_2(t)\}$, $v_2(t) = \max\{t_2, e_3(t), e_4(t)\}$,
$v_3(t) = \max\{t_2, e_5(t)\}$, $v_4(t) = e_0(t)$, $v_5(t) = t_2$.

(Constr-3) Further, define the system $(w_1, \ldots, w_4)$ of functions over $\mathcal{F}(R^6_+, R_+)$, as, for each $t = (t_1, \ldots, t_6) \in R^6_+$,

(w-def) $w_1(t) = \max\{t_2, e_1(t)\}$, $w_2(t) = \max\{t_2, t_6\}$,
$w_3(t) = \max\{t_2, e_5(t)\}$, $w_4(t) = t_2$.

(Constr-4) Then, define the couple (z_1, z_2) of functions over $\mathcal{F}(R_+^6, R_+)$, as, for each $t = (t_1, \ldots, t_6) \in R_+^6$,

$z_1(t) = \min\{t_1, t_3, t_4, t_6\}$, $z_2(t) = t_1 t_3 t_4 t_6$.

(Constr-5) Finally, define the couple $(\top, *)$ of binary relations on R_+ as

$(a \top b = \max\{a, b\}, a * b = a + |b - a|; a, b \in R_+)$; as precise,

$(\top)$ and $(*)$ are strongly regular and $a \top b \leq a * b$, $\forall a, b \in R_+$.

It is our aim to discuss of to what extent are the triples (u, v, w, z), where

$u \in \{u_1, u_2, u_3\}$, $v \in \{v_1, v_2, v_3, v_4, v_5\}$, $w \in \{w_1, w_2, w_3, w_4\}$, $z \in \{z_1, z_2\}$,

compatible with the regularity conditions encountered in the couple of fixed point statements above. The response to this question is to be precise as below.

(R-1) Let $Y_0 = (y_n)$ be a (Ba-o-f) sequence, and $(r_n := d(y_n, y_{n+1}); n \geq 0)$, $(p_n := d(y_n, y_{n+2}))$, stand for the attached telescopic sequences. Further, let $(A_n := (r_n, r_n, p_n, 0, r_{n+1}, r_{n+1}); n \geq 0)$ denote the vectorial sequence generated by these. We have

$(\forall n)$: $e_0(A_n) = r_n * r_{n+1}$, $e_1(A_n) = r_{n+1}$, $e_2(A_n) = 0$,
$e_3(A_n) = r_n \top r_{n+1}$, $e_4(A_n) = (1/2)p_n \leq r_n \top r_{n+1}$, $e_5(A_n) = r_{n+1}$,

and this, by definition, gives

$(\forall n)$: $u_1(A_n) = r_{n+1}$, $u_2(A_n) = r_{n+1}$, $u_3(A_n) = r_{n+1}$,
$v_1(A_n) = r_n \top r_{n+1}$, $v_2(A_n) = r_n \top r_{n+1}$, $v_3(A_n) = r_n \top r_{n+1}$,
$v_4(A_n) = r_n * r_{n+1}$, $v_5(A_n) = r_n$,
$w_1(A_n) = r_n \top r_{n+1}$, $w_2(A_n) = r_n \top r_{n+1}$, $w_3(A_n) = r_n \top r_{n+1}$, $w_4(A_n) = r_n$,
$z_1(A_n) = 0$, $z_2(A_n) = 0$,

proving that

(u_i, v_j, w_k, z_h) is strongly iterative, for each
$i \in \{1, 2, 3\}$, $j \in \{1, 2, 3, 4\}$, $k \in \{1, 2, 3, 4\}$, $h \in \{1, 2\}$,

(u_i, v_j, w_k, z_h) is weakly iterative, for each $i \in \{1,2,3\}$, $j \in \{1,2,3,4,5\}$, $k \in \{1,2,3,4\}$, $h \in \{1,2\}$.

(R-2) Let $Y_0 = (y_n)$ be a (Ba-o-f) sequence; and (ξ, η, ζ) be a triple of functions in $\mathcal{F}(R_+)$. Remember that (u, v, w, z) is called *right asymptotic of type* $(\xi, \eta, \zeta, 0)$ at $[Y_0]$, when

> $\forall$ sequence $(t^n = (t_1^n, \ldots, t_6^n); n \geq 0)$ in $\mathcal{Q}(\leq; B_3 > 0; [Y_0])$, and $\forall b > 0$:
>
> $(t^n; n \geq 0)$ has the right property at $(0, b, b, b, b, 0)$ (in the sense: $t_i^n \to 0$, $\forall i \in \{1,6\}$, $t_j^n \to b+$, $\forall j \in \{2,3,4,5\}$) implies $u(t^n) \to \xi(b)+$, $v(t^n) \to \eta(b)+$, $w(t^n) \to \zeta(b)+$, $z(t^n) \to 0$.

To verify this upon our constructions, let the sequence $(t^n; n \geq 0)$ in $\mathcal{Q}(\leq; B_3 > 0; [Y_0])$, and the point $b > 0$ be as before. We have (as $n \to \infty$)

$$e_0(t^n) \to b,\ e_1(t^n) \to 0,\ e_2(t^n) \to b,\ e_3(t^n) \to 0,\ e_4(t^n) \to b,\ e_5(t^n) \to 0,$$

and this, by definition, gives (again under $n \to \infty$)

$$u_1(t^n) \to b+,\ u_2(t^n) \to b+,\ u_3(t^n) \to b+,$$
$$v_1(t^n) \to b+,\ v_2(t^n) \to b+,\ v_3(t^n) \to b+,\ v_4(t^n) \to b+,\ v_5(t^n) \to b+,$$
$$w_1(t^n) \to b+,\ w_2(t^n) \to b+,\ w_3(t^n) \to b+,\ w_4(t^n) \to b+,$$
$$z_1(t^n) \to 0,\ z_2(t^n) \to 0,$$

proving that

> (u_i, v_j, w_k, z_h) is right asymptotic of type $(I, I, I, 0)$ at $[Y_0]$, for all $i \in \{1,2,3\}$, $j \in \{1,2,3,4,5\}$, $k \in \{1,2,3,4\}$, $h \in \{1,2\}$.

(R-3) Let $X_0 = (x_n)$ be a (Ba-o-f) sequence, and $Z_0 = (z_n)$ be the associated limit sequence. Then, let (ξ, η, ζ) be a triple of functions in $\mathcal{F}(R_+)$. Remember that (u, v, w, z) is called *bilateral asymptotic of type* $(\xi, \eta, \zeta, 0)$ at $[X_0] \cup [Z_0]$, when

> $\forall$ sequence $(t^n = (t_1^n, \ldots, t_6^n); n \geq 0)$ in $\mathcal{Q}(\leq; B_3 > 0; [X_0] \cup [Z_0])$, and $\forall b > 0$: $(t^n; n \geq 0)$ has the bilateral property at $(0, b, b, b, b, 0)$

(in the sense: $t_i^n \to 0, \forall i \in \{1,6\}, t_j^n \to b, \forall j \in \{2,3,4,5\}$) implies $u(t^n) \to \xi(b)$, $v(t^n) \to \eta(b)$, $w(t^n) \to \zeta(b)$, $z(t^n) \to 0$.

To verify this upon our constructions, let the sequence $(t^n; n \geq 0)$ in $\mathcal{Q}(\leq; B_3 > 0; [X_0] \cup [Z_0])$, and the point $b > 0$ be as before. We have (as $n \to \infty$)

$e_0(t^n) \to b$, $e_1(t^n) \to 0$, $e_2(t^n) \to b$, $e_3(t^n) \to 0$, $e_4(t^n) \to b$, $e_5(t^n) \to 0$,

and this, by definition, gives (again under $n \to \infty$)

$u_1(t^n) \to b$, $u_2(t^n) \to b$, $u_3(t^n) \to b$,

$v_1(t^n) \to b$, $v_2(t^n) \to b$, $v_3(t^n) \to b$, $v_4(t^n) \to b$, $v_5(t^n) \to b$,

$w_1(t^n) \to b$, $w_2(t^n) \to b$, $w_3(t^n) \to b$, $w_4(t^n) \to b$, $z_1(t^n) \to 0$, $z_2(t^n) \to 0$,

proving that

(u_i, v_j, w_k, z_h) is bilateral asymptotic of type $(I, I, I, 0)$ at $[X_0] \cup [Z_0]$,

for all $i \in \{1,2,3\}$, $j \in \{1,2,3,4,5\}$, $k \in \{1,2,3,4\}$, $h \in \{1,2\}$.

(R-4) Let $X_0 = (x_n)$ be a (Ba-o-f) sequence, and (ξ, η, ζ) be a triple of functions in $\mathcal{F}(R_+)$. Remember that (u, v, w, z) is called *point asymptotic of type* $(\xi, \eta, \zeta, 0)$ at $[[X_0]]$, when

$\forall$ sequence $(t^n = (t_1^n, \ldots, t_6^n); n \geq 0)$ in $\mathcal{Q}(\leq; B_3 > 0; [[X_0]])$, and $\forall b > 0$:

$(t^n = (t_1^n, \ldots, t_6^n); n \geq 0)$ has the point property at $(0,0,b,0,b,b)$ (in the sense: $t_i^n \to 0$, $\forall i \in \{1,2,4\}$, $t_j^n \to b$, $\forall j \in \{3,5,6\}$, and $t_6^n = b$, $\forall n$) implies $u(t^n) \to \xi(b)$, $v(t^n) \to \eta(b)$, $w(t^n) \to \zeta(b)$, $z(t^n) \to 0$.

To verify this upon our constructions, let the sequence $(t^n; n \geq 0)$ in $\mathcal{Q}(\leq; B_3 > 0; [[X_0]])$, and the point $b > 0$ be as before. We have (as $n \to \infty$)

$e_0(t^n) \to b$, $e_1(t^n) \to b$, $e_2(t^n) \to 0$, $e_3(t^n) \to b$,

$e_4(t^n) \to b/2$, $(e_5(t^n))$ is not convergent,

and this, by definition, gives (again under $n \to \infty$)

$u_1(t^n) \to b$, $u_2(t^n) \to b$, $u_3(t^n) \to b$,
$v_1(t^n) \to b$, $v_2(t^n) \to b$, $(v_3(t^n))$ is not convergent,
$v_4(t^n) \to b$, $v_5(t^n) \to 0$,
$w_1(t^n) \to b$, $w_2(t^n) \to b$, $(w_3(t^n))$ is not convergent, $w_4(t^n) \to 0$,
$z_1(t^n) \to 0$, $z_2(t^n) \to 0$,

proving that

$(\forall i \in \{1,2,3\}, \forall j \in \{1,2,4,5\}), \forall k \in \{1,2,4\}), \forall h \in \{1,2\})$:
(u_i, v_j, w_k, z_h) is point asymptotic of type $(I, \eta_j, \zeta_k, 0)$,
where $(\eta_j = I; j \in \{1,2,4\})$, $\eta_5 = 0$, $(\zeta_k = I; k \in \{1,2\})$, $\zeta_4 = 0$;
$\forall i \in \{1,2,3\}, \forall h \in \{1,2\})$:
(u_i, v_j, w_k, z_h) is not point asymptotic when $j \in \{3\}$ and/or $k \in \{3\}$.

(R-5) Finally, remember that the quadruple $(u, v, w.z)$ is invariant, provided

$(\forall a > 0)$: $u(A) = v(A) = w(A) = a$, $z(A) = 0$, where $A = (0, a, a, a, a, 0)$.

To verify this upon our constructions, we have

$e_0(A) = a$, $e_1(A) = 0$, $e_2(A) = a$, $e_3(A) = 0$, $e_4(A) = a$, $e_5(A) = 0$,

and this, by definition, gives

$u_1(A) = a$, $u_2(A) = a$, $u_3(A) = a$,
$v_1(A) = a$, $v_2(A) = a$, $v_3(A) = a$, $v_4(A) = a$, $v_5(A) = a$,
$w_1(A) = a$, $w_2(A) = a$, $w_3(A) = a$, $w_4(A) = a$,
$z_1(A) = 0$, $z_2(A) = 0$,

proving that

(u_i, v_j, w_k, z_h) is invariant,
for all $i \in \{1,2,3\}$, $j \in \{1,2,3,4,5\}$, $k \in \{1,2,3,4\}$, $h \in \{1,2\}$.

As a direct consequence of these facts and Rhoades–Dutta–Choudhury principles we just exposed, one may now derive a couple of fixed point statements with methodological value.

The former of these, starting from Rhoades–Dutta–Choudhury principle for general functions (RDC-gen) is based on a general framework upon ψ.

Theorem 25. *Let the selfmap T be taken so as*

(71-i) *T is $(\leq, \mathcal{R}; \psi, \varphi, \chi; u_i, v_j, w_k, z_h)$-contractive or, equivalently, $(\leq; \mathcal{R}; \Upsilon)$-contractive,*

where the relation $\mathcal{R}$ over X, the triple (ψ, φ, χ) of functions over $\mathcal{F}(R_+, R)$, the quadruple (u_i, v_j, w_k, z_h) of functions over $\mathcal{F}(R_+^6, R_+)$ fulfill (for the moment)

(71-ii) *$\mathcal{R}$ is B_1-admissible and φ is strictly positive ($\varphi(R_+^0) \subseteq R_+^0$)*

(71-iii) *χ is null right continuous at origin ($\chi(0+0) = 0 = \chi(0)$)*

(71-iv) *$i \in \{1, 2, 3\}$, $j \in \{1, 2, 3, 4\}$, $k \in \{1, 2, 3, 4\}$, $h \in \{1, 2\}$*

(71-v) *$(\psi, \varphi; I, I, I)$ is non right oscillating.*

Further, let X be (Ba-o-f,d)-complete; and $X_0 = (x_n)$ be a (Ba-o-f) sequence. Then,

(71-a) *$X_0 = (x_n)$ is semi-Picard (modulo $(\leq, \mathcal{R}; T)$)*

(71-b) *$X_0 = (x_n)$ is Picard (modulo $(\leq, \mathcal{R}; T)$)*

(71-c) *$X_0 = (x_n)$ is strongly Picard (modulo $(\leq, \mathcal{R}; T)$), whenever T is (Ba-o-f,d)-continuous*

(71-d) *$X_0 = (x_n)$ is Bellman Picard (modulo $(\leq, \mathcal{R}; T)$), provided $(\leq)$ is (Ba-o-f,d)-selfclosed and (in addition to the general conditions)*

$j \in \{1, 2, 4\}$, $k \in \{1, 2, 4\}$ and $(\psi, \varphi; I, \eta_j, \zeta_k)$ is non point oscillating, where $(\eta_j = I; j \in \{1, 2, 4\})$, $(\zeta_k = I; k \in \{1, 2\})$, $\zeta_4 = 0$

(71-e) *$X_0 = (x_n)$ is almost Bellman Picard (modulo $(\leq, \mathcal{R}; T)$), provided $(\leq)$ is (Ba-o-f,d)-sup-selfclosed, $X_0 = (x_n)$ is regular, and*

(in addition to the imposed general conditions)

$(\psi, \varphi; I, I, I)$ *is non point oscillating*

(71-f) T *is fix-*$(\leq)$*-asingleton (hence, fix-*$(\leq)$*-singleton, under any of these extra requirements) when, in addition to the conditions above,*

$\mathcal{R}$ *is* B_2*-admissible:* $(B_2 > 0) \subseteq \mathcal{R}$.

The latter of these, starting from Rhoades–Dutta–Choudhury principle for increasing functions (RDC-incr) is based on an increasing condition upon ψ.

Theorem 26. *Let the selfmap* T *be taken so as*

(72-i) T *is* $(\leq, \mathcal{R}; \psi, \varphi, \chi; u_i, v_j, w_k, z_h)$*-contractive*

or, equivalently, $(\leq; \mathcal{R}; \Upsilon)$*-contractive,*

where the relation $\mathcal{R}$ *over* X*, the triple* (ψ, φ, χ) *of functions over* $\mathcal{F}(R_+, R)$*, and the quadruple* (u, v_j, w_k, z) *of functions over* $\mathcal{F}(R^6_+, R_+)$ *fulfill (for the moment)*

(72-ii) ψ *is increasing and* φ *is strictly positive*

(72-iii) χ *is null right continuous at origin* $(\chi(0+0) = 0 = \chi(0))$

(72-iv) $i \in \{1, 2, 3\}$, $j \in \{1, 2, 3, 4, 5\}$, $k \in \{1, 2, 3, 4\}$, $h \in \{1, 2\}$

(72-v) $(\psi, \varphi; I, I, I)$ *is non right oscillating.*

Further, let X *be (Ba-o-f,d)-complete; and* $X_0 = (x_n)$ *be a (Ba-o-f) sequence. Then,*

(72-a) $X_0 = (x_n)$ *is semi-Picard (modulo* $(\leq, \mathcal{R}; T)$*)*

(72-b) $X_0 = (x_n)$ *is Picard (modulo* $(\leq, \mathcal{R}; T)$*)*

(72-c) $X_0 = (x_n)$ *is strongly Picard (modulo* $(\leq, \mathcal{R}; T)$*), whenever* T *is (Ba-o-f,d)-continuous*

(72-d) $X_0 = (x_n)$ *is Bellman Picard (modulo* $(\leq, \mathcal{R}; T)$*), provided* $(\leq)$ *is (Ba-o-f,d)-selfclosed and (in addition to the general conditions)*

$j \in \{1, 2, 4, 5\}$, $k \in \{1, 2, 4\}$ *and* $(\psi, \varphi; I, \eta_j, \zeta_k)$ *is non point oscillating, where* $(\eta_j = I; j \in \{1, 2, 4\})$, $\eta_5 = 0$, $(\zeta_k = I; k \in \{1, 2\})$, $\zeta_4 = 0$

(72-e) *$X_0 = (x_n)$ is almost Bellman Picard (modulo $(\leq, \mathcal{R}; T)$), provided $(\leq)$ is (Ba-o-f,d)-sup-selfclosed, $X_0 = (x_n)$ is regular, and (in addition to the general conditions)*

$(\psi, \varphi; I, I, I)$ *is non point oscillating*

(72-f) *T is fix-$(\leq)$-asingleton (hence, fix-$(\leq)$-singleton, under any of these extra requirements) when, in addition to the conditions above,*

$\mathcal{R}$ *is B_2-admissible:* $(B_2 > 0) \subseteq \mathcal{R}$.

Two basic particular cases of this result are to be discussed.

(Part-Case-I) [$\psi(.)$ = increasing, $\chi(.)$ is nonzero].
As precise in a previous place, the regularity of d-convergent sequence $X_0 = (x_n)$ is a consequence of

T is $(\leq)$-increasing and $(\leq)$ is (Ba-o-f,d)-sup-selfclosed.

In this case, under

$\chi(.)$ is linear ($\chi(t) = \lambda t$, $t \in R_+$, for some $\lambda \geq 0$)

the almost Bellman part of Theorem 26 yields the main result in Mustafa *et al.* [35] that may be viewed as an extension of some altering related statements due to Luong and Thuan [30].

(Part-Case-II) [$\psi(.)$ = increasing, $\chi(.)$ is identically zero].
Clearly, the (Ba-o-f,d) completeness upon X is assured when [X is d-complete] Note that, in such a case, the strong part of Theorem 26 includes in a direct way the related statement in Jaggi [24] and Arshad *et al.* [4]; see also Chouhan *et al.* [16]. At the same time, the almost Bellman part of Theorem 26 yields the main result in Gupta *et al.* [19]. proved via different methods. Moreover, as precise, the included result extends the statement in Luong and Thuan [30]; which, in turn, includes the one in Harjani *et al.* [21]; hence, so does the underlying variant of Theorem 26.

Finally, by the same completeness condition as before, the Bellman part of Theorem 26 includes the related statements in Chandok *et al.* [14], Vujaković *et al.* [63], and Choudhury *et al.* [15], proved via direct methods. Further, under the same framework, the obtained part includes as well a lot of *rational* type fixed point results in Altun and Simsek [3], Cabrera *et al.* [11], Kumam *et al.* [28], or Can and Thuan [12]; see also Nashine *et al.* [37].

(Part-Case-III) [$\psi(.)$ = arbitrary].
As precise, the (Ba-o-f,d) completeness upon X is holding when [X is d-complete]. In this case, Theorem 25 may be viewed as a partial extension of a related statement in Proinov [41], which, in turn, appears as a generalization of the result in Xue [65].

Finally, concerning further enlargements of these facts to the relational metric setting, the following observations are to be made.

(Obs-1) Due to the very specific nature of involved elements, the almost Bellman portion of Theorems 25 and 26 cannot be extended to (genuine) relational metric spaces.

(Obs-2) On the contrary, an extension of strongly/Bellman portion of Theorems 25 and 26 to relational metric spaces is possible, under the lines in Turinici [59]. This tells us that the results in Chakraborty *et al.* [13] are available in this way.

Further aspects concerning such an extension will be delineated elsewhere.

References

[1] Akkouchi, M. Common fixed points for weakly compatible maps satisfying implicit relations without continuity, *Dem. Math.*, 44, 151–158 (2011).

[2] Alam, A., Imdad, M. and Arif, M. Observations on relation-theoretic coincidence theorems under Boyd-Wong type nonlinear contractions, *Fixed Point Th. Appl.*, 2019, 2019:6.

[3] Altun, I. and Simsek, H. Some fixed point theorems on ordered metric spaces, *Fixed Point Th. Appl.*, 2010, 621469.

[4] Arshad, M., Karapinar, E. and Ahmad, J. Some unique fixed point theorems for rational contractions in partially ordered metric spaces, *J. Ineq. Appl.*, 2013, 2013:248.

[5] Banach, S. Sur les opérations dans les ensembles abstraits et leur application aux équations intégrales, *Fund. Math.*, 3, 133–181 (1922).

[6] Beg, I. and Butt, A.R. Common fixed point for generalized set valued contractions satisfying an implicit relation in partially ordered metric spaces, *Math. Commun.*, 15, 65–76 (2010).

[7] Berinde, V. and Vetro, F. Common fixed points of mappings satisfying implicit contractive conditions, *Fixed Point Th. Appl.*, 2012, 2012:105.

[8] Bernays, P. A system of axiomatic set theory: Part III. Infinity and enumerability analysis, *J. Symbolic Logic*, 7, 65–89 (1942).

[9] Berzig, M. Coincidence and common fixed point results on metric spaces endowed with an arbitrary binary relation and applications, *J. Fixed Point Th. Appl.*, 12, 221–238 (2013).

[10] Boyd, D.W. and Wong, J.S.W. On nonlinear contractions, *Proc. Amer. Math. Soc.*, 20, 458–464 (1969).

[11] Cabrera, I., Harjani, J. and Sadarangani, K. A fixed point theorem for contractions of rational type in partially ordered metric spaces, *Ann. Univ. Ferrara*, 59, 251–258 (2013).

[12] Can, N.V. and Thuan, N.X. Fixed point theorem for generalized weak contractions involving rational expressions, *Open J. Math. Modelling*, 1, 29–33 (2013).

[13] Chakraborty, P., Choudhury, B.S. and De La Sen, M. Relation theoretic fixed point theorems for generalized weakly contractive mappings, *Symmetry*, 2020, 12, 29.

[14] Chandok, S., Choudhury, B.S. and Metiya, N. Fixed point results in ordered metric spaces for rational type expressions with auxiliary functions, *J. Egyptian Math. Soc.*, 23, 95–101 (2015).

[15] Choudhury, B.S., Konar, P., Rhoades, B.E. and Metiya, N. Fixed point theorems for generalized weakly contractive mapping, *Nonlin. Anal.*, 74, 2116–2126 (2011).

[16] Chouhan, S., Tiwari, A. and Solanki, M. Some fixed point result in metric spaces for rational expression, *Math. Theory Modelling*, 4, 77–84 (2014).

[17] Cohen, P.J. Set Theory and the Continuum Hypothesis, Benjamin, New York (1966).

[18] Dutta, P.N. and Choudhury, B.S. A generalisation of contraction principle in metric spaces, *Fixed Point Th. Appl.*, 2008, 406368.

[19] Gupta, V., Mani, N. and Jindal, J. Some new fixed point theorems for generalized weak contraction in partially ordered metric spaces, *Comp. Math. Methods*, 2020, e1115.

[20] Halmos, P.R. *Naive Set Theory,* Van Nostrand Reinhold Co., New York (1960).

[21] Harjani, J., Lopez, B. and Sadarangani, K. A fixed point theorem for mappings satisfying a contractive condition of rational type on a partially ordered metric space, *Abstr. Appl. Anal.*, 2010, 190701.

[22] Hitzler, P. Generalized metrics and topology in logic programming semantics, PhD Thesis, Natl. Univ. Ireland, Univ. College Cork (2001).

[23] Jachymski, J. The contraction principle for mappings on a metric space with a graph, *Proc. Amer. Math. Soc.*, 136, 1359–1373 (2008).

[24] Jaggi, D.S. Some unique fixed point theorems, *Indian J. Pure Appl. Math.*, 8, 223–230 (1977).

[25] Kasahara, S. On some generalizations of the Banach contraction theorem, *Publ. Res. Inst. Math. Sci. Kyoto Univ.*, 12, 427–437 (1976).

[26] Khan, M.S., Swaleh, M. and Sessa, S. Fixed point theorems by altering distances between the points, *Bull. Austral. Math. Soc.*, 30, 1–9 (1984).

[27] Krasnoselskii, M.A. and Stetsenko, V.Ya. Towards a theory of equations with concave operators (Russian), *Sib. Math. Zh.*, 10, 565–572 (1969).

[28] Kumam, P., Rouzkard, F., Imdad, M. and Gopal, D. Fixed point theorems on ordered metric spaces through a rational contraction, *Abstr. Appl. Anal.*, 2013, 206515.

[29] Leader, S. Fixed points for general contractions in metric spaces, *Math. Japonica*, 24, 17–24 (1979).

[30] Luong, N.V. and Thuan, N.X. Fixed point theorem for generalized weak contractions satisfying rational expressions in ordered metric spaces, *Fixed Point Th. Appl.*, 2011, 2011:46.

[31] Matkowski, J. Integrable solutions of functional equations, *Dissert. Math.*, 127, *Polish Sci. Publ.*, Warsaw (1975).

[32] Meir, A. and Keeler, E. A theorem on contraction mappings, *J. Math. Anal. Appl.*, 28, 326–329 (1969).

[33] Moore, G.H. *Zermelo's Axiom of Choice: Its Origin, Development and Influence,* Springer, New York (1982).

[34] Moskhovakis, Y. Notes on Set Theory, Springer, New York (2006).

[35] Mustafa, Z., Karapinar, E. and Aydi, H. A discussion on generalized almost contractions via rational expressions in partially ordered metric spaces, *J. Ineq. Appl.*, 2014, 2014:219.

[36] Nadler S.B. Jr. Multi-valued contraction mappings, *Pacific J. Math.*, 30, 475–488 (1969).

[37] Nashine, H.K., Kadelburg, Z. and Kumam, P. Implicit-relation-type cyclic contractive mappings and applications to integral equations, *Abstr. Appl. Anal.*, 2012, 386253.

[38] Nieto, J.J. and Rodriguez-Lopez, R. Contractive mapping theorems in partially ordered sets and applications to ordinary differential equations, *Order*, 22, 223–239 (2005).

[39] Pasicki, L. A strong fixed point theorem, *Topology Appl.*, 282, 107–130 (2020).

[40] Popa, V. Some fixed point theorems for compatible mappings satisfying an implicit relation, *Dem. Math.*, 32, 157–163 (1999).

[41] Proinov, P.D. Fixed point theorems for generalized contractive mappings in metric spaces, *J. Fixed Point Th. Appl.*, 2020, 22:21.

[42] Ran, A.C.M. and Reurings, M.C. A fixed point theorem in partially ordered sets and some applications to matrix equations, *Proc. Amer. Math. Soc.*, 132, 1435–1443 (2004).

[43] Reich, S. Fixed points of contractive functions, *Boll. Un. Mat. Ital.*, 5, 26–42 (1972).

[44] Rhoades, B.E. A comparison of various definitions of contractive mappings, *Trans. Amer. Math. Soc.*, 226, 257–290 (1977).

[45] Rhoades, B.E. Some theorems on weakly contractive maps, Nonlin. Anal., 47, 2683–2693 (2001).

[46] Roldán, A.F. and Shahzad, N. From graphical metric spaces to fixed point theory in binary related distance spaces, *Filomat*, 31, 3209–3231 (2017).

[47] Rus, I.A. *Generalized Contractions and Applications,* Cluj University Press, Cluj-Napoca (2001).

[48] Samet, B. and Turinici, M. Fixed point theorems on a metric space endowed with an arbitrary binary relation and applications, *Commun. Math. Anal.*, 13, 82–97 (2012).

[49] Samet, B., Vetro, C. and Yazidi, H. A fixed point theorem for a Meir-Keeler type contraction through rational expression, *J. Nonlinear Sci. Appl.*, 6, 162–169 (2013).

[50] Schechter, E. *Handbook of Analysis and its Foundation,* Academic Press, New York (1997).

[51] Sierpinski, W. *Cardinal and Ordinal Numbers,* Polish Sci. Publ., Warsaw (1965).

[52] Tarski, A. Axiomatic and algebraic aspects of two theorems on sums of cardinals, *Fund. Math.*, 35, 79–104 (1948).

[53] Turinici, M. Fixed points of implicit contraction mappings, *An. Şt. Univ. "Al. I. Cuza" Iaşi* (S I-a, Mat), 22, 177–180 (1976).

[54] Turinici, M. Fixed points of implicit contractions via Cantor's intersection theorem, *Bul. Inst. Polit. Iaşi* (Sect I: Mat., Mec. Teor., Fiz.), 26(30), 65–68 (1980).

[55] Turinici, M. Fixed points for monotone iteratively local contractions, *Dem. Math.*, 19, 171–180 (1986).

[56] Turinici, M. Abstract comparison principles and multivariable Gronwall-Bellman inequalities, *J. Math. Anal. Appl.*, 117, 100–127 (1986).

[57] Turinici, M. Function pseudometric VP and applications, Bul. Inst. Polit. Iaşi (Sect.: Mat., Mec. Teor., Fiz.), 53(57), 393–411 (2007).

[58] Turinici, M. Ran-Reurings theorems in ordered metric spaces, *J. Indian Math. Soc.*, 78, 207–214 (2011).

[59] Turinici, M. Implicit contractive maps in ordered metric spaces, in: Th.M. Rassias and L. Toth, (eds.), *Topics in Mathematical Analysis and Applications*, pp. 715–746, Springer International Publishing, Switzerland (2014).

[60] Turinici, M. Contractive maps in locally transitive relational metric spaces, *The Sci. World J.*, 2014, 169358.

[61] Turinici, M. *Selected Topics in Metrical Fixed Point Theory* (Revised Ed.), Pim Editorial House, Iaşi (2017).

[62] Turinici, M. Meir-Keeler sequential contractions and applications, D. Andrica and Th.M. Rassias, (eds.), *Differential and Integral Inequalities*, pp. 761–829, Springer International Publishing, Switzerland (2019).

[63] Vujaković, J., Mitrović, S., Pavlović, M. and Radenović, S. On recent results concerning F-contraction in generalized metric spaces, *Mathematics*, 2020, 8, 767.

[64] Wolk, E.S. On the principle of dependent choices and some forms of Zorn's lemma, *Canad. Math. Bull.*, 26, 365–367 (1983).

[65] Xue, Z. Fixed point theorems for generalized weakly contractive mappings, *Bull. Austral. Math. Soc.*, 93, 321–329 (2016).

https://doi.org/10.1142/9789811293009_0038

Chapter 38

Equivalent Properties of Two Kinds of Hardy-Type Integral Inequalities with the General Non-homogeneous Kernel in the Whole Plane

Bicheng Yang

School of Mathematics, Guangdong University of Education, Guangzhou, Guangdong 510303, P.R. China
bcyang@gdei.edu.cn
bcyang818@163.com

By means of the way of real analysis and the weight functions, a few equivalent conditions of two kinds of Hardy-type integral inequalities with the general non-homogeneous kernel in the whole plane are obtained. The constant factors are proved to be the best possible. As applications, a few equivalent conditions of two kinds of Hardy-type integral inequalities with the general homogeneous kernel in the whole plane are deduced. We also consider the operator expressions, a few particular cases and examples.

1. Introduction

Assuming that $f(x), g(y) \geq 0$, $0 < \int_0^\infty f^2(x)dx < \infty$, and $0 < \int_0^\infty g^2(y)dy < \infty$, we have the following well-known Hilbert's integral inequality (see [1]):

$$\int_0^\infty \int_0^\infty \frac{f(x)g(y)}{x+y} dxdy < \pi \left(\int_0^\infty f^2(x)dx \int_0^\infty g^2(y)dy \right)^{\frac{1}{2}}, \quad (1)$$

with the best possible constant factor π. Recently, by means of the weight functions, a lot of extensions of (1) were given by two books (see [2,3]). Some Hilbert-type inequalities with the homogenous kernels and non-homogenous kernels were provided by [4–9]. In 2017, Hong [10] also gave a equivalent condition between a Hilbert-type inequalities with the homogenous kernel and some parameters. Some other kinds of Hilbert-type inequalities were obtained by [11–16]. Most of them are built in the quarter plane of the first quadrant.

Using the way of real analysis, in 2007, Yang [17] gave a Hilbert-type integral inequality in the whole plane as follows:

$$\int_{-\infty}^\infty \int_{-\infty}^\infty \frac{f(x)g(y)}{(1+e^{x+y})^\lambda} dxdy$$
$$< B\left(\frac{\lambda}{2}, \frac{\lambda}{2}\right) \left(\int_{-\infty}^\infty e^{-\lambda x} f^2(x)dx \int_{-\infty}^\infty e^{-\lambda y} g^2(y)dy \right)^{\frac{1}{2}}, \quad (2)$$

with the best possible constant factor $B(\frac{\lambda}{2}, \frac{\lambda}{2})(\lambda > 0,\ B(u,v)$ is the beta function) (see [44]). He *et al.* [18–43] proved some new Hilbert-type integral inequalities in the whole plane with the best possible constant factors.

In this chapter, by means of the way of real analysis and the weight functions, a few equivalent conditions of two kinds of Hardy-type integral inequalities with the general nonhomogeneous kernel in the whole plane are obtained. The constant factors are proved to be the best possible. As applications, a few equivalent conditions of two kinds of Hardy-type integral inequalities with the general homogeneous kernel in the whole plane are deduced. We also consider the operator expressions, a few particular cases and examples.

2. Two Lemmas

In the following, we make appointment that $p > 1, \frac{1}{p} + \frac{1}{q} = 1, \sigma_1, \sigma \in \mathbf{R} = (-\infty, \infty)$, $h(u)$ is a non-negative measurable function in $\mathbf{R}$

such that

$$K^{(1)}(\sigma) := \int_0^1 (h(-u) + h(u))u^{\sigma-1}du, \tag{3}$$

$$K^{(2)}(\sigma) := \int_1^\infty (h(-u) + h(u))u^{\sigma-1}du. \tag{4}$$

Lemma 1. *If $K^{(1)}(\sigma) > 0$, there exists a constant M_1 such that for any non-negative measurable functions $f(x)$ and $g(y)$ in* $\mathbf{R}$, *the following inequality*

$$\int_{-\infty}^{\infty} g(y)\left(\int_{-\frac{1}{|y|}}^{\frac{1}{|y|}} h(xy)f(x)dx\right)dy$$
$$\leq M_1\left[\int_{-\infty}^{\infty} |x|^{p(1-\sigma)-1}f^p(x)dx\right]^{\frac{1}{p}}\left[\int_{-\infty}^{\infty} |y|^{q(1-\sigma_1)-1}g^q(y)dy\right]^{\frac{1}{q}} \tag{5}$$

holds true, then we have $\sigma_1 = \sigma$ and $K^{(1)}(\sigma) \leq M_1 < \infty$.

Proof. Since $K^{(1)}(\sigma) > 0$, it follows that $h(-u) + h(u) > 0$ a.e. in an interval $(a, b)(\subset (0, 1))$. □

If $\sigma_1 > \sigma$, then for $n \geq \frac{1}{\sigma_1-\sigma}$ $(n \in \mathbf{N})$, we set the following two functions:

$$f_n(x) := \begin{cases} |x|^{\sigma+\frac{1}{pn}-1}, 0 < |x| \leq 1 \\ 0, |x| > 1 \end{cases}, g_n(y) := \begin{cases} 0, 0 < |y| < 1 \\ |y|^{\sigma_1-\frac{1}{qn}-1}, y \geq 1 \end{cases}$$

and find

$$J_1 := \left[\int_{-\infty}^{\infty} |x|^{p(1-\sigma)-1}f_n^p(x)dx\right]^{\frac{1}{p}}\left[\int_{-\infty}^{\infty} |y|^{q(1-\sigma_1)-1}g_n^q(y)dy\right]^{\frac{1}{q}}$$
$$= \left[\int_{-1}^{1} |x|^{p(1-\sigma)-1}|x|^{p(\sigma+\frac{1}{pn}-1)}dx\right]^{\frac{1}{p}}$$
$$\times\left[\int_{\{y;|y|\geq 1\}} |y|^{q(1-\sigma_1)-1}|y|^{q(\sigma_1-\frac{1}{qn}-1)}dy\right]^{\frac{1}{q}}$$
$$= \left(2\int_0^1 x^{\frac{1}{n}-1}dx\right)^{\frac{1}{p}}\left(2\int_1^\infty y^{-\frac{1}{n}-1}dy\right)^{\frac{1}{q}} = 2n.$$

We obtain

$$I_1 := \int_{-\infty}^{\infty} g_n(y)\left(\int_{-\frac{1}{|y|}}^{\frac{1}{|y|}} h(xy)f_n(x)dx\right)dy$$
$$= \int_{-\infty}^{-1}\left(\int_{\frac{1}{y}}^{\frac{-1}{y}} h(xy)|x|^{\sigma+\frac{1}{pn}-1}dx\right)(-y)^{\sigma_1-\frac{1}{qn}-1}dy$$
$$+\int_{1}^{\infty}\left(\int_{\frac{-1}{y}}^{\frac{1}{y}} h(xy)|x|^{\sigma+\frac{1}{pn}-1}dx\right)y^{\sigma_1-\frac{1}{qn}-1}dy$$
$$= \int_{1}^{\infty}\left[\int_{\frac{-1}{y}}^{\frac{1}{y}}(h(-xy)+h(xy))|x|^{\sigma+\frac{1}{pn}-1}dx\right]y^{\sigma_1-\frac{1}{qn}-1}dy$$
$$= \int_{1}^{\infty}\left[\int_{-1}^{1}(h(-u)+h(u))|u|^{(\sigma+\frac{1}{pn})-1}du\right]y^{(\sigma_1-\sigma)-\frac{1}{n}-1}dy, \quad (6)$$

and then by (5), we have

$$2K^{(1)}\left(\sigma+\frac{1}{pn}\right)\int_{1}^{\infty} y^{(\sigma_1-\sigma)-\frac{1}{n}-1}dy$$
$$= \left[\int_{-1}^{1}(h(-u)+h(u))|u|^{(\sigma+\frac{1}{pn})-1}du\right]\int_{1}^{\infty} y^{(\sigma_1-\sigma)-\frac{1}{n}-1}dy$$
$$= I_1 \le M_1 J_1 = 2M_1 n < \infty. \quad (7)$$

Since $(\sigma_1-\sigma)-\frac{1}{n} \ge 0$, it follows that $\int_1^\infty y^{(\sigma_1-\sigma)-\frac{1}{n}-1}dy = \infty$. By (7), in view of $h(-u)+h(u) > 0$ a.e. in $(a,b)(\subset(0,1))$, namely,

$$K^{(1)}\left(\sigma+\frac{1}{pn}\right) = \int_{0}^{1}(h(-u)+h(u))u^{\sigma+\frac{1}{pn}-1}du > 0,$$

we have $\infty \le 2M_1 n < \infty$, which is a contradiction.

If $\sigma_1 < \sigma$, then for $n \ge \frac{1}{\sigma-\sigma_1}$ $(n \in \mathbf{N})$, we set the following two functions:

$$\widetilde{f}_n(x) := \begin{cases} 0, 0 < |x| < 1 \\ |x|^{\sigma-\frac{1}{pn}-1}, |x| \ge 1 \end{cases}, \widetilde{g}_n(y) := \begin{cases} |y|^{\sigma_1+\frac{1}{qn}-1}, 0 < |y| \le 1 \\ 0, |y| > 1 \end{cases}$$

and find

$$\begin{aligned}\widetilde{J}_1 &:= \left[\int_{-\infty}^{\infty} |x|^{p(1-\sigma)-1} \widetilde{f}_n^p(x)dx\right]^{\frac{1}{p}} \left[\int_{-\infty}^{\infty} y^{q(1-\sigma_1)-1}\widetilde{g}_n^q(y)dy\right]^{\frac{1}{q}} \\ &= \left[2\int_1^{\infty} x^{p(1-\sigma)-1}x^{p(\sigma-\frac{1}{pn}-1)}dx\right]^{\frac{1}{p}} \\ &\quad \times \left[2\int_0^1 y^{q(1-\sigma_1)-1}y^{q(\sigma_1+\frac{1}{qn}-1)}dy\right]^{\frac{1}{q}} \\ &= \left(2\int_1^{\infty} x^{-\frac{1}{n}-1}dx\right)^{\frac{1}{p}} \left(2\int_0^1 y^{\frac{1}{n}-1}dy\right)^{\frac{1}{q}} = 2n.\end{aligned}$$

We obtain

$$\begin{aligned}\widetilde{I}_1 &:= \int_{-\infty}^{\infty} \widetilde{f}_n(x) \left(\int_{-\frac{1}{|x|}}^{\frac{1}{|x|}} h(xy)\widetilde{g}_n(y)dy\right) dx \\ &= \int_{-\infty}^{-1} \left(\int_{\frac{1}{x}}^{\frac{-1}{x}} h(xy)|y|^{\sigma_1+\frac{1}{qn}-1}dy\right)(-x)^{\sigma-\frac{1}{pn}-1}dx \\ &\quad + \int_1^{\infty} \left(\int_{-\frac{1}{x}}^{\frac{1}{x}} h(xy)|y|^{\sigma_1+\frac{1}{qn}-1}dy\right) x^{\sigma-\frac{1}{pn}-1}dx \\ &= \int_1^{\infty} \left[\int_{-\frac{1}{x}}^{\frac{1}{x}} (h(-xy)+h(xy))|y|^{\sigma_1+\frac{1}{qn}-1}dy\right] x^{\sigma-\frac{1}{pn}-1}dx \\ &= \int_1^{\infty} \left[\int_{-1}^{1} (h(-u)+h(u))|u|^{\sigma_1+\frac{1}{qn}-1}du\right] x^{(\sigma-\sigma_1)-\frac{1}{n}-1}dx, \quad (8)\end{aligned}$$

and then by Fubini theorem (cf. [46]) and (5), it follows that

$$\begin{aligned}&2K^{(1)}\left(\sigma_1+\frac{1}{qn}\right)\int_1^{\infty} x^{(\sigma-\sigma_1)-\frac{1}{n}-1}dx \\ &= \left[\int_{-1}^{1} (h(-u)+h(u))|u|^{\sigma_1+\frac{1}{qn}-1}du\right]\int_1^{\infty} x^{(\sigma-\sigma_1)-\frac{1}{n}-1}dx \\ &= \widetilde{I}_1 = \int_{-\infty}^{\infty} \widetilde{g}_n(y)\left(\int_{\frac{-1}{|y|}}^{\frac{1}{|y|}} h(xy)\widetilde{f}_n(x)dx\right)dy \le M_1\widetilde{J}_1 = 2M_1 n.\end{aligned} \quad (9)$$

Since $(\sigma-\sigma_1)-\frac{1}{n}\geq 0$, it follows that $\int_1^\infty x^{(\sigma-\sigma_1)-\frac{1}{n}-1}dx=\infty$. By (9), in view of $h(-u)+h(u)>0$ a.e. in and interval $(a,b)\subset(0,1)$, we find that $K^{(1)}(\sigma_1+\frac{1}{qn})>0$, and then $\infty\leq 2M_1n<\infty$, which is a contradiction.

Hence, we conclude that $\sigma_1=\sigma$.

For $\sigma_1=\sigma$, we reduce (9) as follows:

$$K^{(1)}\left(\sigma_1+\frac{1}{qn}\right)=\int_0^1(h(-u)+h(u))u^{\sigma_1+\frac{1}{qn}-1}du\leq M_1.$$

Since $\{(h(-u)+h(u))u^{\sigma+\frac{1}{qn}-1}\}_{n=1}^\infty$ is increasing in $(0,1)$, by Levi theorem (cf. [46]), we have

$$\begin{aligned}K^{(1)}(\sigma)&=\int_0^1\lim_{n\to\infty}(h(-u)+h(u))u^{\sigma+\frac{1}{qn}-1}du\\&=\lim_{n\to\infty}\int_0^1(h(-u)+h(u))u^{\sigma+\frac{1}{qn}-1}du\leq M_1<\infty.\end{aligned}$$

The lemma is proved. □

Lemma 2. *If $K^{(2)}(\sigma)>0$, there exists a constant M_2 such that for any non-negative measurable functions $f(x)$ and $g(y)$ in $\mathbf{R}$, the following inequality*

$$\begin{aligned}&\int_{-\infty}^\infty g(y)\left(\int_{\{x;|x|\geq\frac{1}{|y|}\}}h(xy)f(x)dx\right)dy\\&\quad\leq M_2\left[\int_{-\infty}^\infty|x|^{p(1-\sigma)-1}f^p(x)dx\right]^{\frac{1}{p}}\left[\int_{-\infty}^\infty|y|^{q(1-\sigma_1)-1}g^q(y)dy\right]^{\frac{1}{q}}\end{aligned}\tag{10}$$

holds true, then we have $\sigma_1=\sigma$ and $K^{(2)}(\sigma)\leq M_2<\infty$.

Proof. Since $K^{(2)}(\sigma)>0$, it follows that $h(-u)+h(u)>0$ a.e. in an interval $(c,d)(\subset[1,\infty))$.

If $\sigma_1 < \sigma$, then for $n \geq \frac{1}{\sigma - \sigma_1}$ $(n \in \mathbf{N})$, we set two functions $\widetilde{f}_n(x)$ and $\widetilde{g}_n(y)$ as in Lemma 1 and find

$$\widetilde{J}_1 = \left[\int_{-\infty}^{\infty} |x|^{p(1-\sigma)-1} \widetilde{f}_n^p(x)dx\right]^{\frac{1}{p}} \left[\int_{-\infty}^{\infty} |y|^{q(1-\sigma_1)-1} \widetilde{g}_n^q(y)dy\right]^{\frac{1}{q}} = 2n.$$

We obtain

$$\begin{aligned}
\widetilde{I}_2 &:= \int_{-\infty}^{\infty} \widetilde{g}_n(y) \left(\int_{\{x;|x|\geq\frac{1}{|y|}\}} h(xy)\widetilde{f}_n(x)dx\right) dy \\
&= \int_{-1}^{0} \left(\int_{\{x;|x|\geq\frac{-1}{y}\}} h(xy)|x|^{\sigma-\frac{1}{pn}-1}dx\right)(-y)^{\sigma_1+\frac{1}{qn}-1}dy \\
&\quad + \int_{0}^{1} \left(\int_{\{x;|x|\geq\frac{1}{y}\}} h(xy)|x|^{\sigma-\frac{1}{pn}-1}dx\right) y^{\sigma_1+\frac{1}{qn}-1}dy \\
&= \int_{0}^{1} \left[\int_{\{x;|x|\geq\frac{1}{y}\}} (h(-xy)+h(xy))|x|^{\sigma-\frac{1}{pn}-1}dx\right] y^{\sigma_1+\frac{1}{qn}-1}dy \\
&= \int_{0}^{1} \left[\int_{\{u;|u|\geq 1\}} (h(-u)+h(u))|u|^{\sigma-\frac{1}{pn}-1}du\right] y^{(\sigma_1-\sigma)+\frac{1}{n}-1}dy,
\end{aligned}$$

and then by (10), it follows that

$$2K^{(2)}\left(\sigma - \frac{1}{pn}\right) \int_0^1 y^{(\sigma_1-\sigma)+\frac{1}{n}-1}dy = \widetilde{I}_2 \leq M_2\widetilde{J}_1 = 2M_2 n. \quad (11)$$

Since $(\sigma_1 - \sigma) + \frac{1}{n} \leq 0$, it follows that $\int_0^1 y^{(\sigma_1-\sigma)+\frac{1}{n}-1}dy = \infty$. By (11), in view of $h(-u) + h(u) > 0$ a.e. in an interval $(c,d)(\subset [1,\infty))$, we find that

$$K^{(2)}\left(\sigma - \frac{1}{pn}\right) = \int_1^{\infty} (h(-u)+h(u))u^{\sigma-\frac{1}{pn}-1}du > 0,$$

and then $\infty \leq 2M_2 n < \infty$, which is a contradiction.

If $\sigma_1 > \sigma$, then for $n \geq \frac{1}{\sigma_1 - \sigma}$ $(n \in \mathbf{N})$, we set two functions $f_n(x)$ and $g_n(y)$ as in Lemma 1 and find

$$J_1 = \left[\int_{-\infty}^{\infty} |x|^{p(1-\sigma)-1} f_n^p(x)dx\right]^{\frac{1}{p}} \left[\int_{-\infty}^{\infty} |y|^{q(1-\sigma_1)-1} g_n^q(y)dy\right]^{\frac{1}{q}} = 2n.$$

We obtain

$$\begin{aligned}
I_2 &:= \int_{-\infty}^{\infty} f_n(x)\left(\int_{\{y;|y|\geq\frac{1}{|x|}\}} h(xy)g_n(y)dy\right)dx \\
&= \int_{-1}^{0}\left(\int_{\{y;|y|\geq\frac{-1}{x}\}} h(xy)|y|^{\sigma_1-\frac{1}{qn}-1}dy\right)(-x)^{\sigma+\frac{1}{pn}-1}dx \\
&\quad + \int_{0}^{1}\left(\int_{\{y;|y|\geq\frac{1}{x}\}} h(xy)|y|^{\sigma_1-\frac{1}{qn}-1}dy\right)x^{\sigma+\frac{1}{pn}-1}dx \\
&= \int_{0}^{1}\left[\int_{\{y;|y|\geq\frac{1}{x}\}} (h(-xy)+h(xy))|y|^{\sigma_1-\frac{1}{qn}-1}dy\right]x^{\sigma+\frac{1}{pn}-1}dx \\
&= \int_{\{u;|u|\geq 1\}} (h(-u)+h(u))|u|^{(\sigma_1-\frac{1}{qn})-1}du \int_0^1 x^{(\sigma-\sigma_1)+\frac{1}{n}-1}dx,
\end{aligned}$$

and then by Fubini theorem (cf. [46]) and (8), it follows that

$$\begin{aligned}
&2K_2\left(\sigma_1 - \frac{1}{qn}\right)\int_0^1 x^{(\sigma-\sigma_1)+\frac{1}{n}-1}dx \\
&= \int_0^1 x^{(\sigma-\sigma_1)+\frac{1}{n}-1}dx \int_{\{u;|u|\geq 1\}} (h(-u)+h(u))|u|^{\sigma_1-\frac{1}{qn}-1}du \\
&= I_2 = \int_0^{\infty} g_n(y)\left(\int_{\{x;|x|\geq\frac{1}{|y|}\}} h(xy)f_n(x)dx\right)dy \leq M_2 J_1 = 2M_2 n.
\end{aligned} \tag{12}$$

Since $(\sigma - \sigma_1) + \frac{1}{n} \leq 0$, it follows that $\int_0^1 x^{(\sigma-\sigma_1)+\frac{1}{n}-1}dx = \infty$. By (12), in view of $(h(-u) + h(u)) > 0$ a.e. in $(c,d)(\subset [1,\infty))$, we find that $K^{(2)}(\sigma_1 - \frac{1}{qn}) > 0$, and then $\infty \leq 2M_2 n < \infty$, which is a contradiction.

Hence, we conclude the fact that $\sigma_1 = \sigma$.

For $\sigma_1 = \sigma$, we reduce (12) as follows:

$$K^{(2)}\left(\sigma - \frac{1}{qn}\right) = \int_1^{\infty} (h(-u) + h(u))u^{\sigma - \frac{1}{qn} - 1} du \leq M_2. \tag{13}$$

Since $\{(h(-u) + h(u))u^{\sigma - \frac{1}{qn} - 1}\}_{n=1}^{\infty}$ is increasing in $[1, \infty)$, still by Levi theorem (cf. [46]), we have

$$\begin{aligned} K^{(2)}(\sigma) &= \int_1^{\infty} \lim_{n\to\infty} (h(-u) + h(u))u^{\sigma - \frac{1}{qn} - 1} du \\ &= \lim_{n\to\infty} \int_1^{\infty} (h(-u) + h(u))u^{\sigma - \frac{1}{qn} - 1} du \leq M_2 < \infty. \end{aligned}$$

The lemma is proved. □

3. Main Results and Somc Corollaries

Theorem 3. *If $K^{(1)}(\sigma) > 0$, then the following conditions are equivalent:*

(i) *There exists a constant M_1 such that for any $f(x) \geq 0$, satisfying $0 < \int_{-\infty}^{\infty} |x|^{p(1-\sigma)-1} f^p(x)dx < \infty$, we have the following first kind of Hardy-type integral inequality with the non-homogeneous kernel:*

$$\begin{aligned} J &:= \left[\int_{-\infty}^{\infty} |y|^{p\sigma_1 - 1} \left(\int_{\frac{-1}{|y|}}^{\frac{1}{|y|}} h(xy)f(x)dx\right)^p dy\right]^{\frac{1}{p}} \\ &< M_1 \left[\int_{-\infty}^{\infty} |x|^{p(1-\sigma)-1} f^p(x)dx\right]^{\frac{1}{p}}. \end{aligned} \tag{14}$$

(ii) *There exists a constant M_1 such that for any $f(x)$, $g(y) \geq 0$, satisfying $0 < \int_{-\infty}^{\infty} |x|^{p(1-\sigma)-1} f^p(x)dx < \infty$, and*

$0 < \int_{-\infty}^{\infty} |y|^{q(1-\sigma_1)-1} g^q(y)dy < \infty$, *we have the following inequality:*

$$I := \int_{-\infty}^{\infty} g(y) \left(\int_{\frac{-1}{|y|}}^{\frac{1}{|y|}} h(xy)f(x)dx \right) dy$$

$$< M_1 \left[\int_{-\infty}^{\infty} |x|^{p(1-\sigma)-1} f^p(x)dx \right]^{\frac{1}{p}} \left[\int_{-\infty}^{\infty} |y|^{q(1-\sigma_1)-1} g^q(y)dy \right]^{\frac{1}{q}}. \tag{15}$$

(iii) $\sigma_1 = \sigma$, *and* $K^{(1)}(\sigma) < \infty$.

If Condition (iii) *holds, then the constant* $M_1 = K^{(1)}(\sigma)$ $(\in \mathbf{R}_+)$ *in* (14) *and* (15) *(for* $\sigma_1 = \sigma$*) is the best possible.*

Proof. (i) => (ii). By Hölder's inequality (cf. [45]), we have

$$I = \int_{-\infty}^{\infty} \left(|y|^{\sigma_1 - \frac{1}{p}} \int_{\frac{-1}{|y|}}^{\frac{1}{|y|}} h(xy)f(x)dx \right) \left(|y|^{\frac{1}{p}-\sigma_1} g(y) \right) dy$$

$$\leq J \left[\int_{-\infty}^{\infty} |y|^{q(1-\sigma_1)-1} g^q(y)dy \right]^{\frac{1}{q}}. \tag{16}$$

Then by (14), we have (15).

(ii) => (iii). By Lemma 1, we have $\sigma_1 = \sigma$, and $K^{(1)}(\sigma) \leq M_1 < \infty$.

(iii) => (i). We obtain the following weight function: For $y \neq 0$,

$$\begin{aligned} \omega_1(\sigma, y) &:= |y|^{\sigma} \int_{\frac{-1}{|y|}}^{\frac{1}{|y|}} h(xy)|x|^{\sigma-1} dx \\ &= |y|^{\sigma} \int_{\frac{-1}{|y|}}^{0} h(xy)(-x)^{\sigma-1} dx + |y|^{\sigma} \int_{0}^{\frac{1}{|y|}} h(xy)x^{\sigma-1} dx \\ &= |y|^{\sigma} \int_{0}^{\frac{1}{|y|}} h(-xy)x^{\sigma-1} dx + |y|^{\sigma} \int_{0}^{\frac{1}{|y|}} h(xy)x^{\sigma-1} dx \\ &= |y|^{\sigma} \int_{0}^{\frac{1}{|y|}} (h(-x|y|) + h(x|y|)x^{\sigma-1} dx \\ &= \int_{0}^{1} (h(-u) + h(u))u^{\sigma-1} du = K^{(1)}(\sigma). \end{aligned} \tag{17}$$

By Hölder's inequality with weight and (17), we have

$$\left(\int_{\frac{-1}{|y|}}^{\frac{1}{|y|}} h(xy)f(x)dx\right)^p$$

$$= \left\{\int_{\frac{-1}{|y|}}^{\frac{1}{|y|}} h(xy)\left[\frac{|y|^{(\sigma-1)/p}}{|x|^{(\sigma-1)/q}}f(x)\right]\left[\frac{|x|^{(\sigma-1)/q}}{|y|^{(\sigma-1)/p}}\right]dx\right\}^p$$

$$\leq \int_{\frac{-1}{|y|}}^{\frac{1}{|y|}} h(xy)\frac{|y|^{\sigma-1}f^p(x)}{|x|^{(\sigma-1)p/q}}dx\left[\int_{\frac{-1}{|y|}}^{\frac{1}{|y|}} h(xy)\frac{|x|^{\sigma-1}}{|y|^{(\sigma-1)q/p}}dx\right]^{p-1}$$

$$= \int_{\frac{-1}{|y|}}^{\frac{1}{|y|}} h(xy)\frac{|y|^{\sigma-1}}{|x|^{(\sigma-1)p/q}}f^p(x)dx\cdot\left[\omega_1(\sigma,y)|y|^{q(1-\sigma)-1}\right]^{p-1}$$

$$= (K^{(1)}(\sigma))^{p-1}|y|^{-p\sigma+1}\int_{\frac{-1}{|y|}}^{\frac{1}{|y|}} h(xy)\frac{|y|^{\sigma-1}}{|x|^{(\sigma-1)p/q}}f^p(x)dx. \quad (18)$$

If (18) takes the form of equality for a $y \in \mathbf{R}\backslash\{0\}$, then (cf. [45]), there exists constants A and B such that they are not both zero and

$$A\frac{|y|^{\sigma-1}}{|x|^{(\sigma-1)p/q}}f^p(x) = B\frac{|x|^{\sigma-1}}{|y|^{(\sigma-1)q/p}} \text{ a.e. in } \mathbf{R}.$$

We suppose that $A \neq 0$ (otherwise $B = A = 0$). It follows that

$$|x|^{p(1-\sigma)-1}f^p(x) = |y|^{q(1-\sigma)}\frac{B}{A|x|} \text{ a.e. in } \mathbf{R},$$

which contradicts the fact that $0 < \int_{-\infty}^{\infty}|x|^{p(1-\sigma)-1}f^p(x)dx < \infty$. Hence, (18) takes the form of strict inequality.

For $\sigma_1 = \sigma$, by above results and Fubini theorem (cf. [46]), we have

$$J < (K^{(1)}(\sigma))^{\frac{1}{q}}\left\{\int_{-\infty}^{\infty}\left[\int_{\frac{-1}{|y|}}^{\frac{1}{|y|}} h(xy)\frac{|y|^{\sigma-1}}{|x|^{(\sigma-1)p/q}}f^p(x)dx\right]dy\right\}^{\frac{1}{p}}$$

$$= (K^{(1)}(\sigma))^{\frac{1}{q}}\left\{\int_{-\infty}^{\infty}\left[\int_{\frac{-1}{|x|}}^{\frac{1}{|x|}} h(xy)\frac{|y|^{\sigma-1}}{|x|^{(\sigma-1)(p-1)}}dy\right]f^p(x)dx\right\}^{\frac{1}{p}}$$

$$= (K^{(1)}(\sigma))^{\frac{1}{q}} \left[\int_{-\infty}^{\infty} \omega_1(\sigma, x)|x|^{p(1-\sigma)-1} f^p(x)dx\right]^{\frac{1}{p}}$$

$$= K^{(1)}(\sigma) \left[\int_{-\infty}^{\infty} |x|^{p(1-\sigma)-1} f^p(x)dx\right]^{\frac{1}{p}}.$$

Setting $M_1 \geq K^{(1)}(\sigma)$, then for $K^{(1)}(\sigma) < \infty$, we have

$$J < K^{(1)}(\sigma) \left[\int_{-\infty}^{\infty} |x|^{p(1-\sigma)-1} f^p(x)dx\right]^{\frac{1}{p}}$$

$$\leq M_1 \left[\int_{-\infty}^{\infty} |x|^{p(1-\sigma)-1} f^p(x)dx\right]^{\frac{1}{p}},$$

namely, (14) follows.

Therefore, Conditions (i)–(iii) are equivalent.

When Condition (iii) follows, if there exists a constant $M_1 \leq K^{(1)}(\sigma)$ such that (15) is valid, then by Lemma 1, we have $K^{(1)}(\sigma) \leq M_1$. It follows that the constant factor $M_1 = K^{(1)}(\sigma)$ in (15) is the best possible.

The constant factor $M_1 = K^{(1)}(\sigma)(\in \mathbf{R}_+)$ in (14) is still the best possible. Otherwise, by (16) (for $\sigma_1 = \sigma$), we can conclude that the constant factor $M_1 = K^{(1)}(\sigma)$ in (15) is not the best possible.

The theorem is proved. □

In particular, for $\sigma = \sigma_1 = \frac{1}{p}$ in Theorem 1, we have the following:

Corollary 4. *If $K^{(1)}(\frac{1}{p}) > 0$, then the following conditions are equivalent:*

(i) There exists a constant M_1 such that for any $f(x) \geq 0$, satisfying $0 < \int_{-\infty}^{\infty} |x|^{p-2} f^p(x)dx < \infty$, we have the following inequality:

$$\left[\int_{-\infty}^{\infty} \left(\int_{\frac{-1}{|y|}}^{\frac{1}{|y|}} h(xy)f(x)dx\right)^p dy\right]^{\frac{1}{p}}$$

$$< M_1 \left(\int_{-\infty}^{\infty} |x|^{p-2} f^p(x)dx\right)^{\frac{1}{p}}. \tag{19}$$

(ii) There exists a constant M_1 such that for any $f(x), g(y) \geq 0$, satisfying $0 < \int_{-\infty}^{\infty} |x|^{p-2} f^p(x)dx < \infty$, and $0 < \int_{-\infty}^{\infty} g^q(y)dy < \infty$, we have the following inequality:

$$\int_{-\infty}^{\infty} g(y)\left(\int_{\frac{-1}{|y|}}^{\frac{1}{|y|}} h(xy)f(x)dx\right)dy$$
$$< M_1\left(\int_{-\infty}^{\infty} |x|^{p-2} f^p(x)dx\right)^{\frac{1}{p}}\left(\int_{-\infty}^{\infty} g^q(y)dy\right)^{\frac{1}{q}}. \quad (20)$$

(iii) $K^{(1)}(\frac{1}{p}) < \infty$.

If Condition (iii) holds, then the constant $M_1 = K^{(1)}(\frac{1}{p}) (\in \mathbf{R}_+)$ in (19) and (20) is the best possible.

Setting $y = \frac{1}{Y}$, $G(Y) = g(\frac{1}{Y})\frac{1}{Y^2}$ in Theorem 1, and then replacing Y by y, we have the following:

Corollary 5. *If $K^{(1)}(\sigma) > 0$, then the following conditions are equivalent:*

(i) There exists a constant M_1 such that for any $f(x) \geq 0$, satisfying $0 < \int_{-\infty}^{\infty} |x|^{p(1-\sigma)-1} f^p(x)dx < \infty$, we have the following inequality follows:

$$\left[\int_{-\infty}^{\infty} |y|^{-p\sigma_1 - 1}\left(\int_{-|y|}^{|y|} h\left(\frac{x}{y}\right) f(x)dx\right)^p dy\right]^{\frac{1}{p}}$$
$$< M_1\left[\int_{-\infty}^{\infty} |x|^{p(1-\sigma)-1} f^p(x)dx\right]^{\frac{1}{p}}. \quad (21)$$

(ii) There exists a constant M_1 such that for any $f(x), G(y) \geq 0$, satisfying $0 < \int_{-\infty}^{\infty} |x|^{p(1-\sigma)-1} f^p(x)dx < \infty$, and $0 < \int_{-\infty}^{\infty} y^{q(1+\sigma_1)-1} G^q(y)dy < \infty$, we have the following inequality:

$$\int_{-\infty}^{\infty} G(y)\left(\int_{-|y|}^{|y|} h\left(\frac{x}{y}\right) f(x)dx\right)dy$$
$$< M_1\left[\int_{-\infty}^{\infty} |x|^{p(1-\sigma)-1} f^p(x)dx\right]^{\frac{1}{p}}\left[\int_{-\infty}^{\infty} |y|^{q(1+\sigma_1)-1} G^q(y)dy\right]^{\frac{1}{q}}. \quad (22)$$

(iii) $\sigma_1 = \sigma$, and $K^{(1)}(\sigma) < \infty$.

If Condition (iii) holds, then the constant $M_1 = K^{(1)}(\sigma)\,(\in \mathbf{R}_+)$ in (21) and (22) (for $\sigma_1 = \sigma$) is the best possible.

Note. $h(\frac{x}{y})$ is a homogeneous function of degree 0, namely, $h(\frac{x}{y}) = k_0(x,y)$.

Setting $h(u) = k_\lambda(u,1)$, where $k_\lambda(x,y)$ is the homogeneous function of degree $-\lambda(\in \mathbf{R})$, for $g(y) = y^\lambda G(y)$ and $\mu = \lambda - \sigma_1$ in Corollary 2, we have the following:

Corollary 6. *If* $k_\lambda^{(1)}(\sigma) := \int_0^1 (k_\lambda(-u,1) + k_\lambda(u,1))u^{\sigma-1}du > 0$, *then the following conditions are equivalent:*

(i) *There exists a constant* M_1, *such that for any* $f(x) \geq 0$, *satisfying* $0 < \int_{-\infty}^{\infty} |x|^{p(1-\sigma)-1} f^p(x)dx < \infty$, *we have the following first kind of Hardy-type integral inequality with the homogeneous kernel:*

$$\left[\int_{-\infty}^{\infty} y^{p\mu-1}\left(\int_{-|y|}^{|y|} k_\lambda(x,y)f(x)dx\right)^p dy\right]^{\frac{1}{p}} < M_1\left[\int_{-\infty}^{\infty} |x\}^{p(1-\sigma)-1} f^p(x)dx\right]^{\frac{1}{p}}. \tag{23}$$

(ii) *There exists a constant* M_1, *such that for any* $f(x), g(y) \geq 0$, *satisfying* $0 < \int_{-\infty}^{\infty} |x|^{p(1-\sigma)-1} f^p(x)dx < \infty$, *and* $0 < \int_{-\infty}^{\infty} |y|^{q(1-\mu)-1} g^q(y)dy < \infty$, *we have the following inequality:*

$$\int_{-\infty}^{\infty} g(y)\left(\int_{-|y|}^{|y|} k_\lambda(x,y)f(x)dx\right)dy < M_1\left[\int_{-\infty}^{\infty} |x|^{p(1-\sigma)-1} f^p(x)dx\right]^{\frac{1}{p}} \times\left[\int_{-\infty}^{\infty} |y|^{q(1-\mu)-1} g^q(y)dy\right]^{\frac{1}{q}}. \tag{24}$$

(iii) $\mu + \sigma = \lambda$, *and* $k_\lambda^{(1)}(\sigma) < \infty$.

If Condition (iii) *holds, then the constant* $M_1 = k_\lambda^{(1)}(\sigma)\,(\in \mathbf{R}_+)$ *in* (23) *and* (24) *is the best possible.*

In particular, for $\lambda = 1, \sigma = \frac{1}{q}, \mu = \frac{1}{p}$ in Corollary 3, we have the following:

Corollary 7. *If $k_1^{(1)}(\frac{1}{q}) > 0$, then the following conditions are equivalent:*

(i) There exists a constant M_1 such that for any $f(x) \geq 0$, satisfying $0 < \int_{-\infty}^{\infty} f^p(x)dx < \infty$, we have the following inequality:

$$\left[\int_{-\infty}^{\infty} \left(\int_{-|y|}^{|y|} k_1(x,y)f(x)dx\right)^p dy\right]^{\frac{1}{p}} < M_1 \left(\int_{-\infty}^{\infty} f^p(x)dx\right)^{\frac{1}{p}}. \tag{25}$$

(ii) There exists a constant M_1 such that for any $f(x), g(y) \geq 0$, satisfying $0 < \int_{-\infty}^{\infty} f^p(x)dx < \infty$, and $0 < \int_{-\infty}^{\infty} g^q(y)dy < \infty$, we have the following inequality:

$$\begin{aligned} I &= \int_{-\infty}^{\infty} g(y) \left(\int_{-|y|}^{|y|} k_1(x,y)f(x)dx\right) dy \\ &< M_1 \left(\int_{-\infty}^{\infty} f^p(x)dx\right)^{\frac{1}{p}} \left(\int_{-\infty}^{\infty} g^q(y)dy\right)^{\frac{1}{q}}. \end{aligned} \tag{26}$$

(iii) $k_1^{(1)}(\frac{1}{q}) < \infty$.

If Condition (iii) holds, then the constant $M_1 = k_1^{(1)}(\frac{1}{q}) \, (\in \mathbf{R}_+)$ in (25) and (26) is the best possible.

Similarly, in view of Lemma 2, we obtain the following weight function: For $y \neq 0$,

$$\begin{aligned} \omega_2(\sigma, y) &:= |y|^\sigma \int_{\{x;|x|\geq\frac{1}{|y|}\}} h(xy)|x|^{\sigma-1}dx \\ &= \int_1^{\infty} (h(-u) + h(u))u^{\sigma-1}du = K^{(2)}(\sigma), \end{aligned}$$

and then in the same way, we have the following:

Theorem 8. *If $K^{(2)}(\sigma) > 0$, then the following conditions are equivalent:*

(i) There exists a constant M_2 such that for any $f(x) \geq 0$, satisfying $0 < \int_{-\infty}^{\infty} |x|^{p(1-\sigma)-1} f^p(x)dx < \infty$, we have the following second kind of Hardy-type integral inequality with the non-homogeneous kernel:

$$\left[\int_{-\infty}^{\infty} y^{p\sigma_1 - 1} \left(\int_{\{x;|x|\geq \frac{1}{|y|}\}} h(xy)f(x)dx\right)^p dy\right]^{\frac{1}{p}} < M_2 \left[\int_{-\infty}^{\infty} |x|^{p(1-\sigma)-1} f^p(x)dx\right]^{\frac{1}{p}}. \tag{27}$$

(ii) There exists a constant M_2 such that for any $f(x), g(y) \geq 0$, satisfying $0 < \int_{-\infty}^{\infty} |x|^{p(1-\sigma)-1} f^p(x)dx < \infty$, and $0 < \int_{-\infty}^{\infty} |y|^{q(1-\sigma_1)-1} g^q(y)dy < \infty$, we have the following inequality:

$$\int_{-\infty}^{\infty} g(y) \left(\int_{\{x;|x|\geq \frac{1}{|y|}\}} h(xy)f(x)dx\right) dy < M_2 \left[\int_{-\infty}^{\infty} |x|^{p(1-\sigma)-1} f^p(x)dx\right]^{\frac{1}{p}} \left[\int_{-\infty}^{\infty} y^{q(1-\sigma_1)-1} g^q(y)dy\right]^{\frac{1}{q}}. \tag{28}$$

(iii) $\sigma_1 = \sigma$, and $K^{(2)}(\sigma) < \infty$.

If Condition (iii) holds, then the constant $M_2 = K^{(2)}(\sigma)$ ($\in \mathbf{R}_+$) in (27) and (28) (for $\sigma_1 = \sigma$) is the best possible.

In particular, for $\sigma = \sigma_1 = \frac{1}{p}$ in Theorem 2, we have the following:

Corollary 9. *If $K^{(2)}(\frac{1}{p}) > 0$, then the following conditions are equivalent:*

(i) *There exists a constant M_2 such that for any $f(x) \geq 0$, satisfying $0 < \int_{-\infty}^{\infty} |x|^{p-2} f^p(x)dx < \infty$, we have the following inequality:*

$$\left[\int_{-\infty}^{\infty} \left(\int_{\{x;|x|\geq \frac{1}{|y|}\}} h(xy)f(x)dx\right)^p dy\right]^{\frac{1}{p}} < M_2 \left(\int_{-\infty}^{\infty} |x|^{p-2} f^p(x)dx\right)^{\frac{1}{p}}. \tag{29}$$

(ii) *There exists a constant M_2 such that for any $f(x), g(y) \geq 0$, satisfying $0 < \int_{-\infty}^{\infty} |x|^{p-2} f^p(x)dx < \infty$, and $0 < \int_{-\infty}^{\infty} g^q(y)dy < \infty$, we have the following inequality:*

$$\int_{-\infty}^{\infty} g(y) \left(\int_{\{x;|x| \geq \frac{1}{|y|}\}} h(xy) f(x) dx \right) dy$$
$$< M_2 \left(\int_{-\infty}^{\infty} |x|^{p-2} f^p(x)dx \right)^{\frac{1}{p}} \left(\int_{-\infty}^{\infty} g^q(y)dy \right)^{\frac{1}{q}}. \quad (30)$$

(iii) $K^{(2)}(\frac{1}{p}) < \infty$.

If Condition (iii) *holds, then the constant $M_2 = K^{(2)}(\frac{1}{p})\ (\in \mathbf{R}_+)$ in* (29) *and* (30) *is the best possible.*

Setting $y = \frac{1}{Y}$, $G(Y) = g(\frac{1}{Y})\frac{1}{Y^2}$ in Theorem 2, and then replacing Y by y, we have the following:

Corollary 10. *If $K^{(2)}(\sigma) > 0$, then the following conditions are equivalent:*

(i) *There exists a constant M_2 such that for any $f(x) \geq 0$, satisfying $0 < \int_{-\infty}^{\infty} |x|^{p(1-\sigma)-1} f^p(x)dx < \infty$, we have the following inequality:*

$$\left[\int_{-\infty}^{\infty} y^{-p\sigma_1 - 1} \left(\int_{\{x;|x| \geq |y|\}} h\left(\frac{x}{y}\right) f(x)dx \right)^p dy \right]^{\frac{1}{p}}$$
$$< M_2 \left[\int_{-\infty}^{\infty} |x|^{p(1-\sigma)-1} f^p(x)dx \right]^{\frac{1}{p}}. \quad (31)$$

(ii) *There exists a constant M_2 such that for any $f(x), G(y) \geq 0$, satisfying $0 < \int_{-\infty}^{\infty} |x|^{p(1-\sigma)-1} f^p(x)dx < \infty$, and $0 < \int_{-\infty}^{\infty} |y|^{q(1+\sigma_1)-1} G^q(y)dy < \infty$, we have the following inequality:*

$$\int_{-\infty}^{\infty} G(y) \left(\int_{\{x;|x| \geq |y|\}} h\left(\frac{x}{y}\right) f(x)dx \right) dy$$
$$< M_2 \left[\int_{-\infty}^{\infty} |x|^{p(1-\sigma)-1} f^p(x)dx \right]^{\frac{1}{p}} \left[\int_{-\infty}^{\infty} |y|^{q(1+\sigma_1)-1} G^q(y)dy \right]^{\frac{1}{q}}. \quad (32)$$

(iii) $\sigma_1 = \sigma$, *and* $K^{(2)}(\sigma) < \infty$.

If Condition (iii) *holds, then the constant* $M_2 = K^{(2)}(\sigma)\,(\in \mathbf{R}_+)$ *in* (31) *and* (32) *(for* $\sigma_1 = \sigma$*) is the best possible.*

Setting $h(u) = k_\lambda(u,1)$, where $k_\lambda(x,y)$ is the homogeneous function of degree $-\lambda(\in \mathbf{R})$, for $g(y) = y^\lambda G(y)$ and $\mu = \lambda - \sigma_1$ in Corollary 6, we have the following:

Corollary 11. *If* $k_\lambda^{(2)}(\sigma) := \int_1^\infty (k_\lambda(-u,1) + k_\lambda(u,1))u^{\sigma-1}du > 0$, *then the following conditions are equivalent:*

(i) *There exists a constant* M_2 *such that for any* $f(x) \geq 0$, *satisfying* $0 < \int_{-\infty}^\infty |x|^{p(1-\sigma)-1} f^p(x)dx < \infty$, *we have the following inequality:*

$$\left[\int_{-\infty}^\infty y^{p\mu-1}\left(\int_{\{x;|x|\geq|y|\}} k_\lambda(x,y)f(x)dx\right)^p dy\right]^{\frac{1}{p}} < M_2\left[\int_{-\infty}^\infty |x|^{p(1-\sigma)-1} f^p(x)dx\right]^{\frac{1}{p}}. \tag{33}$$

(ii) *There exists a constant* M_2, *such that for any* $f(x), g(y) \geq 0$, *satisfying* $0 < \int_{-\infty}^\infty |x|^{p(1-\sigma)-1} f^p(x)dx < \infty$, *and* $0 < \int_{-\infty}^\infty |y|^{q(1-\mu)-1} g^q(y)dy < \infty$, *we have the following inequality:*

$$\int_{-\infty}^\infty g(y)\left(\int_{\{x;|x|\geq|y|\}} k_\lambda(x,y)f(x)dx\right)dy < M_2\left[\int_{-\infty}^\infty |x|^{p(1-\sigma)-1} f^p(x)dx\right]^{\frac{1}{p}} \times \left[\int_{-\infty}^\infty |y|^{q(1-\mu)-1} g^q(y)dy\right]^{\frac{1}{q}}. \tag{34}$$

(iii) $\mu + \sigma = \lambda$, *and* $k_\lambda^{(2)}(\sigma) < \infty$.

If Condition (iii) *holds, then the constant* $M_2 = k_\lambda^{(2)}(\sigma)\,(\in \mathbf{R}_+)$ *in* (33) *and* (34) *is the best possible.*

In particular, for $\lambda = 1, \sigma = \frac{1}{q}, \mu = \frac{1}{p}$ in Corollary 7, we have the following:

Corollary 12. *If $k_1^{(2)}(\frac{1}{q}) > 0$, then the following conditions are equivalent:*

(i) *There exists a constant M_2 such that for any $f(x) \geq 0$, satisfying $0 < \int_{-\infty}^{\infty} f^p(x)dx < \infty$, we have the following inequality:*

$$\left[\int_{-\infty}^{\infty}\left(\int_{\{x;|x|\geq|y|\}} k_1(x,y)f(x)dx\right)^p dy\right]^{\frac{1}{p}} < M_2\left(\int_{-\infty}^{\infty} f^p(x)dx\right)^{\frac{1}{p}}. \tag{35}$$

(ii) *There exists a constant M_2 such that for any $f(x), g(y) \geq 0$, satisfying $0 < \int_{-\infty}^{\infty} f^p(x)dx < \infty$, and $0 < \int_{-\infty}^{\infty} g^q(y)dy < \infty$, we have the following inequality:*

$$I = \int_{-\infty}^{\infty} g(y)\left(\int_{\{x;|x|\geq|y|\}} k_1(x,y)f(x)dx\right)dy < M_2\left(\int_{-\infty}^{\infty} f^p(x)dx\right)^{\frac{1}{p}}\left(\int_{-\infty}^{\infty} g^q(y)dy\right)^{\frac{1}{q}}. \tag{36}$$

(iii) $k_1^{(2)}(\frac{1}{q}) < \infty$.

If Condition (iii) *holds, then the constant $M_2 = k_1^{(2)}(\frac{1}{q})\,(\in \mathbf{R}_+)$ in* (35) *and* (36) *is the best possible.*

4. Operator Expressions and Examples

For $\mu = \lambda - \sigma$, we set the following functions: $\varphi(x) := |x|^{p(1-\sigma)-1}, \psi(y) := |y|^{q(1-\sigma)-1}, \phi(y) := |y|^{q(1-\mu)-1}$, wherefrom, $\psi^{1-p}(y) = |y|^{p\sigma-1}, \phi^{1-p}(y) = |y|^{p\mu-1}$ $(x, y \in \mathbf{R})$. Define the following real normed linear spaces:

$$L_{p,\varphi}(\mathbf{R}) := \left\{f : ||f||_{p,\varphi} := \left(\int_{-\infty}^{\infty} \varphi(x)|f(x)|^p dx\right)^{\frac{1}{p}} < \infty\right\},$$

wherefrom,

$$L_{q,\psi}(\mathbf{R}) = \left\{ g : ||g||_{q,\psi} := \left(\int_{-\infty}^{\infty} \psi(y)|g(y)|^q dy \right)^{\frac{1}{q}} < \infty \right\},$$

$$L_{q,\phi}(\mathbf{R}) = \left\{ g : ||g||_{q,\phi} := \left(\int_{-\infty}^{\infty} \phi(y)|g(y)|^q dy \right)^{\frac{1}{q}} < \infty \right\},$$

$$L_{p,\psi^{1-p}}(\mathbf{R}) = \left\{ h : ||h||_{p,\psi^{1-p}} = \left(\int_{-\infty}^{\infty} \psi^{1-p}(y)|h(y)|^p dy \right)^{\frac{1}{p}} < \infty \right\},$$

$$L_{q,\phi^{1-p}}(\mathbf{R}) = \left\{ h : ||h||_{p,\phi^{1-p}} = \left(\int_{-\infty}^{\infty} \phi^{1-p}(y)|h(y)|^p dy \right)^{\frac{1}{p}} < \infty \right\}.$$

(a) In view of Theorem 1 (for $\sigma_1 = \sigma$), for $f \in L_{p,\varphi}(\mathbf{R})$, setting

$$h_1(y) := \int_{\frac{-1}{|y|}}^{\frac{1}{|y|}} h(xy)f(x)dx \ (y \in \mathbf{R}),$$

by (14), we have

$$||h_1||_{p,\psi^{1-p}} = \left[\int_{-\infty}^{\infty} \psi^{1-p}(y)h_1^p(y)dy \right]^{\frac{1}{p}} < M_1||f||_{p,\varphi} < \infty. \quad (37)$$

Definition 13. Define a first kind of Hardy-type integral operator with the non-homogeneous kernel $T_1^{(1)} : L_{p,\varphi}(\mathbf{R}) \to L_{p,\psi^{1-p}}(\mathbf{R})$ as follows: For any $f \in L_{p,\varphi}(\mathbf{R})$, there exists a unique representation $T_1^{(1)}f = h_1 \in L_{p,\psi^{1-p}}(\mathbf{R})$, satisfying for any $y \in \mathbf{R}, T_1^{(1)}f(y) = h_1(y)$.

In view of (37), it follows that $||T_1^{(1)}f||_{p,\psi^{1-p}} = ||h_1||_{p,\psi^{1-p}} \leq M_1||f||_{p,\varphi}$, and then the operator $T_1^{(1)}$ is bounded satisfying

$$||T_1^{(1)}|| = \sup_{f(\neq 0) \in L_{p,\varphi}(\mathbf{R})} \frac{||T_1^{(1)}f||_{p,\psi^{1-p}}}{||f||_{p,\varphi}} \leq M_1.$$

If we define the formal inner product of $T_1^{(1)}f$ and g as follows:

$$(T_1^{(1)}f, g) := \int_{-\infty}^{\infty} \left(\int_{\frac{-1}{|y|}}^{\frac{1}{|y|}} h(xy)f(x)dx \right) g(y)dy,$$

then we can rewrite Theorem 1 (for $\sigma_1 = \sigma$) as follows:

Theorem 14. *The following conditions are equivalent:*

(i) There exists a constant M_1 such that for any $f(x) \geq 0, f \in L_{p,\varphi}(\mathbf{R})$, $||f||_{p,\varphi} > 0$, we have the following inequality:

$$||T_1^{(1)} f||_{p,\psi^{1-p}} < M_1 ||f||_{p,\varphi}. \tag{38}$$

(ii) There exists a constant M_1 such that for any $f(x), g(y) \geq 0, f \in L_{p,\varphi}(\mathbf{R}), g \in L_{q,\psi}(\mathbf{R}), ||f||_{p,\varphi}, ||g||_{q,\psi} > 0$, we have the following inequality:

$$(T_1^{(1)} f, g) < M_1 ||f||_{p,\varphi} ||g||_{q,\psi}. \tag{39}$$

(iii) $K^{(1)}(\sigma) < \infty$.

If Condition (iii) follows, then we have $||T_1^{(1)}|| = K^{(1)}(\sigma) \leq M_1$.

(b) In view of Corollary 3 ($\mu + \sigma = \lambda$), for $f \in L_{p,\varphi}(\mathbf{R})$, setting

$$h_2(y) := \int_{-|y|}^{|y|} k_\lambda(x, y) f(x) dx \ (y \in \mathbf{R}),$$

by (23), we have

$$||h_2||_{p,\phi^{1-p}} = \left[\int_{-\infty}^{\infty} \phi^{1-p}(y) h_2^p(y) dy\right]^{\frac{1}{p}} < M_1 ||f||_{p,\varphi} < \infty. \tag{40}$$

Definition 15. Define a first kind of Hardy-type integral operator with the homogeneous kernel $T_1^{(2)} : L_{p,\varphi}(\mathbf{R}) \to L_{p,\phi^{1-p}}(\mathbf{R})$ as follows: For any $f \in L_{p,\varphi}(\mathbf{R})$, there exists a unique representation $T_1^{(2)} f = h_2 \in L_{p,\phi^{1-p}}(\mathbf{R})$, satisfying for any $y \in \mathbf{R}, T_1^{(2)} f(y) = h_2(y)$.

In view of (40), it follows that $||T_1^{(2)} f||_{p,\phi^{1-p}} = ||h_2||_{p,\phi^{1-p}} \leq M_1 ||f||_{p,\varphi}$, and then the operator $T_1^{(2)}$ is bounded satisfying

$$||T_1^{(2)}|| = \sup_{f(\neq 0) \in L_{p,\varphi}(\mathbf{R})} \frac{||T_1^{(2)} f||_{p,\phi^{1-p}}}{||f||_{p,\varphi}} \leq M_1.$$

If we define the formal inner product of $T_1^{(2)} f$ and g as follows:

$$(T_1^{(2)} f, g) := \int_{-\infty}^{\infty} \left(\int_{-|y|}^{|y|} k_\lambda(x, y) f(x) dx\right) g(y) dy,$$

then we can rewrite Corollary 3 (for $\mu + \sigma = \lambda$) as follows:

Corollary 16. *The following conditions are equivalent:*

(i) *There exists a constant* M_1 *such that for any* $f(x) \geq 0, f \in L_{p,\varphi}(\mathbf{R})$, $||f||_{p,\varphi} > 0$, *we have the following inequality:*

$$||T_1^{(2)} f||_{p,\phi^{1-p}} < M_1 ||f||_{p,\varphi}. \tag{41}$$

(ii) *There exists a constant* M_1 *such that for any* $f(x), g(y) \geq 0, f \in L_{p,\varphi}(\mathbf{R}), g \in L_{q,\phi}(\mathbf{R}), ||f||_{p,\varphi}, ||g||_{q,\phi} > 0$, *we have the following inequality:*

$$(T_1^{(2)} f, g) < M_1 ||f||_{p,\varphi} ||g||_{q,\phi}. \tag{42}$$

(iii) $k_\lambda^{(1)}(\sigma) < \infty$.

If Condition (iii) *follows, then we have* $||T_1^{(2)}|| = k_\lambda^{(1)}(\sigma) \leq M_1$.

(c) In view of Theorem 2 (for $\sigma_1 = \sigma$), for $f \in L_{p,\varphi}(\mathbf{R})$, setting

$$H_1(y) := \int_{\{x;|x| \geq \frac{1}{|y|}\}} h(xy) f(x) dx \ (y \in \mathbf{R}),$$

by (27), we have

$$||H_1||_{p,\psi^{1-p}} = \left[\int_{-\infty}^{\infty} \psi^{1-p}(y) H_1^p(y) dy \right]^{\frac{1}{p}} < M_2 ||f||_{p,\varphi} < \infty. \tag{43}$$

Definition 17. Define a second kind of Hardy-type integral operator with the nonhomogeneous kernel $T_2^{(1)} : L_{p,\varphi}(\mathbf{R}) \to L_{p,\psi^{1-p}}(\mathbf{R})$ as follows: For any $f \in L_{p,\varphi}(\mathbf{R})$, there exists a unique representation $T_2^{(1)} f = H_1 \in L_{p,\psi^{1-p}}(\mathbf{R})$, satisfying for any $y \in \mathbf{R}, T_2^{(1)} f(y) = H_1(y)$.

In view of (43), it follows that $||T_2^{(1)} f||_{p,\psi^{1-p}} = ||H_1||_{p,\psi^{1-p}} \leq M_2 ||f||_{p,\varphi}$, and then the operator $T_2^{(1)}$ is bounded satisfying

$$||T_2^{(1)}|| = \sup_{f(\neq 0) \in L_{p,\varphi}(\mathbf{R})} \frac{||T_2^{(1)} f||_{p,\psi^{1-p}}}{||f||_{p,\varphi}} \leq M_2.$$

If we define the formal inner product of $T_2^{(1)} f$ and g as follows:

$$(T_2^{(1)} f, g) := \int_{-\infty}^{\infty} \left(\int_{\{x;|x| \geq \frac{1}{|y|}\}} h(xy) f(x) dx \right) g(y) dy,$$

then we can rewrite Theorem 2 (for $\sigma_1 = \sigma$) as follows:

Theorem 18. *The following conditions are equivalent:*

(i) *There exists a constant* M_2 *such that for any* $f(x) \geq 0, f \in L_{p,\varphi}(\mathbf{R})$, $||f||_{p,\varphi} > 0$, *we have the following inequality:*

$$||T_2^{(1)} f||_{p,\psi^{1-p}} < M_2 ||f||_{p,\varphi}. \tag{44}$$

(ii) *There exists a constant* M_2, *such that for any* $f(x), g(y) \geq 0, f \in L_{p,\varphi}(\mathbf{R}), g \in L_{q,\psi}(\mathbf{R}), ||f||_{p,\varphi}, ||g||_{q,\psi} > 0$, *we have the following inequality:*

$$(T_2^{(1)} f, g) < M_2 ||f||_{p,\varphi} ||g||_{q,\psi}. \tag{45}$$

(iii) $K^{(2)}(\sigma) < \infty$.

If Condition (iii) *follows, then we have* $||T_2^{(1)}|| = K^{(2)}(\sigma) \leq M$.

(d) In view of Corollary 7 ($\mu + \sigma = \lambda$), for $f \in L_{p,\varphi}(\mathbf{R})$, setting

$$H_2(y) := \int_{\{x;|x|\geq|y|\}} k_\lambda(x,y) f(x) dx \ (y \in \mathbf{R}),$$

by (33), we have

$$||H_2||_{p,\phi^{1-p}} = \left[\int_{-\infty}^{\infty} \phi^{1-p}(y) H_2^p(y) dy\right]^{\frac{1}{p}} < M_2 ||f||_{p,\varphi} < \infty. \tag{46}$$

Definition 19. Define a second kind of Hardy-type integral operator with the homogeneous kernel $T_2^{(2)} : L_{p,\varphi}(\mathbf{R}) \to L_{p,\phi^{1-p}}(\mathbf{R})$ as follows: For any $f \in L_{p,\varphi}(\mathbf{R})$, there exists a unique representation $T_2^{(2)} f = H_2 \in L_{p,\phi^{1-p}}(\mathbf{R})$, satisfying for any $y \in \mathbf{R}, T_2^{(2)} f(y) = H_2(y)$.

In view of (46), it follows that $||T_2^{(2)} f||_{p,\phi^{1-p}} = ||H_2||_{p,\phi^{1-p}} \leq M_2 ||f||_{p,\varphi}$, and then the operator $T_2^{(2)}$ is bounded satisfying

$$||T_2^{(2)}|| = \sup_{f(\neq 0) \in L_{p,\varphi}(\mathbf{R})} \frac{||T_2^{(2)} f||_{p,\phi^{1-p}}}{||f||_{p,\varphi}} \leq M_2.$$

If we define the formal inner product of $T_1^{(2)} f$ and g as follows:

$$(T_2^{(2)} f, g) := \int_{-\infty}^{\infty} \left(\int_{\{x;|x|\geq|y|\}} k_\lambda(x,y) f(x) dx \right) g(y) dy,$$

then we can rewrite Corollary 3 (for $\mu + \sigma = \lambda$) as follows:

Corollary 20. *The following conditions are equivalent:*

(i) *There exists a constant* M_2 *such that for any* $f(x) \geq 0, f \in L_{p,\varphi}(\mathbf{R})$, $||f||_{p,\varphi} > 0$, *we have the following inequality:*

$$||T_2^{(2)} f||_{p,\phi^{1-p}} < M_2||f||_{p,\varphi}. \tag{47}$$

(ii) *There exists a constant* M_2 *such that for any* $f(x), g(y) \geq 0, f \in L_{p,\varphi}(\mathbf{R}), g \in L_{q,\phi}(\mathbf{R}), ||f||_{p,\varphi}, ||g||_{q,\phi} > 0$, *we have the following inequality:*

$$(T_2^{(2)} f, g) < M_2||f||_{p,\varphi}||g||_{q,\phi}. \tag{48}$$

(iii) $k_\lambda^{(2)}(\sigma) < \infty$.

If Condition (iii) *follows, then we have* $||T_2^{(2)}|| = k_\lambda^{(2)}(\sigma) \leq M_2$.

Example 21. Setting $h(xy) = \frac{|\ln|xy||^\beta}{(\max\{|xy|,1\})^{\lambda-1}|xy-1|}$, and

$$k_\lambda(x, y) = \frac{|\ln|x/y||^\beta}{(\max\{|x|, |y|\})^{\lambda-1}|x - y|} \; (x, y \in \mathbf{R}),$$

for $\beta > 0, \sigma > 0, \sigma + \mu = \lambda$, it follows that $h(u) = \frac{|\ln|u||^\beta}{(\max\{|u|,1\})^{\lambda-1}|u-1|}$,

$$\begin{aligned}
K^{(1)}(\sigma) = k_\lambda^{(1)}(\sigma) &= \int_0^1 \frac{|\ln u|^\beta u^{\sigma-1}}{(\max\{u, 1\})^{\lambda-1}} \left(\frac{1}{u+1} + \frac{1}{|u-1|}\right) du \\
&= \int_0^1 (-\ln u)^\beta \left(\frac{1}{u+1} + \frac{1}{1-u}\right) u^{\sigma-1} du \\
&= 2\int_0^1 (-\ln u)^\beta \frac{u^{\sigma-1}}{1-u^2} du = 2\int_0^1 (-\ln u)^\beta \sum_{k=0}^\infty u^{2k+\sigma-1} du.
\end{aligned}$$

By Lebesgue term by term integration theorem (cf. [46]), we have

$$\begin{aligned}
K^{(1)}(\sigma) = k_\lambda^{(1)}(\sigma) &= 2\sum_{k=0}^\infty \int_0^1 (-\ln u)^\beta u^{2k+\sigma-1} du \\
&= 2\sum_{k=0}^\infty \frac{1}{(2k+\sigma)^{\beta+1}} \int_0^\infty v^\beta e^{-v} dv \\
&= \frac{\Gamma(\beta+1)}{2^\beta} \zeta\left(\beta + 1, \frac{\sigma}{2}\right) \in \mathbf{R}_+,
\end{aligned}$$

where $\zeta(s,a) = \sum_{k=0}^{\infty} \frac{1}{(k+a)^s} (\mathrm{Re}\, s > 1; a > 0)$ is the extended Riemann zeta function ($\zeta(s,1) = \sum_{k=1}^{\infty} \frac{1}{k^s} (\mathrm{Re}\, s > 1)$ is the Riemann zeta function) (cf. [44]).

Then by Theorem 3 and Corollary 9, we have

$$||T_1^{(1)}|| = ||T_1^{(2)}|| = \frac{\Gamma(\beta+1)}{2^\beta} \zeta\left(\beta+1, \frac{\sigma}{2}\right). \tag{49}$$

In the same way, for $\beta > 0, \mu > 0, \sigma + \mu = \lambda$, we find

$$\begin{aligned} K^{(2)}(\sigma) = k_\lambda^{(2)}(\sigma) &= \int_1^\infty \frac{|\ln u|^\beta u^{\sigma-1}}{(\max\{u,1\})^{\lambda-1}} \left(\frac{1}{u+1} + \frac{1}{|u-1|}\right) du \\ &= \int_0^1 \frac{|\ln u|^\beta u^{\mu-1}}{(\max\{u,1\})^{\lambda-1}} \left(\frac{1}{u+1} + \frac{1}{|u-1|}\right) du \\ &= \frac{\Gamma(\beta+1)}{2^\beta} \zeta\left(\beta+1, \frac{\mu}{2}\right) \in \mathbf{R}_+. \end{aligned}$$

Then by Theorem 4 and Corollary 10, we have

$$||T_2^{(1)}|| = ||T_2^{(2)}|| = \frac{\Gamma(\beta+1)}{2^\beta} \zeta\left(\beta+1, \frac{\mu}{2}\right). \tag{50}$$

Example 22. Setting $h(xy) = \frac{1}{|xy-1|^\lambda}, k_\lambda(x,y) = \frac{1}{|x-y|^\lambda}$ $(x, y \in \mathbf{R})$, for $\sigma > 0, \sigma + \mu = \lambda < 1$, it follows that

$$\begin{aligned} K^{(1)}(\sigma) = k_\lambda^{(1)}(\sigma) &= \int_0^1 \left(\frac{1}{(u+1)^\lambda} + \frac{1}{|u-1|^\lambda}\right) u^{\sigma-1} du \\ &= \int_0^1 \frac{u^{\sigma-1}}{(u+1)^\lambda} du + \int_0^1 \frac{1}{(1-u)^\lambda} u^{\sigma-1} du \\ &= \sum_{k=0}^{\infty} \binom{-\lambda}{k} \frac{1}{k+\sigma} + B(1-\lambda, \sigma) \in \mathbf{R}_+. \end{aligned}$$

Then by Theorem 3 and Corollary 9, we have

$$||T_1^{(1)}|| = ||T_1^{(2)}|| = \sum_{k=0}^{\infty} \binom{-\lambda}{k} \frac{1}{k+\sigma} + B(1-\lambda, \sigma). \tag{51}$$

In the same way, for $\mu > 0, \sigma + \mu = \lambda < 1$, we find

$$K^{(2)}(\sigma) = k_\lambda^{(2)}(\sigma) = \int_1^\infty \left(\frac{1}{(u+1)^\lambda} + \frac{1}{|u-1|^\lambda}\right) u^{\sigma-1} du$$
$$= \int_0^1 \left(\frac{1}{(u+1)^\lambda} + \frac{1}{|u-1|^\lambda}\right) u^{\mu-1} du$$
$$= \sum_{k=0}^\infty \frac{\binom{-\lambda}{k}}{k+\mu} + B(1-\lambda, \mu) \in \mathbf{R}_+.$$

Then by Theorem 4 and Corollary 10, we have

$$||T_2^{(1)}|| = ||T_2^{(2)}|| = \sum_{k=0}^\infty \frac{\binom{-\lambda}{k}}{k+\mu} + B(1-\lambda, \mu). \tag{52}$$

Example 23. Setting $h(xy) = \frac{|\ln|xy||^\beta}{(\min\{|xy|,1\})^{\lambda-1}|xy-1|}$, and

$$k_\lambda(x, y) = \frac{|\ln|x/y||^\beta}{(\min\{|x|, |y|\})^{\lambda-1}|x-y|} \; (x, y \in \mathbf{R}),$$

for $\beta > 0, \lambda = \sigma + \mu, \mu < 1$, it follows that $h(u) = \frac{|\ln|u||^\beta}{(\min\{|u|,1\})^{\lambda-1}|u-1|}$, and

$$K^{(1)}(\sigma) = k_\lambda^{(1)}(\sigma) = \int_0^1 \frac{|\ln u|^\beta u^{\sigma-1}}{(\min\{u,1\})^{\lambda-1}} \left(\frac{1}{u+1} + \frac{1}{|u-1|}\right) du$$
$$= 2\int_0^1 (-\ln u)^\beta \frac{u^{\sigma-\lambda}}{1-u^2} du = 2\int_0^1 (-\ln u)^\beta \sum_{k=0}^\infty u^{2k+\sigma-\lambda} du.$$

By Lebesgue term by term integration theorem (cf. [46]), we have

$$K^{(1)}(\sigma) = k_\lambda^{(1)}(\sigma) = 2\sum_{k=0}^\infty \int_0^1 (-\ln u)^\beta u^{2k+\sigma-\lambda} du$$
$$= 2\sum_{k=0}^\infty \frac{1}{(2k+\sigma-\lambda+1)^{\beta+1}} \int_0^\infty v^\beta e^{-v} dv$$
$$= \frac{\Gamma(\beta+1)}{2^\beta} \zeta\left(\beta+1, \frac{\sigma-\lambda+1}{2}\right) \in \mathbf{R}_+.$$

Then by Theorem 3 and Corollary 9, we have

$$||T_1^{(1)}|| = ||T_1^{(2)}|| = \frac{\Gamma(\beta+1)}{2^\beta}\zeta\left(\beta+1, \frac{\sigma-\lambda+1}{2}\right). \tag{53}$$

In the same way, for $\beta > 0, \lambda = \sigma + \mu, \sigma < 1$, we find

$$\begin{aligned} K^{(2)}(\sigma) = k_\lambda^{(2)}(\sigma) &= \int_1^\infty \frac{|\ln u|^\beta u^{\sigma-1}}{(\min\{u,1\})^{\lambda-1}}\left(\frac{1}{u+1} + \frac{1}{|u-1|}\right) du \\ &= \int_0^1 \frac{|\ln u|^\beta u^{\mu-1}}{(\min\{u,1\})^{\lambda-1}}\left(\frac{1}{u+1} + \frac{1}{|u-1|}\right) du \\ &= \frac{\Gamma(\beta+1)}{2^\beta}\zeta\left(\beta+1, \frac{\mu-\lambda+1}{2}\right) \in \mathbf{R}_+. \end{aligned}$$

Then by Theorem 4 and Corollary 10, we have

$$||T_2^{(1)}|| = ||T_2^{(2)}|| = \frac{\Gamma(\beta+1)}{2^\beta}\zeta\left(\beta+1, \frac{\mu-\lambda+1}{2}\right). \tag{54}$$

Acknowledgments

This work is supported by the National Natural Science Foundation (No. 61772140) and characteristic innovation project of Guangdong Provincial colleges and universities in 2020 (No. 2020KTSCX088). We are grateful for this help.

References

[1] Hardy, G.H., Littlewood, J.E. and Pólya, G. *Inequalities*, Cambridge University Press, Cambridge, USA (1934).

[2] Yang, B.C. *The Norm of Operator and Hilbert-Type Inequalities*, Science Press, Beijing, China (2009).

[3] Yang, B.C. *Hilbert-Type Integral Inequalities*, Bentham Science Publishers Ltd., The United Arab Emirates (2009).

[4] Yang, B.C. On the norm of an integral operator and applications, *J. Math. Anal. Appl.*, 321, 182–192 (2006).

[5] Xu, J.S. Hardy-Hilbert's inequalities with two parameters, *Advances in Mathematics*, 36(2), 63–76.

[6] Yang, B.C. On the norm of a Hilbert's type linear operator and applications, *J. Math. Anal. Appl.*, 325, 529–541 (2007).

[7] Xin, D.M. A Hilbert-type integral inequality with the homogeneous kernel of zero degree, *Mathematical Theory and Applications*, 30(2), 70–74 (2010).

[8] Yang, B.C. A Hilbert-type integral inequality with the homogenous kernel of degree 0, *Journal of Shandong University* (Natural Science), 45(2), 103–106 (2010).

[9] Debnath, L. and Yang, B.C. Recent developments of Hilbert-type discrete and integral inequalities with applications, *International Journal of Mathematics and Mathematical Sciences*, 2012, 871845, 29.

[10] Hong, Y. On the structure character of Hilbert's type integral inequality with homogeneous kernal and applications, *Journal of Jilin University* (Science Edition), 55(2), 189–194 (2017).

[11] Rassias, M.Th. and Yang, B.C. On half-discrete Hilbert's inequality. *Applied Mathematics and Computation*, 220, 75–93 (2013).

[12] Yang, B.C. and Krnic, M. A half-discrete Hilbert-type inequality with a general homogeneous kernel of degree 0. *Journal of Mathematical Inequalities*, 6(3), 401–417 (2012).

[13] Rassias, Th.M. and Yang, B.C. A multidimensional half — discrete Hilbert — type inequality and the Riemann zeta function, *Applied Mathematics and Computation*, 225, 263–277 (2013).

[14] Rassias, M.Th. and Yang, B.C. On a multidimensional half — discrete Hilbert — type inequality related to the hyperbolic cotangent function. *Applied Mathematics and Computation*, 242, 800–813 (2013).

[15] Rassias, M.Th. and Yang, B.C. A multidimensional Hilbert — type integral inequality related to the Riemann zeta function, in: N.J. Daras, (ed.), *Applications of Mathematics and Informatics in Science and Engineering*, Springer, New York, 417–433 (2014).

[16] Chen, Q. and Yang, B.C. A survey on the study of Hilbert-type inequalities. *Journal of Inequalities and Applications*, 2015, 302 (2015).

[17] Yang, B.C. A new Hilbert-type integral inequality, *Soochow Journal of Mathematics*, 33(4), 849–859 (2007).

[18] He, B. and Yang, B.C. On a Hilbert-type integral inequality with the homogeneous kernel of 0-degree and the hypergeometrc function, Mathematics in Practice and Theory, 40(18), 105–211 (2010).

[19] Yang, B.C. A new Hilbert-type integral inequality with some parameters, *Journal of Jilin University* (Science Edition), 46(6), 1085–1090 (2008).

[20] Yang, B.C. A Hilbert-type integral inequality with a non-homogeneous kernel, *Journal of Xiamen University* (Natural Science), 48(2), 165–169 (2008).

[21] Zeng, Z. and Xie, Z.T. On a new Hilbert-type integral inequality with the homogeneous kernel of degree 0 and the integral in whole plane, *Journal of Inequalities and Applications*, 2010, 256796, 9.

[22] Wang, A.Z. and Yang, B.C. A new Hilbert-type integral inequality in whole plane with the non-homogeneous kernel, *Journal of Inequalities and Applications*, 2011, 123 (2011).

[23] Xin, D.M. and Yang, B.C. A Hilbert-type integral inequality in whole plane with the homogeneous kernel of degree -2, *Journal of Inequalities and Applications*, 2011, 401428, 11.

[24] He, B. and Yang, B.C. On an inequality concerning a non-homogeneous kernel and the hypergeometric function, *Tamsul Oxford Journal of Information and Mathematical Sciences*, 27(1), 75–88 (2011).

[25] Yang, B. A reverse Hilbert-type integral inequality with a non-homogeneous kernel, *Journal of Jilin University* (Science Edition), 49(3), 437–441 (2011).

[26] Xie, Z.T., Zeng, Z. and Sun, Y.F. A new Hilbert-type inequality with the homogeneous kernel of degree -2-. *Advances and Applications in Mathematical Sciences*, 12(7), 391–401 (2013).

[27] Huang, Q.L., Wu, S.H. and Yang, B.C. Parameterized Hilbert-type integral inequalities in the whole plane, *The Scientific World Journal*, 2014, 169061, 8.

[28] Zhen, Z., Raja, K., Gandhi, R. and Xie, Z.T. A new Hilbert-type inequality with the homogeneous kernel of degree -2 and with the integral, *Bulletin of Mathematical Sciences & Applications*, 3(1), 11–20 (2014).

[29] Rassias, M.Th. and Yang, B.C. A Hilbert — type integral inequality in the whole plane related to the hyper geometric function and the beta function, *Journal of Mathematical Analysis and Applications*, 428(2), 1286–1308 (2015).

[30] Huang, X.Y., Cao, J.F., He, B. and Yang, B.C. Hilbert-type and Hardy-type integral inequalities with operator expressions and the best constants in the whole plane. *Journal of Inequalities and Applications,* 2015, 129 (2015).

[31] Gu1, Z.H, and Yang, B.C. A Hilbert-type integral inequality in the whole plane with a non-homogeneous kernel and a few parameters. *Journal of Inequalities and Applications*, 2015, 314 (2015).

[32] Rassias, M.Th. and Yang, B.C. On a Hilbert-Type integral inequality related to the extended hurwitz zeta function in the whole plane. *Acta Appl. Math.* https://doi.org/10.1007/s10440-018-0195-9.

[33] He, L.P., Li, Y. and Yang, B.C. An extended Hilbert's integral inequality in the whole plane with parameters. *Journal of Inequalities and Applications,* 2018, 216 (2018).

[34] Xin, D.M., Yang, B.C. and Wang, A.Z. Equivalent property of a Hilbert-type integral inequality related to the beta function in the whole plane. *Journal of Function Spaces*, 2018, 2691816, 8.

[35] Zhong, Y.R. Huang, M.F. and Yang, B.C. A Hilbert-type integral inequality in the whole plane related to the kernel of exponent function. *Journal of Inequalities and Applications,* 2018, 234 (2018).

[36] Rassias, M.Th. and Yang, B.C. On a Hilbert-type integral inequality in the whole plane related to the extended Riemann zeta function. *Complex Analysis and Operator Theory* (2018). https://doi.org/10.1007/s11785-018-0830-5.

[37] Rassias, M.Th. and Yang, B.C. On a Hilbert-type integral inequality in the whole plane related to the extended Riemann zeta function. Published in Mathematical Analysis and Applications (eds.), M.Th. Rassias, P.M. Pardalos, Springer, (2019).

[38] Rassias, M.Th., Yang, B.C. and Raigorodskii, A. On Hardy-type integral inequality in the whole plane related to the extended Hurwitz-zeta fanction. *Journal of Inequalities and Applications*, 2020, 94 (2020).

[39] Rassias, M.Th., Yang, B.C. and Raigorodskii. A. On the reverse Hardy-type integral inequalities in the whole plane with the extended Riemann-Zeta function. *Journal of Mathematical Inequalities*, 14(2), 525–546 (2020).

[40] Wang, A.Z. and Yang, B.C. Equivalent statements of a Hilbert-type integral inequality with the extended Hurwitz zeta function in the whole plane. *Journal of Mathematical Inequalities*, 14(4), 1039–1054 (2020).

[41] Rassias, M.Th., Yang, B.C. and Raigorodskii, A. A Hilbert-type integral inequality in the whole plane related to the Arc tangent function. *Symmetry*, 13, 351 (2021). https://doi.org/10.3390/sym13020351.

[42] Yang, B.C. On a Hilbert-type integral inequality in the whole plane. Published in Applications of Nonlinear Analysis (Ed.), M.Th. Rassias, Springer, (2018).

[43] Yang, B.C., Xin, D.M. and Wang, A.Z. Equivalent property of a Hilbert-type integral inequality related to the beta function in the whole plane. Published in Recent Studies in Mathematics and Computer Science, Vol. 2 (eds.), Manuel Alberto M. Ferreira, Book Publisher International, (2020).

[44] Wang, Z.Q. and Guo, D.R. *Introduction to Special Functions*, Science Press, Beijing, China (1979).

[45] Kuang, J.C. *Applied Inequalities.* Shangdong Science and Technology Press, Jinan, China (2004).

[46] Kuang, J.C. *Introduction to Real Analysis.* Human Education Press, Changsha, China (1996).

https://doi.org/10.1142/9789811293009_bmatter

Index

A

B

S

T

U

V

W

Z

www.ingramcontent.com/pod-product-compliance
Lightning Source LLC
LaVergne TN
LVHW012052011026
849162LV00018B/164